T0186896

SNOW ENGINEERING: RECENT ADVANCES

PROCEEDINGS OF THE THIRD INTERNATIONAL CONFERENCE ON SNOW
ENGINEERING/SENDAI/JAPAN/26 - 31 MAY 1996

Snow Engineering:
Recent Advances

Edited by

MASANORI IZUMI
Tohoku University, Japan

TSUTOMU NAKAMURA
Iwate University, Japan

RONALD L.SACK
University of Oklahoma, USA

A.A.BALKEMA/ROTTERDAM/BROOKFIELD/1997

Photo cover: Tsukasa Tomabechi

The texts of the various papers in this volume were set individually by typists under the supervision of each of the authors concerned.

Authorization to photocopy items for internal or personal use, or the internal or personal use of specific clients, is granted by A.A. Balkema, Rotterdam, provided that the base fee of US$1.50 per copy, plus US$0.10 per page is paid directly to Copyright Clearance Center, 222 Rosewood Drive, Danvers, MA 01923, USA. For those organizations that have been granted a photocopy license by CCC, a separate system of payment has been arranged. The fee code for users of the Transactional Reporting Service is: 90 5410 865 7/97 US$1.50 + US$0.10.

Published by
A.A. Balkema, P.O. Box 1675, 3000 BR Rotterdam, Netherlands (Fax: +31.10.4135947)
A.A. Balkema Publishers, Old Post Road, Brookfield, VT 05036-9704, USA (Fax: +1.802.2763837)

ISBN 90 5410 865 7
© 1997 A.A. Balkema, Rotterdam
Printed in the Netherlands

Snow Engineering: Recent Advances, Izumi, Nakamura & Sack (eds) © 1997 Balkema, Rotterdam. ISBN 90 5410 865 7

Table of contents

2 Building and construction engineering

3 Infrastructure and transportation

4 Housing and residential planning

5 Development strategy in snow countries

Perspectives

Snow Engineering: Recent Advances, Izumi, Nakamura & Sack (eds) © 1997 Balkema, Rotterdam. ISBN 90 5410 865 7

Foreword

People living in heavy snow regions potentially suffer from failure and damage of buildings, snow avalanche disasters, traffic and transport difficulties, and other various economic and psychological burdens. To overcome these problems and to realize more comfortable living conditions in snow regions, a close cooperation among scientists, engineers, governments and private sectors is essential. For this purpose, it is useful to promote exchange of informations and experiences to stimulate economic and technical progress in this area.

The third International Conference on Snow Engineering (ICSE-3) was held in Sendai, Japan from May 26 to 31, 1996. This is the third of a series held every four years started in 1988. Both of the previous conferences chaired by Prof. Ronald L. Sack took place in Santa Barbara, California, USA. Those proceedings were published as special reports of Cold Regions Research and Engineering Laboratory.

The objective of this Conference was to provide a forum for engineers, architects and scientists to discuss a broad range of research and design methods for various problems related to snow engineering. Specialists in building and civil engineering, architecture, traffic and transport engineering, environmental engineering, energy engineering, urban planning, and regional development as well as snow scientists were brought together for this Conference.

The Conference began with 'Snow Forum in Sendai '96', as the 8th Sendai International Academic Symposium. Four invited lectures were given in this Forum and those lecture notes are included in this volume as principal papers. Participants of ICSE-3 came from Austria, Canada, China including Taiwan, Finland, France, Italy, Japan, Norway, Russia, Switzerland, and the USA.

Prof. Hirozo Mihashi of Tohoku University served as Secretary General of the Conference Organizing Committee of ICSE-3. The City of Sendai and Intelligent Cosmos Academic Foundation also served as sponsoring agencies. The Conference was held under the auspices of 16 professional organizations. Financial cooperation was provided by Argos Co. Ltd, Association for Technological Progress of Snow Disaster Prevention, Ebara Co., Fujikura Co. Ltd, Japanese Society of Snow and Ice, Niigata Engineering Co. Ltd, Nippon Steel Metal Products Co. Ltd, SECOM Science and Technology Foundation, Sendai Doro Engineering Co., Shinko Kenzai Co. Ltd, Tohoku Construction Association and Tokamachi City.

The technical sessions, extending over five days were in five thematic areas as follows:

1. Snow Technology and Science;
2. Building and Construction Engineering;
3. Infrastructure and Transportation;
4. Housing and Residential Planning; and
5. Development Strategy in Snow Countries.

The final session entitled 'Perspective' was an open discussion format summarizing the session contents and stimulated participants to discuss past, present and future trends in the field.

At the conclusion of the Conference, the advisory committee, chaired by Prof. Masanori Izumi of Tohoku University/Tohoku University of Art & Design, Japan, accepted the invitation of Prof. Kristoffer Apeland, Oslo School of Architecture, Norway, to hold the Fourth International Conference on Snow Engineering in Norway in the year 2000.

Further detailed information about the International Conference on Snow Engineering is available through Internet:

WWW home page: http://www.ta.chiba-u.ac.jp /ICSE/

These proceedings contain 115 papers, including four invited principal papers. All of the papers, which were presented at the conference, were reviewed and edited after the conference. It is our great pleasure to thank Prof. Hirozo Mihashi and Mr Tatsuo Sasaki of Tohoku University, and Prof. Jiro Suzuya and Mr Toshikazu Nozawa of Tohoku Institute of Technology for their enormous efforts. We could not have finished the editing process without the personal engagement.

In closing, we wish to express our thanks to the authors for their excellent contributions to this volume.

October 1996

Masanori Izumi
Tsutomu Nakamura
Ronald L. Sack

Organization

CONFERENCE ADVISORY COMMITTEE

K.Apeland, Oslo School of Architecture, Norway (Co-Chairman)
J.E.Cermak, Colorado State University, USA
M.Gränzer, Landesstelle für Bautechnik, Germany
H.Gulvanessian, Building Research Establishment, UK
P.A.Irwin, Rowan, Williams, Davies and Irwin, Inc., Canada
N.Isyumov, University of Western Ontario, Canada
M.Izumi, Tohoku University/Tohoku University Art & Design, Japan (Chairman)
S.Kobayashi, Niigata University, Japan
I.Mackinlay, Ian Mackinlay Architecture, USA
H.Mihashi, Tohoku University, Japan (Secretary General)
T.Nakamura, Iwate University, Japan
M.O'Rourke, Rensselaer Polytechnical Institute, USA
V.D.Raizer, Russian Academy of Architecture and Construction, Russia
J.Raoul, Ministère des Transports, France
R.L.Sack, University of Oklahoma, USA (Co-Chairman)
M.Salgo, Engineering Foundation, USA
R.Sandvik, Norwegian Council for Building Standardization, Norway
L.Sanpaolesi, University di Pisa, Italy
U.Stiefel, Gruner Ltd. Consulting Engineers, Switzerland
J.Suzuya, Tohoku Institute of Technology, Japan
W.Tobiasson, CRREL, USA

CONFERENCE ORGANIZATION COMMITTEE

T. Fukuda, Tohoku University
N. Hayakawa, Nagaoka University of Technology
M. Higashiura, Shinjo Institute of Snow & Ice Studies, NIED
T. Ito, Akita National College of Technology
M. Izumi, Tohoku University/Tohoku University Art & Design (Chairman)
O. Joh, Hokkaido University
S. Kobayashi, Niigata University
H. Mihashi, Tohoku University (Secretary General)
T. Nakamura, Iwate University (Co-Chairman)
I. Sakamoto, University of Tokyo
J. Suzuya, Tohoku Institute of Technology
K. Tsusima, Toyama University
T. Umemura, Nagaoka University of Technology
M. Watanabe, Hachinohe Institute of Technology

EDITORIAL BOARD

N. Hayakawa, Nagaoka University of Technology
M. Izumi, Tohoku University/Tohoku University Art & Design (Chairman)
S. Kobayashi, Niigata University
H. Mihashi, Tohoku University (Secretary)
T. Nakamura, Iwate University (Co-Chairman)
R. L. Sack, University of Oklahoma (Co-Chairman)
T. Takahashi, Chiba University
W. Tobiasson, CRREL
E. Wright, CRREL

PAPERS COMMITTEE

N. Hayakawa, Nagaoka University of Technology
O. Joh, Hokkaido University
S. Kobayashi, Niigata University (Chairman)
T. Nakamura, Iwate University
N. Shuto, Tohoku University

Principal lectures

Snow Engineering: Recent Advances, Izumi, Nakamura & Sack (eds) © 1997 Balkema, Rotterdam. ISBN 90 5410 865 7

Perspectives on the science engineering effects of snow

Ronald L. Sack
School of Civil Engineering and Environmental Science, University of Oklahoma, Norman, Okla., USA (Presently: Division of Civil and Mechanical Systems, NSF, Arlington, Va., USA)

ABSTRACT: Uncertainty associated with snow loads can be determined through measurements, simulation and analytical procedures. The science used to define snow behavior is transferred into engineering practice through building standards and codes; this process varies from nation to nation. Snow science research and engineering, plus standards and codes, are described for North America.

1 Introduction

The effects of snow on the natural and engineered environments can be significant. Spectacular events such as avalanches and heavy snow falls can devastate entire villages, destroy buildings and bring transportation systems and daily routines to a halt. In the U.S. on January 26, 1996 the jet stream brought frigid air from the North, mixed it with warm moist air from the South, and produced two feet of snow in the Nation's Capitol. Some have dubbed this event the "blizzard of 1996;" it was the largest storm since 1922. The blizzard closed down U.S. Government operations for one week. This disruption, coupled with an on-going budget impasse, is estimated by some to have put various agencies six to eight months behind schedule. In addition to lost time and revenues, plus inconvenience and personal injury, the direct costs of the storm are associated with snow removal, an impaired transportation system and structural damage. While the ancillary impact of snow can be extreme, this paper will focus upon issues associated with the effects of snow on buildings.

1.1 Uncertainty in Structural Design

Loads imposed upon structures by wind, waves, earthquakes and snow are generally unknown. When designing for snow, we are dealing with uncertainty. We deceive ourselves by assuming that the design load is the maximum of a set of recorded values. The largest snow load imposed during the life cycle of the structure might have not been recorded at the time of design. Almost all loads, including the weights of building materials, contain inherent randomness.

Even if loads were deterministic, structural design would still involve uncertainty because of variabilities in the member properties. We have variations in material strengths, dimensions of members, placement of supports, in addition to discrepancies between the engineer's design and the contractor's product.

Laws governing early buildings focused upon structural integrity and made no effort to define extreme forces. Hammurabi, King of Babylon, in 2200 B.C. passed the first law concerning building safety. The penalties for structural failure were severe. If a building collapsed and killed the owner, the builder would be put to death. If the owner's son were killed in a collapse, the builder's son would be slain. If a collapse killed a slave, the builder was obliged to provide another slave to the owner. Also, the builder was required to rebuild the house if it suffered any damage during use.

In 1804, Napoleon established a code mandating imprisonment for the builder and architect if, within the first 10 years, the building serviceability were impaired because of damage caused by foundation failure or poor workmanship.

The emperor of northern China, Che Tsung, in 1097 A.D. revised an earlier "Building Standard" that was never printed. Li Chieh produced a 1,078-page manuscript of the standard in 1100, which was lost. A second edition of Li's standard was produced in 1145 when the capital was moved to the south and the emperor wanted to standardize the new buildings.

This standard describes construction practices in great detail.

In the early twentieth century, most United States building codes defined deterministic design load values. However, the American National Standard Institute's (ANSI) A58.1-1972 (1972) (herein referred to as ANSI72), introduced the probability distributions for the annual extreme fastest mile wind speed and the annual extreme ground snow load. These are also used in the current ASCE Standard (ASCE 7-95). Also, design codes for steel, concrete, and wood were initially written as though the resistance of the members were calculable as a deterministic value. In contrast to most material codes, the 1977 American Concrete Institute Standard 318-77 acknowledged that the probability of obtaining concrete with a strength less than the 28-day strength was less than 10 percent.

In the U.S. we are now beginning to treat loads and resistances as independent quantities; this process was accelerated by a key publication by Ellingwood, et al, (1980). Furthermore, under the guidance of Dr. Ellingwood, the ANSI Standard A58.1-1982 (ANSI82) treated loads as probabilistic and independent of structural resistance. The United States model building codes are emulating the approach of the ANSI Standard. The nine Eurocodes are being developed for concrete steel, composite steel and concrete, timber, masonry, and aluminum, as well as for loads and earthquake resistance.

With probability-based loads and strengths, we need not use crude measures such as central factor of safety to measure reliability. For cases in which the variables are normally distributed, and the relationships between load and resistance are linear, first order, second moment methods give reliability indices. Advanced and approximate methods can be used to account for nonlinearities and to handle cases in which the random variables are not normally distributed.

Generally, we can assume that limit states design is used by the structural engineer to determine either ultimate or serviceability limit states. Probability-based material codes have emerged and are written and generally followed for building design.

1.2 Probability-based load criteria

The objective of structural engineering is to produce a system that will function in a prescribed manner and provide for the safety and welfare of the occupants and the general public. Preliminary member sizing and first analysis is typically followed by additional design-analysis cycles to obtain an acceptable structure, wherein factored resistance is greater than or equal to factored loads. That is, for the case in which the total load is a linear combination of individual loads,

$$\phi R_n \geq \Sigma \, \gamma_i Q_i \qquad (1)$$

where R_n = nominal resistance, ϕ = the resistance factor; Q_i = the applied load; and γ_i = the load factor. The left side of Eq. (1) symbolizes the factored structural resistance, and the right side represents factored load effects.

The structural resistances characterize the capacities of members made from concrete (reinforced and prestressed), metal (steel and aluminum), masonry, and timber that are subjected to tension, compression, and bending. The resistance factor is usually less than one and accounts for: variability in material properties; variability in member dimensions; variability due to modeling error; mode of failure (e.g., brittle versus ductile); importance of member in the system; and the engineer's familiarity with the design method. The load and resistance factor design characterized by Eq. (1) mandates an understanding of the statistical properties of structural resistance.

Load and resistance are not deterministic; both contain uncertainty but exhibit statistical regularity. Consequently, absolute reliability does not exist. We treat load and resistance as random variables with known statistical information for describing their probability laws, and classical reliability theory provides the concepts for describing structural reliability. The reliability index, β (varying from approximately 2 to 8), is a measure of relative reliability stemming from data describing probability distributions and statistics of the resistance, loads and load effects.

Reliability indices are implicit in existing structures built using codes that predate load and resistance factor design. By studying the statistical properties of those earlier resistances and loads, the β's can be deduced and used as target values to calibrate new codes. Calculated reliabilities depend upon the material, type of member and the ratio of applied load to dead load. By examining current practice, Ellingwood, et al (1980) found existing reliabilities for the following load combinations: 3.0 for D + L and D + S; 2.5 for D + L + W; and 1.75 for D + L + E. Where D=dead load, L=live load, S=snow load, W=wind load, and E=earthquake.

Based on the observed β levels, Ellingwood, et al (1980) recommends the following load factors:

1. 1.4D
2. $1.2(D + F + T) + 1.6(L + H) + 0.5(L_r \text{ or } S \text{ or } R)$
3. $1.2D + 1.6(L_r \text{ or } S \text{ or } R) + (0.5L \text{ or } 0.8W)$
4. $1.2D + 1.3W + 0.5L + 0.5(L_r \text{ or } S \text{ or } R)$ (2)
5. $1.2D + 1.0E + 0.5L \text{ or } 0.2S$
6. $0.9D + (1.3W \text{ or } 1.0E)$

where D=dead load; E=earthquake load; F=load due to fluids with well-defined pressures and maximum heights; F_a=flood load; H=load due to the weight and lateral pressure of soil and water in soil; L=live load; L_r=roof live load; R=rain load; S=snow load; T=self-straining force; and W=wind load. See ASCE 7-95 . The load combinations acknowledge the unlikely simultaneous occurrence of wind, snow, earthquake, and live loads. The small load factors for combining these loads implies that factored arbitrary-point-in-time load is less than the nominal load.

Ellingwood, et al (1980) suggest that groups writing material specifications either: choose target β's and select consistent load factors; or calculate β's resulting from a selection of φ's.

Loads imposed by self-weight of building materials and also building contents can be described statistically by measuring load magnitude only. Whereas wind and snow are examples of loads that must be described using data obtained at prescribed periodic time intervals (e.g., once each hour or once per month). Dynamic loads (e.g., earthquake) must be measured during an occurrence. Following is a discussion of static loads imposed by snow, which requires statistical descriptions of ground snow and methodology to convert ground loads to those on buildings.

2 Ground Snow Loads

Snow dictates building design criteria in many locations. To establish roof snow loads, the designer must determine the ground snow load for the site and understand the depth, density, and distribution of the snow.

Snow accumulation on a roof is influenced by the ground snow load. The Soil Conservation Service (SCS) and the National Weather Service (NWS) are the two principal agencies gathering ground snow data in the United States. The NWS makes daily snow load and depth measurements at 204 so-called first-order stations, and daily snow depths are recorded at approximately 9,200 additional locations. In the Western United States, the SCS makes monthly measurements of depth and water equivalent for the accumulated snow. The NWS stations are typically

located adjacent to towns and cities; whereas, the SCS sites are in the remote high mountainous areas. The NWS stations are near the majority of the building activity; thus, the construction industry could potentially make use of these data, but snow depths alone do not yield design loads. In some of the Western States the SCS stations vastly outnumber NWS locations. For the 1990 code for Canada, Newark used data from 1618 Atmospheric Environment Service (AES) stations with daily and/or month-end snow depths. Newark used additional snow water equivalent measurements from 293 British Columbia Ministry of Environment snow courses (i.e., a series of observations at one site) (Newark 1984).

We encounter a difference in the temporal content of the SCS and NWS data (as well as for the AES and climatological network data from Canada). The NWS daily measurements reveal small changes due to deposition and ablation (i.e., net volumetric decrease); whereas, the monthly SCS data do not reflect these changes. Typically, the NWS quantities peak during January and February; whereas, snowpacks in mountainous areas, as characterized by the SCS data, maximize in March and April. Boyd corrected the two sets of Canadian measurements to a common basis (Boyd 1961), and Newark further refined Boyd's so-called melt index (Newark 1984).

Most of the early analyses of ground snow loads used the maximum of the annual values for design: Structural Engineers' Association of Northern California (SEAONC64), and also Colorado (SEAC84), Oregon (SEAO71), and Arizona (SEAA81). The 1941 National Building Code of Canada used the sum of the average snow falls in January, February, and March over a number of years and added the maximum 24-hour rainfall occurring during these months.

Today the annual maxima are extrapolated beyond the historical period of observation using the Frechet (type II), log Pearson type III, Gumbel (type I) or lognormal cumulative probability distribution function (cdf) as a model. The parameters describing the cdf are determined from the data at a given site. Since the cdf extrapolates extreme values from the historical data, it is imperative that the correct model be chosen by examining the data using measures such as the Chi-square test of fit, the Kologorov-Smirnov test, or probability plot correlation coefficients. For example, predicting the annual extreme water equivalents at first-order NWS sites from the Dakotas to the East coast requires both Gumbel and lognormal distributions (Ellingwood and Redfield 1983).

5

Europe and Canada use the Gumbel distribution (Newark 1984); whereas, in the United States, ASCE 7-95 used the lognormal model. In the Western United States, the log Pearson type III distribution is used in Idaho (Sack and Sheikh-Taheri 1986), Montana (Stenberg and Videon 1978), and Washington (SEAW81). Annual probabilities of exceedance, ranging from 0.01 to 0.04, are used in the United States, but attempts are being made to standardize on 0.02. The American Society of Civil Engineers prescribes a 50-year mean recurrence interval (mri), and the states of Idaho, Montana, and Washington have also adopted this value. Canada uses a 30-year mri (Newark, et al 1989; NBCC 95). The mri is the reciprocal of the annual probability of being exceeded (e.g., a 50-year mri corresponds to an annual probability of being exceeded of 0.02).

We must estimate the snow density to use the snow depths recorded at many stations; these sites are typically located near populous areas and constitute a potentially useful data base. For many years, beginning in 1953, Canada has used a constant specific gravity of 0.192 for all locations and added the maximum 24-hour rain occurring during the winter months (Boyd 1961). The assumption of constant density does not acknowledge that snow deposition and density are dependent upon regional climatology. Newark (1984) pointed out that snow density at mountainous sites can be associated with forest type. This rationale gives mean specific gravities of 0.190 to 0.390 for the non-melt period of the year and 0.240 to 0.430 during the spring-melt interval. For mountainous areas of British Columbia, Yukon, and foothills of Alberta the NBCC (1995) assumes that snow loads above critical elevations increase linearly or quadratically with elevation.

The methodology used by ASCE 7-95 for the United States involves plotting the 50-year mri ground depths against the 50-year mri ground loads for the 204 first-order NWS stations. The resulting nonlinear regression curve relating these extreme values was used to predict ground snow loads for the 9,200 NWS stations where only depths are measured. A fourth approach is reflected in the Colorado study (SEAC84); wherein, a power law regression was applied to the snow course data from 128 stations in the state to relate snow load and depth. A similar approach was used for the 1986 study of Idaho; the depth-snow load relation was obtained using bilinear regression and data from 3,000 Western SCS stations with over five years of record (Sack and Sheikh-Taheri 1986).

The country-wide ASCE 7-95 ground snow load map displays zone intervals of $5lb/ft^2$. The zones represent 50-year mri values obtained using a lognormal distribution to model the loads from the 204 first order NWS stations and depths from the 9,200 NWS synoptic stations. The latter were converted to loads using equivalent densities calculated from the 204 first order stations. In certain areas, the snow loads shown are not appropriate for unusual locations such as high country, and some locales may have extreme variations in snow deposition. As a result, building associations, local jurisdictions, and entire state areas have initiated and published their own snow load studies (Brown 1979; Meehan 1979; Placer 1985; Sack and Sheikh-Taheri 1986; Stenberg and Videon 1978; SEAA81, SEAONC64, SEAC84, SEAO71, SEAW81). These local studies are referenced in ASCE 7-95, but not all use a 50-year mri. Also, these analyses use various annual extremes and different extreme value statistical distributions.

The map for the 1990 NBCC shows zones with varying ranges of ground snow load expressed in kPa (Newark 1984). In addition, Newark presents maps of melt index (to convert month-end data to daily data), average seasonal snowpack density, and maximum 24-hour rainfall. This Canadian study uses a Gumbel distribution and a 30-year mri.

3. Flat Roof Snow Loads

Exposure of the roof to wind and sun, thermal losses from the building, roof geometry, roof cladding, and obstructions on and around the roof significantly influence roof snow loads. The 1941 edition of NBCC mandated that

$$p_r = C\, p_g \qquad\qquad (3)$$

where p_g is the ground snow load, C is a dimensionless coefficient that depends upon roof environment and geometry, and p_r is the structural design load for the roof. Canada initiated a country-wide survey of snow loads on roofs in 1956 (Peter et al 1963) and found that the basic roof snow load for a flat roof in a location sheltered from the wind is typically 80% of the ground snow load (i.e., C = 0.80); this information was incorporated into the 1960 NBCC. The survey also indicated that the basic snow load coefficient can be reduced by 25% where the roof is fully exposed to the wind (this reduction was introduced in the 1965 NBCC). The recommendations of ANSI72 were similar to those of the 1965 NBCC.

In 1978, the snow load subcommittee of ANSI recommended the following for the contiguous United States and Alaska, respectively (O'Rourke, Koch, and Redfield 1983):

$$p_f = 0.7\, C_e C_t I p_g \qquad (4a)$$

$$p_f = 0.6\, C_e C_t I p_g \qquad (4b)$$

in which p_f is the flat roof snow load; C is a dimensionless exposure factor; C_t is a dimensionless thermal factor; and I is a dimensionless importance factor that converts the ground snow load to a mri different from 50-year. The coefficient of 0.7 and 0.6 in Eqs. (4a) and (4b) stem from the analysis of the data base of the Cold Regions Research Engineering Laboratory (CRREL). Using measured values for the roof load and the associated ground load, O'Rourke found that the expected value for the conversion factor p_f/p_g was $0.47 C_e C_t$. There was, however, a fair amount of scatter of data points about the expected value, which was modeled as an error term with a lognormal distribution. Considering the variability of both the annual maximum ground load and the conversion factor, O'Rourke and Stiefel (1983) determined that the 50-year mri roof load was equal to the 50-year mri ground times $0.606 C_e C_t$. However, O'Rourke used values of C_e ranging from 1.32 (sheltered) to 0.95 (windswept); whereas, the corresponding ANSI82 values were 1.2 (sheltered) and 0.8 (windswept). Hence the 0.70 factor for the contiguous U.S. (see Eq. (4a)) is a reasonably accurate simplification. ASCE 7-95 uses values of C_e ranging between 1.3 and 0.8 for the continental U.S., with terrain categories that are identical to those used for wind.

4 Sloped Roof Snow Loads

Using field observations (Lutes 1971; Schriever 1967), experience, and judgement, the distributions of snow on shed, gable, and arched roof shapes were obtained and included in Supplement No. 3 to the 1965 National Building Code of Canada. Values of C in Eq. (3) for these same basic building configurations were subsequently recommended by ANSI72. Canada made minor changes in their 1970 and 1975 code editions.

The 1941 NBCC allows a slope reduction factor (C_s) to be multiplied by the flat roof snow load. Roof slopes between 20° and 62.9° are reduced as shown by line C in Fig. 1, while slopes in excess of the upper limit are considered free of snow and slopes less than

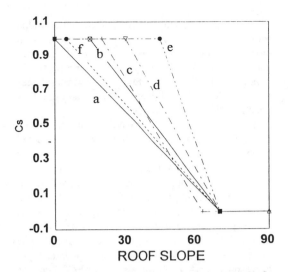

Fig. 1. Slope reduction factors for NBCC41 (line c), NBCC85 (line d), ANSI82 (lines b, d, and e), ASCE/ANSI 7-88 (lines a, b, d, and e) and ASCE 7-95 (lines f, b, d and e)

20° have the flat roof load. This slope reduction factor was changed in the 1960 edition, and NBCC85 prescribed those same slope reduction factors for simple shed and gable roofs sheltered from the wind (see line D in Fig. 1). The values of C_s are to be used in the following equation from NBCC85

$$p_s = C_b\, C_w\, C_s\, C_a\, p_g \qquad (5)$$

where p_s = the sloped roof snow load; C_b = the basic roof snow load factor of 0.8; C_w = the wind exposure factor (i.e., 1.0 for normal sitting and 0.75 for a building exposed to the wind); and C_a = the accumulation factor, which accounts for effects such as roof geometry and sliding snow. To avoid ambiguity within this paper, some symbols different from those in NBCC85 are used. NBCC85 notes that snow can slide from sloped slippery roofs (e.g., glass and metal) for roof slope, a > 15°. Studies in Canada of sloped roof behavior (Taylor 1985) have been conducted.

ANSI82 indicated that if the roof has sufficient slope, the flat roof snow load may be reduced as follows:

$$p_s = C_s p_f \qquad (6)$$

where p_s = the sloped roof snow load, and C_s = a dimensionless coefficient dependent upon roof slope,

7

roofing materials, and thermal characteristics. Slope reduction factors for unobstructed roofs with sufficient room at the eave line to shed snow are shown in Fig. 1. Line B applies to warm ($C_t = 1.0$) slippery surfaces; line D is for all other warm surfaces and cold ($C_t > 1.0$) slippery surfaces; and line E is used for all other cold surfaces.

The American Society of Civil Engineers (ASCE) made substantive changes to C_s for unobstructed slippery surfaces. Based on recent research (Sack et al 1987; Sack 1988), the suggested slope reduction factors for the 1988 ASCE standard (herein referred to as ASCE/ANSI88) in Fig. 1 are: line A for warm (C_t=1.0) slippery surfaces; line B for cold (C>1.0) slippery surfaces; and for all other surfaces lines D and E for warm and cold roofs, respectively. The latter two C_s relationships are identical to corresponding ANSI82 values. The ASCE 7-95 has abandoned line A and substituted line F for warm (C_t= 1.0) slippery surfaces and placed a number of restrictions on the use.

4.1 Sloped roof studies

Snow sliding from sloped roofs provides some interesting insights as to how behavior can be studied experimentally and explained analytically. We know that snow slides off unobstructed slippery sloped roofs; nonetheless, most building codes and standards require slippery sloped roofs to be designed for excessively large snow loads. Sack et al (1987) and Sack (1988) developed a statistical model of snow accumulation for unobstructed slippery roofs to show that sliding at the roof-snow interface is initiated by either a reduction in the resisting forces or shear failure; the latter involves a reduction in the shear resistance of the snowpack. Temperature, temperature-time, and precipitation were used as independent variables to develop the conditional probabilities for sliding from field and laboratory data. Roof snow loads for a site in deep-snow country were simulated using existing meteorological data to yield probabilities of occurrence for temperature and precipitation. A flat-roof building with four different thermal conditions was monitored as a control for the sloped roofs. Also, Sack and Giever (1990) developed probabilistic models from laboratory and field data to predict roof snow loads for cold slippery gabled roofs. The models accounted for the frequency and amount of snow sliding during an event by using actual measured probabilities. One model was based on temperature and precipitation and the other on degree-hours and precipitation. Field data were obtained from three gabled roof structures with slopes of 10°, 30°, and 45° during one winter. In addition, artifical snow was deposited on small, gabled model roofs in the laboratory. Measured laboratory roof loads were related to prototype snow loads through similitude (Giever and Sack 1990). Using simulation techniques, with the probabilistic models and 35 years of meteorological data from the field site, 50-yr mean recurrence-interval roof snow loads were calculated and compared to published design standards.

5 Nonuniform snow loads

Snow desposition is affected by wind speed and direction, terrain relief upwind and around the structure, air temperature, humidity, snow desposition rate, and building geometry (Isyumov and Davenport 1974). These effects influence the exposure factors in Eqs. 4a, 4b, and 5, as well as the slope-reduction factors.

5.1 Unbalanced loads

Wind blowing normal to the ridge of a structure will create an area of aerodynamic shade on the leeward surface, yielding nonuniform snow deposition and transport of existing snow from the windward roof surface. Unbalanced roof snow loads may also result from sliding snow.

The initial data for unbalanced snow distributions came from the Canadian survey of roof snow loads of 1956 through 1967 (Lutes 1971; Schriever 1967). Additional work has been addressed to large multi-level roofs and curved roofs (Taylor 1980). The design criteria resulting from the early Canadian case studies were introduced into the 1965 edition of NBCC; in general, for gable, arch, and curved roofs all snow is removed from one side, with loading on the leeward side. Additional snow accumulations in valley areas of roofs are also prescribed. ANSI72 contained recommendations similar to NBCC, and ANSI82 suggested a unique set of unbalanced loads.

5.2 Drifts on lower roofs and adjacent structures

Strong winds transport snow so that drifts form on roofs at abrupt changes in roof geometry and around obstructions. Drifts on multi-level roofs can constitute loads many times the ground snow load; they are cited as one of the primary causes of structural failure due to snow in the midwestern and eastern United States.

NBCC85 recommended a triangular drift on lower

roofs with a maximum load near the higher roof equal to the unit weight of snow (2.4 kN/m³) times the difference in roof elevation (m). The drift surcharge plus the balanced load is limited to $3p_g$. The base of the drift is twice the difference in roof elevations. For upper roofs less than 15 m long, the designer can reduce the drift. NBCC85 mandated that drift loads be considered when the upper and lower roofs are contiguous or separated by less than 5 m. The value of C_w in Eq. 3 is 1.0 for a distance of 10 times the difference in roof heights downwind from the elevation change.

ASCE/ANSI88 proposed new drift provisions based upon a study of approximately 350 drift snow load case histories gathered from the technical literature and insurance company failure investigations. Multiple regression analysis indicated that drift surcharge height is a function of: (a) length of the upper and lower roofs (i.e., the sources of snow); (b) the ground snow load (i.e., consistency of snow in the vicinity of the building); and (c) the difference in height of the roofs (i.e., the space available for drift formation; O'Rourke et al. 1985). The current ASCE 7-95 following drift design criteria emerged from that study (O'Rourke and Wood 1986; O'Rourke et al. 1986). The triangular snow drift surcharge load (to be superimposed on the balanced roof snow load) has a maximum height, h_d(ft), of

$$h_d = 0.43(L_u)^{1/3}(p_g+10)^{1/4} - 1.5 \qquad (7)$$

where L_u, the length of the upper roof, should be taken as not less than 25 ft. nor greater than 600 ft. The density of the drift, D (lb/ft³), is

$$D = 0.13p_g + 14 \leq 35\ lb/cu\ ft. \qquad (8)$$

The extra snow load at the top of the drift equals $h_d D$, and the total load there equals the drift load plus the balanced roof load (p_s). The maximum height of the drift must not exceed $(h_r - h_b)$, where h_r is the difference in height of the two roofs and h_b is the depth of the uniform snow deposition. O'Rourke suggests that if p_g is less than 10 lb/sq ft or if $(h_r - h_b)/h_b$ is less than 0.20, drift loads need not be considered. The drift surcharge load diminishes to zero at $4h_d$ (for most drifts) from the change in roof elevation.

The drift load on a lower roof within 20 ft of a higher structure should be determined by the method described earlier, except that the maximum intensity of the drift load is to be reduced by the factor (20 - s)/20 to account for the horizontal separation, s(ft), between the buildings.

ASCE/ANSI88 and ASCE 7-95 proposed adaptations of Eqs. (7) and (8) with a drift length of $4h_d$. On-going research in Canada (Irwin et al 1995) is showing the utility of a combined experiemental-analytical technique called the finite area element method to predict nonuniform snow distribution.

6 Summary and Conclusions

Historical information, research results and codes/standards provide the basis for snow load design practices in North America. In most locations, the engineer, contractor and builder are all guided by appropriate building codes and/or standards that have legal status. For buildings, we begin by knowing how much snow is on the ground and then determine how much and what distribution the snow will have on the roof. Building parameters such as siting, wind environment, roof configuation, roof surface and building heat loss all influence the ultimate applied loading on the structure. Snow research must play a significant role in establishing the significance of each of the effects on roof snow loads. Various tools are available to study these effects: partial scale studies in the laboratory and also in nature; full-scale observations in nature; and mathematical simulation. Ultimately, the research findings must stand the scrutiny of engineers, scientists, politicians and commercial interests in order to be included in building standards and codes. In North America we have three model building codes in the United States, and Canada has their National Building Code. An example of code unification is demonstrated by the Eurocode written by the various jurisdictions of the ECC.

7 References

American National Standard - Minimum Design Loads for Buildings and Other Structures. (1972) A58.1-1972, American National Standards Inst., New York, NY.

American National Standard - Minimum Design Loads for Buildings and Other Structures. (1982) A58.1-1982, American National Standards Inst., New York, NY.

American Society of Civil Engineers - Minimum Design Loads for Buildings and Other Structures. (1990) ANSI/ASCE 7-88, New York, NY,

American Society of Civil Engineers - Minimum Design Loads for Buildings and Other Structures. (1996) ASCE 7-95, New York, NY.

Boyd, D.W. (1961) Maximum snow depths and snow loads on roofs in Canada, in *Proc. 29th Annu. Meet. West. Snow Conf.* pp. 6-16. (Res. Pap. 142, Div. Build. Res. NRC 6312, NRCC, Ottawa, Ont., pp. 6-16).

Brown, J.W. (1979) An approach to snow load evaluation, in *Proc. 38th Annu. Meet. West. Snow Conf.*

Ellingwood, B., Galambos, T.V., MacGregor, J.G. and Cornell, C.A. (1980), *Development of a Probability Based Load Criterion for American National Standard A58,* NBS Special Publication 577, Washington, D.C.

Ellingwood, B. and Redfield, R. (1983) Ground snow loads for structural design. *J. Struct. Engrg.,* ASCE, Vol. 109 (4), 950-964.

Elliott, M. (1981) *Snow Load Data for Arizona.* Structural Engineers Association of Arizona, Tucson, AZ.

Isyumov, N. and Davenport, A.G. (1974) A probabilistic approach to the prediction of snow loads. *Can. J. Civ. Eng.,* Vol. 1, 28-49.

Giever, F.M. and Sack, R.L. (1990), Similitude considerations for roof snow loads, *Cold Regions Science and Technology,* Vol. 19, 59-71.

Irwin, P.A., Gamble, S.L. and Taylor, D.A. (1995) Effects of roof size, heat transfer, and climate on snow loads: studies for the 1995 NBC, *Can. J. Civ. Eng.,* Vol. 22, No.4, 770-784.

Lutes, D.A. and Schriever, W.R. (1971) Snow accumulations in Canada: Case histories: II. *Tech. Paper 339,* Div. Build. Res., NRCC No. 11915, Nat. Res. Counc. Can., Ottawa, Ont., 1-17.

Meehan, J.F. (1979) Snow loads and roof failures, in *Proc. Struct. Eng. Ass. Calif.,* 38th Ann. Conv.

National Building Code of Canada 1985, (Supplement). (1985) Part 4 NRCC No. 23178, Assoc. Comm. Nat. Bldg. Code, National Research Council of Canada, Ottawa, Ont.

National Building Code of Canada 1995, National Research Council of Canada, Ottawa, Ont.

Newark, M.J. (1984) A new look at ground snow loads in Canada. *Proc. 41st East. Snow. Conf.,* New Carrolton, MD.

Newark, M.J., Welsh, L.E., Morris, R.J. and Dnes, W.V., (1989) Revised Ground Snow Loads for the 1990 NBC of Canada. *Can. J. Civ. Eng.,* Vol 16, No. 3.

O'Rourke, M.J., Koch, P. and Redfield, R. (1983) Analysis of roof snow load case studies, uniform loads. *CRREL Rept.* 83-1, Hanover, NH.

O'Rourke, M.J. and Stiefel, U. (1983) Roof snow loads for structural design. *J. Struct. Engrg.,* ASCE, Vol. 109 (7), 1527-1537.

O'Rourke, M.J., Speck, R.S., Jr., and Stiefel, U. (1985) "Drift snow loads on multilevel roofs." *J. Struc. Engrg.,* ASCE, Vol. 111(2), 290-306.

O'Rourke, M.J., and Wood, E. (1986) "Improved relationship for drift loads on building," *Can. J. Civ. Engrg.,* 13(6), 647-652.

O'Rourke, M. J., Tobiasson, W., and Wood, E. (1986). "Proposed code provisions for drifted snow loads." *J. Struct. Engrg.,* ASCE, 112(9), 2080-2108.

Peter, B.B.W., Dalgliesh, W.A. and Schriever, W.R. (1963) Variation of snow loads on roofs. *Trans. Eng. Inst. Can.* Vol. 6, No. A-1, 1-11.

Placer County, Building Division (1985) *Placer County Code.* Ch. 4, Sec. 4.20(v), "Snow Load Design," Auburn, CA.

Sack, R.L. and Sheikh-Taheri, A. (1986) *Ground and Roof Snow Loads for Idaho.* Dept. of Civil Eng., Univ. of Idaho, Moscow, ID.

Sack, R.L. and Arnholtz, D. and Haldeman, J.S. (1987) Sloped roof snow loads using simulation. *J. Struct. Engrg.,* ASCE, Vol. 113(8), 1820-1833.

Sack, R.L. (1988) Snow loads on sloped roofs. *J. Struct. Eng.,* ASCE. Vol. 114(3), 501-517.

Sack, R. L. And Giever, P.M. (1990) Predicting roof snow loads on Gabled Structures, *J. Struct. Eng.,* ASCE, Vol. 116, No. 10, 2763-2779.

Schriever, W.R., Faucher, Y. and Lutes, D.A. (1967) Snow accumulations in Canada: Case histories: I. *Tech. Pap. 237,* Div. Build, Res. NRC No. 9287, NRCC, Ottawa, Ont., pp. 1-29.

Stenberg, P. and Videon, F. (1978) Recommended snow loads for Montana structures. Dept. of Civ. Eng./Eng. Mech. Montana State Univ., Bozeman, MT.

Structural Engineers Association of Colorado (1984) Snow load design data for Colorado. Boulder, CO.

Structural Engineers Association of Northern California (1964) Snow load design data for the Lake Tahoe Area.

Structural Engineers Association of Oregon (1971) Snow load analysis for Oregon. Portland, OR.

Structural Engineers Association of Washington (1981) Snow load analysis for Washington. Seattle, WA.

Taylor, D.A. (1980) Roof snow loads in Canada. *Can. J. Civ. Engrg.* Vol. 7(1), 1-18.

Taylor, D.A. (1985) Snow loads on sloping roofs: Two pilot studies in the Ottawa area. *Can J. Civ. Engrg.* Vol. 12(2), 334-343.

Snow Engineering: Recent Advances, Izumi, Nakamura & Sack (eds) © 1997 Balkema, Rotterdam. ISBN 90 5410 865 7

Building design in snow country: Past, present and future

Masanori Izumi
Department of Architecture and Building Engineering, Tohoku University, Sendai & Tohoku University of Arts and Design, Yamagata & Izumi Research Institute of Shimizu Corporation, Tokyo, Japan

ABSTRACT: Snow is an indispensable water resource for Japan and remains in place over substantial periods. For areas, however, with heavy snow can be troublesome because it impedes economic activities and demands extra financial outlays. Snow has been regarded as architecturally undesirable. First, in terms of dynamics, snow applies loads to building, and sometimes may cause damage by its added depression pressure. Many wooden and steel-framed gymnasiums collapsed during *38 & 56 Gosetu* (the heavy snowfalls of 1963 and 1981).

In this paper, it is presented from a historical view point how Japanese people have been living in snow country. Then building design aspects in snow engineering are shown and finally the economy in snow country is discussed.

1 INTRODUCTION

A passage from "*Tsurezuregusa* (Japanese essays)", written long ago by Kenko Hoshi (the monk Kenko), is almost always quoted regarding the design concept of dwelling houses in Japan. It reads, " A house should be built with summer living in mind; people can live any place in winter." Of course, Kenko Hoshi probably did not know of the conditions in areas such as Hokkaido, where winters are severe and intensely cold. He must have assumed that he would always be able to somehow cope with winter in Japan, warming himself with a *hibachi* (brazier) in his house and was unable to tolerate the uncomfortably high temperatures and high humidity of summer.

Certainly, average winter temperatures in many Japanese cities, excluding those in Hokkaido, are higher than 0℃, unlike those of many cities in northern Europe and the former U.S.S.R.. Also, the area of heavy snowfall on the Sea of Japan side of the country is separated by a boundary of spinal mountains, from the area on the Pacific Ocean side where it is dry in winter. The difference in winter weather conditions between these two regions, which are so near to one another, can be considered a very rare phenomenon not present anywhere else in the world(Fig. 1). The amount of snowfall on the Japan Sea side is far above the world average; for instance,

world meteorological records show that the world's deepest annual snow cover is 11.82 m on Mt. Ibuki, and the world's greatest snowfall in a single day was 2.1 m on Mt. Seki, both of which are in Japan. This means that our narrow land of Japan is divided into two completely different areas by the existence of winter snow. The expression, " After I came out of the tunnel, there was the snow country, before me," from the novel, "*Yukiguni* (Snow Country)," written by Yasunari Kawabata, conveys the impressions of people visiting Niigata from the Kanto plain where there is not much snowfall. In contrast, when I start from Yamagata city, pass through the Sasaya Tunnel, and enter Sendai, I experience the pleasant sensation of the weather with low clouds and constantly-falling snow, changing to fine weather in the blink of an eye, and I understand the burdens of "snow country." But when summer arrives, the distinctive features of snow country disappear almost completely, except in Hokkaido, and high temperatures and humidity predominate everywhere in Japan. For example, Yamagata holds the record for highest air temperature in Japan; climatic conditions there make Kenko Hoshi's theory acceptable.

Japan is an island country where precipitation is generally high, and it has abundant grass fields and forests. Therefore, its buildings were developed on the basis of wooden structures with grass roofing. The reasons why there was little use of stones and

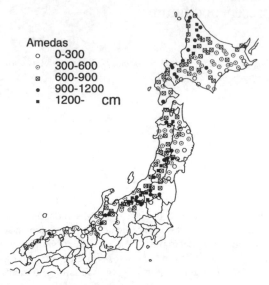

Fig.1 Max. Ground Snow Depth of Normal Year.

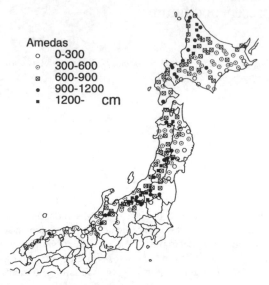

Amedas
o   0-300
⊙   300-600
⊠   600-900
•   900-1200
▪   1200-   cm

temperature. However, snow would also produce loads on a pit dwelling and destroy it; therefore snow removal work has probably been needed ever since ancient days.

Elevated floor type buildings coexisted the pit dwellings. Elevated floor type buildings used for grain storage included the attachment of *nezumi gaeshi* (rodent barriers) to each post of the building. Elevated buildings were also used as a watchtowers. One theory says that the area where pit dwellings originated and the area where elevated floor type buildings originated may be different; that is, the former type is from northern Japan and the latter originates from the south. The elevated floor type building had good ventilation since the roof and outer walls were separated from the beginning. This probably provided a comfortable living space in hot and humid summers and the pit dwellings proved effective in smoking out mosquitoes, etc. At first, the earthen floors were covered by grass or straw. Later, when lumber-processing methods had advanced, logs or boards were laid where grass and straw had previously been used. Later still, these logs or boards were suspended to form an elevated floor that created a moisture barrier, and interior walls were erected. This structure developed into the farmhouse structure, which has both an earthen floor and a wooden floor, as well as a large interior space. The large space under the roof was utilized for sericulture. The earthen floor area was used as a kitchen with the installation of a kamado (wood-burning kitchen stove), and it was also used as the location for various types of work and for storage space. The earthen floor was an important space, particularly in snow country where people were often snowed in during winter. An *irori* (open hearth) was often set up on the wooden floor section. Like a bonfire in a pit dwelling, this was not effective for heating, although smoke from the *irori* had the effect of preserving the thatch roofing(Fig. 2). Due to the difficulty of obtaining grass in towns, wooden slab roofing was used. The roof pitch of the wooden-slab roof was lower, the roof structure was rather simple and stones were placed on the roof to make it able to withstand strong winds.

Japan was greatly influenced by the great teachers of Buddhism who came from the Asian continent and became naturalized in Japan. The introduction and development of architectural techniques advanced through the construction of temples and shrines. For example, roofing tiles gradually came to be used for ordinary buildings, instead of the grass roofing and wooden-slab roofing, both of which had no

bricks included the ease with which grass and wood could be obtained and that, as the climate in Japan is relatively warm, people could tolerate drafts, even in winter. Also, Japan is prone to earthquakes, thus making structures with layers of stones or bricks dangerous. Japanese philosophy has always been more concerned with the vicissitudes of life than with immortality, in tune with nature and the differences found in each of the four seasons. Also, due to the conditions of this island country, the conquerors of Japan probably had little opportunity to show their power by building gigantic and gorgeous structures after conquering vast areas, as did the empires of ancient Egypt and Rome. Except for Nintoku-ryo (Tomb of Emperor Nintoku), structures in Japan were relatively small-sized wooden structures, based upon restrictions dictated by the size of trees.

Pit dwellings using vertical wooden posts sunk into the ground, which were used in Japan beginning in the Jomon period, were in close contact with the ground, so we can assume that these dwellings were relatively cool in summer and warm in winter. Although Jomon people had hearths inside their dwellings, which were probably able to provide warmth with the radiant heat from a fire, the air temperature in the sections where people would have been sitting or sleeping inside such a dwelling did not rise all that much because of the ventilation needed to discharge the smoke. Snow has a good adiabatic effect, so I assume that snow filled the gaps of these dwellings and was useful in maintaining the interior

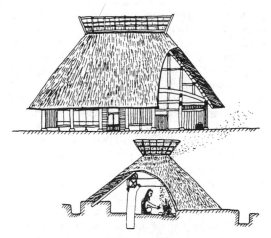

Fig.2 Pit(*Tate-ana*) Dwelling and Framhouse.

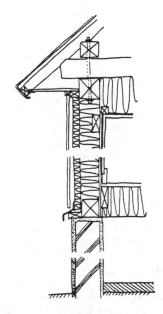

Fig.4 Use of Insulating Materials for Houses.

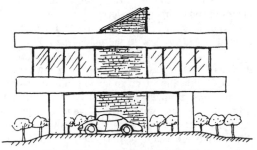

Fig.3 Modern Residential House.

resistance to fire. In the Edo period, a kind of module called *kiwari* (a ratio of dimensions and arrangement of lumber) was developed, standardization was advanced, a division of the specialties regarding production was ahead of that found elsewhere in the world and prefabricated production of fittings, tatami mats and furniture, etc., was practiced.

The opening of Japan that accompanied the Meiji Restoration had a profound impact on architecture. Various building materials, including iron, steel, concrete, glass plates and bricks, etc., were used in addition to wood, paper, grass, earth and roof tiles. The use, scale, design, techniques and construction methods thus changed remarkably(Fig. 3).

Since the beginning of the second half of the 20th century, Japan made great use of its advantages as an island country, and it has grown into a rich and advanced nation from the viewpoint of its economy, importing raw materials and exporting products. Therefore, the lifestyle of the Japanese people

gradually changed from simply working for survival, to one of living to enjoy life. The method of heating also changed to a method of raising the temperature of an entire room using abundant imported energy, and the number of cases of cerebral apoplexy decreased. Concurrent improvements in the medical system in Japan have dramatically lengthened the average life span of Japanese people.

Prefabrication techniques for housing have advanced, and the strength of prefabricated houses, for which a study of their earthquake-resistant characteristics has been thoroughly made, was verified by the earthquake that occurred in southern Hyogo prefecture in 1995. The use of air-conditioning systems in summer has gradually spread to where they can now be found in each home, and there is a continuing trend in housing that emphasizes comfortable living spaces, isolating the house from nature by making its structure airtight, and insulating the entire house(Fig. 4).

The strength of buildings is gradually being increased, so people in snow country can make it through the winter without having to remove snow from their roofs. Snow melting facilities have been also developed and the people living in snow country are now possible to liberate themselves from snow removal work in principle. However, the disadvantages of living in snow country have not yet been completely eliminated.

2 ARCHITECTURE, ENVIRONMENT AND THE TIMES

The statement, "Architecture is a container for the lives of human beings," is frequently made in conversations on architecture. These words effectively express the fact that when the lifestyles of human beings change, architecture must also change. I have been lecturing on architecture using the above-mentioned words with the additional comment that, "Architecture is, in essence, system engineering" (Izumi, 1989).

The goal of architecture is to create a safe, comfortable and convenient space which also appeals to people's aesthetic sensibility, by combining the various elements of structural materials, finishing materials, equipment, fittings and furniture. I consider it to be appropriate to refer to architecture as "typical systems engineering" (Fig. 5). Also, architecture itself must be beautiful and must simultaneously be in harmony with the surrounding environment. If possible, it must also improve the environment itself. It is often said that Japanese

A system is a sub-system of its upper system, consists of more than two sub-systems and/or elements, has one or more aims, and changes as the time goes on.

Examples: from upper to lower systems Cosmos, Earth, Asia, Japan, City, House, Room, Furniture, Metal-parts, Bolt & Nut(Either each bolt or nut can't be a system)

Fig.5 System(Izumi, 1989).

people have superior sensitivity, that they embrace nature and live in harmony with it, but actually this sentiment is only expressed in a few structures, represented by, for example, the *Katsura Rikyu* (Imperial Villa). There are almost no such structures in present-day Japanese cities. On the contrary, I think that the Japanese cannot avoid being criticized for making aggregations of buildings which are assembled together in a state of disorder and confusion, and in the process destroying the natural environment. Architects, planners, legislators and people ordering the construction of buildings - all of them - are responsible for this present situation.

First in the list of problems is that many architects seem to be attempting to generate appeal for their work by developing an "original" style which varies only slightly from that of others. Various awards have been presented for architectural works, but there is no award for "architecture which has no distinction and does not stand out," that is, for architecture that blends in and complements its environment.

Even so, architecture is now, at last, finally being evaluated with a view to the landscape. The goal, "First, to satisfy the present necessity," is often one that is set by the parties which draw up and execute policies, with almost no judgment coming from a comprehensive viewpoint that takes the future into account. For example, in the event of a natural disaster, it has been shown that vacant land areas within a city are useful for emergency relief activities. However, the government has been applying political pressure and increasing the tax rate on farmland within cities to encourage farmers to sell their farmland, ensuring an ample supply of building sites within the cities. The administrative side should instead protect farmers living within a city and treat them well to effectively ensure the presence of green zones that are useful not only for disaster response efforts, but also for environmental conservation. This is one way to rescue many Japanese cities, which have few parks and green zones. In addition, it may be possible to reduce the taxes of a person who owns vacant land in a city, even though that person is not a farmer, on the condition that the owner allow temporary use of the vacant land when a disaster has occurred. At present, however, the government seems to be charging somewhat higher taxes on vacant land. Of course, when a city has green zones, the area of the city will be expanded, and decreases in the efficiency of traffic, lifelines, etc., could result. But these problems could be solved by taking measures such as the formation of a group of high-rise buildings in the central area of the city that takes the environment into consideration, altering the present condition of a mixed bag of low-rise buildings and environmental damage. Finally, there are some problems with the parties which order the construction of buildings. They force architects and builders to construct the maximum size building legally permissible on each small lot, thus making conservation of the surrounding environment impossible. In other words, in order to create a beautiful city with a nice environment, the understanding and cooperation of all citizens and indications as to the intentions of the architects and the administrators are needed. However, as a basis for this, education is needed, and it will take some time before we can see the actual effect of these measures.

Sendai belongs to that category of cities in Japan that are beautiful and possess a good environment. However, I feel that, with the increase in its population, it seems to be losing its advantages. This may be an unavoidable trend by now, but I feel that Sendai should try to avoid practicing small-scale, Tokyo-like development.

3 SNOW AND ARCHITECTURE

If architecture is the container for human living, it is natural to expect the buildings in snow country on the Sea of Japan side of the country to be quite different from those on the Pacific Ocean side. Certainly, when we examine in detail the buildings in snow country, unique contrivances used only in that environment can be seen in many parts of buildings. Generally speaking, however, I can say that the buildings on either side of Japan do not differ greatly. Nonetheless, there are some aspects of architecture in snow country worth mentioning.

The distinctive features of the buildings in snow country can be seen principally in the construction of their roofs and entrances. Steep roof pitches are incorporated into the design so that falling snow will slide off the roofs. However, no one can predict when snow will slide off a roof and it is dangerous in most cases if a large amount of snow slides down suddenly. Therefore, safe entrances and exits are required.

In case of gable such as *gassho zukuri* ("A"-shaped frame structures), it is possible to place the entrance/exit on the gable side(*tsuma-iri*). If the gable is designed to face toward the roadside in urban areas where house lots are small, snow often crashes into a neighbor's lot, and sometimes damages adjacent houses, trees, etc. Also, where there is enough ground space, deep snow accumulated on the ground reaches the snow on the roof, interfering with the sliding of snow off the roof. Meanwhile, the weight

of accumulated snow consequently results in sedimentation pressure on the roof due to sedimentation pressure. Therefore, it is necessary to speedily remove the snow accumulated around the dwellings. Snow removal is more easily accomplished if there are ponds and ditches to which the snow can be removed. In order to avoid damage to neighboring lots caused by snow sliding off roofs, many houses have *hira-iri* (entrances facing at a right angle to the gable). The entrance/exit is on the side, along the direction of the cross-beam(Fig.6). In these houses, measures are taken to protect pedestrians from falling snow, including: (1) installing snow guards to stop snow from falling off (accompanied by occasional snow-removal work); (2) installing protective structures, such as covered alleys, and arcades; (3) designing roofs so snow will not slide off; and (4) safely melting the snow.

Recently, however, many cities which have no nearby vacant lots or river basins have been having difficulty in finding snow dumping sites, and the number of such cases has been increasing markedly. Also, it is not easy to gather people to do snow removal work. Therefore, I can foresee some changes of lifestyle that will include leaving snow on roofs and not removing it all. When the roof is simply horizontal, a snow cornice will grow at the eaves. It is dangerous if this snow cornice falls off. Therefore, a roof type in which the cross section of the building and roof forms the letter "M" was developed and has been widely used. With this type of roof, vertical conduits are installed on the lowest part at the center of the roof, and snow is melted around the conduits by heat produced inside the building. The roof is designed so that snow on the roof collects in the conduits along the circumference of the roof. If the roof is located in a place higher than that of surrounding objects and strong winds are present, snow is blown away by the wind and not much snow accumulates on a horizontal roof. However, if trees and buildings in the location surrounding the horizontal roof are higher than the roof, drifted snow accumulates on the horizontal roof (Fig. 7).

From the structural viewpoint previously mentioned, it is now quite possible to construct ordinary buildings which can withstand snow accumulation of an amount that is the same as the maximum recorded snow depth for each city. But problems occur when, from the viewpoint of function, we need a wide space without pillars, and when we cannot design building structures with sufficient durability due to severe economic constraints. In 1963 and 1981, the gymnasiums of some schools collapsed because the structures and

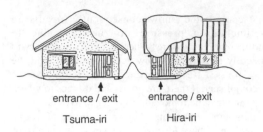

entrance / exit   entrance / exit

Tsuma-iri   Hira-iri

Fig.6   *Tsuma-iri* and *Hira-iri*.

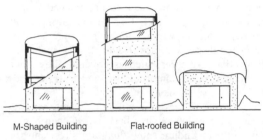

M-Shaped Building   Flat-roofed Building

Fig.7   M-shaped Building & Flat-Roofed Buildings.

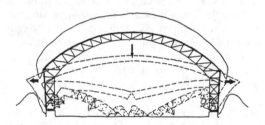

Fig.8   Collapse of Arch-Roofed Gymnasium.

frames of these buildings were designed on the basis of accumulated snow estimates (called "accumulated snow load for design") which were calculated to have not a particularly large value on the assumption that snow removal would be regularly done. When heavy snow fell, however, everyone was busy removing snow from their own homes and all roads were buried by snow, so these schools were unable to gather enough people to remove the snow from the roofs of the gymnasiums (Fig.8). This heavy snowfall began at the end of 1980, and continued into 1981: Not only gymnasiums, but also steel transmission towers in Miyagi Prefecture collapsed because of the heavy snowfall, which had a high moisture content (Izumi, 1991).

In snow country, it is possible to design buildings with wide interior spaces using no (internal) pillars, but the size of the wall pillars and beams becomes

large, giving a dull and heavy impression, and we can hardly expect agreement from designers who rather want to design structures that appear to be lightweight.

I will now mention a few more details regarding buildings in snow country. When snow falls in a very cold area, it consists of dry, non-sticky, fine particles, but once the snow has accumulated on the ground, it changes in quality, developing a sticky, taffy-like character and moving very slowly on inclined surfaces. When water is present between a sheet metal roof and snow, the snow may slide forcefully. If an eave gutter protrudes slightly, the gutter is destroyed by snow, whether the snow oozes down slowly like soft candy or falls rapidly. Therefore, the details of the eaves must be different from those used outside snow country. Many buildings are constructed without the installation of horizontal eave gutters, but if this is done, water dropping from the roof and splashing on the ground makes walls dirty and can damage them by accelerating freezing damage in winter, and rotting or rusting them in summer (Fig.9).

When snow has accumulated on a roof and the outdoor air temperature has dropped below 0℃, the underside of the roof, where the attic is, is warmed by the building's interior heat, melting the snow on the roof with the exception of that on the eaves. The snow and ice on the eaves function like a dam and stores water, causing the phenomenon of roof leaks, called *sugamore* in Japanese. Double roofing, heat insulation of the ceiling, or the melting of the snow on eaves is needed to prevent *sugamore* (Fig. 10).

There is seldom any problem when the roof is simply covered with snow, but a cold area has both the problems of freezing and melting water. Therefore a crack or other damage to an outer wall has occurred, it needs to be fixed immediately. Also, the bottom of the foundation must be placed below the frost line of the ground; otherwise the foundation will be destroyed by ground freezing. In addition, construction of a deep and durable foundation is effective against earthquakes. Of course, in cold areas, heat insulation on wall and ceiling surfaces and double-glazed sashes are effective for the efficient utilization of heat energy. Recently, plastic and wooden sashes have been marketed instead of thermaly conductive aluminum sashes.

Houses in areas with deep snow are partially buried by snow. If such snow is left as it is, the wall is pushed by the lateral pressure of the snow and becomes soggy, so people pile up firewood on the outside of wall surfaces or install fences covering the walls in order to protect them. Since trees are also

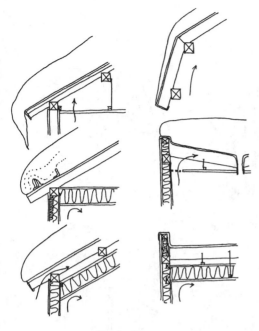

Fig.9 Eaves.

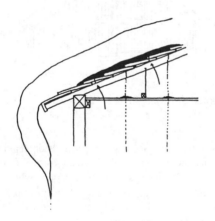

Fig.10 *Sugamore* (roof leaks).

broken by censiting snow people set supports to reinforce and protect them. Various types of labor peculiar to snow country and unrelated to productivity, including the removal of snow from roofs, are needed. When I consider these alone, living in snow country can be said to be disadvantageous, as compared to living outside snow country (Fig. 11).

4 TOWNS AND CITIES IN SNOW COUNTRY

In the old days, I assume that each house in a farm village was prepared to survive on its own for a while should it become isolated by frequent heavy snowfalls, with its people utilizing horse-drawn sleighs as their means of transportation. But, in towns and cities, people's means of making a living was probably achieved through interchanges with other people and, therefore, ensuring clear passage became a top priority. *Gangi-michi* (covered alley roads) are one example of passages, through which tpeople can safely pass, even when it is snowing (Fig.12). But *gangi-michi* were not developed in some towns, even in heavy snowfall areas, and their use was considered to be dependent upon the way people lived in winter. At present, snow is removed from major roads, and winter living in snow country has become remarkably convenient.

In my opinion the ideal for snow country involves the construction of high-rise buildings, the shortening of movement distances of people and vehicles and, if possible, the covering of the lines of travel with arcades and atria to eliminate traffic obstruction by snow. Also, we should always be able to ensure a reliable means of transportation that is not easily affected by snow, as is the Shinkansen Railway.

5 BUILDING STRUCTURE DESIGN

Buildings are designed to safely bear loads and other external forces acting upon them (Fig.13). When we design buildings, we do not know how great a load and what kinds of forces will be applied to these buildings in the future. We are forced to use estimates. At the time of estimation, the life span of a building - how many years the building will be used - is factored into the estimate. When the life span of a building is longer, there is a higher possibility of the building being subjected to heavy snowfalls and earthquakes. Also, for instance, a building which is expected to be usable for 70 years may be damaged by the type of disaster which occurs only once in several hundred years. The recent earthquake in southern Hyogo Prefecture is one such case. Therefore, a building designer adopts a stochastic concept for the assumption of the external force and load for a design, estimates what level of probability is safe (Fig.14), and considers how to design a building which will not collapse, even if it is damaged by rarely occurring phenomena (e.g., very heavy snow, earthquakes, etc.), not factored into the

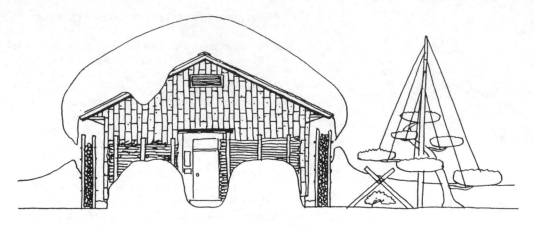

Fig.11 Snow Protect Fences.

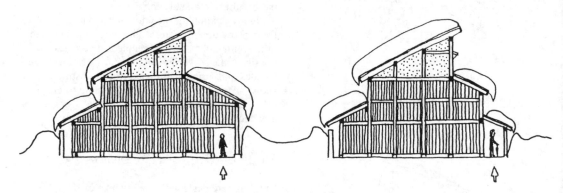

Fig.12 *Gangi - michi* (Roofed Street).

probability estimate. A designer tries to create a "tenacious or ductile" building, a building which is extremely durable and will not collapse, even if great deformation of the building has occurred. In contrast to this, the characteristic of a building collapsing when just a small deformation has occurred, is called brittleness.

Data regarding heavy snow can generally be obtained in the form of the annual maximum depth of accumulated snow at each observation station, and the characteristics of the observation area are shown in the statistical data. When these statistical data are expressed by a function called the "extreme value distribution function," it is then possible to estimate maximum values for very rare phenomena (for instance, rare heavy snowfall occurring once in a hundred years), which is actually not included in the observation data.(Fig. 15). Also, the estimated snow load value is obtained from the results of the average snow density for each depth of the entire snow layer and obtained from different research data. Values of $0.2 kgf/m^2/cm$ when the snow depth is less than 1 m and $0.3 kgf/m^2/cm$ when the snow depth is 1 m or higher, are shown as very rough average snow densities.

The load of snow accumulated on a roof varies according to the shape of the roof and the direction from which the wind blows when the snow is falling(Fig.16); the standard value of snow load shown is then interpolated and used in the design.

There is a movement to use the same data internationally, but, strictly speaking, snow load will vary depending on roofing materials and snow quality, etc., so we should understand that the international standard for snow load shows only a general procedure to determine the snow load value.

In addition, we sometimes need to consider sow loads in combination with wind and earthquakes,

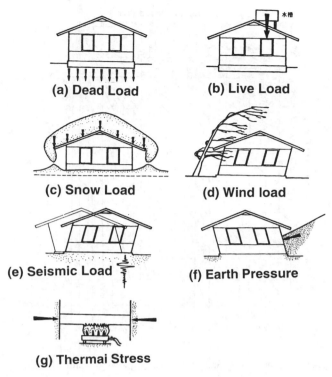

(a) Dead Load   (b) Live Load
(c) Snow Load   (d) Wind load
(e) Seismic Load   (f) Earth Pressure
(g) Thermai Stress

Fig.13 Forces Loaded on Building.

lateral pressure and the densification of snow. In and falls from the building (Fig. 17).

Avalanches have enormous power, and most of dwelling houses are powerless against avalanches. Therefore it is recommended to be sure not to construct them in an avalanche path.

In Japanese design code, loads and external forces are classified into two kinds, short term and long term. The resistance of materials against loads and external forces of a short period of time, such as during an earthquake, is considered to be higher than that against external forces and loads applied for a long duration. The subject of whether snow load is a long-term load or short-term load, has been discussed often for many years. Because the appearance of the difference between a long term load and a short-term load varies depending on the construction method by which the materials used as the members of the framework are assembled, we cannot determine these loads simply.

The effect of snow load decreases as the building weight becomes heavier, and as the number of stories of the building increases. A high-rise building is generally a snow-resistant building. Except for protruding sections such as balconies (Fig.18), I can say that heavy buildings such as reinforced concrete structures are generally strongly resistant to snow. On the contrary, relatively light-weight buildings, such as those of steel frame construction and wooden construction, have weak resistance to snow, and among these, the previously mentioned building structure which has a wide space with no pillars, is considered to be particularly weak with regard to snow load.

A gymnasium needs not only a wide floor space, but also a high ceiling. Therefore many gymnasiums have mountain-shaped or arch-shaped (dome-shaped) roofs. In addition, designers avoid joining the two bottom ends of the roof using beams (because the beams become obstacles). Thus these bottom sections of the roof are separated by the snow load on the roof, and in many cases this results in the collapse of the gymnasium. Usually, it is necessary to take measures to increase rigidity in the orthogonal direction of the beam called the *kaze-uke-bari* (wind-load resisting beam), or to set up buttresses on the outside of the building (Fig.19 and refer Fig.8).

Among the loads and external forces acting upon buildings, snow loads are considered to be loads with "good character" after dead loads (i.e., fixed loads)

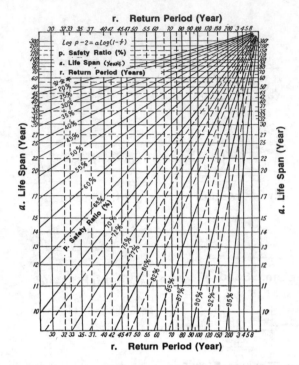

r. Return Period (Year)

r. Return Period (Year)

Fig.14   Relation among Safty-Ratio, Life-Span and Return-Period.

and live loads. This is because snow loads usually act in the vertical direction, and it takes some time for the snow load to become a large load (in other words, until snow accumulation is deep). Also before a heavy snowfall, a weather warning is usually issued. Also, in many cases, the building gives advance warning by making noises and by deformation, often giving residents time to get out of the building before it collapses. The force which has the worst character is seismic force, where, suddenly, without warning, a very great seismic force acts three-dimensionally on the building.

6 SNOW AND THE ECONOMY

There are many large cities in snowy regions, for example, Moscow in Russia and Sapporo in Japan When a city has become as big as one of these cities, the city becomes more active and the effect of snow becomes less, but generally, there is a tendency for population density to decrease as the accumulated snow becomes deeper (Fig. 20). However, the desirability of living in such areas depends on one's taste. For instance, according to the "Index of Easy

and Comfortable Living" announced by the Economic Planning Agency in 1996, among the 47 prefectures in Japan, Fukui Prefecture was in first place and Toyama Prefecture was 2nd; both are in snow country. Eight of the top ten prefectures are in snow country. However, Miyagi Prefecture, which was ranked 41st and Aomori Prefecture which was 42nd (both in snow country), and Saitama Prefecture, which was ranked lowest and many other prefectures would probably object to this ranking. This index shows the tendency to assign low evaluation points to prefectures with large populations and where the population is increasing.

Therefore, I'd like to take a general view of economy and people's lives in the past, present and future, with particular attention being paid to snow country.

In the Edo (Tokugawa shogunate) period, in which feudalism established itself and matured, Japan was divided into many small countries, each governed by a feudal lord, and each with a separate economic environment. Each feudal clan fostered and protected its own industries and, as a result, the disadvantages of snow country were eased. Industries utilizing the abundant cold water of snow

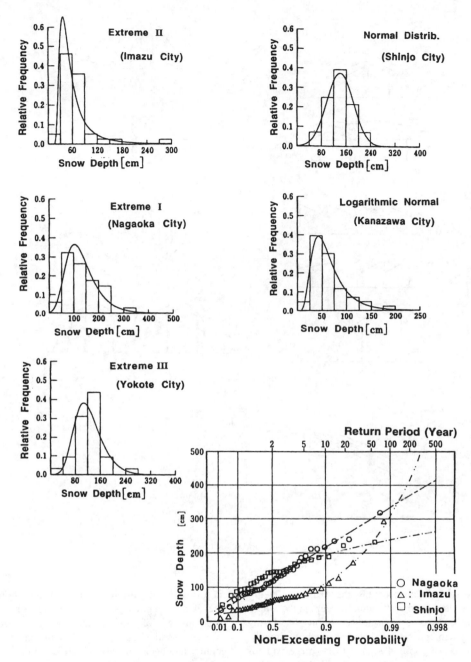

Fig.15   Statistical Date of Snow Depth and Estimation of Return Period (Izumi, 1989).

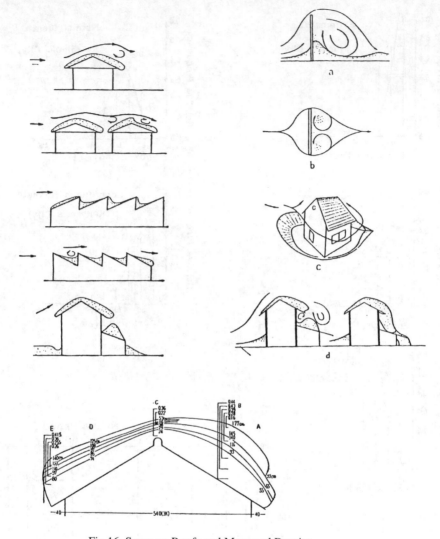

Fig.16 Snow on Roofs and Measured Density.

country, such as paper-making, the textile and dyeing industries and handicrafts which occupied the ample time available to people who were snowed in during the long winters, were developed in snow country, and processed foods designed to be preserved for long periods of time, such as pickled vegetables, were developed. On the other hand, agriculture in "snow country" was favored by its hot summers.

After the Meiji Restoration, however, Japan was strongly affected by capitalism as a result of the industrial revolution overseas. Each industry finally introduced a modern shift work system to increase the effectiveness of its investment in equipment, and

factories went for long, continuous hours of operation, regardless of the season. The disadvantages of snow country were then sharply exposed, and young people began leaving snow country and became seasonal workers in non-snowy areas. Investment in snow country where industries were hardly developed at all, decreased, except in industries seeking the power of cheap labor, and the differential between snow country and non snow country, gradually expanded.

In addition, with the recent advances in the mechanization of agriculture, the entire younger generation has been absorbed by secondary and

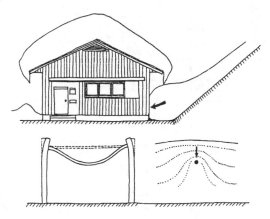

Fig.17 Lateral Pressure and Sedimentation Pressure.

Fig.18 Roofs & Balconies are often broken by Snow, even they are of RC.

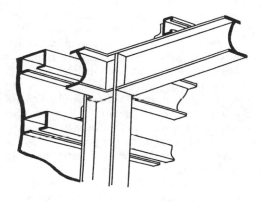

Fig.19 Buttresses or "Kazeuke bari" are essential to mountain or Dome shaped gymnasiums.

tertiary industries, and we have started to feel a sense of crisis in the agriculture, forestry and marine product industries, which are now supported by the elderly.

Now, Japan finds itself in the midst of the "great Heisei depression," and the secondary industries of which Japan was so proud have been cornered and are facing tough times due to attacks from both advanced nations and developing countries. The U.S.A. has been restricting the import of Japanese products in the order of textiles, iron, machinery, automobiles, and electrical and electronic products. Developing countries are now taking the offensive by beginning to export products made using simple technology. They are now occupying markets previously occupied by Japan. Japan cannot resist their cheap labor resources, and we are now barely remaining in the field of industries which require advanced technology. It is only a question of time before this situation reaches a crisis point. Japan is also receiving pressure on its primary industries, which are not Japan's strong point, for its primary industries are weak because Japan has few resources and little energy of its own. It is obvious that, in the future, Japan will not easily be able to get the food which it is presently purchasing from overseas using foreign currency obtained from exports related to secondary industries. Nevertheless, Japan is moving in such a direction that it can be pushed into a situation where it will be far from self-supporting and self-sufficient. The future of Japan will not be easy. Unfortunately, Japan is rapidly becoming a society of the aged, in which a majority of the population will eventually be elderly people who are exclusively consumers, and who will be dependent upon a youthful minority with high productivity. This situation is lowering productivity in Japan, resulting in loss of its competitive edge against foreign nations.

Japan has achieved rapid growth, which has been said to be a menace to the world economy, and the assets accumulated during this rapid growth have contributed greatly to the world in the form of financial aid and investment capital flowing out of Japan. At present, interest rates are being kept at a low level due to the economic depression in Japan and pressure from the U.S.A., China and other debtor nations. This is one reason why the aged cannot live on the interest they receive, and this makes the problems of the aged more difficult. There is a possibility that debtor nations will print large quantities of money with the expectation that inflation will occur and thus escape from their debts, or that debtor nations could fall into a situation in which they cannot repay their debts. As a consequence, Japan would lose most of its accumulated assets. There is also anxiety about the many disgraceful affairs occurring inside and outside of banks in Japan, which has greatly lowered confidence in them.

In addition, there is the crucial matter of the Japanese government deficit becoming so huge that it dwarfs that of other nations. The Japanese

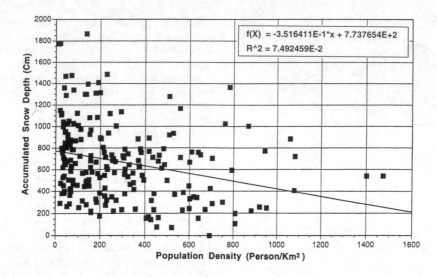

Fig.20   Relation between Snow Depth and Population Density.

Government is now unable to execute necessary policies because it has no money. Presently, the Japanese Government debt is said to be ¥443 trillion, which is 90% of Japan's GDP of ¥496 trillion. Japanese people are already struggling under heavy taxes, so it is impossible to implement a drastic tax increase, but enormous amounts of funds will be needed in the future for economy-boosting public works projects and social welfare work for the aged. Therefore, the Japanese Government needs to make efforts to curtail expenditures, but this will not be easy. In the past, the Government separated the "red ink" Japanese National Railways from government enterprises and thus was Japan Railway Co.(JR) founded. JR has been achieving good results in its own way, but it is still quite far from the situation of completely repaying the debts that ballooned during the Japanese National Railway period. In theory, it is obvious that it is better to carry out deregulation and decrease administrative work and the number of civil servants, but, when we consider the influence of deregulation, new employment for former civil servants and the existence of fields in which the numbers of civil servants cannot be reduced, etc., things get quite complicated.   If the Japanese Government were to act to reduce the number of the members of the National Diet and regional assemblies by half as a starting point for the formation of a small scale government, that would probably be rejected by the Diet and assemblies. (Related to this, it is said that many omissions of work, in violation of regulations, were discovered in buildings seriously damaged by the Southern Hyogo Prefecture Earthquake that occurred in 1995, but, if there had been no regulations, there is a possibility that the damage could have been far more serious.) Public opinion regarding the Japanese bureaucracy has become lower recently, but it was once said that Japan was supported by an excellent bureaucracy. People's eyes are apt to be dazzled by imminent matters, and they do not show much interest in matters for which a decent plan is to be carried out step by step, based upon future prospects. If the civil servants carry out deregulation, eliminate authority, and are devoted to their original work of serving people, an organization will be established which will make it possible in the future to execute plans established in the interest of the people. Basically, this may prove to be the job of politicians and the mass media, although they are still immature, so we may need more time.

Japan is now enduring severe economic conditions. I can say that this is because Japan is at a definite turning point. Ancient people in primitive communal societies lived long ago by hunting animals and collecting food, before they learned farming and stock breeding, which can be considered the "primary industrial revolution." This revolution caused wide differences to crop up among people, between poor and rich and strong and weak, and society changed to become constituted of the nobility, the common people, and slaves. After this, social mechanisms improved and a feudalistic society was born in which industries were protected by the guild system and, simultaneously, their development stopped. However, mass production became possible in the course of the "secondary industrial revolution," and capitalism began. Capitalism led to a number of problems, including the distance it created between capitalists and labor. Socialism created a nobility out of the administrators, which caused the producers to lose their enthusiasm. Socialism is now near collapse, groping toward the formation of a new society in the course of a "tertiary industrial revolution."

Capitalistic society is basically ruled by the law of competition, with each enterprise attempting to produce larger quantities of better and cheaper products, and thus eliminate other enterprises in the same trade. Therefore, individual enterprises have not paid much attention to concerns other than production. Consequently, we have witnessed the wholesale destruction of the natural environment. Resources, energy and labor power have been used wastefully, environmental pollution has become a serious problem, waste has accumulated, and, with the steady increase in the world's population, we are beginning to see the crisis in light of survival of the human species. Thus far, advanced nations have established social infrastructures, have stopped the increase of their populations, and are carrying out preparations to change from consumption oriented societies to recycling societies, making preparations for intelligent living as the result of the "tertiary industrial revolution." On the other hand, developing countries are involved in many big projects, including the establishment of infrastructure, improvement of people's living standards, and the prevention of population growth. Achieving the survival of their people and heading off starvation is the biggest concern for some developing nations, and these countries want to improve their living standards as much as possible, even it means wasteful consumption of their resources and increased pollution. From their point of view, it is natural for developing countries to consider it too selfish that advanced nations, which have wasted enormous amounts of resources and severely polluted the environment in the past, are now forcing the developing countries to stop wasting resources and polluting the environment.

As a result of this so-called "south-north problem" (northern [advanced] nations versus southern [developing] nations), humans will delay in adjusting their direction and getting on the right track for survival. They will find themselves in a more serious crisis.

The "tertiary industrial revolution" may satisfy the materialistic demands of human living to a certain degree, and may give more mental satisfaction than is possible now. Japan itself cannot be said to be an advanced nation, considering the condition of its social infrastructure, but it is now possible to bring Japan up to the level of other advanced nations. Japan's problems are that its self-supporting levels for energy, resources and food are lower than those of other advanced nations, and Japan has already devoted itself to maintaining its own state on the basis of foreign trade. Japan needs to prepare to increase its self-supporting capabilities while avoiding pressure from overseas, which may be contrary to the trend of the worldwide-scale division of work. The land of Japan at present does not have much utilizable area as compared with its population of 120 million people, and then Japan is too small for self-sufficiency. Perhaps it will be possible to improve this condition gradually, by making as much use of the wide sea area surrounding Japan and by further utilization of Japan's land area. When the territorial waters within its 200 nautical mile zone are taken into consideration, Japan is not a small country from the viewpoint of area. Utilization of additional land area will be done mainly in snow country and in the Tohoku Region where development has been delayed since the Meiji Restoration. However development in these regions will be different from development accompanied by the destruction of nature. A way of development which co-exists with nature will be implemented, while controlling nature to a certain degree. Like the rice paddies in Japan, the land has been developed and people are getting the things they need from it, even though they have created a "new nature." In other words, people have built up a natural ecosystem different from that of a simple natural reed field. During the "tertiary industrial revolution," which is presently under way, primary industries will first be supported by secondary and tertiary industries, secondary industry will be renovated by tertiary industry, and human life will advance to become full of intelligence and abundant individuality. The development of recycling technology and recyclable energy will be advanced, and waste and pollution will be drastically reduced. The "south and north problem" is a serious matter, but if northern nations steadily provide assistance and aid to the southern nations, we can assume that we will somehow be able to stop this situation before it becomes uncontrollable and results in the destruction of the human species.

7 CONCLUSION

I have stated my opinions regarding human living from the present to the near future in relation to snow. The "great Heisei depression" is different from the valley of an economic cycle and cannot just simply be described from the viewpoint of economic theory. It is related to the industrial revolution, and it has a very profound influence on Japan, but I consider it actually to be a good thing for snow country, because we can expect the future development of snow country to be different from development there in the past.

REFERENCES

Izumi, M. 1989. Structural mechanics for building energees, II, 292p., Baifukan.
Izumi, M. 1991. Design wind and snow load, 145-151, Shokokusha.

Snow Engineering: Recent Advances, Izumi, Nakamura & Sack (eds) © 1997 Balkema, Rotterdam. ISBN 90 5410 865 7

The removal and the melting of snow on roads

Tadashi Fukuda
Department of Infrastructure Planning, Tohoku University, Sendai, Japan

ABSTRACT: The effects of snow on roads are extremely significant. For such a highly urbanized city as Sendai, the effects of snow coverage on road traffic are extremely significant and the outlawing of studded tries called for more through snow removal and melting efforts.

Snow removal and melting tasks consume a great deal of time and money. The extensive sprinkling of deicing agents also adversely affects the natural environment. In response to these problems, technologies are being developed for snow removal and melting. For the protection of the environment, the city of Sendai has been continuously conducting research on the environmental effects of deicing agents.

1 INTRODUCTION

Snow has a great impact on road traffic conditions. It is no exaggeration to say that the long-time backwardness of the Tohoku region was largely due to the hindrance of traffic caused by snow. Even today, snow continues to interfere with road traffic in the Tohoku region.

Weather conditions in Sendai are relatively moderate, compared to those of many other cities in the Tohoku region, and the amount of snowfall is minimal. But, in Sendai, where the process of urbanization is relatively advanced, the impact of snowfall on road traffic conditions is great. For example, the studded tires which were once used on vehicles in winter caused dust pollution. The movement to eliminate studded tires began in Sendai in 1981, a turning point that eventually led to the restriction of studded tires throughout Japan (Miyagi Prefecture, 1993).

In this paper, I'll present some innovative technologies for snow removal and snow melting with regard to roads, paying particular attention to the problem of snow in Sendai.

2 PROBLEMS RELATED TO THE ELIMINATION OF STUDDED TIRES

In winter, an air mass cooled on the Asian continent crosses the Sea of Japan and flows into the Tohoku region. This cold air mass is blocked by the Ou Mountains, bringing heavy snowfall to the Sea of Japan side of the Tohoku region. However, the Pacific Ocean side of Tohoku has a relatively large number of days of fine weather in winter. The distance between Yamagata and Sendai is about 60 km, but the meteorological conditions of Yamagata are very different from those of Sendai. For example, the maximum snow depth recorded this winter in the area along National Highway No. 48, which connects the two cities, was 66 cm (February 3) in Yamagata city, and 135 cm (February 7) in Sekiyama on the prefectural border. However, only 11 cm was recorded in Sendai (January 10). As mentioned above, when driving along roads in the Tohoku region, road surface conditions change drastically after only driving a short distance.

The Tohoku region is mountainous; therefore all roads passing through this area are characterized by a series of many curves, and steep gradients. Road conditions are particularly hazardous in winter. Due to such conditions, studded tires were first used beginning around the 1960s, and the percentage of vehicles equipped with studded tires increased very rapidly from about 1980 onward. As a result, a pollution problem developed that was caused by dust consisting of pavement particles that had been scraped off by studded tires in urban areas where road surface conditions were mostly dry, as in Sendai (Fig. 1).

A movement to eliminate the use of studded tires began. Miyagi Prefecture established its "Studded

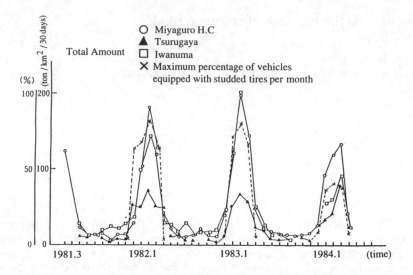

Fig.1   Variation of poluution (total amount of pavement particles in the air).

Tire Measure Ordinance" in 1985, and the National Diet established the "Law Concerning the Prevention of the Occurrence of Studded Tire Dust" in 1990. The use of studded tires was basically prohibited by this law and the ordinance, and the environmental problem caused by studded tires was eliminated. However, the problems related to the technology needed for the management of road traffic after the prohibition of studded tires still were in need of attention.

3 DEICING AGENT

1) Mechanism of snow melting

Deicing agents, such as sodium chloride and calcium chloride, have the characteristic of lowering the freezing point when they are in solution. We can use this characteristic of the agents to melt snow and ice on road surfaces. Fig. 2 shows the relationship between concentration and freezing temperature for solutions made with these deicing agents (Japan Construction Mechanization Association, 1993). As shown in the diagram, when the concentration of the solution is increased, its freezing temperature drops, but when the concentration of the solution has increased to a point greater than that equivalent to the lowest freezing temperature, the freezing temperature characteristically rises again.

Upon sprinkling a solution of deicing agent on a road before snowfall, even if the air temperature is

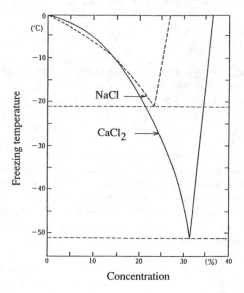

Fig.2   Relationship between concentraition of deicing agent solution and freezing temperature.

below 0℃, the snow that falls will melt until thefreezing temperature of the solution is reached. If theair temperature continues to drop, some freezing results, the concentration of the residual solution increases, and its freezing temperature falls yet further. As a result, complete freezing of the road surface can be prevented. When we use deicing agent by sprinkling it before a snowfall, it is called an anti-icing agent. The method of using deicing agents as anti-icing agents is said to be particularly effective in (very) cold areas. However, it is difficult to obtain accurate snowfall predictions that are useful for managing roads. If the anti-icing agents have been sprinkled and snow does not fall, it will be a wasteful expenditure and will also increase the amount of anti-icing agents used, polluting the natural environment in the area along the road. For these reasons, the technique of scattering deicing agents after snow has fallen to melt snow by lowering the freezing temperature on the road surface is the one most generally adopted.

2) Methods of scattering deicing agents

Rock salt is the main deicing agent used in North America and Europe. The sodium chloride used in Japan is an imported sun dried salt. Calcium chloride, also used, is a product manufactured in chemical plants. There are various methods of using these deicing agents, such as the method of scattering solid deicing agent as is, the method of sprinkling deicing agent in the form of a solution, and the method of sprinkling the solid agent and its solution simultaneously (wet salt method).

The amount of deicing salt used generally varies from 20 to 30 g/m^2 according to weather conditions. Recently, however, there has been a tendency to decrease the amount of deicing salt sprinkled, due to environmental concerns. The price of the deicing salts made from the above-mentioned chlorides used in Japan is much more expensive than that of the rock salt used in North America and Europe. For instance, if we sprinkle calcium chloride at the rate of 30g/m^2 (35% solution), the cost per unit area is about ¥4/m^2; therefore, treating a 10-m-wide, 2-lane road for a distance of 100 km costs approximately ¥4 million.

3) Environmental impact of deicing agents

There are some reports that deicing salts using chlorides have caused salt damage to structures along roadsides, corrosion of vehicles, and affected the natural environment along the roads (Fukuda, 1986). When the protection of the natural environment is the primary concern, deicing agents such as urea, CMA, etc., should be used. The immediate efficacy of urea is inferior to that of chlorides, but it encourages the growth of vegetation, and urea causes less rusting of structures than chlorides do. Road conditions can be improved by using increased amounts of deicing agents, but, on the other hand, such practices are likely to have a greater adverse impact on the natural environment. Therefore, the quantity of deicing agent used must be limited to within a range that does not affect the natural environment.

4) Environmental impact study in Sendai City

Sendai City has been continuously studying the effects of deicing agents on areas along arterial roads, on soil, vegetation and other living things, and on river water quality since 1983. The following provides a generalization of the results of the study to date (Sendai City, 1994).
(1) The effects of deicing agents on soil quality have yet to be confirmed.
(2) The discoloration of the leaves of some plants that appeared to have been in direct contact with scattered deicing agent was detected, and the germination rates of some plants slowed. However, no plants were confirmed to have withered as a result of using deicing agents.
(3) The impact of deicing agents on aquatic plants has yet to be confirmed.
(4) Some corrosion of guard rails, etc., was confirmed, but whether or not this was caused by a deicing agent alone, has not been confirmed.

4 NEW TECHNOLOGIES FOR THE MANAGEMENT OF ROAD TRAFFIC IN WINTER

1) Deicing agents that are gentle to the environment

At the end of the 1970s, the U.S.A. required the use of a deicing agent gentle to the environment. It developed CMA (calcium magnesium acetate), which is an excellent anticorrosive with other advantages, including its improvement of soil conditions and encouragement of vegetation growth. Its principal disadvantages are that its lowest freezing point is somewhat high (-10℃), it is fairly expensive and it has an odor.

A number of deicing agents using potassium acetate and calcium magnesium acetate, etc., have been developed recently, because such acetate compounds have a generally low toxicity and are gentle to the environment. In particular, its lowest freezing point of potassium acetate is low, and it is considered to have the same snow-melting ability as calcium chloride.

2) Freezing resistant pavement

This pavement uses an asphalt mixture containing deicing agent capsules or powdered deicing agent. When the surface of a road covered by this pavement is worn with the traffic of vehicles, the deicing agent is released, thus preventing the surface of the pavement from freezing due to snow. The level of snow-melting effectiveness of this treated pavement is only enough to delay the freezing of the road surface, weaken the adhesiveness of snow that has become a sheet of ice on the surface of the pavement and ease the work of snow removal machines. In other words, we cannot expect high levels of effectiveness from this pavement.

There is one type of anti-icing pavement which is quite flexible. That flexibility is obtained through the addition of rubber chips several mm in size to the surfacing material of the asphalt pavement. The resulting deformation of the pavement surface
is expected to have the effect of physically breaking down ice when a vehicle runs along a road paved with such a material. Also, ice does not adhere easily to rubber.

3) Road heating

One method involves the use of a heat source installed inside the pavement to melt accumulated snow on the surface of a road.

The exothermic wire method uses electricity (Fig. 3). The method of conducting heat from a heat source (e.g., a hot water spring) using of a heat pump

is another alternative.

Generally, road heating involves high construction and maintenance costs; therefore, it is used in locations where traffic are particularly busy and where the road grade is relatively steep.

4) Rough surface forming machine for frozen road surfaces

When snow on a road surface has frozen and solidified, it is slippery and very dangerous. When a deicing agent is sprinkled on such a frozen road surface, a water film is created, sometimes making it all the more slippery. The rough surface forming machine has been developed as a measure to cope with such frozen road surfaces.

The rough surface forming machine creates grooves in the ice on the frozen road surface by tapping it using chain hammers made of carbon steel, or with the application of pressure using a toothed steel roller (Fig. 4). When a deicing agent is sprinkled on the road surface after grooving has been completed, elimination of ice from the road surface can be effectively accomplished.

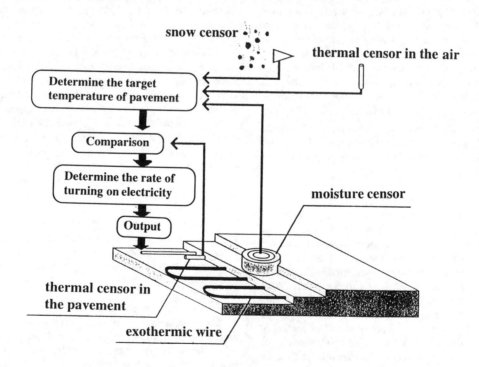

Fig.3 Road heating system with exothermic wire.

Fig.4 Rough surface forming machine on the frozen road surface.

5) Snowfall forecasting system

If the application of a deicng agent is done before snowfall as an anti-icing measure, the quantity of deicing agent needed is considered to be drastically less than that needed if it is applied after a snowfall. Therefore, if advance application of a deicing agent is possible, it will not only curtail the cost of road management in winter, but also prevent traffic jams when snow has fallen. It will also reduce environmental pollution. To operate with this strategy, however, we need to improve the snowfall forecasting system.

Sapporo city is using a winter road traffic information system in order to more effectively manage snow removal operations and road heating, and the snowfall forecasting system is utilized as a subsystem. The goal of this subsystem is to output a short-term forecast every 30 minutes and ensure a constant output that identifies the area where snowfall can be expected and the amount of snowfall in 1 km mesh units, using exclusive meteorological radar for these short-term forecasts.

The system also forecasts snowfall amounts, air temperature, wind velocity and other factors regarding expected weather conditions 12 hours later, based upon data obtained from the Meteorological Agency (Mayors' Council of Northern Cities, 1994).

5 CONCLUSION

Measures for snow on road in winter in conformity with the actual conditions found in each area are needed. For example, the ability of drivers to cope with snowy roads differs greatly depending upon whether each driver is familiar with driving on snowy roads. From the viewpoint of people living in areas where there is heavy snowfall, Sendai, where there is not much snowfall, appears to not have any a snow problem. However, the process of urbanization has advanced to a great degree and intra-city traffic patterns are relatively complex in Sendai. In addition, people in Sendai are not all that familiar with driving on snowy roads and whenever there is even the lightest of snowfalls, thorough snow removal is demanded.

It should be noted that such thorough snow removal will consequently increase the financial burden on the citizens of Sendai. Also, there is concern that excessive snow melting measures may have a bad effect on the natural environment. In other words, we must be aware that snow removal and snow melting to ensure the safety of roads are measures that are not without complications, including their impact on the environment and the problem of the financial burden that they create.

Finally, I would ask every driver to observe good manners when driving in winter. For instance, the installation of winter tires is common sense for any driver, and each driver needs to be wise enough to put on tire chains when road surface conditions make them necessary. The carelessness of only one driver can cause great hindrance to the flow of traffic. Also, I want each driver to cooperate with snow removal and snow melting workers by driving slowly. In particular, illegal parking on roads in winter is an irresponsible and prohibited action, as it interferes with snow removal and snow melting work that urgently needs to be done.

REFERENCES

Fukuda T., 1986. Problems of the Management of Roads Using Deicing agents in Winter. Proc. Second Snow Engineering Symposium, JSSE.

Japan Construction Mechanization Association, 1993. Road Snow Removal Handbook.

Mayors' Council of Northern Cities, 1994. Technological precedents for the management of road in winter. Winter Urban Environmental Problems Research Group.

Miyagi Prefecture, 1993. Clean Roads and Sky since Studded Tires have Disappeared: Progress in the past 10 years.

Sendai City, 1994. Investigation of environmental effects of deicing agents - Secular summary report, Road Division, Construction Bureau.

Snow Engineering: Recent Advances, Izumi, Nakamura & Sack (eds) © 1997 Balkema, Rotterdam. ISBN 90 5410 865 7

Snow becomes a splendid cooler – Exciting challenge in the town of Funagata

Masayoshi Kobiyama
Muroran Institute of Technology, Japan

ABSTRACT: An air-conditioning system using storage snow was developed and installed in a building. This is a simple system that fully utilized the positive characteristics of snow: for example, it lends us refreshing, gentle feelings, and that it absorbs dust and harmful gases. This system is also a low-cost technology that would meet today's strict economic requirements.

In this paper, the outline of this air-conditioning system and test results measured in a pilot project are presented.

1 INTRODUCTION

About 50 % of Japan is snow country, and in summer the weather in snow country is hot. In this experiment, we have developed a practical system for what is called "air-conditioning with snow," using snow accumulated during winter, saved until summer and designed for medium-sized needs. In this snow air-conditioning system, a flow of warm air is transported through holes bored into the snow to cool the air directly. The following results can be expected when adopting the use of this snow-air direct heat exchange system.

(1)Cool air at a stabilized temperature can be obtained throughout the entire period of use of this system.

(2)By ensuring direct contact between the air and the snow, contaminants such as dust, ammonia gas, etc., can be removed by absorption and through their being dissolved in the low-temperature melted water on the surface of the snow, thus achieving a high-grade filter effect.

(3)The structure of the system can be made quite easily.

(4)The snow storage room of this system can be also used as a snow dumping site in winter, so we can expect not only an energy-saving and environmental conservation effect, but also an economic effect.

Testing of the snow-air direct heat exchange air-conditioning system, revealed it to have the excellent performance characteristics mentioned above, and proved it to be useful. Testing were carried out in Funagata, Yamagata Prefecture, from early spring of 1995, and the practical-use testing of this system was completed by midsummer. The following provides system details and major results obtained from the practical-use tests.

2 SYSTEM

Fig. 1 shows the snow air-conditioning system installed in the Agriculture, Forestry and Fishery Practices Building in Funagata. Tables 1-3 show the plan and various design specifications of the system. We designed the 120-m^3 snow storage room as a semi-underground unit utilizing the existing topography, and we constructed the room together with a preparation room and a machine room. Cool air to the air-conditioned area (two training rooms, A and B, with identical floor areas and a total floor area of 54 m^2) is transported by way subterranean air ducts. Warm air is transferred into the snow storage room and distributed to 20 pipes (inner diameter: 80.7 mm, length: 320 mm) installed on the ceiling of the snow storage room, forced to flow through the holes bored in the snow, and thus cooled. The cool air then passes through a gap (175 mm) between the floor and an expanded metal plate at the bottom of the

Table 1 Design & air-conditioning load.

	Air-conditioning area m²	Air-conditioning load kcal/m²h kcal/h	Air-conditioning hour load rate, hours	Annual total air-conditioning load Mcal [tons of snow]
Training rooms A, B	27x2=54	140 x54=7,560	400 x0.5=200	1512 [18.9]

Table 2 Quantity of stored snow and details.

Air-conditioning load [18.9/(π/4)]	24.1 tons
Transport loss [18.9 x 10 %]	1.9 tons
Open air loss [18.9 x 30 %]	5.7 tons
Spontaneous melting	24.5 tons
Reserve snow	3.2 tons
Quantity of snow initially stored	59.4 tons

Table 3 Snow storage room.

Capacity	6m x 5m x 4m(depth) = 120m³
Density of snow stored	0.55 ton/m³
Space fill rate	90%
Volume of snow and ice initially stored	120 x 0.55 x 0.9 = 59.4 tons

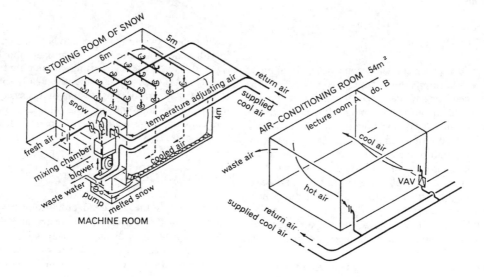

Fig. 1 Overall view of snow air-conditioning system.

snow. The air is transported through the system using of an air blower.

This "snow-air direct heat exchange air-conditioning system" can control both the temperature and humidity of the cool air to be supplied. The facility in which we carried out this verification test did not require intensive air-conditioning; therefore we decided to control only the temperature of the air-conditioned area. Adjustments to the temperature of each training room are accomplished by way of a variable air volume (VAV) controller, and adjustment of the total air volume was controlled by the rotation speed of the air blower according to air pressure detected. The temperature of the cooled air that was to be supplied to the air-conditioning rooms was pre-set by way of mixing the cool air from the snow storage room with a portion of warm circulating air. Adjustment of the air volume was accomplished by adjusting the opening of the motor dampers (MD) installed on the bypass ducts which branched off from the circulating air ducts.

3 PRELIMINARY TEST

3.1 *Circumstances prior to preliminary testing, and test method*

We filled the snow storage room with snow on February 24, 1995, where it remained untouched for approximately four months. We confirmed the condition of the stored snow on June 15 and 16 prior to starting the practical operation of the snow air-conditioning system on July 10 to test the system and examine whether it had a practical application.

On June 15, we measured the quantity of snow remaining and examined its condition. Later, we bored holes in the snow, and applied a heat load to the air that was to be circulated by using a jet burner in the area to be air-conditioned (training room A only). The diameter of each snow hole was around 20 cm, and we measured the gas absorbability on the snow surface. On the 16th, we confirmed the action of the system and measured the effect of the air-conditioning.

We set the cool air temperature at the exit point of the machine room at a fixed temperature, 14℃, and opened the divider between training rooms A and B in a series of like experiments.

3.2 *Change in the quantity of the stored snow*

On February 24, 1995, using tire shovels and a rotary snowplow, we loaded snow into the storage room until it touched the ceiling of the tank. The average density of the stored snow was 5.02 ton/m³.

The average snow density on June 15, when we measured the remaining snow and ice, was 5.83 ton/m³, due to the compacting action of the snow's own weight. Fig. 2 shows a comparison of the remaining rate of snow as theoretically estimated, with the actual measured value of remaining snow.

The actual quantity of snow that melted was larger by about 2.5% (1.5 tons) than the estimated value,

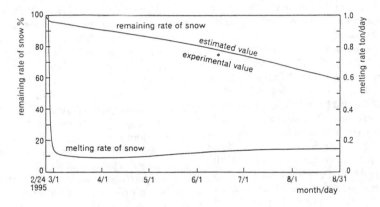

Fig. 2 Estimated and measured values of snow remaining rate.

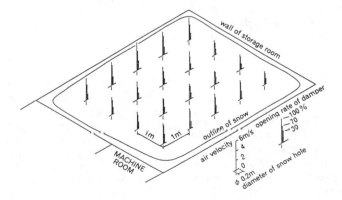

Fig. 3 Distribution of outlet air velocity from the feed pipes in the snow storage room.

35

although it was within the allowable range. The heat transfer design of the snow storage room was generally good, so we judged that snow storage room to be satisfactory for practical use. The average gap between the ceiling and the top surface of the snow was 870 mm; Fig. 3 also shows the average vertical gap between the snow and the wall surfaces of the snow storage tank.

3.3 Distribution of air through the warm air feed pipes and the condition of the holes bored in the snow

Generally speaking, it is difficult to distribute air uniformly, and it is also difficult to bore holes of same shape and dimensions in snow. In order to achieve a uniformity of air distribution, we arranged to install the main pipes of the circulation air ducts for the snow storage room in positions nearly symmetrical with those of the cold air inlet ports installed.

Fig. 3 shows the outlet air velocity for each feed pipe. We set the opening of the circulation air side damper at zero, and varied conditions by changing the opening of the snow storage room side damper. For 100 % air capacity, we opened the snow storage room side damper 100 %. The average diameter of the bored holes is also shown in Fig. 3. Each hole was made with a funnel shape from the top surface to a depth about 1 m down, and below this level each hole was bored as a straight tube. The surface at each hole undulated almost systematically with a pitch of about 100 - 150 mm. The dimensions and shape of the holes were not uniform due to the effects of the initial boring operation, and but refrained from attempting to reshape them in consideration of operational control in a practical use. We observed considerable fluctuations in the outlet air velocity at each pipe outlet due to the influence of the dimensions and shapes of holes, in addition to the effect of the unevenness of distribution of the piping, but noted quite sufficient levels of uniformity with the most frequently used low flow rate.

3.4 Air-conditioning test

We checked the temperature control system and measured the effect of air-conditioning on June 16. It was mostly rainy and the outdoor air temperature was rather low(17.5℃). In consideration of the low temperatures we heated the circulating air using a jet burner (installed in training room A only) to establish a heat load, set the temperature of training room at the fixed temperature of 27℃, and carried out automatic operation of the system (called "automatic operation tests with heat addition"). In the following test, we did not heat the circulating air, but fixed the set temperature of the training room at 18℃, and carried out automatic operation of the system (a "automatic operation test"). The amount of outdoor air bleeded in these tests was zero. In the tests that took place after 18:00, we carried out a self operation test, and simultaneously bleeded outdoor air in an amount of 15 % of the circulating air (a "automatic operation test with bleeding of outdoor air) .

Fig. 4 shows the temperature results of the air-conditioning tests described above and Fig. 5 provides the humidity results.

In the operation test that involved the application of heat, conducted until shortly after 14:00, the heat loads in training room A and training room B were different, and the room temperatures of the two rooms had a large difference in distribution. We observed a difference in the air temperatures at the exits of the two training rooms, but the temperature at the VAV control detector was kept set at 27℃. After this test was completed, the heat loads in training rooms A and B became approximately even in the self operation test, and the temperature at the VAV control detector gradually approached the set temperature of 18℃. The cold air temperature at the outlet from the machine room, that is, the VAV outlet temperature, was controlled to a fixed temperature of 14℃, regardless of fluctuations in temperature of the circulating air. Also, the cold air temperature at the outlet of the snow storage room was low, as we had estimated, and remained stable regardless of fluctuations in the temperature of the circulating air.

In addition, beginning at 18:00, we carried out the automatic operation test with bleeding outdoor air when rain was falling, while the VAV outlet temperature remained stabilized. As shown in Fig. 5, the system did not take a conditions with fog in the rooms.

3.5 Gas absorbability at snow-air containing part

We supplied ammonia gas from the inlet port in the training room, and measured the concentrations of ammonia gas at the outlet of the feed pipe in the snow storage room and at the entrance of the cold air outlet pipe, to measure the gas absorption ability of the bored snow holes. We set the concentration of ammonia gas supplied at the inlet port at 5-30 ppm. The gas absorbability of the snow holes was about 50 %, so we confirmed a high gas-absorption filter effect. However, we did not carry out the forced

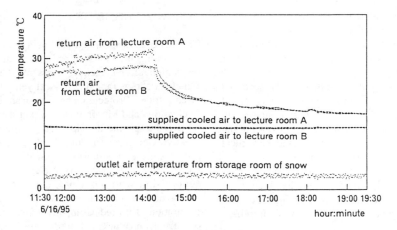

Fig. 4 System operation results (temperature).

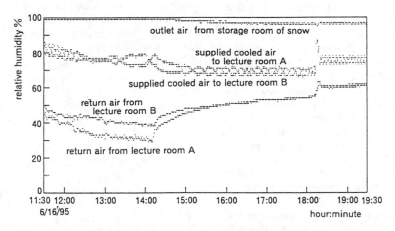

Fig. 5 System operation results (humidity).

heating of air as we did in test done on June 15, and the temperature of the circulating air was low, approximately 15 ℃. Therefore, the rate of melting at the snow surface was lower than that during practical use, and accordingly, the value representing the gas-absorption ability also seemed to be lower than it actually would be in times of practical application.

Here, it is enough to mention that the gas-absorption ability of the snow surface accompanying the melting of the snow surface is high. In the future, we will perform a more accurate evaluation of the gas absorbability and conduct tests using various gases in accordance with the conditions found in the course of practical application of the system, in order to clarify the special characteristics and distinctive features peculiar to this system. Among these are the fact that

the area of snow and air in contact increases as the system is used, and that its capacity for gas-absorption likewise increases.

4 PRACTICAL-USE TESTS

The following summarizes by the practical-use tests carried out from early July to the end of August.

(1) Both the temperature of the cold air supplied and the training room temperatures coincided with the set values, and the temperature control system operated as expected. The humidity inside the training rooms reached a comfortable level of approximately 65%. In addition, the filtering effect of the snow surface received favorable comment, and it

proved to be particularly effective in the removal of cigarette smoke from the room environment.

(2) Temperature at the snow storage room outlet was stabilized at 4-7℃ over the entire period of practical-use testing.

When the melting of the snow had advanced to a certain degree, the sectional shape of each bored snow hole stabilized into a funnel shape, as mentioned in Section 3.3, and the progression of the snow melting process advanced parallel to the vertical while this funnel shape was maintained.

The constancy of the temperature at the outlet of snow storage room is one of the excellent features of this snow air-conditioning system, making it possible for it to be operated with an even more simplified control system.

(3)When we finished using this system the gaps between the snow edges and the side walls was about 35-45 cm at places other than the snow loading inlet, and judged to be close to 50 cm that we had initially estimated. The gap near the snow loading inlet was about 55 cm, so we will need to increase snow density by more thoroughly compacting the snow.

(4) We detected more water than estimated in the mixing box where advance adjustments of the temperature of the cold air to be circulated are made to put it in accordance with the temperature selected, which is accomplished through mixing the warm circulating air from the training rooms with the cold air from the snow storage room. It appears that it will be necessary to install a water eliminator or to adopt another mixing approach at the design stage as needed.

(5) The blower noise level is not particularly low, but it did not interfere with conversations. It may be necessary to adopt dampening device in the design stage if lower noise levels are desired.

5 CONCLUSION

The "air-snow direct heat exchange air-conditioning system" was shown to produce performance levels of a degree we estimated, according to the results of tests conducted from February 24 to June 16 in 1995. We began actual operation of the air-conditioning system on July 10. In the course of this actual operation of the air-conditioning system, we also carried out practical-use testing and produced satisfactory results throughout the summer, confirming that this system can fulfill its function.

ACKNOWLEDGMENTS

The testing was carried out using the snow air-conditioning facility designed and manufactured by the National Land Agency, Yamagata Prefecture and Funagata-cho as their 1994 fiscal year Heavy Snow Area Model Project. The Muroran Institute of Technology and Muroran Heat Pipe Research Group participated in the development of the system and Hokuyu Construction Consultants Co., Ltd. carried out design and supervision work. The Mami Construction & Design Office also provided supervision and control, and the Marumitsu Construction Co., Ltd. and Marusa Housing Equipment Limited Company participated in the construction work. Finally, the Development Division of Sanki Industry Co., Ltd. and the Plant Business Division of Satake Works Co., Ltd. cooperated in providing basic concepts, design, instrumentation and measurement. We wish to express our sincere gratitude to all of the concerned parties mentioned above.

1 Snow technology and science

Snow Engineering: Recent Advances, Izumi, Nakamura & Sack (eds) © 1997 Balkema, Rotterdam. ISBN 90 5410 865 7

Equitemperature metamorphism of snow

Robert L. Brown
College of Graduate Studies, Montana State University, Bozeman, Mont., USA

Michael Q. Edens
Civil Engineering Department, Montana State University, Bozeman, Mont., USA

Michael Barber
Engineering Technology Department, MSU-Northern, Havre, Mont., USA

Atsushi Sato
Shinjo Branch of Snow and Ice Studies, NIED, Yamagata, Japan

ABSTRACT: A mixture theory is developed to analyze the equitemperature metamorphism of snow. The formulation was used to analyze the time dependent changes in grain size distribution and intergranular bonding in the absence of a temperature gradient. Results show that the rate at which this type of metamorphism proceeds is intricately connected to the initial microstructure of the material.

1.0 Introduction

Equitemperature metamorphism is a process which occurs continuously in snow, regardless of whether or not mechanical loads or a temperature gradient is present. The vapor in the pore space will tend to migrate toward the regions with ice surfaces with lower equilibrium vapor pressures. Such surfaces, which we will refer to as low energy surfaces are usually associated with large grains that have large radii of curvature. Necks also in effect have surfaces with large curvatures, since one of the principal radii of curvature is negative (Figure 1). Vapor tends to sublimate off small grains with small radii of curvature and diffuse to larger grains or necks, where it condenses. As a result, small grains and surfaces with small radii of curvature lose mass, while those with large radii tend to gain mass. Due to this process, intergranular bonding increases, and the mean grain size tends to increase as the small grains are sacrificed.

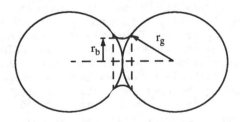

FIGURE 1. Basic neck geometry

Temperature gradient metamorphism occurs only when a temperature gradient is present in the material. Mass is transferred from the warmer regions of the material to the colder regions. This occurs because the vapor pressure decreases as the temperature decreases. Therefore a temperature gradient produces a vapor pressure gradient, resulting in a flow of vapor toward the colder regions within the snow mass. This is supported by the sublimation of mass off of the warmer ice grains, diffusion of vapor toward the colder ice grains, and subsequent condensation of vapor onto the colder ice grains

Even with small-to-moderate temperature gradients, the effects of a temperature gradient will overwhelm the process of equitemperature metamorphism. Equitemperature metamorphism is a very slow process often requiring times on the order of years to produce measurable changes in microstructure. Temperature gradient metamorphism, on the other hand, can require only a few days to produce significant effects on the physical and mechanical properties of snow. Studies by de Quervain (1963), Akitaya (1974), Bradley (1967), Bradley et al. (1977), Colbeck (1980), Adams and Brown (1989, 1990), and many others attest to the rapid changes that can occur when snow is subjected to gradients larger than 10 °C/m. The review paper by Colbeck (1980) provides a historical perspective of past studies of metamorphism of snow.

A number of important situations exist in which temperature gradients are either missing or are very small. One such case involves the slow consolidation of Antarctic firn into ice. Once the firn is buried under a few meters of firn, temperature gradients have been reduced to insignificant levels, and the only important processes affecting the material are mechanical loads produced by the overburden and the process of equitemperature metamorphism. It often requires times on the order of a century for the firn to consolidate into ice, and equitemperature metamorphism is a significant factor in this process. Other situations where equitemperature metamorphism may play a role is in the stability of snow road beds and snow beds upon which temporary structures are built. These regions within a snow mass usually have strong variations in density due to snow processing or consolidation to form the beds. Strong variations in density may cause gradients in surface energy and therefore be responsible for movement of mass from low density areas to high density areas. This is form of equitemperature metamorphism.

2.0 Description of Theory

In this paper, a mixture theory is used to model equitemperature metamorphism. The formulation utilizes a description of the material microstructure to determine the rate at which metamorphism proceeds. The material is modeled as a mixture of vapor occupying the pore spaces, dry air occupying the pore space, ice grains, and ice necks which interconnect the ice grains. The grains and necks are broken into a number of separate constituents. Each grain constituent is assumed to have a mean grain size $r_{i\alpha}$ and an initial mass density $\rho_{i\alpha}$, where the subscript "i" indicates this is an ice constituent, and "α" indicates which one it is ($\alpha = 1,...N$). The neck constituents likewise have an initial mass density $\rho_{i\alpha}$ and an initial bond radius $r_{b\alpha}$. The vapor and air phases also have initial mass densities ρ_v and ρ_a. The time-dependent changes in grain size distribution and neck size distribution are calculated. While this formulation does not include the process of densification due to overburden, continuing studies are now under way to add this to the model. Future studies will also include the effects of layering and temperature gradients.

The details of the mixture theory cannot be presented here. The reader is referred to the papers by Adams and Brown (1990, 1991) and Hansen (1991) for details regarding the development of similar mixture theories. Here we discuss only briefly the results relating to the interchange of mass between the various constituents.

The mixture theory makes use of the principles of balance of mass, linear momentum, angular momentum, and energy. In addition the second law of thermodynamics is used to place restrictions upon the various constitutive relations. This includes the mass interactions $\hat{c}_v$, $\hat{c}_{i\alpha}$, $\hat{c}_a$, which represent the rate at which each constituent is interchanging mass with the others by means of phase changes. The second law can be used to find the following forms for the mass interactions for the ice phases and the vapor phase:

$$\hat{c}_{i\alpha} = m_{\alpha v}(\Delta_v - \Delta_{i\alpha}), \ (\alpha = 1, ..., n)$$

$$\hat{c}_v = -\sum_{\alpha = 1}^{n} m_{\alpha v}(\Delta_v - \Delta_{i\alpha}) \tag{1}$$

where

$$\Delta_{i\alpha} = e_{i\alpha} - \mu_{i\alpha}$$
$$\Delta_v = e_v - \mu_v \tag{2}$$

$\Delta_{i\alpha}$ and Δ_v will be referred to as the potential differences. The term $\mu_{i\alpha}$ is the chemical potential which is related to the Helmholtz free energy $\psi_{i\alpha}$ by the following relationship:

$$\mu_{i\alpha} = \frac{\partial}{\partial \rho_{i\alpha}}(\rho_{i\alpha}\psi_{i\alpha}) \tag{3}$$

The term $e_{i\alpha}$ is the internal energy for each ice constituent. The vapor phase has analogous variables.

The mass supply can therefore be seen to be determined directly from the free energy. The expressions $\Delta_{i\alpha}$ and Δ_v represent the differences between the internal energy and the chemical potential. This difference then governs the rate at which mass exchanges take place. The above results, while somewhat surprising, are direct results of the restrictions imposed by the second law of thermodynamics and therefore must be observed. As will be seen later, this variable is partially determined by the surface energy of each constituent.

By examining Equation (1), one can see that each ice constituent exchanges mass only with the vapor phase, while the vapor phase can exchange mass simultaneously with all ice constituents. This assumption is a realistic one, since mass is sublimated off an ice surface into the pore space occupied by the vapor, while other grains may be acquiring mass by having vapor condensed onto their ice surfaces. It is the vapor phase, therefore, that indirectly provides for mass leaving one ice grain and eventually being deposited upon another ice surface. This representation of the mass interaction process precludes transfer of ice from one grain directly to another by volume diffusion or surface diffusion through the necks from one grain to another. Maeno (1983) has demonstrated that vapor diffusion dominates under most situations.

The energy exchange is assumed to have the form:

$$\hat{e}_{i\alpha} = K_{i\alpha}(\theta_v - \theta_{i\alpha})$$
$$\hat{e}_v = \sum_{\alpha = 1}^{n} -K_{i\alpha}(\theta_v - \theta_{i\alpha}) \tag{4}$$

where $\theta_{i\alpha}$ and θ_v are the absolute temperatures for the ice and vapor phases. This is a form similar to that used earlier by Adams and Brown (1990) and is essentially Newton's law of cooling and will account for the interchange of thermal energy between constituents. Here, we neglect the direct interchange of energy between ice grains and assume energy is interchanged directly between the ice constituents and the vapor phase. This is reasonable if the intergranular bonding is typical for snow of low-to-medium density. One can verify that the second law is satisfied by this expression.

The balance equations reduce to the following expressions:

Balance of mass:

$$\frac{\partial \rho_{i\alpha}}{\partial t} = m_{\alpha v}(\Delta_v - \Delta_{i\alpha}), \alpha = 1, \dots n$$
$$\frac{\partial \rho_v}{\partial t} = -\sum_{\alpha = 1}^{n} m_{\alpha v}(\Delta_v - \Delta_{i\alpha}) \tag{5}$$

Balance of energy:

$$\rho_{i\alpha}\frac{de_{i\alpha}}{dt} = K_{i\alpha}(\theta_v - \theta_{i\alpha}), \alpha = 1, \dots n$$
$$\rho_v\frac{de_v}{dt} = \sum_{\alpha = 1}^{n} -K_{i\alpha}(\theta_v - \theta_{i\alpha}) \tag{6}$$

Balance of linear momentum does not need to be considered, since all velocities vanish for the case of a homogeneous material with no loads or temperature gradients applied to the material. Each particle in the vapor phase does have a nonzero velocity vector, but the bulk material does not in the sense that the averaged velocity vectors of the particles within any differential volume element will sum to zero, giving a zero macroscopically measured velocity.

3.0 The Microstructure

The effect of surface tension on the material stress and deformation will first be considered. It is well understood that the surface tension in the ice grains will attempt to compress the material to a higher density to reduce the surface energy within the material. We start by considering the case when the strain is maintained at zero and consider the stress that must be developed within the material to keep the strain tensor $E_{i\alpha} = 0$. If there are no deformations, temperature changes or mass exchanges the stress tensor can be shown to be:

$$T_{i\alpha} = \sigma_{0\alpha}I \tag{7}$$

Figure 2 illustrates conceptually the internal stresses within the ice grains that will be developed due to the presence of the surface tension on the surface of each grain.

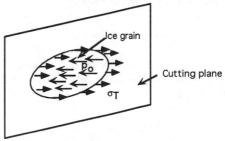

FIGURE 2. Ice grain inner surface exposed by a cutting plane through the material

If the internal stress must balance the surface tension σ_T, we must have for the internal stress $p_{o\alpha}$ within each ice grain:

$$p_{o\alpha} = \frac{2\sigma_T}{r_{i\alpha}} \tag{8}$$

Since the surface fractions on any plane cut through a material is proportional to the mass fractions if the material is homogeneous and isotropic, we can then verify that $\sigma_{o\alpha} = (\rho_{i\alpha}/\rho)p_{o\alpha}$ so that:

$$\sigma_{o\alpha} = \frac{\rho_{i\alpha}}{\rho}\frac{2\sigma_T}{r_{i\alpha}} \tag{9}$$

This then gives the partial stress of the ice constituent, i.e., the contribution to the total stress developed within the snow mixture.

Now determine how the surface energy of each constituent determines the rate in which they exchange mass with the vapor phase. We assume that there are no point sources of mass within the mixture. Therefore, for the N ice constituents and the vapor phase, we must have:

$$\sum_{\alpha=1}^{N} \hat{c}_{i\alpha} + \hat{c}_{v} = 0 \tag{10}$$

We can verify that:

$$\hat{c}_{i\alpha} = m_{\alpha v}(\Delta_v - \Delta_{i\alpha}), \ (\alpha = 1, \ldots, n)$$

$$\hat{c}_v = -\sum_{\alpha=1}^{n} m_{\alpha v}(\Delta_v - \Delta_{i\alpha}) \tag{11}$$

The free energy, entropy, chemical potential and internal energy are all reckoned per unit mass, whereas the surface energy $\sigma_{i\alpha}$ is reckoned per unit area of ice surface. Let $\Sigma_{i\alpha}$ be the surface energy per unit mass. Then we can write:

$$\sigma_s(4\pi r^2) = \Sigma_s\left(\frac{4}{3}\pi r^3 \gamma_i\right) \tag{12}$$

where γ_i is the mass density of ice. One can show after some algebra that in the reference configuration we have

$$(\Delta_v - \Delta_{i\alpha})_R = \frac{\sigma_s}{4\gamma_i r_{i\alpha}} \tag{13}$$

for the difference between the potential differences for the vapor and the αth ice constituent. Therefore, even in the reference configuration, we can have mass exchanges, since their potential differences are not necessarily equal. What will happen is that when we start with an initial state consisting of a collection of ice grains of different sizes and a vapor phase, the ice grains will lose mass to the vapor until its energy begins to exceed the energies of some of the ice grains. Then the vapor will begin to supply mass to those grains with the lower energies, while the grains with the higher energies will continue to lose mass to the vapor. The vapor will reach a stable configuration with an energy level where it is continuously acquiring mass from some grains while at the same time giving up mass to the grains.

Now characterize the geometry of the necks. To do this consider two ice grains in direct contact with each other, as shown in Figure 3. In this figure, the neck will have two surface curvatures. r_b is the positive radius of curvature describing the bond

radius. Typically this will be anywhere between ten percent and seventy percent of the grain radius. The other curvature, r_c, is the concave curvature seen in Figure 3 and forms due to either pressure sintering, vapor diffusion to the neck or diffusion of ice molecules through the surface or volume of the ice grains to the necked region. The concave radius, r_c, will always represent negative curvature but is given a positive algebraic value in this work. Therefore the mean curvature of the neck is given by the following relation, where the minus sign reflects the negative curvature.

$$\frac{2}{r_n} = \frac{1}{r_b} - \frac{1}{r_c} \tag{14}$$

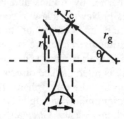

FIGURE 3. Neck geometry showing the negative curvature, r_c and other dimensions of the neck, r_b, l, and r_g.

From the above figure, the following relations can be found.

$$l = 2r_g\left[1 - \frac{r_g}{r_c + r_g}\right] \tag{15}$$

$$r_c = \frac{r_b^2}{2 \cdot (r_g - r_b)} \tag{16}$$

The surface area of a neck will be approximated by the relation:

$$\Sigma_n = 2\pi r_b l \tag{17}$$

An important factor that determines how rapidly the various ice constituents interchange mass is the free surface areas $S_{i\alpha}$, since available free surface area is needed to facilitate the sublimation and condensation of vapor. These values are determined by the density and the radii of curvature. Approximate values can be found by the relations:

$$S_{i\alpha} = \frac{3\rho_{i\alpha}}{\gamma_i r_{i\alpha}}, \quad \text{ice grains}$$

$$S_{i\alpha} = \frac{2\rho_{i\alpha}}{\gamma_i r_{i\alpha}}, \quad \text{necks} \tag{18}$$

The term γ_i is the density of ice.

4.0 Examples Of Equitemperature Metamorphism

4.1 Five Grain Sizes With Equal Number Densities

In this example we consider the metamorphism of snow consisting of a distribution of initial grain sizes in which the grain size distribution varies over more than an order of magnitude, from 2.5×10^{-4} m to 5.0×10^{-3} m. In this case the intergranular bonding is not included in the analysis, as we are just concerned about the affect to the ice grains on each other and the interchange of mass between the different ice grains. The ice grains are broken into five constituents, with mean grain sizes of 0.25, 0.5, 1.0, 2.5, and 5.0 mm radii of curvature. The temperature is assumed to be -10 °C.

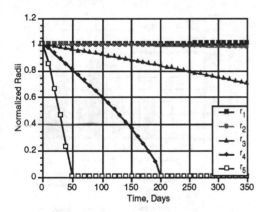

FIGURE 4. Variation of radii when all constituents have equal number densities.

In this example we assume the number density (number of particles per unit volume) is the same for all constituents. As a consequence, the mass density increases with grain size. The initial densities are shown in Table 1. In addition, the free surface area per unit volume also increases with grain size. To have snow with a density approaching 250

kg/m^3 and consisting of these grain sizes occurring in equal number densities, the number density of all constituents would equal approximately 420,000 particles per cubic meter. Table 1 contains information about these parameters.

TABLE 1. Microstructure for Example 4.1.

Constituent No.	Initial Grain size mm	Initial Density kg/m^3	Initial Surface Area, m^{-1}	Number Density, m^{-3}
1	5.00	200	1,310	420,000
2	2.50	25	327	420,000
3	1.00	1.60	52.3	420,000
4	0.50	0.20	13.1	420,000
5	0.25	0.0025	3.27	420,000

As can be seen in figure 5, the largest grains (constituent #1) grow at the expense of all of the smaller ones (constituents #2,3,4,5). The smaller sized constituents have significantly smaller mass densities and smaller total surface areas. As a consequence the constituent with the largest surface area per unit volume (constituent #1) dominates the mass exchange process, and all constituents give up mass to this one constituent.

4.2 Five Grain Sizes With Equal Mass Densities

If on the other hand, we had assumed that all of the constituents have the same initial mass density rather that the same number density, the result is substantially different. Assume the five constituents each have a density of 50 kg/m^3. In this case, the free surface areas of the constituents increase as the grain sizes decrease, so that the smallest grain-sized constituent has substantially more surface area than the largest grain-sized constituent.

A significantly different patter emerges, as may be seen by comparing Figures 4 and 5. The constituent #4 actually gains mass by interacting with constituent #5 until that constituent is essentially depleted. These two constituents interact in such a manner, since their surface areas per unit volume are much larger than that of the other constituents. After constituent #5 is almost depleted, constituent #4 begins to lose mass to constituent #3 because it has more surface area that either constituents #1 or #2. While the larger constituents do gain mass, the activity between the smaller grains is much more substantial, since they have considerably more surface area to support mass exchange.

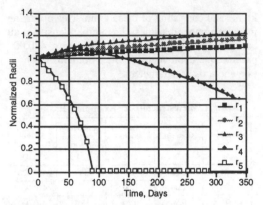

FIGURE 5. Variation of radii when all constituents have equal mass densities.

TABLE 2. Microstructure for Example 4.2

Constituent No.	Initial Grain size mm	Initial Density kg/m³	Initial Surface Area, m⁻¹	Number Density, m⁻³
1	5.00	50.0	327	1.0×10^6
2	2.50	50.0	654	8.3×10^6
3	1.00	50.0	1,640	1.3×10^8
4	0.50	50.0	3,270	1.0×10^9
5	0.25	50.0	6,540	8.3×10^9

In actuality, neither of these models could be considered to be very realistic. Until image analysis systems can determine actual gain size distributions, the two examples are primarily of value in expanding our understanding of the physics involved in metamorphism processes. A more realistic model might be to assume a normal distribution of grain sizes centered about the mean grain size.

4.3 Effect of Grain Size Distribution.

In the previous examples the snow had a maximum grain size of about 5 mm, which makes it large-grained snow. We now consider what would happen if the distribution of grain sizes was altered by making each constituent smaller by a factor of 10, so that the grain size for the five constituents would range from 0.025 mm to as large as 0.5 mm.

Snow with this size distribution will be termed fine-grained. Obviously the free surface area available for mass exchanges is increased substantially, and hence one would expect the redistribution of grain sizes to proceed much more rapidly in fine-grained snow. For this example, we assume that

each constituent has equal initial densities of 50 kg/m³ as in Example 4.2. Each constituent in the fine-grained snow will have ten times the free surface area as each corresponding constituent for the large-grained snow discussed in Example 4.2.

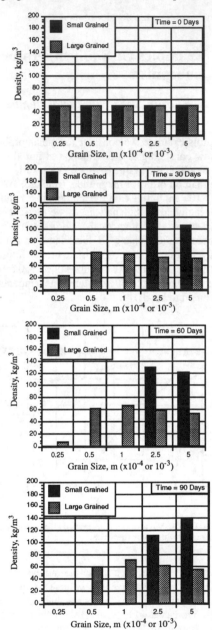

FIGURE 6. Effect of initial grain size

Figure 6 illustrates the relative rate of metamorphism for the large-grained snow and the fine-grained snow. The figure depicts the density distributions for the two snows at given times. The horizontal axis represents the initial grain sizes, though it must be kept in mind that the initial dimensions for the two snows differ by a factor of ten. The first figure shows the distribution at the initial time when all constituents have densities of 50 kg/m^3. By 30 days, the two smallest constituents in the fine-grained snow have completely disappeared. In contrast to this, the large-grained snow has gone though a relatively minor change in grain size distribution. The remaining two parts of the figure are for 60 and 90 days.

4.4 Snow With Intergranular Necks

We consider now an extension of the first example in which we assume a distribution of both grain sizes and neck sizes in which the necks are taken to be uniformly distributed in number density and have dimensions which initially are ten percent of the corresponding grain sizes. Repeating the calculations would produce grain size variations which are nearly identical to what is given in Figure 4. The presence of the necks connecting the grains has very little effect upon the changes in the grain sizes. This is due to the fact that the free surface area of the necks are so small compared to the ice grains that their effect is insignificant. The necks, however do experience significant growth.

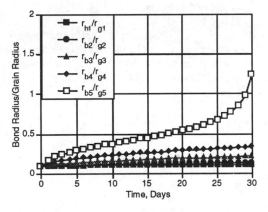

FIGURE 7. Growth of bond size relative to the grain size.

Figure 7 demonstrates the time-wise variation of the bonds in the material. The bond radii were assumed to initially be one tenth the initial radii of the ice grains. As can be seen, all of the bonds grow fairly rapidly, with the smallest bonds growing fastest, since they will have the most negative neck radii of curvature r_n given by Equation 14. The effective neck radius r_n will always be larger than the radius of the grain to which it is attached. The neck will therefore continue to grow relative to the grain size, although the rate will decrease as the ratio r_n/r_g exceeds the value of 0.65.

5.0 Conclusions

In this paper, equitemperature metamorphism was evaluated by modeling snow as a mixture of constituents with different grain sizes. The mixture theory utilizes the material microstructure to determine just how the constituents will exchange mass. The rate of mass exchange is determined by differences between the values of the potential differences $\Delta_{i\alpha} = e_{i\alpha} - \mu_{i\alpha}$ and $\Delta_v = e_v - \mu_v$. These potential terms for the ice constituents and the vapor contain terms which reflect the surface energy of the material, so that high energy surfaces will lose mass to the vapor. The vapor is seen to gain mass from high energy surfaces and to lose it to low energy surface with large radii of curvatures. The results of the mixture theory were determined by utilizing the second law to place restrictions upon these processes.

In one example the snow was assumed to be represented by a distribution of five grain sizes ranging form 0.25 mm to 5.0 mm with all five constituents having equal particle number densities (number of grains per unit volume). In the second example the same grain size distribution was used but with each constituent having the same mass density. Neither of these distributions could be considered to be very realistic, but these distributions were chosen to demonstrate the effect of microstructure on the process of equitemperature metamorphism. The calculated results for both examples show that the mean grain size increases with time but that rates are different. For the first example, the two smallest grain sizes (0.25 mm and 0.5 mm) are rapidly eliminated, since their mass densities are quite small. Both results do confirm, however, that the process is a slow one requiring years to produce a uniformly sized grain structure.

In a third example, the effect of grain size was investigated. Snow with a grain size distribution of 0.25 mm to 5.0 mm was evaluated and compared to snow with a grain size distribution of 0.025 mm to 0.5 mm. Both samples had a grain size distribution with the constituents having equal initial mass densities. The small-grained snow, of course, had ten times the surface area per unit volume and hence was able to experience a much faster rate of metamorphism. Within thirty days the two smallest sized constituents (0.025 mm and 0.05 mm) were completely eliminated, whereas for the large-grained snow, only a modest decrease in density for the smallest constituent could be observed, while the second smallest constituent had experienced only a very small increase in density. By 90 days only the two largest sized constituents were left for the small-grained snow. For the large-grained snow, only the smallest constituent had been depleted at 90 days.

In the final example, the snow was modeled by a grain size distribution ranging from 0.25 mm to 5.0 mm with equal number densities. In addition a distribution of necks with five bond sizes ranging from 0.025 mm to 0.5 mm was added as five additional constituents. The results show that the presence of the necks have little effect on the growth or decrease of the grains, since all but the smallest sized grains have significantly larger surface areas and mass densities than the necks. The bonds are all seen to grow, since their effective surface curvatures are negative, thereby giving them low energy surfaces which will collect mass from the vapor phase. The rate of sintering is seen to decrease as the neck sizes grow. The smaller necks are seen to sinter more rapidly than the larger necks.

The results of this study indicate that the use of mixture theories which include the microstructure of the grains and the connecting necks can be of use of characterizing metamorphism of snow. However, more precise descriptions of the initial grain size and bond size distributions are needed in order to use these theories, since the microstructure apparently affects the metamorphism process. With the recent improvements of stereological theory or the use of sieving methods to obtain grain size distributions, such information on grain size distributions may soon be available. This data could then be used to more accurately validate metamorphism models such as the one presented in this paper.

Acknowledgment: The research reported here has been funded by the terrestrial Sciences Program of the Army Research Office. the authors wish to express their appreciation to the /Army Research Office for this support.

6.0 References

Adams, E. E. and R. L. Brown, 1989. A continuum theory for snow as a multiphase mixture, *Journal of Multiphase Flow*, Vol. 15, pp. 553-572.

Adams, E. E. and R. L. Brown, 1990. A study of heat and mass transport in non homogeneous snow cover, *Journal of Continuum Mechanics and Thermodynamics*, Vol. 2, pp. 31-63.

Akitaya, E., 1974. Studies of depth hoar, *Low Temperature Science, Vol. A*, No. 26, pp. 1-67.

Bradley, C. C., 1967. Metamorphism of dry and wet snow, *Proceedings of the USDA Western Snow Conference.*

Colbeck, S. C., 1980. Thermodynamics of snow metamorphism due to variations in curvature, *Journal of Glaciology*, Vol. 26, No. 94, pp. 291-301.

de Quervain, M. R., 1963. On the metamorphism of snow, Ice and Snow: Properties, Processes, and Applications: *Proceedings of a Conference held at Massachusetts Institute of Technology*, Cambridge Press, pp. 203-226.

Hansen, A. C. (1991). reexamining some basic definitions of modern mixture theory, *International Journal of Multiphase Flow*, Vol. 27, No. 12, pp. 1531-1534.

Maeno, N. and T. Ebinuma, 1983. Pressure sintering of ice and its implication to densification of snow at polar glaciers and ice sheets, *Journal of Physical Chemistry*, Vol. 87, No. 21, pp. 4103-4110.

Snow Engineering: Recent Advances, Izumi, Nakamura & Sack (eds) © 1997 Balkema, Rotterdam. ISBN 90 5410 865 7

Heat and mass transport in snow under a temperature gradient

S.A. Sokratov & N. Maeno
Institute of Low Temperature Science, Hokkaido University, Sapporo, Japan

ABSTRACT: An experiment on heat and mass transfer in snow under a temperature gradient was carried out in the laboratory. Opposite ends of snow samples were kept at different temperatures for prolonged periods of time. The time to achieve a quasi–steady temperature distribution varied from about 10 hours to about 40 hours depending on the sample length. It was found that quasi–steady temperature distributions were not linear but that the temperature gradient decreased in the direction of heat flow. This result suggests that heat transport cannot be explained as a pure conduction mechanism in snow having uniform thermal conductivity, and that the contributions of water vapor transport and phase change can be important to different degrees in snow.

1 INTRODUCTION

Heat and mass transport in snow is one of the most essential processes characterizing a variety of snow properties, which play important roles in snow technology and science. Among various physical quantities the effective thermal conductivity k_e [$W \cdot m^{-1} \cdot K^{-1}$] and effective water vapor diffusivity D_e [$m^2 \cdot s^{-1}$] of snow are the most important and useful.

However the obtained values of k_e and D_e in previous investigations vary in wide ranges. It is usually accepted that k_e is almost independent of temperature (only 0.17 % change per °C) and increases with increasing density (0.02 – 0.07 $W \cdot m^{-1} \cdot K^{-1}$ at 100 $kg \cdot m^{-3}$ and 0.35 – 0.7 $W \cdot m^{-1} \cdot K^{-1}$ at 400 $kg \cdot m^{-3}$ (Maeno and Kuroda 1986; Fukusako 1990)), but the physical understanding of the heat transfer mechanism is not complete (Yosida 1955; Giddings and LaChapelle 1962; Yen 1965). Most investigators agree that the value of D_e is about 5 times larger than that in air. Observed numerical values are 0.65 $m^2 \cdot s^{-1}$ (Yen 1963) – 0.85 $m^2 \cdot s^{-1}$ (Yosida 1950), but the conclusions given by various authors are not in agreement (Voitkovskiy et al. 1988; Colbeck 1993).

The poor agreement in published data forces to analyze how the values of k_e and D_e were obtained. The first step in common way of experimental determination of these coefficients was to consider heat and mass fluxes in snow as independent from each other. Then the Fourier's and Fick's Laws could be easily used for final calculations on base of obtained temperature and density data. An assumption was made almost in all the works —both the effective heat conductivity and effective water vapor diffusivity were considered as uniform through snow. A similar assumption was usually made in theoretical modeling of heat and mass transfer.

The measurable data for such experiments could be temperature distributions in snow and density change with time. Use of the above assumption forces to accept the linear temperature distribution in snow in conditions of steady state, or at least negligibly small difference from linear temperature distribution sometimes used in theoretical modeling (Albert and McGilvary 1992) and evaporation condensation processes in the space between neighboring grains (Colbeck 1993). Such conditions for condensation–evaporation processes mean no density variations caused by water vapor flux.

The measured temperature distributions and density variations in snow show that the temperature gradient could vary considerably through snow samples (Kondrat'eva 1954; Voitkovskiy et al. 1988) and that the density change took place (Yosida et al. 1955; Voitkovskiy et al. 1988). There are two explanations for the non-linearity of temperature distribution in snow except possible experimental errors — densification of snow related with recrystallization of snow (Kondrat'eva 1954), and air flux in snow caused by the pressure difference over snow surface (Colbeck 1989).

The purpose of the present study was to make a systematic investigation of heat and mass transport in snow and to construct a model for the process. This paper gives preliminary results of an experiment in which time variations of temperature distributions in snow were measured.

2 EXPERIMENTAL APPARATUS AND PROCEDURE

Heat and mass transfer was studied by measuring temperature distributions after sudden heating of one end of a snow sample and keeping the applied temperature difference for a prolonged period of time. The density change in snow was also measured by weighing thin slabs of snow cut perpendicular to the direction of heat and water vapor flow after the experiment.

Several runs of the experiment were done. Samples were natural blocks of snow with a density of about $350 \ \mathrm{kg \cdot m^{-3}}$, screened compacted snow with density about $500 \ \mathrm{kg \cdot m^{-3}}$ and, screened new snow with a density of about $200 \ \mathrm{kg \cdot m^{-3}}$.

The blocks of naturally compacted snow were cut from bigger pieces of snow to have dimensions available for experimental set–up. Screened snow samples were made by sifting naturally compacted snow through 0.84 mm sieve. New snow was sifted through 1.19 mm. Sifting was done into plastic boxes which were set aside for one day during which sintering took place, forming blocks of snow. Then the plastic walls perpendicular to the direction of heat flow was removed on one end of each sample.

The experimental set–up was similar in principle to that used by Yosida et al. (1955) and Voitkovskiy et al. (1988), but several developments have been done. The horizontal plan of the set–up is shown in Figure 1. It consisted of a temperature–controlled plywood box in a cold room at -19 °C, in which four snow samples with lengths of 10, 20, 30 and 40 cm were placed. The cross section of each snow sample perpendicular to the direction of heat flow was 18 cm×18 cm. One end of each sample (without wall) was in contact with a brass tank filled with kerosene and the opposite end was exposed to the surrounding cold air in the plywood box. Actually a thick metal plate was applied on the cold end to decrease possible influence of temperature variations in the plywood box. All the other sides of the snow samples were thermally insulated with 5 cm thick foam plastic plates. The temperature inside the plywood box was kept constant by two fans and a thermostatically controlled electric heater.

The brass tank was connected by thermoinsulated tubes with a thermostat situated

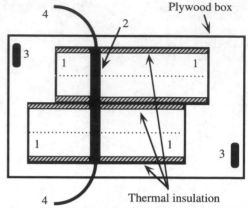

Figure 1   Horizontal plan of the experimental device in a cold room
1 — four snow samples of different lengths; 2 — brass tank; 3 — fans; 4 — thermoinsulated tubes circulating kerosene; dotted lines — positions of thermocouples.

outside the cold room. The thermostat could control the temperature of the circulating kerosene within about ±0.5 °C. An additional brass tank of the same dimensions as that in contact with the snow samples was situated in the coldroom and connected to the same thermostat. The kerosene was circulated through this tank for one day before the experiment to stabilize the kerosene temperature.

Temperatures in the snow samples were measured by 0.2 mm copper–constantan thermocouples installed with minimal snow disturbances along the central horizontal axes parallel to heat flow. Five to nine thermocouples per sample, depending its length, were inserted into the snow through the thermal insulation. Small vertical holes were drilled to the depth of 9 cm in each sample. The temperature data were recorded by a data logger situated outside the coldroom. The logger was manipulated by a computer and the data were stored on floppy disks. The temperature measurements were made every 1 minute during the first day of the first experimental run and every 10 minutes thereafter because temperature variations, by then, were within the measurements accuracy of ±0.1 °C.

The experimental procedure was as follows. After four snow samples were set in the experimental apparatus and the temperature became uniform in the whole set–up (it took about one day), kerosene circulating in a warmer tank was connected to the brass tank contacting the four samples in plywood box. Then the temperatures inside each snow sample were measured for about two weeks.

After the experiment the snow samples were cut perpendicular to the direction of heat flow and the density of each of the 1 – 2 cm thick slabs was measured. Sample cutting was done by tensioned nichrome wires heated electrically. Initial density had been measured before the experiment by weighing the sample or several snow slabs from similar sample prepared by the same way. Thin sections of samples were also prepared for microscopic observations, but the results of the density and thin section analyses are not included in this paper.

3 EXPERIMENTAL RESULTS

An example of temperature data obtained is shown in Figure 2 for a 20 cm sample. The initial density was $350 \ kg \cdot m^{-3}$. In this run when the temperature of the whole system became -15 °C, kerosene at temperature -1 °C was circulated. It is seen that temperatures in the snow sample increased with time and reached a steady state in roughly 10 hours. The steady state lasted until the end of the experimental run (14 days).

Figure 3 shows the temperature distributions in the same snow sample at several measurement times from the beginning of -1 °C kerosene circulation. As already noted in Figure 2, the temperature change after about 10 hours was very slow, and practically no variations were measured after 17 hours. Thus a thermal steady state had been attained in the sample and the heat flux might be the same at any points in the sample. The steady state

temperature distribution was not linear as it might be in materials with uniform thermal conductivity, but gave a curved trend. We call such a steady state temperature distribution the *quasi–steady distribution* to distinguish it from the linear one.

Quasi–steady temperature distributions for various length samples are shown in Figure 4. For this experimental run (naturally compacted snow with a density of about $350 \ kg \cdot m^{-3}$) the time from the beginning of the experiment to when temperature distributions could be considered to the quasi–steady state was 12 hours for the 10 cm long snow sample, 17 hours for the 20 cm sample, 21 hours for the 30 cm sample and 39 hours for the 40 cm sample. The bars show the range of temperature variation during the quasi–steady period of measurements. It is recognized that all the four quasi–steady temperature distributions are not linear but convex toward the warmer side. The result suggests that the thermal steady state in snow cannot be explained by thermal conduction only because the difference is much higher than that which can be explained by the temperature dependence of thermal conductivity of ice. Thus the phase change and water vapor transport in snow might be important in the establishment of the steady temperature distribution in snow.

In all other experimental runs at different temperature gradients and at densities from $200 \ kg \cdot m^{-3}$ to $500 \ kg \cdot m^{-3}$, the general character of temperature distributions was similar to that described above.

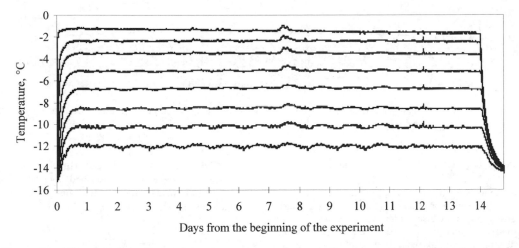

Figure 2 Time variation of temperature in a snow sample (naturally compacted snow, 20 cm length, density $350 \ kg \cdot m^{-3}$)

The positions of temperature measurements are (from top to bottom): 0, 1, 3, 5, 8, 11, 15 and 19 cm from the warm end.

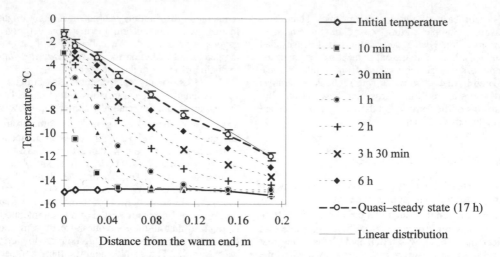

Figure 3   Time variation of temperature distribution in a snow sample (the same run as Figure 2)

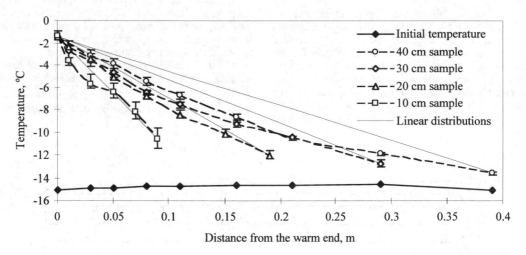

Figure 4   Quasi–steady temperature distributions for various length samples (naturally compacted snow, density 350 kg·m⁻³)

4 DISCUSSION AND CONCLUSIONS

Our experiment showed that when the ends of a snow sample were kept at different temperatures, a quasi–steady temperature distribution was established within a relatively short period of time and did not change noticeably during the next two weeks. The quasi–steady temperature gradients were not uniform but diminished toward the colder end.

We assume that the heat flux (q) at the quasi–steady state can be described by the local temperature gradients, and then the effective thermal conductivity of snow (k_e) can be defined as:

$$q = -k_e \frac{dT}{dy}, \qquad (1)$$

where T is the temperature and y is the coordinate.

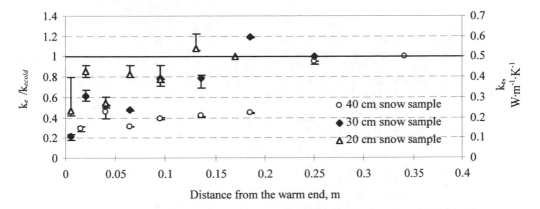

Figure 5   Normalized effective thermal conductivity for various length samples (naturally compacted snow, density 350 kg·m^{-3})

k_e — effective thermal conductivity; k_{ecold} — k_e at the cold end.

Equation 1 is often written in the form (Colbeck 1993):

$$q = -k_{sn}\frac{dT}{dy} + F\left(L + c_p\left(T_0 - T\right)\right), \qquad (2)$$

where k_{sn} is the pure thermal conductivity of snow and the second term is the heat due to water vapor transfer in snow: F — water vapor flux, L — latent heat of sublimation, c_p — heat capacity of water vapor, T_0 — temperature of the warm end of snow. As water vapor flux is also dependent on the temperature gradient, k_e includes all heats transported by conduction, water vapor diffusion and evaporation–condensation.

Values of k_e at each point were calculated from the temperature distributions in Figure 4, normalized by those at the colder end (k_{ecold}), and plotted against the distance in Figure 5. Though the data scatter is appreciable the values of normalized k_e are mostly smaller than unity and increase towards the colder parts. The increase is too large to be explained by the larger thermal conductivity of ice at lower temperatures: the thermal conductivity of ice at -15 °C is not more than 4 % larger than that at -1 °C. The additional axis was given to show rough magnitudes of the effective heat conductivity of snow samples.

As the structure and thermal conductivity of the initial snow sample before the experiment were uniform, the obtained difference in k_e at the warmer and colder parts implies that some structural and physical processes were in progress. The growth of bonds between ice grains due to sintering, which will increase the value of k_e, is faster at the warmer parts of samples. The second term in Equation 2 decreases towards colder parts. These considerations lead to some other mechanism of mass and heat transfer. The probable effect is heat transport through evaporation–condensation between ice grains. The density measurements and structure inspection of samples after the experiment are in progress to give evidence of the effect suggested above, and the results will be combined with the present measurement to give a more detailed description of heat and mass transport in snow.

ACKNOWLEDGMENTS

The study was partly supported by the special fund for Scientific Research of the Ministry of Education, Science and Culture, Japan. The study of one of the authors (S. A. Sokratov, Institute of Geography, Russian Academy of Science, Moscow, Russia) in Japan was made possible by the Japanese Government (Monbusho) Scholarship. The authors are grateful to the following individuals for various assistance: M. Arakawa, H. Narita, K. Nishimura of Institute of Low Temperature Science, Hokkaido University; S. Fukusako of Department of Mechanical Engineering, Hokkaido University. Finally the authors are grateful to W. Tobiasson, CRREL and R. Decker, University of Utah for their useful comments given to the manuscript.

REFERENCES

Albert, M.R. and McGilvary W.R. 1992. Thermal effects due to airflow and vapor transport in dry snow. *J. Glaciol.* 38 (129): 273–281.

Colbeck, S.C. 1989. Air movement in snow due to windpumping. *J. Glaciol.* 35 (120): 209–213.

Colbeck, S.C. 1993. The vapor diffusion coefficient for snow. *Water Resources Research.* 29. 1: 109–115.

Fukusako, S. 1990. Thermophysical properties of ice, snow, and sea ice. *International Journal of Thermophysics.* 11. 2: 353–372.

Giddings, J. and LaChapelle, E. 1962. The formation rate of depth hoar. *J. Geophys. Res.* 67. 6: 2377–2383.

Kondrat'eva, A.S. 1954. Thermal conductivity of the snow cover and physical processes caused by the temperature gradient. *SIPRE, Translations* . 22. 13 pp.

Maeno, N. and Kuroda, T. 1986. *Fundamental Glaciology I. Structural and Physical Properties of Snow and Ice.* Tokyo: Kokon Shoin. 208 pp.

Voitkovskiy, K.F., Golubev V.N., Sazonov A.V. and Sokratov S.A. 1988. New data on diffusion coefficient of water vapor in snow. *Data on Glaciological Studies.* 63: 76–81.

Yen, Y.-C. 1963. Heat transfer by vapor transfer in ventilated snow. *J. Geophys. Res.* 68. 4: 1093–1101.

Yen, Y.-C. 1965. Effective thermal conductivity and water vapor diffusivity of naturally compacted snow. *J. Geophys. Res.* 70. 8: 1821–1825.

Yosida, Z. 1950. Heat transfer by water vapour in a snow cover. *Cont. Inst. Low Temp. Sci.* 5: 93–100.

Yosida, Z. and Colleagues 1955. Physical studies of deposited snow. I. Thermal properties. *Cont. Inst. Low Temp. Sci.* 7: 19–74.

Snow Engineering: Recent Advances, Izumi, Nakamura & Sack (eds) © 1997 Balkema, Rotterdam. ISBN 90 5410 865 7

Numerical study of a snow wind scoop

S. Kawakami, T. Uematsu, T. Kobayashi & Y. Kaneda
JWA, Sapporo, Japan

ABSTRACT : A three-dimensional simulation method for blowing snow was developed by Uematsu and others (1991). Using this simulation method for blowing snow and snowdrift, a wind scoop in snow around a hut was studied. The snow wind scoop was observed by setting a hut in an open snow field. The simulation model reconstructing the wind scoop digitally from meteorological data. The calculated snow depth by the simulation method was compared with the observed snow depth. The computational results showe a good agreement with the observed results. In order to investigate the influence of parameters for the formation of the wind scoop, the model was run by changing some parameters such as wind speed, wind direction and threshold friction velocity. As a result, the wind scoop cannot be created for wind speeds larger than about 20 m/s and for a threshold friction velocity smaller than about 0.1 cm/s.

1 INTRODUCTION

In snowy regions, "wind scoops" can be seen frequently. The definition is not clear, however Seligman (1980) described it as follows:

"Eddies often assume a cockscrew-like shape to leeward on each side of an obstruction, one on one side revolving right-handsided and the other left-handed. The line between the two eddies is one of deposition, so that rock and the like are frequently found with twin depressions separated by a strip of deposited snow."

Figure 1 shows the sketch of a wind scoop made by Seligman (1980).

Some investigators became interested in the wind scoop, and had observations of it and described it. For example, Kobayashi (1979) investigated the flow dynamics that caused the wind scoop around two huts by using an ultrasonic anemometer and discussed the formation of wind scoops. Although he studied them in detail, it is difficult to know the formation mechanism of them exactly, due to their complexity of the wind scoops and lack of adequate tools for their investigation.

In order to overcome the shortcomings of field observations, we have developed a three-dimensional numerical simulation model of snowdrift in a previous paper (Uematsu and others (1991)) and described primarily the wind scoop around a hut.

In this paper we show the results of observations and numerical simulation of the wind scoop around a hut and discuss the mechanism of its formation.

2 OBSERVATION

The experiment was done by building a 1.8 x 1.8 x 2.4 m hut on flat ground (figure 2). Snow depths around the hut were measured on 2/2 and 3/12 in 1991 and the weather was observed at the same time.

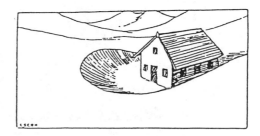

Figure 1. A wind scoop around an Alpine hut sketched by Seligman (1980).

Figure 3 shows a photograph of the snow surface around the hut taken on 3/12. The wind scoop is evident. Since no wind scoop was present on 2/2, it developed in the period from 2/2 to 3/12.

Figure 4 shows the official meteorological data of the Japan Meteorological Agency at the nearest observation station, Teine-Yamaguchi AMeDAS (Automated Meteorological Data Acquisition System) point from 1991/2/2 to 1991/3/12. The storm occurred from 3/6 to 3/7 and the second largest wind speed as well as precipitation were observed on 3/7. The largest wind speed was observed on 2/17, however, the precipitation on that day was small and therefore it seems that no wind scoop was developed on that day. On 2/16 the largest precipitation was observed, but the wind speed on that day was not strong enough to develop a wind scoop.

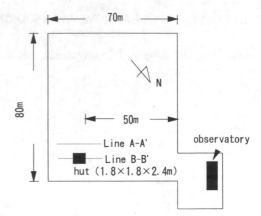

Figure 2. Observation site.

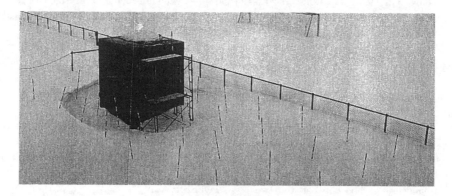

Figure 3. Photograph of a wind scoop around a hut on 12 March 1991. The size of hut is 1.8 x 1.8 x 2.4 m.

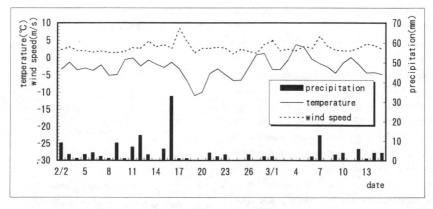

Figure 4. Meteorological data at the nearest official observation station from 2 Feb. 1991 to 12 Mar. 1991.

Figure 5 shows the observed snow depth on lines A-A', B-B' (Figure 2). These cross sections of snow cover show clearly the wind scoop .

3 SIMULATION

3.1 Simulation method
3.1.1 Wind-velocity and friction velocity

The friction velocity, u_*, which mainly controls snowdrifting (Uematsu and others 1991) , can be obtained from a velocity profile.

A numerical simulation method developed by Arisawa (1987) is applied to obtain a three-dimensional velocity field. Arisawa's method is a control-volume method developed by Patanker's algorithm (Patanker, 1981). The Navier-Stokes equations with boundary conditions are numerically solved. The Navier-Stokes equations in the three-dimensional case are expressed as:

$$\frac{\partial(\rho u)}{\partial t} + \frac{\partial(\rho uu)}{\partial x} + \frac{\partial(\rho uv)}{\partial y} + \frac{\partial(\rho uw)}{\partial z} =$$

$$\frac{\partial}{\partial x}(\rho K_m \frac{\partial u}{\partial x}) + \frac{\partial}{\partial y}(\rho K_m \frac{\partial u}{\partial y}) + \frac{\partial}{\partial z}(\rho K_m \frac{\partial u}{\partial z}) - \frac{\partial P}{\partial x}$$

(1)

$$\frac{\partial(\rho v)}{\partial t} + \frac{\partial(\rho vu)}{\partial x} + \frac{\partial(\rho vv)}{\partial y} + \frac{\partial(\rho vw)}{\partial z} =$$

$$\frac{\partial}{\partial x}(\rho K_m \frac{\partial v}{\partial x}) + \frac{\partial}{\partial y}(\rho K_m \frac{\partial v}{\partial y}) + \frac{\partial}{\partial z}(\rho K_m \frac{\partial v}{\partial z}) - \frac{\partial P}{\partial y}$$

(2)

$$\frac{\partial(\rho w)}{\partial t} + \frac{\partial(\rho wu)}{\partial x} + \frac{\partial(\rho wv)}{\partial y} + \frac{\partial(\rho ww)}{\partial z} =$$

$$\frac{\partial}{\partial x}(\rho K_m \frac{\partial w}{\partial x}) + \frac{\partial}{\partial y}(\rho K_m \frac{\partial w}{\partial y}) + \frac{\partial}{\partial z}(\rho K_m \frac{\partial w}{\partial z}) - \frac{\partial P}{\partial z} - \rho g$$

(3)

$$\frac{\partial(\rho)}{\partial t} + \frac{\partial(\rho u)}{\partial x} + \frac{\partial(\rho v)}{\partial y} + \frac{\partial(\rho w)}{\partial z} = 0$$

(4)

where x, y and z are the Cartesian coordinates, and u, v and w are longitudinal, lateral and vertical wind speeds, respectively. p is pressure, ρ is air density

and K_m is the eddy diffusivity of momentum.

We used the mixing length model to calculate the eddy diffusivity, K_m, since it does not require much computer time. The equation used is as follows:

$$K_m = l^2 \sqrt{\left(\frac{\partial u}{\partial z}\right)^2 + \left(\frac{\partial v}{\partial z}\right)^2}$$

(5)

and

$$l = \kappa r$$

(6)

where l is the mixing length, κ the on Karman constant (=0.4), and r the distance from the nearest obstruction.

3.1.2 Simulation of snowdrift transport

Snow transport is possible by saltation, by suspension or by creep. As mentioned above, only saltation and suspension are taken into consideration in our simulation.

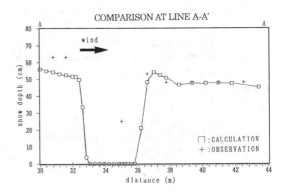

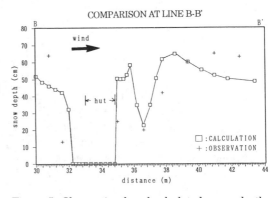

Figure 5. Observational and calculated snow depth comparisons on lines A-A' and B-B'.

a. Simulation of saltation

The snowdrift transport rate, q (more correctly called the vertically integrated snowdrift flux) is expressed by

$$q = \int_0^h (u_s \cdot \Phi)dz \tag{7}$$

where u_s is the mean horizontal particle velocity, Φ the snowdrift density, and h the height of the saltation layer. Many empirical formulae have been developed for the snowdrift transport rate, q, in the two-dimensional case. Iversen et al. (1980) expressed the snowdrift transport rate as a function of friction velocity

$$q = c \left(\frac{\rho}{g}\right) \frac{|w_f|}{u_{*t}} u_*^2 \, (u_{*t} - u_*) \tag{8}$$

where ρ is the fluid density, g the acceleration due to gravity, w_f the snow-particle terminal fall speed, and u_{*t} the threshold value of friction velocity.

The distribution of the snowdrift-transport rate, q(x), at any particular point and time, is computed by applying Equation with the distribution of friction velocity, $u_*(x)$. The constant, c, is obtained from the experimental data. This value is similar to that used by Schmidt (1982).

b. Simulation of suspension

Suspension is considered in this model by using the following diffusion equation:

$$\frac{\partial(\Phi)}{\partial t} + \frac{\partial(\Phi u)}{\partial x} + \frac{\partial(\Phi v)}{\partial y} + \frac{\partial(\Phi w)}{\partial z} =$$

$$\frac{\partial}{\partial x}\left(K_s\frac{\partial\Phi}{\partial x}\right) + \frac{\partial}{\partial y}\left(K_s\frac{\partial\Phi}{\partial y}\right) + \frac{\partial}{\partial z}\left(K_s\frac{\partial\Phi}{\partial z}\right) - \frac{\partial(w_f\Phi)}{\partial z} \tag{9}$$

where Φ and K_s are the snowdrift density and the eddy diffusion coefficient of snowdrift density and w_f is the snowfall velocity. The eddy diffusion coefficient of snowdrift density is assumed equal to that of momentum, $K_s = K_m$. The transport velocity of snow particles is assumed equal to the wind velocity.

3.2 Simulation conditions

The simulation model developed can be applied to

Table 1. Input data during the blowing snow period.

Wind speed	7.9 m/s
Wind direction	NNW NW WNW W
Precipitation	1.5 mm/h
Threshold friction speed	0.25 m/s
Blowing snow time	94 hours
	NNW:33 hours, NW:9 hours
	WNW:26 hours, W:26 hours

any meteorological conditions by changing the input values of wind speed, wind direction, precipitation, and threshold shear stress. The model is now applied to reconstruct the observed wind scoop on 3/12. For the input data the mean values in the storm period were used. Table 1 shows the input data for simulation. Blowing snow is assumed to occur when the wind speed is stronger than 6 m/s. The mean wind speed and direction during the blowing snow event were used in the simulation. Precipitation for the simulation was obtained by dividing the total precipitation between 2/2 and 3/12 to each blowing snow event. Threshold wind speed was not observed during this period. The value 0.25 m/s, which was assumed by Kobayashi's equation (Kobayashi1979),was used in the simulation. The simulation model calculates snowdrift density and snowdrift rate, and consequently the snow depth around the hut can be calculated by multiplying the snowdrift rate by the period of the blowing snow event.

When the meteorological conditions in one blowing snow event change, the snow depth is calculated for

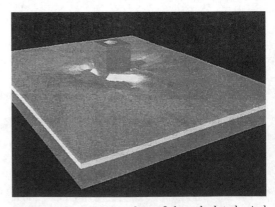

Figure 6. Computer graphics of the calculated wind scoop.

each subevent and integrated. To obtain the overall snow depth distribution, the calculated results for each blowing snow event are integrated.

3.3 Simulation results

Figure 6 shows a computer graphics display of snow depth calculated by the simulation model. Comparing Figure 6 with Figure 3, the simulated and observed wind scoops are remarkably similar. The snow accumulated on the roof and attached to the wall are similar also.

Calculated snow depths on lines A-A' and B-B' are also shown in Figure 5. Comparing calculated and observed snow depths, the calculated results show a good agreement with the observed field data. However, the wind scoop on A-A' is wider in the observed. This discrepancy could be due to the small threshold

friction velocity adapted.

The snow depth distribution of a wind scoop is very sensitive to the threshold friction velocity as mentioned later and we do not have adequate information for the threshold friction velocity. No peak occured immediately leeward of the hut on the line B-B'. This is due to the influence of the steps as seen in Figure 3.

Figures 7a-d show the computed wind field at 0.5 m above ground surface for NNW, NW, WNW, and W wind directions. The arrows indicate the direction and magunitude of the wind. The strong wind is clearly seen at the side of the hut and the upward flow can been seen windward of the hut. Comparing the four cases it is found that vortexes exist at the windward corners of the hut in all cases but each case has different points of vortexes and scales of vortexes. The

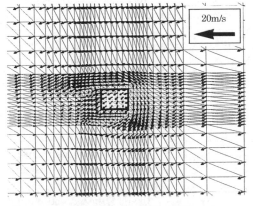

Figure 7-a. Computed wind field at 0.5 m above the ground surface.(Wind direction NNW)

Figure 7-b. Computed wind field at 0.5 m above the ground surface.(Wind direction NW)

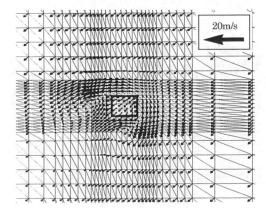

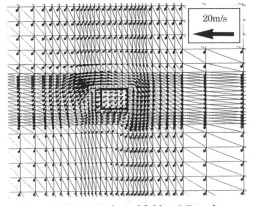

Figure 7-c. Computed wind field at 0.5 m above the ground surface.(Wind direction WNW)

Figure 7-d. Computed wind field at 0.5 m above the ground surface.(Wind direction W)

wind speed above the roof of the hut for the case of wind direction NNW. Most of these phenomena show a good agreement with our experiences.

4 DISCUSSION

We have shown that the simulation model can reconstruct the observed wind scoop adequately. In this section we will show the results of sensitivity tests of the model. The tested parameters are wind speed, wind direction and threshold friction speed.

Table 2 shows the values used in the tests.

In each case except for case 1 only are parameter is changed from case 1. The test results compared are maximum width and maximum length of the wind scoop. The width is defined as the distance along the direction from east to west. The length is defined as the distance along the direction from north to south.

Figures 8a-g show the calculated snow depth distributions. According to Figure 8, no wind scoop exits in case 2 and case 6 and the wind direction influences the shape of the wind scoop and snow accumulation on the roof of the hut remarkably.

Table 3 shows the results of the sensitivity test.

Comparing case 1 with case 2, a critical wind speed for the formation of a wind scoop is found. Namely, when the wind speed at the height of 10 m is 10 m/s , the wind scoop is formed, it however, when the wind speed is 20 m/s, most of the snow is blown away and the wind scoop is not formed.

Comparing cases 1,3,4 and 5, the influence of wind direction on the wind scoop is found. When the wind direction was normal (case1), the width of the wind scoop is the largest of the four cases. On the other hand, when the wind direction was 67.5 degree from normal, the length at the wind scoop is largest in four cases. Comparing case 1, 6 and 7, the critical threshold friction velocity for the formation of a wind scoop is found to be in the range from 0.2 m/s to 0.15 m/s.

In the sensitivity test only one parameter in three parameters was changed at a time. Uematsu (1993) suggested that threshold friction velocity and snowfall velocity can be calculated by the method of White (1940). His calculation results showed that the threshold friction velocity var from 0.17 m/s to 0.29 m/s in the range of snow particle diameter from 0.1 to 0.3 mm. In this test, we calculated for friction

Table 2. Sensitivity test data.

	Wind speed	Wind direction	Threshold friction speed
case 1	10 m/s	normal (0)	0.2 m/s
case 2	20 m/s	normal (0)	0.2 m/s
case 3	10 m/s	+22.5	0.2 m/s
case 4	10 m/s	+45	0.2 m/s
case 5	10 m/s	+67.5	0.2 m/s
case 6	10 m/s	normal (0)	0.1 m/s
case 7	10 m/s	normal (0)	0.3 m/s

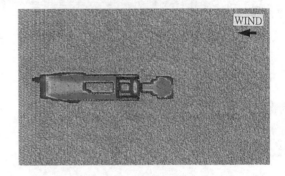

Figure 8-a case 1. Wind scoop pattern.

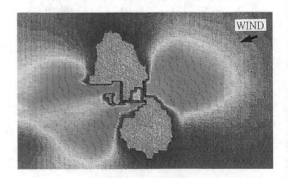

Figure 8-b case 2. Wind scoop pattern.

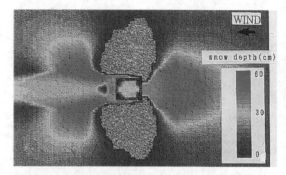

Figure 8-c case 3. Wind scoop pattern.

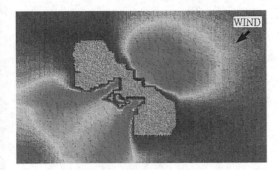

Figure 8-d case 4. Wind scoop pattern.

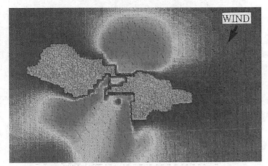

Figure 8-e case 5. Wind scoop pattern.

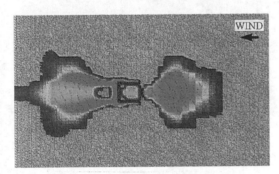

Figure 8-f case 6. Wind scoop pattern.

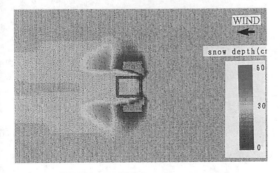

Figure 8-g case 7. Wind scoop pattern.

Table 3. Results of sensitivity test.

	width		length	
	right(m)	left(m)	right(m)	left(m)
case 1	6.3	6.3	4.6	5.0
case 2	∞	∞	∞	∞
case 3	5.6	5.3	4.2	4.8
case 4	5.2	5.2	3.5	5.0
case 5	5.2	5.2	6.1	6.8
case 6	∞	∞	∞	∞
case 7	1.6	1.6	2.2	2.2

∞:Shows no wind scoop

velocities of 0.1, 0.2, and 0.3 m/s all three of which correspond to the snow particle diameter range mentioned above. Consequently, there is a critical value of snow particle diameter in snowdrift formation when wind speed is assumed constant.

5 CONCLUSION

Using the simulation model for blowing snow and snowdrift, the snow wind scoop around a hut was studied. The following results were obtained:

The snow wind scoop was observed by setting a hut in an open snow field. The wind scoop was calculated by a simulation model using meteorological data. The calculated snow depth distribution by the simulation model showed good agreement with that observed around the hut.

In the sensitivity test the parameters of wind speed, wind direction and threshold friction velocity were changed to investigate the influence of each parameter on the resulting wind scoop. This test showed that the wind scoop cannot be formed when the wind speed is above a critical value or when threshold friction velocity is below a critical value.

The critical wind speed for the formation of a wind scoop was found to be between 10 m/s and 20 m/s at a height of 10 m . The critical friction velocity was found to be in the range of 0.1 m/s to 0.2 m/s. According to White (1940) this corresponds to, a snow psrticle diameter smaller than 0.1 mm for blowing snow phenomenumon.

References

Anno, Y. 1985. Development of a snowdrift wind tunnel. Cold Reg. Sci. Technol.,10(2), pp.153-161.

Arisawa, Y. 1987. Two-dimensional numerical experiments on the mechanism of long-rang etransport of air pollution. Ph.M. thesis. University of Tsukuba.

Hatano, T. and T. Imai 1969. On the function of the slat snow fence.Seppyo, 31, 2, pp37-43. (In Japanese)

Irwin, P.A. 1983. Application of snow-simulation model tests to planning and design. Proc. East Snow Conf., 28, pp.118-130.

Iversen, J.D., R. Greeley, B.R. White, and J.B. Pollack 1980. Eolian erosion of the Martian surface, Part 1; Erosion rate similitude. Icarus, 26,3, pp.321-331.

Kawamura,S. 1982. Earth and sand hydraulics. Morikita Shuppan. (In Japanese)

Kikuchi, T. 1981. Studies on aerodynamic surface roughness associated with drifting snow. Mem. Fac. Sci. Kochi Univ., 2, Ser. B, pp13-37.

Kobayashi, D. 1972. Studies of snow transport in low-level drifting snow. Contrib. Inst. Low Temp. Sci., Ser. A, 24.

Kobayashi, S. 1979. Studies on interaction between wind and dry snow surface. Contrib.Inst. Low Temp. Sci., Ser. A, 29.

Orlanski, I. 1976. A simple boundary condition for unbounded hyperbolic flows. J. Comput. Phys.,21, pp.251-269.

Patanker, S.V. 1981. Numerical heat transfer and fluid flow. Hemisphere Pub.193pp.

Schmidt, R.A. 1982. Properties of blowing snow. Reviews of Geophysics and Space Physics, 20,pp.39-44.

Seligman, G. 1980. Snow structure and ski field. I.G.S. 555pp.

Tabler, R.D. 1988. Snow fence handbook. Tabler Association.

Tabler, R.D. 1974. New engineering criteria for snow fence systems. Transp. Res. Rec. 506, pp.65-78.

Takeuchi,M., K. Ishimoto, T. Nohara, and Y. Fukuzawa 1984. Study of snow fence. Yuki to Dohro, 1,pp 96-100. (in Japanese)

Takeuchi, M. 1989. Snow-collection mechanisms and the capacities of snow fences. Annals of Glaciology, 13, pp 248-251.

Uematsu, T., Y. Kaneda, K. Takeuchi, T. Nakata, and M. Yukumi 1989. Numerical simulation of snowdrift development. Annals of Glaciology, 13, pp.265-267.

Uematsu, T., T. Nakata, K. Takeuchi. Y. Arisawa and Y.Kaneda 1991.Three dimensional numerical simulation of snowdrift. Cold Reg. Sci. Technol., 20(1), 65-73.

Uematsu, T. 1993.Numerical study on snow transport and drift formation. Annals of Glaciology, 18, pp.135-141.

Uematsu, T. 1993.Numerical study on snow transport and drift formation. Annals of Glaciology, 18, pp.135-141.

White, C.M. 1940.The equilibrium of grains on the bed of a stream. Proc. R. Soc. London,Ser. A, 174(958),322-338.

Wipperman, F.K. and G. Gross 1985. The wind-induced shaping and migration of an isolated dune. A numerical experiment. Boundary Layer Meteorol., 36, pp.319-334.

Snow Engineering: Recent Advances, Izumi, Nakamura & Sack (eds) © 1997 Balkema, Rotterdam. ISBN 90 5410 865 7

Avalanche experiments with styrene foam particles

Yasuaki Nohguchi
Nagaoka Institute of Snow & Ice Studies, NIED, Niigata, Japan

ABSTRACT: A sufficiently developed natural avalanche always has a head at the front end and a tail at the rear end. To experimentally simulate avalanches with such a head-tail structure, avalanche experiments were carried out using styrene foam particles (1.5-2.0 mm in diameter) as a model material for snow. In these experiments, the styrene foam particle (SFP) avalanches flow down half pipe shaped chutes with diameters of 50, 75, 100, 125 and 150 mm and a length of 2 m to examine the similarity of the model. From these experiments, it was found that the similarity of the head-tail structure relates to the chute length L, the acceleration of gravity g and the terminal velocity of the granular avalanche on the chute V_e through a dimensionless number V_e^2/Lg. When $V_e^2/Lg \ll 1$, the head-tail structure is formed. This condition corresponds to that when the avalanche motion is almost steady. Therefore, a light granular material or a long chute is favorable for the formation of the head-tail structure because a light granular avalanche has a low terminal velocity. This means that a light granular material such as styrene foam is good for model experiments of avalanches on a reduced scale.

1 INTRODUCTION

To make clear the dynamics of snow avalanches, various kinds of model experiments have been carried out instead of direct observations of actual snow avalanches. Such a model study has the advantage that experiments can be repeated many times under the same conditions. In general, the model experiments for avalanches are carried out on a reduced scale to obtain the data safely. For a model experiment on a reduced scale it is necessary to find a similarity law between real avalanches and the reduced scale experiments.

One of the most impressive properties of the shape of large natural avalanches is that they have a head-tail structure like a tear drop as is well known in gravity currents. In model experiments using plastic beads by Plüss et al.(1987), the head was not formed at the front of the model avalanches. Savage and Nohguchi (1988), Nohguchi et al.(1989) and Hutter and Nohguchi(1990) examined the behavior of granular avalanches by qualitative analysis of the similarity solutions. As a result, a body of granular material always spreads on a even slope, but a head-tail structure could not be formed using their theoretical frame work. Water tank experiments have also been used to study powder snow avalanches. In water tank experiments, heads similar to gravity currents always appear.

Recently, model experiments of granular avalanches using light materials such as table tennis balls have been continued (Nohguchi,1994, Kosugi et al.,1995, Nohguchi, 1996). A head-tail structure is formed on a

about 20 m long model. This paper is a continuation of such granular model experiments. Its purpose is to show that granular avalanches on an about 2 m long model using styrene foam particles have a head-tail structure. This paper discusses what governs the similarity.

2 EXPERIMENT

The styrene foam particles used to model snow are spherical and range in diameter from 1.5 mm to 2.0

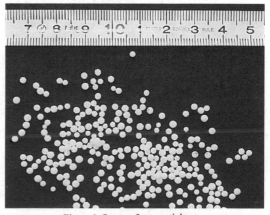

Figure 1. Styrene foam particles.

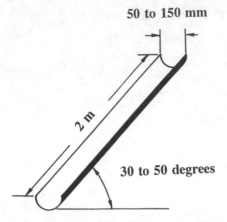

50 to 150 mm

2 m

30 to 50 degrees

Figure 2. Half pipe chute used in the experiments.

from rest, and the motion of the model avalanches was recorded by a high speed video camera. Avalanches were repeated several times under the same condition with changing the position of the video camera.

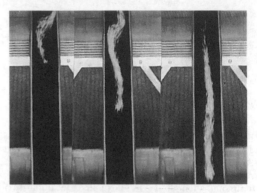

Figure 3. A typical example of a SFP avalanche.

Table 1. Conditions of each experiment.

No.	Diameter of chute ϕ m	Length of chute L m	Inclination degree	Number of particles	Number of runs
1	0.050	2.0	40	2,500	3
2	0.050	2.0	40	1,900	3
3	0.050	2.0	40	1,200	3
4	0.050	2.0	40	600	3
5	0.050	2.0	40	300	3
6	0.050	2.0	40	150	3
7	0.050	2.0	40	75	3
8	0.050	2.0	40	38	3
9	0.050	2.0	40	20	3
10	0.050	2.0	40	1	5
11	0.050	2.0	30	2,500	3
12	0.050	2.0	30	1,900	3
13	0.050	2.0	30	1,200	3
14	0.050	2.0	30	600	3
15	0.050	2.0	30	300	3
16	0.050	2.0	30	150	4
17	0.050	2.0	30	1	6
18	0.050	2.0	50	2,500	1
19	0.050	2.0	50	1,900	3
20	0.050	2.0	50	1,200	3
21	0.050	2.0	50	600	3
22	0.050	2.0	50	300	3
23	0.050	2.0	50	150	3
24	0.050	2.0	45	600	25
25	0.075	2.0	45	2,500	20
26	0.100	2.0	45	5,000	15
27	0.125	2.0	45	10,000	5
28	0.150	2.0	45	18,000	4

mm as shown in Figure 1. The bulk density, average mass per a particle and the terminal velocity of an isolated particle falling in air are 16.8 kg/m³, 2.0 x 10⁻⁷ kg, 1.5-2.0 x 10⁻³ m and 0.5 m/s respectively. The half pipe chute used for experiments is illustrated in Figure 2.

Under 28 different conditions for the experiment as summarized in Table 1, SFP avalanches were released

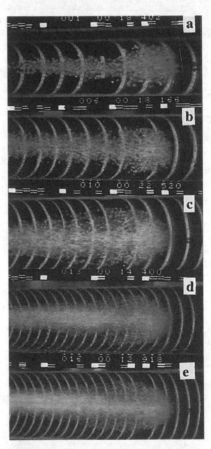

Figure 4. Front end of SFP avalanches at the dimensionless distance $L/\phi = 6$; a)$L = 50$ mm, $N = 600$, b)$L = 75$ mm, $N = 2500$, c)$L = 100$ mm, $N = 5000$, d)$L = 125$ mm, $N = 10000$, e)$L = 150$ mm, $N = 18000$.

3 RESULTS

In all SFP avalanches, a head was formed at the front of the flow and the rear looked like a tail as seen in Figure 3. The fronts of 5 different SFP avalanches are shown in Figure 4.

The front velocity of SFP avalanches increases to a steady state value as the avalanche flows down the chute (Fig. 5). This result shows that the 2 m slope length used in the experiments was sufficient to achieve a steady motion.

The effect of the number of particles on front velocity is shown in Figure 6. The velocity is the average value between dimensionless distances (L/ϕ) of 10 and 30. Therefore it can be considered as terminal one. The front velocity increases with increasing number of the particles.

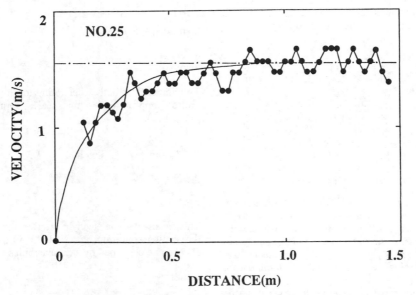

Figure 5. Variation of front velocity with flow distance.

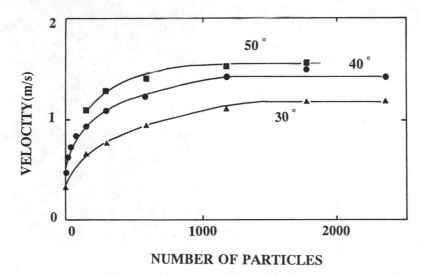

Figure 6. Relation between terminal velocity at front and number of particles.

4 SIMILARITY OF HEAD-TAIL STRUCTURE

As shown in Figure 6, a bigger group of particles has a faster terminal velocity. Thus means the front group always must be the biggest group of particles when the terminal velocity is reached. A smaller group is overtaken by a bigger group. As a result, the concentration of particles occures at the front end, which becomes a head (Fig.7a). On the other hand, if a small group releases later, the distance between the small group and the main body increases with time. As a result, at a rear end, the group of the particles is increasingly extended and becomes a tail (Fig.7b). Thus, the head-tail structure is formed (Fig.7c).

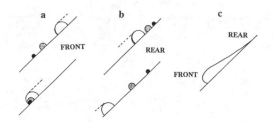

Figure 7. Schematic diagrams of the head-tail structure formation.

The appearance of the head-tail structure indicates that the terminal velocity of the group has been reached; in other words, the motion is steady under the balance between the force of gravity and forces resisting. Air drag is one of the mechanisms that defines the terminal velocity. Air drag is dominant for light materials.

Phenomena which accelerate to reach a steady motion are, in general, characterized by the terminal velocity, V_e, the acceleration of gravity, g, and a system size such as a slope length, L. Using these parameters, we derive a similarity law for a simple model of a granular avalanche.

The equation of motion of a group of granular materials is as follows;

$$m\,dv/dt = mg(\sin\theta - \mu\cos\theta) - \alpha v \qquad (1)$$
$$dx/dt = v \qquad (2)$$

where m and θ are the mass and inclination angle of chute respectively, and μ and α are the coefficients of the resisting force.

To make Eqs.(1) and (2) dimensionless, let velocity, v, distance, x, and time, t, be

$$v = Vv, \quad x = Lx, \quad t = Tt, \qquad (3)$$

where V and T are the scales for velocity and time respectively.

Then Eqs.(1) and (2) are

$$dv/dt = Tg^*/V - (Tg^*/V_e)v$$
$$= Tg^*/V(1 - (V/V_e)v) \qquad (4)$$
$$dx/dt = (TV/L)v \qquad (5)$$

where

$$mg^*/\alpha = V_e, \quad g^* = g(\sin\theta - \mu\cos\theta). \qquad (6)$$

If

$$T = V_e/g^* = T_c, \qquad (7)$$
$$L = V_e^2/g^* = L_c, \qquad (8)$$
$$V = V_e, \qquad (9)$$

then Eqs.(4) and (5) are

$$dv/dt = 1 - v \qquad (10)$$
$$dx/dt = v. \qquad (11)$$

This means that both accelerating motion and steady motion appear under the scales represented by Eqs.(7)-(9).

If

$$V = V_e \text{ and } L \gg L_c, \text{ that is, } T \gg T_c, \qquad (12)$$

then Eq.(4) is

$$v = 1. \qquad (13)$$

This means that the steady motion is dominant under the scaling condition represented by Eq.(12), which also can be described by a dimensionless number as

$$V_e^2/Lg^* \ll 1. \qquad (14)$$

Therefore Eq.(14) or Eq.(12) governs the formation of the head-tail structure. V_e^2/Lg^* is a kind of Froude number where V and L are the terminal velocity and slope length, respectively.

If

$$V \ll V_e, \qquad (15)$$

then

$$dv/dt = 1. \qquad (16)$$

In this case, only acceleration is dominant; therefore, the head-tail structure cannot be formed.

In Table 2, typical cases in natural avalanches and granular avalanches are summarized. TTB avalanches (Kosugi 1995) on 25 to 100 m long slopes and SFP avalanches on 1 to 4 m long slopes are similar to natural powder snow avalanches on slopes a few kilometers long.

Table 2. Comparison of various kinds of avalanches.

Avalanche	Speed V_e m/s	$T_c=V_e/g^{3)}$ sec	$L_c=V_e^2/g$ m	Slope length V_e^2/Lg m	
SFP[1]	1	0.1	0.1	1	0.1
SFP	2	0.2	0.4	4	0.1
TTB[2]	5	0.5	2.5	25	0.1
TTB	10	1.0	10.0	100	0.1
Flow	20	2.0	40.0	400	0.1
Powder snow	50	5.0	250.0	2500	0.1

1)Styrene foam particle.
2)Table tennis ball.
3)Here, $g*$ is considered as g.

5 DEPENDENCE OF TERMINAL VELOCITY ON THE NUMBER OF PARTICLES

As shown in Figure 6, V_e is an increasing function of the number of particles, N. Here we will derive the function to satisfy the similarity characterized by dimensionless number V_e^2/Lg. when two model experiments of different sizes (Fig.8) are carried out using the same material.

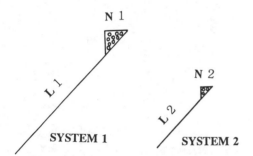

Figure 8. Two similar systems of different size.

For simplicity, we assume that the function can be represented by a power law as

$$V_e \propto N^\alpha. \tag{17}$$

Then the similarity of systems 1 and 2 is

$$N_1^{2\alpha}/L_1 g = N_2^{2\alpha}/L_2 g. \tag{18}$$

Therefore,

$$(N_1/N_2)^{2\alpha} = L_1/L_2. \tag{19}$$

If the systems are 3-dimensional, the ratio N_1/N_2 must be

$$N_1/N_2 = (L_1/L_2)^3. \tag{20}$$

From Eqs.(19) and (20)

$$\alpha = 1/6. \tag{21}$$

For 2-dimensional systems

$$\alpha = 1/4. \tag{22}$$

Figure 9 shows the relation between the front velocity and the number of particles for 5 similar systems with 5 different scales as summarized in Table 3. In this case we can see the dependence is obeying to this law.

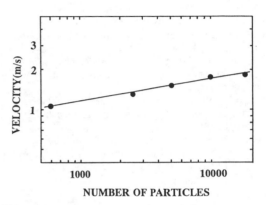

Figure 9. Relation between the front velocity and the number of particles. The line represents the theoretical one for $\alpha = 1/6$.

Table 3. 5 similar systems with 5 different scales.

NO.	ϕ m	ϕ / ϕ_{24}	L m	L/L_{24}	N	$(N/N_{24})^{1/3}$
24	0.050	1.0	0.30	1.0	600	1.0
25	0.075	1.5	0.45	1.5	2500	1.6
26	0.100	2.0	0.60	2.0	5000	2.0
27	0.125	2.5	0.75	2.5	10000	2.6
28	0.150	3.0	0.90	3.0	18000	3.1

CONCLUDING REMARKS

From a theoretical analysis, we found that the similarity of the head-tail structure relates to the chute length L, the acceleration of gravity g and the terminal velocity of the granular avalanche on the chute V_e through a dimensionless number V_e^2/Lg. When $V_e^2/Lg \ll 1$, the head-tail structure is formed. This condition corresponds to avalanche motion that is almost steady; therefore, a light granular material or a long chute is favorable for the formation of the head-tail structure because a light granular avalanche has a low terminal velocity. This means that a light granular material such as styrene foam particles is good for model

67

experiments of avalanches on a reduced scale.

REFERENCES

Hutter, K and Y. Nohguchi. 1990. Similarity solutions for a Voellmy model of snow avalanches with finite mass. *Acta Mechanica* 82, 99-127.

Kosugi, K et al. 1995. Table tennis ball avalanche experiments. ISSW'94 Proceedings, 636-642.

Nohguchi, Y., K. Hutter and S. B. Savage. 1989. Similarity solutions for granular avalanches of finite mass with variable bed friction. *Continuum Mechanics and Thermodynamics* 1, 239-265.

Nohguchi, Y. 1994. Study on avalanches by table tennis balls (in Japanese). *Science and Technology for Disaster prevention* 72, 40-45.

Nohguchi, Y. 1996. Similarity of avalanche experiments by light particles. Interpraevent 1996 Proceedings, 2, 147-156.

Plüss, Ch. 1987. Experiments on granular avalanches. Diplomarbeit, Abt X, Eidg. Techn. Hochshule, Zürich, 113 pp.

Savage, S. B. and Y. Nohguchi. 1988. Similarity solutions for avalanches of granular materials down curved beds. *Acta Mechanica* 75, 153-174.

Snow Engineering: Recent Advances, Izumi, Nakamura & Sack (eds) © 1997 Balkema, Rotterdam. ISBN 90 5410 865 7

Development of the nowcasting map of the risk of dry snow avalanches

Koyuru Iwanami, Yasuaki Nohguchi, Yutaka Yamada & Takashi Ikarashi
Nagaoka Institute of Snow and Ice Studies, NIED, Niigata, Japan

Tsutomu Nakamura
Nagaoka Institute of Snow and Ice Studies, NIED, Niigata, Japan (Presently: Faculty of Agriculture, Iwate University, Morioka, Japan)

ABSTRACT: The method of nowcasting the risk of dry snow avalanches in two to three hours was devised for 3-km × 3-km meshes into which the entire Niigata Prefecture was divided, and a prototype of the system for the calculation and display of the nowcasting maps was developed. In this study, the objective of nowcasting was limited to the risk of large-scale dry snow avalanches that occur when it is snowing heavily under the condition of a prevailing winter monsoon. The system can display distributions of various basic data such as the number of snow avalanches in the past. It can also calculate and display hazard maps that show the potential risk of occurrence of dry snow avalanches, nowcasting maps of snowfall amount, and nowcasting maps of the risk of dry snow avalanches.

1 INTRODUCTION

Heavy snowfalls and the many snow avalanches that inflicted damage have occurred in Niigata Prefecture, Japan (Nakamura et al., 1987). So it is important for the promotion of the district to prevent disasters by snow avalanches. Snow avalanches that have occurred in Hokuriku district, Japan, can be divided broadly into two categories. One is the wet snow avalanche that occurs when the air temperature rises after a series of snowfalls or in early spring. The other is the dry snow avalanche (hereafter, referred to as DSA) that occurs during heavy snow storms. It is known that large-scale DSAs that tend to cause significant damage occur when new snow deposition exceeds ten centimeters (for example, Nohguchi, 1992). A typical large-scale DSA occurred January 26, 1986 at Maseguchi, Nou-town in Niigata Prefecture, resulting with thirteen fatalities (Kobayashi, 1986; Yamada and Ikarashi, 1987). Such large-scale DSAs have not happened since 1986, because for several years we had warm, dry winter seasons. However, the number of disasters by snow avalanches in ski areas and mountainous regions has increased recently (Ikarashi, 1994).

The purpose of this study is the development of a method of nowcasting (comprehension of the present condition and very short-range forecast) of the risk of DSAs for all of Niigata Prefecture with two-three hour advance notice. The construction of the system for real-time data acquisition and distribution is out of the scope of this study.

2 OUTLINE OF THE NOWCASTING SYSTEM

The prototype of the nowcasting system was constructed on an engineering workstation. It has mainly four functions and all operations can be done by selecting items and typing in parameters on hierarchical menu windows. The first function is the display of the key maps and distributions of various basic data on avalanches, radar echoes, and surface meteorological observations. It is possible to superimpose information about topography, land use, railroads, roads, boundary lines of municipalities processed using Digital National Land Information on these distribution maps.

The snowfall information was derived from C-band weather radar data that were observed with the Yakushi-dake Radar, Ministry of Construction located in the western part of Nagaoka-city (37° 28'N, 138° 43'E). Transmitted/received frequency, transmitted peak power, observation range and range resolution of the radar are 5,330 MHz, 250 kW, 120 km and 3 km, respectively. We can obtain snowfall information covering all of Niigata Prefecture using PPI (Plan Position Indicator) data at three elevation angles every five minutes.

Surface observations of depth of snow cover and newly fallen snow and air temperature have been made at about ninety points once a day at 09:00 JST as a project of Niigata Prefecture since the 1969/70 winter season. We can use results that have been published with the same kind of data observed in the prefecture by the Automated Meteorological Data Acquisition System (AMeDAS) by Japan Meteorological Agency.

69

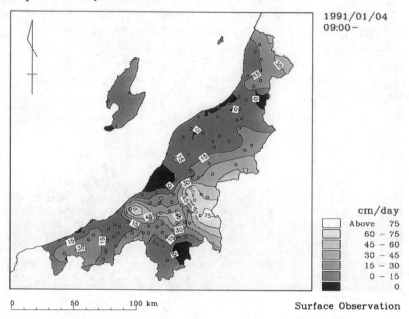

Depth of Daily Snowfall

1991/01/04
09:00~

cm/day
Above	75
60	– 75
45	– 60
30	– 45
15	– 30
0	– 15
	0

0   50   100 km

Surface Observation

Fig. 1. Distribution of daily snowfall observed at 09:00 JST on January 5, 1991. The small open circles and rectangles show the observation points by Niigata Prefecture and Japan Meteorological Agency, respectively.

Figure 1 shows distribution of daily snowfall from January 4 to 5, 1991 by both contours and shading.

The second function of the system is the calculation and display of the hazard maps showing the distributions of the potential risk of DSAs. The calculation and display of the nowcasting maps of snowfall amount is the third function of the system. Finally the system can calculate and display the nowcasting maps of the risk of DSAs. The risk of DSAs is obtained by the combination of the hazard map and the time series of snowfall information derived from radar data with the nowcasting map of snowfall amount. These functions are described in the next section in detail.

The composite of two data of different time or kinds, several simple calculations between data of different time, and the individual display for four divided areas of the prefecture are also available by the system.

3 METHOD OF NOWCASTING OF THE RISK OF DSA

3.1 *Hazard maps of DSA*

The data on topography (mean value, maximum difference and standard deviation of the altitude), land use and the number of disastrous snow avalanches in the past in each 3 km × 3 km mesh were prepared in order to estimate the potential risk of occurrence of DSAs for each mesh. The distribution map of the potential risk of DSAs is referred to as the "hazard map" in the developed system. The land use was classified into six kinds: fields, forests, urban areas, riverbeds and lakes, wasteland and others. The percentage of each kind of land use in each mesh was used as the factor for the estimate of the potential risk of DSAs. The contribution to the potential risk of occurrence of DSAs can be given for each factor and calculated for each mesh by the following relation:

$$PR = PRda \times (PRma + PRsa + PRlu + PRad); \qquad (1)$$

where *PR* shows the potential risk of occurrence of DSAs for each mesh, and *PRda*, *PRma*, *PRsa* indicates the contribution of maximum difference, mean value and standard deviation of the altitude in each mesh to the potential risk. The contribution of land use and the number of snow avalanche disasters in the past is denoted by *PRlu* and *PRad*, respectively. The contribution of each factor to the potential risk is set up by a kind of linear function with weight. The method of assigning a contribution to the potential risk of each factor is one object of this study.

Figure 2 shows the hazard map by considering only the maximum difference and mean value of the altitude in each mesh. It is felt that the maximum difference indicates the existence of large slopes

70

where large DSAs may occur whereas mean altitude may indicate the potential for dry snow. The potential risk of DSAs is expressed in arbitrary units in the figure.

3.2 *Nowcasting maps of snowfall amount*

The relationship, $Z = 0.636S^{0.996}$ between the radar reflectivity, $10\log Z$ dBZ and snowfall intensity, S in cm/hr was obtained using radar data above the surface observation points and observed depths of newly fallen snow on the ground for 24 hours from January 4 to 5 in 1991 (Iwanami et al., 1994). The reflectivity data of the Yakushi-dake Radar were converted to snowfall amount using the empirical relationship. Time series of hourly snowfall depths covering all of Niigata Prefecture were also used for nowcasting the risk of DSAs.

The advection velocity of radar echoes (snowfall clouds) was calculated from the best matching position obtained from the maximum cross-correlation of two sequential echo patterns at fifteen minute intervals (Asuma et al., 1984a). Then the snowfall amount was predicted within two to three hours in advance by the method of kinematic linear extrapolation of two sequential radar echoes described in Asuma et al (1984b,c).

Figure 3 shows a nowcasting snowfall amount

calculated using two PPI data observed at 06:45 and 07:00 JST on January 5, 1991. In this case, radar echoes moved eastward so fast that it was difficult to estimate the depth of hourly snowfall three hours after the start of the prediction.

3.3 *Nowcasting maps of the risk of DSA*

The occurrence of large-scale DSAs is closely related to very recent snowfall patterns because new snow has a low density and low strength while older snow layers are generally denser and stronger. Also the continuance of snowfall tends to increase the risk of a DSA, but snow cover stabilizes by densification and sintering after snowfall ceases, hereby decreasing the DSA hazard. We assumed therefore that the risk of DSAs can be obtained by the integration of a time series of snowfall amounts weighted by an exponentially decreasing time function (Nohguchi, 1994) based on the idea of the N-index proposed by Nohguchi (1992). The N-index can express lightness and weakness of snow cover by integration of the depth of the snow cover layer weighted by the inverse of snow density. The time series of snowfall amounts covering all of Niigata Prefecture were derived from weather radar data as described previously.

The risk of DSAs, $K(N)$, N hours after the start of nowcasting is calculated by the following equations.

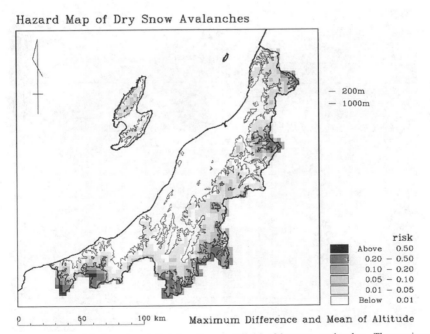

Fig.2. An example of the hazard maps that show the potential risk of dry snow avalanches. The maximum difference and mean of the altitude in meshes were considered. The risk is expressed in arbitrary units. The contour lines of 200 and 1,000 m are indicated by thin and slightly thick solid lines, respectively.

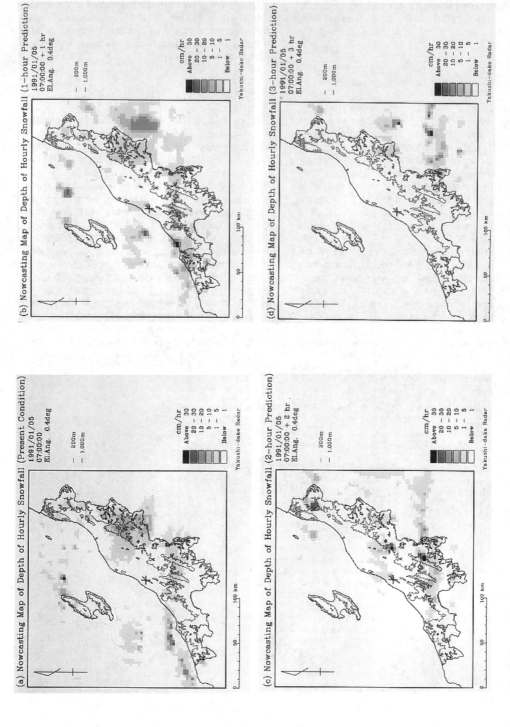

Fig. 3. An example of nowcasting maps of hourly snowfall depths predicted by radar data at 06:45 and 07:00 JST on January 5, 1991. (a) shows the present condition, and (b), (c) and (d) show the distribution of hourly snowfall depths at 08:00, 09:00 and 10:00 JST on January 5, 1991, respectively. The location of the Yakushi-dake Radar is indicated by a cross.

$$K(0)=\sum_{T=0}^{W}H(-T)G(T) \qquad (2)$$

$$K(1)=F(1)G(0)+\sum_{T=0}^{W}H(-T)G(T+1) \qquad (3)$$

$$K(2)=F(2)G(0)+F(1)G(1)+\sum_{T=0}^{W}H(-T)G(T+2) \quad (4)$$

$$K(3)=F(3)G(0)+F(2)G(1)+F(1)G(2)$$
$$+\sum_{T=0}^{W}(-T)G(T+3), \qquad (5)$$

where $H(-T)$ and $F(N)$ are the depth of hourly snowfall T hours ($T = 0, 1, 2, \cdots, W-1, W$) before, and predicted hourly snowfall depth N hours after the start of nowcasting, respectively. $G(T)$ is the weight function calculated by the following two expressions:

$$G(T)=\exp(-T/\tau) \qquad (6)$$

$$G(T)=\begin{cases} 1 & \text{when } 0 \leqq T \leqq \tau \\ 0 & \text{when } \tau < T. \end{cases} \qquad (7)$$

The time constant, τ in hours and the influence time, W in hours are parameters that can be set up. K, H, F are calculated for each mesh.

Furthermore we assumed that the risk of DSAs can be estimated by multiplication of the potential risk in the hazard map and the risk converted from time series of hourly snowfall depth by the method described above. Figure 4 shows a result of the nowcasting of the risk of DSAs. They were calculated using the hazard map shown in Fig. 2, the nowcasting map of snowfall amount shown in Figure 3 and the time series of hourly snowfall derived from the radar data. The exponential weight function and parameters of $\tau = W = 20$ hours were used in the calculation. The distribution of the risk of DSAs can be obtained for all of Niigata Prefecture with 3 km resolution as shown in Figure 4. However, the risk has not yet been normalized.

The temporal changes of the risk of DSAs from 07:00 to 08:00 JST and 08:00 to 09:00 JST shown in Figure 4 are indicated in Figure 5. Plus and minus numbers indicate increasing and decreasing risk in the figure, respectively. The risk of a DSA changed with the movement of snowfall clouds.

4 CONCLUSIONS AND FUTURE DIRECTIONS

The purpose of this study was development of the nowcasting method of the risk of large-scale DSAs that are a result of heavy snowfall and which may cause heavy damage. This method for nowcasting the risk of DSAs two to three hours in advance was devised for 3-km $\times$ 3-km meshes for the entire Niigata Prefecture. A prototype of a system for calculating and displaying the nowcasting maps was developed.

The system can display distributions of various basic data such as the number of snow avalanches in the past, and calculate and display hazard maps that show the potential risk of DSAs. It can also produce nowcasting maps of snowfall amount and the risk of DSAs. This method uses weather radar data to obtain continuous snowfall information covering a wide area with fine resolution in time and space, and estimates the risk of DSAs by the conversion of a time series of snowfall amount based on the idea of N-index (Nohguchi, 1992).

We still have several points to resolve. One of them is verification of the results of nowcasting by the system. Comparisons of the results of nowcasting to real DSAs have not been made because large DSAs have not been reported during our research since warm winters with little snow have occurred. The development of methods to monitor snow avalanches by remote sensing techniques covering wide areas with satellite-borne sensors is expected. It will become possible to normalize the value of the risk of DSAs by storing information about DSAs in wide area. The relationship between a time series of newly fallen snow depths and occurrence of DSAs should be also studied.

There is room for improvement in each function. Improvement of the relationship between radar reflectivity and snowfall amount, taking orographic effects and development of snow clouds into consideration are expected. The contribution of each factor to the potential risk of DSAs should be examined. In this study nowcasting was limited to DSAs that occur when it was snowing heavily. It seems that we have to also consider the formation of weak layers in snow cover since they are especially important for DSAs in mountainous regions.

ACKNOWLEDGEMENTS

The authors wish to thank the Shinano River Construction Office, the Hokuriku Regional Construction Bureau of the Ministry of Construction for providing the Yakushi-dake radar data. The authors are deeply indebted to Dr. Yoshio Asuma, Hokkaido University for his considerable assistance with the computer programs for nowcast mapping of snowfall amount. The authors wish to thank Prof. Shun'ichi Kobayashi, Niigata University, Prof. Norio Hayakawa, Nagaoka University of Technology, Niigata Local Meteorological Observatory, Japan Meteorological Agency, Hokuriku Regional Construction Bureau of Ministry of Construction, and Niigata Prefecture for their support. We used the Digital National Land Information of coastline and altitude and others made by the Geophysical Survey Institute, Ministry of Construction to draw various maps in this study.

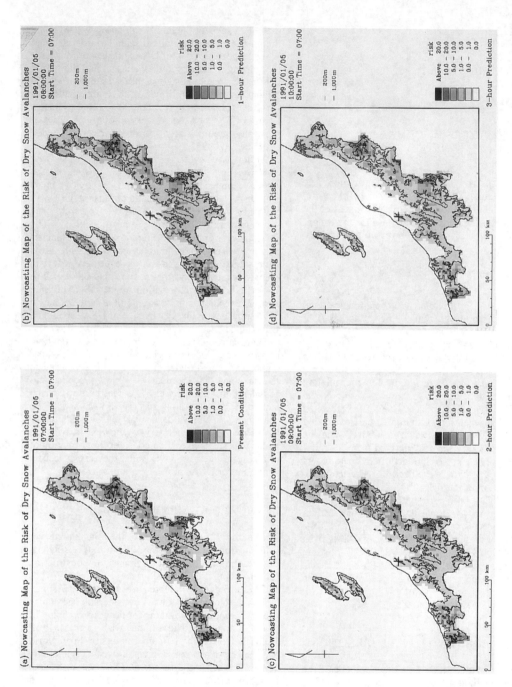

Fig. 4. An example of nowcasting maps of the risk of dry snow avalanches obtained at 07:00 JST on January 5, 1991. (a) shows the present condition, and (b), (c) and (d) show distribution of the risk of dry snow avalanches at 08:00, 09:00 and 10:00 JST on January 5, 1991, respectively. The risk is expressed in arbitrary units.

74

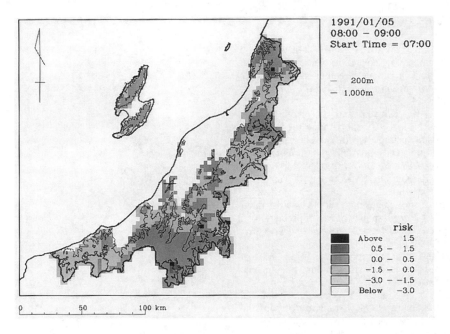

(a) Temporal Change of the Risk of Dry Snow Avalanches

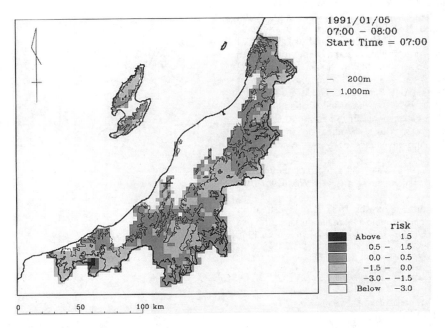

(b) Temporal Change of the Risk of Dry Snow Avalanches

Fig. 5. Temporal changes of the risk of dry snow avalanches shown in Fig. 4. Parts (a) and (b) indicate the temporal changes from 07:00 to 08:00 JST and from 08:00 to 09:00 JST on January 5, 1991, respectively.

REFERENCES

Asuma, Y., K. Kikuchi and H. Kon, 1984a: A method for estimating the advection velocity of radar echoes using a simple weather radar system. *Geophysical Bulletin of Hokkaido University*, Vol.44, 23-34.

Asuma, Y., K. Kikuchi and H. Kon, 1984b: Experiments for a very-short-range prediction of snowfall using a simple weather radar systems Part 1. −Outline and possibility−. *Geophysical Bulletin of Hokkaido University*, Vol.44, 35-51.

Asuma, Y., K. Kikuchi, H. Kon and T. Taniguchi, 1984c: Experiments for a very-short-range prediction of snowfall using a simple weather radar systems Part 2. −Examples of actual prediction−. *Geophysical Bulletin of Hokkaido University*, Vol.44, 53-65.

Ikarashi, T., 1994: Disasters that injured and lost human bodies caused by snow and ice phenomena. *Gekkan Shobo*, Vol.16, 96-102.

Iwanami, K, Y. Nohguchi, Y. Yamada, T. Ikarashi and T. Nakamura, 1994: Estimation of snowfall amount from radar data. Proceedings of Annual Meeting of the Japanese Society of Snow and Ice in 1994, 7.

Kobayashi, S., 1986: On the powder snow avalanche, which occurred in Maseguchi, Nou-machi, Niigata Prefecture, 1986. *Seppyo*, Vol.48, 87-91.

Nakamura, T., H. Nakamura, O. Abe, M. Higashiura, N. Numano, H. Yuuki, S. Kutsuzawa, Y. Yamada and T. Ikarashi, 1987: Prefectural distribution of disastrous snow avalanches and prediction of them in the northern parts of Japan. International Symposium on Avalanche Formation, Movement and Effects, IAHS Publication No.162, 639-646.

Nohguchi, Y., 1992: On the N-index of snow cover. Report of a Grant-in-Aid for Scientific Research on Priority Areas, "Basic Studies on the Internal Structure and Dynamics of Snow Avalanches", 03201106, 54-58.

Nohguchi, Y, 1994: Values like depth of snow cover and newly fallen snow. *Seppyo Hokushin'etsu*, No.12, 38.

Yamada, Y. and T. Ikarashi, 1987: The investigation of the Nou catastrophic avalanche disaster. *Review of Research for Disaster Prevention*, No.117, pp.35.

Snow Engineering: Recent Advances, Izumi, Nakamura & Sack (eds) © 1997 Balkema, Rotterdam. ISBN 90 5410 865 7

Slip tests between the surfaces of snow/ice and some kinds of shoes

Toshiichi Kobayashi, Yasuaki Nohguchi, Katsuhisa Kawashima & Takashi Ikarashi
Nagaoka Institute of Snow and Ice Studies, NIED, Niigata, Japan

Tsutomu Nakamura
Faculty of Agriculture, Iwate University, Morioka, Japan

Kaoru Horiguchi & Yukiko Mizuno
Institute of Low Temperature Science, Hokkaido University, Sapporo, Japan

ABSTRACT : This paper describes slipperiness for pedestrians on a snowy or icy surface of a road. In order to investigate the coefficients of static friction between the soles of some kinds of shoes and snowy or icy surfaces, slip tests were made in a coldroom with the examinee wearing various shoes. As a result, the coefficients of static friction for four kinds of shoes, i.e., low shoes with flat soles, low shoes with notched soft soles, winter high shoes and rubber boots were calculated. In addition , it was found that sprinkling sand on an icy surface was effective in increasing the coefficient of static friction μ_s below 0 °C.

1 INTRODUCTION

In areas such as Tokyo with little snow, many pedestrians slip on snowy roads and are injured whenever the roads are covered with snow (Figure 1). On the other hand, pedestrians who live in snowy and cold areas ought to be accustomed to walking safely on snowy roads. In Japan, the use of studded tires was prohibited by law in April 1991. Since April 1992, anyone who infringes on the law has been fined up to

¥100,000. Very slippery frozen road surfaces called "Tsuru-tsuru-romen" have appeared frequently in winter in Hokkaido, especially in Sapporo. At the same time, the number of pedestrians that slipped on such roads and were injured, and traffic accidents have increased (Figure 2). We measured in a coldroom the slip factor of shoes worn by pedestrians on snow or ice. The anti-slip effect of sand sprinkled on the surface of ice was also investigated.

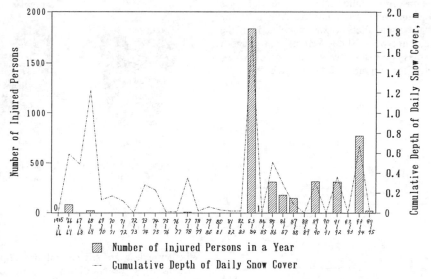

Figure 1. Relation between the number of injured persons and the cumulative depth of daily snow cover in Tokyo.

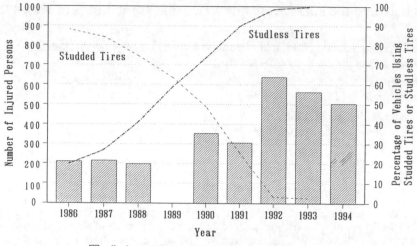

Figure 2. Relation between the number of injured persons
and the percentage of vehicles using studded tires
or studless tires in Sapporo.

2 EXPERIMENTAL METHOD

In Figure 3, a 1 m long, 0.55 m wide and 0.1 m thick block of ice was set horizontally in a coldroom. A load cell was fixed to the wall of the coldroom and wire was extended from the end of the load cell. A handle was fixed to the another end of the wire. An examinee who wore the test shoes stood on the ice, and while crouching, pulled the handle until he began to slip on the ice. The maximum static frictional force which was generated at that time was recorded. After that, the coefficient of static friction was calculated. In the experiment, four types of shoes were used ; low shoes with flat soles, low shoes with notched soft soles, winter high shoes and rubber boots (Figure 4).

Low Shoes with Flat Soles

Low Shoes with Notched
Soft Soles

Winter High Shoes

Rubber Boots

Figure 4. Soles of four pairs of shoes
used in the experiment.

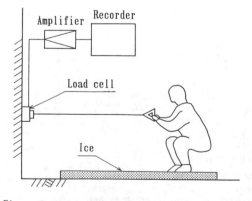

Figure 3. Measuring system for static friction.

3 RESULTS

Whether pedestrians slip and fall on a slippery road is influenced by the coefficient of static friction μ_s. Thus the slipperiness for pedestrians on a snowy or icy road was estimated by μ_s.

3.1 *Temperature dependence on* μ_s

In order to investigate whether ice temperature influenced μ_s, slip tests were made on ice at temperatures from -0.4℃ to -5.3℃. The results of these tests are shown in Figure 5. The coefficient of friction μ_s, increased linearly with decreasing ice temperature. In other words, pedestrians are most apt to slip on ice at about 0℃. Thus all the slip tests were made at a temperature 0℃.

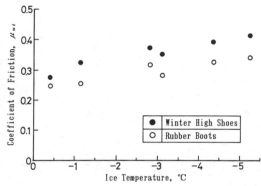

Figure 5. Relation between ice temperature and μ_{si}.

3.2 *Coefficient of friction* μ_s, *on an icy surface*

The coefficient of static friction on ice μ_{si} for low shoes with flat soles was the lowest of the four kinds of shoes (Figure 6). Low shoes with notched soft soles were not as slippery as winter high shoes and rubber boots (Figure 6). That was due to the soft soles and the sharp-edged notches in them. The shoes "sank into" the icy surface, gripping it.

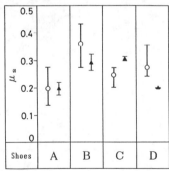

O : μ_s on icy surface,
▲ : μ_s on thin snowy surface with icy base.

A : Low shoes with flat soles,
B : Low shoes with notched soft soles,
C : Winter high shoes,
D : Rubber boots.

Figure 6. Coefficients of static friction on icy surface or thin snowy surface with icy base.

3.3 *Coefficient of friction on a thin snowy surface with an icy base*

It is well-known that pedestrians are apt to slip on a thin layer of newly deposited snow on ice. Thus we conducted slip tests on such a new snow surface. The snow was deposited through a sieve with 1 mm openings. The coefficients of friction of the low shoes with notched soft soles and rubber boots decreased compared with that for the icy surface (Figure 6). That was because the slip did not occur between the soles and the snow but between the snow and the ice. The smallest coefficient of friction on the snow was for rubber boots (Figure 6).

3.4 *Coefficient of friction on ice with sand*

Sand is sometimes used for preventing pedestrians from slipping on an icy road. Slip tests were conducted on ice with sand which was sprinkled through a sieve with 1 mm openings. The coefficient of static friction μ_{ss} for the four kinds of shoes increased to more than 0.5 (Table 1).

Table 1. Coefficient of friction.

Shoes	μ_{ss}	μ_{ss}/μ_{si}
A	0. 582	1. 60
B	0. 524	1. 24
C	0. 578	1. 59
D	0. 511	1. 38

A : Low shoes with flat soles,
B : Low shoes with notched soft soles,
C : Winter high shoes,
D : Rubber boots.

4 DISCUSSION

According to Brungraber (1976), for the longest stride likely to be taken when walking the required minimum static coefficient of friction should be 0.5. The static coefficient of friction μ_{ss} on ice with sand for all four types of shoes tested was more than 0.5 as shown in Table 1. Therefore, sprinkling sand on a slippery icy road surface at 0℃ is quite effective, particularly for low shoes with flat soles and winter high shoes. The relation between the improvement in coefficient of static friction and the ground contact area of shoes is shown in Figure 7.

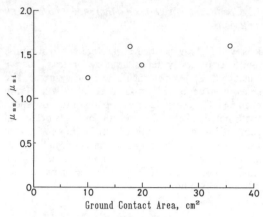

Figure 7. Relation between μ_{ss}/μ_{si} and the ground contact area.

5 CONCLUSIONS

In this study, the slipperiness for pedestrians on snowy or icy surfaces was investigated by making slip tests in a coldroom. The results of the tests are as follows:

(1)When a pedestrian was wearing winter high shoes or rubber boots on an icy road, he was more apt to slip at a temperature 0℃than -5℃.

(2)When a pedestrian was walking on an icy road wearing low shoes, a slip was much less likely with notched soft soles than with flat soles.

(3)When fresh snow covers the ice, a pedestrian wearing low shoes with notched soft soles has a greater tendency to slip than on an icy road.

(4)A pedestrian wearing rubber boots was as apt to slip as if wearing low shoes with flat soles on a thin snowy surface with icy base.

(5)Sprinkling sand on an icy road is quite effective in increasing walking safety for pedestrians.

REFERENCES

Brungraber, R. J. 1976. An overview of floor slip-resistance research with annotated bibliography. *NBS Technical Note.* 895:7-8.

Hara, F. , Kawabata, T. & Kobayashi, H. 1990. Study of pedestrian injuries in winter in Sapporo. *Proceedings of ' 90 Cold Region Technology Conference.* 151-157.

Ito, H. 1994. About spreading antifreeze chemicals in Sapporo. *Snow and Its Measures ' 94 ～ ' 95 (Yuki to taisaku).* Economic Research Association. 50-53.

Kobayashi, T. , Nohguchi, Y. & Kawashima, K. 1995. Fundamental studies on pedestrians falling down on snow and ice road. *Proceedings of ' 95 Cold Region Technology Conference.* 476-479.

Sapporo Fire Bureau. 1996. Report of occurrence of pedestrians falling down on snowy roads.

Takamori, M. , Takagi, H. & Onuma, H. 1994. Actual situation of pedestrians fall down walking on road in winter and study of countermeasure against snowy and icy side walk surface. *Proceedings of ' 94 Cold Region Technology Conference.* 45-50.

Investigations on avalanche disasters and related weather situations

Takeshi Ito
Department of Urban and Environmental Studies, Akita National College of Technology, Japan

Takeshi Hasegawa
Department of General Science, Akita National College of Technology, Japan

ABSTRACT: This research is intended to review the avalanche disasters, and weather situations in the Tohoku District, northwestern part of Japan. From the avalanche disasters analysis, it was found that there are several specific distribution patterns of atmospheric pressure that are prone to cause snow avalanches. Each pattern and its characteristic will be mentioned for each prefecture of this area. Next, a potential index for avalanches was investigated using daily snow cover depth. The proposed index was confirmed as a reasonable method for predicting surface layer avalanches.

1 ANALYTICAL RESULTS OF SNOW AVALANCHES

From the beginning of the 20th century up to 1995, deaths and missing people caused by snow avalanches in the Tohoku District, Japan are estimated approximately to 1,300. This accounts for a quarter percent of those killed and missing in all avalanche disasters in Japan. Figure 1 shows the studied area. Avalanches tend to appear under the specific weather situations in this area. We found that there are about seven weather types which may bring avalanches.

2 WEATHER SITUATIONS RELATED TO AVALANCHES

From the analytical results, northwest monsoons (Type I) and the distribution of troughs of low atmospheric pressure over the Sea of Japan (Type II) are main triggers in Aomori, Akita, and Yamagata prefectures, which front the Sea of Japan, and the Aizu District of Fukushima prefecture. Figure 2(a) shows the typical Type I when historical heavy snowfall recorded on January 26, 1974 and Figure 2(b) shows the series of daily traces of cyclone centers (x marks) and -40 ℃

isothermal contours at 500hpa plane in those days. The cyclone centers traced at north side a little compared with normal years, and a very strong cold vortex which produced heavy snow was tightly performed .

On the other hand, Taiwan cyclones (Type III)coming on the Japanese archipelago cause frequently avalanches in the Pacific Ocean side, for

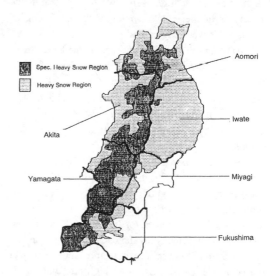

Figure 1. Special heavy snow region and heavy snow region in the Tohoku District.

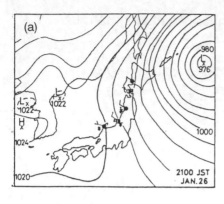

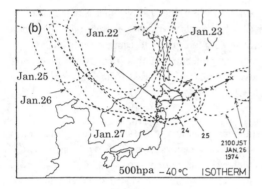

Figure 2. (a)Example of TypeI observed on Jan.26, 1974 and (b) daily traces of-40℃ contours.

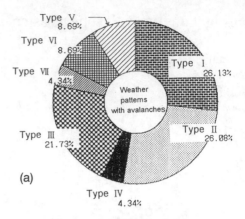

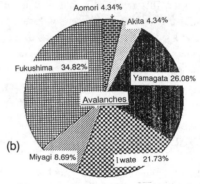

Figure 3. (a)Weather patterns with avalanches, and (b)avalanche occurrence rates of 6 prefectures.

example, in the Iwate prefecture and the east side of Aomori prefecture. This weather pattern tends to appear at the end of winter or the beginning of spring with wet heavy snow.

Another weather situations related to snow avalanches are the two cyclones (Type Ⅳ) coming parallel side by side on the Japanese archipelago like a sandwich, migratory anticyclones (Type Ⅴ), abnormal high temperature (Type Ⅵ) and abnormal low temperature (Type Ⅶ) situations.

The occurrence rate of these weather patterns during 1976 to 1990 in the Tohoku District are shown in the Figure 3(a). Figure 3(b) also represents the occurrence rate of avalanche disasters of each prefecture. In which, Type Ⅰ and Type Ⅱ cause many avalanches especially in the Aizu District, Fukushima prefecture and Yamagata prefecture. Type Ⅳ is not so high rate, but this is a very severe pattern especially in the Pacific Ocean side because of its unseasonable appearance.

3 AVALANCHE DISASTERS AND MAIN AVALANCHE TYPES

Investigating about old avalanche disasters in Akita prefecture, fifty avalanche events were recorded since 1900 to 1995. Figure 4 shows the avalanche accident points. The distribution concentrates fairy toward the special heavy snow regions. According to the snow disaster records, most of huge avalanches have been occurred in heavy snow fall years, for example, 1918, 1929, 1963, and 1974. The tendencies of their occurrence conditions in this area are shown in Figure 5. These data were all observed in 1973-1974 winter, although sample is limited, however, we can point out roughly as one of regularity of occurrences that wet point-starting full-depth avalanche tends to appear under the snow cover depth 100 cm, and dry slab full-depth avalanche tends to appear over the temperature of -1.5 ℃ .

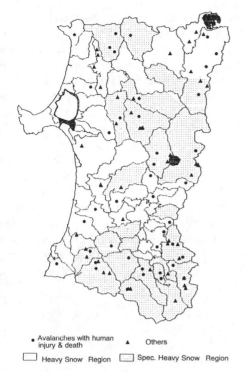

- Avalanches with human injury & death　　▲ Others

☐ Heavy Snow Region　☒ Spec. Heavy Snow Region

Figure 4. Avalanche accidents occurred during 1900-1995.

4　AVALANCHE POTENTIAL INDEX

From the view point of daily snow cover depth changes, avalanche frequently occurs from certain depth. Most of huge avalanches with human accidents appeared at snow cover depth from the vicinity of 200 cm and over, therefore, to find avalanche regional potential index, we tried to make current potential index using the prior snow cover depth of 1 day, 3 days and 5 days. Surface avalanches tend to appear at snow cover depth from more than 200 cm and a slope angle more than 30 degrees in this area.

According to our investigation of an index of increase and decrease of daily snow cover depth changes, surface layer avalanches will occur when the index reaches to certain value. This potential index consists of accumulation variations of increase and decrease of daily snow cover depth as shown in the following example:

$$H_5 = 5H_i - (H_{i-1} + H_{i-2} + H_{i-3} + H_{i-4} + H_{i-5})$$

If we take data of snow cover depth of going back five days (H_5), and if the current time's

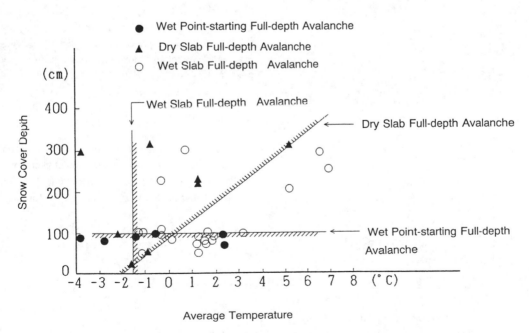

Figure 5. Avalanche patterns observed in 1973/1974 winter in Akita prefecture.

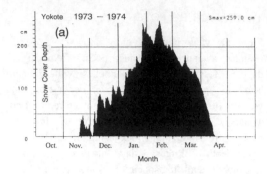

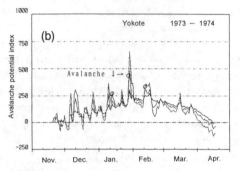

Figure 6. (a)Daily snow cover depth and (b)potential index for surface layer avalanches.

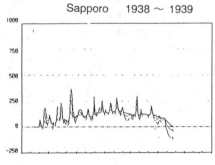

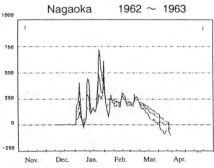

Figure 7. Patterns of potential index by districts.

(i-th) rises to more than 150 suddenly, a surface layer avalanche will occur near the observation point, especially in the Tohoku District. If it rises to 200, avalanche will occure at anywhere in this area. Figure 6 shows a seasonal case of 1973 to 1974 winter's time series variance of the index at Yokote city, Akita prefecture using 1 day, 3 days and 5 days back data, respectively. Circle marks in the figure show when surface layer avalanches appeared. Figure 7 shows the another typical seasonal examples of index variations of Sapporo in Hokkaido District, and Nagaoka in Hokuriku District. Taking note of the index variations, it fluctuates very widely in case of Nagaoka city, on the contrary, the fluctuation is not so widely in Sapporo. In this way, the index value varies a little, depending on the individual regional snow pack quality.

CONCLUSIONS

Weather patterns with snow avalanches were pointed out that there are seven types in the Tohoku District, Japan. In which, northwest monsoons (Type I), the distribution of troughs of low atmospheric pressure over the Sea of Japan (Type II) and Taiwan cyclones (Type III) are main triggers in this area. From the analytical results of daily snow cover depth changes, avalanche potential index was found as 150 and over for this area.

REFERENCES

ITO,T.1992. Investigations on snow disasters and development of a disaster potential index, *Proceedings of the Second International Conference on Snow Engineering*, 147-156.

Snow Engineering: Recent Advances, Izumi, Nakamura & Sack (eds) © 1997 Balkema, Rotterdam. ISBN 90 5410 865 7

Study of avalanches in the Tianshan Mountains, Xinjiang, China

Jiaqi Qiu, Junrong Xu & Fengqing Jiang
Xinjiang Institute of Geography, Chinese Academy of Sciences, Urumqi, People's Republic of China

Osamu Abe & Atsushi Sato
Shinjo Branch of Snow and Ice Studies, NIED, Yamagata, Japan

Yasuaki Nohguchi
Nagaoka Institute of Snow and Ice Studies, NIED, Niigata, Japan

Tsutomu Nakamura
Nagaoka Institute of Snow and Ice Studies, NIED, Niigata, Japan (Presently: Iwate University, Morioka, Japan)

ABSTRACT: In order to investigate avalanche internal dynamic structures, a joint research project was carried out in the western area of the Tianshan Mountains, Xinjiang, China. A medium-size avalanche path has been instrumented with six pressure transducers and two load cells to measure avalanche velocities and impact pressures in the interior of avalanches. During the 1995-1996 winter, the data were recorded of an artificially released avalanche passing by the instruments. The impact velocities were measured to be 6.9 m/s by image analysis and 7.1 m/s by the lag time of a pair of first impact waves. It was also found that the avalanche velocity almost reached a terminal velocity, and the shape of dense flow declined towards downstream.

1 INTRODUCTION

Avalanches have been studied for many years in the western area of the Tianshan Mountains with continental climate conditions, which are related to the avalanche hazards and their control along some mountain highways(Hu & Jiang 1989, Qiu 1995). The internal dynamic structures of avalanches have been investigated worldwide (Schaerer et al. 1980, Shimizu et al. 1980, Gubler, 1987, Kawada et al. 1989, Nishimura et al. 1993, Dent et al. 1994), but had not been studied substantially in the Tianshan Mountains until a new research project between China and Japan began in 1995. Some avalanche defense structures were destroyed in the past because they were designed when relatively little is known about their dynamic features in this area. In recent years, the construction works, including highways have increased rapidly in the Tianshan Mountains. It is getting more and more important to study the internal dynamic structures of avalanches with their impact pressure being considered as the main research target. In January 1996, an avalanche observation of the internal dynamic structures was carried out by means of artificial release. The focus of this paper is a preliminary report of some research results.

2 OBSERVATION SITE

The study area is located on the western slope of the

Aiken Pass upstream of the Gongnaisi River in the western area of the Tianshan Mountains. The Tianshan Station for Snowcover and Avalanche Research (TSSAR) is located in this area (at 84°24'E, 43°16' N, elevation 1,776 m), Xinjiang Institute of Geography, Chinese Academy of Sciences. The predominant wind in the study area is from the west. The annual average precipitation is 837.3 mm at TSSAR. Its no. 3 avalanche path was selected as the

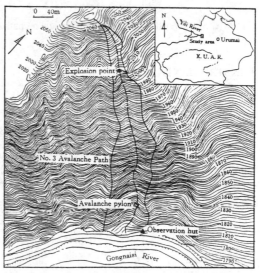

Figure 1. No. 3 avalanche path of TSSAR.

experiment path for this project (Fig. 1). It is 100 m west of the station. The Yili-Yanqi Highway crosses the lower end of the avalanche path. Avalanches occur here very frequently, 35 were registered during the period from 1968 to 1985. In snowy years, many avalanches occur each winter. Favorable conditions are available for recording of frequent avalanche events. This path is a southeasterly facing gully with a steep slope, a medium size, a total length of about 480 m, and elevations ranging from 1,770 to 2,072 m (Fig. 2). The dip angles of the gully are variable, their maximum, average and minimum being 51°, 41° and 29°. The gully bed surface is undulatory due to terraces built for avalanche control. Two sides of the gully are outlined by bedrock, which cause its lines of maximum depth and width to meander.

Figure 3. Avalanche experiment pylon.

1(A,B), 2(C,D), 3(E,F) and two load cells; LC-1 and 2, with a 0.12 m diameter pressure plate, were mounted vertically to the impact face of the pylon at different heights. The rear transducer of each pair of MPTs was placed 0.5 m immediately downstream of the front transducer. The transducers were mounted this way to determine the velocity of avalanches (Table 1). The capacities of the transducer and load cell are 981 kPa and 9,806 N. Three pairs of transducers arranged vertically determine velocities as a function of height.

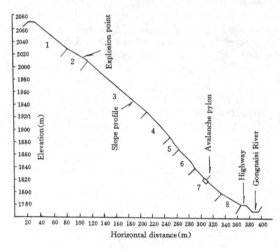

Figure 2. Slope profile along the no.3 avalanche path.

3 INSTRUMENTATION

3.1 Avalanche experiment pylon

A pylon was installed in the center of the no. 3 avalanche path at the lower end of the gully, which is the impact face. The pylon consists of one primary member made of steel pipe with a diameter of 190 mm, standing 5 m perpendicular to the slope, and three steel braces downslope (Fig. 3). The base plate of the pylon was secured with common reinforced concrete (see Clayton et al. 1992).

3.2 Impact pressure sensors

Three pairs of miniature pressure transducers; MPT-

Table 1. Vertical arrangement of MPT and LC accessaries on pylon.

MPT & LC		VH (m)	Arm		CH no.
			L(m)	ΔL(m)	
MPT-1	A	3.45	0.213	0.500	1
	B	3.25	0.713	0.000	2
LC-1		2.75	0.200	0.513	3
MPT-2	C	2.45	0.213	0.500	4
	D	2.25	0.713	0.000	5
LC-2		1.75	0.200	0.513	6
MPT-3	E	1.45	0.213	0.500	7
	F	1.25	0.713	0.000	8

3.3 *Recording equipment*

The avalanche pressure signals were transmitted to an automatic recorder through 8 armored cables and two strain amplifiers. The strain amplifier is a 4-channel subminiaturized DC amplifier, the recorder is an 8-channel thermal dot recorder employing a 9-inch electroluminescence touch panel display. These instruments were located in a hut on the slope and supplied with electric power from a storage battery. Two video cameras were employed to monitor the avalanche motion.

4 ANALYSES OF EXPERIMENTAL DATA

4.1 *Operational problems*

A dry-snow avalanche was artificially released using an explosive charge in the snow of the avalanche starting zone (elevation 2,010 m). The snow depth in the starting zone was about 1.5 m. The avalanche followed along a dashed line path of Figure 1, and terminated at the highway after running about 370 m(Fig. 4). The observations yielded impact pressures of the avalanche at different heights for its entire duration.

4.2 *Snow condition*

Snow profiles and avalanche debris were measured. The data on 17 Jan.,1996 shown by the international classification (ICSI of IASH 1990) in Figure 5 is typical of snowcover in the study area. The snowcover here is characterized by its low temperature, steep temperature gradient, low density, low strength, medium depth of maximum snowcover and thick layer of depth hoar. The snowcover had a mean density of 160 kg/m³.

The avalanche debris was like a long dike. Its front section was deposited at the highway with a length of 6 m, width from 2.6 to 12.1 m and thickness ranging from 0.52 to 1.14 m; its rear section accumulated on the slope with a length of 8.35 m.

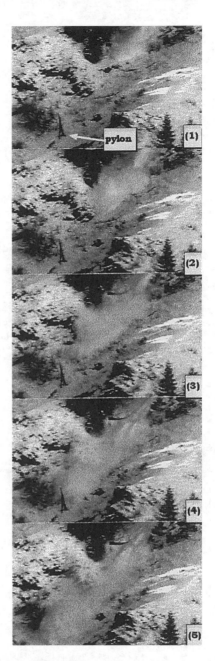

Figure 4. The avalanche filmed every 1 s on 18 Jan., 1996.

The deposited snow, which was completely mixed, had a mean density of 390 kg/m³ and a temperature of -11.2 °C.

Table 2. Numerical description of the avalanche.

CH no.	Height (m)	Time (s)	P_max (kPa)
6	1.75	1.596	27.8
7	1.45	2.287	88.7
8	1.25	1.948	84.2

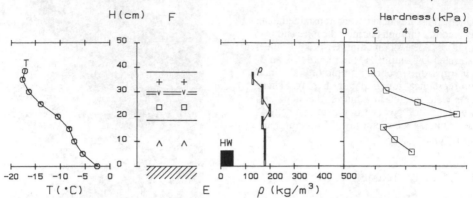

Figure 5. Snow profile beside the meteorological platform of TSSAR.

4.3 *Impact pressure*

The impact pressure waves within the avalanche were obtained as a function of height. Three impact waves for the lower sensors are shown in Figure 6, but the other five waves are not shown because of low level signals. The maximum impact pressure of 88.7 kPa appeared at a height of 1.45 m (CH 7) as shown in Table 2. The impact signals of the avalanche appear frequently at the lowest sensor (CH 8). The impact signals obtained by the load cell with a pressure plate (CH 6) have smooth shapes for the large impact area.

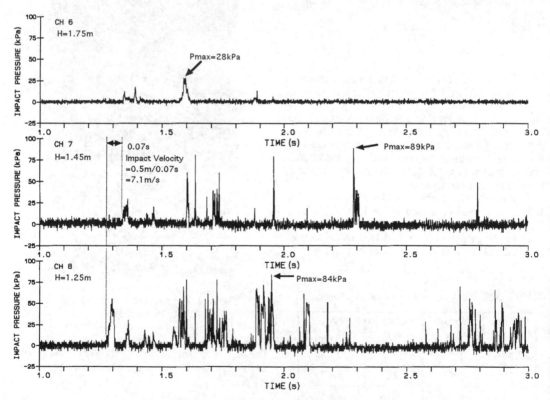

Figure 6. Impact pressure recording of the avalanche.

4.4 *Avalanche velocity*

From the record of the impact pressure waves and video images of the avalanche, some avalanche behavior can be deduced as follows. The avalanche front velocities were measured by means of two methods, (1) interpretation of multiple video images and (2) timing of impact pressure waves of the avalanche. According to successive video images of the avalanche passing by the pylon, the velocity was determined to be 6.9 m/s. As shown in Figure 4, the avalanche almost reached a terminal velocity, because the front velocities between the two images does not increase. On the other hand, on the basis of the lag time of the first signals of the avalanche passing by impact pressure sensors MPT-3-E and F and their difference in arm lengths, the velocity was determined to be 7.1 m/s. The velocities determined by the two methods are very close. Other signals of the impact waves between CH 7 and CH 8 in this avalanche have no good correlation, so the impact velocities could not be calculated for them.

4.5 *Internal structure*

A dry-snow avalanche mainly consists of two parts: dense flows and snow clouds (Perla & Martinelli 1976). In this experiment, the maximum height of the snow clouds is estimated to about 5.0 m, because the snow clouds of the moving avalanche reached to the top of the pylon as shown in Figure 4. The dense flow was outlined by the impact pressure waves. To define the shape of the dense flow, the impact waves were treated as follows. First, the impact waves of CH 6 and CH 7 were shifted forward to the left due to the delay time Δt of the impact wave of CH 8. Here, Δt can be calculated as,

$$\Delta t = \Delta l / V. \tag{1}$$

Where Δl is the distances between each two sensors (CH 6 - CH8 and CH7 - CH 8, see Table 1) and V is the velocity along the slope. Here, V was assumed to constant at 7.1 m/s. These procedures are for the

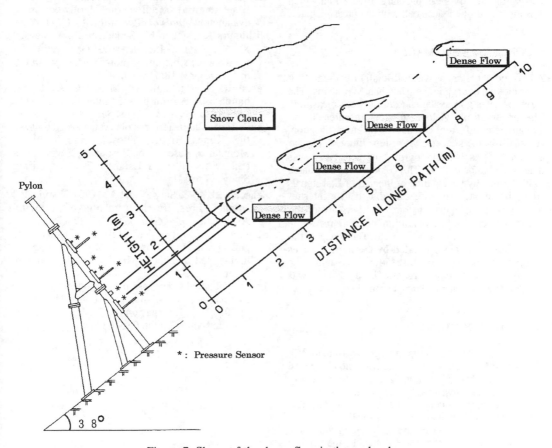

Figure 7. Shape of the dense flow in the avalanche.

purpose of examining readings taken at the same time. The periods of the impact waves are shown in Figure 7 as a function of height. The dense flow in the avalanche is declined towards downstream. For example, the declination angles for the leading edges of the avalanche was estimated about 90 and 115 degrees for the first and second parts of the dense flow. The shape of the dense flow describes a well developed type of avalanche on a steep slope by model experiments (personal communication from Nohguchi Y.).

The impact pressure is usually estimated by a dimensional analysis as follows:

$$P = k \rho V^2 \qquad (2)$$

where P is the impact pressure, and k is the resistant factor. k could be used to be 3 for a peak of snow block impact (Abe, 1996). In this case, V was assumed 7.1 m/s, so the internal density could be estimated from the maximum pressure. Using equation (2) the internal density was estimated to be 184 kg/m^3 for the peak pressure of 27.8 kPa which was measured by the load cell with the pressure plate.

5 CONCLUDING REMARKS

A dry-snow avalanche was artificially released using an explosive charge in the Tianshan Mountains. The snow profiles and avalanche debris were measured. The observations yielded impact pressures of the avalanche at different heights for its entire duration. According to successive video images of the avalanche passing the pylon, the velocity was determined to be 6.9 m/s. The velocity was calculated to be 7.1 m/s by the lag time of the pair of first signals from the impact pressure transducers.The internal density was estimated to be 184 kg/m^3 from the impact velocity and the maximum pressure measured by a load cell with a pressure plate. The shape of the dense flow also was estimated by the impact waves and sensor heights. The dense flow in the avalanche declined towards downstream. The avalanche was small, and almost reached its a terminal velocity.

ACKNOWLEDGEMENTS

We are grateful to the Xinjiang Highway Bureau, Yili Highway-Maintaining Sector General, and Gongnaisi and Nalati Highway-Maintaining Sectors for their help in closing the highway as the avalanche was released and in clearing the deposited avalanche snow. The authors also wish to thank Dr. Masao Higashiura and K.C. Wong Education Foundation, Hong Kong for their assistance. This study was funded by the Science and Technology Agency, Japan and the Expert Committee for Basic Studies on Cryosphere Dynamic Change, CAS, China.

REFERENCES

Abe O. 1996. Experimental study on the avalanche impact forces. Ph.D. thesis, Niigata University.

Clayton A., Decker R., Richardson C.R. & Abe O. 1992. Installation design of the avalanche impact pylon facility, Alta, Utah. ISSW'92 Proc.:182-190.

Dent J.D., Adams E.E., Bailey I.J. & Schmidt D.S. 1994. Velocity and mass transport measurements in Snow Avalanche. ISSW'94 Proc.:347-359.

Gubler H. 1987. Measurements and modelling of snow avalanche speeds. IAHS Publ. 162:405-420.

Hu R. & Jiang F. 1989. Avalanche and Prevention in the Tianshan Mountains (Zhongguo Tianshan Xuebeng Yu Zhili). Beijing: The People's Traffic Press.

Kawada K., Nishimura K. & Maeno N. 1989. Experimental studies on a powder-snow avalanche. Annals of Glaciology. 13:129-134.

Nishimura K., Maeno N., Sandersen F., Kristensen K., Norem H. & Lied K. 1993. Observations of the dynamic structure of snow avalanches. Annals of Glaciology. 18:313-316.

Perlra R. & Martinelli Jr. M. 1976. Avalanche handbook. Agriculture Handbook 489, U.S. Dept. of Agriculture, Forest Service.

Qiu J. 1995. Characteristics of Avalanche Activity in the Kunes River Valley of the Tianshan Mountains. Research on Resources, Environment and Oasis in Arid Land. 197-204. Beijing: Science Press.

Schaerer, P. A. & Salway, A. A. 1980. Seismic and Impact Pressure Monitoring of Flowing Avalanches. J. of Glaciology. 26: 179-187.

Shimizu H., Hujioka T., Akitaya E., Narita H., Nakagawa M. & Kawada K. 1980. A study on high-speed avalanches in the Kurobe Canyon, Japan. J. of Glaciology 26:141-151.

The International Commission on Snow and Ice of the International Association of Scientific Hydrology 1990. The International Classification for Seasonal Snow on the Ground.

Snow Engineering: Recent Advances, Izumi, Nakamura & Sack (eds) © 1997 Balkema, Rotterdam. ISBN 90 5410 865 7

Statistics on avalanche accidents in the central part of Japan (1900-1989)

K. Izumi & S. Kobayashi
Research Institute for Hazards, Niigata University, Japan

K. Yano
Faculty of Science, Yamagata University, Japan

Y. Endo, Y. Ohzeki & S. Watanabe
Tohkamachi Experiment Station, FFPRI, Niigata, Japan

ABSTRACT: Data on avalanche accidents occurring in Niigata and neighboring prefectures (Nagano, Gunma, Fukushima and Yamagata) were collected mainly from local newspaper articles and statistically analyzed for the past ninety years (1900-1989). In these prefectures, more than 3300 avalanche accidents and 2300 avalanche fatalities were compiled and distributed into categories of damage and by prefecture. Avalanche accidents related to leisure activities, such as mountaineering and skiing, have increased recently.

1 INTRODUCTION

The heaviest snow areas of Japan, where annual mean maximum snow depth is more than 2 m, are distributed mainly around the border between Niigata and neighboring prefectures: Yamagata, Fukushima, Gunma and Nagano, in the central part of Japan (Figure 1). Snow depths sometimes reach more than 4 or 5 m and most avalanche accidents in Japan have occurred in this region.

Nakamura et al. (1987) surveyed avalanche accidents occurred in the northern part of Japan from 1926 to 1981 through local newspapers. This survey covers Hokkaido Island as well as Aomori, Iwate, Akita, Yamagata, Miyagi, Fukushima and Niigata prefectures. But there is no report on avalanche accidents occurred in the heaviest snow region of Japan over a long period of time.

Therefore, data on avalanche accidents occurring in Niigata and neighboring prefectures were collected for the past ninety years (1900-1989), and statistically analyzed to understand the characteristics and time-variations of avalanche accidents in this region.

2 DATA COLLECTION

Articles that describe avalanche accidents were collected mainly from local newspapers: Niigata Nippo (Niigata), Yamagata Shinbun (Yamagata), Fukushima Minpo (Fukushima), Johmo Shinbun (Gunma) and Shinano-mainichi Shinbun (Nagano), for the past ninety years (1900-1989). The local newspaper is the only record possible to trace avalanche accidents back about one hundred years in Japan. A database for

Figure 1. Distribution of annual mean maximum snow depth more than 2 m in the central part of Japan.

Table 1. Accidents and casualties due to avalanches occurred in five prefectures from 1900 to 1989.

Prefecture	Niigata	Yamagata	Fukushima	Gunma	Nagano	Total
Number of avalanche accidents	1541	738	538	86	448	3351
(Average per winter)	17.1	8.2	6.0	1.0	5.0	37.2
Number of avalanche accidents with casualty	487	233	114	63	233	1130
Number of avalanche fatalities	1074	508	181	155	403	2321
(Average per winter)	11.9	5.6	2.0	1.9	4.5	25.8
(Average per fatal avalanche)	2.8	2.7	1.8	3.1	2.1	2.5
Maximum number of fatalities due to one avalanche accident	158 (1918)	154 (1918)	8 (1934)	42 (1936)	12 (1982)	

avalanche accidents was prepared from these articles then analyzed statistically.

3 AVALANCHE ACCIDENTS AND CASUALTIES BY PREFECTURE

More than 3300 avalanche accidents and 2300 avalanche fatalities are compiled for the past ninety years and distributed by prefecture (Table 1). The number of avalanche accidents in Niigata reaches 1541, which is nearly half of the total accidents. In each prefecture, with the exception of Gunma, avalanche accidents occur at an average frequency of 5 or more per winter. The average number of accidents in this region is about 37 per winter.

More than 1100 avalanche accidents, about one-third of the total, are accompanied by casualties. The total of avalanche fatalities is 2321 and the average is about 26 per winter in this region. In Niigata, 1074 fatalities, which is about half of the total, are recorded during the ninety years.

In each prefecture the average fatalities per fatal avalanche is about 2 or 3. The maximum number of fatalities due to one avalanche accident in this region is 158, which was recorded at Mitsumata village, Niigata,

in 1918 and is the worst record in Japan. A loose snow avalanche from dry new snow hit most of the village and killed 158 people.

4 TIME-VARIATION OF AVALANCHE ACCIDENTS

The avalanche accidents in this region were summed up every decade and distributed by prefecture (Figure 2). They increased rapidly during the early 1900s with the economic growth of Japan, but decreased markedly in the 1940s. After the 1950s they increased again and became stable in the 1970s and 1980s. In the 1930s and 1960s they had peak values because these decades had several heavy snowfall winters. The remarkable decrease in the 1940s is considered to be due to the decrease of traffic and working in mountainous areas under the influence of the Second World War and its aftermath.

5 AVALANCHE ACCIDENTS AND FATALITIES BY CATEGORY

The avalanche accidents and fatalities of this region

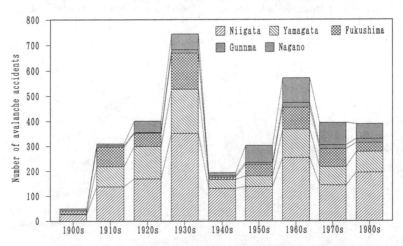

Figure 2. Decade-variation of avalanche accidents distributed by prefectures.

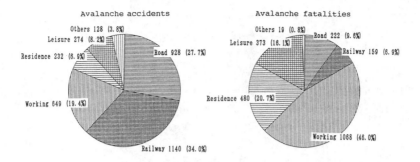

Figure 3. Avalanche accidents and fatalities of five prefectures distributed into six damage categories.

were distributed into the following six damage categories: road, railway, working (in mountainous areas), residence, leisure and others (Figure 3). As shown in Figure 3, the railway has the largest number of avalanche accidents and the accidents concerning traffic (road and railway) occupy about 60 % of the total. But the fatalities concerning traffic are not great compared with the number of accidents. This is because the majority of traffic accidents are caused only by depositing of avalanche snow on the road or railway. On the other hand, working, residence and leisure have many avalanche fatalities in comparison with the number of accidents. Working has the largest number of avalanche fatalities, which is 46% of the total.

5.1 Avalanche accidents concerning traffic

The avalanche accidents related to traffic were distributed into four subjects (road traffic, vehicle, pedestrian and railway), and summed up every decade (Figure 4). The majority of them before the 1950s occurred on railways. The majority of them after the 1960s occurred on roads. This is because of the

development of countermeasures for avalanches along railways and the progress of motorization and snow removal for roads in mountainous areas. The increase of avalanche accidents involving vehicles on roads is of special concern. On the other hand, avalanche accidents involving pedestrians have decreased recently. The maximum number of fatalities due to one traffic accident is 90, which was recorded on the Hokuriku-line railway, Niigata prefecture, in 1922. A wet full-depth avalanche hit a train carrying workers for snow removal and killed 90 persons. But we have had almost no fatalities due to avalanche accidents related to railway since the 1950s.

5.2 Avalanche accidents concerning working

The avalanche accidents related to working in mountainous areas were divided into six activities (mining, hydroelectric power generation, charcoal burning, timber producing, construction & snow removal and others) and summed up every decade as shown in Figure 5. The avalanche accidents related to charcoal burning were frequent until the 1950s, but have

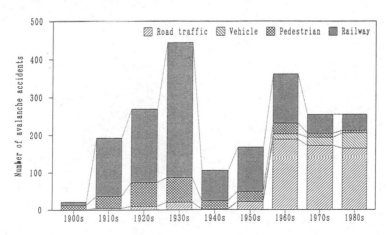

Figure 4. Decade-variation of avalanche accidents related to traffic distributed into four subjects.

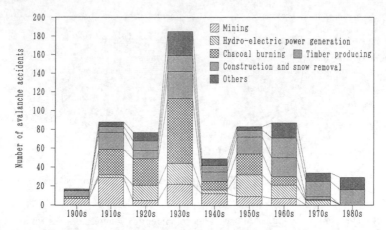

Figure 5. Decade-variations of avalanche accidents related to working divided into six activities.

almost ceased since the 1970s. This is because charcoal production has rapidly declined due to the increased use of oil, gas and electricity as a heat source. The avalanche accidents related to mining, hydroelectric power generation and timber production have decreased because of the decline of domestic production or the result of no further development. Accordingly, the avalanche accidents related to working have recently decreased as a whole as shown in Figure 5. Instead, the rate of avalanche accidents for construction and snow removal has increased recently. This tendency is expected to continue because work related to winter construction and snow removal is increasing in mountainous areas.

The maximum number of fatalities due to one avalanche accident related to working is 154, which was recorded at Ohtori mine, Yamagata prefecture, in 1918 as shown in Table 1. A loose snow avalanche from dry new snow hit the mine community of Ohtori and killed many miners and their families.

6 AVALANCHE ACCIDENTS CONCERNING LEISURE

Avalanche accidents concerning leisure were summed up every decade and distributed into five activities: mountaineering, ski area, ski tour, ski training and others (Figure 6). In Japan, winter leisure activities have become popular after the Second World War. Avalanche accidents related to leisure, therefore, have increased since the 1950s. The rate of avalanche accidents during mountaineering is still high among winter leisure activities, but the rate of avalanche accidents at ski areas has increased remarkably since the 1960s. This increase corresponds to the development of large-scale ski areas at high elevations since the 1960s. Considering the recent trend in skiing, the avalanche accidents related to ski activities are expected to increase.

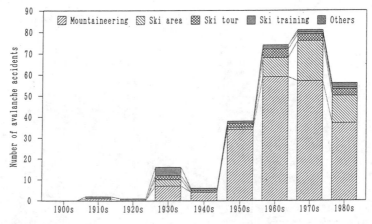

Figure 6. Decade-variation of avalanche accidents related to leisure distributed into five activities.

6.1 Monthly frequency of avalanche accidents

Monthly frequency of avalanche accidents related to leisure is shown in Figure 7 in comparison with that of the other avalanche accidents. As seen in the figure, the maximum frequency of the other avalanche accidents is found in February when snow depth usually reaches the maximum. On the contrary, the avalanche accidents related to leisure have three peaks of frequency in January, March and May. This is because these months have successive holidays when a lot of people tend to enter mountainous areas for leisure activities.

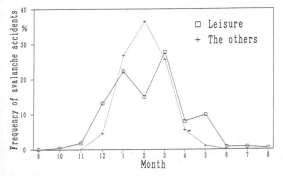

Figure 7. Monthly frequency of avalanche accidents related to leisure and the others.

6.2 Comparison of fatalities in Nagano and Niigata prefectures

The avalanche accidents related to leisure have occurred mainly in Nagano and Gunma prefectures, which have high mountain ranges. Meanwhile, the other avalanche accidents have occurred mainly in Niigata, Yamagata and Fukushima prefectures. Now we choose Nagano and Niigata prefectures as the former type and the latter type, respectively, and compare avalanche fatalities of these two prefectures.

The avalanche fatalities in Nagano and Niigata prefectures were distributed into six subjects: road, railway, working, residence, leisure and others (Figure 8).

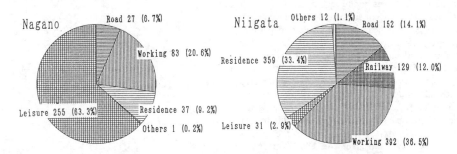

Figure 8. Avalanche fatalities in Nagano and Niigata distributed into six subjects.

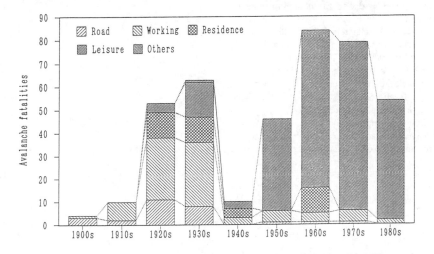

Figure 9. Decade-variation of avalanche fatalities in Nagano distributed into five subjects.

In Niigata, working and residence occupies about 70% of the total fatalities. On the contrary, in Nagano the leisure occupies more than 60% of the total fatalities. There are no railway fatalities in Nagano.

The avalanche fatalities in Nagano were summed up every decade and distributed into above categories (Figure 9). As seen in this figure, before the 1940s, a lot of fatalities in Nagano related to working and residence as well as in Niigata. Since the 1950s leisure accounts for more than 80% of avalanche fatalities. This corresponds to the fact that, after the Second World War, winter leisure activities have become popular, and avalanche accidents concerning leisure have increased as mentioned above.

7 CONCLUDING REMARKS

Data on avalanche accidents in Niigata and neighboring prefectures of the central part of Japan were collected mainly from local newspaper articles and statistically analyzed for the past ninety years (1900-1989). The following results were obtained :

More than 3300 avalanche accidents and 2300 avalanche fatalities have occurred in these prefectures. More than 1100 avalanches killed or injured people. Among these prefectures Niigata has the largest number of avalanche accidents and fatalities, which correspond to about half of the total. In the 1930s and 1960s avalanche accidents occurred frequently due to repeated heavy snowfall winters in this region.

Avalanche accidents related to traffic (railway and road) account for about 60% of the total for the ninety years. The majority of them before the 1950s occurred on railways, and after the 1960s on roads. Working in mountainous areas accounts for about half of the avalanche fatalities for the ninety years. But avalanche accidents and fatalities related to working have decreased recently due to the changing industrial structure of Japan.

Avalanche accidents related to leisure activities, such as mountaineering and skiing, have increased rapidly after the Second World War. The remarkable increase since the 1960s, at ski areas corresponds to the development of large-scale ski areas at high elevations. Their monthly frequencies peak in January, March and May, which have holidays convenient for leisure activity. They occur mainly in Nagano and Gunma which have high mountain ranges. In Nagano since the 1950s more than 80% of avalanche fatalities relate to leisure.

This research was focused on the heaviest snow region of Japan, Niigata and neighboring prefectures. We intend to extend this research to the whole of Japan.

REFERENCE

Nakamura, T., H. Nakamura, O. Abe, M. Higashiura, N. Numano, H. Yuuki, S. Kutsuzawa, Y. Yamada & T. Ikarashi 1987. Prefectural distribution of disastrous snow avalanches and prediction of them in the northern parts of Japan. *Avalanche Formation, Movement and Effects*, IAHS Publ. no. 162: 639-646.

Snow Engineering: Recent Advances, Izumi, Nakamura & Sack (eds) © 1997 Balkema, Rotterdam. ISBN 90 5410 865 7

Performance testing and effectiveness of avalanche blaster Gaz.Ex in Japan's central mountains

Isao Kamiishi
ARGOS Co. Ltd, Arai, Niigata, Japan

Norio Hayakawa & Yusuke Fukushima
Nagaoka University of Technology, Niigata, Japan

Kunio Kawada
Toyama University, Japan

Masanori Yamada
Arai Resort Co. Ltd, Niigata, Japan

ABSTRACT: Performance of an avalanche inducing device, Gaz.Ex, in Japan's central mountains is reported, and this device has a good probability of inducing an avalanche. The best condition among others to induce an avalanche is found to be heavy snowfall. Of a large-scale avalanche released with use of the Gaz.Ex, the dynamics of avalanche movement is simulated with a set of ordinary differential equations and showed a good agreement with the observed data in terms of the avalanche velocity.

1. INTRODUCTION

Generally, avalanche control technology in Japan relies heavily on anti-avalanche structure. Inducing artificial avalanche is seldom practiced expect during the spring season. In recent years, however, large-scale development of the mountainous area has been attempted with the result of increased danger of avalanches encountered in high mountains.

An avalanche blasting device Gaz.Ex, originally developed in France, has been widely used in a number of countries. In most cases, however, this apparatus has been used in mountains where winter is very cold and snow dry. This study intends to establish the effectiveness of Gaz.Ex. in Japan's central mountains where winter weather is rather mild and snow is wet. The Gaz.Ex. device to be tested was installed in the Slope of Arai Ski Resort on the Ougenashi mountain where elevation is 1,200 m and slope is about 40 degree (Figure 1). This Gaz.Ex was installed in the winter of 1992 and has been used ever since. Moreover, various aspects of its performance have been under study.

In this paper, performance of Gaz.Ex. is investigated to show its effectiveness. Among many studied items so far, numerical simulation of avalanche movement induced by Gaz.Ex. is then presented.

2. PERFORMANCE OF Gaz.Ex. FIRING.

Gaz.Ex. has been operated by the personnel of the Arai Ski Resort almost daily and record of the occurrence of avalanches together with the weather and snow conditions has been taken. For two winter seasons of 1993-94 and 1994-95, Gaz.Ex has been fired for 47 days and avalanches have been successfully induced for 25 days, with more than

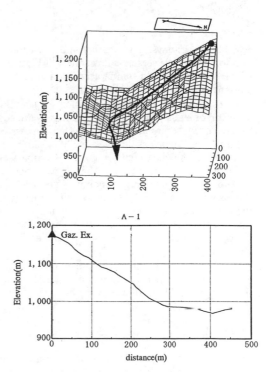

Figure 1: Perspective view and longitudinal profile of Gaz.Ex. testing field.

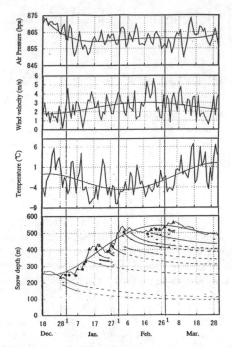

Figure 2: Snow and weather at Ougenashi Mt. in 1994-1995 (Filled circles indicate Gaz.Ex. induced avalanches).

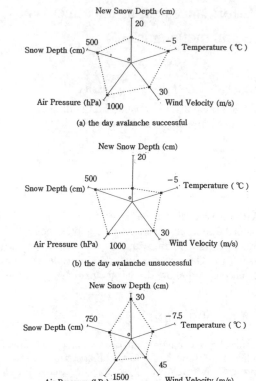

(a) the day avalanche successful

(b) the day avalanche unsuccessful

(c) the previous day avalanche successful

Figure 3: Weather conditions at Gaz.Ex. firing.

50% of success days. The weather and snow condition for 1994-1995 season is illustrated in Figure 2, which shows the mild temperature and enormous amount of snow on this mountain.

To examine the favorable condition for successful firing of avalanches, weather conditions are collected and data are averaged for either success or failure in inducing avalanches. The results are plotted in radial diagrams in Figure 3. Comparison of these three figures indicates that an avalanche is best induced when a large amount of snow fell on the previous day (Gaz.Ex. has been fired mostly in the early morning) and avalanche inducing is not successful when there is little new snow. Other weather conditions do not seem to matter, judging from this diagram.

It may be safely concluded that avalanche inducing by Gaz.Ex. is successful in this kind of mountain if it is fired immediately after heavy snowfall.

3.LARGE-SCALE AVALANCHE AND SIMULATION OF AVALANCHE DYNAMICS.

As long as practice is firing of Gaz.Ex. on a daily basis, large-scale avalanches do not seem to occur often. There are, however, times when a large-

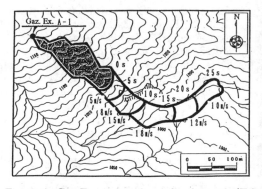

Figure 4: Gaz.Ex. induced anvalanche path (Feb. 17, 1992) (Velocity and lapse time indicated).

scale avalanche is induced. This happens under a combination of favorable conditions, notably existence of a weak layer within the snow. Such an avalanche was observed on February 27, 1992, with width of avalanche 100 m, run-out distance of 600 m and depth of 1.2 m. Its movement was recorded

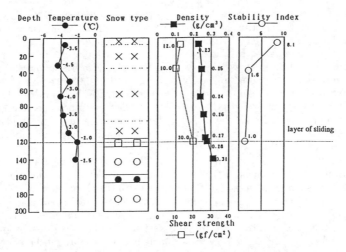

Figure 5: Snow layer profile at Gaz.Ex. station.

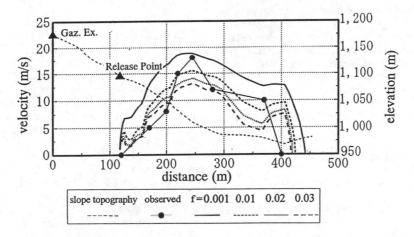

Figure 6: Velocity of avalanche front.

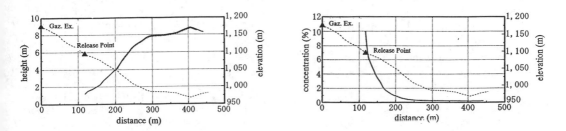

Figure 7: Results of numerical simulation (a) Height of avalanche and (b) Snow concentration.

by a video camera and is reproduced in Figure 4. The total volume of snow moved is estimated as 24,000 cubic meters. A snow pit observation around this avalanche is illustrated in Figure 5, which shows existence of a 1.20 m thick layer lying over bottom layer, with a thin fragile layer of solid-type depth hoar between. In Figure 5, stability index is calculated with shear strength divided by the gravity force in the direction of the slope. The figure shows that the upper layer is at the brink of sliding at the fragile layer.

Figure 4 tests enebles one to the study on the dynamics of avalanche movement. Density of snow and visible observations indicate that this avalanche was a flow-type avalanche and not a powder-snow type avalanche. Powder snow avalanche movement was analyzed by Fukushima and Parker (1990). Here, analysis of flow-type avalanche is formulated and applied to this record. The major difference between flow-type avalanche and powder snow is that the latter does not involve much turbulent motion. Otherwise the fundamental equations to govern the avalanche motion are written in a similar fashion. They are the equation of continuity of air inside an avalanche, conservative equation of snow particles, momentum equation and the equation to define the location of the avalanche. They are written as follows:

(1) Equation of continuity for air contained in the avalanche:

$$d(\rho A)/dt = \rho_a P E U \qquad (1)$$

(2) Equation for continuity of snow:

$$d(CA)/dt = 0 \qquad (2)$$

(3) Conservation equation for momentum:

$$\frac{d(\rho U A)}{dt} = (\rho - \rho_a)gA\sin\theta - \frac{1}{2}\rho_a c_D U^2 h$$

$$- \frac{1}{2}\rho_a f_i U^2 P - \frac{1}{2}\rho_b f_b U^2 B \qquad (3)$$

(4) Equation for avalanche posotion.

$$ds/dt = U \qquad (4)$$

where ρ and ρ_b are the avarage and near bed density of avalanche, respectively, A is the cross-sectional area, t is time, ρ_a is the density of air, P is the length of avalanche-air interface, E is the entrainment coefficient, U is the avalanche velocity, C is the volumetric concentration of snow particles in an avalanche, θ is the slope angle, c_D is the drag coefficient, h is the height of the avalanche,

f_i, f_b are the friction coefficients, B is the length of the avalanche-earth interface and S is the distance along the slope.

The calculation is run with a drag coefficient of $c_D = 0.6$, and a friction coefficient f_i, f_b changing from 0.001 to 0.03. The result for avalanche velocity is illustrated in Figure 6. This figure shows that the avalanche front is accelerated at the distance 120 to 250 m, where slope angle is very steep, and is decelerated to stop at mid slope around 250 to 400 m away.

The calculation is insensitive for change of friction coefficient within this range and agrees very well with the observed data. Figure 7(a) is the calculation result for the avalanche height, in which the initial condition is set as 1.2 m from the observation. This figure shows the growth of avalanche height along the avalanche course. Figure 7(b) is also gives the calculated result for snow particle concentration of the avalanche. It shows that the concentration decreases rapidly after initial rise.

4. CONCLUSIONS

An avalanche blaster Gaz.Ex. is being used in one of Japan's central mountains, the Ougenashi Mountain where the winter weather is mild and snow is wet and heavy. The record of its daily use shows that the success of releasing an avalanche reaches 50%. Further analysis of the firing date shows that avalanche release is most successful after a heavy snowfall, irrespective of other weather and snow conditions.

Large avalanches can be released under a combination of favorable conditions, notably existence of a fragile layer within the snow. Such a case is reported and snow-pit data show the existence of such a fragile layer and that the upper layer is on the brink of sliding. Using the data of avalanche movement, avalanche movement dynamics is analyzed formulating a set of ordinary differential equations, integrating them numerically. The calculation results show good agreement with the observed data in terms of avalanche velocity.

REFERENCES

Fukushima, Y. & Parker, G. 1990. Numerical simulation of powder snow avalanches, *J. Glaciology*, 123.

Snow Engineering: Recent Advances, Izumi, Nakamura & Sack (eds) © 1997 Balkema, Rotterdam. ISBN 90 5410 865 7

Experience with snow measurement devices in avalanche research

Peter Höller
Institute for Torrent and Avalanche Research, Federal Forest Research Center Innsbruck, Austria

ABSTRACT: The paper describes some different snow measurement devices we used in avalanche research, especially for the project "Avalanche Formation in Mountain Forests." Moreover, the experience we made with the different systems is presented.

1 INTRODUCTION

Several years ago our institute started with detailed investigations concerning the structure of snow cover and the phenomenon of snow gliding in mountain forests. The purpose of this study is to find out if avalanches are possible in openings and low density forests. As the necessary condition for snow slab release is the existence of a weak layer within the snowpack (Gubler & Salm 1992), we have to concentrate on snow cover and snow gliding investigations in different parts of mountain forests (dense forest, low density forest, opening in a forest, etc.).

The objectives of the study can be summarized as the following questions:
- Does structure of snow cover (snow gliding) depend on canopy density, particularly stand density?
- Is there a minimum extension of gaps and openings so that weak layers are possible within the forest snowpack?
Results of this study can be found in other papers (Höller 1995); this paper deals with the different snow measurement devices we used in the last years.

2 THE SNOW PARAMETERS

Besides measuring different meteorological parameters such as air temperature, relative humidity, wind speed, wind direction, solar radiation and radiation balance, it was necessary to register typical snow parameters, too.

The first and perhaps most important parameter is the total depth of snow cover (HS). Although the depth of new snowfall (HN) is more relevant for snow slab release, it is possible by continuous measurement of HS to calculate HN for a given period. HN is mainly responsible for additional stress on the present snow cover.

In order to get more information on snow metamorphism at different sites of mountain forests (dense stands, low density stands, openings in a forest, etc.) we are also measuring snow temperature, particularly temperature gradient. Especially the formation of depth hoar depends on high temperature gradients within the snow cover; the higher the gradient, the faster the growth rates. So, if there will be found higher temperature gradients in the snowpack of gaps or openings (or in forests with a low canopy density), this will result in an increasing of depth hoar crystals. Thus exists one of the necessary conditions (i. e., a weak layer) for the formation of slab avalanches.

To understand the formation of surface hoar it is necessary to record snow surface temperature, too (including air temperature and rel. humidity). Surface hoar is one of the most common weak layers within the snowpack and arises during clear nights as a result of cooling the snow cover by long wave radiation. Surface hoar regularly will be found on open fields, but may exist in low density forests, too. If so this would be an additional condition for snow slab release in mountain forests.

Last but not least we also have to take into account the snow gliding phenomenon. Observations of the last years showed that snow gliding (snow gliding is a downhill motion of snow cover on the ground) is very frequent in low density forests. The snow movement sometimes results in small glide avalanches. The aim of these investigations is the continuous recording of glide rates (particularly glide velocities) in different parts of mountain forests.

3 INSTRUMENTS AND SENSORS

To measure total depth of snow cover we tested two types of ultrasonic depth gauges (Fig.1). This system was developed several years ago (Gubler 1981) and is regularly used in avalanche research and avalanche warning. It measures the time needed by an ultrasonic signal transmitted by the sensor above the snow surface and reflected from the surface back to sensor.

It is important to know that speed of sound in air depends on air temperature (331.4 m/s at 0°C, increasing with higher temperatures) so it is necessary to take into account this parameter in the final calculations.

Our ultrasonic depth sensors come from two different companies. All types of sensors include temperature sensors, which means that it is not necessary to compensate the measured values with a special software. The measuring range of two types is between 0 m and 10 m. Within periods of heavy snowfall we observed that the ultrasonic signal was reflected by falling snowflakes and not at the snow surface; this results in a number of errors. A similar problem regularly occured in periods with a snow surface of light powder snow; so the signal was not reflected at the surface, but in a more

deeper layer of the snowpack. To prevent such problems we also used a new sensor type with a maximum distance of 20 m. Our measurements showed that there were no failing values with these more powerful types of ultrasonic device.

The snow temperature measuring device was designed at our institute (Lindner 1992). We used a white painted pole of polyacetal (a special synthetic material) to minimize heat conductivity and warming of the pole as a result of solar radiation. To measure snow temperature in different depths of the snowpack (temperature gradient), we mounted platinum RTDs every 20 cm on the pole (Fig. 2). The sensor is a platinum element of a characterstic resistance (100 ohms at 0°C). To reduce resistance of the cable we used 4 -wire connections. In that case the sensor is connected to a constant current; thus a voltage drop can be measured at the datalogger. Our platinum elements meet the German Industry Standard (DIN). The sensors are fitted into small tubes so that they cannot be destroyed by snow load or creep and glide processes. To minimize warming of the tubes they are covered with a white film.

The experience with the equipment was good. Only the sensors lying next to the snow surface sometimes were influenced by solar radiation; data of this

Fig. 1: Snow depth gauge, infrared thermometer and different radiation instruments.

Fig. 2: Snow temperature measuring device.

sensors cannot be used for further processing. An example for snow temperature measurements is shown in Appendix A.

As I have mentioned just before it is very difficult to measure surface temperature of snow because of solar radiation. We chose an infrared thermometer (Fig.1). This type of thermometer (Tränkler 1990) is based on the fact that each body transmits electromagnetic radiation. Snow is almost a black body (in the infrared wave length); the emissivity is between 0.82 and 0.99 (Sellers 1965). The maximum radiation of snow (temperature -10°C) will be found near to the wave length of 11 μm (law of Wien). The lens system of the sensor consists of a concave mirror; it works within the wavelength of 8 μm to 14 μm. While the sensor worked well down to -30° C, we had problems with the electronic equipment, because it was not compensated for temperatures below the freezing point. After doing some modifications and additional tests we verified the data with calculated surface temperatures from radiation measurements (by using the Stefan-Boltzmann equation). The results of measured and calculated snow surface temperatures are shown in Appendix B.

For the snow gliding measurements we used glide shoes designed by In Der Gand (1954). The recording system was developed by Lindner (1992), and two reading units were recently used. The first one is equipped with microswitches, which transmits impulses to the datalogger every time when a glide shoe moves more than 2 mm. For the second system we used special potentiometers; thus parameters are stored in the form of resistance values. We found that the recording system based on impulse count works well if slow movements have to be registered only. Otherwise it is necessary to use dataloggers with very short recording intervals (50 ms). If using potentiometers one can register fast gliding movements, too (for example, glide avalanches).

4 DATALOGGERS, TRANSFERRING OF DATA

As some of our study fields are situated in high alpine regions and power supply is not available, we were forced to use solar energy and storage batteries. Therefore it is impossible to use devices with a high power consumption (for example, we cannot have a heating for the radiation sensors). Regularly we had no problems with our solar energy equipment, but sometimes (especially if the solar panels were covered with snow over a couple of days) we observed power failure so that the loggers didn't record actual data (only stored data were preserved). In general we can say that the dataloggers worked well during the different winter periods (except in case of power failure, see above).

All data are measured in intervals of 15 min or 30 min, except snow gliding, which is registered anytime if the glide rate exceeds 2 mm. The memory capacity of our loggers is between 40 kB and 170 kB and depends on logger type; so it is (theoretically) possible to have measurements over two and more months without any reading. The question of how to transfer data for further processing was not really a great problem. We discussed two variants, via cellular phone and modem, or recording data by a transportable PC. While the first possibility is more comfortable, the second one is cheaper. As it is necessary to go to the field every two weeks (measuring snow profiles, checking sensors, loggers and other devices...) we decided for the cheaper solution.

5 CONCLUSIONS

To summarize our experience with snow measurment devices, we can say that the different sensors (partly modified by our technicians) and loggers worked well during the last winter periods.

Although the complete device is working automatically it is necessary to check the equipment regularly. Only personal and regular checks assure trouble-free running and avoids questionable data.

REFERENCES

Gubler, H. 1981. An inexpensive remote snow-depth gauge based on ultrasonic wave reflection from the snow surface. *Journal of Glaciology*. Vol.27, No. 95:157- 163.

Gubler, H. & B. Salm 1992. Bildung von Schnee-brettlawinen. *Proc.Int. Symposium Schiberg-steigen, Rudolfshütte, April 1992*, 5-15.

Höller, P. 1995. Snow gliding in subalpine forests: *Proc. XX-IUFRO World Congress, Finnland, August 1995, Technical Session on Natural Disasters in Mountainous Areas*, 217-221.

In Der Gand, H. 1954. Beitrag zum Problem des Gleitens der Schneedecke auf dem Untergrund. *Winterbericht d. Eidgen. Inst. f. Schnee- u. Lawinenforschung* 17.

Lindner, S. 1992. Entwicklung einer meteorolo-gischen und nivologischen Meßstation für das Projekt „Lawinenbildung im Schutzwald". *Lawinenkundliche Informationsblätter, FBVA- Wien/Innsbruck* 3.

Sellers, W. 1965. *Physical Climatology*. The University of Chicago Press, Chicago & London

Tränkler H.-R., 1990 *Taschenbuch der Meßtechnik*. Oldenbourg Verlag München Wien.

Appendix A: Example for snow temperature measurements, study field Kaserstattalm winter 1994-95

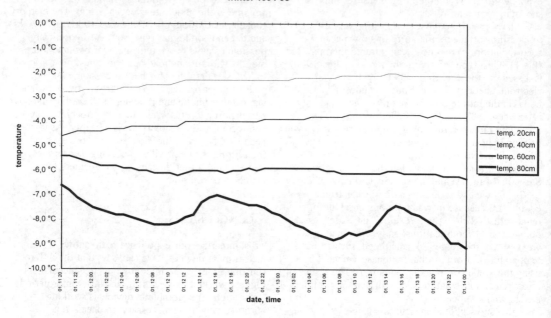

FBVA 1996 Lindner

Appendix B: Example for snow surface temperature measurements, study field Kaserstattalm winter 1994-95

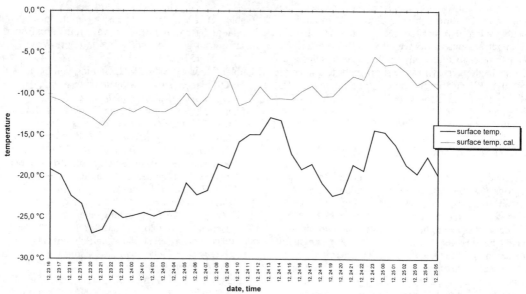

FBVA 1996 Lindner

Snow Engineering: Recent Advances, Izumi, Nakamura & Sack (eds) © 1997 Balkema, Rotterdam. ISBN 90 5410 865 7

A measuring system for snow depth profiles in Maglev guideway using the light cross section method

Katsuhisa Kawashima, Shigehiro Iikura, Toru Endo, Toshishige Fujii & Toshiaki Imai
Railway Technical Research Institute, Tokyo, Japan

Tatsuhide Nakane
Kokusai Kogyo Co., Ltd, Tokyo, Japan

ABSTRACT: Aiming at controlling the running of Maglev (magnetically levitated) vehicles under snowfall, a measuring system for profiling snow depths in a U-shaped Maglev guideway by image processing was developed using the light cross section method. This system, composed of control and data analysis devices, a CCD-camera unit and a semiconductor laser scanning unit, measures the snow depth at 80 points in 2 minutes for every snow cover profile 4.6 m in width. The field tests revealed that the system was capable of measuring the profiles with an accuracy of ±10 mm except when the snow cover was saturated with water, irrespective of the measuring conditions such as the weather and the type of snow.

1 INTRODUCTION

The superconducting magnetically levitated system (Maglev) has been studied at the Railway Technical Research Institute, Japan, since 1970, and running tests of the test vehicles have been continued for more than 15 years. In 1990, the development of the system was authorized as a national project. It is now in final confirmation situation for practical use (Nakashima 1994).

Maglev vehicles run on the U-shaped guideway at high speed, levitating as high as 10 cm above it. If the snow accumulates on the guideway exceeding the levitation height, it may be highly resistant and obstructive to the running of Maglev vehicles, in which case proper countermeasures must be taken. Partly because the cross section of the guideway has projections such as sidewalls and tracks, and partly because the erosion-deposition phenomena take place at the snow surface in a strong wind, an uneven surface structure of deposited snow often occurs in the guideway. Thus, the evaluation of safety based on the measurement of snow depth profiles is necessary for the running of Maglev vehicles.

This paper describes the principles, construction and implementational aspects of a measuring system for snow depth profiles which we have developed using the light cross section method.

2 BASIC PRINCIPLES

The light cross section method employed in this study can be regarded as a 2-dimensional expansion of an optical range finder. It is sometimes used for the measurement of surface roughness and recognition of polyhedrons (Shirai 1971, Takeshita et al. 1993). This is the first attempt to apply the method to the measurement of snow. Although the slit beam was used as a light source in most cases up to now, we tried to employ a scanning spot beam to function in the harsh measuring environment.

2.1 Determination of snow depth

Suppose that a laser beam is vertically irradiated from the point S to the snow cover having a thickness of H_S and a picture of a projected bright spot P on the snow surface is taken diagonally by a CCD-camera at a distance d from a laser beam source, as shown in Figure 1(a). The points O, C are the foot of the perpendicular from the point S to the reference plane (horizontal plane) and the principal point of the lens, respectively. If the CCD-camera is installed so that the optical axis of the lens conforms to the line CO, the points $R(0,0)$, $Q(x_i, y_i)$ represent the center of the image plane (the origin of the coordinate axes on the image plane) and the image of the point P on the image plane. In such a case, the snow depth can be calculated by means of trigonometry and is expressed as a function of y_i:

$$H_S = \frac{y_i H}{(f \cos\theta + y_i \sin\theta)\sin\theta}, \tag{1}$$

where H is the distance between the point C and the reference plane, f is the focal length of the lens, and

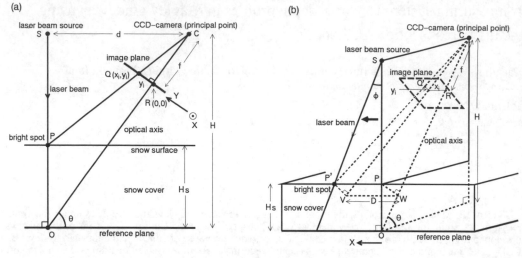

Figure 1. Schemes for measuring snow depth profiles using the light cross section method.

$\theta = \tan^{-1}(H/d)$.

2.2 Determination of measuring position

If pictures of a projected bright spot on the snow surface are taken by a CCD-camera at regular intervals while the laser beam scans in the direction of the X-axis, as shown in Figure 1(b), we can also obtain values of snow depth by use of Equation (1). The measuring position, however, differs depending on the snow depth even if the beam angle ϕ is the same. Let the points P' and Q' be the center of the projected bright spot on the snow surface and the image of the point P', respectively. The distance between the line SO and the point P', which is equal to the length of the segments PP' and WV, can be defined as a measuring position, and its value D is calculated by means of trigonometry:

$$D = \frac{x_i}{f}\left(\frac{H}{\sin\theta} - H_s \sin\theta\right). \qquad (2)$$

3 CONSTRUCTION OF SYSTEM AND MEASURING PROCEDURE

The system is composed of control and data analysis devices, a CCD-camera unit and a semiconductor laser scanning unit, as shown in Figure 2. The snow cover on the Maglev guideway is scanned by the collimated laser beam (wavelength, 830 nm; output, 170 mW), by $0.72°$ beam angle at a time, with a pulse motor controlled by a personal computer. At every position of the projected bright spot on the snow surface, ten pictures taken by a CCD-camera with a wide-angle lens (f=6.5 mm) and 768×493 picture elements are transmitted to the video tracker (a real-time tracking and coordinate measuring device) so that ten coordinates of the spot center on the image plane are obtained by means of binary image processing in an instant and sent to the personal computer to calculate the snow depth and measuring position using Equations (1) and (2). Then, the computer directs the pulse motor to rotate the laser beam by $0.72°$ and the next measurement commences. After a series of measurements along the cross section of the guideway are finished, the storage of data and the display of snow depth profile on CRT are done. We adopt the mean values of ten measurements as measured results. It takes 2 minutes to complete the measurement of a profile 4.6 m in width at 80 points.

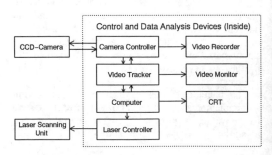

Figure 2. Block diagram of the measuring system for snow depth profiles in a Maglev guideway.

front view

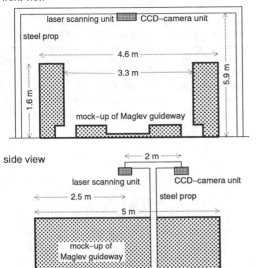

side view

Figure 3. Configuration of the CCD-camera and laser scanning units on the mock-up of the Maglev guideway.

4 EXPERIMENTAL MEASUREMENT

The CCD-camera and laser scanning units were made drip-proof and installed outdoors at the Shiozawa Snow Testing Station, 2 m apart, 5.9 m above a mock-up of the Maglev guideway, as shown in Figure 3, to verify implementational aspects and the accuracy of the system using natural snow.

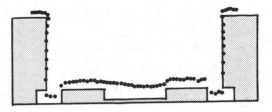

Figure 4. Example of measured snow depth profile in the Maglev guideway during snowfall.

4.1 Implementational aspects

An example of a snow depth profile in the Maglev guideway measured by the system during snowfall is shown in Figure 4. Weather conditions such as snowfall and rain were found not to exert an influence on the measurements. The system can detect not only the snow surface but also the inner faces of vertical sidewalls, indicating that the detection of snow accretion on the inner side faces of the guideway is also possible.

Each measured point in Figure 4 is the mean value of ten measurements as mentioned before. Although the large scatter in the data of ten measurements was seen in the case of dry snow and wet snow with low water content, as shown in Figure 5(a), each measured point distributes both above and below the snow surface uniformly to result in the approximate agreement between the mean value and height of snow. Contrary to this, the laser beam is transmitted into the snow in the case of water-saturated snow, so that all the measured points exist below the snow surface (Figure 5(b)).

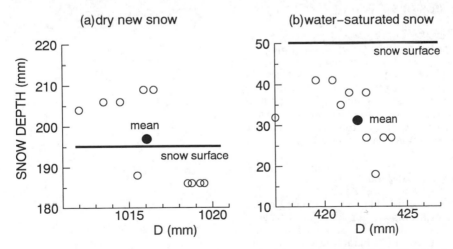

Figure 5. Distributions of original data (open circles), together with their mean values (solid circles) and heights of snow surface measured by a ruler (solid lines).

107

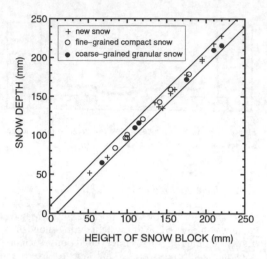

Figure 6. Comparison between height of snow block and snow depth measured by the system.

4.2 Accuracy

Accuracy was verified by measuring snow blocks of known height with a fixed laser beam. The snows used were new snow, fine-grained compact snow and coarse-grained granular snow. The snow depth measured by the system is shown in Figure 6, along with the height of snow block measured precisely with a steel ruler. The differences between them do not exceed 10 mm, regardless of snow type. Thus, this system was capable of assuring accuracy within ±10 mm.

5 CONCLUSIONS

A measuring system for snow depth profiles in a Maglev guideway was developed using the light cross section method. The measurement accuracy of the system was ±10 mm, which suffices for practical use, showing that this method is available for the measurement of snow. This system is believed to be applicable to other situations such as railway tracks, roads and roofs as a monitoring system for the prevention of snow damage.

ACKNOWLEDGMENTS

The authors would like to acknowledge the assistance of Messrs. T. Suzuki and T. Tomita of OKK Inc. for developing the software of the system and Mr. M. Ozawa of Kokusai Kogyo Co., Ltd. for measurements at the Shiozawa Snow Testing Station. This work was financially supported by the Japan Ministry of Transportation.

REFERENCES

Nakashima, H., 1994: The status of Maglev system development. *Quarterly Report of RTRI*, 35(3), 145-149.

Shirai, Y., 1971: Recognition of polyhedrons with a range finder. *Bulletin of the Electrotechnical Laboratory*, 35(3), 290-296.

Takeshita, K., K., Takagi and Y., Narita, 1993: Method of rail profile measurement with light section method. *RTRI Report*, 7(11), 47-54. (in Japanese with English abstract.)

Snow Engineering: Recent Advances, Izumi, Nakamura & Sack (eds)© 1997 Balkema, Rotterdam. ISBN 90 5410 865 7

Areal observation of blowing snow using a telemetering network of snow particle counters

Kenji Kosugi, Takeshi Sato & Masao Higashiura
Shinjo Branch of Snow and Ice Studies, NIED, Yamagata, Japan

ABSTRACT: This paper describes a new system for areal observation of blowing snow. A telemetering system of snow particle counters (SPCs), the photoelectronic instruments measuring the number and size of blowing snow particles, was developed in the Ishikari Plain, Hokkaido, Japan. Data measured at two observation sites were automatically transmitted every one minute to the main station. The results of an observation in January 1994 showed an ability of the system for areal observation of blowing snow.

1 INTRODUCTION

Blowing snow often causes serious problems in snowy regions, such as sudden obstruction to visibility and formation of snowdrifts on roads. It is necessary to clarify the nature of blowing snow to make an efficient forecast and countermeasures for it.

Many measurements have been conducted on blowing snow at a single observation site using trenches or blowing snow particle collectors (e.g., Budd et al. 1966, Kobayashi 1972, Kobayashi 1978, Takeuchi 1980). The results are summarized as functions of the wind velocities and other physical parameters. These studies revealed characteristics of blowing snow averaged for several minutes or more, because it takes several minutes to complete a measurement in these kind of observations.

Kobayashi (1972), Takeuchi (1980) and Higashiura et al. (1993) measured the mass of snowdrift or mass of transported snow particles with trenches or snow collectors as a function of horizontal distance. The results showed areal variations of the intensity of blowing snow in a distance range from tens meters to some kilometers averaged over some tens minutes to some hours. However, it is difficult to carry out simultaneous measurements at distant sites continuously in higher time resolution with these methods.

Recently, photoelectronic instruments for measurements of blowing snow, called snow particle counters (SPCs), were developed (Schmidt 1977, Gubler 1981, Brown and Pomeroy 1989). A SPC produces voltage pulses for each snow particle

passing through its sampling area. Schmidt (1982) and Schmidt (1986) used SPCs to measure vertical profiles of blowing snow. After that, Kimura (1991) and Sato et al. (1993) improved SPCs and confirmed a sufficient performance of the SPCs in the mass flux measurements of blowing snow.

In the present study, surface observations of blowing snow are carried out using a telemetering network of SPCs to investigate the characteristics of areal variations of blowing snow. This paper reports the outline of the observation system and some preliminary analysis of a blowing snow event observed in January 1994. This study is a part of the research project "Study on areal prediction techniques of drifting snow and development of warning system", conducted by the National Research Institute for Earth Science and Disaster Prevention. The purpose of the project is to clarify the whole structure of snowstorms and the characteristics of temporal and spatial variations of blowing snow with a combination of radar and surface observations toward the development of warning system of blowing snow (Higashiura et al. 1994).

2 OBSERVATION

2.1 *Observation area*

The observations were conducted in the northern part of the Ishikari Plain in Hokkaido, Japan (Figure 1). In 1994, two SPC observation sites were made in Maeda, Sapporo. The direction connecting these sites is roughly in parallel with the northwesterly

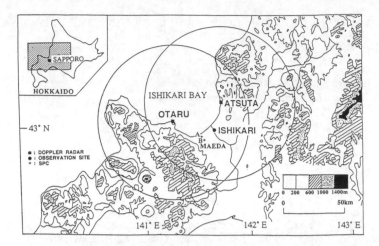

Figure 1. Topography of the Ishikari Plain and observation sites. A and B indicate the SPC observation sites. The solid dot indicates the Ishikari Meteorological Station. The large circles show the observation scope of the Doppler radars.

wind, which is the prevailing wind in winter of this area. The sites were 2 km apart from each other.

The main station was established in Ishikari. Detailed meteorological observations, including blowing snow measurements, were carried out in this site. Some of the results are reported by Sato and Higashiura (1996). All the data from the SPC observation sites were collected at this station.

2.2 Observation system

Figure 2 shows a schematic illustration of the observation system of this study. The SPC sensors of the sites were mounted at a height of 0.5 m above the surface of the snow cover. The signals of snow particles from the sensors were processed every 10 s and transmitted every 1 min. to the Ishikari Meteorological Station via telephone lines. The data were collected by a multiplexer and stored by a personal computer on magneto-optical disks.

3 RESULTS AND SUMMARY

In January 1994, a continuous observation was carried out. Here an event of blowing snow is analyzed. The event took place from 20:32 to 21: 37 on Jan. 27. Figure 3 shows meteorological conditions around the event at a station of the Automated Meteorological Data Acquisition System near the SPC observation sites. Westerly winds of 4–5 m/s blew and the air temperature was −8 °C during the event.

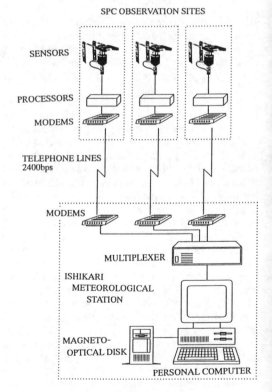

Figure 2. Schematic illustration of the observation system.

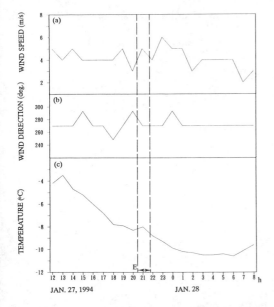

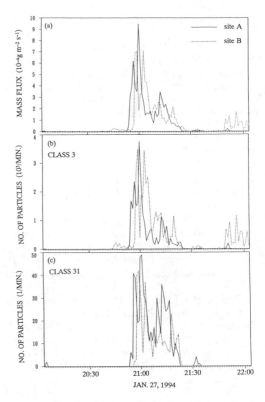

Figure 3. Meteorological conditions of the SPC observation sites; (a): wind speed, (b): wind direction, (c): temperature. E indicates the period of the blowing snow event analyzed in this paper.

Figure 4. Mass flux (a) and number of particles passing the sampling area of the SPCs, 2 mm x 25 mm, in the diameter classes 3 (115–130μm) (b) and 31 (535–550μm) (c) during the blowing snow event. Solid lines and dashed lines indicate SPC observation sites A and B, respectively.

Figure 4 shows mass flux and number of particles in diameter classes 3 (115–130μm) and 31 (535–550μm) at sites A and B during the blowing snow event. It is found that the intensity of blowing snow sometimes changed by an order of magnitude in a few minutes. Particles in diameter classes 3 and 31 are those of blowing snow and precipitating snow, respectively. As shown in the figures, blowing snow occurred with snow precipitation during the event. Relations are observed between number of particles in diameter class 31 and that in diameter class 3 or mass flux. It seems that the blowing snow intensity depended mainly on the intensity of snow precipitation during the event.

A correlation is also seen between values of site A and those of site B in Figure 4. Figure 5 shows cross–correlation coefficients between values of sites A and B as a function of time lag. A time lag of 2 or 3 min. gives the largest correlation coefficients. The maximum cross–correlations are significantly large. This probably implies the snow blowing area moved from site A to site B because of the advection of the clouds feeding the snow precipitation.

As a summary, it is shown that the telemetering network of SPCs has a practical ability for areal observation of blowing snow. It will be a useful tool to study the characteristics of blowing snow distribu-

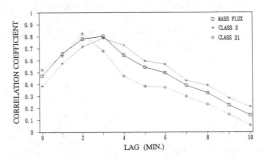

Figure 5. Cross–correlation coefficients of mass flux and number of particles between SPC observation sites A and B.

tion when the area of the network is spread. The number of the observation site has been increased to 6 by January 1996. Analyses in combination with radar sounding of snow clouds has also been started.

111

ACKNOWLEDGEMENTS

We would like to thank the Hokkaido Development Bureau for permitting us to construct the main station in an experimental field of the Bureau. Thanks are also due to Dr. Ishimoto of the Civil Engineering Research Institute, Hokkaido Development Bureau, for help with the observation.

REFERENCES

Brown, T. & J.W. Pomeroy 1989. A blowing snow particle detector. *Cold Reg. Sci. Technol.* 16, 167–174.

Budd, W.F., W.R.J. Dingle & U. Radok 1966. The byrd snow drift project: outline and basic results. In M.J. Rubin (ed.), *Studies in Antarctic Meteorology (Antarctic Research Series*, vol. 9): 71–134. Washington, D.C.: American Geophysical Union.

Gubler, H. 1981. An electronic remote snow–drift gauge. *J. Glaciol.* 27: 164–174.

Higashiura, M., T. Sato, A. Sato, T. Kimura, M. Maki, S. Nakai, H. Nakamura & T. Yagi 1993. Areal investigation of drifting snow on Tsugaru Plain, Japan. *Ann. Glaciol.* 18: 155–160.

Higashiura, M., T. Sato, T. Kimura, M. Maki, & K. Iwanami 1994. Toward the development of areal warning system of blowing snow. *Proc. International Snow Science Workshop '94*: 203–210.

Kimura, T. 1991. Measurements of drifting snow particles. *J. Geogr.* 100: 250–263. (In Japanese with English summary)

Kobayashi, D. 1972. Studies of snow transport in low–level drifting snow. *Contributions from the Institute of Low Temperature Science, Ser. A* 24: 1–58.

Kobayashi, S. 1978. Snow transport by katabatic winds in Mizuho Camp area, East Antarctica. *J. Meteorol. Soc. Japan* 56: 130–139.

Sato, T., T. Kimura, T. Ishimaru & T. Maruyama 1993. Field test of a new snow–particle counter (SPC) system. *Ann. Glaciol.* 18: 149–154.

Sato, T. & M. Higashiura 1996. Characteristics of blowing snow fluctuation. *Proc. third ICSE (this issue)*.

Schmidt, R.A. 1977. A system that measures blowing snow. *U.S. Department of Agriculture, Forest Service Research Paper* RM–194.

Schmidt, R.A. 1982. Vertical profiles of wind speed, snow concentration, and humidity in blowing snow. *Boundary–Layer Meteorol.* 23: 223–246.

Schmidt, R.A. 1986. Transport rate of drifting snow and the mean wind speed profile. *Boundary–Layer Meteorol.* 34: 213–241.

Takeuchi, M. 1980. Vertical profiles and horizontal increase of drift snow transport, *J. Glaciol.* 26: 481–492.

Snow Engineering: Recent Advances, Izumi, Nakamura & Sack (eds)© 1997 Balkema, Rotterdam. ISBN 90 5410 865 7

The temporal and spatial distribution of snowmelt in the subalpine region, Mt. Hachimantai

A. Kojima, T. Ohta, T. Hashimoto & K. Suzuki
Faculty of Agriculture, Iwate University, Morioka, Japan

ABSTRACT: When water vapor pressure in subalpine regions becomes to be larger than 6.11 hPa in the snow-melt season, this feature effects on the temporal and spatial distribution of snowmelt. In this case, the percentage of net radiation for snowmelt become small because of the growth of latent heat flux. The spatial distribution of snowmelt becomes to be characterized by the spatial distribution not only of net radiation but of wind speed. The fluctuations in the energy for snowmelt are largely defined by the fluctuations in wind speed.

1. INTRODUCTION

The prediction of snowmelt rate is desirable in developing water management and preventing disasters caused by snowmelt water, especially in subalpine regions where accumulated snow represents much of the annual runoff. The features of snowmelt in this region in Japan are as follows: 1) Because of the great amount of snow, snowmelt continues from spring to summer and involves not only seasonal variations but also fluctuations of meteorological elements. 2) Snowmelt occurs in complex terrain and various altitudes simultaneously. Therefore it is important to clarify the effects of the temporal and spatial distribution of meteorological conditions on snowmelt. Although Ohta *et al*. (1994) analyzed the temporal distribution of snowmelt at Mt. Hachimantai, this paper discusses it with new data and adds the analysis of it's spatial distribution.

1.1 *Energy balance on the snow surface*

The energy balance on the snow surface is written as

$$Q_{SM} = Q_{NR} + Q_H + Q_E + Q_R + Q_B \qquad (1)$$

where Q_{SM} is the energy for snowmelt, and Q_{NR}, Q_H and Q_E are the net radiative, sensible and latent energy fluxes, respectively. The flux of heat from rain Q_R and the heat conduction Q_B can be disregarded because these are negligible. The trends of meteorological features during snowmelt at the regions can be assumed as follows. The albedo of snow surface α has a decreased trend because of the rise of the snow

density ρ_s and the accumulation of dirt on the surface. The daily sum of global solar radiation on a clear day $I\downarrow$ must increase from spring to summer. The global solar radiation condition of each day R that is expressed as dividing $I^\downarrow$ into the daily sum of solar radiation at the surface $S^\downarrow$ fluctuates every day. The daily sum of the downward longwave radiation $L^\downarrow$, the daily average of air temperature T_a, water vapor pressure e_a and wind speed U_a have not only a seasonal change but also a fluctuation range. Using these elements, we may write the ranges of Q_{NR}, Q_H and Q_E approximately as

$$Q_{NR}(t) = (0.5 + R')\{1 - \alpha(t)\}I^\downarrow(t) + \overline{L}^\downarrow(t) \pm L^{\downarrow\prime} - L^\uparrow \qquad (2)$$

$$Q_H(t) = A_H \{\overline{T}_a(t) \pm T_a' - 0\} \{\overline{U}_a(t) \pm U_a'\} \qquad (3)$$

$$Q_E(t) = A_E \{\overline{e}_a(t) \pm e_a' - 6.11\} \{\overline{U}_a(t) \pm U_a'\} \qquad (4)$$

$$A_H = C_H \rho C_P \qquad (5)$$

$$A_E = l C_E \rho \frac{0.622}{P} \qquad (6)$$

where t represents the number of days. The functions with a bar and a prime show the average trends and the maximum fluctuation ranges from thier average, respectively. C_P, ρ, l and P are the specific heat of air at constant pressure, the air density, the latent heat of vaporization and the air pressure, respectively. C_H and C_E are the bulk coefficients for sensible and latent heat, respectively. We predict that

$L^{\downarrow}$, $\overline{T}_a$ and $\overline{e}_a$ are increasing functions in the snowmelt season and that $\overline{U}_a$ has a small seasonal variation. We assumed for simplifing the equations that the snow surface temperature is 0℃ and $L^{\uparrow}$ is constant.

1.2 The Average seasonal variation of temporal and spatial distribution of Q_{SM}

The Q_{SM}'s average variation $\overline{Q}_{SM}$ can be written as

$$\overline{Q}_{SM}(t) = \overline{Q}_{NR}(t) + \overline{A}_{HE}(t)\,\overline{U}_a(t) \tag{7}$$

where

$$\overline{A}_{HE}(t) = A_H\,\overline{T}_a(t) + A_E\,\{\overline{e}_a(t) - 6.11\}. \tag{8}$$

When T_a and e_a are low, the percentage of Q_{NR} for Q_{SM} is high (e.g., Ohta et.al., 1994). According to (7), there is a possibility that the percentage of $\overline{Q}_H + \overline{Q}_E$ for $\overline{Q}_{SM}$ may compete with the one of $\overline{Q}_{NR}$ at the later period of the thaw because $\overline{A}_{HE}$ increases. Assuming that T_a, e_a and α do not differ between sites close to each other, the average trend of the difference of Q_{SM} between sites is written as

$$\overline{\delta_T Q}_{SM}(t) = \overline{\delta_T Q}_{NR}(t) + \overline{A}_{HE}(t)\,\overline{\delta_T U}_a(t) \tag{9}$$

where symbols with δ_T indicate differences between sites. On the other hand, using symbols *low* and *high* for the signs of low and high altitudes, the average trend of the difference of Q_{SM} between altitudes $\overline{\delta_A Q}_{SM}$, that is calculated by subtracting $\overline{Q}_{SM}$ at a high altitude from the one at a low altitude, can be shown as

$$\overline{\delta_A Q}_{SM}(t) = \overline{\delta_A Q}_{NR}(t) + D(t)\,\overline{U}_{a\cdot low}(t)$$
$$+ \overline{A}_{HE\cdot high}(t)\,\overline{\delta_A U}_a(t) \tag{10}$$

where $\overline{\delta_A Q}_{NR}$, D, B and $\overline{\delta_A U}_a$ were defined as

$$\overline{\delta_A Q}_{NR}(t) = \overline{Q}_{NR\cdot low}(t) - \overline{Q}_{NR\cdot high}(t) \tag{11}$$

$$D(t) = A_H\,\{\overline{T}_{a\cdot low}(t) - \overline{T}_{a\cdot high}(t)\}$$
$$+ A_{E\cdot low}\,\{\overline{e}_{a\cdot low}(t) - B\,\overline{e}_{a\cdot high}(t) - 6.11\,(1-B)\} \tag{12}$$

$$P_{low} = B\,P_{high} \tag{13}$$

$$\overline{\delta_A U}_a(t) = \overline{U}_{a\cdot low}(t) - \overline{U}_{a\cdot high}(t). \tag{14}$$

If the general conditions that ρ_S and $\overline{T}_a$ at a low

altitude are larger than at a high altitude and the distribution of $\overline{e}_a$ is dominated by the distribution of $\overline{T}_a$, both $\overline{\delta_A Q}_{NR}$ and D are always positive and D has no obvious seasonal change. Given that $\overline{Q}_{H\cdot low} + \overline{Q}_{E\cdot low}$ and $-\overline{\delta_A U}_a$ are positive, the third term on the right side of (10) is always negative and decreases with the melt season. According to (9) and (10), when $\overline{A}_{HE}$ gets to be large, $\overline{\delta_T Q}_{SM}$ and $\overline{\delta_A Q}_{SM}$ may be dominated by not only $\overline{\delta_T Q}_{NR}$ and $\overline{\delta_A Q}_{NR}$ but also $\overline{\delta_T(Q_H + Q_E)}$ and $\overline{\delta_A(Q_H + Q_E)}$. Therefore, it is relevant to make clear the seasonal variation of $\overline{A}_{HE}$ and measure its effects on the spatial distribution of $\overline{Q}_{SM}$.

1.3 The seasonal variation of the Q_{SM}'s fluctuation

It is necessary to find the main elements for the fluctuation of Q_{SM} in order to find the elements that are needed precisely for the snowmelt's computing. Hence, meteorological elements that affects the Q_{SM}'s fluctuation and its seasonal variation have to be analyzed. If F_{NR}, F_H and F_E indicate the maximum fluctuations of Q_{NR}, Q_H and Q_E respectively, they can be shown as the difference between the maximum and minimum value of Q_{NR}, Q_H and Q_E, i.e.,

$$F_{NR}(t) = 2R'\{1 - \alpha(t)\}\,I^{\downarrow}(t) + 2L^{\downarrow\prime} \tag{15}$$

$$F_H(t) = 2A_H\,\{\overline{T}_a(t)\,U_a' + T_a'\,\overline{U}_a(t)\} \tag{16}$$

$$F_E(t)\begin{cases} = 2A_E\,\left[e_a'\,\overline{U}_a(t) + U_a'\,\{-\overline{e}_a(t) - 6.11\}\right] \\ \qquad\qquad (when\ \overline{e}_a(t) + e_a' < 6.11) \\ = 2A_E\,\left[e_a'\,\overline{U}_a(t) + U_a'\,\{\overline{e}_a(t) - 6.11\}\right] \\ \qquad\qquad (when\ \overline{e}_a(t) - e_a' \geq 6.11) \end{cases} \tag{17}$$

According to (15), (16) and (17), if the meteorological fluctuation ranges are constant, it can be thought that F_{NR}, F_H and F_E increase as it gets closer to summer and then the fluctuation of Q_{SM}, that is described as F_{SM}, increases. It is noticeable that F_E has a sudden increase which can be predicted by the change of e_a's condition, and which may change the main meteorological element that affects the Q_{SM}'s fluctuation.

2. STUDY AREA AND METHOD

At Mt. Hachimantai, in the northern part of Japan, the field study was done from May to June of 1992 -1994. The points of the measurement were located at a flat

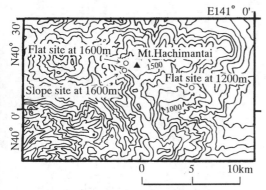

Fig.1 Location of study area.

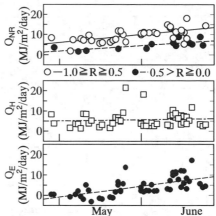

$O\ -1.0 \geqq R \geqq 0.5$ $\bullet\ -0.5 > R \geqq 0.0$

Fig.2 Variation of the daily snowmelt energy at flat site.

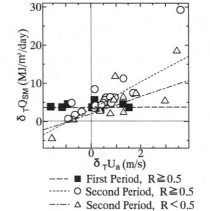

$\delta_T U_a$ (m/s)

--- ■ First Period, $R \geqq 0.5$
...... O Second Period, $R \geqq 0.5$
—·— △ Second Period, $R < 0.5$

Fig.3 Relation between $\delta_T U_a$ and $\delta_T Q_{SM}$.

and a sloping site that inclines 30° at a 1600 m. The aspect of sloping site was south-southwest (Figure 1). The meteorological elements that were needed for calculating the heat budget and for estimating the observed daily snowmelt rates $Q_{SM \cdot obs}$ were measured at the 1600 m. $Q_{SM \cdot obs}$ is indicated by (18), i.e.,

$$Q_{SM \cdot obs} = \rho_S\, h - E \qquad (18)$$

where h is the daily change in snow depth. E is the daily evaporation rate. At a 1200 m point, U_a, ρ_s and h were measured throughout the period. For computing Q_{SM} at the 1200 m when observation is held at the 1600 m, the other elements were predicted with the data at 1600 m. Q_{SM} at the 1200 m and 1600 m points was calculated by a heat balance method (Kondo and Yamazaki, 1990) using the meteorological elements based on 1-hour averaged data. The estimation of daily sums of Q_{SM} reproduced $Q_{SM \cdot obs}$.

3. RESULTS AND DISCUSSION

The daily average data of meteorological elements at the sites had the temporal and spatial variations that accorded with the same trends as in section 1. U_a had no seasonal conspicuous variation. There were no obvious differences of T_a and e_a between the flat and slope site. $U_{a \cdot high}$ was almost always greater than $U_{a \cdot low}$. In the following, the period when e_a was smaller than 6.11 hPa, i.e., until 21 May, is named "the first period" and the period after that is named "the second period."

3.1 The trends of $\overline{Q}_{SM}$, $\overline{\delta_T Q}_{SM}$ and $\overline{\delta_A Q}_{SM}$

The trends of Q_H and Q_E suggest that the increase of $\overline{A}_{HE}$ was caused by the growth of $\overline{e}_a$ (Figure 2). The trend of $\overline{Q}_E$ shows larger growth than $\overline{Q}_{NR}$ when

$1.0 \geqq R \geqq 0.5$. Figure 3 shows the relationship between $\delta_T U_a$ and $\delta_T Q_{SM}$ that are the subtractions of Q_{SM} and U_a at the flat site from at the sloping site. In the first period, $\delta_T Q_{SM}$ was defined by $\delta_T Q_{NR}$. In the second period, $\delta_T Q_{SM}$ was also dominated by $\delta_T U_a$ and $\delta_T Q_{NR}$ was offset by $\delta_T (Q_H + Q_E)$ on some days. It is deduced that in the second period, larger snowmelt can more likely occur at a north aspect site than at a south site depending on $\delta_T U_a$. Figure 4 shows that when e_a was more than 6.11 hPa, $Q_{H \cdot high} + Q_{E \cdot high}$ was greater than $Q_{H \cdot low} + Q_{E \cdot low}$ especially when $\delta_A U_a$ was large. Hence, there is a possibility that $\overline{\delta_A Q}_{SM}$ becomes negative at the second period and decreases continuously along with season with the general condition $U_{a \cdot high} > U_{a \cdot low}$.

115

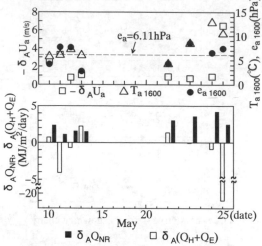

Fig.4 Relation between $\delta_A Q_{NR}$, $\delta_A (Q_H + Q_E)$ and meteorological conditions.

3.2 Seasonal variation of F_{SM}

When the contribution rate of F_{NR} and $F_H + F_E$ on F_{SM} are shown as C_{NR} and C_{TUR}, they are defined as

$$C_{NR}(t) = \frac{Q_{NR}(t) - \overline{Q}_{NR}(t)}{Q_{SM}(t) - \overline{Q}_{SM}(t)} \quad (19)$$

$$C_{TUR}(t) = \frac{\{Q_H(t) + Q_E(t)\} - \{\overline{Q}_H(t) + \overline{Q}_E(t)\}}{Q_{SM}(t) - \overline{Q}_{SM}(t)} \quad (20)$$

where $\overline{Q}_{SM}$, $\overline{Q}_{NR}$, $\overline{Q}_H$ and $\overline{Q}_E$ were found by calculating the regression line of Q_{SM}, Q_{NR}, Q_H and Q_E. In Figure 5, there is no obvious difference between C_{NR} and C_{TUR} in the first period and C_{TUR} is greater than C_{NR} in the second period. This result demonstrates that the sudden increase trend of F_E had an effect on F_{SM} in the second period.

For estimating the effect of U_a' on $F_H + F_E$, the

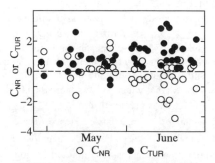

Fig.5 Seasonal variations of C_{NR} and C_{TUR}.

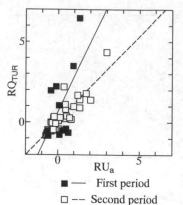

Fig.6 Relation between RU_a and RQ_{TUR}.

fluctuation rates of $Q_H + Q_E$ and U_a for 2 consecutive days are defined as RQ_{TUR} and RU_a, i.e.,

$$RQ_{TUR}(t) = \frac{\{Q_H(t) + Q_E(t)\} - \{Q_H(t-1) - Q_E(t-1)\}}{Q_H(t-1) + Q_E(t-1)} \quad (21)$$

$$RU_a(t) = \frac{U_a(t) - U_a(t-1)}{U_a(t-1)}. \quad (22)$$

Figure 6 derives $F_H + F_E$ was determined by U_a' more dominantly than T_a' and e_a' in the second period.

4 CONCLUSIONS

In the case where e_a increases from under to above 6.11 hPa, $\overline{A}_{HE}$ clearly increases. This phenomenon has influences on the features of the temporal and spatial distribution of Q_{SM}, i.e., the increase of Q_{SM} results from not only Q_{NR} but also Q_E. In the second period, the Q_{SM}'s spatial distribution is dominated by not only the Q_{NR}'s distribution but also the distribution of U_a. At the same time, F_{SM} consists of the higher level of F_E compared with F_{NR} and F_E is made by the U_a's fluctuation. Hence, if we have to compute an areal snowmelt under this situation, the estimate of U_a's distribution becomes important.

REFERENCES

Kondo, J. & T.Yamazaki. 1990. A prediction model for snow melt, snow surface temperature and freezing depth using a heat balance method. *J. Appl. Meteor.* 29: 375-384.

Ohta, T, A.Kojima, K.Suzuki. 1994. Heat balance analysis on snowmelt rates in the subalpine region, Mt. Hachimantai. *Proceedings of the International Symposium on Forest Hydrology, Tokyo, Japan:*193-200.

Snow Engineering: Recent Advances, Izumi, Nakamura & Sack (eds) © 1997 Balkema, Rotterdam. ISBN 90 5410 865 7

Penetration tests of snow covering a pit

Yasuaki Nohguchi, Takashi Ikarashi & Koyuru Iwanami
Nagaoka Institute of Snow and Ice Studies, NIED, Niigata, Japan

Tsutomu Nakamura
Faculty of Agriculture, Iwate University, Morioka, Japan

ABSTRACT:Accidents related to a small river or channel covered with snow frequently have occurred recently in Japan. To know the safe thickness of snow covering a channel we carried out pit observations of snow using three types of penetration tests: Canadian hardness tests, Rammsonde hardness tests and a man direct penetration tests. We propose a simple model to roughly estimate the safe thickness. The result obtained from this model coincides with that of the direct man penetration test, and it was found that the dangerous thickness for wet granular snow in Niigata prefecture is 10 cm.

1 INTRODUCTION

Recently, accidents related to a small river or channel covered with snow frequently have occurred in Japan (Ikarashi 1993 and Numano 1993). Such snow sometimes behaves itself like a kind of trap. In this paper, the strength of the snow covering a pit will be reported based on hardness tests of snow.

2 OBSERVATION OF STRENGTH OF SNOW COVERING A SMALL CHANNEL

To examine the snow covering a small channel, pit observations were carried out at Yunotani Village, Niigata prefecture, from Feb. 1995 to Apr. 1995. In these observations, vertical profiles of Canadian hardness and Rammsonde hardness were measured in both the snow covering the small channel and that on the ground surface near the channel. Moreover, penetration tests of a man into the the channel were carried out by gradually removing snow from snow surface as shown in Figure 1.

3 RESULTS OF OBSERVATIONS

In general, the strength of the snow covering a channel

Figure 1. A man penetration test on snow covering a channel.

Figure 2. A vertical section of snow covering a channel. A water drainage channel is formed at the center of the snow bridge.

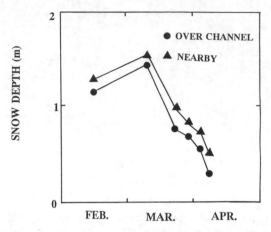

Figure 3. Comparison of the snow depth over the channel and that over the ground nearby.

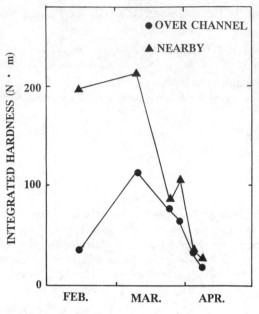

Figure 5. Comparison of the integrated rammsonde hardnesses on depth over the channel and that over the ground nearby.

4 ESTIMATION OF PENETRATION DEPTH

To roughly estimate the stress at the penetration we proposed an effective area of the load on a snow surface as shown in Figure 6. Then the stress on a layer, f, is represented by

$$f = Wg / \pi (R + L)^2 \qquad (1)$$

where L is the depth of that layer, g is the acceleration of gravity, W is the weight of the man and R is the radius of the circle whose area is equivalent to the contact area between the man and the snow surface. This stress rapidly decreases with increasing depth. Therefore, the man penetrates into the snow near the surface, but the penetration will stop at the depth where the strength of snow is larger than the penetration stress. When the stress exceeds the hardness of the snow for every depth, the man will penetrate through the whole snowcover and fall into the channel (Fig. 7).

In this rough estimation, complete penetration of the snow occurs when the thickness of the snow is about 10 cm. This result is consistent with that of the man penetration tests.

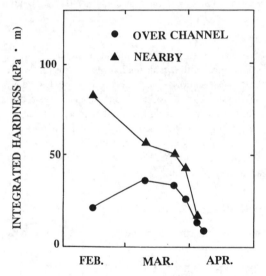

Figure 4. Comparison of the integrated canadian hardnesses on depth over the channel and that over the ground nearby.

are lower than those of the natural snowcover because a water drainage channel was always formed at the center of the covering snow Figure 2. Figures 3-5 show comparisons of snow depth, the integrated Canadian hardness and Rammsonde hardness between the snow over the channel and that over the ground surface nearby the channel. Snow near the bottom of the snow over the channel was always wet granular snow.

In man penetration tests, when the thickness of the remaining snow layer was between 10 and 20 cm, a man weighing about 60 kg penetrated into the channel.

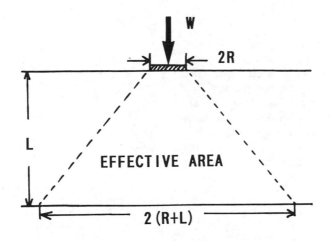

Figure 6. Effective area of penetration stress.

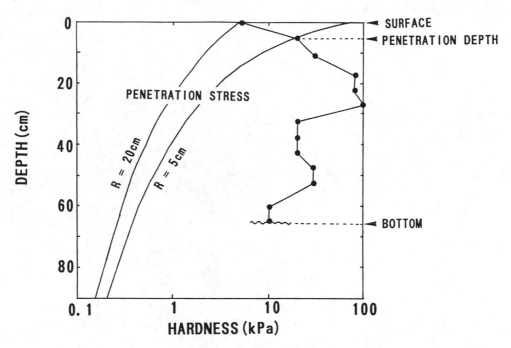

Figure 7. Penetration stress (W:60kg, R:5-20cm) and vertical profile of Canadian hardness on March 28th, 1995. The penetration depth from snow surface for R=5 cm is 7 cm. If the upper 50 cm of snow is removed, complete penetration will occur for R=5 cm.

5 CONCLUSION

To know the safe thickness of snow covering a channel we carried out pit observations of snow and 3 types of penetration tests: Canadian hardness test, Rammsonde hardness test and a man direct penetration tests. We propose a simple model to roughly estimate the safe thickness (i.e., the dangerous thickness). The result obtained from our model coincides with that of the direct man penetration test, and it was found that the dangerous thickness for wet granular snow in Niigata prefecture is 10 cm.

REFERENCES

Ikarashi, T. 1993. Casualties due to snow and ice phenomena (in Japanese). *Gekkan Shobo*, 16, 95-102.

Numano, N. 1993. Long-term trend of human body snow damage and its social background: -An examination of two prefectures, Yamagata and Niigata-. *J. Japanese Soc. Snow and Ice*, 55, 317-326.

Snow Engineering: Recent Advances, Izumi, Nakamura & Sack (eds) © 1997 Balkema, Rotterdam. ISBN 90 5410 865 7

On a relation between probability occurrence of solid precipitation and ground air temperature

Tatsuo Hasemi
Science of Snow and Ice Co., Ltd, Tokyo, Japan

ABSTRACT: Relations between probability occurrence of solid precipitation and ground air temperature for winter precipitation under two types of surface pressure patterns, winter monsoon and extratropical cyclone, were analyzed using data from meteorological observatories in Japan. Temperature is related to the elevation of the observatory, are as follows:

(winter monsoon type) $T_{50} = 4.0 - 0.44 \ln(Z)$
(extratropical cyclone type) $T_{50} = 2.2 - 0.21 \ln(Z)$

where T_{50} is the ground air temperature (°C) corresponding to 50% probability occurrence and Z is the elevation of the observatory (m). A possible reason for the observed T_{50} decrease with elevation, is that precipitation particles are smaller at high elevations. Observed differences in T_{50} for the winter monsoon type and the extratropical cyclone type are attributed to differences in the heat content or other conditions in the upper layers of the atmosphere where falling snow particles melt. Using the results of this analysis, the type of precipitation, either snow or rain, can be forecasted more precisely. Also the amount of new snow deposition or falling snow which contributes to development of snow accretion can be estimated more reasonably.

1 INTRODUCTION

The effect of precipitation on life and industrial activity and the preparation for it are much different when it is solid than when it is liquid. If precipitation is snow, it may remain much longer as snow accumulates. Occasionally it should be removed quickly from roads and roofs. Conversely, if it is rain, it will soon run off and one should take care for another circumstances such as overflow of rivers and floods. Forecast of precipitation type is important especially to prevent hazards from occurring.

In winter in Japan, except for mountainous areas of high altitude, precipitation frequently falls under near critical conditions of snow or rain. At present, the critical condition is estimated with ground air temperature (observed at 1.5m height above surface). If the ground air temperature is lower than T_c, critical temperature, then precipitation may be snow. The value of T_c has been experimentally determined to be constant at 2°C to 3°C in rather large areas regardless of synoptical meteorological conditions or local conditions such as altitude. Vertical distribution of air temperature of the upper atmosphere near the ground is related to the melting of snow particles as they fall to the ground. Winter precipitation in Japan mainly occurs under two types of surface pressure patterns;

winter monsoons and extratropical cyclones. Lapse rate is generally larger for the former condition than for the latter, and T_c will be different for monsoons and cyclones.

In this study, relations between probability occurrence of solid precipitation (snow, sleet) and ground air temperature for winter precipitation under winter monsoons and extratropical cyclones, were analyzed.

2 RESULTS AND DISCUSSION

2.1 T_{50} as a representative index of the trend of the relation

Figure 1 shows relations between probability occurrence of solid precipitation (probability, hereafter) and ground air temperature at 6 observatories. Probability decreases with temperature but the rate of decrease differs for each observatory. Probability decreases more slowly for the winter monsoon (monsoon) type than for the extratropical cyclone (cyclone) type. As a representative index of these trends, T_{50}, ground air temperature (°C), corresponding to 50% probability, is introduced. Usually T_{50} is used as the critical temperature in forecasting precipitation type.

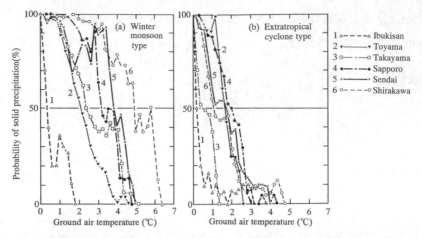

Figure 1　Relations between probability occurrence of solid precipitation and ground air temperature.
(a) monsoon type　(b) cyclone type.

2.2　T_{50} as a reflection of difference in synoptical and local meteorological condition

Figure 2 shows the distributions of T_{50} at each observatory in Japan for monsoon type (a) and cyclone type (b). Range of variety in T_{50} for the monsoon type is 0.3 to 5.0°C which is about 2 times larger than that for the cyclone type, which is 0.2 to 2.4°C. Differences in T_{50} are considered to reflect differences in synoptical and local meteorological conditions at each place. As one feature of the distribution, T_{50} is small at high elevation for both types and rather large at observatories which are far from the windward Japan sea coast for the monsoon type.

Figure 3 shows the relation between T_{50} and the elevation of the observatory. For the monsoon type (Fig.3(a)) solid circles are for observatories where winter precipitation is mainly by monsoon and the snow is rather heavy. At these observatories, which are named on the figure, T_{50} can be represented by the following relation:

$$T_{50} = 4.0 - 0.44 \, ln\,(Z) \quad \text{(winter monsoon type).}$$

For the cyclone type, almost all observatories are in the same slope and T_{50} is represented as follows:

$$T_{50} = 2.2 - 0.21 \, ln\,(Z) \quad \text{(extratropical cyclone type)}$$

where Z is the elevation of the observatory (m).　A

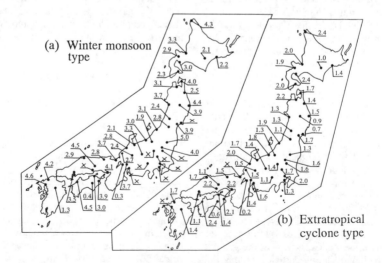

Figure 2　Distribution of T_{50}.　(a) monsoon type　(b) cyclone type.

122

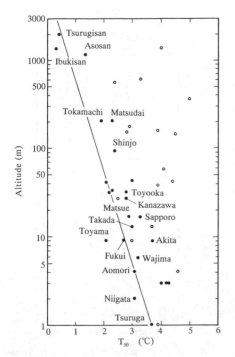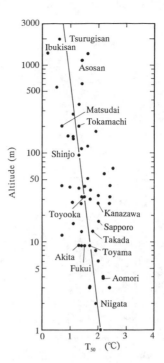

Figure 3   Relation between T_{50} and elevation.   (a) monsoon type   (b) cyclone type.

possible reason for the observed T_{50} decrease with elevation is that precipitation particles are smaller at high elevations.

Observed differences in T_{50} for monsoon type and cyclone type are thought to be attributed to differences in the heat content of the upper layers of atmosphere where falling snow particles melt. Because the lapse rate is usually larger for monsoon types than for cyclone types, the heat content of the upper atmosphere from the ground to 0°C level is larger for monsoon types than for cyclone types for the same ground air temperature. Figure 4 shows the relationship between the ground air temperature, T_g and the heat content of the upper atmosphere from the ground up to the 0°C level, Q, calculated from the following equation:

$$Q = \bar{C}_v \cdot \bar{\rho} \int_0^{H_0} T(h)dh \qquad (1)$$

where,

$\bar{C}_v$, $\bar{\rho}$: mean specific heat and mean density of the atmosphere from the ground up to the 0°C level (0.173 calg⁻¹K⁻¹, 0.12 kgm⁻³) respectively,

T(h): air temperature (°C),

h : height (m),

H_0: height of 0°C (m).

Symbols in the figure show the type of precipitation, and the relation of Q and T_g at these positions. The solid curve shows the inclination and T_{50} with vertical line.

Other conditions, such as relative humidity or precipitation intensity, will affect the melting of snow particles (Matuo *et al.*, 1982). Melting will delay in drier air because snow particles cool by evaporation. Open circles in Fig.3 shows that precipitation by monsoon falls in drier conditions at these observatories, which are far from the windward Japan seacoast. However, forecast accuracy cannot be expected to rise by introducing relative humidity as a second parameter when forecasting precipitation type because relative humidity itself is difficult to forecast. There is still uncertainty of the same order as using only ground air temperature.

2.3   *Critical temperature, T_{50}, in forecasting precipitation type*

Using the results of this analysis, the type of precipitation, either snow or rain, can be forecasted more precisely. Itoh (1944) indicated that T_c is 2°C in the Pacific Ocean coastal area and 3°C in the Japan sea coastal area. Considering that winter precipitation in the Japan Sea coastal area is mostly by monsoon and in Pacific coastal area, mostly by cyclone, T_c, as indicated by Itoh, are mean values of T_{50} used in this study. This study clarified the effect of locality and increased the accuracy of forecasting precipitation type.

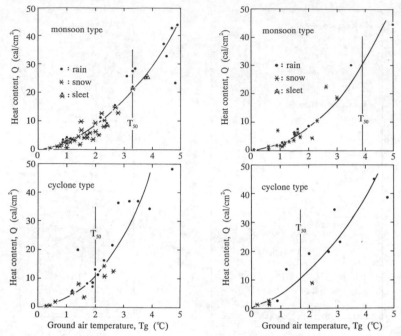

Figure 4   Relation between the ground air temperature T_g and the heat content of the upper atmosphere from the ground up to the 0℃ level, Q.   (a)Sapporo   (b)Sendai.

Also the amount of additional snow cover or the amount of falling snow, which contributes to the development of snow accretion, *etc.*, can be estimated more reasonably.

REFERENCES

Hasemi, T.   1991.   Relations between probability occurrence of solid precipitation and ground air temperature(1) - On the locality of the relation and possibility of prediction of precipitation type. (in Japanese), *Seppyou*. 53: 33-43.

Itoh, H.   1944.   Rain and air temperature. (in Japanese) *Tenkiyohou Shiryo Shoroku* (Abstract of weather forecasting data). Meteorological Agency of Japan.

Matsuo, T. & Y.Sasyo   1982.   Melting of snow pellets in the atmosphere. *Papers in Meteorology and Geophysics* . 33: 55-64.

Snow Engineering: Recent Advances, Izumi, Nakamura & Sack (eds)© 1997 Balkema, Rotterdam. ISBN 90 5410 865 7

Are neural networks a distinct tool for prediction and estimate of newly fallen snow?

Yutaka Yamada & Koyuru Iwanami
Nagaoka Institute of Snow and Ice Studies, NIED, Niigata, Japan

Naoya Maeda & Michihiro Amenomori
Department of Informational Science, Faculty of Science, Hirosaki University, Aomori, Japan

Munemi Michigami
Department of Physics, Faculty of Science, Hirosaki University, Aomori, Japan

ABSTRACT: We tried to predict and to estimate the depth and water equivalent of daily new snowfall by using the neural network method based upon the back-propagation law. To increase the accuracy, the effective input meteorological factors were, first, selected by statistical means, then the selected factors were applied to the neural network. The result of a neuro-computing estimate by this cooperative procedure was as well as a statistical multiple regression analysis.

1 INTRODUCTION

New snow depth is an important quantity for snow removal operations on roads, snow avalanche prediction and others. However, the present Japanese automatic meteorological data acquisition system (AMeDAS) measures only total snow depth every hour. Of more than 1,300 AMeDAS stations, about two hundred are equipped with a supersonic snow depth gage.

It would be very useful if we could use the AMeDAS data to get new additional snow information. Therefore, we have studied the prediction and estimate of new snow depth (Yamada and Ikarashi, 1992; Yamada, 1993a, b; Amenomori and Hashiyama, 1993).

Here, we apply neural networks based upon the back-propagation law to the prediction of depth of the daily newly fallen snow from remote aerological data of the previous day, and to estimate depth and water equivalent of the daily newly fallen snow from the twenty-four hours' data similar to the AMeDAS data up to the estimate time, nine o'clock. We expected that the flexibility of the neural network algorithm could allow data analyses of highly nonlinear data such as snowfall phenomena.

2 NEURAL NETWORK

An artificial neural network consists of three layers: the input layer, hidden layer, and output layer. Each layer has a set of processing nodes, and its node is fed to all other nodes of the next forward or back layer via weights. We used the back-propagation method (Rumerhart, Hinton and Williams, 1986), a kind of least mean square algorithm, as a learning law whose objective is to find a set of adjusted weights whose predicted or estimated output and the actual output or observed value resemble one another.

3 INPUT DATA AND STRUCTURE OF NEURAL NETWORKS

3.1 *Input data*

The target of prediction is daily new snow depth, and those of the estimate are depth and water equivalent of daily new snow. These newly fallen snow data were measured routinely and manually on snow boards once a day at nine o'clock in the Nagaoka Institute of Snow and Ice Studies (Yamada et al., 1995). New snow depth was measured on a snow board (45x45 cm²) having a scale in the center.

For the prediction of new snow depth, aerological data at twenty-one o'clock on the previous day at Vladvostok, USSR, were collected as input valuables from weather charts. Seven aerological variables having high correlation with the observed depth of daily newly fallen snow were used in the neural networks: the three altitudes of 500, 700 and 850 hPa, two air temperatures of 500 and 700 hPa and the two layer thicknesses of 500-700 hPa and 700-800 hPa.

For the estimate of new snow depth, we selected six ground meteorological variables from the data observed by the Nagaoka Institute of Snow and Ice Studies: precipitation, the summation of the positive increment of hourly snow depth, air temperature, wind speed, humidity, and the daily difference of snow depth in the previous day. Five meteorological variables excluding the last difference of snow depth measured between the standard time and the estimated time in the previous day were used for estimating the water equivalent of new snow. These data are all twenty-four hour means or summations at the estimate time.

3.2 *Structure of neural networks*

The structure of neural networks used in this study were three-layer back-propagation networks (Table 1). The number of nodes in the input layer is the number of variables. Computations become very intensive as the number of input variables and hidden layers increase. The output layer in these three cases has only one node and the number of nodes in the hidden layer was selected empirically considering the results of former experience (Amenomori and Hashiyama, 1993).

Table 1. Structure of neural networks (NH: new snow depth; NHW: water equivalent of new snow).

	Input layer	Hidden layer	Output layer
Prediction of NH	7	8	1
Estimate of NH	6	14	1
Estimate of NHW	5	12	1

4 RESULTS AND DISCUSSION

We let the above three neural networks learn from fifty thousand to two hundred and fifty thousand times. As there was no significant difference between correlation coefficient in this range of learning times, neural networks discussed here are mainly for fifty thousand learnings.

The correlation coefficient of predicted or estimated value with observed value for both the depth and water equivalent of new snow is shown in Table 2. The correlation coefficient by the multiple regression analysis is slightly larger than the correlation coefficient by the neural networks, for both cases. On the other hand, the correlation coefficient estimated by neural network with multiple regression for new snow depth was largest, 0.981 and that for water equivalent of new snow was 0.952. These high values of correlation coefficient mean that the standard error of both methods is approximately the same.

Table 2. Correlation coefficients.

	Predicted by Neural Network	Estimated by Neural Network	Estimated by Multiple Regression
Observed NH	0.649	0.969	0.977
Observed NHW	—	0.881	0.891

4.1 *Estimate of depth and water equivalent of new snow*

After the neural network learned 356 days' data during five winter seasons from 1982/83 to 1986/87, we estimated 57 days' new snow depth for January and February in 1988. Figures 1 and 2 are the results of this estimate of new snow depth after fifty thousand learning times by the neural network. Estimates by multiple regression analysis are also shown in those figures. A comparison of these estimated values and the observed new snow depth is shown as time series in Figure 1 and as correlation with observed value in Figure 2.

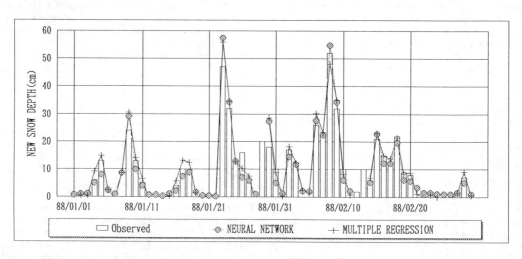

Figure 1. Time series of observed new snow depths together with estimates by the neural network and by multiple regression.

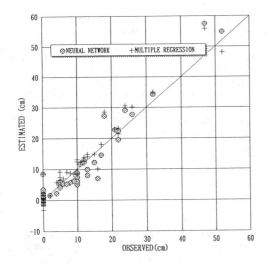

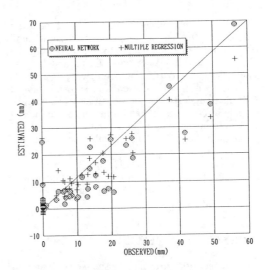

Figure 2. Correlation of observed new snow depth with that estimated by the neural network and by multiple regression.

Fig.3. Correlation of observed water equivalents of new snow with that estimated by the neural network and by multiple regression.

Both plots fit observed new snow depths well overall and scattering of data is small. The correlation coefficient at the estimating phase of the neural network was 0.969, that of multiple regression was 0.977. For the same data period and learning times, the correlation coefficient of water equivalent of new snow was 0.881 and 0.891, respectively (Fig. 3 and Table 2).

4.2 Prediction of new snow depth

For prediction, the neural network learned fifty thousand times from 168 data from January and February during three winter seasons from 1982 to 1984. The correlation coefficient with the observed value during January and February in 1987 was 0.649.

5 DISCUSSION

The correlation of new snow depths estimated by the neural networks with observed values was slightly smaller than that obtained by multiple regression analysis.

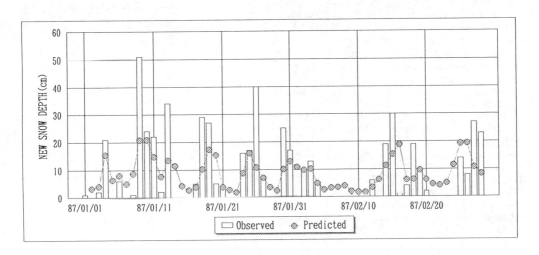

Figure 4. Time series of observed new snow depth together with predicted by neural networks.

But, the correlation coefficient, 0.970, of the neural network in the learning stage for new snow depth estimate was larger than that of the multiple regression analysis in the fitting stage of the linear model, 0.964. The order of the correlation coefficients reverses the order of the estimating phase that is shown in table 2. This fact suggests that the neural network is highly flexible to complex data and it needs a longer period of data to improve estimate accuracy.

. For prediction of new snow depth, this new method was not as good as the estimate of new snow depth. However, it can be improved further by selecting the more appropriate aerological parameters such as wind speed (Asuma and Kikuchi 1988, Ishihara et al. 1965, Rikiishi and Yamada 1993).

6 CONCLUDING REMARKS

Neural networks are models similar to traditional statistical models but not inferential statistical models. Their major objective is to minimize a difference between predicted values and target values no matter what it takes. Yet, experts in this field have asserted that neural networks are an analytical tool distinct from traditional models.

The conclusion of this study was that the accuracy of the neural networks applied for estimating new snow depth proved to be no better than the multiple regression analysis or a traditional statistical model. However, the high correlation coefficient of the neural networks in the learning stage suggests the possibility of high accuracy.

REFERENCES

Amenomori, M. and H. Hashiyama(1993) Prediction of daily snowfall depth obtained from neuro-computing method (in Japanese), *J. of Japanese Soc.of Snow and Ice*, Vol. 55, No. 3, 191-196.

Asuma,Y. and K.Kikuchi(1988) Characteristics of radar echoes by the precipitating snow clouds over the Ishikari Bay, Hokkaido, Japan -Comparison with aerological data- (in Japanese with English abstract), *Geophysical Bulletin of Hokkaido University*, Sapporo, Japan, No. 50, 1-13.

Ishihara, K. and others(1965) Snowfall Prediction for small area based on multiple correlation regression equation (in Japanese with English abstract), *Reports of Cooperative Research for Disaster Prevention*, No. 2, 41-58.

Rikiishi,K. and M.Yamada(1993)Prediction formula of daily new snowfall depth in Aomori City using aerological data (in Japanese), *Bulletin of Annual Report of the Cold Regions Meteorology Laboratory*, Dep. of Sci., Hirosaki Univ. Vol.5, 1-14.

Rumerhart, D.E., Hinton, G.E. and R.J. Williams (1986) Leaning representation by back-propagating errors, *Nature*, 323, 533-536.

Yamada, Y. (1990) Distribution of new snow density in Japan, *Workshop on Atmospheric icing on structures*, Tokyo, October 1990, B5-2, 1-7.

Yamada, Y and T. Ikarashi (1992) Simulation on Depth of Newly Fallen Snow Based on AMeDAS Data, *Second International Conference on Snow Engineering*, Santa Barbara, California, June 1992, 169-178.

Yamada, Y.(1993a) Estimation Method of Newly fallen snow depth based on hourly meteorological data: Raingage method and multiple regression method (in Japanese with English abstract), *Rep. of National Research Institute for Earth Science and Disaster Prevention*, No.52, 69-80.

Yamada, Y.(1993b) Frequency Distribution and Mean of newly Fallen Snow Density at the Twenty-two Cities along the Coast of the Japan Sea (in Japanese with English abstract), *Rep. of National Research Institute for Earth Science and Disaster Prevention*, No.52, 51-67.

Yamada, Y. et al. Eds. (1995) Data on snow cover for 30 years in Nagaoka (in Japanese with English summary), *Technical Note of the National research Institute for Earth Sciences and Disaster Prevention*, 162, pp.250.

Snow Engineering: Recent Advances, Izumi, Nakamura & Sack (eds) © 1997 Balkema, Rotterdam. ISBN 90 5410 865 7

Characteristics of blowing snow fluctuation

T. Sato & M. Higashiura
Shinjo Branch of Snow and Ice Studies, NIED, Yamagata, Japan

ABSTRACT: The characteristics of blowing snow fluctuation were investigated based on field observations. Spectral analysis revealed that mass flux correlates well with wind speed fluctuations for periods longer than about 30 sec and vice versa. Due to this nature of blowing snow fluctuation, the 1-sec averaged mass flux values are scattered against wind speed, and the shape of the size distribution of blowing snow particles detected for a 1-sec period varies second by second. The probability distribution of the mass flux averaged over a 1-sec period can be approximated by a log-normal distribution. This can be used as a statistical method to estimate the magnitude of mass flux fluctuation from wind speed.

1 INTRODUCTION

Heavy blowing snow can cause snow disasters, since it forms snowdrifts on roads and reduces visibility. Recently in Japan, highway networks have been developed even in snowy regions, and the visibility reduction due to blowing snow is a major cause of traffic accidents. The magnitude of blowing snow depends both on meteorological conditions such as wind speed and air temperature and on the conditions of snow cover such as snow type. Although previous researchers on blowing snow have revealed its mechanism and its dependence on the above conditions (e.g., Budd 1966, Schmidt 1982), they were concerned with the mean state of blowing snow and mass flux values averaged over a proper period of time.

The intensity of blowing snow changes rapidly and visibility varies as a result. It is important to clarify the fluctuation of blowing snow to understand its mechanism deeply, as well as for taking measures to cope with blowing snow disasters. The purpose of this study is to investigate the characteristics of blowing snow fluctuation based on field observations.

This study is a part of the research project "Study of areal prediction techniques of drifting snow and development of a warning system" conducted by the National Research Institute for Earth Science and Disaster Prevention (NIED), in which blowing snow will be predicted from upper air sounding data, wind speed and snowfall intensity, with a Doppler radar system. The result of this study will be incorporated into the algorithm to predict minimum visibility associated with blowing snow fluctuation.

2 INSTRUMENTS AND OBSERVATIONS

Blowing snow fluctuation was measured with two types of snow-particle counters (SPC), originally developed by Schmidt (1977). The two types used were improved in the Shinjo Branch of Snow and Ice Studies, NIED (Kimura 1991, Sato et al. 1993), which detect photoelectrically the sizes and numbers of blowing snow particles passing the sampling volume (25 mm wide, 2 mm high and 0.5 mm deep). The output of the first type (Model SPC-S4) is a voltage proportional to the mass flux of blowing snow. The output voltage changes stepwise at an interval of 1 sec. The SPC-S4 detects precisely particles with diameters from 50 to 300 μm. The second type (Model SPC-S7) almost linearly divides the diameter range from 70 to 550 μm into 32 classes and counts the particle number of each class every second. The data are directly transferred to a personal computer.

For the measurement of wind speed fluctuation, a small ultrasonic anemometer was fixed near the SPC. The ultrasonic anemometer measures the orthogonal three components of fluctuating wind with frequencies up to 10 Hz. The wind speed was calculated from the horizontal two components.

The mass flux at $z=1$ m was measured with the SPC-S4 in the Tsugaru Plain, Aomori Prefecture, during the winter of 1992. The ground surface was covered with new snow. The mean wind speed at $z=1.1$ m was 7 to 10 m/s and the air temperature was -7°C. Particle counts at $z=0.5$ m were obtained with

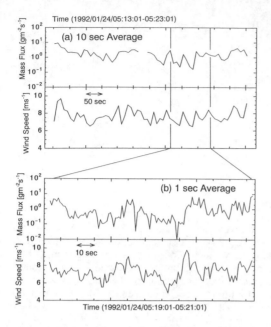

Figure 1. Time series of the mass flux of blowing snow and wind speed. (a):10-sec averaged values, (b):1-sec averaged values.

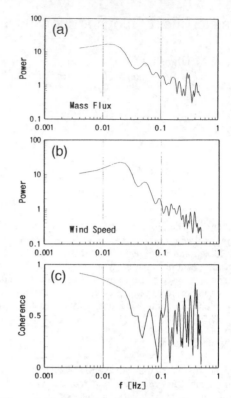

Figure 2. Power spectral densities of mass flux (a) and of wind speed (b) normalized with the respective variances. (c) is the coherence between mass flux and wind speed.

the SPC-S7 in the Ishikari Plain, Hokkaido, during the winter of 1995. There was a thin new snow layer on the hard snow cover. The mean wind speed at $z=0.6$ m was 8 m/s and the air temperature was -3°C. Both plains are located in the northern part of Japan and sometimes suffer from heavy blowing snow in winter.

The analysis was performed of the blowing snow in the suspension layer under stationary conditions.

3 RESULTS

3.1 *Mass flux fluctuation*

Figure 1 is a time series of the mass flux of blowing snow and wind speed, where 10-sec averaged values are plotted in (a) and 1-sec averaged values in (b). The details of fluctuating mass flux and wind speed do not correlate to each other.

For a quantitative discussion on the correlation between mass flux and wind speed, spectral analysis was performed of the 256 one-second averaged data. The power spectral density of mass flux is shown in Figure 2(a) and that of wind speed is shown in Figure 2(b), where the power spectral densities are normalized with respective variances. Both spectra have peaks at almost the same frequency. The slope of the mass flux spectrum in the high frequency region is gentler than that of wind speed. Figure 2(c) is the coherence between mass flux and wind speed.

The coherence is low in the frequency region higher than 0.03 Hz. This means that mass flux correlates well with the wind speed fluctuation of a period longer than about 30 sec and vice versa.

3.2 *Wind speed dependence of mass flux*

The relationship between mass flux and wind speed is shown in Figure 3, where the averaging times of the plotted data are 1 sec (a), 10 sec (b) and 30 sec (c). The sampling duration for the three cases is the same and is about 31 minutes. In the case of 1-sec averaged data, mass flux values are scattered against wind speed. The range of the scattering is about ±1 order. Thus the 1-sec averaged mass flux cannot be related to wind speed. For the longer averaging time, the scattering decreases and one-to-one correspondence between mass flux and wind speed can be found. These results are consistent with the low coherence between mass flux and wind speed in the high frequency region as shown before.

The relationship is approximated by the power law. For the wind speed between 7 and 11 m/s, mass flux is proportional to $U^{4.7}$, and the exponent is

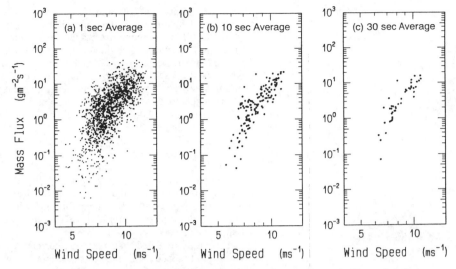

Figure 3. Relationship between mass flux and wind speed. Averaging times of the plotted data are 1 sec (a), 10 sec (b) and 30 sec (c).

1995/01/25 07:36:36-07:36:45

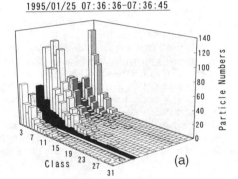

(a)

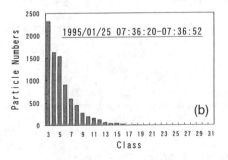

1995/01/25 07:36:20-07:36:52

(b)

Figure 4. Examples of the size distribution of blowing snow particles. (a):Ten consecutive distributions, each of which corresponds to a 1-sec period, (b):distribution for a 32-sec period. The class in the abscissa is almost linearly related to the particle diameter.

greater than 4.7 when the wind speed is below 7 m/s.

3.3 Size distribution of blowing snow particles

Figure 4(a) represents a time series of the size distribution of blowing snow particles. Ten consecutive distributions are illustrated and each one corresponds to the particles detected for a 1-sec period. The class in the abscissa is almost linearly related to the particle diameter. Class 3 corresponds to the diameter of 115-130 μm and Class 31 to 535-550 μm. Although the number of the small particles is generally great, the shape of the distribution varies second by second and the distribution is not smooth.

On the other hand, the size distribution of the particles detected for a 32-sec period, which includes the 10-sec period shown above, is shown in Figure 4(b). The particle number monotonously decreases as the diameter increases. The present distribution in Figure 4(b) is similar to the Gamma distribution, which was used to fit the size distribution by Schmidt (1982).

Figure 4 suggests that the coherence between the number of blowing snow particles and wind speed will also have a similar tendency as the coherence between mass flux and wind speed.

3.4 Probability distribution of mass flux

The above-mentioned results lead to the conclusion that mass flux cannot be estimated from wind speed if they are not averaged over a long enough time period, for example, 30 sec or more. So, the probability distribution of 1-sec averaged mass flux

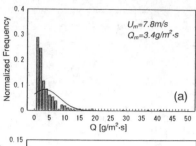

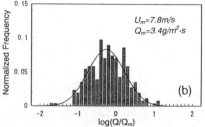

Figure 5. Normalized probability distribution of the 1-sec averaged mass flux Q (a) and $\log(Q/Q_m)$ (b) during a 256-sec period, where Q_m is the arithmetic mean and solid lines are the Gaussian distributions.

was investigated to find the statistical characteristics of blowing snow fluctuation. The normalized probability distribution of the mass flux Q during a 256-sec period is plotted in Figure 5(a). The solid line in the figure is the Gaussian distribution determined from the arithmetic mean Q_m and the standard deviation of the mass flux. The probability distribution is skewed and the probability is relatively large for a mass flux smaller than the arithmetic mean.

The same data are plotted in Figure 5(b) against the logarithm of the mass flux normalized by Q_m. The solid line is the Gaussian distribution determined from the arithmetic mean and the standard deviation of $\log(Q/Q_m)$. The mean value of $\log(Q/Q_m)$, which corresponds to the center of the Gaussian distribution, is smaller than zero because the geometrical mean of Q is smaller that its arithmetic mean. Figure 5(b) indicates that the probability distribution of the mass flux averaged over a 1-sec period can be approximated by a lognormal distribution. Based on this finding, it becomes possible to estimate the magnitude of blowing snow fluctuation from mean wind speed.

4 DISCUSSION AND CONCLUSIONS

In the suspension layer of blowing snow, the particles are transferred upward by atmospheric turbulence. According to similarity theory for the atmospheric surface layer, where blowing snow occurs, the cospectrum between the longitudinal and vertical components of wind is described as a function of stability and nondimensional frequency fz/U, where f is the frequency, z is the height and U is the mean wind speed. The integral of the cospectrum over the whole frequency yields the momentum flux, which is the key parameter of the structure of the turbulent surface layer. For example, the integration of the cospectrum obtained by McBean and Miyake (1972) from fz/U=0.005 to infinity yields about 90% of the momentum flux under a neutral condition. Substituting U=8 m/s and z=1 m into fz/U=0.005 as the typical conditions for the cases analyzed, f=0.04 Hz is obtained. Therefore, most of the momentum transfer is maintained by the eddies whose time scale is less than 25 sec and the turbulent transfer is regarded to be in equilibrium with the mean wind speed of 8 m/s if the averaging time is longer than 25 sec.

This study revealed that mass flux correlates well with wind speed and size distribution becomes smooth when the averaging time is longer than about 30 sec. Considering the discussion on the critical time scale of the turbulent transfer in the atmospheric surface layer, the present findings suggest that the blowing snow in the suspension layer is also in equilibrium with the mean wind speed through the turbulent transfer mechanism of the particles, if the averaging time is longer than the critical time scale of about 30 sec.

The statistical method should be effective to estimate the magnitude of blowing snow fluctuation with a period less than 30 sec. This study indicates that the probability distribution of such high frequency mass flux fluctuation can be approximated by a log-normal distribution.

Further research should be performed under stronger blowing snow conditions and similar analysis of the particle number fluctuation of various sizes will be useful.

REFERENCES

Budd, W.F. 1966. The drifting of nonuniform snow particles. *Antarctic Res.* Ser.9: 59-70.
Kimura, T. 1991. Measurements of drifting snow particles. *J. Geogr.* 100: 250-263. (In Japanese with English summary)
McBean, G.A. & M.Miyake 1972. Turbulent transfer mechanisms in the atmospheric surface layer. *Quart. J. R. Met. Soc.* 98: 383-398.
Sato, T, T.Kimura, T.Ishimaru & T.Maruyama 1993. Field test of a new snow-particle counter (SPC) system. *Ann. Glaciol.* 18: 149-154.
Schmidt, R.A. 1977. A system that measures blowing snow. *U.S. Dep. Agric. For. Serv. Res. Pap.* RM-194.
Schmidt, R.A. 1982. Vertical profiles of wind speed, snow concentration, and humidity in blowing snow. *Boundary Layer Met.* 23: 223-246.

Snow Engineering: Recent Advances, Izumi, Nakamura & Sack (eds) © 1997 Balkema, Rotterdam. ISBN 90 5410 865 7

Application of a random walk model to blowing snow

Takamitsu Sato, Takahiko Uematsu & Yasuhiro Kaneda
JWA, Sapporo, Japan

ABSTRACT : The random walk method has been applied to calculate a snow particle's movement. This model, which was introduced by Etling et al.(1986), was used in the wind field calculation using a k-ε turbulence model. The model can show the behavior of a single snow particle, and then we can analyze the relation of saltation and suspension which has not been spotlighted until now. The computed snow drift density against height shows good agreement with observed data. This model visualizes a snow particle shift from saltation to suspension or vice versa. This study suggests that this shift can be originated by turbulent diffusion mainly. After some improvement, this model will be a useful method for forecasting visibility in blowing snow.

1 INTRODUCTION

In snowy country, snow depth is found to be very variable, whether it be in the city, or the mountains. The main process that causes spatially varying snow depth is redistribution of snow by wind.

Recently, computational and physical modeling efforts have attempted to describe the physical process associated with snow transport. Decker and Brown (1985) utilized modern mixture theory to study blowing snow in mountainous terrain. Uematsu et al (1989) developed a two-dimensional finite-element model of snowdrift development and Uematsu et al (1991) modeled snowdrift around three-dimensional obstructions. To date our knowledge about blowing snow has been developed by dividing snow movement into three types, saltation, suspension and creep. Although it seems that a snow particle shifts from saltation to suspension or vice versa, the saltation and suspension mechanisms are separately taken into consideration separately in these models.

On the other hand, Etling et al (1986) applied a random walk model to turbulent diffusion in complex terrain. The trajectories of particle movement of air pollution are computed with a random walk model and air pollution concentration is computed.

We present a computational model by using a random walk model. In this model the particle's movement is not classified. This model displays trajectories of the snow particle's movement and then we analyze the relation of saltation and suspension which has not been spotlighted until now.

2 THE FORMULATION OF THE MODEL

2.1 *Random Walk Model*

For the random walk simulations we follow the technique described by Etling et al (1986). In the Etling's technique, the inertial force is ignored to compute the trajectories of small particles such as air pollution. Although it seems that the inertial force affects trajectories of large particles such as snow, the Etling's technique is adopted.

In this simulation particle positions are tracked using the Lagrangian relations :

$$x_i(t + \Delta t) = x_i(t) + (u_i + u_{fi})\Delta t + u_i{}'\Delta t \qquad (1)$$

where x_i is in Cartesian coordinates (i=1 being streamwise, i=2 being upwards in the two-dimensional case), t is the time increment and Δt the time step. The mean wind field u_i is assumed to be stationary and has been obtained from a numerical model which is described in the next section. u_{fi} is the falling velocity of a snow particle. The velocity fluctuation $u_i{}'$ is obtained for each time step from a Markov chain simulation after :

$$u_i{}'(t + \Delta t) = u_i{}'(t)R + (1 - R^2)^{1/2}\sigma_i\Omega \qquad (2)$$

where Ω is random number from the Gaussian distribution with zero mean and unit variance, and σ_i is the variance of the velocity fluctuation. An exponential form is taken for the Lagrangian autocorrelation function R, where for simplicity, the same function is supposed to be valid for both velocity components

$$R(\Delta t) = \exp(-\Delta t / \tau) \qquad (3)$$

where τ is the Lagrangian time scale. In the surface layer, τ and σ_i relate to the friction velocity u_* by :

$$\tau = 0.59l / u_* \qquad (4a)$$
$$\sigma_1 = 2.3u_* \qquad (4b)$$
$$\sigma_2 = 1.3u_* \qquad (4c)$$
$$l = \kappa z \qquad (4d)$$

where l is a mixing length, and κ the von Karman's constant. The constants in (4a-c) are taken from Panofsky et al.(1977) for neutral stratification.

In an air flow over complex terrain, since the friction velocity is no longer a constant, it is replaced by the turbulent kinetic energy k, which may be related to the friction velocity by :

$$u_*^2 = c_e k \qquad (5)$$

where c_e=0.16 according to Panofsky et al.(1977). The turbulent kinetic energy has been obtained from a numerical model which is described in the next section.

We assume that a snow particle which comes into collision with a wall surface, such as the ground or an obstruction, reflects as a mirror, and we control the intensity of the reflection with the restitution coefficient (Fig.1). We simply select 1.0 for the restitution coefficient at all surfaces.

2.2 Turbulence model

In the case of homogeneous and flat terrain, the mean wind field and the friction velocity are simply computed with the logarithmic law.

On the other hand, in the case of complex terrain, the mean wind field and the turbulent kinetic energy are computed with the Reynolds equations and the k-ε turbulence model which are expressed as :

$$\frac{\partial u_i}{\partial x_i} = 0 \qquad (6)$$

$$\frac{\partial u_i}{\partial t} + \frac{\partial u_i u_j}{\partial x_i} = -\frac{1}{\rho}\frac{\partial p}{\partial x_i} + \frac{\partial}{\partial x_i}\left(-\overline{u_i{}'u_j{}'}\right) \qquad (7)$$

$$-\overline{u_i{}'u_j{}'} = \nu_t\left(\frac{\partial u_i}{\partial x_j} + \frac{\partial u_j}{\partial x_i}\right) - \frac{2}{3}k\delta_{ij} \qquad (8)$$

$$\nu_t = C_\mu\frac{k^2}{\varepsilon} \qquad (9)$$

$$\frac{\partial k}{\partial t} + \frac{\partial u_i k}{\partial x_i} = \frac{\partial}{\partial x_i}\left(\frac{\nu_t}{\sigma_k}\frac{\partial k}{\partial x_i}\right) + \nu_t S_{ij} - \varepsilon \qquad (10)$$

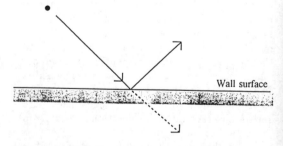

Figure 1. Sketch for the wall surface boundary condition for a snow particle. It is assumed that a snow particle reflects as mirror at the wall surface.

$$\frac{\partial \varepsilon}{\partial t} + \frac{\partial u_i \varepsilon}{\partial x_i} = \frac{\partial}{\partial x_i}\left(\frac{v_t}{\sigma_\varepsilon}\frac{\partial \varepsilon}{\partial x_i}\right) + \frac{\varepsilon}{k}\left(C_{\varepsilon1}v_t S_{ij} - C_{\varepsilon2}\varepsilon\right) \quad (11)$$

$$S_{ij} = \left(\frac{\partial u_i}{\partial x_j} + \frac{\partial u_j}{\partial x_i}\right)\frac{\partial u_i}{\partial x_j} \quad (12)$$

Here p is the pressure and ε is the dissipation rate of turbulent kinetic energy. The empirical constants in these equations are based on the work of Laundar and Spalding (1974).

$$(C_\mu, \sigma_k, \sigma_\varepsilon, C_{\varepsilon1}, C_{\varepsilon2}) = (0.09, 1.0, 1.3, 1.44, 1.92) \quad (13)$$

In summary, the procedure of our simulation is adopted as following steps :
(1) computation of mean wind field and turbulent kinetic energy by the log law or turbulence model;

(2) computation of trajectories of snow particles released at a source by random walk the model;
(3) evaluation of snow drift density by counting the number of particles in each box of the numerical grid.

3 RESULTS

In the following sections we show the results of computation and comparison with observation. We are interested in the shift of a snow particle's movement, namely from saltation to suspension or from suspension to saltation. We applied this model to a flat terrain.

Figure 2 shows some examples of one snow particle's trajectory. In Figure 2A, one particle jumps up to 30 cm height in one moment and rolls on in the other movement. Figure 2B shows an example of one particle which jumps up to 30 cm height in one movement and flies

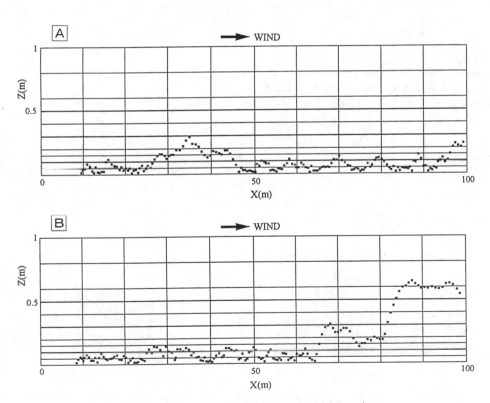

Figure 2. Some examples of one snow particle's trajectory.

over 30 cm height.

Figure 3 shows an example of 500 snow particles' trajectories. This figure shows that most particles move under 20 cm and the number of the other particles is few. We have the concept that snow transport is classified into saltation, suspension or creep. We can see three movement in these figure clearly. To date our knowledge about blowing snow was developed by dividing snow movement to three types, saltation, suspension and creep, and studying each movement. For example, Kobayashi (1972) reported that the mass flux of snow particles in saltation is about 80% of the total mass flux. However these figures show that it is difficult to classify the movement because one particle changes one type of movement to another frequently. Since the turbulence is considered in this model, the shift of the classified movement may be mainly originated by the turbulent diffusion.

Figure 4 shows the computed snow drift density against height above surface. This density is normalized by the surface flux at the source area. This profile shows a good agreement with the observation of Takeuchi et al (1975).

Finally, Figure 5 shows many particles' trajectories in the "back step flow." This flow and turbulent field is obtained with the Reynolds equations and k-ε turbulence model.

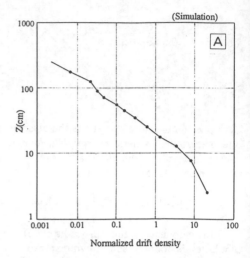

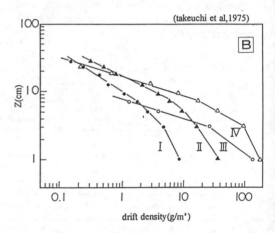

Figure 4. Snow drift density against height above surface. (A:computed, B:observed by Takeuchi et al [1975]).

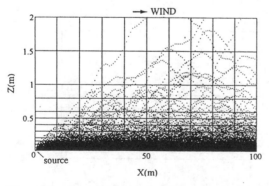

Figure 3. An example of 500 snow particles' trajectories.

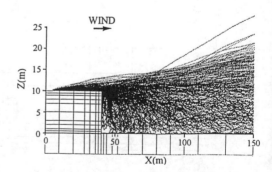

Figure 5. Many particles' trajectories in the back step flow.

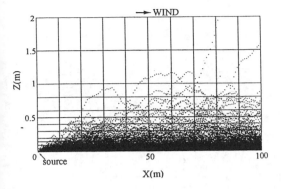

Figure 6. An example of 500 snow particles' trajectories (restitution coefficient=0.2).

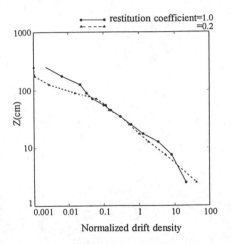

Figure 7. An example of 500 snow particles' trajectories. (solid line: restitution coefficient=1.0, dashed line:0.2)

4 DISCUSSION

In this paper we propose the random walk model for blowing snow. Now this model has many problems when we compare it to actual phenomena. For example snow particles are assumed to move at the same motion as the wind; however actual snow particles cannot move so. Actual snow particles have mass and volume, therefore the inertial force affects the motion of particles. But snow particles have only mass in this model, exactly only falling velocity. The second problem is that of source. Namely how many of particles are released in a condition, for example, a wind speed? This

problem is very difficult. The third problem is that of restitution coefficient. In this model it is equal to 1.0, but in actuality it has various values.

Here we will discuss the problem of the restitution coefficient. Figure 6 shows 500 particles' trajectories and Figure 7 shows the normalized snowdrift density. In these figures, the restitution coefficient is equal to 0.2. In this case, the drift density above 1 m height is smaller than the coefficient equals 1.0. Therefor, it seems that the restitution coefficient controls the number of particles which fly high.

5 CONCLUSION

A random walk model has been developed for blowing snow. The simulated flow is determined as follows :
(1) computation of the mean wind field and turbulent kinetic energy by the log law or turbulence model;
(2) computation of trajectories of snow particles released at a source by the random walk model;
(3) evaluation of snow drift density by counting the number of particles in each box of the numerical grid.

This model visualizes that a snow particle can shift from saltation to suspension and another from suspension to saltation. This study suggests that shift can be originated by turbulent diffusion mainly. After some improvement, this model will be useful method for forecasting the visibility at blowing snow.

REFERENCES

Decker, R. and Brown R.L. 1985. Two Dimensional Solutions for a Turbulent Continuum Theory for the Atmospheric Mixture of Snow and Air. *Annals of Glaciology* 6, 53-58.

Etling, D., Preuss, J. and Wamser, M. 1986. Application of a Random Walk Model to Turbulent Diffusion in Complex Terrain. *Atmospheric Environment*, 20, 741-747.

Kobayashi, D. 1972. Studies on Interaction between Wind and Dry Snow Surface. *Contrib. Inst. Low Temp. Sci. Hokkaido Univ.*, Ser.A, 24.

Laundar, B.E. and Spalding, D.B. 1974. The Numerical Computation of Turbulent Flows.

Comput. Methods Appl. Mech. Eng., 3, 269-289.

Panofsky, P.A., Tennekes, H., Lenschow, D.H. and Wyngaard, J.C. 1977. The Characteristics of Turbulent Velocity Components in the surface layer under convective conditions. *Boundary-Layer Met.* 11, 355-361.

Takeuchi, M., Ishimoto, K. and Nohara, T. 1975. A Study of Drift Snow Transport. *Seppyo*, 37, 8-15.(in Japanese)

Uematsu, T., Kaneda, Y., Takeuchi, K., Nakata, T. and Yukumi, M. 1989. Numerical Simulation of Snowdrift Development. *Annals of Glaciology*, 13, 265-268.

Uematsu, T., Nakata, T., Takeuchi, Y., Arisawa, Y., and Kaneda, Y. 1991. Three-Dimensional Numerical Simulation of Snowdrift. *Cold Reg. Sci. Technol.*, 20, 65-73.

Snow Engineering: Recent Advances, Izumi, Nakamura & Sack (eds) © 1997 Balkema, Rotterdam. ISBN 90 5410 865 7

Experiment of metamorphism using 'standard' snow

Atsushi Sato
Shinjo Branch of Snow and Ice Studies, NIED, Yamagata, Japan

Mike Q. Edens & Robert L. Brown
Department of Civil Engineering, College of Engineering, Montana State University, Bozeman, Mont., USA

ABSTRACT: A sintering experiment was done on "standard snow", which is composed of ice spheres. New methods were adopted for making the standard snow and analyzing surface sections. A snowmaking machine using a dry-ice tunnel as a cooling source was used to produce the snow. The preparation of the surface section was done with using dimethyl phthalate, and the surface sections were analyzed with a computer software system developed recently by the research team. Image analysis shows that mean grain size increases with time while the surface volume ratio and the diameter of the smallest grains both decrease with time.

1 INTRODUCTION

The physical properties of deposited snow are important factors in disasters and snow engineering. They are known to be strongly dependent on the microstructure as well as snow density. For example, thermal properties are shown to be strongly dependent on the bonding of grains of snow (Adams and Sato 1993, Sato and Adams 1995). Microstructure refers to such factors as grain size, grain shape, intergranular bond size, three-dimensional coordination number, and even the statistical distribution of these factors. Mostly, metamorphism governs the evolution of microstructure. For prediction of microstructure, it is quite important to understand the exact process and mechanisms of metamorphism in snow. In the discussion of natural snow, there are always difficulties due to complexity and varieties of microstructure in both time and space.

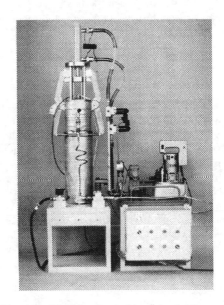

Figure 1. Machine for making "standard snow".

2 STANDARD SNOW

To conduct an observational study of snow metamorphism, we prepared "standard snow" which is an assembly of small ice spheres. There are many ways to produce ice spheres, such as using liquid nitrogen, cold alcohol or kerosene. A newly designed snow machine consists of dry ice (solid carbon dioxide) and a spray nozzle shown as Figure 1. The machine first makes dry ice of a ring shape at the top of the machine. Then the machine pushes down dry ice ring and makes another, finally producing a tunnel of dry ice.

Water is sprayed at the top of the tunnel. Water droplets are cooled as they travel down the tunnel and freeze into ice spheres with an average diameter of 25 micro meters.

3 EXPERIMENT

Observation of microstructure of snow is very important to understand physical and mechanical properties (Kinisita and Wakahama, 1959). To avoid poisonous chemicals as a filling material for void space of the snow, some trials were done by Watanabe (1974) and Perla (1982). This time we adopted dimethyl phthalate, which has a +2 ℃ as melting point, supercools easily but requires considerable time to freeze at -5 ℃. Several 5.0 x 5.0 x 2.5 cm³ metal containers were placed at the bottom of the snow machine, and standard snow was deposited naturally in the metal containers. The snow filled containers was put in plastic bags to avoid sublimation, and stored in coldrooms at -5 ℃, -10 ℃, -15 ℃ and -20 ℃. The snow specimens were frozen with dimethyl phthalate at a predetermined time intervals of 0 min, 20 min, 1 hr, 3 hr, 6 hr, 12 hr, 24 hr, 2 days, 1 week, 2 weeks and 5 weeks for different stages of sintering.

3.1 Surface section

A mixture of standard snow and dimethyl phthalate becomes sufficiently hard to plane after freezing. A smooth surface was made with a sharpened blade. The surface was then polished with a paper towel. A small amount of carbon powder (lamp oil) was painted on the specimen surface. Polishing again with the paper towel high-lighted the ice which became dark since the powder adhered to it. The dimethyl phthalate remained white, since the powder did not adhere to it.

3.2 Image analysis and results

Each specimen was placed on a microscope, and its surface was observed by a CCD camera (charge coupled device) mounted on the microscope. The camera was connected to a Macintosh computer and the surface image of the specimen was taken and digitally recorded as image files which were then stored on an optical

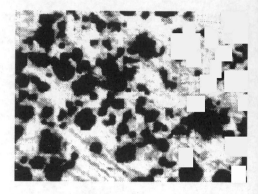

Figure 2. Original surface section image of standard snow aged 20 minutes.

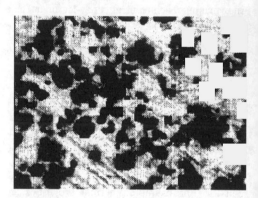

Figure 3. Edge sharpened image, modified from Figure 2.

Figure 4. Final image of Figure 2 as a binary

storage device. The surface section images are usually not very clear, as can be seen in Figure 2. These images were then subjected to an image cleaning process. Figure 3 shows one of the sharpened images. The sharpened image is then converted into a binary image (Figure 4).

4 DISCUSSION AND CONCLUSION

Ice grain growth was known to occur rapidly in the water or water saturated snow (Wakahama 1965, Raymond and Tusima 1979). In Figure 5, it is shown that the radius of ice spheres grows rapidly by dry sintering. Our experiment shows same grain growth in the dry snow without liquid water. The phenomena of sacrifice of small grains are seen in the same figure, where small grains experience decreases with time whereas large grain size grow. The mean grain size also shows growth. This was predicted by our former work (Brown et al. 1993).

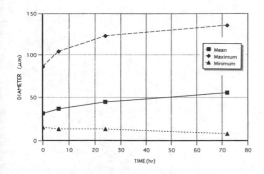

Figure 5. Changes in grain diameter with age.

Modern constitutive theories of snow cover uses microstructural parameters, such as bond radius, neck length as well as grain sizes. In order to reveal these parameters, the images of surface section of metamorphosed standard snow have been analyzed by stereology software (Edens and Brown 1991). Intergranular bonds are found and displayed, which are defined according to the criterion of Kry (1975). Figure 6 is an example of the section images showing the recognized intergranular bonds by the method. Using these microstructural parameters, a statistical analysis is being conducted to show the change of parameters important for characterizing the dry metamorphism of snow .

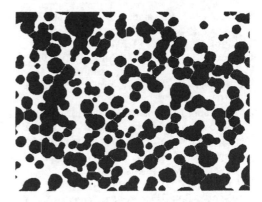

Figure 6. Image of surface section after stereology software has recognized intergranular bonds.

ACKNOWLEDGEMENTS

The authors would like to thank the Science Technology Agency of Japan and the Department of Civil Engineering of Montana State University for support of this work, and Drs. M.Matsuda, K.Yano and the Tecwart Group of Companies for discussion and help in making the new snow machine.

REFERENCES

Adams, E.E. and A. Sato. 1993. Model for effective thermal conductivity of a dry snow cover composed of uniform ice spheres. *Annals of Glaciology* 18: 300-304.

Brown, R.L., M. Q. Edens and A. Sato. 1993. A study of metamorphism of fine-grained snow due to surface curvature differences. *Annals of Glaciology* 19: 69-76.

Edens, M.Q. And R.L.Brown. 1991. Changes in Microstructure of snow under large deformations. *Journal of Glaciology*, Vol.37, No.126: 193-202.

Kinosita, S. and G.Wakahama. 1959. Thin sections of deposited snow made by the use of aniline. *Low Temperature Science, Ser.A*, 18: 77-96. (in Japanese with English summary).

Kry, P.R. 1975. Quantitative stereological analysis of grain bonds in snow. *Journal of Glaciology*, Vol.14, No.72: 467-477.

Perla, R. 1982. Preparation of section planes in snow specimens. *Journal of Glaciology*, Vol.28, No.98: 199-204.

Raymond, C. and K.Tusima. 1979. Grain coarsening of water saturated snow. *Journal of Glaciology*, Vol.22, No.86: 83-105.

Sato, A and E.E.Adams. 1995. Influence of microstructure on effective thermal conductivity of snow cover composed of ice spheres. *Seppyo*, 57, 2: 133-140. (in Japanese with English summary).

Wakahama, G. 1965. Metamorphism of wet snow. *Low Temperature Science, Ser.A*, 23: 51-66. (in Japanese with English summary).

Watanabe, Z. 1974. A model of making thin sections of deposited snow by mixed solution. *Seppyo*, 36, 3: 1-4. (in Japanese With English summary).

Snow Engineering: Recent Advances, Izumi, Nakamura & Sack (eds) © 1997 Balkema, Rotterdam. ISBN 90 5410 865 7

A method for estimating snow weight in mountainous areas

Masujiro Shimizu & Hideomi Nakamura
Nagaoka Institute of Snow and Ice Studies, NIED, Niigata, Japan

Osamu Abe
Shinjo Branch of Snow and Ice Studies, NIED, Yamagata, Japan

ABSTRACT: We estimated snow weight in a mountainous area using precipitation at Takada and the difference in air temperature between Takada and the mountainous area. Estimating snow weight is compared direct measurement data at Okutadami.

1 INTRODUCTION

It is important to know the amount of snow cover in mountainous areas for securing water resources and avoiding snow damage. The National Research Institute for Earth Science and Disaster Prevention (NIED) constructed seven snow stations six years ago, in the mountains of Japan. They are not numerous enough to estimate the amount of snow cover all over Japan.

The climate of the snowy Hokuriku region in Japan is rather mild, with the average air temperature only a few degrees above freezing from December to March. Then, the types of precipitation is rain or snow, but it is almost always snow then in the high mountainous areas of Niigata prefecture.

Air temperatures are almost always above 0 ℃ in the Hokuriku area in winter at the low plain near the coast. Snow cover in the low elevation areas are not used for water resources because the snow cover melt before the time it is needed. However, there is snow in the mountainous area after snow in the low elevation area has melted. Mountain snow continues to melt from spring to summer.

Variations in the amount of snow with elevation have been investigated by Yamada (1978), and Ishizaka (1995) has estimated snow types in Japan. However, the method for estimating snow weights in mountainous areas has not been investigated in Japan.

We have developed a method to estimate the amount of snow cover in the mountains where there is no station.

2 THE CLIMATE OF HOKURIKU REGION IN JAPAN

The distribution of average air temperature in Hokuriku from December 1992 to March 1993 is shown in Figure 1. Moving from the coast to mountainous area, the air temperature becomes lower. The distribution of total precipitation in the Hokuriku area is shown in Figure 2. Total precipitation in this period along the coast was about 500 mm, mountain precipitation exceeded

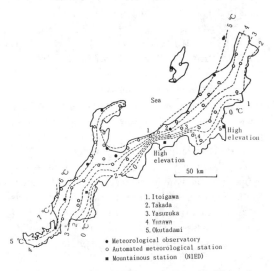

Figure 1. Average air temperature in the Hokuriku region from December 1992 to March 1993.

1100 mm. The tendency is heavy precipitation in the mountainous area, and light precipitation in plain near the coast.

The precipitation is about the same at Takada and other the stations omitting Okutadami. The range of average air temperatures from December to March is from 0 to 5 ℃ at four stations, omitting Okutadami. Differences of air temperature between Takada and the other stations are greatly influenced by the amount of snow cover at each station.

Figure 2.　Total precipitation in the Hokuriku region from December 1992 to March.1993.

3　THE　THRESHOLD　VALUE　OF　AIR　TEMPERATURE FOR RAIN OR SNOW

We investigated the threshold value of air temperature relative to type of precipitation, rain or snow, using daily average air temperatures, precipitation and weight of new snow which were calculated assuming a new snow density of 100 kg/m³. The relationship between daily average air temperature and precipitation type is shown in Figure 3. In Figure 3, example total precipitation at 4 ℃ is accumulated precipitation when daily average air temperature less than 4 ℃. Snow type precipitation is also accumulated snow type precipitation. Total precipitation and weight of New Snow are plotted against daily average air temperature. The line of snow does not increase when daily average air temperature are warmer than 2.7 ℃. This means

that precipitation is rain if the daily average air temperature is higher than 2.7 ℃, and the precipitation is snow if the temperature is lower than 0.7 ℃. In the region where temperatures are higher than 0.7℃ and lower than 2.7 ℃, we cannot distinguish the type of precipitation from air temperatures only. We define the threshold value of air temperature as 1.6 ℃. It is the temperature at which the total precipitation line is crossed by extension of the flat part of the snow line. We consider that this threshold value is statistically suited during snow fall season.

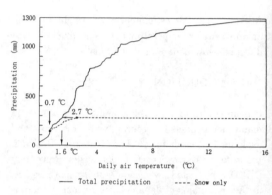

Figure 3. Relationship between daily air temperature and precipitation from December 1992 to March 1993.

4　DIFFERENCE　Of　AIR　TEMPERATURE IN　NIIGATA　PREFECTURE

Observation stations used in this paper are listed in Table 1. Daily average air temperature in the Hokuriku region at low elevation, during winter season are between -5 ℃ and 10 ℃. Frequency distributions of daily air temperature at four meteorological agency observation stations and at Okutadami Maruyama station in Niigata prefecture are shown in Figure. 4. Value at 1℃ is the number of

Table 1. Observation stations used.

Observation station	Latitude	longitude	Altitude
Takada	37° 06. 3'	138° 15. 0'	13 m
Itoigawa	37 04. 9	137 51. 9	10
Yasuzuka	37 06. 4	138 27. 9	135
Yuzawa	36 56. 1	138 49. 2	340
Shiozawa	37 02. 2	138 51. 0	195
Okutadami dam site	37 09. 4	139 14. 9	770
Okutadami Maruyam	37 09. 4	138 13. 6	1200

days when the daily average air temperature is from 0 ℃ to 1 ℃. Different air temperature between Takada and Itoigawa is about 1 ℃. Also the differences of air temperature between Takada and other stations are about 2, 2.5 and 8 ℃. But we do not find out accurate differences of air temperature between Takada and other stations in Figure 4. We investigated the accumulated distributions of air temperature at five stations. The five distributions are shown in Figure 5. There are no days lower than 1 ℃ at Itoigawa and almost no days lower than 1 ℃ at Takada. Difference air temperature at 0.5 ratio between Takada and Itoigawa, Yasuzuka, Yuzawa and Okutadami are 0.9, -2.1, -2.7 and -8.2 ℃. Even if the precipitation were distributed the same at all stations, the amounts of snow at them differ, because of ratio of snow to total precipitation differs at each station.

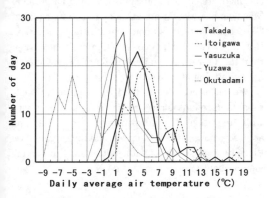

Figure 4. Frequency distribution daily average air temperature from December 1992 to March 1993.

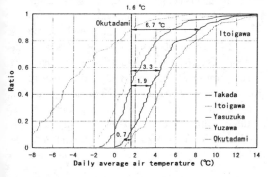

Figure 5. Accumulated distribution of air temperatures at five observation stations from December 1992 to March 1993.

5 CHARACTERERISTICS OF PRECIPITATION IN THE HOKURIHU AREA

Precipitation, which is predominant from December to the middle of February and ranges from 100 to 500 mm/month is caused by the strong winter monsoon. It is not always in the solid phase but also liquid. Accumulated precipitation at five observatories in Niigata prefecture is shown in Figure 6. Whenever a cold air mass comes to Japan, there is precipitation which various among stations. Accumulated precipitation during from December to March is 1700 mm at Okutadami dam site. Precipitation at other station ranges from 1000 to 1300 mm. Precipitation from February to March at Yuzawa is more than at Takada, this is also the case at Okutadami dam site.

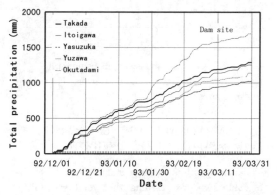

Figure 6. Accumulated precipitation at five observation stations from December 1992 to March 1993.

6 ESTIMATING SNOW WEIGHT AT FOUR STATION

Snow weights at each station were estimated using precipitation at Takada and the difference in air temperature between Takada and each station shown in Figure 5. The daily average air temperature at Takada is 3.5℃ when the temperature at Yasuzuka is 1.6 ℃. So the type of precipitation at Yasuzuka can be distinguished by this temperature (3.5℃) at Takada. We assume that total precipitation is the same at each station, then estimate solid precipitation at Yasuzuka. The snow weights are 780 kg/m^2 at Yusuzuka, 880 kg/m^2 at Yuzawa, and 1040 kg/m^2 at Okutadami. These values did not include the amount of snowmelt. Snow scarcely melts in the mountainous area at an altitude of 1200 m until early spring.

Snow weights are measured at Okutadami by a snow weight gauge and a snow sampler. But precipitation at each station is not equal to that at Takada in Figure 6. Thus the estimates of solid precipitation must be corrected by the total precipitation ratio between Takada and each station. Correcting solid precipitation for differences in total precipitation, the result is 161 kg/m² at Itoigawa, 712 kg/m² at Yuzawa, 1496 kg/m² at Okutadami Maruyama.

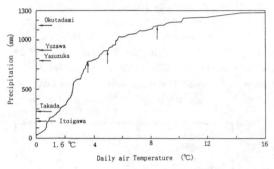

Figure 7. Estimating snow weights using total precipitation at Takada and converted threshold temperatures.

7 COMPARISION OF ESTIMATES TO MEASURED SNOW WEIGHTS

The snow depth at Yuzawa is almost equal to that at Shiozawa. We assume that total snowfall at Yuzawa is equal to that at Shiozawa. We compared our estimated snow weights at Itoigawa and etc, to total snowfall. Snowfalls at Takada, Itoigawa and Yuzawa are shown in Figure 7. Accumulated snowfall at Yuzawa is more than at Takada or Itoigawa,

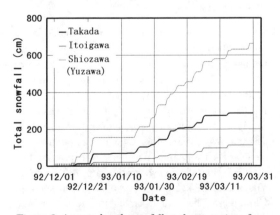

Figure 8. Accumulated snowfall at three stations from December 1992 to March 1993.

because precipitation at Yuzawa is solid when precipitation at Takada and Itoigawa is liquid.

The density of daily snowfall is measured by the Nagaoka Institute of snow and ice studies. The average density of daily snowfall is about 100 kg/m³. Thus we assumed that snowfall density is 100 kg/m³. Total snowfall is 664 mm at Yuzawa using data at Shiozawa, and 113 mm at Itoigawa in Figure 8. We estimated that solid precipitation is 712 mm at Yuzawa and 161 mm at Itoigawa. The difference between observation and estimation is 48 mm at Yuzawa, and the error is 7%. The difference between observation and estimation is 48 mm, and the error is 42 % at Itoigawa. Solid precipitation is estimated at 1496 mm at Okutadami. The maximum snow cover measured using a sampler was 2000 mm. The difference between direct measurement and estimation is 504 mm, and the error is 25 %. We assumed that the precipitation at Okutadami Maruyama station at an altitude of 1200 m, equal at Okutadami dam site station at an altitude of 770 m. But the elevation difference is 530 m. It is consider that the precipitation at Okutadami Maruyama is more than that at dam site.

We developed the method for estimating snow weights in mountainous areas. Precipitation various with elevation is indispensable for estimating accuracy snow weights. Since We will investigate total Precipitation various with elevation in many mountainous areas in the future.

REFERENCES

Ishizaka, M. 1995. Climatic Division of Snow Cover Environments in Japan Using the Mesh Climatic Data. Seppyo, 57. (in Japanese)

Yamada, T., Nishimura, H., Suizu, S. and Wakahama, G. 1978. Distribution and Process of Snow on the West Slope of Mt. Asahidake, Hokkaido. Low Temperature Science. Ser. A , 37. (in Japanese)

Kumagaya, M., Kobayashi ,T., Kimura, T., Shimizu, M., Yamada, Y., Ikarashi, T., and Nohguchi, Y. 1987. Data on Snow Cover in Nagaoka (11) (November 1986 - April 1987). Review of Research for Disaster Prevention. No.120. (in Japanese)

Snow Engineering: Recent Advances, Izumi, Nakamura & Sack (eds)© 1997 Balkema, Rotterdam. ISBN 90 5410 865 7

A few considerations concerning acid snow and acid rain on the Japan Sea Coast

Koichi Katoh & Yoji Taguchi
Department of Chemistry and Chemical Engineering, Niigata University, Japan

Kiyomichi Aoyama
The Research Institute for Hazards in Snowy Areas, Niigata University, Japan

Jiro Endo
Department of Agricultural Engineering, Niigata University, Japan

ABSTRACT : Acid rain damage to forests is a major ecological problem around the world. Recently, particular attention has been drawn to acid snow damage to the environment along the Japan Sea Coast. The acid snow and rain are now acidifying the ground and soil, and the acidified soil is affecting the forest, woods, plants, etc. The degree of acidification of soil and the concentration of some elements leached with a mixed solution of nitric and sulfuric acids were examined. Many examples of buildings damaged by acid snow and acid rain can be observed. The corrosion process by acid is especially effective on marble structures. In Eastern Europe this results from the low quality coal that people have used for electric and heating system for many years.

1. Introduction

Recently, particular attention has been paid to acid snow damage to the environment in the world. The study of acid snow is important as well as acid rain. About ten years ago, it was found that forests were damaged by acid rain. By monitoring program in forest plots in Sweden, O.Westling *et. al.* (1995) have been investigated in 30 sites with deposition of strong acids. From the data, they found out that abnormal internal circulation of nutrient imbalance or deficiency could be indicated by their original calculation method. In Dutch, W. De Vries *et. al.* (1995) studied about effects of acid deposition on forest ecosystems. In past ten years, many researchers, including these of local and national governments, have developed much time to studying the acid rain and related environmental problems. This paper is a study of chemical characteristics of acid deposition and its effects on acidity of soil.

2. Experimental

All samples of snow and rain are gathered by co-workers who live in the cities along the Japan sea coast. The chemical components, pH, or conductivities of samples were measured as soon as possible. Anion concentrations in them were determined by ion chromatography.

3. Results and discussion

3.1 pH values

Snow, and rain samples had a pH between 4.3 and 6.8.The average pH of melted snow was 5.3 and the average of 10 rain samples was 5.6. It seems that snow can easily catch more pollutants such as nitrogen oxides and sulfur dioxide in the air than rain. This is why melted snow shows lower pH than rain does. The ratios of acid snow and normal snow are shown in Figure 1 as well as the ratios for rain.

According to Figure 1, the percentage of acidic rain is 50% ; however, that is 77.8% in fresh snow. The average values of nitrate ions, and sulfate ions in fresh snow are 1.5 ppm, and 5.2 ppm, respectively. This means that fresh snow can easily catch pollutants

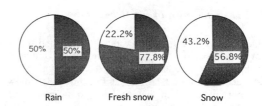

Figure 1 Acidic percents in rain and snow.
■: Acidic, □: Normal

of nitrogen oxides and sulfur dioxide in air. There is no large difference in pH between north and south portions of Japan Island. Shimane prefecture lies at the southern part of Japan, and Akita prefecture lies at the northern part of Japan used for this study. The pH values of snow and rain from south to north were 5.8 (Shimane prefecture), 5.3 (Fukui), 4.5 (Ishikawa), 5.9 (Toyama), 5.4 (Niigata), and 5.1 (Akita). The average value of pH in Niigata prefecture was 5.6. In the case of Niigata city, it was 5.4.

3.2 *Component Ratios of Anions*

Piled snow was artificially polluted by exhaust gas of cars or smoke of manufacturers. Then the ratios of chloride ion, sulfate ion and nitrate ion in rain and fresh snow are compared. Figure 2 shows the component ratios in normal rain and normal fresh snow. The percentage of chloride ions contained in the rain was 63.6%, but 72.8% in fresh snow. The amount of chloride ions was relatively large in each case. Sulfate ions were almost at the same order in both cases. The percentage of nitrate ions was 13.7% in rain, and 6.1% in fresh snow. While, from Figure 3, the chloride ion in fresh snow was 75.0%, but only 48.5% in rain. The nitric acid ion was about 17.3% in rain, but only 5.5% in fresh snow. There are large differences between rain and fresh snow.

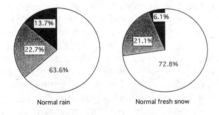

Figure 2 The component ratios of Cl ion, SO4 ion, and NO3 ion in normal rain and fresh snow(over pH 5.6).

□ : Cl⁻ion, ■ : SO4^{2-}ion, ■ : NO3⁻ ion

3.3 *The First Water of The Melted Snow*

During snow melt, the snow repeats thaw and freeze cycles. As the snow melts, impurities contained inside it are gradually squeezed out towards the surface. Therefore the first water of the melted snow is more acidic than that of later snow melt. The component ratio of anions in the first water of the melted snow is shown in Figure 4. The pH value of this water was 5.4 and that of normal melted water was 7.6. The component ratios in both cases show a big difference, but the concentrations of chloride ions

in the normal water of melted snow was 2.61 ppm and in the first water of the melted snow, 2.27 ppm. According to Figure 4(b), the nitrate ion was absent in normal water of melted snow in this case. The acidic water of the melted snow started to flow down in early spring. That water is indispensable to certain animals and plants.

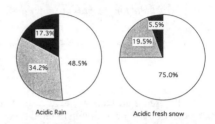

Acidic Rain   Acidic fresh snow

Figure 3 The component ratios of Cl ion, SO4 ion, and NO3 ion in acidic rain and fresh snow(under pH 5.6).

□ : Cl⁻ ion, ■ : SO4^{2-} ion, ■ : NO3⁻ ion

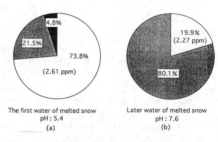

The first water of melted snow   Later water of melted snow
pH : 5.4   pH : 7.6
(a)   (b)

Figure 4 The component ratios of Cl ion, SO4 ion, and NO3 ion in water of the melted snow.

□ : Cl⁻ ion, ■ : SO4^{2-} ion, ■ : NO3⁻ ion

3.4 *Multiple Regression Analysis*

In order to find the effect of anions of acid snow or acid rain on pH, the pH value was taken as a criterion variable and anions of chloride ions, sulfate ions, and nitrate ions were taken as prediction variables. According to the results, the following equations were obtained:

For normal rain (13 samples);

$$pH = -0.061pCl + 0.077pSO_4 - 0.125pNO_3$$
    Level of significance: pCl 5%
    Coefficient of determination: 0.422

For acid snow (13 samples);

$$pH = -1.94pCl + 13.81pSO_4 - 0.726pNO_3$$
    Level of significance: pCl 5%, pSO4 1%
    Coefficient of determination: 0.327

148

3.5 *Comparison of pH(H₂O) and pH(KCl)*

The relations of pH values of the soil sample measured after shaking with deionized water (pH(H₂O)), potassium chloride solution (pH(KCl)), and acidic solution (pH(acidic)) were depicted in Figure 5. pH values on the abscissa indicate the pH(H₂O) measured using the water, while the pH values on the ordinate are pH(KCl) and pH(acidic). The solid lines were drawn by the least squares method. In the relationship between pH(KCl) and pH(H₂O) the line has a slope of about 1.0 but is deviating from a broken line, which means that the pH values on the abscissa are equal to those on the ordinate. The deviating width from the broken line is read off about 0.8 in pH described in Doshitsu Kogakkai (1979). It is generally believed that potassium ions can exchange with hydrogen ions on the soil surface. Larger amounts of hydrogen ions exchanged suggests that the soil had already been covered with larger amounts of hydrogen ions and had been acidified.

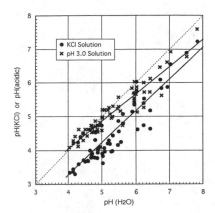

Figure 5 Comparison of soil pH values using
three solutions (H₂O, KCl and
pH = 3.0).

3.6 *Acidification of Soil Samples*

It is difficult to evaluate the degree of acidification of soil by acid snow and acid rain now, except to say that 22 samples (36% of all samples) have already been acidified. It is also impossible to point out where the acidified soil mostly exists, which kind of trees can grow on the acidified soil, where on mountains the soils mostly exist , which soils are biased on the sea coast or on inland areas. We can find acidified soils every where.

3.7 *Concentration of Phosphorus Leached*

The concentration of phosphor leached from the soil sample using a pH 3.0 solution is shown in Figure 6. Many concentrations measured showed less than 0.5 mg/l and they are observed at a lower pH. When the concentration of phosphorus is less than 0.5 mg/l, we might say the soil as a poor phosphorus soil, or an acidified soil. The basis for the tentative limiting value of 0.5 mg/l has not yet been defined, but studies by Murashige and Skoog (1962), Sansei Chosa Kenkyu Kai (1993) and Yoshida and Kawahata (1988) support the value of 0.5 mg/l .

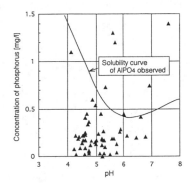

Figure 6 Concentration of phosphorus leached
from soil using a pH 3.0 solution.

4. CONCLUSION

According to experiments and observations, the following conclusions were obtained.
1. The ratio of acidic to normal in rain, piled snow, and fresh snow shows a higher percentage of acidic ratio in fresh snow than in rain. Comparing the component ratio, there is a large difference between the first water of melted snow and normal water of melted snow.
2. From the result of multiple regression analysis, sulfate ion shows a 1% level of significance in acidic rain.
3. From the measurement of soil pH it was suggested that 36% of all the samples gathered have already been acidified. The same soil sample showing lower pH also showed a lower concentration of phosphorus, less than 0.5 mg/l, when leached by an acidic solution of pH 3.0. This value may be considered to be a standard for judgment of a phosphorus-free soil.

REFERENCES

O. Westling, H.Hultberg and G.Malm 1995. *Dev. Plant Soil Sci.* 1995,62: 639-47

W. De Vries, E.E.J.M.Leeters, and V.M.A. Hendricks 1995. *Water, Air, Soil Pollut.* 85 (3): 1063-8

Doshitsu Koggakai, Ed.1979. Doshitsu Shiken Hou: 218-219

T. Murashige and F. Skoog 1962. *Physiol. Plant,* vol. 15: 473-497

Sanseiu Chosa Kenkyu Kai, Ed. 1993. Sanseiu Chosa Hou, Gyousei: 268

M. Yoshida and Y. Kawahata: *Dohi Shi* (Journal Soils & Fertilizers, Japan),Vol.59, 1988. 413-415

Snow Engineering: Recent Advances, Izumi, Nakamura & Sack (eds) © 1997 Balkema, Rotterdam. ISBN 90 5410 865 7

Sulfur isotope ratios of sulphate in wet deposits in Japan

Fumitaka Yanagisawa
Department of Earth and Environmental Sciences, Faculty of Science, Yamagata University, Japan

Akira Ueda
Mitsubishi Materials Corporation, Central Research Institute, Omiya, Japan

ABSTRACT:Sulfur isotope ratios of sulfate in wet deposits in Japan were measured in an attempt to determine the sources of sulfuroxides. Wet deposits collected every month from 1991 to 1996 at Yamagata, Shinjo, Yonezawa, Ohta, Sakata, Okushiri Island, Rishiri Island, Ohfunato and Tagajo. Sulfur isotope ratios of non-sea salt sulfate were low in summer and high in winter. The sulfur isotope ratios of non-sea salt sulfate ranged from -12 to +18‰. The sulfur isotope ratio of non sea salt sulfate in winter agreed with sulfur isotope values of sulfur-oxides exhausted from Asian anthropogenic activites.

1 INTRODUCTION

Much attention in being denoted to earth environmental problems. Acidification of rain and snow is one of the most serious problems in the world. Sulfur is deposited from the atmosphere as sulfate. Many studies suggest that wet deposits contain acid with mainly sulfate ion and anthropogenic sources are the major sources of sulfate. In Japan, the northwest monsoon dominates in winter and the deposition rate of non-sea salt sulfate increases in winter. Sulfur isotope ratios have been used as natural tracers of sources (Nakai and Jensen 1967). In this paper, sulfur isotope ratios of sulfate in wet deposits in Japan were measured in an attempt to determine the sources of sulfate.

2 SAMPLING

Wet deposits collected every month from 1991 to 1996 at Yamagata, Shinjo, Yonezawa and Ohta, which are located 80 to 100 km inland from the coast of the Japan Sea, at Sakata, Okushiri Island and Rishiri Island, facing the Japan Sea, and at Ohfunato and Tagajo facing the Pacific Ocean (Fig.1). Rains

Figure 1. Location of sampling stations.

and snows were collected using 10-liter polyethylene bottle, connected to a polyethylene funnel (23 cm in diameter) placed on the top of a building.

3 EXPERIMENT

3.1 Chemical analysis

The water was filtered through a 0.45 μ m

151

membrane filter and then analyzed for Na^+, K^+, Ca^{2+} and Mg^{2+} by inductively coupled plasma, for Cl^-, SO_4^{2-}, NO_3^- by ionchromatography, and for NH_4^+ by indophenol colorimetry. Electroconductivity and pH were measured by an electroconductivity meter and a pH meter with glass electrode.

3.2 Sulfur isotope analysis

Sulfate was precipitated as $BaSO_4$. The thermal decomposition method was used to prepare SO_2 gas for sulfur isotopic analysis (Yanagisawa and Sakai 1983). A finely ground mixture of $BaSO_4$ (about 5 mg), V_2O_5 (about 50 mg) and SiO_2 (about 50 mg) was placed at the bottom of a silica extraction glass tube. The tube was heated at $450°C$ for about 30 minutes in open air to remove any organic contaminants. The tube was connected to the preparation vacuum line. This silica glass tube was then heated by an electric furnace. The evolution of SO_2 started at $600°C$ and maximized at about $670°C$. In order to complete the thermal decomposition, the mixture must be heated to $950°C$.

The sulfur isotope ratio of SO_2 gas was measured by mass spectrometer. The isotopic ratios of sulfur were expressed as $\delta^{34}Smes$ notation, defined by:

$$\delta^{34}Smes = \{((^{34}S/^{32}S)mes/(^{34}S/^{32}S)std)-1\} \times 1000 \tag{1}$$

which expressed the parts-per-thousand deviation (permil : ‰) of the isotope ratio of the sample, $(^{34}S/^{32}S)$ mes, from that of the standard, $(^{34}S/^{32}S)$ std. The standard used for sulfur was the troilite (FeS) from the Canyon Diablo Meteorite. The precision was about ± 0.2 permil.

Sulfate ions from sources other than sea salt were contributing to the deposition. The concentration of non sea salt sulfate ($SO_4^{2-}nss$) was calculated by the following equation:

$$SO_4^{2-}nss = SO_4^{2-}mes - (SO_4^{2-}/Na^+)sea \times Na^+mes \tag{2}$$

where $SO_4^{2-}mes$ and Na^+mes were the concentrations

of sulfate and sodium. Sulfate and sodium weight ratio of seawater, $(SO_4^{2-}/Na^+)sea$, was 0.252.

The sulfur isotope ratio of sea salt sulfate ($\delta^{34}Ssea$) was 20.3‰. The sulfur isotope ratio of non sea salt sulfate ($\delta^{34}Snss$) was calculated by the following equation:

$$\delta^{34}Snss = \{ \delta^{34}Smes \times SO_4^{2-}mes - \delta^{34}Ssea \\ \times (SO_4^{2-}mes - SO_4^{2-}nss) \} / SO_4^{2-}nss \tag{3}$$

where $\delta^{34}Smes$ was the sulfur isotope ratio of sulfate in wet deposits.

4 RESULTS AND DISCUSSION

Sulfur isotope ratios of sulfate were low in summer and high in winter, linked with the contribution of sea salt. The $\delta^{34}Smes$ values in atmospheric deposits ranged from -10 to +18‰, which was less than those of seawater sulfate (20.3‰). This result suggests that the main sources of sulfate in rain and snow of these areas did not come from sea spray sulfate. On the other hand, the sulfur isotope ratios of non-sea salt sulfate ($\delta^{34}Snss$) ranged from -12 to +18‰ and showed seasonal variation with a increase in winter.

Japan is divided into three regions according to features of seasonal variation of $\delta^{34}Snss$: A, B and C (Fig. 2). The A region, including Hokkaido and the Japan Sea coastal area of northern Japan, is characterized by high values and wide dispersion (Fig. 3). The B region, including the Pacific Ocean coastal area of northern Japan, has high values and wide dispersion, although sulfur isotope ratio of non sea salt sulfate in summer is represented by low values (Fig. 4). The C region, including southern Japan, is characterized by low values and narrow dispersion (Fig. 5).

The $\delta^{34}Snss$ in winter agreed with sulfur isotope values of sulfur-oxides exhausted from Asian anthropogenic activites. In Japan, the northwest monsoon dominates in winter and deposition rate of non-sea salt sulfate was increased in winter. We

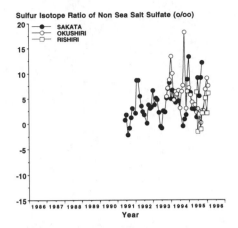

Figure 2. Seasonal variation of sulfur isotope ratios of non-sea sulfate (δ^{34}Snss) in wet deposits at Sakata, Okushiri Island and Rishiri Island.

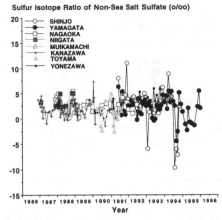

Figure 4. Seasonal variation of sulfur isotope ratios of non-sea sulfate (δ^{34}Snss) in wet deposits at Shinjo, Yamagata (Obinata and Yanagisawa 1996), Yonezawa (Yanagisawa et al. 1994), Niigata (Ohizumi et al. 1994), Nagaoka (Ohizumi et al. 1994), Muikamachi (Ohizumi et al. 1994), Kanazawa (Kitamura et al. 1993) and Toyama (Satake et al. 1991).

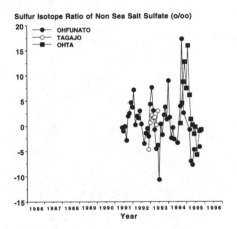

Figure 3. Seasonal variation of sulfur isotope ratios of non-sea sulfate (δ^{34}Snss) in wet deposits at Ohfunato, Tagajo and Ohta.

conclude that sulfate in winter contains sulfur-oxides from Asian human activities.

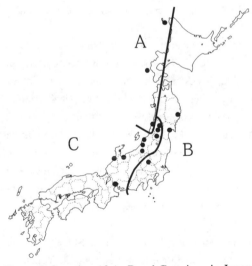

Figure 5. Location of A, B and C regions in Japan.

AKNOWLEDGEMENTS

The authors express sincere appreciation to Ryuuichi Matsuura, Yutaka Obinata, Midori Sasaki, Satoko Kikuchi, Yoshitoshi Sugawara, Hiroaki Sugasawa, Hirokazu Chiba, Junichi Shida, Kunio Sasaki, Mitsumasa Yanagisawa, Yutaka Takigami, Midori Furukawa, Hiromi Kitagawa who assisted with sample collection. The authors wish to thank Professor Hitoshi Sakai, Yamagata University, for his encouragement for this work.

This study was financially supported by the Grant in Aid for Scientific Research, No.07640650 from the Ministry of Education, Science and Culture.

REFERENCES

Kitamura, M., Sugiyama, M., Ohhashi, T. and Nakai, N. 1993. An estimation of the origin of sulfate ion in rain water in view of sulfur isotopic variations., *Geochemistry*, 27, 2, 109-118.

Nakai, N. and Jensen, M. 1967. Sources of atmospheric sulfur compounds., *Geochemical Journal*, 1, 199-210.

Obinata, Y. and Yanagisawa, F. 1996. Sulfur isotope ratio of sulfate in wet and dry deposits., *Sea*, Special Edition, in press.

Ohashi, T. (1988) *Master Course Thesis of Nagoya University*.

Ohizumi, T. Fukuzaki, N. and Kusakabe, M. 1994. Contribution of various sulfur sources to atmospheric sulfate deposition in Niigata Prefecture, Japan., *Journal of the Chemicall Society of Japan*, 9, 822-827.

Satake, H., Yamane, R . and Ichikawa, T . 1992. Abstract of The Geochemical Society of Japan. Seki, K., Maeda, K., Nagamine, Y., Ikeda, Y., Kusakabe, M. and Yoshioka, K. 1994. *Abstract of The Geochemical Society of Japan*.

Yanagisawa, F. and Sakai, H. 1983. Thermal decomposition of barium sulfate - vanadium pentaoxide - silica glass mixtures for preparation of sulfur dioxide in sulfur isotope ratio measurements., *Analytical Chemistry*, 55, 985-987.

Yanagisawa, F., Ueda, A. and Shida, J. 1994. Sources of sulfate in wet deposits in Yonezawa, Yamagata Prefecture, Japan., *Journal of The Japan Society of Analytical Chemistry*, 43, 947-952.

Snow Engineering: Recent Advances, Izumi, Nakamura & Sack (eds) © 1997 Balkema, Rotterdam. ISBN 90 5410 865 7

A database for climatic environments of snowy areas in Japan

Masaaki Ishizaka
Toyama Science Museum, Japan

Toshikazu Kobayashi
Toyama Computer Vocational School, Japan

ABSTRACT: A database for climatic environments in snowy areas in Japan was constructed from the mesh climatic data of the winter season, topographical data at every 1-km square area and categories for indicating the characters of snow cover, that is wet snow, dry snow, depth hoar and intermediate snow. Using the database we could obtain a digital cartographic map showing that "heavy and wet snow " areas were distributed in Hokuriku District and southward, which is one of the typical features of snowy areas in Japan where winter monsoons cause heavy snowfall in a temperate climate. The relationships between climatic monthly maximum snow depth and average air temperature were also discussed according to predominant snow type. A new general climatic condition determined where depth hoar is likely to develop in the snow cover.

1 INTRODUCTION

Topographical features of Japan make the coastal regions facing the Japan Sea, heavy snowfall areas, spanning a wide range of latitudes from 35° to 45° N and different climatic zones. The climatic and meteorological conditions in winter season affect characteristics of snow cover, causing local features of snow cover to change with latitude (Nakamura 1979). Pit-wall observations of snow at fixed points and many snow surveys have supported these facts.

There have been many attempts to clarify the regional characteristics of snow cover. Akitaya and Endo (1980,1982) clarify the regions in which depth hoar is expected to develop in Hokkaido Island and present an expression for deciding whether the depth hoar is present. Izumi and Akitaya (1986) prove the expression applicable to Honshu Island. Kawashima et al. (1994) investigate the snow cover at plain areas along the Japan sea coast, dividing them into 11 regions, and discuss the relationship between the predominant type of metamorphism and meteorological conditions. Their division of snow cover environments based on metamorphism of snow well explains the regional variation of snow type qualitatively and quantitatively. But it is difficult for his division to be applied to other snowy areas because the index for predominant metamorphism introduced are not easily obtained at any area in Japan.

Ishizaka (1995), one of the authors, presents more general criteria of climatic division of snowy areas and applied them to every 1-km square area in Japan

using mesh climatic data developed by the Japan Meteorological Agency (JMA).

In the present work a database for snow cover environments at any 1-km square area in Japan was constructed, including the categories of the climatic division introduced by Ishizaka, climatic data in the winter season and topographical data. Using the database we could obtain a digital cartographic map of "heavy and wet snow " regions. This and other results on the climatic relationship between air temperature and snow depth in snowy areas, prove the database useful for holistic understanding of climatic characteristics of snow cover environments in Japan. New general climatic criteria for indicating the depth hoar region are also discussed.

2 METHODOLOGY OF CLIMATIC DIVISION

Snow cover environments were classified into four categories according to predominant snow type (Table 1,Ishizaka 1995). In the wet snow region, where the climatic monthly mean temperature in January (CMTJ) is higher than 0.3 ℃, every layer of deposited snow is wet due to percolation of snow meltwater throughout the winter season and the proportion of coarse-grained granular snow in snow layers is large. The colder regions, where CMTJ is lower than -1.1 ℃, are divided into two snow types; the dry snow region and the depth hoar region. In the depth hoar region, where depth hoar would be expected to develop, air temperature is low and snow

Table 1. Climatic division of snow cover environments.

Category for division	Climatic conditions	Characteristics of snow cover
wet snow region	T>0.3℃	Every layer of deposited snow is wet throughout the winter season. Granular snow(wet grains) is likely to be developed.
intermediate snow region	0.3℃≤T≤-1.1 ℃	transient charaveristics between wet type and dry type
dry snow region	T<-1.1℃	Deposited snow is dry at least mid winter. Fine grained compacted snow is likely to be developed.
depth hoar region	T<-1.1℃ and R<8 kg	low temperature and a little snow region Temperature gradient in snow layers induces depth hoar. Rammsonde Hardness is small.
		R is calcurated with equation writen as $R=0.339(H_1/\overline{T})^{1.21} \cdot (H_2)^{0.21}$ where T is average of the absolute values of the monthly average air temperature during January and February, H_1 is the average of monthly maximun snow depth during January and February and H_2 is that during February and March.

T is monthly average temperature in January

depth is small. In the dry snow regions, where snow layer is dry at least in midwinter and fine-grained compact snow is likely to develop. The index for dividing theses regions is Rammsonde hardness, which is estimated from climatic values.

An intermediate snow type region with CMTJ ranging from -1.0 to 0.2 ℃ was introduced because the criterion between wet and dry regions is not explicit. Snow cover in this region has a medium character between wet and dry snow and changes to each other snow types in temporal meteorological conditions.

The climatic data used in this division were the mesh climatic data developed by JMA. JMA calculated climatic values at every 1-km square area in Japan with multiple regression formula which reflected the relationship between topographical condition and climate. Adapting the criterion of climatic division (Table 1) to all the mesh climatic data, we could determine which snow type region every square area of Japan belongs to.

3 DATABASE

A database was constructed from snow type, climatic data in the winter season and topographic data. Climatic data used were monthly mean, minimum and maximum temperatures, monthly maximum snow depth and monthly precipitation during winter (from December to March). The database did not include data from the southern part of Japan, that is Kantoh, Tohkai, Kinki, Sikoku Island, Kyusyu Island and southward, because the monthly values of maximum snow depth had not been calculated by JMA. Areas where the monthly maximum snow depth was less than 10 cm in any month between December and March were also eliminated from the database. Topographical data were average altitudes of every 1-km square area of digital cartographic information that had been produced by the Japan Geographical Survey Institute .

The structure of this database is shown in Table 2. The first element ("code") indicates the geographical point corresponding to every square mesh.

Table 2. Data structure.

Code	monthly average temperature				monthly maximum temperature				monthly minimum temperature				monthly maximum snow depth				precipitation amounts				altitude	snow type
	Dec.	Jan.	Feb.	Mar.	Dec.	Jan.	Feb.	Mar.	Dec.	Jan.	Feb.	Mar.	Dec.	Jan.	Feb.	Mar.	Dec.	Jan.	Feb.	Mar.		
55370000	5.2	1.9	2.3	5.3	1.6	-1.3	-1.4	0.8	8.8	5.1	5.9	9.7	20	67	46	19	341	304	206	166	33	W
55370001	5.2	1.9	2.3	5.2	1.6	-1.3	-1.4	0.7	8.8	5.1	6.0	9.7	20	72	49	18	337	298	207	165	33	W
55370002	5.1	1.9	2.2	5.1	1.5	-1.4	-1.6	0.5	8.7	5.1	5.9	9.6	19	76	48	17	331	301	209	163	36	W
55370003	5.1	1.9	2.2	5.1	1.5	-1.3	-1.5	0.5	8.7	5.0	5.9	9.6	19	80	51	19	316	308	220	160	47	W
55370004	5.0	1.8	2.1	4.6	1.3	-1.5	-1.7	0.1	8.6	5.0	5.9	9.1	19	84	52	20	308	312	234	158	63	W

4 RESULTS AND DISCUSSION

4.1 *Characteristic snow cover environments of heavy snow cover areas in Japan*

Querying the database we can derive information about the character of snow cover in heavy snow areas. Figure 1 shows the snow types in the heavy snow cover regions where the monthly maximum snow depth in any month during December and March are more than 1 m and the altitude are less than 500 m. The latter condition was introduced for clarifying the snow cover environments where human activities exist to some extent in winter season. It is found that the snow type in Niigata prefecture and southward is almost all wet or intermediate and

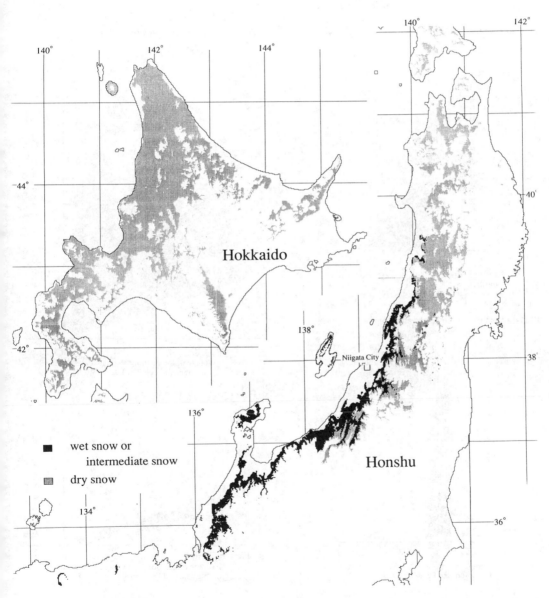

Figure 1. The digital cartographic map of the character of snow in heavy snow cover areas in Japan, where climatic monthly maximum snow depths are greater than 1 m and altitudes are less than 500 m.

that northward it is dry. This result agrees with the fact that people living in Niigata prefecture generally talk about the character of deposited snow as being "wet and heavy". In the wet and intermediate snow regions snow cover is governed by melt metamorphism and its density is relatively large. Therefore it is hard to remove snow, and destruction of roofs and other structures by weight of snow often happens. On the other hand in the dry snow area the density of the snow cover is relatively low but the snow is hard to melt and human activities are influenced for a longer snow cover period.

4.2 Relationship between average air temperature and snow depth

Figure 2 shows the relationship between the average of monthly snow depth during January and February and the monthly average temperature according to

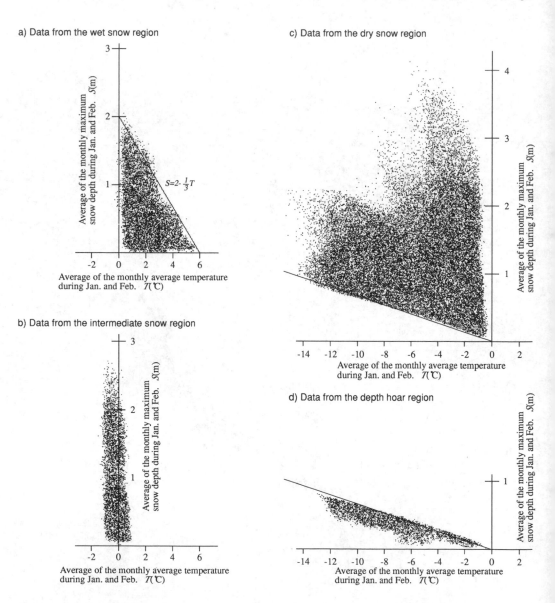

Figure 2. The relationship between the average monthly maximum snow depth during January and February, S and the monthly average temperature, T.

snow type. In this examination places where altitudes are greater than 1 km were eliminated, because insufficient climatic data were available at such points.

In the wet snow region, the snow depth decreases as the air temperature increases. This indicates that accumulation of snow is controlled by a melting process in warmer snowy areas. Almost all wet snow region data exist below the line drawn from 6 ℃ on the abscissa to 2 m on the ordinate as shown in Figure 2a.

4.3 A new criterion distinguishing the depth hoar region from the dry snow region

Comparing (c) with (d) in Figure 2 we found that data from two different regions were distributed on two different areas on the graph. The equation of the line that divides two groups is as follows,

$$S=-0.066T \qquad (1)$$

where S (m) was the average maximum snow depth during January and February and T (℃) was the monthly average temperature in the same period.

To obtain the coefficient of the equation (1) we chose data from the regions where calculated Rammsonde hardness R values (Table 1) was between 7 and 9. The values of chosen points are shown to scatter around the boundary (Figure 3). The coefficient was obtained by averaging the values, S/T of these points.

The inverse of the absolute value of the coefficient, 15℃/m has a physical meaning, that is, it corresponds to the temperature gradient of deposited snow. Akitaya and Endo (1977) introduced the mean temperature gradient, by dividing the mean air temperature in January by the mean snow depth, for indicating depth hoar snow formation. Kawshima et al. (1994) also present the "temperature-gradient index" for indicating the degree of temperature-gradient metamorphism. That is obtained by dividing the mean air temperature from beginning of snow deposition to any day by the mean snow depth during the same period. His index is sensitive to temporal meteorological conditions.

The temperature gradient developed here has a general and climatic meaning. The climatic temperature gradient which distinguishes the depth hoar region from the dry snow region gives better criterion than the presumed Rammsonde hardness climatic division (Table 1).

5 SUMMERY AND CONCLUDING REMARKS

The database was constructed from climatic data, topographical data and the character of snow cover that were introduced for division according to predominant snow type. From the database we derived detailed information about the characteristics of snowy areas in Japan. For example we used digital cartography to map predominant snow types in heavy snow regions, where altitudes are not so high and human activities are expected.

We also examined the relationship between the monthly maximum snow depth and the average temperature of snowy areas according to the snow type, then developed the general and climatic criterion to divide the depth hoar region and the dry snow region.

ACKNOWLEDGMENTS

The authors wish to express their thanks to the Japan Meteorological Agency for providing the mesh climatic data and the Japan Geographical survey Institute for providing digital cartographic information. The authors would like to express their thanks to Professor S. Kobayashi and Dr.K.Izumi, Niigata University, for their helpful suggestions. This work was partly supported by a grant of the Ministry of Education, Science and Culture, Japan.

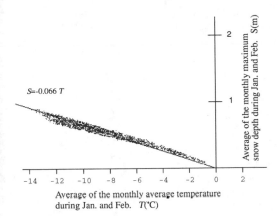

Figure 3. S-T relationship for boundary points where the presumed Rammsonde hardness R (Table 1) satisfy the condition, $7<R<9$.

REFERENCES

Akitaya, E. and Endo, Y.,1977. Regional Characteristics of snow cover in Ishikari and Sorachi District, Hokkaido, Low Temp. Sci. A35,105-115(In Japanese).

Akitaya, E. and Endo, Y.,1980. Regional Characteristics of snow cover in Hokkaido, Low Temp. Sci.
A39, 55-61(In Japanese).

Akitaya, E. and Endo, Y.,1982. Regional Characteristics of snow cover in Hokkaido, Res. Report
Grant-in-Aid for special project of Hokkaido University, 1979-81,1-17 (In Japanese).

Ishizaka M.,1995. Climatic division of snow cover
environments in Japan using the mesh climatic
data, Seppyo. J.Jpn. Soc. Snow Ice, 57, 23-34(In
Japanese).

Izumi K. and Akitaya E., 1986. The distribution of
depth hoar in Honshu, Japan, Seppyo. J.Jpn. Soc.
Snow Ice,48, 199-206 (In Japanese).

Kawashima K., T. Yamada and G. Wakahama,
1994. Regional division of snow-depositional environments and metamorphism of snow cover in
plain areas along the Japan Sea coast, IAHS PUbl,
223, 187-196.

Nakamura T.,1979. Local features of snow cover of
Japan in January, Reprt. National Research
Centr.for Disaster Prevention, 22, 175-177.

Snow Engineering: Recent Advances, Izumi, Nakamura & Sack (eds) © 1997 Balkema, Rotterdam. ISBN 90 5410 865 7

Dielectric properties of wet snow in microwave region

Waichi Nakata & Michiya Suzuki
Department of Electronics and Information Engineering, Aomori University, Japan

ABSTRACT: In order to study the dielectric behavior of wet snow, an experiment using a mixture of glass beads and water was conducted in the 9 to 12.4-GHz range. Glass beads of three different diameters were used to study the effect of particle size on dielectric characteristics. The depolarization factors of the water were estimated by minimizing the error between the measured permittivities and those calculated from the three phase Polder-Van Santen model. Hence, a transition of shape factors from needle-like at lower water content to oblately spheroidal at high content was verified, which is similar to the results obtained for natural snow by Hallikeinenn *et al.*(1986). Larger particle diameters showed larger increases in dielectric loss factor (imaginary part of the relative permittivity) for changes in water content. Changes in dielectric constant (real part of the relative permittivity) due to the glass particle size were minor.

1 INTRODUCTION

Electromagnetically, a snow medium is a three-component dielectric mixture consisting of air, ice particles, and water. Several dielectric mixing models for snow have been proposed. Hallikeinen *et al.* (1986) showed in their extensive investigation that the two-phase and three- phase Polder-Van Santen models describe the dielectric behavior of wet snow well for data measured in the 3 to 37-GHz range, assuming that the shape of the ice particles is spherical, and the depolarization factor of the water is nonsymmetric. In this paper, we study the dielectric properties of a wet snow model through an experiment using glass beads of different sizes. One of the objects in this study was to confirm the possibility of using glass beads in an indoor experiment concerned with the microwave scattering characteristics of wet snow.

2 MEASUREMENT SYSTEM

We used the free-space transmission technique to measure the permittivity ($\varepsilon = \varepsilon' - j\varepsilon''$) of glass bead samples. The block diagram of the measurement system is shown in figure 1. The glass bead sample to be measured is placed on the extruded polystyrene sample holder between the transmitting and receiving horn antennas. The sample is about 30 cm square and the antennas are separated by 136 cm. In the measurement of transmission coefficient T_s, both the magnitude $|T_s|$ and the phase ϕ_s are obtained by reading S_{21} scattering parameter data of the network-analyzer.

The transmission coefficient T_s for a snow sample of length L is given by

$$T_s = |T_s| \, e^{j\phi_s} = \frac{(1 - R^2) \, e^{-(\alpha + j\beta)L}}{1 - R^2 e^{-2(\alpha + j\beta)L}} \quad (1)$$

where α, β are the attenuation constant and the phase factor. The reflection constant R is given in terms of Z_o, the characteristic impedance of the free space, and Z, the characteristic impedance of the sample

$$R = \frac{Z - Z_o}{Z + Z_o} \quad (2)$$

These impedances are given by

$$Z = \frac{2\pi\eta_0}{\lambda_0} \cdot \frac{\beta(1+j\alpha/\beta)}{\alpha^2+\beta^2} \tag{3}$$

$$Z_0 = \eta_0 \tag{4}$$

where λ_0 is the wavelength in the free space and $\eta_0 = (\mu/\varepsilon)^{1/2}$ is the intrinsic impedance of free space. From the given measured values of $|T_s|$ and ϕ_s, the quantities α and β can be determined using an appropriate optimization method (direct research method, gradient method, etc.). Thus, the dielectric constant ε' and the dielectric loss factor ε'' can be computed by the following equations:

$$\varepsilon' = \left(\frac{\lambda_0}{2\pi}\right)^2 \cdot \left(\beta^2 - \alpha^2\right) \tag{5}$$

$$\varepsilon'' = \left(\frac{\lambda_0}{2\pi}\right)^2 \cdot (2\alpha\beta) \tag{6}$$

3 THREE PHASE POLDER-VAN SANTEN MIXING FORMULA

The three phase Polder-Van Santen (PVS) mixing formula for wet snow is given by

$$\varepsilon_{ws} = 1 + \frac{1}{3}\varepsilon_{ws}\, m_i\, (\varepsilon_i - 1)\sum_{j=1}^{3}\left[\varepsilon_{ws} + (\varepsilon_i - \varepsilon_{ws})A_{ij}\right]^{-1}$$
$$+ \frac{1}{3}\varepsilon_{ws}\, m_w\, (\varepsilon_w - 1)\sum_{j=1}^{3}\left[\varepsilon_{ws} + (\varepsilon_w - \varepsilon_{ws})A_{wj}\right]^{-1} \tag{7}$$

where ε_{ws}, ε_i, ε_w are the permittivities of wet snow, ice particles and liquid, respectively, m_i, m_w are the volume fractions of ice and water, and A_{ij}, A_{wj} are the depolarization factors of the ice and water inclusions. A_{ij} and A_{wj} depend on the shapes of ice particles and water. In the experiment using glass beads as the snow model, ε_i, m_i may be replaced by the permittivity of glass ε_g and the volume fraction of glass m_g. In our analysis, the shape of the glass particles is assumed to be spherical ($A_{ij}=1/3$) and the shape of water inclusions, nonsymmetrical ($A_{w1} \neq A_{w2} \neq A_{w3}$) according to Hallikainenn et al. (1986).

4 EXPERIMENTS AND RESULTS

Before the experiment, measurements of polytetrafluoroethylene (Teflon) and pure water were conducted to evaluate system error. It was found that the calculated errors of the relative permittivity of these samples were within 5% over the 9 to 12-GHz range. The physical parameters of the three glass bead samples (#1~#3) used in the experiment are listed in Table 1. A glass bead sample of known weight W_g was first dried sufficiently and a predetermined amount of distilled water (weight W_w) was added to it. The combination was mixed well and then placed on the sample holder. The volumetric moisture is given by $m_w = \rho_s W_w / W_g$, where ρ_s is the density of the glass beads. All the measurements were conducted at $20\pm2\,°C$. Figure 2 shows measurement results of dielectric constant ε' and dielectric loss factor ε'' of sample #2 in the 9 to 12.4-GHz

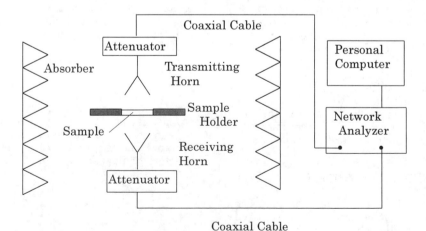

Figure 1. Measurement system.

Table. 1. Physical parameters of glass bead samples.

	Sample #1	Sample #2	Sample #3
Particle diameter (mm)	0.35~0.5	0.71~0.99	0.991~1.397
Density ρ_s (g/cm³)	1.48	1.48	1.49
Volume fraction of glass m_g (%)	59	59	60

range. Figure 3 shows the depolarization factors A_{w1}, A_{w2} and A_{w3} for water content optimized by fitting a three-phase PVS mixing model to the measured values of ε', ε''. As shown in this figure, the water inclusion in the sample appears needle-like in shape (since $A_{w1}=0.0$, $A_{w2}=A_{w3}=0.5$ for perfect needles) for water contents below $m_w=3$ % and becomes disk-like for $m_w>3$ % (since $A_{w1}=A_{w2}=0.0$, $A_{w3}=1.0$ for disks). This change in shape is observed for natural snow and considered to be due to the transition from the pendular regime to the funicular

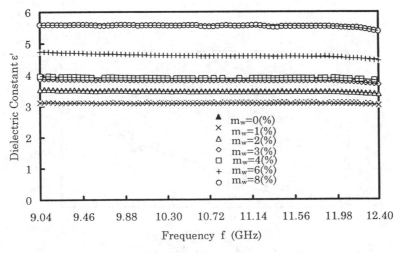

(a) Dielectric constant ε'.

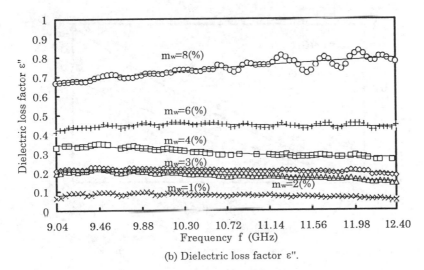

(b) Dielectric loss factor ε''.

Figure 2. Relative permittivity of sample #2.

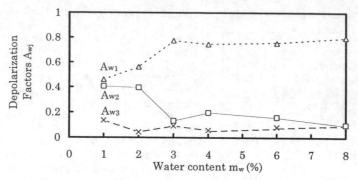

Figure 3. Depolarization factors for water inclusions in sample #2, optimized by fitting the three phase Polder-Van Santen model to the measured values of ε', ε".

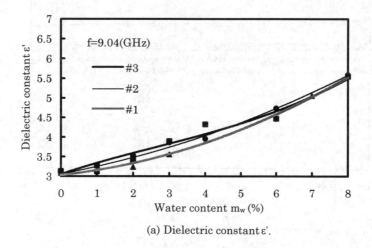

(a) Dielectric constant ε'.

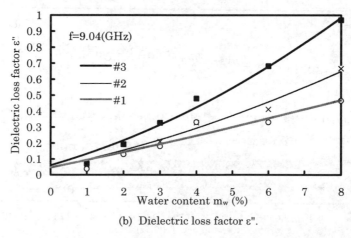

(b) Dielectric loss factor ε".

Figure. 4 Comparison of relative permittivities of three samples at 9.04 GHz.

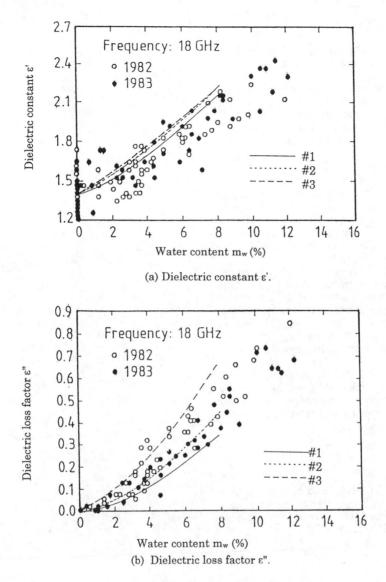

(a) Dielectric constant ε'.

(b) Dielectric loss factor ε''.

Figure 5. Comparison between the model predictions and reported data of natural snow at 18GHz.

regime. Samples #1, #3 also showed a similar tendency. Figure 4 shows comparisons of ε', ε'' for the three samples at 9.04 GHz. In this figure, it can be seen that the samples with larger particles have a larger change in ε'' with water content. Differences in ε' due to particle size are not clear. The difference in ε'' can be attributed to the difference in the shape and size of water captured among the glass particles, i.e, the difference in the depolarization factors of water. In Figure 5 the predicted curves of ε', ε'' calculated using the depolarization factors based on this experiment are compared with reported measurements of natural snow (snow particle size from 0.5 to 1.5 mm) at 18 GHz by Hallikainenn *et al.* (1986). From these figures, it can be seen that the predicted values of ε' are slightly higher than the measured data; however, on the whole the curves of ε'' cover the data with that of sample #2 showing the best fit. Thus, it can be supposed that the dielectric loss factor ε'' measured in natural snow includes a distribution error due to ice particle size variety.

5 CONCLUSIONS

The following conclusions were drawn from the experiments in the 9 to 12.4-GHz range using glass beads.

(1) A three-phase PVS model was applied to the measured ε', ε'' of samples to estimate the depolarization factors of water inclusion. As a result, a transition in shape of water particles from being needle-like to disk-like was verified in the experiment, which is similar to results reported for natural snow.

(2) Larger particles showed larger change in ε'' with changes in water content, due to the difference in the shape and size of water captured among the glass particles. However, differences in ε' due to the particle size were minimal.

(3) The predicted curves of ε', ε'' using depolarization factors based on this experiment were compared witn reported measurements of natural snow at 18 GHz. On the whole, the predicted curves described the changes in dielectric properties of snow with changing water content.

Thus it was verified that the mixture of glass beads and water may be used as a snow model for an indoor experiment concerning with the microwave scattering problem of wet snow.

REFERENCE

Hallikainen,M.,F.T.Ulaby, and M.Abdelrazik .1986. Dielectric Properties of Snow in the 3 to 37 GHz range. IEEE Trans.Antenna Propag, vol.AP-34, No.11, Nov.

Snow Engineering: Recent Advances, Izumi, Nakamura & Sack (eds) © 1997 Balkema, Rotterdam. ISBN 90 5410 865 7

Granulation of snow

Yasuaki Nohguchi, Tosiiti Kobayasi & Koyuru Iwanami
Nagaoka Institute of Snow and Ice Studies, NIED, Niigata, Japan

Kouichi Nishimura
Institute of Low Temperature Science, Hokkaido University, Sapporo, Japan

Atsushi Sato
Shinjo Branch of Snow and Ice Studies, NIED, Yamagata, Japan

ABSTRACT: Snow avalanches and blowing snow activate themselves by granulation. Dry snow avalanches and blowing snow are kinds of pulverizing granulations. In sufficiently developed wet snow avalanches, the snow sometimes changes into rounded lumps like snowballs. This process is a kind of tumbling granulation. To simulate such granulations in a small test, we rotated or shook a small polyethylene bucket. In this test, we succeeded in formating snowballs by tumbling granulation. It was found that small grain size and low density of snow are favorable for tumbling granulation (i.e., snowball formation).

1 INTRODUCTION

The making of particles is called granulation. The size of particles or lumps of snow in motion is an important snow engineering issue since it relates to fluidization.

Natural snowcovers are frequently disturbed by artificial movement such as snow removal and by natural phenomena such as avalanches and blowing snow. These disturbances change the snow. For example, dry snow sometimes becomes small ice particles by pulverizing, and wet snow becomes large lumps by tumbling because of its adhesion.

In cold regions where the snow is always dry except during the snow melting season, pulverizing is the predominant snow granulation process. In the Hokuriku district of Japan, where wet snow can exist throughout the winter, the process of the granulation is not simple. Granulation makes snow a kind of fluid.

In this paper, granulation in nature (i.e., snow avalanches and blowing snow) is classified from the viewpoint of snow conveyance and compared to artificial granulation. Second, a simple test method for granulation is introduced using a polyethylene bucket.

2 GRANULATION OF SNOW IN NATURE

2.1 Granulation by avalanche

A snow avalanche is a kind of natural snow conveyance by gravity. Avalanches are classified into two types in the sense of granulation: dry snow avalanches and wet snow avalanches.

In the case of dry snow avalanches, the snow cover is broken and pulverized by movement, becoming

Figure 1. Snowballs in a wet snow avalanche.

smaller snow lumps or ice particles. Powder snow avalanches are one of the most typical phenomena of pulverizing granulation in nature.

On the other hand, after wet snow avalanches, a lot of snowballs can be found in the debris (Figure 1). Wet snow on a slope become snowballs by tumbling, increasing the fluidity of the snow. Therefore, wet snow avalanches can be considered a kind of tumbling granulations.

2.2 Granulation by wind

Blowing snow is a kind of natural snow conveyance by the force of air. It is a kind of pulverizing granulation.

Figure 2. Snow roller formation as a tumbling granulation.

Snow rollers on a plain in wet snow (Figure 2) are also a kind of natural snow conveyance by air force and can be considered as a kind of tumbling granulation.

2.3 Granulation by snow removal and conveyance

Granulation occurs in snow disturbed by a rotary blower or a snowplow. The disturbed snow is similar to debris snow after an avalanche (Figure 3), so granulation of snow naturally occurs also during snow removal.

To convey wet snow using a pipeline, it is necessary to artificially make snowballs by compression (Kobayashi 1991 and 1993) to increase the fluidity of that snow.

Figure 3. Snow granulated by a snowplow.

3 TUMBLING GRANULATION TEST IN A POLYETHYLENE BUCKET

Tumbling granulation similar to that in a snow avalanche or blowing snow can be created in a small bucket.

As shown in Figure 4, we can simulate tumbling for wet snow and pulverizing for dry snow by rotating or

Figure 4. A simple granulation test using a polyethylene bucket.

Table 1. Summary of bucket granulation tests.

No.	Snow Type	Year	Location	Density kg/m3	Grain Size mm	Snow Temp. ℃	Test Temp. ℃	Ball Size cm
1	NS	1981-96	Nagaoka	<100	<0.5	0	0<	3~8
2	NS	1992-93	Alta	40~100	<0.5	-9~-4	+20	3~5
3	GR	1993	Alta	220	(2-4)	-2	+20	3~5
4	GS	1993	Akiyamago	380	2-4	0	0<	*
5	AFS	1994	Shinjo	170	0.03	-7	+15	0.5~3
6	DH	1996	Tianshan	180	10	-12	+4	*

* Snowballs were not formed by the bucket granulation test.

shaking a polyetylene bucket containing snow. Results of this bucket granulation test for some snow types are discussed below and summarized in Table 1.

Figure 5. Dry new snow after a granulation test.

Figure 6. Wet new snow after a granulation test.

3.1 *New snow (NS)*

Figure 5 shows a granulation test result for dry new snow below 0° C. The snow looks like powder as a result of pulverizing. When that snow is tested above 0° C, it becomes balls as shown in Figure 6 once the snow temperature reaches the freezing point.

3.2 *Granular snow (GS)*

Figure 7 shows wet granular snow (No.4 in Table 1) above 0° C after a granulation test. No snowballs were formed, though a few large and loose lumps appeared in the bucket. The loose lumps were easily broken by shaking the bucket.

3.3 *Graupel (GR)*

In a granulation test above the freezing point of dry

Figure 7. Wet granular snow after a granulation test.

graupel 2-4 mm in diameter (No.3 in Table 1) snowballs (30-50 mm in diameter) were formed.

3.4 *Artificial fine snow (AFS)*

Figure 8 shows the result of a granulation test of artificial fine snow whose grain size is 30 μ m (Sato 1996). The fine grains became small snowballs 5 to 30 mm in diameter. These granulated balls were much smaller than those from new snow (No.1 and 2) and from graupel (No.3). This suggests that the size of granulated snow depends on the grain size before granulation.

Figure 8. Artificial fine snow (Sato, 1996) after a granulation test.

3.5 *Depth hoar (DH)*

In the test of depth hoar above the freezing point in the Tianshan Mountains of China (No.6 in Table 1), no snowballs were formed. The initial grain size of this snow was extremely large in comparison with new snow. Such large grains do not granulate. Small grains and low density are favorable for tumbling granulation, that is, snowball formation.

4 CONCLUSION

To simulate snow granulation in a small test, we rotated and shook a small polyethylene bucket. Some kinds of snow formed snowballs by tumbling granulation. As a result, it was found that small grain size and low density snow are favorable for tumbling granulation, that is, snowball formation.

169

REFERENCES

Kobayashi, T. 1991. Experimental studies of
  pneumatic dispatch systems of snow. *Proc.
  JUWSLDPC* 1991, 161-170.
Kobayashi, T. 1993. Engineering studies on pneumatic
  conveying systems of snow. *CRREL Special Report*,
  92-27, 275-286.
Sato, A. 1996. Metamorphism of artificial fine snow
  and physical modeling. *Proc. 3rd International
  Conference on Snow Engineering.*

Snow Engineering: Recent Advances, Izumi, Nakamura & Sack (eds) © 1997 Balkema, Rotterdam. ISBN 90 5410 865 7

Chemical composition of snow and rime at Mt. Zao, Yamagata Prefecture, Japan

Fumitaka Yanagisawa
Department of Earth and Environmental Sciences, Faculty of Science, Yamagata University, Japan

Katsutoshi Yano
Department of Physics, Faculty of Science, Yamagata University, Japan

ABSTRACTS: The Ice-Monsters (Jyuhyo) at Mt. Zao, Yamagata Prefecture, Japan, resulted from a repeated accretion of rime and snow in winter. Snow and rime samples were collected every winter from 1993 to 1996 at Zao which is about 1680 m high. Chemical compositions of soluble substances, such as sodium ion, chloride ion, potassium ion and magnesium ion from sea spray salt, non-sea salt sulfate ion and nitrate ion from human activity, non-sea salt calcium ion from dust along road and ammonium ion from biological activity, and pH, were measured to reveal atmospheric chemistry. Ion concentrations of rime and snow increased with velocity of the northwest winter Siberia monsoon. The pH values were associated with the concentration of non-sea salt sulfate and nitrate, which means that snow and rime were made acid by sulfate and nitrate exhausted from anthropogenic activites. Non-sea salt sulfate and nitrate ratio of snow, (nss-SO_4^{2-}/NO_3^-), were larger than unity and were identical to those of rime. It is suggested that non-sea salt sulfate and nitrate in snow and rime contains sulfur-oxides and nitrogen-oxides from human activity of the Asian continent.

1 INTRODUCTION

Rime and snow samples were collected in the Kyushu mountainous regions and soluble substances were measured (Nagafuchi et al. 1993; Nagafuchi et al. 1993). It was found that ion concentrations in rime samples were higher than those of snow and that pH of rime was lower than that of snow. It was suggested that rime reflects local atmosphere situation better than snow. Therefore, rime will be a useful indicator for evaluation of atmospheric environments.

The Ice-Monsters (Jyuhyo) at Mt. Zao, Yamagata Prefecture, Japan, resulted from a repeated accreation of rime and snow in winter (Yano 1986; 1991). For implication of the transport of soil particles (Kosa) from the Asian continent to Japan,

chemical composition of insoluble substances in the rime at Mt. Zao were determined (Uchiyama et al. 1991; Uchiyama et al. 1992). These results suggested that the Kosa particles were transported to Japan irrespective of the apparent occurrence of the Kosa phenomenon.

Rime (about 1500 m above the sea) was formed by the northwest Siberia monsoon (Yano 1986; Yano 1994) . On the other hand, snow was formed in clouds several thousand meters high. Snow and rime samples were collected every winter from 1993 to 1996 at Mt. Zao. Chemical composition of soluble substances, such as sodium ion, chloride ion, potassium ion and magnesium ion from sea spray salt, non-sea salt sulfate ion and nitrate ion from human activity, non-sea salt calcium from dust along road and ammonium ion from biological activity, and pH, were measured to reveal the atmospheric chemistry (Yanagisawa et al. 1996).

2 SAMPLING

Snow and rime samples were collected every winter from 1993 to 1996 at Mt. Zao is about 1680 m high (Fig. 1).

Figure 1. Location of sampling stations.

3 EXPERIMENTS

The solutions were filtered through 0.45 μ m membrane filter and then analyzed for sodium ion, potassium ion, calcium ion and magnesium ion by inductively coupled plasma, for chloride ion, sulfate ion and nitrate ion by ionchromatography, and for ammonium ion by indophenol colorimetry. Electroconductivity and pH were measured by an electroconductivity meter and a pH meter with glass electrode. Sulfate and calcium ions from sources other than sea salt were contributing to the deposition. Therefore, non-sea salt sulfate (SO_4^{2-}nss) and non-sea salt calcium (Ca^{2+}nss) were calculated as following equations:

$$SO_4^{2-}nss = SO_4^{2-}mes - (SO_4^{2-}/Na^+)sea \times Na^+mes \quad (1)$$

$$Ca^{2+}nss = Ca^{2+}mes - (Ca^{2+}/Na^+)sea \times Na^+mes \quad (2)$$

where SO_4^{2-}mes, Na^+mes and Ca^{2+}mes were the concentration of sulfate, that of sodium and that of calcium, respectively. Sulfate and sodium weight ratio of seawater, (SO_4^{2-}/Na^+)sea, and calcium and sodium weight ratio of seawater, (Ca^{2+}/Na^+)sea were 0.252 and 0.0384, respectively.

4 RESULTS AND DISCUSSION

Ion concentrations of rime and snow increased with the velocity of the northwest winter Siberia monsoon. Electroconductivities of snow were, always lower than those of rime (Fig. 2). On the other hand, the pH of rime was lower than that of snow (Fig. 3).

The pH values were associated with the concentration of non sea salt sulfate and nitrate (Fig.4). This indicates that the snow and rime were made acid by sulfate and nitrate produced by anthropogenic activites. Non-sea salt sulfate and

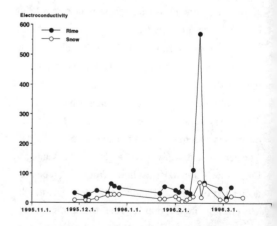

Figure 2. Characteristics of electroconductivity (μ s/cm) in rime and snow as a function of time.

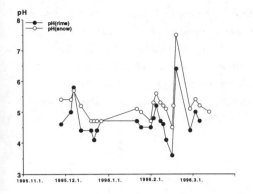

Figure 3. Characteristics of pH in rime and snow as a function of time.

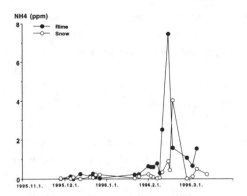

Figure 6. Characteristics of concentration of ammonium ion (ppm) in rime and snow as a function of time.

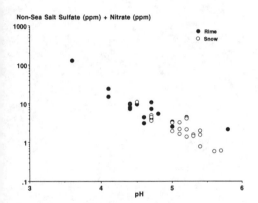

Figure 4. Relation between pH and non-sea salt sulfate plus nitrate.

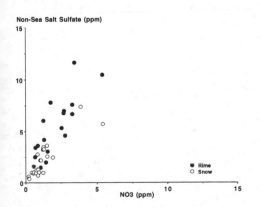

Figure 5. Relation between non-sea salt sulfate (nss-SO_4^{2-}) and nitrate (NO_3^-) in snow and rime.

nitrate ratio of snow (nss-SO_4^{2-}/NO_3^-), were larger than unity and were identical to those of rime (Fig. 5). This suggests that non-sea salt sulfate ion and nitrate ion in snow and rime contains sulfur-oxides and nitrogen-oxides from human activity on the Asian continent.

Ammonium ion is contaminant from biological activity. Concentration of ammonium ion was also increased in strong wind (Fig. 6). It means that ammonium ion also affected from Asian Continent.

CONCLUSION

(1) Ion concentrations of rime and snow increased with velocity of the northwest winter Siberia monsoon.

(2) The pH values were associated with the concentration of non-sea salt sulfate and nitrate, which means that snow and rime were made acid by sulfate and nitrate exhausted from anthropogenic activites.

(3) Non-sea salt sulfate and nitrate ratio of snow, (nss-SO_4^{2-}/NO_3^-), were larger than unity and were identical to those of rime. It is suggested that non-sea salt sulfate and nitrate in snow and rime contains sulfur-oxides and nitrogen-oxides from human activity of the Asian continent.

173

(4) Concentration of ammonium ion was also increased in strong wind. It means that ammonium ion also affected from Asian Continent.

ACKNOWLEDGEMENTS

The authors express sincere appreciation to Kiyoshi Sato, Tomoya Yoshida and Hiroyuki Abe who assisted with sample collection. The authors wish to thank Professor Hitoshi Sakai, Yamagata University, for his encouragement for this work. This study was financially supported by the Grant in Aid for Scientific Research, No.07640650 from the Ministry of Education, Science and Culture.

REFERENCES

Nagafuchi, O., Tagami, S., Ishibashi, T., Murakami, K. and Suda, R. (1993) Evaluation of atmospheric environment by soluble components in rime., *Geochemistry*, 27, 65-72.

Nagafuchi, O., Suda, R ., Ishibashi, T., Murakami, K. and Shimohara, T. (1993) Analysis of long-range transported air pollutions - Origin of acid aerosol in rime found at Kyushu mountainous regions, Japan., *Journal of the Chemical Society of Japan*, 6,788-792.

Ohta, S. (1990) Atmospheric aerosols., *Chemistry of Atmosphere*, 123-145.

Uchiyama, M., Mizuochi, M., Yano, K. and Fukayama, T. (1991) Chemical composition of insoluble substances in the rime sampled at Mt. Zaoh., *Journal of the Chemical Society of Japan*, 5, 517-519.

Uchiyama, M., Fukayama,T ., Mizuochi, M. and Yano, K. (1992) Chemical composition of the winter precipitations of Mt Zaoh; Implication of the transport of soil particles from the Asian continent to Japan., *J. Aerosol Res.*, 7, 1, 44-53.

Yanagisawa, F., Nakagawa, N. and Yano, K. (1996) Chemical composition of snow and rime at Mt. Zao, Yamagata Prefecture, Japan., *Journal of the Japanese Society of Snow and Ice*, (submitted).

Yano, K. (1986) Studies of icing and ice-snow accretion on Abies mariesii., *Bull. Yamagata Univ., (Natural Science)*, 11, 3, 227-247.

Yano, K. (1991) Ice monster (Jyuhyo) in Mt. Zao., *Aerosol Study*, 6, 1, 57-63.

Snow Engineering: Recent Advances, Izumi, Nakamura & Sack (eds) © 1997 Balkema, Rotterdam. ISBN 90 5410 865 7

Study on snow particle changes

Katsutoshi Yano, Hiroyuki Abe & Hiroki Tanaka
Department of Physics, Faculty of Science, Yamagata University, Japan

Atsushi Sato
Shinjo Branch of Snow and Ice Studies, NIED, Yamagata, Japan

Abstract: An observation was conducted on the changes in snow particles with time due to metamorphism. The proposed method is capable of easy investigation of snow particles using a simplified snow particle photographing system. The change in their forms was measured 1-hour, 25-hours, 50-hours, 100-hours under a microscope. Digital image analysis was performed to compare the correlation between the area and the circumferential length of the snow particles with those of round particles. The results show that the area of the snow particles increased with time, and that the snow particles changes their shapes from irregular to round.

1. INTRODUCTION

Dry metamorphism has been studied recently theoretically by Brown et al (1993) as well as wet metamorphism. It is important to observe and analyze the change of snow particle shape or microstructure. The Aniline method is commonly used to understand the time varying change in snow particles, though a special experimental environment is required. The proposed method is capable of easy investigation of snow-particle changes, using a simplified snow particle photographing system.

2. EXPERIMENT

Figure 1 shows the system which uses a personal computer to analyze images of snow particles photographed in the field. The snow sample were taken from compacted snow, new

personal computer

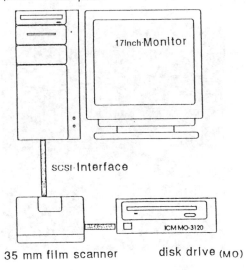

Figure 1. Snow particle photographing system

snow of Mt. Zao, coarse grained snow stored in a low temperature laboratory at 0 ℃ for 15 days. Particles of the compacted snow cover

175

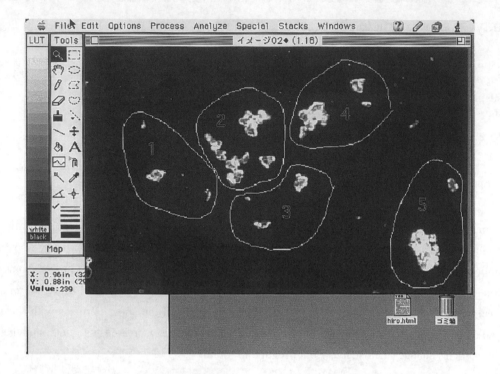

Figure 2.   NIH Image

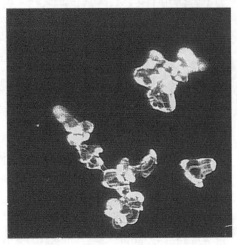

Before binaly process   After binary process(inversed)

Figure 3.   Image-processed pictures of snow particles

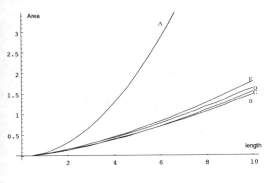

A Circle $\quad y = \left(\dfrac{1}{4\pi}\right)x^2 \cong 0.080x^2 \qquad (1)$

B 0 hours $\quad y = 0.050x^{1.49} \qquad (2)$

C 25 hours $\quad y = 0.056x^{1.42} \qquad (3)$

D 50 hours $\quad y = 0.058x^{1.49} \qquad (4)$

E 100 hours $\quad y = 0.062x^{1.42} \qquad (5)$

Figure 4.　Correlation and graph.

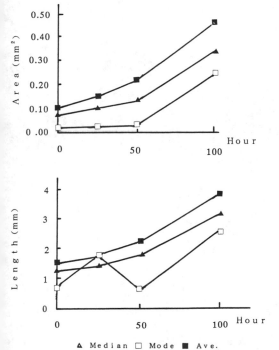

▲ Median ☐ Mode ■ Ave.

Figure 5.　Change in area and circumferential lenbth
of snow particle with time

were inserted into the microscopic photographing system, which was maintained at 0 ℃. Changes in their form were observed at 1 hour, 25 hours, 50 hours, and 100 hours after insertion. We used a film scanner to process the photographs. The digital images of the snow particles were analyzed using photographic analysis software on a personal computer. Figures 2 and 3 showed typical digital images.

3. RESULT AND DISCUSSION

These digital images was then used to compare the correlation between the projected area and the circumferential length of the snow particles with those of round particles. The results show that the projected area of the snow particles increased with time, and that the snow particles changes their shapes from irregular to round (Figure 4).

Figure 5 confirms also the growth of the snow particles with time.

REFERENCES

Brown, R.L., M.Q.Edens and A.Sato. 1993. A study of metamorphism of fine-grained snow due to surface curvature differences. *Annals of Glaciology* 19. 69-76.

Yano, K . Abe H. and Tanaka H. 1996: Study on Snow Particle Change, *Tohoku Journal of Natural Disaster Science*, Vol. 32, 53-60.

Snow Engineering: Recent Advances, Izumi, Nakamura & Sack (eds) © 1997 Balkema, Rotterdam. ISBN 90 5410 865 7

Study of the thermodynamics of Juhyo (ice-monsters)

Katsutoshi Yano, Masaaki Tosabayashi & Hiroo Onda
Department of Physics, Faculty of Science, Yamagata University, Japan

Atsushi Sato
Shinjo Branch of Snow and Ice Studies, NIED, Yamagata, Japan

Shun'ichi Kobayashi
Research Institute for Hazards in Snowy Areas, Niigata University, Japan

Abstract: Many studies have been made on snow cover heat conductivity, but few deal with icing. Heat conductivity of the ice monsters at Mt. Zao was measured in two directions, along the growth direction and perpendicular to the growth direction. From the measured data, the growth surface of the Juhyo (ice-monsters) and its internal structure indicated different heat conductivity. Similar to the snow cover, the heat conductivity of Juhyo increased with an increase in density, to a value several times larger than that of snow cover. The change in the surface temperature of the Juhyo was studied using a thermo-tracer.

1. INTRODUCTION: Juhyo (ice-monsters) are the product of a peculiar icing phenomenon which is seen only in certain mountain areas of the Tohoku District of Japan, including Mt. Zao, Mt. Azuma, Mt. Hachimantai, and Mt. Hakkoda (Figure 1). A commonly accepted explanation for this phenomenon is that supercooled water droplets act as a bind that pastes snowflakes on Abies mariestii (so-called Aomori fir). The ice-snow accretion on Mt. Zao results from both icing and snow accretion and is a complicated compound phenomenon which requires wide-ranging experiments and observations to be explained and overcome. Many studies have been made on snow cover heat conductivity, but few deal with icing. The heat conductivity of snow cover along the horizontal direction is larger than that perpendicular to it due to strong combination of snow particles horizontally (Yamada et al, 1974). Since Juhyo grows to windward, their structure is anisotropic. It is expected that a strong anisotropy exists in the heat conductivity of Juhyo for the different directions.

Figure 1   Juhyo i (ce-monsters) in the Mt. Zao

2. EXPERIMENT

A field experiment was carried out on the evaluation of the heat conductivity of such icing, Juhyo. Heat conductivity of Juhyo at Mt. Zao was measured in the growth direction and perpendicular to the growth direction. Figure 2 shows the

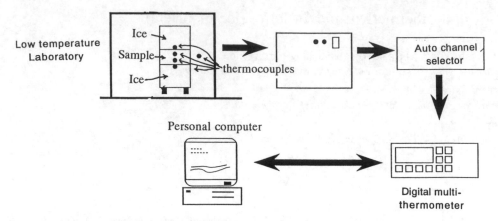

Figure 2 Measuring system for heat conductivity

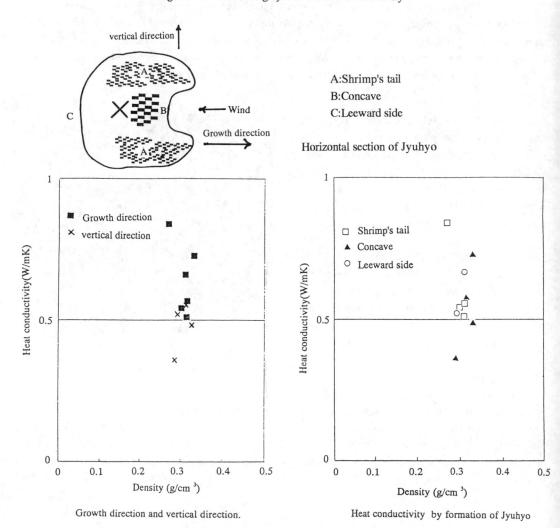

A:Shrimp's tail
B:Concave
C:Leeward side

Horizontal section of Jyuhyo

Growth direction and vertical direction.

Heat conductivity by formation of Jyuhyo

Figure 3 Heat conductivity and density of Jyuhyo

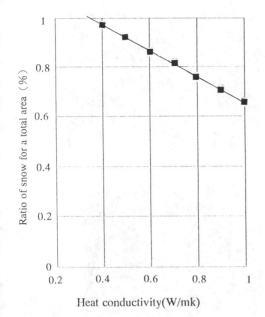

Figure 4 The ratio that heat conductivity and hold snow of Jyuhyo

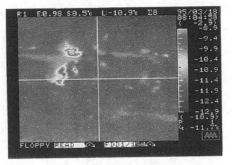

Thermal image(1)
Am 8:05 , 18 march. 1995 Mt. ZAO

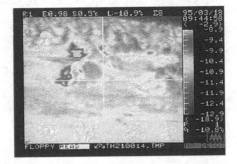

Thermal image(2)
Am 9:35

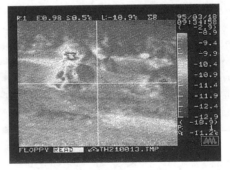

Thermal image(3)
Am 9:45

Figure 5   Thermal images taken at   Mt. Zao

measuring system used for the observations. The sample of snow of Juhyo is sandwiched between ice blocks and surrounded with an insulation of 10 cm thick.   Temperatures of the sample and ice blocks were measured with thermocouples.   We put this sample in the measurement device and placed cryostats above and below to maintain temperatures. Temperature of cryostat established approximately - 8 ℃ at bottom and -10℃ top.

To do the heat conductivity analysis, at first applied Fourier's law of heat conduction.   Heat loss through the insulation from the sides walls of the specimen was found to be small through the temperature gradient measurement. The estimation of vertical heat flow through the specimen was done with the temperature gradient of the ice blocks.

3. RESULT AND DISCUSSION

The measured data of the growth surface of the Juhyo (Ice-monsters) and its internal structure indicated different heat conductivity.   Similar to the snow cover, the heat conductivity of Juhyo

increased with increasing density, but the value was several times larger than that of snow cover (Fig. 3). Figure 4 shows the ratio that heat conductivity and hold snow. The ratio that snow hold is large and heat conductivity becomes small. In the midwinter, the ratio snow accretion and icing were 7:3. This value almost agrees with the value that Ogasawara

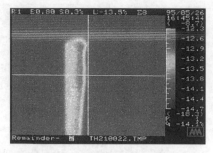
Thermal images (45 min. after)

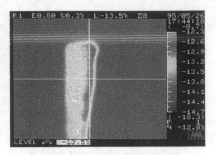
Thermal images (105 min. after)

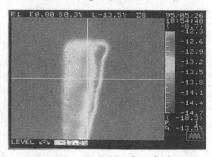
Thermal images(175 min. after)

Icing at the end of the experiment

Figure 6   Thermal images taken at Mt. Zao

(1968) estimated. We can estimate the ratio of ice to snow of Juhyo each year from this fact.

The change in the surface temperature of the Juhyo was studied using a thermo-tracer. Direct observation of the heat distribution of Juhyo showed that the temperature on the growth surface of an ice formation was higher than on the neighboring surface due to the latent heat of freezing (Figure 5). Icing model experiments were performed using a wind tunnel in a low temperature laboratory.

The small wind tunnel is 30 cm wide and 150 cm long. Three of its four sides are plywood and one side is transparent plastic, which enables us to observe the icing and ice-snow accretion processes, and take photographs (Figure 6). Numerical analysis of the data clarified heat movement during ice formation.

4. CONCLUSION

Heat conductivity of Juhyo was measured and showed higher value compare to snow cover of same density range. Anisotropy of heat conductivity was also found to be high conductivity

for the growth direction than vertical direction, that is, perpendicular to the growth direction.

Observation with a thermo-tracer were conducted both at Mt. Zao and in the small wind tunnel located in a cold room. These observations revealed the temperature distribution of the surface of Juhyo, high temperature at growth surface at the windward of Juhyo.

References

Ogasawara, K. 1968. Jyuhyo in Mt. Zao,*Chiri* (Geography in Japanese): 13, 86.

Yamada, T., T.Hasemi, K.Izumi and A. Sato. 1974: On the anisotropy of the snow cover structure and its velocity of elastic wave and heat conductivity (in Japanese with English summary). *Low Temperature Science,A*. 32. 71-80.

Yano, K.1986.Ice-monsters(Jyuhyo) in Mt.Zao、 *Bulletin of the Yamagata University, Yamagata,J apan (Natural Science)*.11:227～247

Yano, K.1989.Ice-monsters(Jyuhyo) in Mt.Zao、 *Society for Snow Engineering*, 5: 2, 23～32

Yano,K.1991.Ice-monsters(Jyuhyo) in Mt.Zao. *Aerosol Research*, 6:57～63

2 Building and construction engineering

Snow Engineering: Recent Advances, Izumi, Nakamura & Sack (eds) © 1997 Balkema, Rotterdam. ISBN 90 5410 865 7

Estimation of ground snow weight based on daily precipitation and daily mean air temperature

Shuji Sakurai
Department of Architecture, Hokkai Gakuen University, Sapporo, Japan

Osamu Joh
Department of Architecture, Hokkaido University, Sapporo, Japan

Takuji Shibata
Department of Architecture, Hokkaido Institute of Technology, Sapporo, Japan

ABSTRACT: This paper presents a simple method for estimating ground snow weight based on standard data taken by meteorological observatories. The ground snow weight (kgf/m^2) on the nth day is considered to be the accumulation of daily precipitation p_i(mm) from the first day (i=1) of a period of continuous snow cover to the (n-1)th day. In this method, daily precipitation p_i (mm) should be adopted when the daily mean air temperature is less than an air temperature limit. The decreasing quantity of snow weight due to snowmelt is neglected for simplicity. In order to verify this method, we used daily precipitation, daily mean air temperature and ground snow weight observed continuously for many years at eight observation sites in heavy snow regions of Japan. Comparison of our estimated values with the observed values of ground snow weight at each site shows that our estimated values trace observed weights fairly well as long as the daily mean air temperature is below +2℃. It is concluded that the proposed method is practical and useful to estimate ground snow weight in heavy snow regions.

1 INTRODUCTION

To determine snow load for structural design, it is particularly important to estimate ground snow weight statistically. Unfortunately, ground snow weight data are limited since such measurements are not included in the standard data taken by meteorological observatories. It would be very convenient for the evaluation of design snow load if the ground snow weight could be estimated using the standard data of meteorological observatories.

The purpose of this paper is to discuss the value of a simple method to estimate ground snow weight using standard data such as daily precipitation and daily mean air temperature observed by meteorological observatories.

2 PROPOSED METHOD

Measurements of daily precipitation appear to be the most important ones which are related to the increasing quantity of ground snow weight during a period of continuous snow cover. We define that the accumulation of daily precipitation p_i (mm) from the first day(i=1) of a period of continuous snow cover to the (n-1)th day is the ground snow weight P_n(kgf/m^2) of the nth day. It can be expressed by the following equation.

$$P_n = \sum_{i=1}^{n-1} p_i$$

In applying the data of daily precipitation to the above equation, it is needed to judge whether the data of daily precipitation can contribute to the increasing of snow weight, in other words whether the data is for rain or snowfall. It is not easy to see how much rain becomes snowfall.

According to the specialized book(1986), the air temperature of about +2℃ near the ground and that of about -6℃ at the height of 1500m are boundary temperatures statistically. Tamura(1990) discussed this problem using the meteorologocal data of Nagaoka and concluded that the air temperature of about +2℃ near the ground can be considered as an air temperature limit. Therefore, an air temperature limit +2℃ is used for the judgment in this paper. Data of daily precipitation should be used in the above equation when the daily mean air temperature is less than the air temperature limit.

As to estimating ground snow weight using measurements of daily precipitation and the air temperature limit, Joh, Shibata et.al discussed in 1987. They considered the accumulated daily precipitation as an increasing quantity and the melting coefficient multiplied by the accumulated daily mean air temperature as main decreasing quantity during a period of snow cover. They showed that there existed an obvious quantitative relation between the estimated value of snow weight and the observed value of that weight based on meteorological data in Sapporo. Kamimura and Uemura(1992) investigated the efficiency of their method using meteorological data in Nagaoka and Tokamachi. Their method is similar to the above method proposed by Joh et al.(1987) with respect to using the accumulated daily precipitation and the melting coefficient multiplied by the accumulated daily mean air temperature. Shinojima and Harada(1992) discussed the efficiency of their method using meteorological data of Tokamachi, in which accumulated daily precipitation, melting coefficient and the accumulated daily mean air temperature were also used.

In this paper, the decreasing quantity of snow weight due to snowmelt is neglected for simplicity mainly because further investigation is necessary to determine the melting coefficient which should be multiplied by the value of accumulated daily mean air temperature.

3 SUMMARY OF DATA

Fortunately, we have obtained data of snow weight measured continuously for many years at eight observation sites in heavy snow regions of Japan. The eight sites are Sapporo(by Hokkaido University), Hirosaki(by Hirosaki University), Kamabuchi(by Yamagata Experimental Site of Tohoku Branch, Forestry and Forest Products Research Institute), Shinjo(by Shinjo Branch of Snow and Ice Studies, National Research Institute for Earth Science and Disaster Prevention), Nagaoka(by Nagaoka Institute of Snow and Ice Studies, National Research Institute for Earth Science and Disaster Prevention), Tokamachi (by Tokamachi Experimental Site and Disaster

Prevention Forest Laboratory, Forestry and Forest Products Research Institute), Joetsu(by Hokuriku National Agricultural Experiment Station) and Shiozawa(by Shiozawa Snow Research Station, Japan Railway Technical Research Institute). They are distributed from 43° north latitude (Sapporo in Hokkaido Prefecture) to 37° north latitude (Shiozawa in Niigata Prefecture) as shown in Figure 1.

A summary of ground snow weights of each site is shown in Table 1. Measurements of snow weight were carried out almost every day at Nagaoka, Tokamachi, Joetsu and Shiozawa, and about every five days to ten days at the others. Though the periods of observation are different, all data are used in the analysis because the reported data of snow weight is very few. On the other hand, measurements of daily precipitation and daily mean air temperature were carried out almost every day at six sites, Kamabuchi, Shinjo, Nagaoka,Tokamachi, Joetsu and Shiozawa. At the two sites of Sapporo and Hirosaki, however, daily precipitation and daily mean air temperature were not measured. Therefore, for the above two sites we used the data taken by meteorological observatories at each city for the lack of daily precipitation and daily mean air temperature.

The average monthly temperatures of January and February at Sapporo are -4.6℃ and -4.0℃, respectively. They are the lowest among the eight sites. Those at Joetsu are both +1.8 ℃ and are the warmest. Thus the meteorological conditions of eight sites cover a wide range.

4 COMPARISON OF THE ESTIMATED VALUE TO THE OBSERVED VALUE

4.1 *Accuracy of estimated value*

As samples of changes in snow weight during a winter, data from eight observation sites are shown in Figures 2(a) to (h). In the figures, the solid lines and the dotted lines indicate the observed values and the estimated values with the above equation for an air temperature limit of +2℃ respectively.

As is evident from these examples we can understand the general tendency that the estimated value by the proposed method traces the increasing process of ground snow weight approximately throughout the period to the peak value.

In order to investigate the efficiency of the proposed method, two topics are chosen :The annual extreme snow weight and the annual maximum of increasing snow accumulation for 7 days, which are the basic value for determining design snow load in AIJ Recommendation for Loads and Buildings (1996). The former topic should be applied when snow load on the roof is not controlled and the latter topic should be applied when controlled.

For examples, at Kamabuchi site shown in Figure 2(c) (1980-1981), the ratio of the estimated value to the observed value is 0.97 (791 kgf/m^2 to 816 kgf/m^2) for the annual extreme snow weight and 1.01(130kgf/m^2 to 129 kgf/m^2) for the annual maximum value of increasing snow weight for 7 days. At Tokamachi site shown in Figures 2(f)(1980-1981), the former value is 0.97 (1529 kgf/m^2 to 1571 kgf/m^2) and the latter value is 0.96 (278 kgf/m^2 to 291 kgf/m^2).

Changes of weights about increasing snow accumulation for 7 days are shown in Figures 3(a) to (h) based on the data of Figures 2(a) to (h). For examples, at Kamabuchi site shown in Figure 3 (c), the estimated annual maximum value(130kgf/m^2) is almost the same with the observed one (129 kgf/m^2) as mentioned above. At Tokamachi site shown in Figures 3(f), the estimated annual maximum value (278kgf/m^2) is similar to the observed one (291 kgf/m^2). As seen in Figures 3(a) to (h), the estimated values approximately agree with the observed values for the annual maximum value of increasing snow accumulation for 7 days.

Figure 4(a) shows the mean values and standard deviations in the ratio of estimated value (P_{max}) to the observed value (W_{max}) for annual extreme snow weight in the period observation. Here, +2℃ is used as an air temperature limit at each site. The mean values are within the range between 1.12(Shiozawa) and 1.28(Kamabuchi) except for Hirosaki, and the standard deviations are between 0.11(Shiozawa) and 0.39(Shinjo) except for Hirosaki. Figure 4(b) also illustrates the amount of scatter in the ratio of estimated value ($P7_{max}$) to the observed value ($W7_{max}$) for the annual maximum weight of increasing snow accumulation for 7 days. The mean values are within the range of 0.78(Shiozawa) to 1.21(Sapporo) except for Hirosaki, and the standard deviations are between 0.14(Tokamachi and Shiozawa) and 0.33(Sapporo) except for Hirosaki. Even though there is an inaccuracy about Hirosaki site, the

proposed method is considered to be accurate enough for practical use to estimate ground snow weights when +2℃ is used as an air temperature limit.

4.2 *Effect of air temperature limit*

In order to verify the effect of air temperature limit on the estimated value, the cases of +3℃ and +4℃ are investigated in addition to +2℃. These results are shown in Figures 5(a) and (b).

Figure 5(a) shows the mean values of P_{max}/W_{max} in the case of +3 ℃ and +4℃. They are in the range of 1.24(Sapporo and Shiozawa) to 1.41 (Shinjo) and 1.27(Sapporo) to 1.50(Kamabuchi) respectively except for Hirosaki. Figure 5(b) also shows the mean values of $P7_{max}/W7_{max}$ in the case of +3℃ and +4℃. They are in the range of 0.78(Shiozawa) to 1.21 (Shinjo) and 0.78(Shiozawa) to 1.24 (Shinjo) respectively except for Hirosaki.

As seen from these results, the estimated value by the proposed method has a tendency to increase as an air temperature limit increases. Consequently an air temperature limit of +2℃ should be adopted to our method rather than +3℃ or +4℃.

5 MEAN RECURRENCE INTERVAL VALUE

Mean recurrence interval(MRI) values are used to generate design snow loads as can be seen in standard specifications such as the ANSI/ASCE Standard (1996), the National Building Code of Canada(1995) and Recommendations for Loads on Buildings of Japan(1996).

In this paper MRI values are calculated using the Type I probability distribution function, judging from results in our papers(1992).

Table 2 compares MRI values calculated by the accumulated daily precipitation to the observed snow weights. The ratios of 50-year and 100-year MRI values of annual extreme snow weights are in the range of 0.97(Shinjo) to 1.15(Joetsu) and 0.95 (Shinjo) to 1.15(Joetsu) respectively except for Hirosaki. The ratios of 50-year and 100-year MRI values of annual maximum of increasing snow accumulation for 7 days are both in the range of 0.87(Tokamachi) to 1.23(Sapporo).

The proposed method can be used to estimate ground snow weights statistically from the first day

Table 1. Summary of ground snow weights.

Observation sites (abbreviation)	Observation years	n	W_{max}(kgf/m²)				$W7_{max}$(kgf/m²)			
			Max.	Min.	m	σ	Max.	Min.	m	σ
Sapporo (Sa)	1964~'91	26	418	162	294	72	92	36	56	14
Hirosaki (Hi)	1965~'86	15	362	61	216	83	75	30	49	15
Kamabuchi (Ka)	1941~'87	42	1071	147	550	204	172	54	97	26
Shinjo (Shn)	1975~'84	10	622	126	407	138	111	56	78	19
Nagaoka (Na)	1984~'91	8	1086	93	583	341	292	50	171	75
Tokamachi (To)	1940~'87	48	1571	233	819	318	331	107	203	60
Joetsu (Jo)	1970~'80	9	721	74	439	190	252	56	186	61
Shiozawa (Shi)	1953~'60	8	964	235	588	237	220	121	176	37

n: no. of years, W_{max}: annual extreme ground snow weight,
$W7_{max}$: annual maximum weight of increasing snow accumulation for 7 days,
Max.: maximum value, Min.: minimum value, m: mean value, σ: standard deviation

Table 2. Ratio of MRI value.

Obsv. sites	P_{MRI}/W_{MRI}		$P7_{MRI}/W7_{MRI}$	
	50 year	100 year	50 year	100 year
Sa	1.11	1.10	1.23	1.23
Hi	1.37	1.36	1.15	1.14
Ka	1.03	1.01	1.03	1.03
Shn	0.97	0.95	1.06	1.05
Na	0.98	0.98	0.97	0.97
To	0.99	0.98	0.87	0.87
Jo	1.15	1.15	0.88	0.89
Shi	1.05	1.05	0.91	0.93

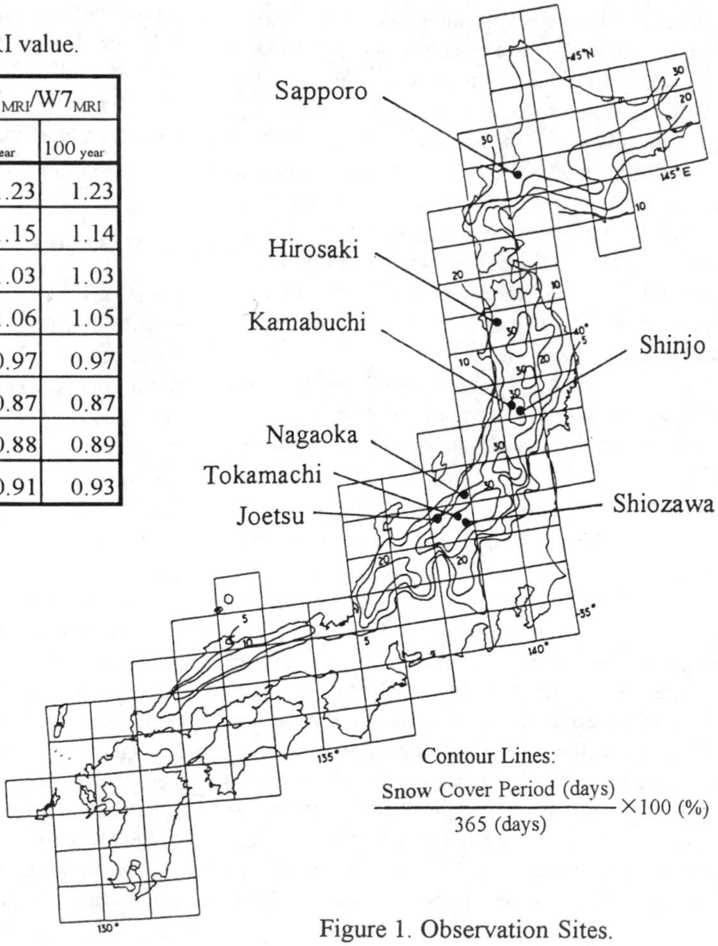

Contour Lines:

$$\frac{\text{Snow Cover Period (days)}}{365 \text{ (days)}} \times 100 \ (\%)$$

Figure 1. Observation Sites.

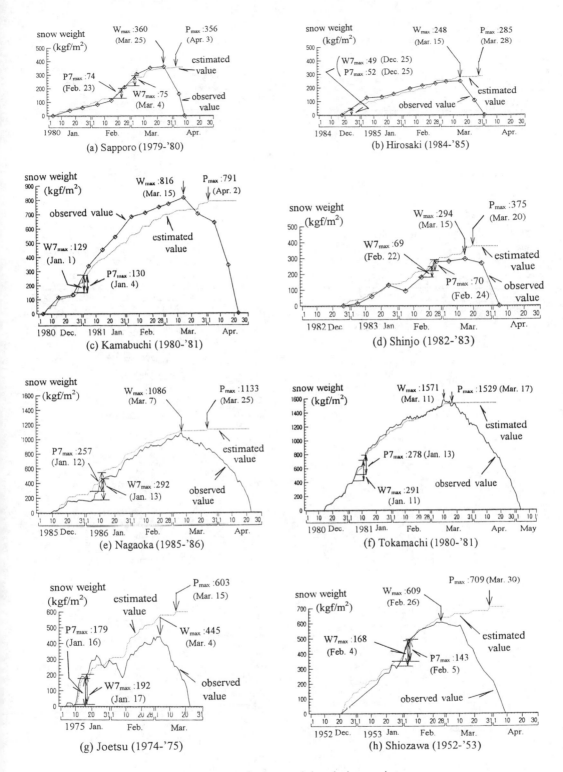

Figure 2. Change of snow weights during a winter.

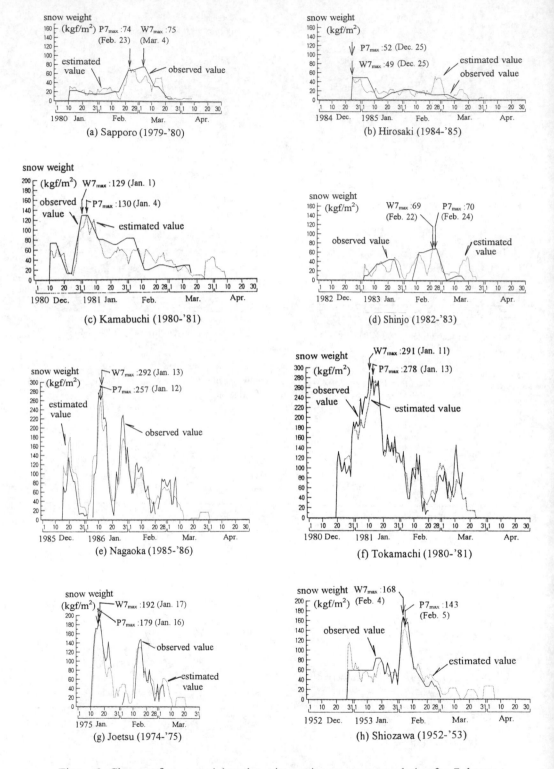

Figure 3. Change of snow weights about increasing snow accumulation for 7 days.

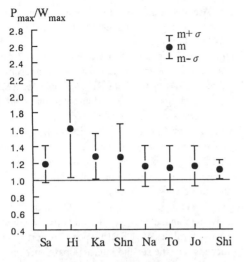

P_{max}/W_{max}

(a) Mean(m) and standard deviation(σ)
for P_{max}/W_{max}.

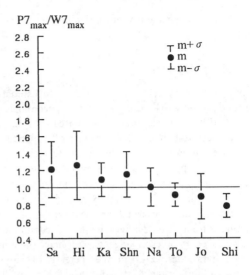

$P7_{max}/W7_{max}$

(b) Mean(m) and standard deviation(σ)
for $P7_{max}/W7_{max}$.

Figure 4. Ratio of the estimated value to the observed value
for an air temperature limit of +2℃.

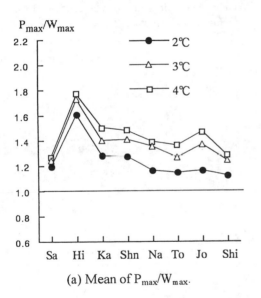

P_{max}/W_{max}

(a) Mean of P_{max}/W_{max}.

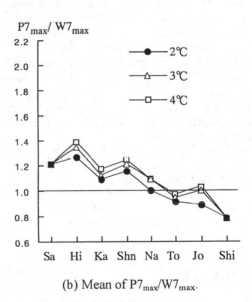

$P7_{max}/W7_{max}$

(b) Mean of $P7_{max}/W7_{max}$.

Figure 5. Ratio of the estimated value to the observed value
for three air temperature limits.

to the annual extreme day of continuous snow cover in heavy snow regions.

6 CONCLUSIONS

This paper presents a simple method to estimate ground snow weight using standard meteorological data. From the analytical results based on data of daily precipitation and daily mean air temperature measured at eight sites in heavy snow regions of Japan, the following conclusions can be drawn:

(1) The analysis shows that the accumulated value of daily precipitation data is closely related to the data of observed snow weight from the first day to the annual extreme day during a period of continuous snow cover. In order to judge whether daily precipitation data contributes to the increase of snow weight, a limit air temperature is introduced for daily mean air temperature. As a result of the investigation three air temperature limits (+2℃, +3℃ and +4℃), +2℃ should be adopted.

(2) As to the mean recurrence interval value of annual extreme weight and annual maximum value of increasing snow accumulation for 7 days, the values based on the estimated snow weight are in reasonable agreement with the observed snow weights

It is noteworthy that the proposed method can evaluate the ground snow weight very simply compared to the case which requires the consideration of snowmelt. The proposed method can provide valuable statistical information to engineers.

ACKNOWLEDGMENTS

The authors are very grateful to the snow-researchers for providing their observed data for long period and their useful comments.

REFERENCES

Handbook of meteorological terms 1986. NHK publication: 160-161. (in Japanese)

Tamura, M. 1990. Snow and rainfall frequencies in nagaoka. Journal of the Japanese society of snow and ice, Vol.52, No.4: 251-257. (in Japanese)

Kaneko, H., Joh, O. & Shibata, T. 1987. Estimation of snow weight on the ground from daily precipitation and temperature data observed in Sapporo city. Architectural Institute of Japan, Summaries of technical papers of annual meeting B(Structures I):1411-1412.(in Japanese)

Kamimura,S. & Uemura, T. 1992. Estimation of daily snow mass on the ground using air temperature and precipitation data", CRREL special report 92-27, proceedings of the second international conference on snow engineering: 157-167.

Shinojima, K. & Harada, H. 1992. Estimation of ground and roof snow weight based on the data of AMeDAS. Japan society for snow engineering, Journal of snow engineering, Vol.8, No.3: 191-204. (in Japanese)

Joh, O., Sakurai, S. & Shibata, T. 1992. Statistical analysis of annual extreme ground snow weights for structural design. CRREL special report 92-27, proceeding of the second international conference on snow engineering: 3-14.

Minimum design loads for buildings and other structures 1993. American society of engineers standard.

National building code of Canada 1995. National research council of Canada.

AIJ recommendations for loads on buildings 1996. Architectural institute of Japan.

Snow Engineering: Recent Advances, Izumi, Nakamura & Sack (eds) © 1997 Balkema, Rotterdam. ISBN 90 5410 865 7

The reduction of earthquake energies on buildings by snow on their roofs

Tsutomu Nakamura
Iwate University, Morioka, Japan

Yasuaki Nohguchi, Toshiichi Kobayashi, Yutaka Yamada & Keiichi Ohtani
Nagaoka Institute of Snow and Ice Studies, NIED, Niigata, Japan

Seitaro Takada
Takada Architecture Corporation, Nagaoka, Japan

ABSTRACT: In Japan today snow remains on the roofs of some wooden houses through out the winter. Japan has many earthquakes. Two types of experiments were conducted to determine the effect of roof snow on the dynamic response of wooden structures. The natural period of experimental structures with snow on their roofs was longer than that of structures without snow. In other words, the structures vibrated more slowly when loaded with snow. Roof snow gives a dynamic lateral force to structures. The attenuation constant for the structure with snow was marginally larger than that for the structure with no snow. A newly defined constant S, the ratio of the damping coefficient against the coefficient of resistance proportional to the relative displacement was 2 to 3 times larger for the structure with roof snow than for the structure without snow. This suggests that roof snow absorbs vibrational energy. This ability may reduce risks when wooden houses are vibrated by earthquakes.

1 INTRODUCTION

More than half the area of the Japanese islands is covered with snow every winter. The depth of snow on the ground exceeds 4 meters in a heavy snowfall winter even in cities with a population of 40,000 (Nakamura, Abe and Takada 1992). A lot of troubles, perhaps disasters, are expected when strong seismic activity occurs when heavy (e.g., one in ten years)snow loads exist on roofs. At 3:39 a.m. on February 2, 1961 a magnitude 5.2 earthquake attacked the suburbs of Nagaoka city, one of the heaviest snowfall cities in Japan. Five persons were killed and 259 houses were seriously damaged (Osawa and Yamamoto 1961). Problems to be expected in a common winter were solved recently, but problems to be expected in a heavy snowfall winter still require attention.In a snowy city we don't have big enough yards to accept snow shovelled off the roof. Therefore, houses with flat or inclined roofs are being designed to keep snow on their roofs. Are houses with heavy roof snow safe in earthquakes?

In past years roof snow load was mainly a staic issue, fundamental to the design of buildings. But in snowy regions, buildings designed to keep snow on their roofs should also consider the dynamic behavior of that roof snow when the buildings were vibrated by an earthquake.

The purpose of this study was to obtain fundamental knowledge of the dynamic behavior of snow on roofs when an earthquake occurs.

2 EXPERIMENTS

Two types of experiments were conducted. The Type A experiment used experimental wooden structures, and the Type B preliminary experiment used a full size wooden house.

2.1 Type A experiment.

This experiment, which might have resulted in some structural fractures, were carried out on three small unoccupied wooden structures as shown in Figures 1 and 2. Each structure was 1.8 m wide x 1.8 m length x 2.4 m high. The slope was 3:10. The roof was of ribbed seam metal roofing with a linear snow stopper.

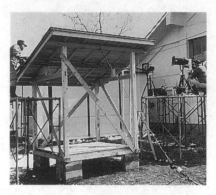

Figure 1　Experimental wooden structure.

Figure 4　Markers in the roof snow.

Figure 2　Experimental wooden structure with snow on its roof.

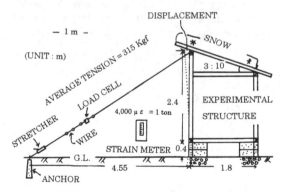

Figure 3　System used to vibrate the experimental structure.

2.1.1.　1988-89 winter experiment.
(1)　Method
　Figure 3 shows the experimental principle used in the 1988-89 winter.Snow was gathered on the ground and placed on the roof, because of the light snowfall that winter.The roof snow depth was 10.3

cm on Roof No.1,15.0 cm on Roof No.2 and 20.0 cm on Roof No.3. The average snow density on Roofs 1, 2 and 3 was 0.42 g/cm^3, 0.46 g/cm^3, 0.59 g/cm^3, respectively.　To vibrate the structure, a wire with a load cell was tightened between the structure and the ground as shown in Fig. 3. After the tension the structure was vibrated by cutting the wire. The average tension was 315 kgf. The unfractured and experimental displacement was 7.5 ± 1.5 mm. Markers were installed to monitor the motion of the roof snow (Fig.4). The high speed (1,000 frames per second) camera shown in Fig.1 recorded the action on videotape. The tape was analyzed and the natural frequency of the structure was calculated.

(2)　Results

　Table 1 summarizes experimental results. Figure 5 shows the relationship between the natural frequency of the structures and the weight of snow on their roofs. The heavier the roof snow, the smaller the natural frequency.

　We can assume that the structure vibrates as an inverted pendulum as shown in Fig.6. The equation of motion can be written as,

$$I \frac{d^2 \theta}{dt^2} + \varepsilon\theta - Mgh \sin\theta = 0 \qquad (1)$$

where I is the moment of inertia, θ is angle of deflection, M is mass of the pendulum, ε is a constant, g is the gravitational acceleration, and h is the length between the mass and the fixed point. If the angle θ is small, then the equation can be written as,

Table 1. Summary of experimental results (Experimental date: February 23, 1989).
* A part of experemental roof No.1 fractured in tension in run 6, so an iron catch(U shaped stopper with sharp tips at the two ends to fasten a joint of two wooden members) was given to the roof , and the experiment was continued.

Run	Name of exp. roof	Displace- ment (mm)	Tension (kg)	Catch	Snow	Remarks
2	3	21	370			Test
		31	400			Test
4	1	8	360	yes	yes	
5	1	7	410	no	yes	
6	1	33	400	no → yes	no	fractured*
7	1	6	300	yes	no	
8	2	10	300	no	yes	
9	2	6-7	295	no	no	
10	3	8-9	290	no	yes	
11	3	6-7	250	no	no	

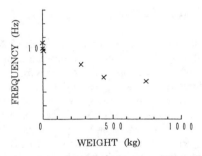

Figure 5   Relationship between the natural frequency of the structure and the weight of snow on its roof.

$$I \frac{d^2 \theta}{dt^2} + (\varepsilon - Mgh) \theta = 0 \qquad (2)$$

If ε > Mgh, the pendulum is stable and the period, T, is written as,

$$T = 2\pi \sqrt{\frac{I}{\varepsilon - Mgh}} \qquad (3)$$

If ε is close to Mgh, the period, T, will increase.

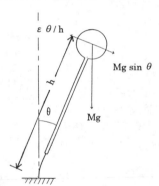

Figure 6   Inverted pendulum.

If the restoring force, $\varepsilon \theta /h$, is larger than the force to destroy the structure, $Mg \sin \theta$, then the structure will not be fractured. As the amount of snow on the roof increases the period, T, increases(i.e., the structure vibrates more slowly.). So, a flexible structure strong enough not to be broken will be safe. Another approach is to use the earthquake to shake the snow off the roof.

When we built a house to conclude this experiment, we had two choices. The first was a stiff house, strong enough to support a heavy roof snow load with the snow dropping off when an earthquake

195

hits. The second was a flexible house which vibrates slowly with heavy snow on the roof, strong enough not to fracture when an earthquake hits.

2.1.2 1989-90 winter experiment.
(1) Method

The method used this winter was the same except that the acceleration was measured, and an experimental wooden structure of a flat roof was used as well as the one of an inclined roof.

Snow was again gathered on the ground and placed on the roof. Figure 7 shows the positions of accelerometers a, b, c and d. Accelerometer a was just below the snow surface, accelerometer b was on the underside of the roof, accelerometer c was on a wooden member between the roof and the floor and accelerometer d was on the floor of the experimental structure.

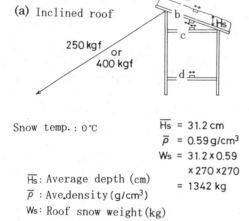

(a) Inclined roof

250 kgf or 400 kgf

Snow temp. : 0 ℃

$\overline{Hs}$ = 31.2 cm
$\overline{\rho}$ = 0.59 g/cm³
Ws = 31.2 × 0.59
      × 270 × 270
   = 1342 kg

$\overline{Hs}$: Average depth (cm)
$\overline{\rho}$: Ave.density (g/cm³)
Ws : Roof snow weight (kg)

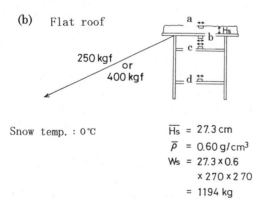

(b) Flat roof

250 kgf or 400 kgf

Snow temp. : 0 ℃

$\overline{Hs}$ = 27.3 cm
$\overline{\rho}$ = 0.60 g/cm³
Ws = 27.3 × 0.6
      × 270 × 270
   = 1194 kg

Figure 7   Position of the accelerometers a, b, c and d.

(2) Results and analyses

Figures 8 and 9 present accelerometer records for the flat roof and inclined roof, respectively. Note the dramatic decrease in accelerations with snow on three roofs. The phase in the vibrational record at a differs from that at the other three points. This is due to the inverted set up of the accelerometer at a. In fact all acceleration records are in the same phase.

For freely damping oscillating motion, the equation of motion can be written as,

$$M\ddot{u} + c\dot{u} + ku = 0 \qquad (4)$$

where M is the mass, u is the displacement, c is the damping coefficient and k is the coefficient of resistance which is proportional to the relative displacement. The equation is rewritten by the use of h, fraction of critical damping (i.e.,damping factor) and ω_0, angular frequency as,

$$\ddot{u} + 2h\omega_0\dot{u} + \omega_0^2 u = 0 \qquad (5)$$

where $k/M = \omega_0^2$, $c/M = 2h\omega_0$.

In equation (4), $M\ddot{u}$ is the inertia term, $c\dot{u}$ is the attenuation term and ku is the restoration term. The fraction of critical damping was obtained from the accelerometer records. Figure 10 shows one example of the decay of acceleration with time. Table 2 summarizes the analytical results of the experiments.

We defined a coefficient S = c/k to quantify the damping effect of snow. We expected that S could indicate the energy absorbed by the snow. The ratio of S_s, with snow, to S_0, without snow, is shown in Table 2. That ratio indicates the magnitude of damping provided by the snow.

2.1.3 1990-91 winter experiment.
(1) Method

The method used this winter was the same as was used during the winter of 1989-90 except that roof snow was that which naturally fell on the roof. Figure 11 shows roof snow and placement of the accelerometer. Figure 12 shows the experimental set up this winter. Only two accelerometers were fixed on the structure and on the snow surface as we learned from the previous experiments that records at those two points are enough for the analysis.

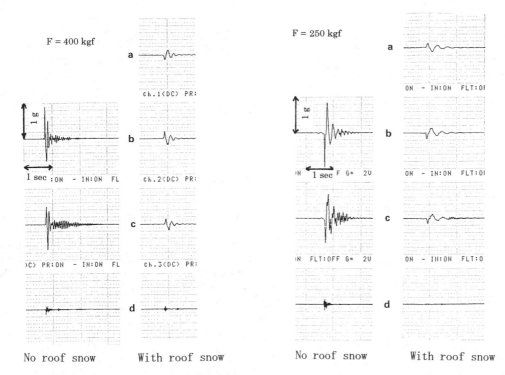

Figure 8. Accelerometer records for the flat roof.

Figure 9. Accelerometer records for the inclined roof.

Table 2. Summary of the vibration experiments of model structures with roof snow(1989-90).

ROOF TYPE	SNOW LOAD (kg)	TENSION (kgf)	PERIOD (sec)	MAX. ACCEL,(G)	ω_0 (1/sec)	h	c/k=S (sec)	S_s/S_0
HORI-ZONTAL	1194	250	0.16	0.10	40	0.17	0.0085	3.4
		400	0.21	0.24	31	0.22	0.0141	3.0
	0	250	0.08	0.50	79	0.10	0.0025	1
		400	0.10	0.90	64	0.15	0.0047	1
IN-CLINED	1342	250	0.38	0.18	17	0.25	0.029	3.1
		400	0.44	0.43	15	0.19	0.025	2.1
	0	250	0.19	0.98	34	0.16	0.0094	1
		400	0.21	1.25	30	0.18	0.012	1

(a)

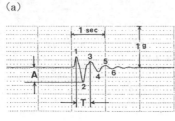

(b)

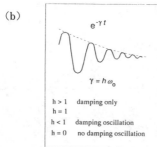

$e^{-\gamma t}$

$\gamma = h\omega_0$

$h > 1$ damping only
$h = 1$
$h < 1$ damping oscillation
$h = 0$ no damping oscillation

(c)

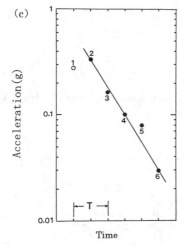

Figure 10. An example of the decay in acceleration with time.

Figure 11. Placing the accelerometer on natural roof snow.

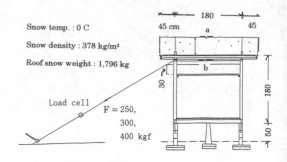

Snow temp. : 0 C

Snow density : 378 kg/m³

Roof snow weight : 1,796 kg

Load cell

$F = 250$,
300,
400 kgf

Figure 12.   Experimental set up in 1990-91 winter.

(2)   Results and analyses

Figure 13 shows examples of roof vibration with and without snow on the roof. Table 3 shows the summary of the experiments carried out on the naturally fallen roof snow.

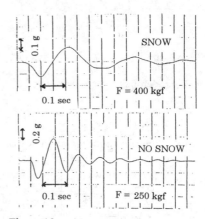

Figure 13.   An example of the records.

(3)   Summary of all Type A experiments.

Table 4 summarizes the results of all Type A experiments. The table shows that the ratio of S_S with snow to S_0 without snow varies from 2.1 to 3.4. This probably means that the snow on the roof is absorbing energy while vibrating. If that is true, roof snow will absorb a part of an earthquake's energy, helping to keep the house from the fracturing.

2.2 Type B experiment.
This was a full size experiment using a type of house Yajirobeh in Japanese(Namamura et. al.

198

Table 3.   Summary of the 1990-91 experiments.   * 1796 kg

	RUN	SNOW	TENSION (kgf)	PREDOMINANT PERIOD, T (sec)	Max. ACCEL (g)	h	$\omega_0 = 2\pi/T = 6.28/T$	c/k = S	S_s/S_o
FLAT ROOF	1	Yes*	300	0.22	0.08	0.20	28.6	0.0140	2.2
	2		400	0.27	0.12	0.17	23.3	0.0146	2.3
	3	No	250	0.10	0.36	0.20	62.8	0.0064	1

Table 4.   Summary of all type A experiments.

SNOW	KIND of ROOF	$\dfrac{S_s}{S_o}$, DEGREE OF ATTENUATION WITH SNOW / DEGREE OF ATTENUATION WITH NO SNOW
ARTIFICIAL SNOW	INCLINED ROOF	3.1
		2.1
	FLAT ROOF	3.4
		3.0
NATURAL SNOW	FLAT ROOF	2.2
		2.3

1992).The structure is shown in Figure 14.The house was pulled by a vehicle for a moment through a wire, then the wire was cut off as was done in Type A experiments.On this experimental day (Jan. 22,1994) roof snow depth was 40 cm at the ridge and 45 cm at the eaves. In the middle of the roof the depth varied from 42 to 48 cm. The average snow density was 0.16 g/cm^3. The total weight of snow on the roof was calculated as 5,570 kgf. The natural oscillation frequency of the house was calculated as 4.5 Hz(i.e., T = 0.22) from other tests. The snow on the roof did not drop off when the wire was cut. The house was vibrated only for a short period as shown in Figure 15. Vibrations started when a portion of the wire was cut through then peaked when the rest of the wire was cut. The house was quite stiff.

3. DISCUSSION AND CONCLUSIONS

With snow on their roofs, the natural(predominant)

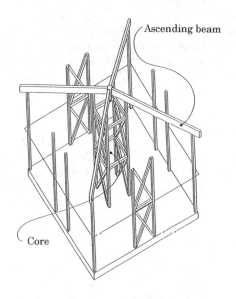

Figure 14.   Structure of the full size house.

Figure 15. Vibration of the full size house. The speedmeter was mounted on the 2nd floor of the house.

period T of the structure became longer. If a house with snow on its roof is flexible and strong enough to vibrate, it will be safe. Coefficient S is defined as the ratio of the damping coefficient, c, to the coefficient of resistance, k, which is proportional to the displacement u. The coefficient S with snow and without snow indicates the damping effect of roof snow. The ratio of S with snow and without snow varied from about 2 to 3. It was also noticed that snow on the roof reduced the amplitude of vibration. These two results show that snow will absorb vibrational energy. It is known that the grain boundaries of polycrystalline ice will absorb vibrational energy at low frequency (Nakamura and Abe 1979) as well as at high frequency(Kuroiwa 1964). Roof snow will be able to absorb vibrational energy as well as polycrystalline ice can. Thus roof snow may help keep houses safe during earthquakes.

Further studies are underway to confirm these findings using continuous vibration experiments. Tests are underway using a concrete block, sand and a vibration machine. We also will study slippage between the roof and the snow as well as slippage within the snow itself.

ACKNOWLEDGEMENTS

The authors express their hearty thanks to Dr. Wayne Tobiasson for his critical readings of this manuscript and to Dr.Vladimir B. Aizen for his technical help to edit this manuscript, respectively.

REFERENCES

Kuroiwa.D. 1964. Internal friction of ice. *Contributions from the Institute of Low Temperature Science, Series A*, No.18:1-62.

Nakamura, T. and O, Abe 1979. A grain-boundary relaxation peak of Antarctic Mizuho ice observed in internal friction measurements at low frequency. *J.Faculty of Sci.,Hokkaido University, Series VII(Geophysics)*,6,1:165-171.

Nakamura,T., O.Abe and S. Takada 1992. Roof snow observation and application to house construction. *CRREL Special Report* 92-27: 81-92.

Osawa, Y. and M. Yamamoto 1961. On the damage to building during the Nagaoka Earthquake of February 2,1961. *Jishinkenkyuushoihou(Bulletin of the Earthquake Research Institute,U. of Tokyo)*,39: 195-208.

Snow Engineering: Recent Advances, Izumi, Nakamura & Sack (eds) © 1997 Balkema, Rotterdam. ISBN 90 5410 865 7

Unbalanced snow loads on gable roofs

Michael J.O'Rourke
Rensselaer Polytechnic Institute, Troy, N.Y., USA

Michael Auren
Friedman and Oppenheimer LLP, New York, N.Y., USA

ABSTRACT: An investigation of unbalanced or drifted snow loading on single gable roofs is presented. A database of 28 gable roof case histories was developed and analyzed to identify trends and key parameters for this type of roof snow drift. The analysis suggests that roof slope, roof width (i.e., ridge to eave distance) and the ground snow load are the key parameters influencing drift size. Furthermore, substantial drifts near the eaves of low sloped roofs (slope less than 15°) are possible, particularly if the width is large and the ground snow load is moderate or low. A number of simple relations were developed which can be used to predict the total drift height. Predicted values from these relations were compared with observed values to verify their accuracy. It is felt that the simple relations developed herein are suitable for use in building codes and load standards.

1 INTRODUCTION

Of particular importance in the design of roof structures for snow is the drift formation process, since the majority of snow-related roof damage is due to drifted snow. Information gathered by insurance companies illustrates this point. Factory Mutual Research (DeAngelis 1993) reported $382.6 million (1990 U.S. dollars) in insured damage attributed to snow drifting for the time period 1977 to 1989. More recently a blizzard in March of 1993 caused in excess of $200 million (1993 U.S. dollars) in roof damage. These March 1993 roof failures also appear to have been due in large part to drifted roof snow.

Three conditions are necessary in order for a drift to form. There must be a snow source to "feed" the drift, some sort of geometric irregularity which (i.e., an abrupt change in roof elevation, change in roof slope, parapets, etc.) where the drift can form, and wind of sufficient speed to allow transport from the source.

A commonly encountered drift, which is the primary focus of this paper, is an unbalanced snow load on a single gable roof. In this case the windward side of the roof is the snow source and the ridge line is the geometric irregularity which results in an area of aerodynamic shade on the leeward side.

Wind with a significant component perpendicular to the roof's ridge line is needed to cause a redeposition of the snow from the windward to the leeward side. The resulting load on the leeward side is usually nonuniform.

This paper presents a brief description of current U.S. design procedures for an unbalanced snow load on a single gable roof, a trend analysis on the gable roof case history database, an empirical relation for estimating gable roof drift loads, and a comparison with observed values. The paper concludes with a discussion of the results.

2 CURRENT U.S. PROCEDURES

Current U.S. procedures for unbalanced snow loads on gable roofs are presented in the American Society of Civil Engineers (ASCE) Standard 7-95 (ASCE 1996). The overall approach is sketched in Figure 1. As can be seen, two separate load cases are considered. The first is the balanced load case which specifies a uniform load of intensity p_s on both sides of the ridge line, while the second load case is an unbalanced load of intensity $1.3 \, p_s/C_e$ on the leeward side only. The coefficient C_e is an exposure factor which varies from 0.7 to 1.3 as a function of terrain topology and roof exposure to wind. The unbalanced load case in the current

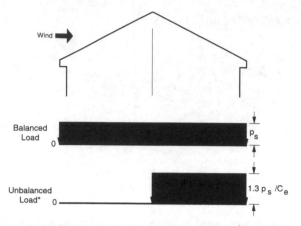

Figure 1 Gable roof snow load cases (from ASCE 1996)

ASCE procedure is required only for gable roofs with slopes from 15 to 70°. In addition, the unbalanced load is not a function of the size of the snow source, that is the size of the windward roof, nor the space available for the drift formation as characterized by the product of the roof slope and the width, αW. As will be shown later, these parameters prove to be significant.

3 TREND ANALYSIS OF GABLE ROOF CASE HISTORY DATABASE

A database of case histories for unbalanced/drift loads on gable roof structures was developed for the purposes of this study. The database contains a total of 28 case histories. These case histories were obtained from the technical literature as compiled in an unpublished study by Kamimura (1994) (18 case histories), as well as a review of insurance company files (10 case histories).

The database contains structure location, case history date, roof geometry, and snow load information. The roof geometry information is the roof slope, α (in degrees from horizontal), the eave height, H, the roof width, W, defined as the horizontal distance from the ridge line to the eave and the roof length, L, defined as the distance from end wall to end wall.

The snow load information includes observed ground snow depths, observed ground snow loads, corresponding ground snow depth and load information for the nearest weather station, as well as roof snow depth and roof snow load for the leeward and windward sides of the roof. A fuller description as well as a listing of the database itself

is contained in O'Rourke and Auren (1995).

Except as noted, in the analysis which follows, the ground snow load and depth values are taken as the average of the observed site and closest weather station values. For case histories where a snow depth was known but not the corresponding load, or vice versa, the following relation from the Cold Regions Research and Engineering Laboratory (CRREL) in Hanover, N.H. (Tobiasson and Greatorex 1996) was used to calculate the missing value

$$p = 1.97 \ d^{1.36} \qquad (1)$$

where p is the snow load (in kN/m^2) and d is the snow depth (in m.). Although this relation was developed specifically for ground snow, it is also used herein for roof snow.

The first item which needed to be established by the trend analysis is the expected spatial distribution of drifting/unbalanced loads on gable roofs. Recall that the current ASCE Standard 7-95 postulates a uniform snow load on the leeward side and no snow on the windward side. The spatial distribution of unbalanced or drifted snow from one of the case history structures is sketched in Figure 2. The observed distributions in the database suggest that the leeward roof load is nonuniform. For example, in some case histories the leeward depth is a maximum near the ridge line, while in others the load is a maximum about halfway between the ridge line and eave. However, for many case histories such as sketched in Figure 2, the leeward load is a maximum near the eave and small near the ridge line.

Irrespective of the exact location of the maximum leeward load, the top of the drift typically

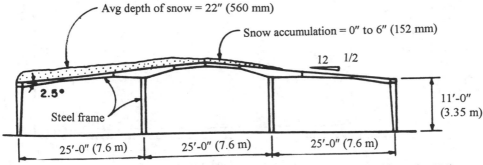

Avg depth of snow = 22" (560 mm)

Snow accumulation = 0" to 6" (152 mm)

12   1/2

2.5°

Steel frame

11'-0"
(3.35 m)

25'-0" (7.6 m)   25'-0" (7.6 m)   25'-0" (7.6 m)

Figure 2   Spatial distribution of drifted snow for case history No. 1 (Chin et al. 1980)

did not extend much above the ridge line. That is, the triangular area above the leeward roof surface and below the ridge line elevation is sheltered from wind (i.e., in the aerodynamic shade region) and hence is the location for potential drift formation. Based upon this concept, it is postulated herein that the drift begins near the ridge line and given a sufficient snow source (e.g., windward roof snow) and wind (both speed and duration) the aerodynamic shade region eventually fills. This leads to a situation where the maximum leeward drift load occurs near the eave. Note that when the wind stops blowing or the snow source is depleted, the drift shape at that point in time becomes the permanent shape.

To determine trends and key parameters, histograms for various parameters in the case history database were developed. In terms of roof slope, 39% of the case histories involving structural failure had a roof slope, α, less than or equal to 15° and hence would not have been considered to be susceptible to a significant unbalanced load condition using the current ASCE Standard 7-95 procedure.

In terms of roof width, W, there are no structural failures in the case history database for roof widths less than 4.6 m (15 ft). Also, the majority of the collapse case histories (88 percent) are for relatively low ground snow loads of 1.44 kN/m^2 (30 psf) or less.

As noted previously, in the current U.S. approach for drifted/unbalanced loads on a gable roof, roof slope, α, is the only geometric parameter of interest. That is, unbalanced loads need not be considered for roof slopes less than 15°, or greater than 70°. However, the concept of drifting in the triangular aerodynamic shade region suggests that the drift load at the eave for a "full" drift is a function of both the roof slope and width.

Figure 3 is a scatter diagram of roof slope and width values for the collapse cases in the database. The lower boundary line is consistent with the concept of the potential drift load being related to the size of the aerodynamic shade region. That is, overload and collapses are possible only when the product αW is reasonably large. Specifically, for small slopes only very large widths could produce a significant overload. Conversely, the aerodynamic shade region and the potential for significant overloads are small for structures with low slope and small roof widths. The database verifies this conjecture in that there are no roof failure case histories in the lower left hand portion of Figure 3.

The influence of roof slope and width on potential drift loads can also be seen by considering a "maximum possible" drift, sketched in Figure 4. In this case the triangular aerodynamic shade region was assumed to be full of snow leading to a total snow depth at the eave equal to αW where the roof slope α is in radians.

The overloads corresponding to this maximum possible or full drift were calculated for various combinations of α, W and the 50 year MRI design ground snow load, $(P_g)_{50}$. Figure 5 presents the overloads as a function of roof slope and width for a $(P_g)_{50}$ of 1.68 kN/m^2 (35 psf). In Figure 5, the ASCE Standard 7-95 density relation is used to convert from depth to load and vice versa, and the produce $C_e C_t I$ was taken as 0.8.

For a roof with a 2% slope, the maximum possible or full gable roof drift leads to 50% overload for a W = 9.1 m (30 ft) and $(P_g)_{50}$ = 0.96 kN/m^2 (20 psf). For the same roof slope, W = 19.8 m (65 ft) and $(P_g)_{50}$ = 3.1 kN/m^2 (65 psf) leads to the same 50% overload. In general, the overload is an increasing function of both W and α, reflecting a larger area available for drift formation. The overload is a decreasing function of the design

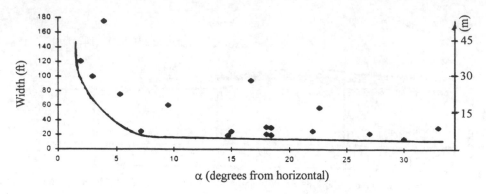

Figure 3   Scattergram of roof width and slope - collapse cases

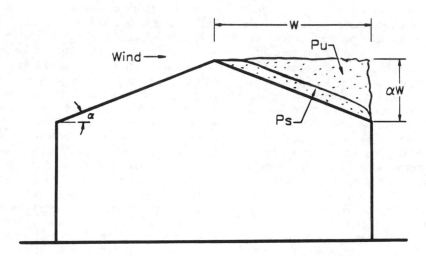

Figure 4   Maximum possible drift

ground snow load, $(P_g)_{50}$. Larger values of $(P_g)_{50}$ result in larger amounts of balanced snow, smaller areas available for drifting, and hence lower overloads. This is compatible with the trend noted previously that drift loads on gable roofs appear to be a low ground snow load phenomenon.

The lower boundary line for the collapse case history scatter diagram (Figure 3) has the same general shape as the overload curve in Figure 5. That is, we have low overloads and no collapses for small values of the product αW. In point of fact, all of the observed collapse case history data points in Figure 3 fall above and to the right of the 75% overload threshold line for a $(P_g)_{50}$ of 0.96 kN/m^2 (20 psf).

Another important characteristic, particularly for the analysis of building frames, is the amount of snow on the windward roof. Recall that the ASCE

Standard 7-95 procedure considers the windward side free from snow. For the case history structures the ratio of windward roof snow load to ground snow load averaged 0.28, while the ratio of windward snow depth to ground snow depth averaged 0.34.

In no instance was the windward snow load or depth greater than its corresponding ground value. In terms of potential code provisions, a windward side snow load for the unbalanced or drifted gable roof load case of 30% of the ground snow load may be appropriate.

4 DRIFT HEIGHT PREDICTION

In order to formulate a reasonable design load incorporating drifting/unbalanced snow loads on

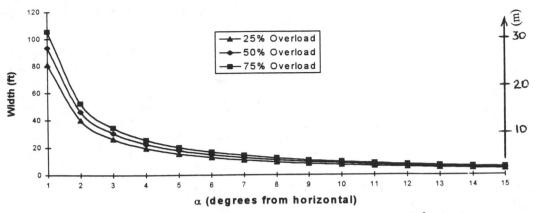

Figure 5 Maximum possible drift overload for $(P_g)_{50} = 1.68$ kN/m^2 (35 psf)

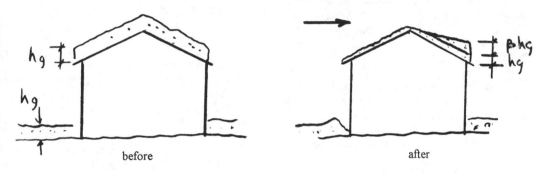

before   after

Figure 6 "Before and after" scenario for snow drift formation on a single gable roof

single gable roofs, an accurate prediction of the drift height must first be obtained. Herein a physical model approach is used. Recall that the size of the potential drift is a function of both the roof slope, α, and the roof width, W. Also recall that the total drift height is limited by the amount of available snow to "feed" the drift. In the extreme case, all of the available snow is redeposited onto the leeward side of the roof, possibly leading to the maximum possible surcharge drift.

Figure 6 presents a sketch of "before and after" for a physical model of gable roof drifting. This model envisions a substantial snow fall with ground snow depth, h_g, followed by strong wind from left to right. In the "before" condition, the balanced load on the roof corresponds to the ground snow load. That is, there is no initial reduction in roof load due to exposure, thermal or slope effects. In the "after" condition, the wind has carried a percentage of the windward side snow across the ridge line, forming a triangular drift on the leeward side. The height of the surcharge drift is βh_g. For the case where the

upwind and downwind widths are equal, the percentage of windward snow which ends up in the drift is then

$$\text{percentage} = \frac{\frac{1}{2} \cdot \beta h_g W}{h_g \cdot W} \times 100 = \frac{\beta}{2} \times 100 \qquad (2)$$

The total height of the lee drift at the eave (i.e., balanced plus surcharge) is then $(1+\beta)h_g$. Based upon the snow source, the maximum total depth is $3h_g$, corresponding to $\beta = 2.0$ or 100 percent of the windward snow ending up in the drift. Based upon geometry (i.e. available space in the aerodynamic shade region) the maximum total height is αW where α is in radians.

That is, the physical model is based upon a percentage, $(\beta/2) \times 100$, of the windward side snow forming the lee side drift. The case history database is used to determine appropriate values for β. Figure 7 is a scatter diagram of observed total eave snow depths against predicted values from the physical model with β equal to 1.4.

205

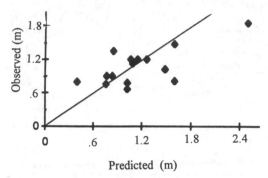

Figure 7. Predicted total drift height with β = 1.4 versus observed total drift height.

From this case history comparison, it appears that β ≅ 1.4 (i.e., roughly 70% of the available snow is deposited in the drift) is a reasonable value in that the 1 on 1 match line in Figure 7 bisects the data. That is, roughly half of the observed heights were larger than the predicted values and vice versa.

Application of the physical model in Figure 6 to the case history database suggests that roughly 70% (β = 1.4) of the windward snow ends up in the leeward drift. However, in relation to potential code provisions it is felt that this percentage should be reduced. In this regard we assume that code provisions for drift loads on gable roofs would be directed at a "50-year Mean Recurrence Interval (MRI)" drift load and the ground snow environment would be characterized by the 50-year MRI ground snow load. Based upon this premise, considerations which logically imply a reduction in β are listed below.

- Due to the source documents used to establish the database (i.e., information from insurance company files, etc.), the drifts are not a random sample of all gable roof drifts. By its nature, it is skewed towards larger drifts and/or those which lead to structural damage.

- The observed ground snow loads for the case history structures were typically somewhat less than the 50-year MRI value for the site in question. The range of the ratio of observed to 50-year MRI ground snow load was between 0.43 and 2.80 with a median of 0.75.

- For some locations, the 50-year MRI ground snow is the result of a series of snowfalls separated in time. Due to thermal effects and crust formation, the "driftable" snow on the

windward roof would be less than the 50-year ground snow value.

- Since wind is needed for drift formation, the joint occurrence of wind and driftable snow is a key parameter. For example, there is a chance that the 50-year ground snow occurs during calm wind conditions. In other words, the annual probability of simultaneous occurrence of the 50-year ground snow and strong winds is less than the annual probability of the 50-year ground snow by itself.

- Wind directionality affects gable roof drifts in that a substantial component perpendicular to the ridge line is required for drift formation. For example, wind parallel to the ridge line is unlikely to generate substantial drifting near the eaves.

The potential impact of various modification or reduction factors for gable roof drifts is shown in Figure 8. This presents the total eave load from the proposed physical model as a function of roof slope and roof width (eave to ridge distance) for design (i.e., 50-year MRI) ground snow loads of 0.96 kN/m² (20 psf) and β equals 0.5. For a given roof width, the horizontal line for small roof slopes is the "balanced" load. For such small roof slopes, the balanced load by itself fills up the aerodynamic shade region (maximum total eave snow depth equal to αW). At large slopes, the horizontal line corresponds to a total eave load of $(1+β)(P_g)_{50}$. For such large slopes, the aerodynamic shade region is large enough to accommodate the specified percentage of windward snow (percentage = (β/2)x100). The inclined line at moderate slopes corresponds to a full aerodynamic shade region due to balance plus some surcharge from the windward side. Based upon our engineering judgment, potential code provisions based upon the physical model, 50 year MRI ground loads and a β in the range of 0.5 to 1.0 seem appropriate.

It is proposed herein that β be based upon the roof's length to width ratio. For roofs which are symmetric about the centerline, the authors propose

$$β = \begin{cases} 0.5 & L/W ≤ 1 \\ 0.333 + 0.167 \, L/W & 1 < L/W < 4 \quad (3) \\ 1.0 & L/W ≥ 4 \end{cases}$$

This increase in β with increasing L/W reflects expected wind directionality effects. That is, one

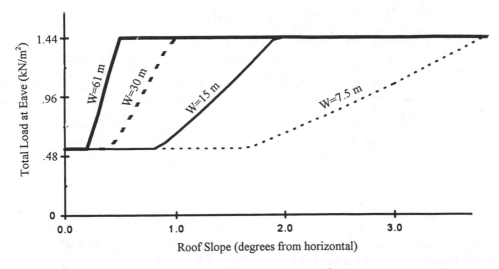

Figure 8 Total load of eave for snow loads of 0.96 kN/m² (20 psf) and β equals 0.5

Table 1 Observed total eave snow load and values predicted by the physical model with P($_g$)$_{50}$ and equation (3)

Case #	L/W	β	$(P_g)_{obs}$ (kN/m²)	$(P_g)_{50}$ (kN/m²)	$(P_r)_{obs}$ (kN/m²)	$(P_r)_{pred}$ (kN/m²)	Observed/ Predicted Roof Ratio	Normalized Ratio
19	12.1	1	1.34	0.48	2.69	0.96	2.81	1.0
20	2.2	0.7	1.37	1.92	2.69	3.25	0.83	–
21	2.9	0.81	0.82	0.96	1.81	1.74	1.04	–
23	2.8	0.8	0.93	0.48	3.16	0.86	3.67	1.90*
26	3.4	0.9	0.40	0.72	1.54	1.37	1.13	–
27	14.2	1	1.51	0.96	2.69	1.92	1.40	0.89
28	5.3	1	0.91	1.92	1.81	3.83	0.47	–

*See text for modified value of 1.24

expects substantial leeward drift loads for roofs with L/W ratios much less than one, only for wind essentially perpendicular to the ridge line. On the other hand, for roofs with L/W much greater than one, substantial drifts could result from a relatively wide range of wind directions (perpendicular to the ridge line or diagonal across the roof).

The potential impact of the relation for β shown in equation (3) is presented in Table 1. This table shows the value of β given by equation (3), the observed and 50-year MRI ground loads for the sites in question, the total eave loads both predicted by the physical model with equation (3) as well as the observed values, and finally two ratios.

The first ratio is simply the observed over predicted total eave load. For some cases, specifically case histories #19, 23 and 27, the overload (observed/predicted roof ratio > 1.0) results in part from the observed ground snow load being larger than the 50-year MRI value for the site. For these three cases, a normalized ratio is also presented. The normalized ratio is the observed/predicted roof ratio divided by the ratio of the observed to 50-year ground load values.

For six of the seven case histories in Table 1 (note Table 1 contains case histories for which all pertinent information was available), the physical model with β given by equation (3) appears appropriate. That is, the observed/predicted roof ratios for case histories with $(P_g)_{obs} < (P_g)_{50}$ or the normalized ratio for cases with $(P_g)_{obs} > (P_g)_{50}$ are less than or close to 1.0 (i.e., 1.0, 0.83, 1.04, 1.13, 0.89 and 0.47).

The one exception is case history #23. For this

case the normalized ratio is 1.9. It is believed that this large ratio is a result of the structure's nonsymmetric roof geometry. The eave to ridge distance of 29 m (95 ft) used in the calculation corresponds to the leeward side where the drift formed and a collapse occurred. However, the eave to ridge distance for the windward side of the roof is 50 m (165 ft). Considering both a larger than expected ground snow load and the asymmetric roof geometry the actual overload is about 24%. Although such a load would overstress structural elements close to the eave, collapse would not be expected.

5 CONCLUSIONS

The purpose of this study was to investigate unbalanced snow loading on single gable roofs. A simple physical model of gable roof drift formation was proposed and used to develop relations for estimating drift size. The drift size was estimated herein by assuming a certain percentage of the windward roof snow eventually ends up in the leeward drift. Based upon the case history database and observed ground snow loads, the appropriate percentage appears to be about 70% ($\beta = 1.4$). The remaining uniform load on the windward side was found to be about 30%. In terms of potential code provisions, a modification factor which accounts for wind directionality, the joint probability of wind and snow and other effects was implemented. For code provisions utilizing the 50-year MRI ground snow load, a β of 0.5 to 1.0 (25 to 50% of the windward snow eventually ends up in the lee side drift) is suggested.

Although the specifics of new code provisions are open to question, the authors feel that the general approach and specifically the physical model proposed herein are proper. Some advantages of the proposed approach are as follows.

The approach is based on an easily understood physical model. The approach utilizes geometric parameters which appear to be important based upon theoretical considerations and failure experience.

The approach discriminates between structures for which gable roof drifts could be substantial (large eave to ridge widths) and those for which drifting is unlikely (low slope and small widths).

Drift loads on asymmetric gable roof structures are easily determined.

6 REFERENCES

American Society of Civil Engineers (1996) *Minimum design loads for buildings and other structures*. ASCE 7-95, New York, NY.

Chin, R. Gouwans, J. and Hanson, M. (1980) *Review of roof failures in the Chicago area under heavy snow loads*. ASCE reprint pp. 80-145.

DeAngelis, Charles A. (1993) letter submitted to ASCE 7 Main Committee regarding Proposed 1995 Revisions of ASCE 7-88.

Kamimura, S. (1994), *Snow accumulations on gable roof case histories*. Unpublished research report, Rensselaer Polytechnic Institute.

O'Rourke, M. and Auren, M. (1995). "Unbalanced snow loads on gable roofs." Report to Metal Building Manufacturers Association, Cleveland, OH, Project Number MBMA 906.

Tobiasson, W. and Greatorex, A. (1996) Data and methodology for conducting site specific snow load case studies for the United States, in *Proceedings Third international snow engineering conference*, Sendai, Japan, 1996.

Snow Engineering: Recent Advances, Izumi, Nakamura & Sack (eds) © 1997 Balkema, Rotterdam. ISBN 90 5410 865 7

Characteristics of roofing materials as related to adfreezing of snow

Toshiyuki Ito & Tsukasa Tomabechi
Department of Architecture, Hokkaido Institute of Technology, Sapporo, Japan

Hirozo Mihashi
Department of Architecture, Tohoku University, Sendai, Japan

ABSTRACT: The adfreeze mechanism at the interface of snow and roofing and the influence of material characteristics on adfreeze strength must be clarified for the purpose of methodically controlling snow sliding on roofs. Experimental results on adfreeze strength showed the following: Roof temperature was mainly influenced by outside temperature and wind velocity, with the result that it varied with location on the roof. This variation influenced the properties of snow and it's adhesion to the roof. The adfreeze strength can be estimated knowing the roughness and water repellency of the roof's surface.

1 INTRODUCTION

Appropriate disposal of snow on roofs is indispensable. Especially, disposal of snow on roofs of large-scale buildings or those with light-permeable roofs is an important problem. On a sloped roof, sliding resistance occurs at the interface of the roofing and the snow(Maeda 1979 ; Watanabe et al.1987a ; Taylor 1985). Adfreeze of snow to roofing at below freezing temperatures creates the greatest sliding resistance (Ueno et al.1987b ; Shimizu et al.1992). Adfreeze prevents sliding of snow,resulting in a surplus load on the roof and a decline in light-permeability of the roof (Tomabechi et al.1991).

Little is known about the adfreeze properties of snow for various types of roofing such as light-permeable membranes and glass. Furthermore, no studies have concretely examined the process of adfreeze of snow to roofs. We related the adfreeze process of snow on a roof to measurements of the surface temperature of the roof. It was necessary to clarify the relationship between the surface properties of materials and the adfreeze properties of various roofing materials.

The purpose of this study was to measure changes in roof surface temperature during natural snowfall and to examine the relationship between adfreeze strength, temperature change and the surface properties of various types of roofing. Surface temperature of the roof during snowfall ia measured and the adfreeze process of the snow on a roof based on the snow conditions on the roof is examined. Three models of adfreeze are presented and the relation between adfreeze strength and the surface property of various types of roofing is clarified.

2 TEMPERATURE OF ROOF AND ADFREEZE PROCESS

2.1 Measurement of roof temperature

To study how roof temperature during snowfall affects adfreeze of snow to roofing materials, a model house was built in Sapporo and roof temperatures were measured during snowfall. The roof of the model house had a slope roof which facilitated regular natural sliding. The temperature of the interior of the model house was maintained at 25 ℃ ~ 28 ℃ with an electric heater. Temperatures were measured at three points on the roof; the ridge, the center, and the eaves and in the attic, indoors, and outdoors. Because roof temperature was influenced by sunlight and wind, solar radiation, average wind velocity, maximum wind velocity.

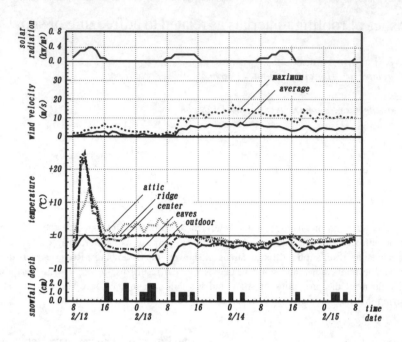

Figure 1. Variations in temperature and meteorological elements the model house
during period of snowfall.

2.2 Weather element about roof temperature

Roof temperature was affected by outdoor temperature, attic temperature, solar radiation,and wind velocity prior to snowfall. Temperature measurements and related meteorological observations during a three day period experiencing snowfall are shown in Figure 1. Roof temperatures at the ridge, the center, and the eaves were influenced by the following meteorological elements:

When there was no snow cover on the roof and there was sunlight before snowfall (12th from 9:00 a.m. to 2:00 p.m.), temperature changes at the three measurement points on the roof were similar. At 3:00 p.m. when the outdoor temperature fell and snowfall began, the temperature at the ridge and the center dropped to about 0 ℃ due to the influence of the attic temperature, but the temperature change at the eaves followed outdoor temperature due to the relative absence of attic temperature influence. Observation of snow conditions on the roof at each point showed that the snow at the ridge and at the center was melting and that there was little adfreezing. The snow on roof at the eaves was melting to a lesser extent and it frozen to the roofing. In this way, the roof temperature

immediately after the start of snowfall was highly related to the adfreeze process which subsequently occurred because it was mainly influenced by the quantity of moisture at the interface of the snow and the roofing materials. As for the attic temperature, it was above freezing at an average wind velocity of about 3m/s. At a wind velocity of 4m/s and above, attic temperature approximated outdoor temperature. Regarding this, the amount of ventilation in the attic increased with wind velocity. When the attic temperature declined in this way, the roof temperature dropped to the freezing point at the ridge and the center, too, an on the 13th ∼ 14th, and the snow on roof was in a state of adfreezing to the roofing. The above-mentioned roof temperature was mainly affected by outdoor temperature, solar radiation and wind velocity and great differences occurred at every part on the roof.

2.3 Condition of snow on roof

When evaluating the condition of snow on a roof based on the change of roof temperatures, the following should be kept in mind. In one case, the temperature at the eaves drops from positive to negative when snow falls. In another case, the temperature there is negative both before and after the start of snowfall. In the first case, adfreeze occurs on the roof surface where water from melted snow is present. In the second case, because the snow on the roof is dry due to its negative temperature, adfreeze is minimal. Conditions at the center and the ridge are similar to those at the eaves, but there are a few differences in temperature changes for several hours after thestart of snowfall. Positive temperatures at the center and ridge for several hours after the start of snowfall result in adfreeze since some snow is melted. Incidentally, the temperature at which adfreeze occurs this case is near 0 ℃.

During adfreeze, the snow at and near the roof surface has a water content which varies according to the melting conditions. There is mainly related to the adfreeze process which occurs on the roof.

2.4 Modeling of the adfreeze process

Three adfreeze processes between roofing and snow, based on our analysis of outdoor temperatures and roof temperatures, are shown in Figure 2. "Moist" adfreeze occurs when the outdoor temperatures is around 0 ℃ and when there is much melting of damp snow on the roof. "Dry" adfreeze occurs when relatively dry snow accumulates on the roof surface under outdoor temperatures of about -5 ℃. "Sub-moist" adfreeze, intermediate between "moist" adfreeze and "dry" adfreeze may also occur.

To remove the snow on a roof regularly by sliding, the adfreeze strength between the snow and the roofing must be evaluated for each adfreeze process shown in Figure 2.

3 ADFREEZE STRENGTH AND SURFACE PROPERTIES OF ROOFING

3.1 Method of measuring for adfreeze strength

Table 1 lists the roofing materials for which adfreeze strength was measured. They were chosen for their durability, light-permeability and other general qualities. The adfreeze strength was

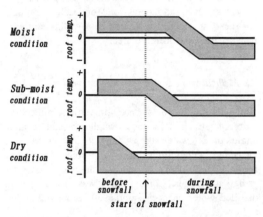

Figure 2. Three adfreeze processes based on roof temperature.

Table 1. Outline of roofing materials for evaluated surface properties.

sample code	roofing type	surface roughness Rz(μ m)
M1	membrane (teflon–coated glass fiber cloth)	21.84
P1	coated steel sheet (polyester resin, glossy)	1.16
P2	coated steel sheet (polyester resin)	2.64
P3	coated steel sheet (fluorine resin)	3.10
S1	stainless steel sheet	3.82
S2	coated stainless steel sheet (fluorine resin)	4.30
Z1	zinc sheet	7.89
G1	float glass sheet	0.03
G2	polished glass sheet	11.63

measured using the horizontal strength. sliding device shown in Figure 3. It delaminates the ice block frozen to the sample in shear and measures adfreeze strength (Jellinek 1959).

We established various ways to freeze ice to the samples. To represent "Moieter vinyl chloride pipe, and frozen. To represent "sub-moist" adfreeze, the same initial steps were done, but then the ice block was removed from the roofing sample. Its underside was sprayed with water then placed back on the roofing sample and allowed to freeze to it. In case of "dry" adfreeze, the ice block was placed on the roofing sample without spraying. The cooling temperature was -10 ℃ and -2 ℃. The cooling time was standardized in all three models as the time required for water to freeze in moist adfreeze. Adfreeze resistance was measured by inflicting a very low-speed horizontal load on each ice block. Adfreeze strength was found by dividing adfreeze resistance by the adfreeze area.

3.2 Method of measuring surface property

We measured two surface properties influencing adfreeze strength; surface roughness and contact angle of a water drop. Surface roughness was measured determining the ten point height of irregularities (Rz) of cross directions at six orthogonal points. Contact angle was measured with distilled water on each sample 60 seconds after fell a drop.

3.3 Moisture quantity of adfreeze interface and adfreeze strength

Test results for the coated steel sheet (sample P1), the float glass sheet (sample G1), and the teflon-coated membrane (sample M1) are shown in Figure 4. The sample represented in the figure is a material in which surface roughness(Rz) and material type differ. At both test temperatures (-10 ℃ and -2 ℃), the adfreeze strength is the greatest for the moist condition, followed by the sub-moist condition, and then the dry condition. The adfreeze strength varies according to the quantity of moisture at the interface and the temperature there.

When considering that the quantity of moisture at the interface influences adfreeze strength, the interface situation of the roofing and the ice block can be represented as in Figure 5. When sufficient moisture exists at an interface in the moist adfreeze

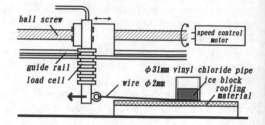

Figure 3. Device for measurement of adfreeze

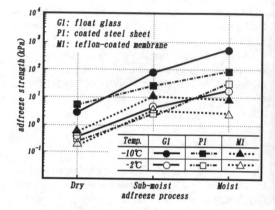

Figure 4. Adfreeze strength in each adfreeze process.

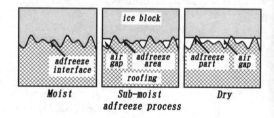

Figure 5. Interface condition of roofing and ice block in each adfreeze process.

situation, the ice block adfreezes according to the surface shape of the roofing material, and adfreeze strength is greatest because the area of adfreeze is biggest. In the dry adfreeze situation, because the adfreeze is partial and limited to the high points on the roofing, in the adfreeze strength is small because the adfreeze area is small. Sub-moist adfreeze is in the intermediate condition.

We found that the adfreeze area and adfreeze strength are influenced by the roughness of the surface and the quantity of moisture present.

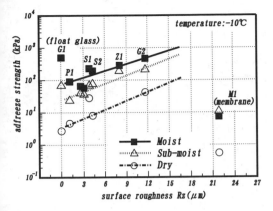

Figure 6. Relation between surface roughness and adfreeze strength.

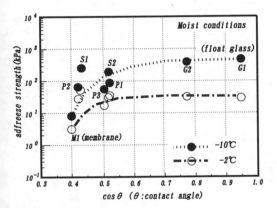

Figure 7. Relation between contact angle and adfreeze strength.

3.4 Relation between surface roughness and adfreeze strength

Figure 6 relates surface roughness (Rz) to adfreeze strength, for all samples except the float glass (sample G1) and the membrane (sample M1), the adfreeze strength increases as the surface roughness(Rz) increase. However, the surface roughness of the teflon-coated membrane is quite large, but the adfreeze strength is quite small. The rounded nature of the membrane's roughness partially explains this result. The float glass had the smallest roughness and the largest adfreeze strength. Surface properties other than roughness explain this tendency.

3.5 Relation between contact angle and adfreeze strength

The contact angle of a water drop is another surface property which can be used to evaluate wetness and adhesion to materials. The contact angle shows the water repellency of the material.

Figure 7 shows the relationship between contact angle and adfreeze strength under moist conditions for various types of roofing. As can be seen, adfreeze strength tends to be great for materials with small contact angles (i.e., large cosines). As shown in Figure 6, the float glass (sample P1) and the membrane (sample M1) are exceptions to the general relationship between surface roughness (Rz) and adfreeze strength. In the float glass, Rz is the smallest, the adfreezing strength is great, and the contact angle is the smallest. In the membrane, Rz is the greatest, adfreeze strength is small, and the contact angle is the greatest.

The reason for these results is as follows. The membrane, which is coated material, is remarkably superior to the other materials in water repellency. When ice adfreezes to it, the interface is not fully filled with water (i.e., there are mainly intervening gaps). We conclude that adfreeze strength to this material is influenced by the small area of actual adfreeze.

The adfreeze strength of roofing materials is varies with the contact angle of a drop and surface roughness(Rz). With knowledge of both the surface roughness and contact angle of a roofing material, the adfreeze strength of ice to roofing can be evaluated.

4 CONCLUSION

To examine the adfreeze of snow on a roof, we measured and analyzed roof temperatures during periods of snowfall, and concluded adfreeze experiments for various roofing materials. The following results were obtained:

1) Roof temperature after the start of snowfall is mainly influenced by meteorological elements and attic temperature.
2) Moreover, the quantity of moisture contained in the snow on a roof varies with temperature and is the main influence on the adfreeze strength of the snow on a roof.
3) The adfreeze strength of ice to roofing can be

213

evaluated using two surface properties, the surface roughness and the contact angle of a drop of water.

REFERENCES

Jellinek H.H.G., 1959, "Adhesive properties of ice," *Journal of Colloid Science*, 14, 268-280

Maeda H., 1979, "Condition of snow sliding on metal sheet roofs," *Journal of Japanese Society of Snow and Ice*, 41-3, 199-204

Shimizu M., and Kimura T., 1992, "Snow adhesion to roof sheets at sub-freezing temperatures," *Journal of Japanese Society of Snow and Ice*, 54-3, 269-275

Taylor D.A.,1985," Snow loads on sloping roofs:two pilot studies in the Ottawa area," *Canadian Journal of Civil Engineering*, 12, 334-343

Tomabechi T., Yamaguchi H., Ito T. and Hoshino M., 1991, "Fundamental Study on Sliding on a Roof," *Journal of Structural and Construction Engineering*, Architectural Institute of Japan, 426, 99-105

Ueno M., Takashima K., Takamura H. and Fukumoto H., 1987b, "Characteristic evaluation of roof sheet for snow sliding," *Journal of Japanese Society of Snowand Ice*, 49-3, 131-137

Watanabe M. and Hirai K., 1987a, "Study on friction between roofing materials and snow on roof," *Journal of Snow Engineering*, 3, 1-11

Snow Engineering: Recent Advances, Izumi, Nakamura & Sack (eds) © 1997 Balkema, Rotterdam. ISBN 90 5410 865 7

Japanese recommendation 1993 for snow loads on buildings

Hirozo Mihashi
Department of Architecture, Tohoku University, Sendai, Japan

Toru Takahashi
Department of Architecture, Chiba University, Japan

ABSTRACT: The basic concept and statistical background of the Architectural Institute of Japan's (AIJ) 1993 recommendation for snow loads on buildings are explained. Snow loads are based on snow depths on the ground with a return period of 100 years. It is recommended that the shape coefficient for complex or large scale roofs should be established by an experimental study with models. Remarkable systems for controlling roof snow have been developed in Japan recently. Where they are used, load reductions are allowed.

1 INTRODUCTION

Since the Japanese islands are surrounded by sea and scattered over a rather wide range of latitudes, snow-fall mechanisms and the snow depth distributions vary from region to region. Hokkaido and northwestern Honshu are covered with heavy snow during each winter. However, areas on the Pacific Ocean side sometimes receive heavy snow at the end of winter due to approaching low pressures in the temperate zone.

In spite of the climate difference from region to region and these two different snowfall mechanisms, the present recommendation uses unified concepts as much as possible in determining snow loads on roofs, on the basis of recent studies. A flowchart of the procedure for determining the snow loads on roofs is shown in Figure 1.1. The main portion of the procedure is subdivided into three parts: setting up the design concept on how to deal with roof snow loads, considering snow conditions at the construction site and considering snow conditions on the roof itself.

First of all, the designer must decide if an active snow control system is to be introduced to limit the snow load on the roof. If an appropriate system is provided, the design snow load on the roof may be reduced. Recent development of control system technologies have been remarkable in Japan, but there have been a lot of accidents in which buildings with control apparatuses have failed. Their reliability needs to be critically assessed, taking human error into account. This new AIJ recommendation is expected to encourage their improvement and use.

Snow loads on roofs vary as a function of the characteristic snow load on the ground, climate (including temperature and wind speed during the winter), roof shape and roofing material, and also from one winter

to another. Most available snow load data are of snow depths on the ground. The AIJ recommendation assumes that snow loads on roofs are proportional to that on the ground. Snow depth on the ground at the construction site is estimated by a statistical treatment and by an interpolation technique. The basic load value in this recommendation is based on the snow depth on the ground with a return period of 100 years. If a building is designed for another return period, a conversion factor is introduced to modify the value. The snow load on the ground is calculated by multiplying the snow depth by an equivalent snow density.

The snow load on the roof is estimated by multiplying the snow load on the ground by a ground-to-roof conversion factor, that is, a shape coefficient. Since the shape coefficient varies with roof shape, temperature, snow type (dry or wet), wind direction and wind speed, it is impossible to determine precisely. Therefore, it is recommended that the shape coefficient for the roof be estimated by an experimental study with models. However, tables in the AIJ recommendation give values for several typical shapes. For a large roof, even if the shape is simple, an experimental study is also suggested because shape coefficients for large roofs usually differ from those of small roofs.

2 SNOW LOADS ON THE GROUND

Because few direct observations of snow loads on roofs exist, snow loads on the ground are used as characteristic values. For ordinary buildings without snow control systems, the annual maximum load for the whole season is used as the characteristic snow load S_0, which is given by Equation (2.1) as the product of factors:

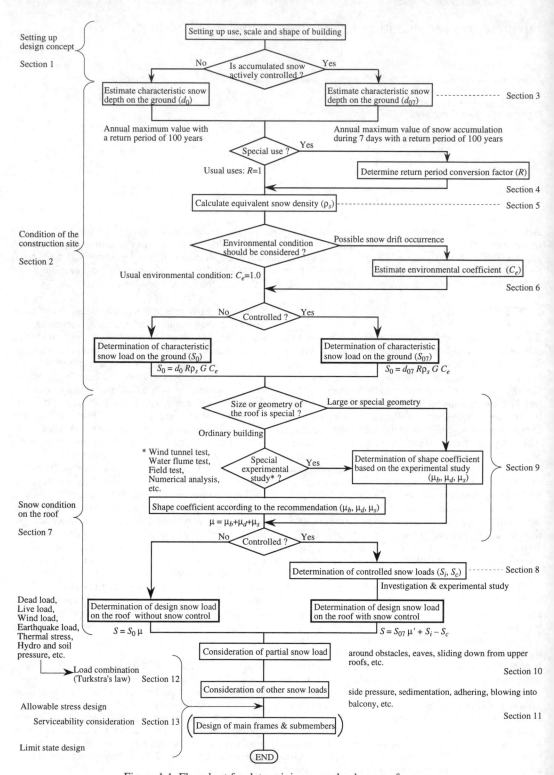

Figure 1.1 Flowchart for determining snow loads on roofs.

$$S_0 = d_0\, R\, \rho_s\, G\, C_e \qquad (2.1)$$

where S_0 = characteristic snow load (N/m^2) on the ground used for design when the snow load on the roof is not controlled; d_0 = characteristic snow depth (m) on the ground when the snow load on the roof is not controlled; R = return period conversion factor; ρ_s = equivalent snow density (kg/m^3); G = gravity acceleration (m/s^2); C_e = environmental coefficient. A heavy short-term snowfall produces the critical design condition for buildings whose snow loads are limited by reliable snow control systems. In such cases, the annual maximum snow load for the whole season is not used. Instead the annual maximum increase in snow depth during 7 days is used. For buildings with reliable control systems, the characteristic snow load on the ground S_{07} is given by Equation (2.2) as the product of several factors:

$$S_{07} = d_{07}\, R\, \rho_s\, G\, C_e \qquad (2.2)$$

where S_{07} = characteristic snow load (N/m^2) on the ground used for design when snow load on the roof is controlled; d_{07} = characteristic snow depth (m) on the ground when the snow load on the roof is controlled. Other factors are as described above for equation (2.1). All are discussed separately in the following sections.

3 BASIC SNOW DEPTH ON THE GROUND

For a roof without a snow control system, the characteristic snow depth on the ground d_0 is defined as the annual maximum value for the whole season (AMD) with a return period of 100 years. For a roof with a reliable snow control system, the basic snow depth on the ground d_{07} is defined as the annual maximum value of snow accumulation during a 7 day period (AMI-7) with a return period of 100 years. The AMI-7 is defined as shown in Figure 3.1. Because of the limited number of meteorological observatories, the basic snow depth at the construction site is estimated from topographical data.

Statistical properties of snow depth at 423 meteorological observing points in Japan were studied to estimate value with a long return period (Izumi *et al.* 1989). Five types of probability distribution functions were applied with two plotting techniques: Hazen and Thomas. Statistical properties in Japan were subdivided into three groups as shown in Figure 3.2. Finally, it was concluded that statistical properties of the AMD and/or AMI could not be explained by any single distribution or plotting technique. As a result, a new method was proposed in which a linear regression analysis was applied only to the maximum 1/3 of all the data plotted on Gumbel probability paper. In the AIJ recommendation, this method was used to estimate statistical values with a return period of 100 years for the AMD and for the AMI-7. Since the Gumbel distribution is given by Equation (3.1), the snow depth on the ground with a return period of r years ($10<r<200$) can be estimated by Equation (3.2).

$$F_X(x) = \exp[-\exp\{-a(x-b)\}] \qquad -\infty<x<\infty \qquad (3.1)$$

where, a = scale parameter; b = position parameter;

$$d = b + 1/a\,\ln(r) \qquad (3.2)$$

To clarify the difference between the AMD, AMI-3, and AMI-7, two contrasting examples plotted on Gumbel probability paper are shown in Figure 3.3. In cold Hokkaido, where snow continues to accumulate for many months, AMD is much larger than AMI-3 and AMI-7. In Gifu, located in the middle of Honshu, values of AMI are closer to values of AMD because this area is warmer, and snow depth increases and decreases rather rapidly. The relationship of AMI-7 to AMD at 423 observatories across Japan is shown in Figure 3.4.

It is usually impossible to obtain meteorological data over a long period at a construction site. Therefore, the AIJ recommendation gives an empirical equation for estimating the snow depth at a location without any observatories. According to previous studies (Takahashi *et al.* 1992), the dominant topographic factors influencing snow depth on the ground are altitude and sea ratio, which is defined as the ratio of sea area to total area around the site. In this recommendation, the annual maximum value of snow depth on the ground is estimated for a given point by:

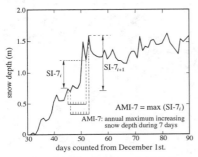

Figure 3.1 Definition of AMI-7.

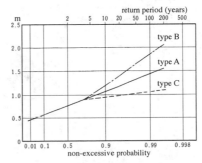

Figure 3.2 Type classification on Gumbel probability paper.

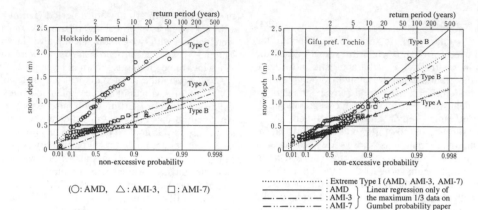

(○: AMD, △: AMI-3, □: AMI-7)

···············: Extreme Type I (AMD, AMI-3, AMI-7)
——————— : AMD   ⎫ Linear regression only of
—·—·—·— : AMI-3 ⎬ the maximum 1/3 data on
—··—··—·· : AMI-7 ⎭ Gumbel probability paper

Figure 3.3 Comparison of AMD, AMI-3 and AMI-7 (Gumbel probability paper, Hazen plot).

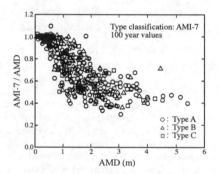

Figure 3.4 Relationship of AMI-7 to AMD across Japan.

$$d = (\alpha \cdot \text{altitude}) + (\beta \cdot \text{sea ratio}) + \gamma \qquad (3.3)$$

where α = coefficient for altitude; β = coefficient for sea ratio; γ = constant. The details are shown by Takahashi et al. (1996). Figure 3.5 shows simply the distribution of 100 year snow depths in Japan. Each

circle is plotted on the meteorological observatory point. Its radius indicates the snow depth.

4 RETURN PERIOD CONVERSION FACTOR

The return period conversion factor (R) is used when the design return period (r) is not 100 years. If the statistical properties follow a Gumbel distribution, Equation (3.1) gives the value of snow depth (d) with an arbitrary return period of r years. The factor R is the ratio of the snow depth (d) to the basic value of 100 years. R is described as a function of the standardized value (y_r):

$$R = b_0 + a_0 y_r \qquad (4.1)$$

where

$$y_r = -\ln[\ln\{r/(r-1)\}] \qquad (4.2)$$

When the return period is longer than about 10 years, the following approximation is available.

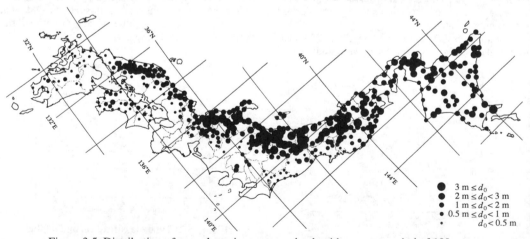

Figure 3.5 Distribution of annual maximum snow depth with a return period of 100 years.

$$y_r = \ln(r) \qquad (4.3)$$

5 EQUIVALENT SNOW DENSITY

Usually snow cover of the annual maximum depth comprises several layers, each of which has accumulated one after another and with varying properties from snowflakes to granular snow. Figure 5.1 shows typical examples of the relation between snow depth and the snow mass as the winter progress. Until the annual maximum snow depth is recorded, the mass increases in proportion to snow depth. After that, the mass still increases and then the maximum value is recorded, even though the depth decreases. Thus, there is usually a time lag between the annual maximum depth and the annual maximum mass recorded, though in most cases the meteorological observation data give only the snow depth. Thus, we need an equivalent snow density, which is the ratio of the annual maximum snow mass to the annual maximum snow depth.

Mihashi and Takahashi (1992) indicate that the formula proposed in ISO/DIS 4355 (1992) may give an overestimation for warm areas and for heavy snow areas where the maximum depth is above 2 meters. Statistical properties of the annual maximum snow depth and the annual maximum snow mass were studied for 12 meteorological observatories in Japan (Joh et al. 1993). Figure 5.2 shows the relation between snow depth and equivalent snow density for return periods of 100 years and 10 years. The relation is described by:

$$\rho_s = 73\sqrt{d} + 237 \qquad (5.1)$$

The equivalent snow density of snow accumulating for 7 days (d_{07}) is expected to be lighter. A significant amount of data have been analysed. It also indicates

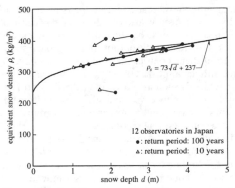

Figure 5.2 Relation between AMD and equivalent snow density (Joh et al. 1993).

the density of new snow is quite variable. Because of the variability the AIJ recommendation uses Equation (5.2) simplified from Equation (5.1), to calculate the snow load from both AMD and AMI-7

$$\rho_s = 73\sqrt{d} + 240 \qquad (5.2)$$

where $d = d_0 \cdot R$ without snow control, or $d = d_{07} \cdot R$ with snow control.

6 ENVIRONMENTAL COEFFICIENT

Snow loads on roofs vary according to various factors due to the environmental conditions. Table 6.1 shows environmental factors influencing the depth of snow on roofs.

Macrofactors are mostly included in the estimated depth of snow on the ground. Microfactors are too variable to consider in determining the design snow load. In this recommendation, only mesofactors are considered by the environmental coefficient (C_e).

The environmental coefficient (C_e) is generally defined as unity. When the snow depth on the ground is estimated to increase locally because of the environmental condition, C_e should be correspondingly larger than the unity.

7 DESIGN SNOW LOAD ON THE ROOF

Generally, design snow loads on roofs are estimated

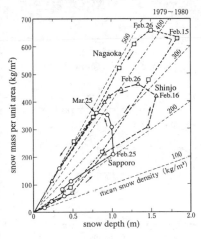

Figure 5.1 Relation between snow depth and snow mass.

Table 6.1 Environmental Factors Influencing the Depth of Snow on Roofs.

Macrofactors	distance from the nearest seaside, altitude, topographical inclination, topographical curvature, etc.
Mesofactors	trees, neighboring buildings, local topography, etc.
Microfactors	thermal condition on the roof surface, roofing materials, orientation of the roof, etc.

219

from ground snow loads. Snow loads on roofs are usually less than snow loads on the ground because of wind, sunshine, and so on. In addition, roof shape and slope influences the ratio of roof snow to ground snow. Therefore, shape coefficients μ and μ' are introduced, taking into account meteorological conditions. Partial snow loads such as snowdrifts caused by projecting structures, snow eaves, and sliding snow from upper roofs at eaves and/or lower levels of multilevel roofs are considered in section 10.

The equations used to determine design snow loads on roofs differ depending on whether or not snow on the roof is limited by snow control operations. If the roof snow is not controlled, the design snow load on the roof S is the characteristic snow load on the ground S_0 multiplied by μ:

$$S = S_0\,\mu \tag{7.1}$$

If the roof snow is controlled, S is determined as:

$$S = S_{07}\,\mu' + S_i - S_c \tag{7.2}$$

In this case, μ' might be different from μ. However, because S_{07} is determined from snow accumulation during 7 days, it is difficult to clarify the difference by wind tunnel tests or other estimation methods. That also depends on the method of snow control. Therefore, in the AIJ recommendation, μ' takes the same value as μ. When the snow load is controlled, S_i is the initial snow load of snow remaining on the roof when another heavy snowfall is expected. S_c is the snow load removed by a device whose performance is guaranteed even during a heavy snowfall. The performance should be estimated by research or experiments and then the value S_c should be determined.

8 CONTROLLED SNOW LOAD

There are three methods of artificially removing snow from a roof: 1. Mechanical snow removal (including snow removal with man power); 2. Positive snow sliding; 3. Snow melting. A combination of methods 2 and 3 is also used. In methods 1 and 2, snow is removed after it accumulates, and it is unclear how much snow has accumulated when removal starts. In method 3, there are two possibilities: all snow is melted and some snow remains. Therefore, design snow load on the roof should be decided by the planner. Controlling capacity and reliability against heavy snowfalls should be considered. When the performance of the snow control device is guaranteed even during heavy snowfalls, the controlled snow load S_c may be reduced from $S_{07}\,\mu'$ as expressed in Equation (7.2).

According to some research, a melting system that melts all the snow on a roof all the time consumes a lot of energy, and a melting system that only works after some snow has accumulated is inefficient because of snow caving between the snow and the roof (Mori-

no *et al.* 1984 and Nishi *et al.* 1986). A new method that consumes little energy (Ohtsuka *et al.* 1990 and Tomabechi *et al.* 1991). creates a thin ice layer which is then melted, inducing snow sliding.

9 SHAPE COEFFICIENT

According to recent work, the occurence of snow drifting on roofs is related to wind speed and temperature as shown in Figure 9.1 (Tomabechi *et al.* 1986 and Izumi *et al.* 1987).

The shape coefficient μ is defined by:

$$\mu = \mu_b + \mu_d + \mu_s \tag{9.1}$$

where, μ_b = basic shape coefficient; μ_d = shape coefficient for irregular distribution caused by snow drift; μ_s = shape coefficient for irregular distribution caused by sliding. The shape coefficient for large or special shaped buildings should be defined after special field research or experiments.

9.1 *Basic shape coefficient*

The basic shape coefficient μ_b is given in Figure 9.2. In the figure, wind speed V (m/s) is the average wind speed in January and February. For intermediate values of V, μ_b should be determined by interpolation.

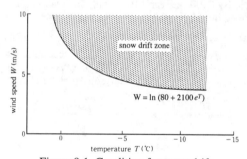

Figure 9.1 Condition for snow drift.

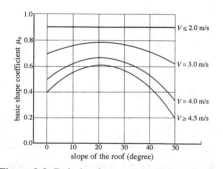

Figure 9.2 Relation between μ_b and roof slope.

Table 9.1 μ_d in the troughs of M-shaped roof, multiple pitched roof and multispan roof

slope of the roof	M-shaped roof and multiple pitched roof				multispan roof			
	average wind speed in Jan. through Feb.				average wind speed in Jan. through Feb.			
	≤ 2 m/s	3 m/s	4 m/s	4.5 m/s ≤	≤ 2 m/s	3 m/s	4 m/s	4.5 m/s ≤
≤ 10°	0	0	0	0.02	0	0	0	0.02
25°	0	0.02	0.14	0.20	0.10	0.22	0.34	0.57
40°	0	0.19	0.37	0.47	0.10	0.29	0.47	0.57
50° ≤	0	0.28	0.56	0.70	0.10	0.38	0.66	0.80

9.2 Shape coefficient μ_d for snow drifts

(1) Shape coefficients for snow drifts in the troughs of M-shaped roofs, multiple pitched roofs and multiple gable roofs are given in Table 9.1. At the ridge, μ_d should be zero. At the halfway point, μ_d is calculated by linear interpolation. For intermediate values of V, μ_d should be determined by interpolation.
(2) For multilevel roofs, distribution of shape coefficient for the lower roof should be determined from Figure 9.3. In the figure, μ_d at point O is given in Table 9.2. For intermediate values of V, μ_d should be determined by interpolation.
(3) If there is a parapet or similar structure at the edge of the roof, μ_d at the edge of the roof is defined by equation (9.2).

$$\mu_d = (h_p - \mu_b d)/d \quad (h_p \geq \mu_b d) \tag{9.2}$$

where, h_p = height of parapet. The shape of the snow drift is given in Figure 9.3. If $h_p < \mu_b d$, μ_d is zero.

9.3 Shape coefficient, μ_s, caused by sliding

Shape coefficient due to sliding on M-shaped roofs, multiple pitched roofs and multigable roofs are determined from either Equation (9.3) or (9.4) according to the roof slope. μ_s is positive in the troughs of these roofs and negative at the ridge. At the halfway point, μ_s is determined by linear interpolation. When the slope of the roof is between those defined in Equation (9.3) and Equation (9.4), μ_s is determined from sliding performance of the roofing materials.

1) When roof slope is smaller than 10 degrees,

$$\mu_s = 0 \tag{9.3}$$

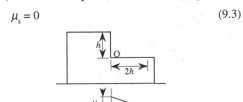

Figure 9.3 μ_d for multilevel roofs

Table 9.2 μ_d for multilevel roofs

average wind speed in Jan. and Feb.	≤ 2 m/s	3 m/s	4 m/s	4.5 m/s ≤
μ_d	0.10	0.30	0.50	0.60

2) When roof slope is larger than 25 degrees,

$$\mu_s = \mu_b \tag{9.4}$$

10 PARTIAL SNOW LOADS ON ROOFS

Snow loads on roofs increase locally because of snow drifts caused by projecting objects like chimneys, ventilators and so on. While this load might not be critical for the main frame, it should be considered when designing secondary members. Such loads are considered to be concentrated loads. To determine their location, wind direction should be considered. Snow eaves are sometimes up to 2 to 3 meters long. Therefore, they should be considered as concentrated loads with a density of about 400 kg/m³. Snow should be prevented from sliding first. However, if it cannot be prevented, the distance and impact of sliding should be considered. The impact force might be twice the gravity force.

11 OTHER SNOW LOADS

Side pressure might be assumed as follows, after Matsushita et al. (1963):

$$q = (15 \sim 20) \, G \, d^2 \tag{11.1}$$

where, q = side pressure [N/m²].
When the building might be buried under snow, snow load caused by sedimentation should be considered because eaves or braces at the outside might be broken.
   When snow adheres to the building or snow covers the building, snow mushrooms at the tops of the projecting objects sometimes grow to over 1 meter, and therefore cannot be neglected. When snow blows into balconies or outside corridors, it accumulates in sheltered zones.

12 LOAD FACTOR FOR LOAD COMBINATION

When Turkstra's rule (Turkstra 1972) is adopted, the load factor for snow load in allowable stress design under wind load, earthquake load, and thermal stress, may be given by the average value of snow load in stochastic processes. Matsushita and Izumi (1957) expressed changing snow loads as:

$$S_t = At^2(P - t) \tag{12.1}$$

$$A = \frac{27S_{max}}{4P^3} \tag{12.2}$$

where, A is a coefficient determined from snow cover duration P and maximum snow load S_{max}. After deriving these equations, they estimate that the snow load combined with earthquake load is lower than 21% of S_{max}. Kanda (1990) assumed that the stochastic process of snow load as rectangular, and estimates its load factor under allowable stress design. When snow cover duration is assumed as three months and average annual maximum snow depth is assumed as a half of the value for a return period of 50 years, the load factor might be 0.2 ~ 0.36. On the other hand, Suzuya et al. (1989) stated that when the real process is replaced by a rectangular process, 70% of real maximum snow depth might be reasonable for a rectangular process. Therefore, considering the uncertainty of the snow covering process, the AIJ recommendation applies a load factor of 0.3 for over three months of snow cover.

REFERENCES

ISO/DIS 4355 1992. Bases for design of structures — Determination of snow loads on roofs.

Izumi, M., Mihashi, H., Sasaki, T., Takahashi, T., Matsumura, T. 1987. Fundamental study on estimation of roof snow loads part 15, Summaries of Technical Papers of Annual Meeting, Vol. B, Architectural Institute of Japan (AIJ), pp.1405 – 1406 (in Japanese).

Izumi, M., Mihashi, H. and Takahashi, T. 1989. Statistical properties of the annual maximum series and a new approach to estimate the extreme values for long return periods, Proc. of 1st. Int. Conf. on Snow Engineering, CRREL Special Report 89-6, pp.25 – 34.

Joh, O., Sakurai, S. 1993. Equivalent snow density in heavy snowing area, Journal of Snow Engineering, Japan Society for Snow Engineering (JSSE), Vol.9, No.2, pp.112 – 114, (in Japanese)

Kanda, J. 1990. Consideration on load combination factor for snow load, Summaries of Technical Papers of Annual Meeting, Vol. B, AIJ, pp.127 – 128

(in Japanese).

Matsushita, S. and Izumi, M. 1957. Loads and actions for buildings part 3, load combination of earthquake and snow, Technical Reports of the AIJ, No.39, pp.1 – 4 (in Japanese).

Matsushita, S., Nakajima, H. and Izumi, M. 1963. A study on side pressure of snow, Transactions of the AIJ, No.89 (Summaries of Technical Papers of Annual Meeting) p.82 (in Japanese).

Mihashi, H., Takahashi, T. 1992. Field Measurement and Characteristic Analyses of Snow Load on Flat Roofs, Proceedings of 2nd. International Conference on Snow Engineering, pp.69 – 80.

Morino, K., Kobayashi, M., Takebayashi, Y., Kawai, M., Kawashima, M. 1984. A snow melting experiment on a pneumatic structure (air supported dome) No.2 a snow melting experiment by a wind box, Summaries of Technical Papers of Annual Meeting, Planning division, AIJ, pp.789 – 790 (in Japanese).

Nishi, Y., Nishikawa, K., Ishii, H. 1986. Analysis of snow melting process of the air supported structure -2, Summaries of Technical Papers of Annual Meeting, Vol. D, AIJ, pp.895 – 896 (in Japanese).

Ohtsuka, K., Joh, O., Homma, Y., Miyagawa, Y., Masumo, T. and Okada, H. 1990. A Study on the removal of snow from membrane structures, Proc. of Membrane Structures Association of Japan, No.4, pp.55 – 68 (in Japanese with English abstract).

Suzuya, J. and Kawana, H. 1989. Continuous process of snow depth and design snow load, Proc. of 5th. Symposium on Snow Engineering, JSSE, pp.81 – 84 (in Japanese).

Takahashi, T., Fukaya, M., Mihashi, H. and Izumi, M. 1992. Influence of altitude and sea area ratio for geographic distribution of snow depth, Summaries of Technical Papers of Annual Meeting, Vol. B, AIJ, pp.219 – 220 (in Japanese).

Takahashi, T., Mihashi, H. and Izumi, M. 1996. Influence of altitude and sea factor on the geographic distribution of snow depths in Japan, Proceedings of 3rd. International Conference on Snow Engineering, paper IIa 01. Rotterdam: Balkema.

Tomabechi, T., Izumi, M. and Endo, A. 1986. A Fundamental study on the evaluation method of roofsnowfall-distributions of buildings, Journal of Structural Engineering, AIJ, Vol.32B, pp.49 – 62 (in Japanese with English abstract).

Tomabechi, T., Yamaguchi, H., Ito, T., Hoshino, M. 1991. Fundamental study on sliding of snow on the roof of membrane structure, Journal of Structural and Construction Engineering, AIJ, No.426, pp.99 – 105 (in Japanese with English abstract).

Turkstra, C.J. 1972. Theory of structural design decisions, Solid Mechanics Study No.2, University of Waterloo, Ontario, Canada.

The new European Code on snow loads

Riccardo Del Corso & Luca Sanpaolesi
*Istituto di Scienza delle Costruzioni, University
of Pisa, Italy*

Manfred Gränzer
Landstelle für Bautechnik, Germany

Haig Gulvanessian
BRE, UK

Joel Raoul
SETRA, France

Rune Sandvik
NBR, Norway

Ulrich Stiefel
Gruner AG, Switzerland

ABSTRACT: Structural Eurocodes have been studied since many years by European Community EEC, as the new rules in the structural engineering field. These codes are nine and regard all the items of interest, from actions, to the reinforced concrete or steel structures, including the earthquake engineering.

In the Eurocode 1, EC1, "Action on Structures", there is a chapter regarding snow loads, which is described in the present paper.

1 PREMISE

The structural Eurocodes are the new European codes in the field of structural engineering. The first studies began more than 10 years ago and were aimed only at the design of concrete and steel structures. Thereafter a very large program was instituted which encompassed the whole of structural engineering, from loads to safety factors, to specific codes for design of the most common types of structures, including earthquake engineering.

There are nine structural Eurocodes and each of them has been studied by a specific Project Team made up of a group of European scientists, charged to produce the code itself, or part of it, with the cooperation of experts from all European countries.

At first the studies of individual topics were began under the aegis of the European Commission. In 1990 works were transferred to and developed by CEN (European Committee for Standardization).

In 1990 a specific Project Team (PT) was formed under the Convenorship of Professor Luca Sanpaolesi and charged with producing the chapter of Eurocode 1 dealing with Snow Loads. The PT was made up of: Luca Sanpaolesi (*University of Pisa - Italy*), Manfred Gränzer (*Landstelle für Bautechnik - Germany*), Haig Gulvanessian (*Building Research Establishment - United Kingdom*), Joel Raoul (*SETRA - France*), Rune Sandvick (*NBR, Norway*) and Ulrich Stiefel (*Gruner AG, Switzerland*). In addition the following

contributed to the research: John Tory (*Building Research Establishment - United Kingdom*), Diana Currie (*Building Research Establishment - United Kingdom*), Riccardo Del Corso (*University of Pisa - Italy*).

The first meeting was held in Pisa on April 25-26, 1991. Several drafts were prepared and subsequently improved based on snow loads studies and observations received from the 18 CEN countries. The final draft, with some matters not fully resolved, was approved on June 30, 1993 during the Berlin meeting of Sub-Commission - 1 Action on Structures of the Technical Committee of CEN - TC/250 Structural Eurocodes.

The code, as approved, is being published by the Standards Organizations of member States as ENV 1991-2-3 Eurocode 1.

2 THE NEW CODE ON SNOW LOADS

In carrying out its work, the PT studied and discussed many specific issues relevant to the various aspects involved in defining snow loads on structures. The scientific criteria followed by the PT in defining the Code has been based on the present state of the art, rather than specific reviews of existing codes. Nevertheless, during the actual drafting, particular attention has been paid to the ISO 4355 (1981), not to introduce its contents into the New Code, but to verify research results with existing ones. The new draft of ISO 4355, dated 1993, has not been taken into account, it being much too complicated.

The contents of Part 2-3 of the EC1: "Snow Loads" is as follows:

Foreword
    Objectives of Eurocodes
    Background of the Eurocode Programme
    Eurocode Programme
    National Application Documents (NAD's)
    Matters specific to this Prestandard
1 General
    1.1 Scope
    1.2 Normative References
    1.3 Distinction between principles and
        application rules
    1.4 Definitions
    1.5 Symbols
2 Classification of actions
3 Design situations
4 Representation of action
    4.1 Nature of the load
    4.2 Modelling of the load
5 Load arrangements
    5.1 Snow loads on roofs
    5.2 Snow overhanging the edge of a roof
    5.3 Snow loads on snowguards and obstacles
    5.4 Snow loads on bridges
6 Snow load on the ground - characteristic values
7 Snow loads shape coefficients
    7.1 General
    7.2 Pitched roofs
    7.3 Cylindrical roofs
    7.4 Abrupt changes of roof heights
    7.5 Drifting at projections and obstructions
ANNEX A: Characteristic values of snow load on ground (informative)
ANNEX B: Snow load shape coefficients for specific climatic regions (Normative)
ANNEX C: Adjiustment of the return period of ground snow load
ANNEX D: Adjustment of the return period of ground snow load (Informative)
ANNEX E: Bulk weight density of snow (Informative).

The fundamental paragraphs of the code discuss snow loads on the ground and snow loads on roofs and the parameters necessary for their determination. Nevertheless, as can be seen, many other topics are treated, such as definition of design situations, ultimate limit state and serviceability limit state; the determination of snow loads on bridges, making distinction between roofed bridges and open ones; snow overhanging the roof and drifting due to obstacles and obstructions, giving quite simple formulas to evaluate snow load to be considered in particular situations.

The Annexes also provide information on some relevant problems related to adjustments to code provisions for different needs, such as adjustment of the return period of ground snow load for temporary structures, modifications to take into account shape coefficients for specific climatic regions or the definition of a simple model for the evaluation of the snow load where there is an abrupt change in roof height.

The determination of roof snow loads has been adopted as a product formula as in the CEN and ISO 4355 formats. Factors are as follows:
- the characteristic ground snow load;
- shape coefficients, which depend on roof type, slope and, in certain cases, dimensions;
- exposure coefficient;
- thermal coefficient.

The first problem encountered in studying snow loads is, consequently, related to climatic conditions and to the need for quantitative definition of ground snow loads. The problem is quite complex and depends on several factors, such as a region's climate, altitude and topographic features, many of which affect the determination of snow loads. This leads to the need for more detailed studies in the field.

The soundest basis for assessing ground snow loads are long term records of snow loads measured at a large number of stations. Such a solid basis is difficult to achieve because of the scarcity, both in frequency and in geographical density, of available data, which have often been collected not with engineering objectives. Obtaining homogeneous measurements taken all over Europe is another problem. Different techniques, such as weighing snow cover and estimating the water equivalents from measurements of snow cover depth have been used. To date, most records are of snow depth.This has lead to the need to describe with empirical formulae, since no physical models exist which would permit this calculation, the correlation between snow depth and snow density. All factors which affect the deposition of snow, such as wind, temperature, rainfall onto snow and the nature of the snow layer need to be considered to be the result of multiple snow events in climates where the snow accumulates over a relatively long period of time. The snow cover can be considered the result of a single snow event, in climates where snow tends to melt completely between successive weather systems.

The ground snow load values to be used in the product formula mentioned above, are to be obtained by a statistical analysis of data. The snow load value for each station of measurement is to be based on an accepted annual probability of excedance, corresponding to a known mean recurrence interval.

Point values must be regionalized to arrive at a geographic representation of the measurement results.

Once the ground snow load is obtained the criterion suggested in the code generates the roof snow load by multiplying by a shape coefficient. It must be said that in CEN standard the exposure coefficient and the thermal coefficient are fixed at unity. The shape coefficient depends on roof shape. It has been determined by empirical research carried out in cold regions. Since results are not directly applicable to all of Europe, it has been necessary to develop empirical formulae, supported by experience and engineering judgement.

3 PROBLEMS TO BE INVESTIGATED

During EC1 work a large number of choices have been made and numerous unresolved problems have been found. At the end of this first stage it was thought fitting to compile the most significant issues encountered and publish them for the sake of interested parties. The aims envisioned by the PT in drafting this document, published as a"Background Document", are as follows:
- illustrate the underlying rationale for and the choices made in Eurocode 1: Snow Loads;
- provide information regarding the basic studies to National Competent Authorities;
- furnish broader guidelines and explanations to designers.

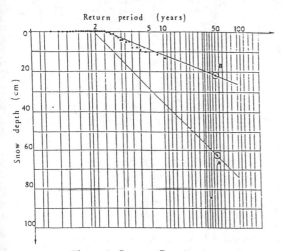

Figure 1. *Snow at Perpignan.*

Each important issue encountered in the EC1 phase is discussed in the Background Document. Information on each issue follows:

3.1 *Statistical analysis of snow loads*

Data records have to be treated with statistical procedures to establish design loads.

Statistical analysis is first applied to the record made at a single station alone. Daily registered snow load values combine to give a record for the whole winter. The values of interest are the yearly maximums. These extreme values, one for each winter, have a statistical distribution, which may be approximated by one of the well-known extreme value distribution functions.

The reliability of statistical analysis depends on the length of records. It has been proved, thanks to a German investigation based on 94-years of snow depth record, that the design value derived from samples of a floating period of 30 consecutive winters are not yet stable, but still influenced by exceptional years. Consequently, for the purpose for CEN Code, in which snow loads are given with a mean recurrence interval of 50 years, a record length of 40 to 50 years have been suggested for the statistical analysis of the collected data.

Compared to imposed or wind loads, snow loads may have a notably higher coefficient of variation. The smallest coefficients are found in mountainous regions where snow falls quite regularly and accumulates during the winter. In many areas, especially in coastal areas and in the southern part of Europe, snowfalls do not occur every year. If many zero values occur, the statistical analysis should lead to unrealistic results. In such cases the analysis should be restricted to the non-zero values only by adjusting of the return period, following procedures in the Annexes of the code.

Another problem which has been encountered is "exceptional snowfalls". These values are so high that they clearly do not fit the distribution calculated when they are discounted. A study, carried out in France, has shown the great influence that these values would produce on the distribution function's parameter if taken into account (Figure 1).

The PT has agreed to deal with exceptional snowfalls separately when determining its implications. A still open problem is the drawing of a European map to define all areas where exceptional snowfalls have to be considered.

3.2 The characteristic snow load and return period

The characteristic ground snow load is based on an annual probability of exceedance of 0.02 (1/50) This corresponds to a mean recurrence interval of 50 years as recommended in the EC1: "Basis of Design". The choice of a 50 year return period avoids inappropriate extrapolation from a data sample which generally covers several decades and represents a good agreement between all the return periods adopted in European country codes.

3.3 Regionalization

All the procedures and problems described above deal with the analysis of snow measurements at a single station, in order to find the characteristic ground snow load for that station. A procedure must be found to arrive at a geographic representation of results covering a region starting from point values obtained at observation stations. The following has been observed:

Mathematical approachs to this problem, through one of several existing methods, can give a continuos best fitting geographic distribution of characteristic snow loads. However, if such an automatic procedure completely ignores the knowledge and the experience of meteorologists it often furnish misleading results.

Sample data and the corresponding characteristic value obtained at a single station are influenced by several factors: orography, frontal waves, presence of great lakes, distance to the sea (macroscale effects); slope and contour of terrain, canopy and crop density (mesoscale effects); surface roughness and presence of obstructions (microscale effects). All these parameters have to be taken into account when making a snow load map.

It has been shown that very important parameters for local snow load variation are altitude, air temperature, solar orientation and wind exposure. In particular, it is often possible to arrive at a quite simple relationship between snow load and altitude alone, determining the "Altitude function".

Large European countries have used this method of zoning in their national codes. This simple procedure is not suitable for all European regions. As shown in a Norwegian study there are areas where snow load does not increase with altitude (Figure 2).

The PT has always aimed at the definition of general rules, applicable in all CEN member states, to achieve a homogeneous framework for determination of design snow load.

In the first phase of this work an attempt was made to collect existing snow load data from several European countries and to develop a new European snow map. Differences among countries in the measuring, collecting and elaborating that data, made it impossible to proceed in this direction. Thus the PT went back to the national codes re-elaborating them to achieve a common level of safety. This resulted in conceptual inconsistencies and differences

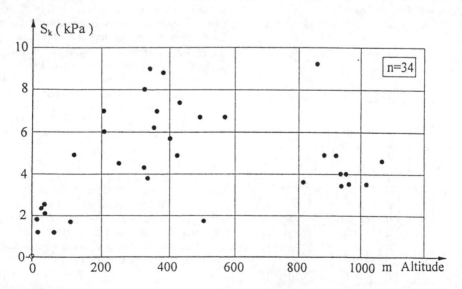

Figure 2. *Fifty year return period snow load for Norway, Hordaland. Poor correlation between snow load and altitude*

at the borderlines between the countries. This inconvenience had to be accepted.

A great research programme would be needed to develop a new European snow load map, based on common and homogeneous techniques all over the Europe. This research would update records of each country and standardise and simplify the application of the Code.

3.4 *Snow load on structures*

The roof snow load, as already mentioned, is normally calculated from the ground snow load by multiplying by conversion factors which account for roof shape, thermal characteristics, exposure and, depending on the code, other influences that may increase or decrease the roof snow load.

The scientific basis underlying determination of the roof coefficients is rather limited and most supporting research has been carried out in cold regions, thus these results are not directly applicable to all of Europe. It has been necessary to develop empirical formulae, supported by experience. Comparison of adopted criteria and parameters for the determination of snow loads on structures in the CEN and ISO 4355 Codes has been very useful (Figure 3).

In the determination of conversion factors there are three main sources of uncertainty: natural uncertainty, statistical and model uncertainty. Natural uncertainties cannot be dealt with;but treatment of statistical and modelling procedure

uncertainties would follow lines similar to that for determining ground snow loads. All the influences that affect ground snow load determination also affect the roof snow load, to which are added additional uncertainties related to the roof itself.

Statistical uncertainties begin with the difficulties of measuring directly the snow load on roofs and to the enormous number of different types of roofs. Although the code attempts to standardize such types, the huge number of existing roof shapes must be underscored.

As for ground snow loads, the problem of translating height into load, until new practical techniques of measurement are established, also exists for the roof snow load. The probability distribution function, or the probability model for analysis of sampled data has been studied only rarely. It is necessary to develop simple models which permit calculation of the design load, with respect to fixed levels of safety. Within a reasonable degree of uncertainty, the selection of two different loading types can be proved: a uniform and an unbalanced distribution of the snow layer.

The substantial lack of scientific knowledge on a probabilistic basis has emerged from elaboration of the shape coefficient within the EC1 work. Only further research will be able to reduce such uncertainties and therefore future efforts must be concentrated on this issue. Specific arguments which could be the object of this research are listed below:
- specific study of the shape coefficients for the more frequent types of roofs;

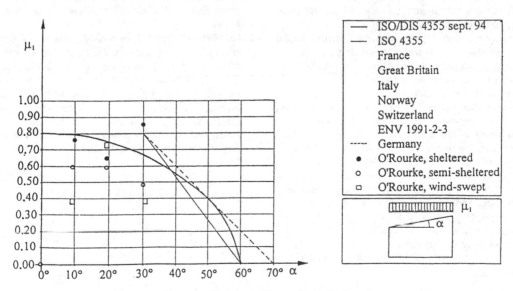

Figure 3. *Shape coefficient μ_1 for mono-pitched roofs*

- probabilistic basis (only with such research will it be possible to provide roof snow loads with a defined mean recurrence interval);
- shape coefficient for regions where single snowfalls create design loads.

3.5 *Design situations*

From the point of view of risk analysis it must be mentioned that the selection of relevant design situations is far more important than trying to develop "precise" partial factors. Therefore it is important to use good engineering judgement in selecting design situations that may occur and for which the design of the structure must be performed with reference to the SLS and ULS.

It has been found that particular attention has to be used: (1) in evaluating snow loads on multi-level roofs where drifts may form; (2) in determining snow distribution on roofs in windy regions where drifting predominates; (3) in establishing snow loads for structures without walls, where high snow loads, acting in conjunction with horizontal wind loads or horizontal earthquake motions may cause failures. For each of the above mentioned problems the code furnish indications, based upon research results or engineering practice.

The combination factors between different loads given in "Basis of Design" have been calibrated using national code values and general reflections. So far, no systematic checks have been performed for snow loads combined with other loads and no investigation has been made to evaluate modifications which may be required for different geographical regions.

In designing for serviceability, the functioning and appearance of a structure and its parts and the comfort of people must be achieved by checking the structure in appropriate load combinations similar to the ultimate limit state ones. The corresponding representative values are obviously dependent on snow dispersion, but are also strongly influenced by the duration of the snow cover on the ground, which depends on the region's climate. Research should be carried out over Europe to determine correction factors to be applied to snow load values for serviceability checks, distinguishing between short-term and long-term effects. Such research, performed only for Switzerland at present, would permit serviceability verifications with special reference to long term effects of great importance to some types of structures, for example, those made of timber.

4 CONCLUSIONS

The new recently published European code is being applied throughout Europe. It is published as ENV, the V indicating a provisional code. Conversion to EN (European Code) will soon begin, taking into account scientific results of recent European research.

Numerous open question persist in many areas. There is no doubt that the ENV-1991-2-3 represents a good base for harmonization of snow loads on structures. More detailed studies and research are needed in order to achieve more faithful standardization.

Other codes on snow loads, such as ISO 4355 (ed. 1993), do not solve the open questions mentioned above, for the extremely non homogeneous region of Europe, which extends from North Cape to Sicily. Climatic conditions in Europe vary widely.

The authors of this paper, PT's members, hope this work will be a useful step towards the enrichment of knowledge on European engineering applications and towards the integration of Europe.

REFERENCES

Del Corso, R. & Sanpaolesi, L. 1990. Stato degli studi per la definizione del carico da neve nelle norme europee. *Proc. CTE 1990*, Bologna, Italy. 123-127.

Del Corso, R. & Sanpaolesi, L. 1991. Il codice europeo sulle azioni e le problematiche del carico neve. *Proc. XIII Conf. CTA*, Abano Terme, Italy. 85-93.

ENV 1991-2-3, 1994. *Eurocode 1: Basis of design and actions on structures. Part 2-3: Snow loads.* CEN European Committee for Standardization, Brussels, Belgium.

Del Corso, R. & Gränzer, M. & Gulvanessian, H. & Raoul, J. & Sandvik, R. & Sanpaolesi, L. & Stiefel, U. 1995. New European code for snow loads: Background Document. *Proc. of the Struc. Eng. Dept. Univ. of Pisa*, N. 264, 1-76.

Snow Engineering: Recent Advances, Izumi, Nakamura & Sack (eds) © 1997 Balkema, Rotterdam. ISBN 90 5410 865 7

Lateral loads on structures due to snow avalanche impacts

Rand Decker
Department of Civil and Environmental Engineering, University of Utah, Salt Lake City, Utah, USA

ABSTRACT: Snow avalanches pose a significant threat to structures built in their paths. This threat is dominated by the lateral loads exerted on the structure by the initial impact of the avalanche, and the subsequent sustained loads associated with the avalanche flowing around the structure. There are few design code guidelines for establishing the lateral load environment on structures which are sited in avalanche paths. Typically, if such guidelines are available, they are in the range of two (2) tonnes per square meter (19.6 to 29.4 kN per square meter, or 410 to 615 lbf per square foot) and are applied to the structure as an equivalent, quasi-static load. This range of estimated lateral load due to snow avalanche impact commonly finds its way into practice. Recent and on-going experimental investigations of full scale avalanche dynamics on the joint Japan - US Alta Avalanche Impact Pylon, amongst others, shows that the peak impulse associated with the initial impact of an avalanche may be as much as an order of magnitude greater than 2 tonnes/meter squared. These peak impulses are very short lived; on the order of a few milliseconds. The sustained loads associated with the passage of the avalanche flow may be as large as 1/2 to 1/3 of the initial peak impulse, and have durations on the order of seconds. There is evidence from both visual observations as well as the impulse record that indicates that avalanche flow is more analogous to a dynamic wave than equilibrium flow. The significance of this is that the dynamic wave speeds will always be greater than the equilibrium flow velocity, and only approach the equilibrium flow velocity as flow durations progress over a relatively long period of time. One consequence of this hydraulic wave model for snow avalanche flow is that the dynamic wave will very quickly move to and become the leading edge of the flow. This will result in large peak impulses associated with the initial arrival of the dynamic wave. These peak impulses will be first order in the mixture density and the square of the wave speed. A dynamic or transient hydraulic wave model for avalanche flow, with resultingly large peak impulses, may require that the basis and methodologies for estimating lateral loads on structures sited in avalanche paths be reconsidered. This is particularly true in the rapidly developing Intermountain West of the U.S., where prior provision for avalanche hazard zones which preclude building are not in effect, and there are insufficient design codes, similar to those available for wind or earthquake, to guide the engineer.

1 INTRODUCTION

There is an extensive body of literature on experimental and analytic modeling efforts in snow avalanche dynamics (Abe 1991, Abe 1994, Brown 1992, Clayton 1992, Kawada 1989, Lang 1980a, Lang 1980b, Lang 1980c, Mears 1987, Mears 1990, McClung 1985, McClung 1990, McClung 1993a, McClung 1993b, Nishimura 1993, Norem 1985, Pedersen 1979, Perla 1980, Schaerer 1980, Shimizu 1980, Voellmy 1964). Many of these investigations were aimed specifically at considerations of or methodologies for estimating engineering parameters associated with avalanche flow, including; runout distances, and the impact pressures and loads imparted on structures as a consequence of avalanche impacts (Abe 1994, Kawada 1989, Lang 1980a, Lang 1980b, Mears 1987, Mears 1990, McClung 1985, Nishimura 1993, Norem 1985, Pedersen 1979, Perla 1980, Schaerer 1980, Shimizu 1980, Voellmy 1964). The earliest works in these areas provided the basis for the very successful European avalanche hazard zoning guidelines.

2 ANALYTIC MODELING AND AVALANCHE IMPACT LOADS

In the classic work; *On the Destructive Forces of Avalanches*, (Voellmy's 1964) recognizes that moving avalanches may be viewed as a homogenous flow amenable to treatment with the model equations of steady, uniform open channel hydraulics. The underlying premise of this model for avalanche flow is that the gravitational potential driving the flow is in balance with the various resisting forces of bed, wall, and internal friction. The net result is a methodology to determine the equilibrium or center-of-mass velocities of the avalanche during its decent, as well as the point at which the avalanche comes to rest or its runout distance. This equilibrium flow model for avalanche dynamics formed the basis for almost all subsequent modeling efforts for the next three decades. The resulting pressures, as a consequence of avalanche impact, where then correlated directly with the stagnation potential of the flow at the equilibrium flow velocities:

$$p_{max} = \rho_{aval} v^2 / 2g \qquad (1)$$

Early, successful efforts in avalanche hazard zoning in Europe used the impact pressure potential of the equilibrium flow velocities from Voellmy's model, applied to various specific sites, to establish the threshold of the "no habitable buildings" or "red" zones. Using the calculated equilibrium flow velocity, the resulting impact pressures were determined and the hazardous footprint of the avalanche path delineated in those areas where the impact pressures would exceeded 2 to 3 tonnes per meter squared (Frutiger 1970, Holler 1990). One consequence of these building restrictions was that the very construction needed to verify the calculations, by virtue of actual avalanche impacts on engineered structures, was precluded. In addition, in the 1950's, when these guidelines were being imposed, 2 to 3 tonnes per square meter was a daunting design load from the standpoint of materials and construction technique.

There is one very serious consequence of this early, Voellmy model based work that is now at issue. The rapid growth in alpine development, especially in the Intermountain West of the U.S., coupled with advances in construction materials and techniques has spurred design and erection of a number of "engineered", habitable dwellings and public buildings which, for lack of any other code or guideline, have been designed for lateral loads due to avalanche impact in the range of 3 to 5 tonnes per square meter. Though corroborated by the equilibrium flow models, these loads are far short of the peak loads found in the experimental record and in certain cases the results have been disastrous (Giraud 1994).

The general premises of Voellmy's analytic model for avalanche flow and the subsequent calculation of engineering products like runout distance and impact

forces were retained in more recent models of avalanche flow, including the popular two parameter model (Perla 1980). These two parameters are, respectively; the coefficient of friction and the ratio of mass to drag. The resulting model output includes the unsteady, uniform equilibrium velocities for the mass center of the avalanche flow.

In as much as the velocity history of the avalanche was the primary goal of the two parameter modeling effort, impact forces were not discussed. However, in subsequent efforts to apply this and other models which evolved from it (McClung 1990) the equivalent, quasistatic impact forces due to snow avalanche impact were determined from the stagnation potential of the equilibrium flow (Mears 1987, McClung 1993). The impact pressures which were determined from these equilibrium velocities were 11 and 5.37 tonnes per meter squared for sites in Japan and Utah, respectively.

In an effort to model phenomena related to viscous effects in avalanche flow, a wide variety of computational solutions of the Navier - Stokes equations, with various models and estimates for the viscosity of flowing snow were attempted (Brown 1992, Lang 1980a, Lang 1980b, Lang 1985, Pedersen 1979). In addition to the equilibrium velocity history of the avalanche, some of these modeling efforts had, as their specific goal, computation of the attendant impact loads (Lang 1980a, Lang 1980b). Impact pressures were determined from the stagnation potential of the normal component of the equilibrium flow velocities.

Unlike the previous steady and unsteady, uniform open channel hydraulic models, the Navier-Stokes derivative models do capture the phenomena of initial peak impact pressures associated with the avalanche impact. This result is ascribed to the shape of the leading edge of the flow at the time of impact, as well as both the variation in the depth of flow about the model obstacle and the reorientation of the initial flow from nearly normal to the obstacle to more nearly parallel to it. In the former case, the shape of the leading edge of the avalanche has been described as a "wave front" (Lang 1980a).

3 THE ALTA AVALANCHE IMPACT PYLON EXPERIMENT

The Alta Avalanche Impact Pylon investigation is an ongoing, joint Japan/U.S. project aimed at furthering our understanding of the basic features of snow avalanche flow. It is operated for the benefit of the snow and avalanche community at large by the Center for Snow Science at Alta (CSSA) and presently has the patronage of Cornell University, the Nagaoka (Japan) Institute for Snow and Ice Studies, the (U.S.) National Cooperative

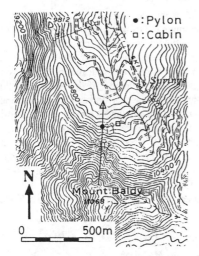

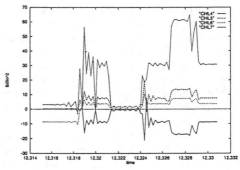

Fig.3. An impulse record

Fig.1. Topographical map around the observation site

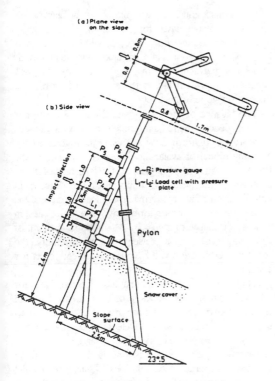

Fig.2. Pylon facility installed at Alta

West Baldy avalanche path, which releases both naturally and under the influence of explosives delivered as an element of the Alta Ski Lift company's ongoing operational avalanche control program. Instrumentation presently includes two load cells with impact plates of 78.5 cm squared (10 cm diameter) and three sets of paired/streamwise lagged pressure transducers with 1 cm squared impact surfaces. Data acquisition is via a PC, running the Labview application and hardware, located in an alpine cabin ~100m laterally from the avalanche path. This alpine station has year round, buried 220 vac power. A more detailed description of the Alta Avalanche Impact Pylon facility may be found elsewhere (Abe 1991).

Figures 1 depicts the physical surroundings of the pylon, while Figure 2 depicts the pylon and its instrumentation layout.

Figure 3 depicts an impulse record typical of the Alta Avalanche Impact Pylon site. The largest impulse records correspond to the lower and upper load cells (channels 4 and 5, respectively) and the upper pair of pressure transducers. Based on cross correlation of the pressure transducer signals, the velocity of this avalanche is ~33 m/s.

Of importance to the discussion of this paper, please note that the impulse record shows two distinct peaks. One at the leading edge of the avalanche and a second shortly thereafter. The peak impulse(s) are on the order of 42 tonnes per square meter (60psi). The sustained loads are about one half of the peak loads and are considerable higher than 2 to 3 tonnes per square meter guideline for equivalent, quasistatic lateral loads due to avalanche impacts.

Highway Research Program (NCHRP), the USDA Forest Service National Avalanche Center, and the University of Utah.

The pylon is an engineered facility (Clayton 1992) consisting of a 6m tall, slope normal, strutted steel structure of circular cross section. It is located in the

4 ADDITIONAL FIELD EXPERIMENTS FOR AVALANCHE IMPACT PRESSURE

There are a variety of field investigations, spanning several decades and using instrumentation that runs from purely mechanical to exotic, that have been carried out to investigate the dynamics of avalanche flow. One element common to many of the experimental investigations of avalanche impact forces is the observation that there are short lived peak impulses (maximum impact pressures) which exceed the sustained loads by a factor of 2 to 5. In a significant number of these investigations these peak impulses are an order of magnitude greater than 3 tonnes per square meter (Kawada 1989, Mears 1990, McClung 1985, Nishimura 1993, Norem 1985). In certain cases there is specific reference to "wave forms", "wave fronts", "surges" and "slugs" in these avalanche flows (Abe 1994, Kawada 1989, McClung 1985, Nishimura 1993, Norem 1985, Schaerer 1980, Shimizu 1980).

There are available explanations for peak pressures or impulses due to avalanche impacts, other than the existence of hydraulic waves in the avalanche flow. However, one very compelling argument for their existence is that in those field investigations where there are multiple sensors distributed on a stream normal line or plane, the peak impulse arrives at all of the sensors simultaneously (Kawada 1989, McClung 1985, Nishimura 1993, Shimizu 1980). This includes the impulse record depicted in Figure 3, obtained from the Alta Avalanche Impact Pylon. Conversely, if the sensors are distributed in the streamwise direction, the peak impulse wave is preserved and arrives at each sensor in turn [Kawada 1989, Shimizu 1980).

In the former case a density fluctuation which spans the dimension of the sensor array (on the order of meters) would produce the same result as a dynamic wave. However, that a density fluctuation could be preserved for several hundred meters during the course of its streamwise transport seems unlikely.

5 AN ARGUMENT FOR AN UNSTEADY, NONUNIFORM HYDRAULIC MODEL OF AVALANCHE FLOW

It is postulated here that dynamic wave growth, as a consequence of the unstable relationship between depth of flow and velocity of any step wise change in depth, is responsible for the peak impulse record found in avalanche field experiments. The hydraulics of these waves has been considered extensively in the analysis of the open channel dam break problems (Henderson 1966).

Hydraulically, any perturbations in depth of flow (or discharge; Q per unit width of flow) will result in a dynamic wave which may travel faster than the equilibrium velocity of the flow. If the equilibrium flow regime is accelerating, the dynamic wave will continue to grow in magnitude and velocity. Theoretically it will grow unbounded. In truth, it will eventually become unstable in wave length to amplitude and crest. Conversely, If the equilibrium flow velocity is constant the dynamic wave will (theoretically) remain constant in both amplitude and velocity. If the equilibrium flow is decelerating the wave will decrease in both amplitude and velocity until it asymptotically approaches both the equilibrium discharge (depth of flow) and velocity.

In the case of avalanche impact pressure, as with the sustained impact pressures associated with equilibrium flow, the peak impulses should correlate with the stagnation potential of the hydraulic *wave speeds*.

Consider the following expressions (Henderson 1966):

$$s_f - s_o + \partial y / \partial x + v / g(\partial v / \partial x) + 1 / g(\partial v / \partial t) = 0 \quad (2)$$
$$\partial Q / \partial x + B \partial y / \partial t = 0 \quad\quad\quad (3)$$
$$c = 1 / B(\partial Q / \partial y) \quad\quad\quad (4)$$

Equations 2 and 3 are recognizable as the energy and continuity balance laws for an incompressible, slope driven, open channel flow. They serve as a set of two independent equations in two unknowns; equilibrium velocity; v and depth of flow; y. The first two term of equation 1 represent the energy slope or potential, and the friction slope respectively.

The balance of the energy and friction slope terms, while neglecting the remain terms, results in the analysis of a steady, uniform flow. This is analogous to the Voellmy model of avalanche flow. The balance of the first four terms, while neglecting the last, results in the analysis of an unsteady, uniform flow. Successful consideration of all five terms would lead to an analysis of an unsteady, non-uniform flow, the only case in which a discharge or step wise perturbation in depth of flow may exist. As a consequence of continuity, the resulting perturbation will have a wave speed; c, as per equation 4.

These systems of equations have been solved for relatively simple hydraulics problems: Channels with simple, known geometric properties and limitations to the ratio of amplitude to wavelength (kinematic wave forms) of the perturbation. The latter allows for the linearization of the expressions and the resulting existence of characteristic solutions. Avalanche channel geometries may not render themselves amenable to simple descriptions. Similarly, the ratio of amplitude to wave length for discharge perturbations in avalanches may be very large and preclude linearization of the problem. However, these same issue are commanding the attention of the unsteady open channel flow community relative to the "near dam" problem and progress is being made by that community.

232

6 CONCLUSIONS

There is an extensive literature on analytic models for the equilibrium flow velocities of avalanches. The resulting impact pressures, as a function of the stagnation potential of the equilibrium flow, may be calculated. These impact pressures have led to a guideline for equivalent, quasistatic lateral loads due to avalanche impact on the order of 2 to 3 tonnes per square meter.

Field investigations of avalanche impact pressures often have peak impulses on the order of tens of tonnes per square meter and sustain loads that are one third to one half the peak values.

The impulse record of many of the field investigations shows that the peak impulses are synchronous on sensors which are widely space about a plane normal to the flow. In addition, some field investigations show that the peak impulse is also coherent relative to sensors that are separated in the streamwise direction. These observations, along with reports of higher than expected (relative to 2 to 3 tonnes per meter squared) amounts of damage and anomalous damage outside the avalanche debris zone lead to the postulate that avalanche flows are capable of generating dynamic waves akin to hydraulic waves in unsteady, non-uniform open channels. The resulting impact pressure from these avalanche would be first order in the stagnation potential of the wave speed.

Building on the analytic modeling of the dam break and near dam break problem by the open channel flow community, along with continued field investigations of avalanche flow, may produce the requisite additional insight needed to continue testing the postulate of dynamic wave forms in snow avalanche flow.

Advances in building technology, along with the growing demand for alpine development in avalanche terrain is creating the temptation to accept, design and build based on the advice that avalanche impacts can be treated as an equivalent, quasistatic lateral load on the order of a few to several tonnes per square meter, well below experimentally measured peak impulses of tens of tonnes per square meter. At present, engineers in the U.S. do not enjoy the same sound, coded guidelines and advice for design and construction in avalanche terrain that they do for similar natural hazard induced loads such as wind or earthquake.

7 REFERENCES

Abe, O., Nakamura, H., Sato, A., Numano, N., and Nakamura, T., 1991, Snow Block Impact Pressures Against a Wall, a Post, and a Disk, *Proceedings of the Japan - U.S. Workshop on Snow Avalanche, Landslide, Debris Flow Prediction and Control.*

Abe, O., Nakamura, T., Nohguchi, Y., Decker, R., Femenias, T., and Howlett, D., 1994, Observations of Snow Avalanches on Dynamic Internal Structures at Alta, Utah, *Snowbird ISSW Proceedings.*

Brown, R., 1992, A Review of Avalanche Dynamics Modeling, *Nagaoka International Symposium on Avalanche Control.*

Clayton, A., Decker, R., Richardson, C., and Abe, O., 1992, Installation Design of the Avalanche Impact Pylon Facility, *Breckenridge ISSW Proceedings.*

Frutiger, H., 1970, The Avalanche Zoning Plan, *Alta Avalanche Study Center Translation #11.*

Giraud, R., 1994, The Allen Residence, a Mountain Dream Home Destroyed by Avalanches, *Snowbird ISSW Proceeding.*

Henderson, F., 1966, Open Channel Flow.

Holler, P., 1990, Avalanche Protection and Avalanche Research in Austria, *Bigfork ISSW Proceedings.*

Kawada, K., Nishimura, K., and Maeno, N., 1989, Experimental Studies on a Powder Avalanche, *Annals of Glaciology, #13.*

Lang, T., and Brown, R., 1980a, Snow Avalanche Impacts on Structures, *J. of Glaciology, Vol. 25, #93.*

Lang, T., and Dent, J., 1980b, Scale Modeling of Snow Avalanche Impact on Structures, *J. of Glaciology, Vol. 26, #94.*

Lang, T., Nakamura, T., Dent, J., and Martinelli, M., 1985, Avalanche Flow Dynamics with Material Locking, *Annals of Glaciology #6.*

Mears, A., 1987, Snow Avalanche Loading Analysis: Catholic Chapel, Alta, Utah, *unpublished.*

Mears, A., 1990, Measurements of Avalanche Loads; East River Avalanche Shed, Colorado, *Bigfork ISSW Proceedings.*

Mears, A., 1992, Snow Avalanche Hazard Analysis for Land Use Planning and Engineering, *Colorado Geological Survey Bulletin #49.*

McClung, D., 1990, A Model for Scaling Avalanche Speeds, *J. of Glaciology, Vol. 36, #123.*

McClung, D., Kobayashi, S., and Izumi, K., 1993, Simulation of a Destructive Avalanche at Maseguchi, Japan, *Annals of Glaciology # 18.*

McClung, D., and Schaerer, P., 1985, Characteristics of Flowing Snow and Avalanche Impact Pressures, *Annals of Glaciology #6.*

McClung, D., and Schaerer, P., 1993, The Avalanche Handbook.

Nishimura, K., Maeno, N., Sandersen, F., Kristensen, K., Norem, H., and Leid, K., 1993, Observation of the Dynamic Structure of Snow Avalanches, *Annals of Glaciology #18.*

Norem, H., Kvisteroy, T., and Eversen, B., 1985, Measurement of Avalanche Speeds and Forces: Instrumentation and Preliminary Results of the Ryggfonn Project, *Annals of Glaciology #6.*

Pedersen, R., Dent, J., and Lang, T., 1979, Forces on Structures Impacted and Enveloped by Avalanches, *J. of Glaciology, Vol. 22, #88.*

Perla, R., Cheng, T., and McClung, D., 1980, A Two Parameter Model of Snow Avalanche Motion, *J.of Glaciology, Vol. 26, #94.*

Schaerer, P., 1992, Suggestions for Snow Research, *CSSA Snow Science; Reflection on the Past, Perspectives on the Future.*

Schaerer, P., and Salway, A., 1980, Seismic and Impact-Pressure Monitoring of Flowing Avalanches, *J. of Glaciology, Vol. 26, #94.*

Shimizu, H., Huzioka, T., Akitaya, E., Narita, H., Nakagawa, M., and Kawada, K., 1980, Study on High Speed Avalanches in the Kurobe Canyon, Japan, *J. of Glaciology, Vol. 26, #94.*

Voellmy, A., 1964, On the Destructive Force of Avalanches, *Alta Avalanche Study Center Translation #2.*

Snow Engineering: Recent Advances, Izumi, Nakamura & Sack (eds)© 1997 Balkema, Rotterdam. ISBN 90 5410 865 7

Snow load carrying capacity of single-layer latticed cylindrical roof structures

Seishi Yamada
Toyohashi University of Technology, Japan

Takashi Taguchi
Yahagi Construction Co., Ltd, Nagoya, Japan

ABSTRACT: Based upon our previous parametric nonlinear elastic and elasto-plastic buckling analyses, the present paper proposes how to estimate the snow load carrying capacity of rigidly jointed single-layer latticed cylindrical roofs. Simple equations and formulae obtained by using the reduced stiffness lower bound concept are presented. For structural design the present procedure is shown to not only have the advantages of allowing safe prediction of snow load carrying capacity but also provides a framework for actually making design decisions.

1 INTRODUCTION

A critical design condition for light-weight cylindrical roof structures is that of providing adequate resistance to buckling under snow loading. It has been widely recognized that under snow loading the nonlinear snap buckling exhibited by even geometrically perfect cylindrical roofs is very complex. The first author and co-workers have engaged in the buckling problems of cylindrical shells and latticed barrel vault roofs during the last decade, and have demonstrated that this class of space frame exhibits all the characteristics associated with thin shells (Yamada 1991). Also the previous studies have indicated (A) the effects of initial geometric imperfections on the buckling load and the buckling mode (Yamada & Croll 1989), (B) the good agreement between the critical load carrying capacity by using the *reduced stiffness method* and the lower bound to imperfection-sensitive, experimental or nonlinear numerical, buckling loads (Yamada & Taguchi 1994 and Yamada et al. 1993), (C) how to apply the reduced stiffness model to a prediction of design-load-carrying capacity for an elastic plastic structures (Yamada 1995a), and (D) the insensitivity of first full plastic lower bounds for imperfect latticed cylindrical

roofs to the irregular distribution of snow load (Yamada & Taguchi 1995). Based upon these mechanical features, the present paper provides an alternative simple procedure to estimate the snow load carrying capacity of single-layer latticed cylindrical roof structures for design.

2 THE PRESENT OBJECTIVE STRUCTURES

The present single-layer latticed cylindrical roof, of central angle ϕ, radius of curvature R, and length in longitudinal direction L, has an equilateral triangle network pattern of constitutive member length l as shown in Figure 1. All the joints are assumed to be rigid or nearly rigid.

The parameter q represents a load carrying capacity of the roof or a uniformly distributed vertical load which is simply modeled on total self-weight and snow loading. For the distributions of snow loading we can use various models as shown in Figure 2. In our previous studies (Yamada 1991 and Yamada & Taguchi 1995) it has been shown that for the buckling problem the uniform external loading system, case-E, provides the most conservative cri-

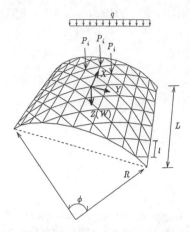

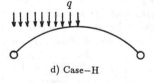

Figure 1. Single layer latticed roof.

terion. The effect of partial distributed loading, for example, half-loading (case-H), on the elastic buckling load carrying capacity is not very large. However, as it has been shown that a considerable stress concentration occurs in the case of partial distributed loading (Yamada & Taguchi 1995), the effect of the partial distributed loading is considered to be included in the estimation of stress concentration failure load carrying capacity q^c.

For boundary conditions, all support points of the present structure are assumed not to be able to move in the vertical direction. For the horizontal translation and/or for the rotation, we can choose a mechanically suitable model out of many; some of which are indicated in Figure 3. The roller support points of R-type in this figure can move in both X- and Y-directions. (Yamada & Taguchi 1994 has used this type.) In the other roller support model, S-type, which is approximate to a so-called

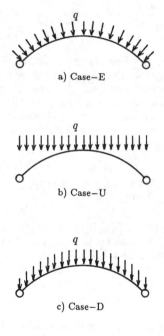

Figure 2. Loading models.

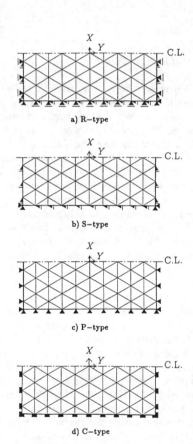

Figure 3. Boundary support models.

"SS3" boundary condition in the continuum shell analysis (Yamada 1991), the support points can only move horizontally in the orthogonal direction to the boundary beams except for the four fixed corner joints. The pinned support points of P-type are restricted against translation in any direction but are free to rotate (Yamada & Taguchi 1994). The clamped support points of C-type are restricted against rotation as well as translation in any direction (Yamada 1995b). It is seldom that the support point condition on the arch edge beams affects the buckling behavior (Yamada 1995b). On the other hand, the restraint of horizontal translation of the support points on the longitudinal edge beams makes the buckling load increase. To account for this, in the present estimation procedure, a "b factor" (see Section 3) is introduced. The effect of the rotation restraint of support points of latticed shells, which is different from that in continuum shells or plates, is negligible (Yamada 1995b). In view of obtaining a conservative criterion, all support points are now defined as rotationally free.

3 HOW TO DETERMINE BUCKLING MODES B_D

Our previous studies have shown that the elastic buckling lower bound, associated with the sensitivity to initial geometric imperfections, is closely related to the circumferential half-wave number n, which is generalized to be as $B \equiv nL/(R\phi)$. Therefore in the present estimation procedure of snow load carrying capacity summarized as shown in Figure 4, we first determine a buckling mode parameter for design, B_D, as follows.

• Calculate the fundamental buckling wave number B_0 from the equation

$$B_0 = K_0 - K_1 \times \log_{10}\beta + K_2 \times (\log_{10}\beta)^2$$

(1)

where β is the geometric parameter (Yamaki 1984 and Yamada 1991) and $\beta = 0.272L^2/(iR)$ for the equilateral triangle network pattern. Now i is the radius gyration of area of the mechanical representative

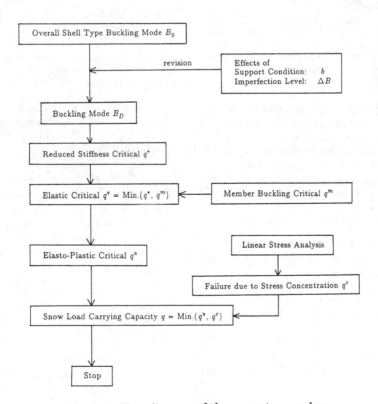

Figure 4. Flow diagram of the present procedure.

of its constitutive members. The coefficients K_0, K_1 and K_2 in Eq.1 are

for $1 \leq \beta < 10^3$;

$$K_0 = 1.029, \quad K_1 = 0.056, \quad K_2 = 0.444 \tag{2a}$$

for $10^3 \leq \beta < 10^5$;

$$K_0 = 10.371, \quad K_1 = 6.131, \quad K_2 = 1.431 \tag{2b}$$

Equation 1 is illustrated in Figure 5 by a solid line which is the exact analytical solution for pressure loaded cylindrical shells.

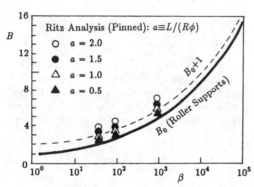

Figure 5. Buckling mode parameter B versus geometric parameter β.

• Determine the boundary factor associated with the condition of support points on the longitudinal edge beams, b;
$b=0$ for horizontally free rollers
$b=1$ for pinned or clamped supports
$0<b<1$ for horizontally elastic supports.
(When determining the value, we would require an additional experiment or numerical analysis for actually used supports.)

• Define the mode number reduction factor ΔB related to the accuracy of production and construction,
$\Delta B=0$ for low level imperfections
$\Delta B=0.5$ for moderate level imperfections
$\Delta B=1.0$ for high level imperfections

• Calculate the mode parameter for overall buckling from

$$B_D = B_0 + bL/(R\phi) - \Delta B \tag{3a}$$
$$\text{and} \quad B_D \geq 1 \tag{3b}$$

In Eq.3b, the condition $B_D = 1$ provides the minimum of the capacity formula, Eq.5, which will be written in the next section. The previous nonlinear finite element results in Yamada & Taguchi (1995) have suggested that Eq.3a the lower bound of B in some ranges of imperfections.

If the rigidity of the joints is doubted, we would need to examine the local (nodal) buckling mode written as

$$B_{local} = L/(\sqrt{3}\, l) \tag{4}$$

Then B_D may be revised, for instance, using a rotational stiffness parameter k $(0<k<1)$, as

$$B_D^r = \text{Min.}\{\ B_D\ , \quad kB_D + (1-k)B_{local}\ \}$$

This revision is on the basis of the reduced stiffness concept in which the smaller wave number (or the longer wave length) provides the more conservative lower bound of elastic buckling load carrying capacity. But, when determining the value of k, we would require an additional experiment or numerical analysis for actually used joints.

4 HOW TO ESTIMATE ELASTIC BUCKLING DESIGN LOAD q^e

Secondly, the elastic buckling snow load carrying capacity for design, q^e, would be determined as follows.

• Calculate the reduced stiffness lower bound to snow load carrying capacity q^* (see Eq.16 in Yamada 1995c) associated with B_D (Eq.3).

$$q^* = q^m \frac{49}{52}\left\{ \frac{l}{L}\left[B_D + \frac{1}{B_D} \right] \right\}^2 \tag{5}$$

where q^m is the member buckling load associated with Euler buckling for a pinned end column, and may be given by considering a pre-buckling uniform membrane stress state (Yamada 1995c) as

$$q^m = \sqrt{3}\ \frac{\pi^2 EI}{R\, l^{\,3}} \tag{6}$$

In Eq.6, EI is the bending stiffness of the constitutive member.

• Determine q^e from Eqs.(5) and (6) as

$$q^e = \text{Min.}(\ q^*\ ,\ q^m\ ) \tag{7}$$

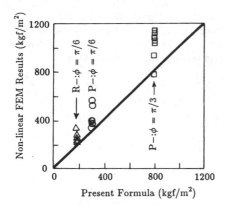

Figure 6. Confirmation of the conservativeness of the present formula, Eq.5.

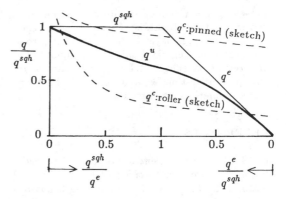

Figure 7. Criterion for each mechanism.

Figure 6 shows the relationship between the nonlinear numerical experiments and the present formula, Eq.5, for imperfect rigidly jointed latticed cylindrical roofs. The present overall buckling formula based upon a thin-walled shallow shell theory is in good agreement with the lower bound to scattered buckling load for imperfect space frames.

5 ELASTIC-PLASTIC BUCKLING DESIGN LOAD q^u

Finally, the elastic-plastic ultimate snow load carrying capacity for design, q^u, would be determined as follows.

• Calculate the plastic squash load, associated with the idealized compressed force of diagonal members, $N_d^F = -qRl/\sqrt{3}$, (by assuming uniform membrane stress state related to that in Eq.6) from the following

$$q^{sqh} = \sqrt{3}\,\frac{A}{Rl}\,\sigma_Y \qquad (8)$$

where $A \equiv$ the cross sectional area of the member and $\sigma_Y \equiv$ the material yield stress.

• Calculate the load carrying capacity q^u using q^e(Eq.7) and q^{sqh}(Eq.8) from the following

$$\frac{q^u}{q^e} + \left[\frac{q^u}{q^{sqh}}\right]^2 = 1 \qquad (9)$$

This elastic-plastic interaction formula is well-known as the "Dunkerley equation" and has been applied to reinforced concrete shell roofs

by IASS (1979) and to space frame design by Kato (1995).

6 EXAMINATION OF SAFETY

The above instability criterion q^u shown in Figure 7 should be compared with a boundary (or corner) zone failure criterion q^c which will be computed from a well-known linear stress analysis (Yamada 1995b) using, for example, the finite element analysis or the continuum shell analogy. Then we can examine whether the following equation

$$\text{Min.}(\,q^u\,,\,q^c\,) \geq \gamma_{self}q_{self} + \gamma_{snow}q_{snow} \qquad (10)$$

is satisfied or not, where q_{self} is self-weight which would be very small in such single-layer space frames, q_{snow} is snow load for design, and γ_{self} and γ_{snow} are the associated load factors which would be involved with the uncertainty of these irregular distributions. For the pinned support boundary, Yamada & Taguchi (1995) have already found $q^u < q^c$. However, for the roller support boundary, the failure criterion q^c, which has been called a "corner failure criterion" in Yamada (1995a), would be much smaller than q^u: for example, his results have shown $0.15 < q^c/q^{sqh} < 0.30$ even at $q^{sqh}/q^e = 1$.

7 CONCLUSIONS

Based upon our previous parametric nonlinear elastic and elasto-plastic buckling analyses, the present paper proposes an alternative new procedure for the estimation of snow load carrying capacity of single-layer latticed cylindrical roof structures having rigid joints. The quality of the results is enhanced by the comparison of the present estimates and some of the published numerical experiments.

REFERENCES

IASS. 1979. *Recommendations for Reinforced Concrete Shell and Folded Plates*, Madrid.

Kato, S. 1995. Elasto-plastic buckling behavior and estimation procedure of the buckling loads by using the generalized slenderness ratio. *Analysis and Design of Cylindrical Latticed Shell Roofs*, Committee of Shell and Spatial Structures, AIJ: 101-110. (in Japanese)

Yamada, S. and Croll, J.G.A. 1989. Buckling behavior of pressure loaded cylindrical panels. *J. Engrg Mech.*, ASCE, 115: 327-344.

Yamada, S. 1991. Relationship between nonlinear numerical experiments and a linear lower bound analysis using finite element method on the overall buckling of reticular partial cylindrical space frames. *Proc. the 4th ICCCBE*: 259-266.

Yamada, S., Uchiyama, K. and Croll, J.G.A. 1993. Theoretical and experimental correlations for pressure buckling of partial cylindrical shells. *Proc. SEIKEN–IASS Int. Symp.*: 151-158.

Yamada, S. and Taguchi, T. 1994. Nonlinear buckling response of single layer latticed barrel vaults. *Spatial, Lattice and Tension Structures*, *Proc. IASS–ASCE Int. Symp.*: 519-528.

Yamada, S. and Taguchi, T. 1995. Elastic buckling behavior of single layer latticed cylindrical roofs under snow loading. *J. of Snow Engrg of Japan*, 11: 11-22. (in Japanese)

Yamada, S. 1995a. Buckling analysis for design of pressurized cylindrical shell panels. *Proc. Int. Conf. on Stability of Struct.*: 9-23.

Yamada, S. 1995b. Linear stress and linear deflection behavior. *Analysis and Design of Cylindrical Latticed Shell Roofs*, Committee of Shell and Spatial Structures, AIJ: 53-60. (in Japanese)

Yamada, S. 1995c. Effects of imperfections. *Analysis and Design of Cylindrical Latticed Shell Roofs*, Committee of Shell and Spatial Structures, AIJ: 93-100. (in Japanese)

Yamaki, N. 1984. *Elastic Stability of Circular Cylindrical Shells*, North-Holland.

Snow Engineering: Recent Advances, Izumi, Nakamura & Sack (eds) © 1997 Balkema, Rotterdam. ISBN 90 5410 865 7

Influence of altitude and sea factor on the geographic distribution of snow depths in Japan

Toru Takahashi
Department of Architecture, Chiba University, Japan

Hirozo Mihashi
Department of Architecture, Tohoku University, Sendai, Japan

Masanori Izumi
Department of Architecture, Tohoku University of Art and Design, Yamagata, Japan

ABSTRACT: Multiple regression analyses were carried out to estimate design snow depths in areas without observatories. Target variables are AMD (Annual Maximum snow Depth) and AMI-7 (Annual Maximum Increase in snow depth during 7 days), which have been observed in observatories of the Japan Meteorological Agency. Independent variables are altitude and sea factor with radius of 20 km and 40 km of each observatory. Altitude generally has a large influence for snow depth, and sea ratio is influential especially for seaside areas.

1 INTRODUCTION

The Architectural Institute of Japan revised its recommendation for loads on buildings for the first time in twelve years in 1993 (AIJ 1993). One of the basic principles of the new recommendation is that basic snow load values for design be 100-year-return-period snow loads, based on depths of snow on the ground.

It is usually impossible to obtain meteorological data over a long period at a construction site. Several studies are available to estimate snow depths using topographical factors (Ishihara et al. 1972, Shibata et al. 1980, Takahashi et al. 1992) or altitude (Sack et al. 1986, Judge 1989, Zuranski 1989, Sandvik 1992). However, studies with many topographical factors are hard to apply for practical design. On the other hand, studies with a few topographical factors have been applied only for small areas in Japan (Nakatao et al. 1975). This study aims to estimate snow depth for the whole of Japan with a few topographical factors, for the sake of practical design.

2 ESTIMATION OF SNOW DEPTH

2.1 *Data on Snow Depth*

The snow depths used in this paper are observed at the 423 observation points of the Japan Meteorological Agency. The period of record is at least 15 years; the maximum is 68 years; and the average is 34 years. Target variables of the analyses are the 100 year mean recurrence interval extreme values of AMD (Annual Maximum snow Depth) and AMI-7 (Annual Maximum Increase in snow depth during 7 days). Extreme values for a mean recurrence interval of 100 years are estimated by the method of the authors (Izumi et al., 1989).

2.2 *Data on Topographical Factors*

According to previous studies (for example, Nakatao et al. 1975, Takahashi et al. 1992), the dominant topographical factors influencing snow depth on the ground are altitude and sea ratio, which is defined as the ratio of sea area to total area around the site, as shown in Figure 1. As shown in Figure 2, snow depth is almost proportional to the altitude where the site is inland and in a small limited area. In this figure, observation points are all located in one small ravine of the Mogami River, and the data, average of annual maximum snow depth, are observed at fields of primary schools or public halls in Nishikawa-cho, Yamagata Prefecture, for ten years. However, the altitude alone is not sufficient to interpolate in general, as shown in Figure 3. Therefore, the authors selected as independent variables, altitude and sea ratio with radius of 20 km and 40 km, of each observation point. Altitude of each observation point was obtained from the Monthly Report for each Prefecture by the Japan Meteorological Agency (JMA 1985). Sea ratio of each observation point was calculated by using digital national land information (Geographical Survey Institute, the Minis-

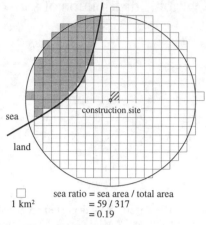

sea

land

□   sea ratio = sea area / total area
1 km²   = 59 / 317
             = 0.19

Figure 1. Definition of sea ratio
(Example: radius of 10 km)

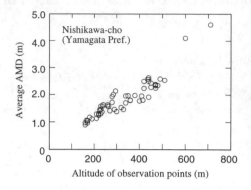

Figure 2. Relation between altitude of observatories
and snow depth for inland sites in Nishikawa-cho.

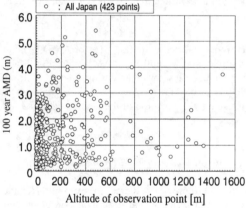

(a) Annual maximum snow depth (AMD)

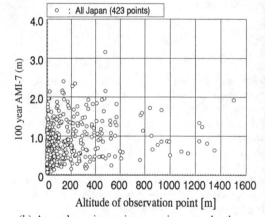

(b) Annual maximum increase in snow depth
during 7 days (AMI-7)

Figure 3. Relation between altitude of observatories and snow depth for all of Japan.

try of Construction 1985). When a designer wants to calculate these factors for a construction site, they can be obtained and calculated easily by using 1/25,000 and 1/100,000 scale topographical maps available at stores.

2.3 Multiple Regression Analyses

To estimate the influence of altitude and sea ratio, multiple regression analyses are carried out carefully. At first, all of Japan area was divided into 8 blocks as shown in Tables 1 and 2, and then they were divided into 40 areas for AMD and 35 areas for AMI-7, by trial and error. The maps for AMD and AMI-7 are shown in Figure 4. The number of divided areas is

about twice that estimated by the authors (Takahashi et al. 1992), which used ten topographical factors. By using a stepwise method, an empirical equation for each area has been given for estimating the snow depth at a location without any observatories. The annual maximum snow depth on the ground is estimated for a given point by:

$$d = (\alpha \cdot \text{altitude}) + (\beta \cdot \text{sea ratio}) + \gamma \qquad (2.1)$$

where d = snow depth; α = partial regression coefficient for altitude; β = partial regression coefficient for sea ratio; γ = constant. The values are given in Tables 1 and 2. In Hokkaido, observation points Shari and Rausu, which are located on the Shiretoko Peninsula, have a different nature from near sites; further, there are only two observation points in the area. Therefore,

242

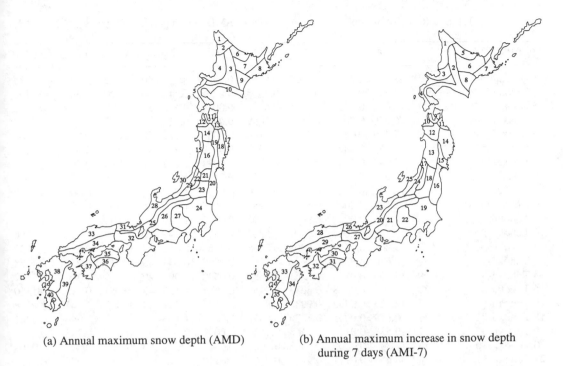

(a) Annual maximum snow depth (AMD)

(b) Annual maximum increase in snow depth during 7 days (AMI-7)

Figure 4. Area divisions used in the multiple regression analyses

they were not included in the analysis. The Tokai region has no observation points, but the equation estimated for the Kanto region might be used in practical design. In some cases, observation points near the boundary were used for both areas. Therefore, the total number of observation points shown in Tables 1 and 2 are more than 423.

3 RESULT OF ANALYSES

3.1 *Reliability of the Analyses*

In tables 1 and 2, results of the F-test, contribution ratio, partial regression coefficients and standardized partial regression coefficients are shown for AMD (Table 1) and AMI-7 (Table 2), respectively. For AMD, except of four inland areas, sea ratio is estimated as influential. In 30 areas, 75 % of the area, the F-test indicates highly significant results with the rejection probability at most 5 %. Sea ratio was judged as influential for AMI-7, except of one inland area. In 27 areas, 77 % of the area, the equations have been judged highly significant since the F-test rejection probability was at most 5 %.

3.2 *Verification of Fitness in the Figure*

Figure 5 shows AMD results for area number 10 (Hokkaido-10), which is located around Funka Bay. The map of observation points is shown in Figure 7. Observation points that are encircled by the sea, like as Erimomisaki, have small snow depths in spite of high altitude. That might be the influence of the warmth of the sea. The AMD estimated by Equation (2.1) almost coincides with the observed data as shown in Figure 5 (b).

Figure 6 shows the result of area number 25 (Chubu Hokuriku-6) for AMI-7. The map of observation points is shown in Figure 8. It is along the seaside of Niigata Prefecture. In Sado Island, the altitude of Aikawa is higher than that of Ryotsu, though, the AMI-7 at Aikawa is smaller than that of Ryotsu. In the flat land region of Niigata Prefecture, the more inland, the deeper snow depth. The equation estimates these properties. On the other hand, for heavy snowy points on the seaside like as Kashiwazaki and Takada, it under estimates. It might be that the tolerance depends on the limited number of topographical factors. In the practical design, designers should use these results with understanding of their tolerance and limitation.

Table 1. Result of multiple regression analysis of AMD, return period: 100 years.

No. Area name	N	F-val.	prob.	R²	Adj-R²	α	β rad.	γ	α (std.)	β (std.)
1 Hokkaido-1	5	3.217	0.2371	0.7629	0.5257	0.1063	3.160 40	-0.892	1.8933	1.2434
2 Hokkaido-2	5	2.539	0.2826	0.7174	0.4348	0.0215	-0.616 20	2.420	0.3663	-0.4999
3 Hokkaido-3	11	12.036	0.0039	0.7506	0.6882	0.0030	9.461 20	1.330	0.6791	0.2562
4 Hokkaido-4	14	16.676	0.0005	0.7520	0.7069	0.0105	0.410 40	1.557	0.8737	0.1283
5 Hokkaido-5	9	7.499	0.0233	0.7143	0.6190	-0.0045	-2.126 20	2.602	-0.0897	-0.8568
6 Hokkaido-6	7	4.953	0.0828	0.7123	0.5685	-0.0079	-3.797 40	3.311	-1.4713	-2.0567
7 Hokkaido-7	8	3.084	0.1341	0.5523	0.3732	0.0021	0.172 20	0.895	0.7965	0.1098
8 Hokkaido-8	6	17.634	0.0219	0.9216	0.8693	0.0111	-1.165 20	1.523	0.3799	-0.7675
9 Hokkaido-9	4	1.373	0.5167	0.7330	0.1990	0.0120	1.060 20	1.205	0.7803	0.5597
10 Hokkaido-10	12	9.189	0.0067	0.6713	0.5982	0.0010	-1.038 20	1.367	0.1005	-0.7878
11 Tohoku-1	5	184.795	0.0054	0.9946	0.9892	0.0005	-1.169 20	2.194	0.0232	-1.0077
12 Tohoku-2	5	3.768	0.2097	0.7903	0.5805	-0.0317	1.296 20	2.427	-1.2925	0.7580
13 Tohoku-3	7	9.800	0.0287	0.8305	0.7458	0.0155	0.612 40	0.370	0.7960	0.2532
14 Tohoku-4	7	8.670	0.0351	0.8126	0.7188	0.0052	0.641 40	1.122	0.9591	0.1168
15 Tohoku-5	9	14.150	0.0054	0.8251	0.7668	0.0342	-2.095 20	1.756	0.4707	-0.5162
16 Tohoku-6	13	8.204	0.0078	0.6213	0.5456	0.0056	1.125 40	1.864	0.8158	0.1550
17 Tohoku-7	5	12.125	0.0762	0.9238	0.8476	-0.0144	5.821 20	-0.857	-1.0615	1.0136
18 Tohoku-8	11	10.679	0.0055	0.7275	0.6594	0.0041	1.162 40	-0.113	1.0404	0.4221
19 Tohoku-9	11	9.588	0.0128	0.5158	0.4620	0.0022	0. -	0.646	0.7182	0.
20 Tohoku-10	18	13.346	0.0005	0.6402	0.5922	0.0021	0.175 40	0.193	0.8468	0.0763
21 Tohoku-11	4	88.658	0.0111	0.9779	0.9669	0.0110	0. -	-0.411	0.9889	0.
22 Tohoku-12	7	2.819	0.1723	0.5850	0.3774	0.0031	-5.297 20	2.801	0.5963	-0.2465
23 Tohoku-13	7	6.830	0.0513	0.7735	0.6603	0.0029	25.552 40	0.376	0.9910	0.3087
24 Kanto-1	16	34.013	0.0001	0.8396	0.8149	0.0006	-0.066 40	0.306	0.8980	-0.0329
25 Chubu Hokuriku-1	25	29.691	0.0001	0.7297	0.7051	0.0058	3.302 40	0.323	0.9036	0.1762
26 Chubu Hokuriku-2	17	78.828	0.0001	0.8401	0.8295	0.0021	0. -	-0.181	0.9166	0.
27 Chubu Hokuriku-3	8	12.046	0.0122	0.8281	0.7594	0.0006	6.964 40	0.126	0.7748	0.3269
28 Chubu Hokuriku-4	27	41.964	0.0001	0.7776	0.7591	0.0039	-2.590 40	3.017	0.4845	-0.5054
29 Chubu Hokuriku-5	9	44.399	0.0003	0.9367	0.9156	0.0111	-1.330 40	2.535	0.8671	-0.1647
30 Chubu Hokuriku-6	12	21.531	0.0004	0.8271	0.7887	0.0058	-3.576 20	2.937	0.0708	-0.9053
31 Kinki-1	8	5.927	0.0479	0.7033	0.5847	0.0084	1.681 40	0.689	0.9306	0.2741
32 Kinki-2	8	6.069	0.0489	0.5029	0.4200	0.0010	0. -	0.227	0.7091	0.
33 Chugoku-1	44	36.590	0.0001	0.6409	0.6234	0.0040	0.772 40	0.290	0.9206	0.1937
34 Chugoku-2	19	17.235	0.0001	0.6830	0.6434	0.0004	-0.228 40	0.370	0.5231	-0.3635
35 Shikoku-1	9	13.612	0.0059	0.8194	0.7592	0.0012	-0.473 20	0.446	0.6347	-0.3729
36 Shikoku-2	7	7.533	0.0440	0.7902	0.6853	0.0004	-0.717 40	0.314	0.5099	-0.4357
37 Shikoku-3	13	13.250	0.0015	0.7260	0.6712	0.0015	-0.770 20	0.536	0.5579	-0.4012
38 Kyushu-1	20	25.145	0.0001	0.7474	0.7176	0.0007	-0.103 20	0.230	0.8166	-0.1237
39 Kyushu-2	7	3.312	0.1418	0.6235	0.4352	0.0003	-0.048 20	0.106	-0.2095	-0.8568
40 Kyushu-3	5	40.537	0.0241	0.9759	0.9518	-0.0001	-0.362 20	0.509	0.8096	-0.2256

3.3 Consideration of Partial Regression Coefficients

In Figure 9, relation of regression coefficients for altitude and sea ratio are shown for AMD (Figure 9 (a)) and AMI-7 (Figure 9 (b)). Generally, coefficients for altitude indicate plus values and those of sea ratio indicate minus values, though these variances are large. The influence of altitude is smaller for AMI-7 than AMD, in general. That means that the increasing snow depth for 7 days is not as varying for a low area as for a mountainous area. This is well known as *Satoyuki* in Japan; which means that sometimes the short term snowfall is heavier in low areas than in mountainous areas.

Figure 10 shows the relation between standardized partial regression coefficients for altitude and those of

No. Area name	N	F-val.	prob.	R^2	Adj-R^2	α	β	rad.	γ	α (std.)	β (std.)
1 Hokkaido-1	14	5.158	0.0263	0.4839	0.3901	0.0032	-0.330	40	1.000	0.3702	-0.3690
2 Hokkaido-2	19	12.196	0.0006	0.6039	0.5544	0.0008	-0.279	40	0.723	0.5145	-0.3744
3 Hokkaido-3	10	2.999	0.1146	0.4615	0.3076	0.0053	-0.449	40	1.030	0.7019	-0.3394
4 Hokkaido-4	14	6.876	0.0116	0.5556	0.4748	0.0012	-0.527	20	1.148	0.3709	-0.4760
5 Hokkaido-5	7	20.114	0.0082	0.9096	0.8643	-0.0024	-1.851	20	1.649	-0.4588	-1.3242
6 Hokkaido-6	8	1.999	0.2301	0.4444	0.2221	0.0009	-0.139	20	0.732	0.5786	-0.1429
7 Hokkaido-7	6	5.727	0.0945	0.7925	0.6541	-0.0021	-0.529	40	1.169	-0.1914	-0.9334
8 Hokkaido-8	5	12.407	0.0746	0.9254	0.8508	0.0019	-0.188	40	1.016	0.8214	-0.2757
9 Tohoku-1	7	3.396	0.1374	0.6294	0.4441	-0.0008	0.333	40	0.721	-0.0999	0.7522
10 Tohoku-2	5	1.887	0.3464	0.6536	0.3072	-0.0153	0.601	20	1.256	-1.1296	0.5542
11 Tohoku-3	10	10.085	0.0087	0.7424	0.6687	0.0036	-0.237	40	0.735	0.8698	-0.1512
12 Tohoku-4	7	7.163	0.0476	0.7817	0.6726	0.0015	-0.193	20	0.845	0.8369	-0.1111
13 Tohoku-5	19	14.869	0.0002	0.6502	0.6064	0.0010	-0.936	20	1.045	0.3060	-0.6037
14 Tohoku-6	19	7.890	0.0041	0.4965	0.4336	0.0023	1.396	40	0.130	0.7686	0.8583
15 Tohoku-7	12	6.729	0.0163	0.5993	0.5102	0.0009	-0.590	40	0.614	0.2325	-0.6278
16 Tohoku-8	19	18.887	0.0001	0.7025	0.6653	0.0015	0.045	40	0.262	0.8543	0.0246
17 Tohoku-9	9	29.703	0.0008	0.9083	0.8777	0.0005	-2.645	20	1.638	0.1707	-0.8349
18 Tohoku-10	12	14.679	0.0015	0.7654	0.7132	0.0009	5.427	40	0.835	0.8128	0.1899
19 Kanto-1	16	35.657	0.0001	0.8458	0.8221	0.0004	-0.085	40	0.321	0.8885	-0.0552
20 Chubu Hokuriku-1	25	22.691	0.0001	0.6735	0.6438	0.0025	1.579	40	0.540	0.8713	0.1866
21 Chubu Hokuriku-2	17	41.089	0.0001	0.7326	0.7147	0.0010	0.	0	0.156	0.8559	0.
22 Chubu Hokuriku-3	8	9.569	0.0195	0.7929	0.7100	0.0002	10.276	40	0.328	0.4284	0.6850
23 Chubu Hokuriku-4	27	25.237	0.0001	0.6777	0.6509	0.0019	-0.704	40	1.642	0.5836	-0.3316
24 Chubu Hokuriku-5	9	6.342	0.0331	0.6789	0.5718	0.0045	0.908	40	1.277	0.9520	0.3047
25 Chubu Hokuriku-6	12	13.356	0.0020	0.7480	0.6920	-0.0012	-2.032	20	2.080	-0.0249	-0.8650
26 Kinki-1	8	10.026	0.0178	0.8004	0.7206	0.0041	1.344	40	0.490	1.0112	0.4855
27 Kinki-2	8	3.892	0.0957	0.6089	0.4525	0.0011	-0.669	40	0.233	0.7657	-0.4393
28 Chugoku-1	44	15.239	0.0001	0.4264	0.3984	0.0013	0.174	40	0.529	0.7210	0.1060
29 Chugoku-2	19	16.516	0.0001	0.6737	0.6329	0.0004	-0.173	20	0.337	0.5788	-0.3383
30 Shikoku-1	9	10.427	0.0112	0.7766	0.7021	0.0010	-0.464	20	0.447	0.5761	-0.4097
31 Shikoku-2	7	7.598	0.0434	0.7916	0.6874	0.0005	-0.368	40	0.197	0.5028	-0.4439
32 Shikoku-3	13	10.330	0.0037	0.6738	0.6086	0.0010	-0.724	20	0.539	0.4514	-0.4759
33 Kyushu-1	20	6.533	0.0079	0.4346	0.3681	0.0003	-0.142	20	0.256	0.5468	-0.2318
34 Kyushu-2	7	3.858	0.1166	0.6586	0.4879	0.0003	-0.048	20	0.106	-0.2009	-0.8774
35 Kyushu-3	5	40.537	0.0241	0.9759	0.9518	-0.0001	-0.368	20	0.508	0.8096	-0.2256

sea ratio. In many areas, the coefficient for altitude is larger than that of sea ratio and it indicates a plus value. In more than half of areas, the coefficient for sea ratio is minus. That means that increasing sea ratio decreases snow depth.

There are some areas whose regression coefficient of altitude is minus, for example, area number 12 and 17 in AMD and area number 9 and 10 in AMI-7. This might be caused by a unique land form in those areas. Let us consider area number 12 in AMD and area number 10 in AMI-7. They are the same area in Aomori Prefecture, as shown in Figure 11. There, observation point Fukaura is located seaside at an altitude of 66 m.

On the other hand point Goshogawara is located inland but its altitude is only 9 m. Values of AMD and AMI-7 of Goshogawara are larger than those of Fukaura. It is the same with area number 10 in AMD as shown in Figure 5. However, more inland and higher points such as Hirosaki and Kuroishi have lower snow depths than Goshogawara. That suggests that this area is a special condition. These points are located in the same basin of the Iwaki River. The reason for such a complicated condition might be the existence of the hilly region of Tsugaru, located east of Goshogawara, or Mount Iwaki, located west of Hirosaki.

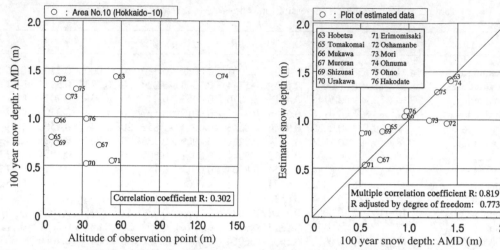

Figure 5. Example results of the multiple resression analysis (AMD).

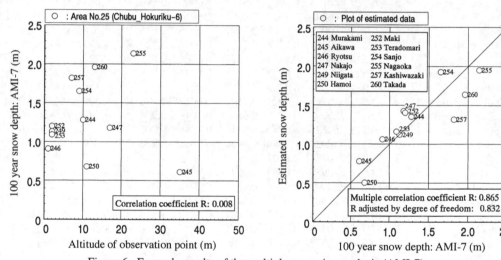

Figure 6. Example results of the multiple resression analysis (AMI-7).

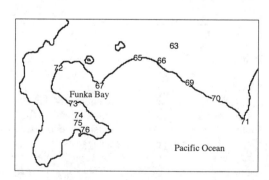

Figure 7. Map of observation points corresponding to Figure 5.

Figure 8. Map of observation points corresponding to Figure 6.

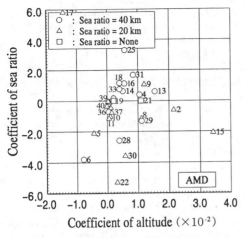

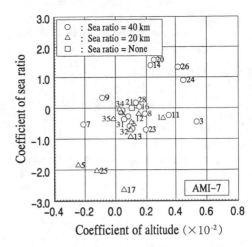

Figure 9. Relation between non-standardized partial regression coefficient of altitude and sea ratio.

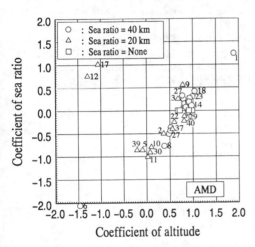

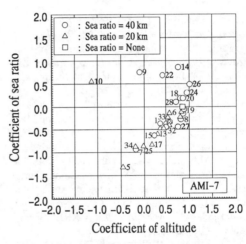

Figure 10. Relation between standardized partial regression coefficient of altitude and sea ratio.

Figure 11. Map of Aomori Prefecture.

4 CONCLUSIONS

For the sake of practical design, multiple regression analyses have been carried out for estimating snow depth on sites without observatories. For more than 75 % of the total area of Japan, equations using a stepwise method were highly significant. The rejection probability was, at most, 5 %. Observation points were locatd up to 1,500 m as shown in Figure 3. The equations given in this paper are applicable only to sites located below an altitude of 1,000 m. In this practical design paper, only altitude and sea ratio are used in the analyses. That means the effects of small hills or inclinations of the land, which are considered by the authors (Takahashi et al. 1992), are neglected.

Designers should establish the design value of snow depth on the ground considering this margin of error.

ACKNOWLEDGMENTS

Data of daily snow depth and the digital national land information used in this paper were provided for the project "Development of Comprehensive Technology on the Construction of Cities to Prevent Snow Disaster," supported by the Japanese Ministry of Construction. The authors wish to acknowledge Mr. T. Murota who made this study possible. The authors also thank Mr. M. Fukaya who assisted with the analyses.

REFERENCES

Architectural Institute of Japan 1993. *AIJ Recommendations for Loads on Buildings*, AIJ, Tokyo.

Geographical Survey Institute, the Ministry of Construction 1985. *Summary of digital national land information*, Japan Map Center, Tokyo. (in Japanese)

Ishihara, K. et al. 1972. Study on Making 4 km Mesh Maps for Annual Maximum Snow Depth and Mean Temperature in Winter - for Niigata Prefecture, *Report of Japan Snow Union* No.120. (in Japanese)

Izumi, M., Mihashi, H. and Takahashi, T. 1989. Statistical Properties of the Annual Maximum Series and a New Approach to Estimate the Extreme Values for Long Return Periods, *First International Conference on Snow Engineering*, CRREL Special Report: 89-6, 25-34.

Izumi, M., Mihashi, H., Takahashi, T. and Ono, M. 1987. An Estimation of Geographical Distribution of Annual Maximum Snow Depth Based on Topographic Factors, *The Architectural Reports of The Tohoku University*, No.26: 75-83. (in Japanese with English abstract)

Izumi, M., Mihashi, H., Takahashi, T. and Yamamoto, M. 1989. Estimation of Geographical Distribution of Annual Maximum Increasing Intensity of Snow Depth and its Relation with Heavy Snowfall Damage *The Architectural Reports of The Tohoku University*, No.28: 29-40. (in Japanese with English abstract)

Japan Meteorological Agency 1985. *Monthly Report for each Prefecture by Japan Meteorological Agency*, Japan Meteorological Agency, Tokyo. (in Japanese)

Judge, C.J. 1989. Developing the Eurocode, *First International Conference on Snow Engineering*, CRREL Special Report 89-6: 403-412.

Nakatao, T., Kitagawa, S. 1976. Linear Expression of Distribution of Snowcover Depth in Fukui Prefecture, *Journal of Snow and Ice*, Vol.37, No.4: 14-21. (in Japanese with English abstract)

Sack, R.L. and Sheikh-Taheri, A. 1986. *Ground and Roof Snow Loads for Idaho*, Univ. of Idaho, Moscow.

Sandvik, R. 1992. Snow Load Variation with Altitude in Norway, *Proceedings of 2nd. International Conference on Snow Engineering*, CRREL Special Report No. 92-27: 15-20.

Shibata, H. and Tanaka, A. 1980. Estimation of the Distribution of the Depth of Snow Cover, *Journal of Meteorological Research*, Vol.32, Nos.1-2: 51-57, Japan Meteorological Agency (in Japanese).

Takahashi, T., Mihashi, H., Izumi, M. 1992. Estimation of Ground Snow Depth Based on Topographic Factors, *Proceedings of 2nd. International Conference on Snow Engineering*, CRREL Special Report No. 92-27: 21–32.

Zuranski, J.A. 1988. Snow Loads on Roofs in East European Standards and Codes of Practice, *First International Conference on Snow Engineering*, CRREL Special Report 89-6: 419-428.

Snow Engineering: Recent Advances, Izumi, Nakamura & Sack (eds) © 1997 Balkema, Rotterdam. ISBN 90 5410 865 7

Database and methodology for conducting site specific snow load case studies for the United States

Wayne Tobiasson & Alan Greatorex
Cold Regions Research and Engineering Laboratory (CRREL), Hanover, N.H., USA

ABSTRACT: We have developed data and a methodology for determining the ground snow load at locations not covered in our ground snow load map of the United States due to extreme local snow load variations in the area. The elevation, the years of record available, the maximum observed value and the "50-year" ground snow load at a number of nearby sites are considered. A plot of elevation vs. load is often helpful.

1 DATABASE AND STATISTICS

Measurements collected by the National Weather Service (NWS) are the largest source of information on snow on the ground in the United States. At 266 "first-order" NWS stations across the nation, both the depth of the snow on the ground and its load (i.e., its "water equivalent") have been measured frequently each winter. At about 11,000 other NWS "co-op" stations only the depth of the snow on the ground has been measured.

At each NWS station we determined the maximum depth of the snow on the ground each winter. At each NWS first-order station we also determined the maximum water equivalent for each concurrent winter. Log-normal extreme value statistics (Ellingwood and Redfield 1983) were then used to estimate the depth of snow on the ground and, where available, the ground snow load (i.e., the water equivalent) having a 2% annual probability of being exceeded (i.e., the 50-year mean recurrence interval value). We did not do statistics for locations having less than 10 years of data or locations with 10 or more years of data but less than 5 years in which snow was observed.

The nonlinear equation of best fit between the 50-year depths and 50-year loads at the 204 first-order stations in the continental United States which met our criteria for analysis was as follows:

$$L = 0.279 \, D^{1.36} \qquad (1)$$

where L = 50-year load in lb/ft^2 and D = 50-year depth in inches. In SI units with the load in kN/m^2 and the depth in meters, the equation becomes $L = 1.97 \, D^{1.36}$.

A separate equation was developed for Alaska from the 20 first-order stations located there.

Our database also contains information from about 3300 additional locations where the water equivalent of snow on the ground is measured several times each winter. Some of those measurements are by companies that generate hydroelectric power, others are by the Corps of Engineers for flood forecasting, but most have been collected by the Natural Resource Conservation Service (NRCS) for similar purposes and for monitoring water supplies. Until recently NRCS was known as the Soil Conservation Service (SCS). Most of these "non-NWS" locations are in high mountain watersheds, not in populated areas.

2 THE NEW SNOW LOAD MAP

The 50-year loads determined at the NWS first-order stations and the 50-year loads at the NWS co-op stations generated by use of Equation 1 were used to construct a new national snow load map, a portion of which is shown in Figure 1. Snow loads are presented as zones. Some zones contain elevation limits above which the zoned value should not be used. These elevation limits, in feet, are shown in parentheses above the zoned value.

In some areas extreme local variations in ground snow loads preclude mapping at this scale. In those areas the map contains the designation "CS" instead of a value. CS indicates that a *case study* is required to establish ground snow loads for locations in this area.

We examined the possibility of integrating the non-NWS information into the new map with the hope of being able to reduce the extent of areas needing case

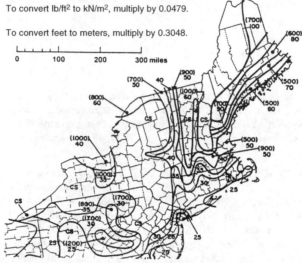

In CS areas, site-specific Case Studies are required to establish ground snow loads. Extreme local variations in ground snow loads in these areas preclude mapping at this scale.

Numbers in parentheses represent the upper elevation limits in feet for the ground snow load values presented below. Site-specific case studies are required to establish ground snow loads at elevations not covered.

To convert lb/ft^2 to kN/m^2, multiply by 0.0479.

To convert feet to meters, multiply by 0.3048.

Figure 1. Northeast portion of the new ground snow load map of the United States. Loads are in lb/ft^2.

studies but determined that little could be gained by doing this. Thus our new map is based only on NWS data. However, all the non-NWS information has been incorporated into our database, making it available for case studies.

The new map was recently published in the 1995 version of the national design load standard, "Minimum Design Loads for Buildings and Other Structures" (ASCE 1995), which is known as ASCE 7-95.

That standard requires that all ground snow loads used in design "be based on an extreme value statistical analysis of data available in the vicinity of the site using a value with a 2% annual probability of being exceeded (50-year mean recurrence interval)." Our data and methodology make this possible anyplace in the United States.

3 "SNOW LOADS FOR THE UNITED STATES"

Our report with the above title (Tobiasson and Greatorex 1996) presents the new snow load map and explains in detail the many steps we took to consider missing and questionable data and to develop the equations used to convert 50-year depths to 50-year loads.

That report contains numerous state maps on which all NWS first-order, NWS co-op and non-NWS stations are located. Figure 2 is one such map containing only NWS stations. Black dots indicate co-op stations and black triangles indicate first-order stations. The station number is presented adjacent to its dot or triangle. All non-NWS stations are presented on separate maps for the 24 states in which such stations exist.

The report also tabulates ground snow load data for all of these stations.

Table 1 shows a small portion of the NWS tabulation for Maine. Water equivalent information for first-order stations is presented in bold type. As with the maps, the NWS and non-NWS tabulations are presented separately.

The state maps contain a latitude and longitude reference grid and county boundaries. With this information a site of interest can be located on a state map. Using the scale shown on that map and a compass, we draw circles with radii of 25 and 50 miles (40 and 80 km) around the site of interest. The number of any NWS first-order station present within the 50-mile (80-km) outer circle is determined and noted on the case

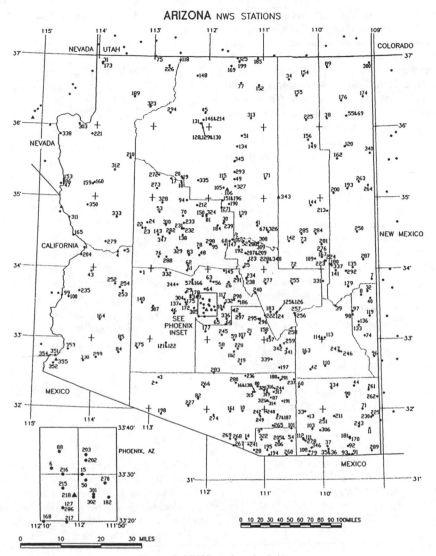

Figure 2. NWS stations in Arizona.

study form. If there is more than one first-order station, the closest is listed first. The scale on the map is used to determine the distance from each site of interest and those distances are noted on the case study form. An example case study form is shown in Figure 3. Space is provided for two lines of information for each first-order station. On the first line, the water equivalent (W.E.) values are presented. The second line presents loads generated by converting the 50-year and maximum observed *depths* measured during the same period to loads using Equation 1 (everywhere but Alaska, where a somewhat different equation is used). The first and second lines of first-order station infor-

mation are labeled (W.E.) and (DEPTH EQ) to show that the first is based on water equivalent measurements and the second on snow depth measurements.

Next the numbers of all NWS co-op stations within the 25-mile (40-km) inner circle are listed on the case study form in order of increasing distance. Distances from the site of interest are determined and noted on the form.

If the circles enter an adjacent state, stations within that state are also needed. To facilitate that, the location of all adjacent-state stations within 25 miles (40 km) of a state's border are also shown on each state map as in Figure 2. Their numbers are not shown. They

251

are obtained from the map of that adjacent state.

The state-by-state, station-by-station tabulation for NWS stations is then consulted for information on each station. That information is added to the case study form. In this way, the elevation, 50-year ground snow load, record maximum load, total years of record and number of years with no snow are transferred to the case study form for each nearby NWS station. For some locations in the tabulation not enough data were available to determine a 50-year ground snow load. Nonetheless, all information available for that location is transferred to the case study form since it is of some value in the analysis.

Then maps in the non-NWS series are examined to determine if there are any non-NWS stations within 25 miles (40 km) of the site of interest. Information on each non-NWS station found is obtained from the non-NWS tabulation and transferred to the case study form.

The new snow load map (Fig. 1) is consulted and the mapped ground snow load and any elevation limitations on it are listed on the case study form. If the case study is being done for a place in a CS zone, as is often the case, CS is listed as the ASCE 7-95 mapped value.

Since we conduct numerous snow load case studies we have computerized the assembly of information on case study forms. We input the name, latitude, longitude and elevation of the site of interest and, with a few manual prods, the form is printed out. Figure 4 is an example. It contains the station name instead of its number. The distance and azimuth of each station from the site of interest are calculated and tabulated automatically. Although the manual and computerized case study forms are slightly different, both contain all the information needed to do the case study.

4 OBTAINING AN ANSWER

Once the case study form is filled in, it is analyzed to obtain a ground snow load for that location.

4.1 NWS first-order stations

For all NWS first-order stations within 50 miles, we compare the (W.E.) values to the (DEPTH EQ) values. When the two ground snow load (P_g) values are about equal, we give Equation 1 credit for doing a good job of converting 50-year depths to 50-year loads in this area. When the values are not close, either the depth or water equivalent measurements are suspect or Equation 1 is not good at predicting loads here. Our investigations convince us to place somewhat more trust in the water equivalent values, but we keep an open mind as we examine the rest of the case study data. When

Table 1. A portion of the NWS station tabulation for Maine. P_g is the ground snow load.

Sta	Elev	P_g	Rec	Years	
			Max	Tot	No
#	ft	lb/ft^2	lb/ft^2		Sno
Maine					
1	470	42	49	10	0
2	600		39	5	0
3	350	67	51	43	0
4	190	69	62	41	0
5	110	46	51	35	0
6	20	61	59	42	0
7	710		31	1	0
8	400		32	2	0
9	600	70	42	11	0
10	1060		48	8	0
11	420	154	44	15	0
12	560	94	101	31	0
13	70	56	45	37	0
14	80	35	25	16	0
15	**620**	**95**	**68**	**34**	**0**
15	620	85	76	43	0
16	1000		38	5	0
17	1000	94	57	23	0
18	360	81	62	42	0
19	400		25	4	0

non-NWS information is available, it is quite helpful in resolving questions about NWS first-order station information since it represents independent water equivalent (i.e., load) measurements, thereby sidestepping any depth-to-load-equation concerns.

We always examine each first-order station's years of record and its record maximum values before deciding how much we trust its 50-year value.

A first-order station only a few miles (kilometers) away from the site of interest is given more weight than one close to 50 miles (80 km) away. Elevation differences are also an important issue and are considered in a similar way.

4.2 NWS co-op stations

This is usually the largest body of information on the case study form. Since it is arranged state-by-state according to distance ("radius" on the form) from the site of interest, the uppermost stations in each state's array are the most valuable. If there are a few stations within 12 to 15 miles (19 to 24 km) with long periods

FORT RICHIE, MARYLAND

Latitude 39° 40'   Longitude 77° 28'   Elevation 1320 ft

Station	#	Radius (mi.)	Elev. (ft)	P_g (lb/ft²)	Record Max. (lb/ft²)	Years of Record Total	Years of Record No Snow
NWS FIRST ORDER							
(W.E.)	202	49	290	23	15	29	1
(DEPTH EQ)	202	49	290	25	19	29	1
MARYLAND	25	1	1610	43	31	24	0
	49	4	910	50	46	35	1
	51	6	720		11	6	1
	52	9	709	25	26	36	0
	30	11	570	40	30	24	0
	64	14	660	25	23	43	0
	19	14	740	58	34	15	0
	59	17	440	22	20	20	1
	58	17	380	27	28	18	0
	56	17	300		26	8	1
	66	18	420	12	8	11	0
	57	19	390	29	16	11	2
	120	21	360	30	35	19	0
	108	21	430	20	25	36	2
	32	23	580		12	9	0
PENNSYLVANIA	297	13	1520	31	28	38	0
	106	14	520	15	16	10	0
	127	17	500	27	20	31	0
	7	19	710	26	20	23	0
	54	20	640	27	23	43	0
	23	22	720	25	19	18	0

ASCE 7-95 mapped value "CS" lb/ft²   Case Study answer 45 lb/ft²

By A. GREATOREX   and W. TOBIASSON   DEC 1995

Comments: ELEVATION PLOT (ATTACHED) POINTS TO 40 lb/ft² BUT LARGE SCATTER ON PLOT AND 43 lb/ft² AT MARYLAND #25, ONE MILE AWAY AND 50 lb/ft² AT MARYLAND #49, FOUR MILES AWAY CAUSE US TO INCREASE TO 45 lb/ft²

Figure 3. "Manual" case study form for Fort Richie, Maryland.

of record, their collective answer often overpowers anything the rest of the co-op stations farther away can contribute. But, at times, the most valuable information lies farther away. For example, the closest three co-op stations on the Figure 4 case study form for Scranton, Pennsylvania, do not have enough information to permit calculation of P_g values. The next two stations have long records (44 and 38 years) but they are at a much higher elevation than Scranton. Finally the sixth and seventh stations, 16 and 17 miles (26 and 27 km) away have relatively long periods of record and elevations about the same as Scranton.

4.3 Non-NWS stations

These records are based on water equivalent (i.e., load) measurements, not depth measurements, so they are somewhat more valuable than those of NWS co-op stations with similar periods of record. However, fewer readings are taken each winter and these stations are frequently at higher elevations than most sites of interest. Nonetheless, some sites of interest are best represented by these stations.

4.4 Elevation vs. P_g plots

Figures 3 and 4 illustrate case studies where the answer does not become self evident after a few minutes of examining the form. In such situations elevation vs. P_g plots are helpful. Often we plot the 6 to 10 nearest stations, make a copy of that plot then add all other stations within 25 miles (40 km) to the plot. The two plots give us an appreciation for the effect of distance.

Figure 5 shows the 10-closest-values plot for

253

SNOW LOAD CASE STUDY FOR

SCRANTON, PENNSYLVANIA

Latitude __41° 24'__   Longitude   __75° 40'__   Elevation   __730 ft__

Station	Radius (mi.)	Azimuth (from site)	Elev. (ft)	P_g (lb/ft²)	Record Max. (lb/ft²)	Years of Record Total	Years of Record No Snow
NWS FIRST ORDER							
WILKES-BARRE-SCRANTON (W.E.)	6	225	930	18	13	37	0
WILKES-BARRE-SCRANTON WSO AP (DEPTH EQ)	6	225	930	27	19	29	0
PENNSYLVANIA							
SCRANTON	1	360	750		19	8	0
AVOCA CAA AP	5	233	920		6	6	0
SCRANTON WB AP	6	225	940		22	9	0
HOLLISTERVILLE	12	94	1370	49	54	44	0
GOULDSBORO	15	125	1890	60	73	38	0
WILKES-BARRE	16	232	580	28	25	38	0
DIXON	17	306	620	31	31	20	0
TOBYHANNA	20	131	1950	43	49	29	0
FRANCIS E WALTER DAM	20	193	1510	52	57	30	0
LAKEVILLE 2 NNE	21	83	1440	68	65	20	0
PAUPACK 2 WNW	22	90	1360	50	57	40	0
HONESDALE 4 NW	23	58	1410	29	46	16	0
PIKES CREEK	25	258	1320	28	20	13	0
HAWLEY 1 S DAM	25	82	1200		54	6	0
PENNSYLVANIA (NON-NWS)							
TOBYHANNA	20	127	2040	49	48	13	0
F. E. WALTER RESERVOIR	20	193	1700	41	30	14	0
PROMPTON-JADWIN RESERVOIR	23	58	1600	41	33	14	0
LONG POND	25	157	1860	75	38	11	1

ASCE 7-95 mapped value __"CS"__ lb/ft²   Case Study answer __30__ lb/ft²

By   __ALAN GREATOREX__   and __WAYNE TOBIASSON__   __DEC.__ 19__95__

Comments: GOOD LONG RECORDS AT SCRANTON BUT W.E. (18 lb/ft²) AND DEPTH EQ (27 lb/ft²)
DO NOT AGREE, PERHAPS EQUATION OVERPREDICTS IN THIS AREA BUT SCRANTON W.E.
DATA COULD BE BAD INSTEAD. PLOT SHOWS AVERAGE VALUE OF 28 lb/ft².
CONSIDERING SCATTER ON PLOT, 30 OR 35 lb/ft² MIGHT BE SELECTED. SINCE
SOME OVERPREDICTION BY EQUATION IS EXPECTED, WE CHOSE 30 lb/ft²

Figure 4. "Computerized" case study form for Scranton, Pennsylvania.

Scranton. Note that two of the values are from non-NWS stations. The least squares line of best fit is shown along with the value of P_g at the elevation of the site of interest. Since it is usually possible to arrive at appropriate answers without having to calculate the least squares best-fit value, we only generate that information in our computerized version.

Such plots not only clarify any elevation effect but they also illustrate the amount of scatter in the local database. Since there is generally noticeable scatter in such records, we usually select a value somewhat above the least squares best-fit value.

Figure 6 shows the all-values-within-25-miles (40 km) plot for Fort Richie. The plot of the nearest eight values generated the same P_g value of 39 lb/ft² (1.87 kN/m²). Considering the scatter of points on the plot and the 50 lb/ft² (2.40 kN/m²) value at Edgemont only 4 miles (6 km) away where a long record is available, we chose 45 lb/ft² (2.16 kN/m²) as our Fort Richie answer. Our concern with the 58 lb/ft² (2.78 kN/m²) at Boonsboro (14 miles or 23 km away) was not great

since it is based on only 15 years of record during which the record maximum value was only 34 lb/ft² (1.63 kN/m²).

Another two individuals examining the Fort Richie case study form (Fig. 4) and elevation plot (Fig. 6) might have settled on 50 lb/ft² (2.40 kN/m²). We would find that hard to argue against, since the data available do not permit loads to be established with great accuracy.

Case study answers are rounded to the nearest 5 lb/ft² (0.24 kN/m²) up to a value of 40 lb/ft² (1.92 kN/m²) and to the nearest 10 lb/ft² (0.48 kN/m²) above 40 lb/ft² (1.92 kN/m²). Table 2 shows the lower and upper limits used for mapped zones and case study answers. For example, if our best estimate of a case study answer is 37 lb/ft² (1.77 kN/m²) we would round that down to $P_g = 35$ lb/ft² (1.68 kN/m²). If our best estimate was 38 lb/ft² (1.82 kN/m²) we would round that up to 40 lb/ft² (1.92 kN/m²).

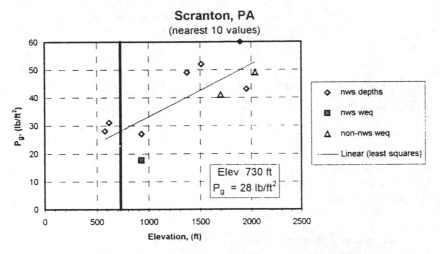

Figure 5. Elevation plot for Scranton, Pennsylvania.

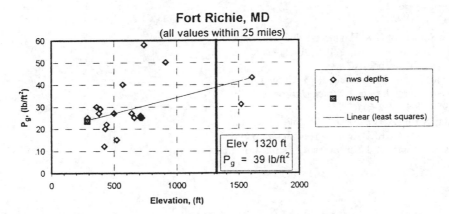

Figure 6. Elevation plot for Fort Richie, Maryland.

5 CONCLUSIONS AND RECOMMENDATIONS

We have conducted over 400 case studies using the data and methodology described in this article. The case study answer is self-evident and uncontroversial after a few minutes of study for some sites of interest, but inconsistencies and considerable variability exist in the data available for many sites. Plots of P_g vs. elevation are quite helpful but, even with them, we often arrive at somewhat different answers. We have always been able to agree upon a single answer after a brief discussion of issues.

We recently initiated cooperative work with structural engineering associations from Maine and New Hampshire. We provided computerized case study forms and snow load vs. elevation plots for many sites

of interest to them and have trained their volunteers to conduct snow load case studies. We are also analyzing the same material and will compare results. If we and they arrive at similar answers, we will have gained confidence in our methodology. If answers do not agree, we hope to incorporate any lessons learned into our report (Tobiasson and Greatorex 1996) before it is published.

Perhaps we will conduct similar studies for other states. Alternatively we may be able to simply pass that task along to others to do it on their own using the information in our report. We may place the database and computerized case study procedure on the worldwide web.

It will be interesting to compare case study results for the Rocky Mountain States with the values devel-

255

Table 2. Upper and lower limits of snow load zones (lb/ft^2).

Upper Limit	2	7	12	17	22	27	32	37
ZONE	*0*	*5*	*10*	*15*	*20*	*25*	*30*	*35*
Lower Limit	0	3	8	13	18	23	28	33
Upper Limit	44	54	64	74	84	94	104	
ZONE	*40*	*50*	*60*	*70*	*80*	*90*	*100*	
Lower Limit	38	45	55	65	75	85	95	

To convert lb/ft^2 to kN/m^2, multiply by 0.0479

oped years ago by several structural engineering groups in those states using a variety of methods. Perhaps all this can lead to a consistent analytical approach for determining ground snow loads for design purposes all across the nation.

Once a large body of case studies accumulates, it should be possible to use them to improve the national map.

6 REFERENCES

American Society of Civil Engineers 1995. Minimum design loads for buildings and other structures. *ASCE Manual 7-95*, New York, NY.

Ellingwood, B. and R. Redfield 1983. Ground snow loads for structural design. *J. Struct. Engrg.*, ASCE, 109(4), 950–964.

Tobiasson, W. and A. Greatorex 1996. Snow loads for the United States. CRREL Report in preparation, Cold Regions Research and Engineering Laboratory, Hanover, NH.

Snow Engineering: Recent Advances, Izumi, Nakamura & Sack (eds) © 1997 Balkema, Rotterdam. ISBN 90 5410 865 7

Analysis of snowfall series observed during long periods at four meteorological stations in the Italian territory

Riccardo Del Corso
Istituto di Scienza delle Costruzioni, University of Pisa, Italy

Luca Mercalli
Idrogeological Department, Regione Piemonte, Italy

ABSTRACT: Snowfalls in the north-west part of Italy have been observed for more than one century at several meteorological stations. Particularly interesting are the series of Bra, Moncalieri and Cuneo, which give, together with 207 years of observations at Turin, useful information about snow loads in this part of Italy. Although the snow data of Florence are typical of another climatic region, they were included in this study because the series is quite long. The data sets were tested for two different probability distribution functions and both the effect of exceptional values and the influence of the period of observation were studied.

1 LONG PERIOD SERIES IN ITALY

Research on ground snow loads for the Italian territory were carried out in the past, making use of the data sets from 105 stations (Sanpaolesi et al., 1983). Since the information available at that time covers periods of about 30 - 40 years, there is great interest in the acquisition of further data, in particular longer periods of record.

In the Italian north-western region some snow data series have been collected for more than a century. Among the meteorological stations, whose daily data have been recovered from the original manuscripts, particularly interesting are those of Turin (45° 03' N, 7° 40' E, 240 m), Moncalieri (44° 59' N, 7° 41' E, 267 m), Bra (44° 42' N, 7° 41' E, 290 m) and Cuneo (44° 23' N, 7° 33' E, 550 m).

The four stations are indicative of different conditions of urbanization, since the town of Turin includes an urban area of about 130 km², Moncalieri is a small town located in the hills to the south of Turin, Bra lies in a little built-up zone, and Cuneo is a town at higher altitude at the margins of the Maritime Alps.

Observations of snowfall in Turin began at the end of the seventeenth century, but only since 1788 are daily records available (Leporati et al., 1993). This series of more than 200 years, probably the longest in the world, was investigated in other works (Del Corso, 1995). The series of Bra (Brizio et al., 1991) and Moncalieri (Di Napoli et al., 1996), even

if shorter, covers a period of 130 years with continuity, while that of Cuneo (Biancotti et al., 1991) (Romano et al., 1994) consists of 91 years.

Taken as a whole, the data collected at these four stations form a significant basis for the north-western part of Italy.

The series of Florence , which covers a period of 130 years, is indicative of a densely urbanized zone in central Italy. The observations were being collected by the Osservatorio Ximeniano, placed in the middle of the town. The observatory, founded in 1756 for astronomical studies, has recorded meteorological data since 1813 (Bravieri et al., 1993) and in particular snowfalls since 1874.

2 THE SNOW DATA

The observations in the stations of Turin, Moncalieri, Bra and Cuneo consist of daily measurements of the amount of the fresh snowfall during the preceding 24 hours. The water equivalent is also measured , by melting the snow.

The data are suitable for estimating the specific weight of the snow, but they do not give direct information about the ground snow load. For this reason, after careful control on the reliability of the data, the water equivalents of snowfall during consecutive days have been summed to give the snow load on the ground.

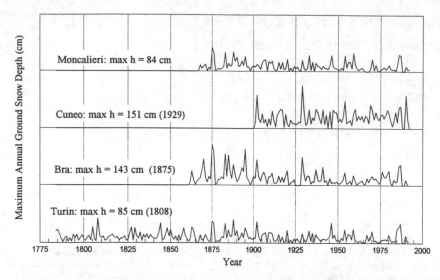

Figure 1 - Maximum Annual Ground Snow Depth for four different stations in the Piemonte Region

The hypothesis of absence of losses due to the melt is reasonable, since the duration of snow on the ground does not exceed a few days, and the occurrence of snowfalls in several consecutive days happens only during the colder months.

When rain is mixed with snow, the water equivalent data do not allow direct calculation of their relative percentages. Herein precipitation is assumed to be snow if its specific weight is lower than 2 kN/m³ , including in this way the effect of rains of moderate intensity, that do not melt the snow cover. A heavier rain is added only if the amount of snow on the ground is so high that the mean specific weight of the whole snow cover results less than 2 kN/m³.

A comparison of the yearly maxima of the snow heights in the stations of Turin, Bra, Moncalieri and Cuneo is shown in Figure 1.

At Florence the thickness of the snow layer has been measured. There the snowfalls are single events, and thus the statistics of the snow heights is the same as that of the snow loads, assuming the density is constant.

3 STATISTICAL ANALYSIS OF THE DATA

The yearly maxima have been obtained from the daily observations, which are water equivalent measurements at Moncalieri, Bra and Cuneo and heights of the snow cover at Florence.

Ordering the yearly maxima x_m in a decreasing manner in such a way that x_1 is the largest value, the probability

$$F(x_m) = 1 - \frac{1}{T_m} \ , \qquad (1)$$

is assigned to each value of x_m , being

$$T_m = \frac{N+1}{m}$$

the return period, N the number of years of observations, m the position index. The pairs $[x_m \ , \ F(x_m)]$ are then plotted on the Gumbel probability paper, to determine how well the Gumbel type I distribution function fits the sample data.

Since each of the four stations presents some years with no snowfall, the cumulative distribution function of the population is of the type

$$F(x) = p + (1 - p) \, F_c(x) \ , \qquad (2)$$

where $F_c(x)$ is the cumulative distribution function for the population consisting of the values different from zero and p is the probability of absence.

The $F_c(x)$ function has been taken in the Gumbel type I form

$$F_c(x) = e^{-e^{-y}} \ , \qquad (3)$$

with $y = \alpha \, (x - u)$.

When one or some yearly maxima x_m are recorded, which are far higher than the others in the series, it can be assumed that such values belong to the sample of a population, defined as the population of exceptional events, while all the others belong to the sample of another population, defined as the population of moderate events. In this case the CDF

Table 1 : Results of the statistical analysis (Gu = Gumbel type I, MG= Mixed Gamma)
Water equivalent (mm) at Bra, Moncalieri, Cuneo - Snow height (cm) at Florence

Bra	1862-1991: 130 years		1862-1906: 45 years		1907-1951: 45 years		1952-1991: 40 years	
mean (x_m)	34.42		39.68		18.72		22.81	
std. dev. (x_m)	28.71		37.04		17.57		20.28	
max (x_m)	149.00		149.00		83.60		86.00	
Fractiles	Gu	MG	Gu	MG	Gu	MG	Gu	MG
T_r= 200 years	139.64	148.00	175.87	157.00	82.76	94.00	96.57	94.00
T_r= 50 years	108.67	135.00	135.79	146.00	64.11	74.00	75.04	84.00
Moncalieri	1867-1994: 128 years		1867-1911: 45 years		1917-1961: 45 years		1962-1994: 33 years	
mean (x_m)	18.70		26.69		16.99		11.29	
std. dev. (x_m)	20.93		27.19		16.49		13.86	
max (x_m)	132.90		132.90		79.50		57.20	
Fractiles	Gu	MG	Gu	MG	Gu	MG	Gu	MG
T_r= 200 years	97.00	128.00	127.00	144.00	77.63	88.00	63.87	60.00
T_r= 50 years	74.00	96.00	97.58	123.00	59.85	71.00	48.47	57.00
Cuneo	1901-1991: 91 years		1901-1945: 45 years		1946-1990: 45 years			
mean (x_m)	40.94		36.29		45.57			
std. dev. (x_m)	27.59		29.42		25.43			
max (x_m)	155.40		155.40		99.40			
Fractiles	Gu	MG	Gu	MG	Gu	MG	Gu	MG
T_r= 200 years	140.89	160.00	144.17	173.00	135.02	121.00		
T_r= 50 years	111.56	118.00	112.39	129.00	108.97	104.00		
Florence	1874-1990: 117 years		1874-1923: 50 years		1924-1953: 45 years		1954-1990: 37 years	
mean (x_m)	5.06		1.44		2.48		3.83	
std. dev. (x_m)	5.59		2.37		5.54		5.81	
max (x_m)	30.00		10.00		30.00		19.00	
Fractiles	Gu	MG	Gu	MG	Gu	MG	Gu	MG
T_r= 200 years	22.51	28.00	11.20	12.00	26.88	44.00	21.17	21.00
T_r= 50 years	16.39	20.00	8.35	10.00	19.26	24.00	20.35	20.00

can be taken in the form proposed by the CIB Commission W81 (1991)

$$F_c(x) = p_1 F_1(x) + p_2 F_2(x) \ , \quad (4)$$

in which

$F_1(x)$ is the probability distribution for the population consisting of moderate events,
$F_2(x)$ is the probability distribution for the population consisting of exceptional events,
p_1 is the probability that a moderate event occurs,
p_2 is the probability that an exceptional event occurs.
The $F_1(x)$ and $F_2(x)$ are chosen as incomplete

Gamma functions $I_1(v_1 x, k_1)$ and $I_2(v_2 x, k_2)$, whose parameters v_i and k_i are calculated with the moment methods:

$$v_i = \frac{\text{mean}(x_i)}{\text{var}(x_i)} \ , \quad k_i = \frac{(\text{mean}(x_i))^2}{\text{var}(x_i)}$$

being mean(x_i) and var(x_i) the mean and the variance of the sample of moderate events (i=1) and of the sample of exceptional events (i=2)

The data were analyzed considering for each station the whole data set and three subsets of about 50 years. The results, reported in the Table 1 and in Figures 2,3,4 and 5, allow comparison of the two cumulative distributions (3) and (4), and evaluation

259

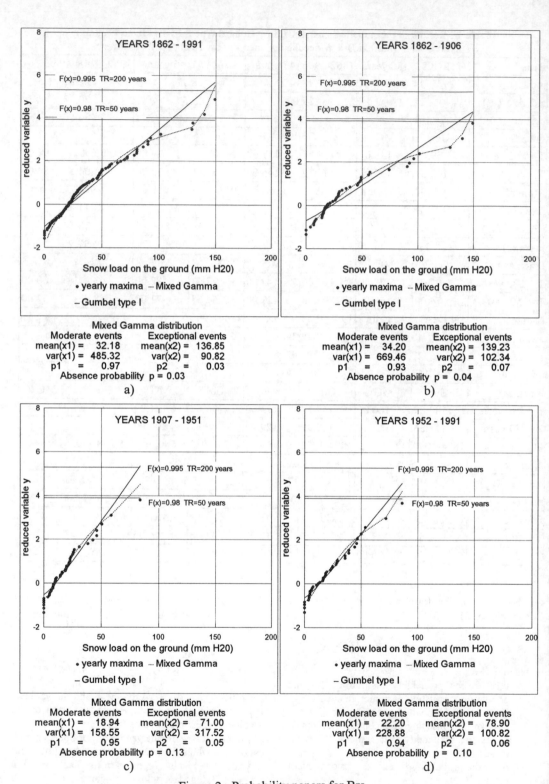

Figure 2 - Probability papers for Bra

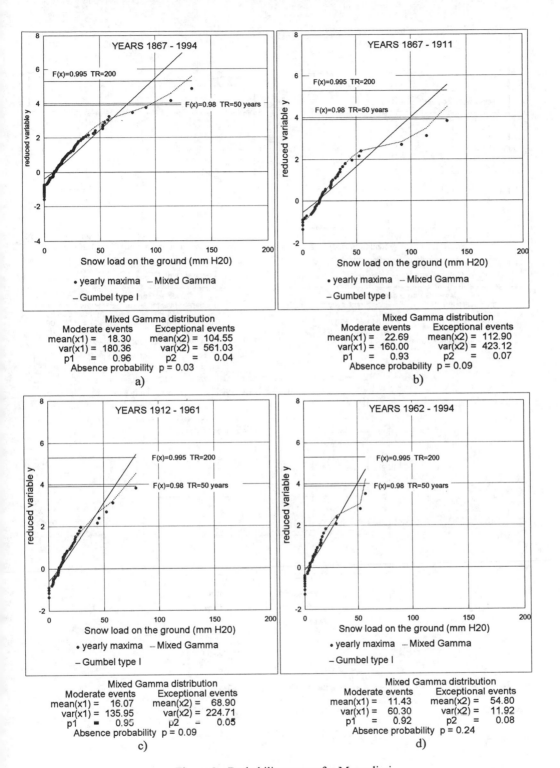

Figure 3 - Probability papers for Moncalieri

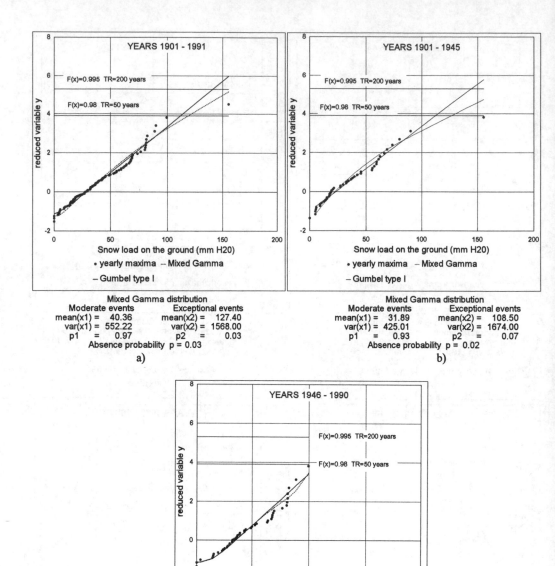

Mixed Gamma distribution
Moderate events   Exceptional events
mean(x1) = 40.36   mean(x2) = 127.40
var(x1) = 552.22   var(x2) = 1568.00
p1 = 0.97   p2 = 0.03
Absence probability p = 0.03
a)

Mixed Gamma distribution
Moderate events   Exceptional events
mean(x1) = 31.89   mean(x2) = 108.50
var(x1) = 425.01   var(x2) = 1674.00
p1 = 0.93   p2 = 0.07
Absence probability p = 0.02
b)

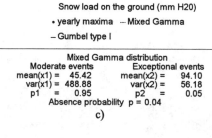

Mixed Gamma distribution
Moderate events   Exceptional events
mean(x1) = 45.42   mean(x2) = 94.10
var(x1) = 488.88   var(x2) = 56.18
p1 = 0.95   p2 = 0.05
Absence probability p = 0.04
c)

Figure 4 - Probability papers for Cuneo

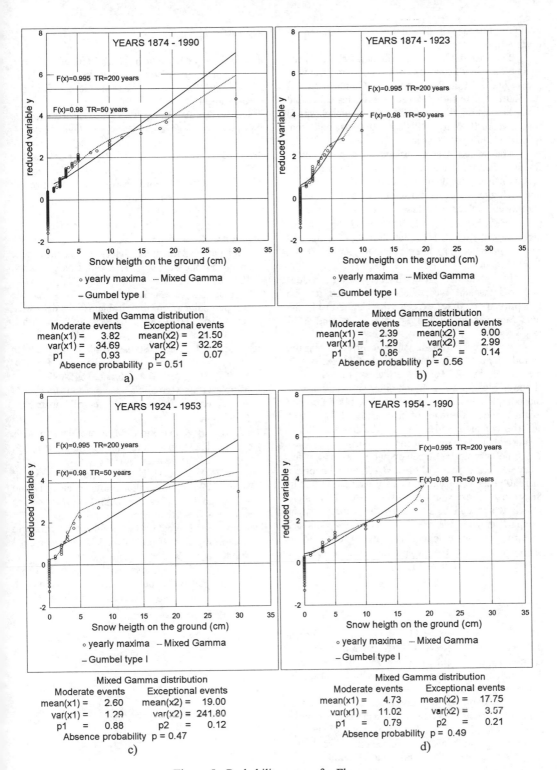

Figure 5 - Probability papers for Florence

of the influence of the length of the period of observation.

The mixed Gamma distribution better fits the sample data in all cases examined and almost always gives higher fractiles than the Gumbel type I distribution. The ratio between the values calculated from the two distributions (Gamma / Gumbel type I) ranges from 0.95 to 1.30 (T_R= 50 years) and from 0.90 to 1.64 (T_R= 200 years).

The mixed Gamma distribution gives lower fractiles than the Gumbel type I distribution for data sets consisting only of moderate values, while it gives higher fractiles for data sets containg exceptional values. The first instance occurs for the Cuneo data for the period 1946 - 1990 (Fig.4c), the second for the whole Mocalieri data set (Fig. 3a) and for the Florence data relevant to the period 1924 - 1953 (Fig. 5c).

In particular at Moncalieri four values occurred which were been considered to be exceptional (they exceed the highest of the moderate values by 128%, 95%, 57% and 36% respectively).

At Florence a single value occurred during the period 1924 - 1953, which exceeds the next one in the series by 275%. Nevertheless the fractiles, although higher, are not much different from those of the complete series. A more remarkable difference exists between the data of the period 1874 - 1923 and the complete series, in that the fractiles of the first are about half those of the second.

It can be inferred that when snowfalls are absent for many years, the period of observation must be longer than 50 years, if reliable results are wanted. In fact in these cases short series are likely to contain single values which can be considered as exceptional in comparison with the other values of the data set (Figure 5c), while longer periods of observation could show that the population consist only of moderate events (Figure 5a).

4 CONCLUSIONS

As a first conclusion of this study, it was observed that there are no essential differencies between the Gumbel type I distribution and the mixed Gamma distribution, unless an exceptional value is present in the series.

The length of the observation period has a limited influence on the results, but it becomes more important for data with a high probability of absence of snow. In such instances the period should be longer than 50 years to obtain reliable results.

REFERENCES

Biancotti, A. & Brancucci, G. & Mercalli, L. 1991. *La serie termopluviometrica di Cuneo (1877-1990)*. Nat. Group of Geogr. Fis. and Geomorph.

Bravieri, D. & Holtz, C. 1993. *L'Osservatorio Ximeniano di Firenze*. Baccini e Baldi, Florence, Italy.

Brizio, D. & Mercalli, L. 1992. Pioggia e neve a Bra, 130 anni di osservazioni (1862-1991). Meteo Bra 1, Quad. Oss. Meteorol. Museo Civico Craveri di Storia Naturale, Bra, Italy.

CIB Commission W81 1991. Actions on structures - Snow loads. *Rep. by CIB Commission W81*

Del Corso, R. & Sanpaolesi, L. 1990. Stato degli studi per la definizione del carico da neve nelle norme europee. *Proc. CTE 1990*, Bologna, Italy. 123-127.

Del Corso, R. & Sanpaolesi, L. 1991. Il codice europeo sulle azioni e le problematiche del carico neve. *Proc. XIII Conf. CTA*, Abano Terme, Italy. 85-93.

Del Corso, R. 1995. Statistics of snow loads in Turin: the effect of exceptional values. *Proc. of the ICASP 7 Conf.*, Paris, France, 645-649.

Di Napoli, G. & Mercalli, L. 1996. *Moncalieri, 130 anni di meteorologia*. Soc. Meteorol. Subalp., Turin, Italy

Ellingwood, B. & O'Rourke, M. Probabilistic models of snow loads on structures. *Structural Safety*. 4: 291-299. Amsterdam: Elsevier

Leporati, E. & Mercalli, L. 1993. Snowfalls series of Turin, 1784 - 1991: climatological analysis and actions on structures. *Int. Symp. on Applied Ice and Snow Res*. Rovaniemi, Finland.

Romano, F. Mercalli, L. 1994. L'osservatorio Meteorologico di Cuneo: dal 1877 sentinella della Granda. *Nimbus* 3.

Sanpaolesi, L. & Del Corso, R. & Ligarò, S. 1983. Analisi statistica dei valori del carico di riferimento al suolo in Italia. Giornale del Genio Civile, Anno 121, 3: 265-278.

Snow Engineering: Recent Advances, Izumi, Nakamura & Sack (eds) © 1997 Balkema, Rotterdam. ISBN 90 5410 865 7

Direct measurement of roof snow load

Teruyoshi Umemura & Tsuyoshi Sasame
Department of Mechanical Engineering, Nagaoka University of Technology, Niigata, Japan

Tadao Fukushima
Daiichi Sokuhan Works Co., Japan

ABSTRACT: In order to reduce the danger and the cost of snow removal on a roof by decreasing the time and frequency of snow removal operations, the possibility of direct measurement of roof snow loads has been studied through measurements of compressive deformation of steel and wooden columns of a garage, and also through the bending deformation of a wooden rafter of an experimental roof. Using the data obtained, the accuracies of the measurements were estimated as 16 kg/m² of roof load in the measurement for the steel column, 98 kg/m² for the wooden column, and 14 kg/m² for the rafter. The measurement of bending angles of a rafter is the most promising method because it appears to be independent of temperature and humidity.

1. INTRODUCTION

For today's inhabitants in the heavy snowfall area of Japan, snow removal from house roofs has become one of their most important problems. Many kinds of devices capable of removing snow without manpower have been developed and sold, but their cost seems too high to be widely accepted.

If an exact roof snow load is continually measured, the time and frequency of snow removal operation can be reduced, along with danger of collapse and cost. Until now, however, no measuring method has succeeded, although several kinds of indirect measuring methods have been investigated (Umemura, 1993).

This paper deals with two kinds of direct measurements due to snow loads: compressive deformation of a column or bending deformation of a beam or a rafter.

2. MEASUREMENT OF COMPRESSIVE DE-FORMATION OF A COLUMN

2.1 Measuring devices

Generally the strain induced in a building column by roof snow loads is so small that it must be amplified for measurements. Two kinds of devices have been assembled for application to existing buildings as shown in Figure 1. The device shown in Figure 1(a) is for a steel column and 1(b) for a wooden column.

For the steel column, shown in Figure 1(a), a steel pipe 2 is fastened with bolts 11 to upper arm 5 which is welded to a column of an existing building. To the lower end, a slide block 8 is bolted. A thin aluminum plate 1 30-mm long is fastened with a bolt 7. The other end of the aluminum plate is fastened to a lower arm plate 6 welded to the column. When compressive force is applied to the column the strain in the 1800 mm of column between the upper and lower arms also occurs in the measuring device. Essentially all of this deformation occurs in the aluminum plate. Hence its strain amplification factor is 1800/30=60. This deformation is measured by strain gauges attached to the aluminum plate.

Figure 1(b) shows the device for a wooden column. Generally a device of this kind is set in a groove cut in an actual column. In Figure 1(b), column 1 is Japanese cedar with a 120 x 120 mm cross section, which is a type of column commonly used in Japan. A steel rod 3 with a 5 mm diameter hangs in the groove 2 (532.5 mm length and 14-mm depth). A potentiometer 4 with a maximum error of 12 μm is attached to the bottom end of the groove, and measures the steel rod displacement, which corresponds to the deformation of 532.5 mm of the column.

2.2 Measurement for steel column

The device was set on a steel column of a garage, which is composed of a 6500 mm x 3600 mm roof of 0.8 mm thick angular wave-shaped (250 mm pitch and 150 mm height) steel plate and 4 supporting columns of 4.5 mm thick square hollow steel (200 x 200 mm section and 3055 mm length). The design load for the roof was 1250 kg/m²,

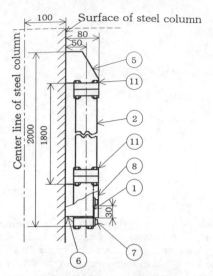

(a) Measuring device for steel columns
(amp. factor = 60)

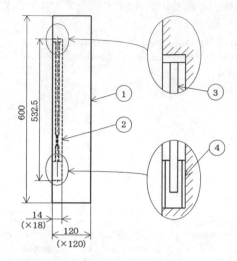

(b) Measuring device for wooden columns

Figure 1. Deformation amplifiers as snow load measuring devices for a streel column (a) and for a wooden column (b). (All dimension in mm)

which is followed by 29.25 t for the whole roof, and 7.31 t for each of the four columns. This causes a 20.34 MPa compressive stress and 99.0 x 10^{-6} strain in each column. Thus the ratio of column strain to snow load is 99 x $10^{-6}/1250 = 7.9$ x 10^{-8} m²/kg. Therefore, the ratio of the amplified strain of the device to a snow load is expected to be 7.9 x 10^{-8} x 60 = 4.8 x 10^{-6} m²/kg.

To test the device, a water pool of 4500 x 2800 mm with 900 mm depth was constructed on the roof of the garage, and water was introduced in it and left for a day. During that time the mass of water was measured with an accumulating flow meter as well as by the device. Results are shown in Figure 2. The introduction of water began at 14:00 hours and stopped at 17:57 hours, when it reached 300 mm depth, that is 3780 kg. The roof was left until 16:30 hours of the next day. Figure 2 shows that the measured strain agrees fairly well with the introduced mass of water. Its fluctuation with air temperature is also evident.

2.3 Measurement for wooden column

The measurement device for a wooden column was loaded by a compression test machine at a constant temperature of 23 °C and a humidity 67 to 75 %. It was observed that the potentiometer reading increases linearly with increasing stress up to the allowable stress of 6.9 MPa, while the slope of the line shows Young's modulus of the wood is 8.2 ± 4 GPa. The maximum error of the potentiometer is 12 μm.

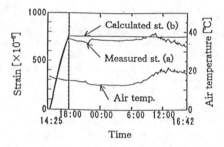

Figure 2. Strain (a) measured with the device for a steel pillar and strain (b) calculated from the mass of water introduced into a pool on the roof, and air temperature. The load on the roof equals 214 x 10^3 x (strain) kg/m².

When the temperature or the humidity varied during the measurements, the potentiometer reading fluctuated considerably. So the effects of temperature and humidity were measured by holding the device for some time in an air-conditioning room at set temperatures and humidities. Such measurements were carried out when the device was new (A) and again, a year later (B).

In order to know the effects of temperature changes, the measurements were taken after the device was held for 3 hours at 50 % humidity and at every 10 °C from 20 °C -30 °C for (A), or 20 °C to -10 °C for (B), and then back to 20 °C. The potentiometer readings were almost linearly dependent on temperature, but the data obtained with decreasing temperatures deviated from those with increasing temperatures, that is to say, a hystelisis was observed as shown in Figure 3. The maximum deviations of hysteresis were 33 μm for (A) and 9 μm for (B). This difference between (A) and (B) is not only due to the difference of the experimental range of temperatures but also the aging of the device.

The effects of humidity were checked in the same manner. The readings showed similar behaviors to Figure 3. The maximum deviations in the hysteresis, 12 μm for (A) and 17 μm for (B).

3. MEASUREMENT OF THE BENDING ANGLE OF A RAFTER

3.1 Measuring apparatus for bending angle

Measurement of bending deformations may be a better way of measuring roof snow loads as it is primarily dependent upon Young's modulus and hardly affected by temperatures and humidity. In particular, the measurement of bending angles of a rafter seems promising, because of the large number of available measuring sensors and the high sensitivity of a rafter to roof loads. A commercially available electronic level gauge was incorporated into the experiments to measure the bending angles. The gauge has a measuring range of ±5 mm/m and a minimum reading of 0.01 mm/m.

3.2 Experimental roof and Application of theory

Two experimental roofs having a pitch of 1/10 and 4/10 were constructed (Figure 4). Slabs of American wood of 100 x 100 mm were used for the posts, poles and tie beams, 60 x 45 mm slabs for the rafters, and 12 mm thick

lauan plywood for the roof panels. And they were assembled with nails and roofed with resin-coated sheet steel. The width of the roof was 1560 mm, having 4 rafters arranged at intervals of 455 mm. A steel box containing the electronic level gauge, i.e., the angle sensor, was fastened to a central rafter at x = 175 mm from the eave as shown in Figure 4(a).

The snow load on the roof, W kg/m², was assembled to be uniform on the roof. The supporting condition of a rafter was supposed to be that of a cantilever. The deformation angle, i (rad), at x were obtained as follows:

$$i = \frac{WB\cos\theta}{6\sum EI}\left(\frac{b}{\cos\theta}\right)^3\left(1-\left(\frac{x}{b}\right)^3\right) \tag{1}$$

where E is Young's modulus, and I the moment of inertia.

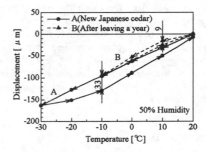

Figure 3. Displacement of the wooden column device due to atmospheric temperature change.

3.3 Experiment

The experimental roof with a 1/10 roof slope was located in the back yard of Nagaoka University of Technology from December 31, 1995 through February 4, 1996.

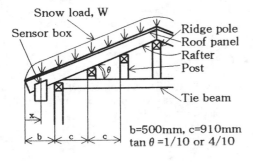

(a) Side view of experimental roof.

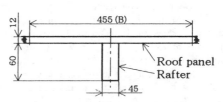

(b) Section of the rafter: B is the tributary load width. (All dimension in mm)

Figure 4. Illustration of experimental roof.

Every day the depth and density of the roof snow were measured and the snow load was calculated (product of depth and density). The observed snow was typical for Nagaoka. The maximum depth was 108 cm and the maximum load was 262 kg/m². Readings of the angle sensor showed no hysteresis by repeated returns to zero. The angle sensor readings as a function of snow loads are shown in Figure 5. Note the linear relation between them: The least square approximation is i = 5.31W μm (the dashed line). The standard deviation of the readings about the least square line was 14.7 kg/m².

Deformation angles were also calculated by equations (1) using a Young's modulas of 2.93 MPa and 6.99 MPa for the roof panel and the rafter, respectively, which were measured by bending tests using those materials. The calculated angles are also shown in Figure 5 as a solid line. Note equation (1) line for a cantilever, i = 5.25W μm, provides well fit to the measured angles.

Angle data from tests with the roof loaded with 15-kg sand bags are also shown in Figure 5 with the hollow circles. There was considerable fluctuation in the measured angle for a given sand bag load. Also the measured angles were substantially less than those predicted by the analytical equation. These effects are thought to be due to the sand bags sliding toward the eave and hence resulting in a non-uniform roof load.

4. DISCUSSION

The accuracy of three measuring methods be compared in this section. We consider the sensitivity of each method first. For the steel column device, the sensitivity of the device is given by 7.9×10^{-8} (ratio of column strain to snow load) $\times 1.8$ (column length for measurement) $= 0.142$ μm/(kg/m²). For the wooden column device, 0.256 μm/(kg/m²) of the sensitivity would be obtained. The sensitivity of the rafter angle sensor is directly obtained from the equation i = 5.31W, as 5.31 μrad/(kg/m²).

Next we consider the maximum errors for each of the three methods. In the steel column device, the maximum deviation of strain in Figure 2 (in difference between measured and calculated strain) was 75×10^{-6} which when multiplied by Young's modulus gives in maximum displacement error of 2.3 μm. Dividing by the sensitivity, we get a maximum measuring error in terms of load of 16.2 kg/m². In the wooden column device, the maximum displacement deviation of 33 μm were observed in the temperature tests for (A) while 12 to 17 μm were observed in the humidity tests. Based on these values we estimate the maximum error in deformation to be 25 μm. Then dividing by the sensitivity, the maximum error of the device in terms of load is 25/0.256 = 98 kg/m². In the rafter angle sensor, the standard deviation of the measured data was 14.7 kg, which includes possible errors in measuring the snow loads. We estimate the error in the angle device by considering the deviation between the

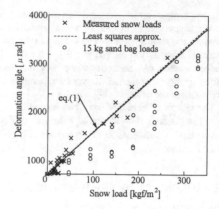

Figure 5. Measured and calculated snow loads.

approximated line (dashed least squares line in Figure 5) and the calculated line (solid of in Figure 5). At a roof load of 1250 kg/m², the deference is 75 μrad. Dividing by the sensitivity, gives the error of the sensor as 75/5.31=14.1 kg/m². This error would be less for a smaller roof load. Also it is affected by neither temperature nor humidity.

5. CONCLUSION

Three methods of direct measurement of roof snow loads were tested and compared from the viewpoint of accuracy for the same roof load conditions of up to 1250 kg/m². The results are summarized as follows:
(1) The method with the strain amplifying device for a steel column had an accuracy of about 16 kg/m². It is only slightly reduced due to temperatures variations.
(2) The similar method for a wooden column had an accuracy of 98 kg/m², owing to temperature and humidity variations.
(3) The method using a deformation angle sensor for a rafter had the estimated accuracy of 14 kg/m² which can be improved in future. Therefore this is the most promising method for measuring roof snow loads.

REFERENCES

Umemura, T. 1993. Evaluations of roof snow loads and snow melting technology. Japan Soc. of Snow & Ice Hokushinetu Branch, 21-96.

Snow Engineering: Recent Advances, Izumi, Nakamura & Sack (eds) © 1997 Balkema, Rotterdam. ISBN 90 5410 865 7

Probabilistic models for snow loads in structural analysis

V. Raizer
Central Research Institute of Building Structures, Moscow, Russia

ABSTRACT: Probabilistic models for snow load are here discussed accordingly to design code making procedures. Collection and analysis of basic metereological information are considered. Maximum values of snow load are investigated. In order to take into account the influence of long-term exploitation process and the analysis of load combination factors, a random process of snow accumulation is analysed. Filter Poison and Markov processes for presentation of snow load are taken into account.

1. INTRODUCTION

A thorough physical analysis must precede statistical and probabilistic models of snow loads. It includes:
— reasons of their occurence and variations connected with nature phenomena;
— space and time features connected with their manifestations and conditions grouped according to territory, shape of building, regime of structural use;
— methods of research helping to estimate homogeneity and reliability of observation and of its results;
— methods of adaptation and regulation of measurements.

On the base of this analysis statistical load models are formulated including description of their general totality and areas of use.

Main physical factors forming a snow load on building are winter atmospheric precipitation, temperature regime at the site, predominant wind direction and geometry of the roof. Snow re-distribution along roofs through wind tranfer is realised after each snowfall.

Snow melt also depends on above mentionned factors and on the thermal resistence of the roof. For regions with continental climate the process of accumulation and melt is illustrated. Yearly fluctuations of snow level are the reason of periodically changing snow load.

2. INTENSITY OF SNOW LOAD

Calculating maximum snow load is necessary in structural analysis for appreciation of an absolute limit state. Intensity of snow load on roof can be presented as a random function of coordinates (x.y), time (t), site (g) and building factors (c):

$$\tilde{S}(x,y,t/g,c) = \tilde{S}_0(t/g)\,\tilde{K}(g,c/S_0)\,\tilde{\mu}(x,y/g,c,S_0) \quad (1)$$

Where S_0 (t/g) is the main climatic parameter of snow load which characterizes a weight of snow (N/m²) accumulated to moment t.

$\tilde{K}(g,c/S_0)$ — transitional function from weight of snow on the ground to accumulation of snow load on the roof.

$\tilde{\mu}(x,y)$ (g,c/S_0) — re-distribution function of snow load along the roof.

Site factors (g) are geographical region, topography, uneveness. These factors have been defined by the inequality of snow accumulation on the ground, wind and temperature — humidity regime, territory, protection of meteorological station and building. *Building factors* (c) are determinated by form, dimentions, thermo-physical properties of structural surface. All these factors together with site factors define snow load on roof.

$\tilde{K}(S,c/S_0)$ and $\tilde{\mu}(y/g,c,S_0)$ functions are very complicated for statistical investigation and probabilistic models.

That is why only random process $\tilde{S}_0(t/g)$ based on observation results will be considered further. There are two kinds of $\tilde{S}_0(t)$ measurements:
— dayly measurements of snow depth;
— ten day period measurements of the depth and density of snow with subsequent calculation of water equivalent.

Several years of observation of $\tilde{S}_0(t)$ over each winter represent rectangular impulse realisation on a regular stream (Figure 1). If parameters of continuity for snow load are not investigated, the process can be presented in the form of a momentary impulse with intermittent time (Figure 2).

Considering features on Figure 1 as conditionaly stationary and taking into account all totality of impulse S_0 as random value, one can hold one-measured (upper index is 0) distribution function $P^0{}_{S_0}(S_0)$ and its parameters.

$$P^0{}_{S_0}(S_0) = 1 - Q^0{}_{S_0}(S_0) = 1 - \Delta t n_{S_0}/t_0 =$$
$$= 1 - n_{S_0}/N = 1 - P^0{}_{S_0\Delta t} \quad (2)$$

where $N = t/\Delta t$ general number of rectangular bound impulses.

n_{S_0} — number of crossings through given level in interval T. This number depends upon number of Δt intervals where $\tilde{S}_0(t) \geq S_0$

$P_{S_0\Delta t}$ — relative frequency of crossings.

$P^0{}_{S_0}(S_0)$ — is one-measured distribution function of momentary $S_0(t)$ values (Figure 1) and in addition to one-measured distribution function of maximums $P^0{}_{S_0\Delta t}(S_0)$ which corresponds to

realisation shown in Figure 2. Mean S_0 and standard deviation σ_s parameters will be:

$$\bar{S}^0_0 = 1/N\sum_{j=1}^{N}S_0\Delta j \quad \sigma^0{}^2_{S_0\Delta} = 1/N - \sum_{j=1}^{N}(S_0\Delta j - \bar{S}^0_\Delta)^2 \quad (3)$$

Since there is a rather large variation of maximum one-year snow load, Som, Gumbel Type 3 distribution can be used (Gumbel, 1967).

$$P(S_{0m}) = \exp(- \exp \alpha - S_{0m}/\beta) \quad (4)$$

Density function will be as following:

$$P(S_{0m}) = 1/\beta\{\exp(- \exp \alpha - S_{0m}/\beta + \alpha - S_{0m}/\beta)\} \quad (5)$$
$$- \infty < \alpha < \infty \qquad \beta > 0$$

α and β parameters can be expressed by mean Som, standard deviation $\sigma^2(S_{0m})$ factors

$$S_{0m} = \alpha + 0{,}577$$

$$\sigma^2(S_{0m}) = 1{,}645 \ \beta^2 \quad (6)$$

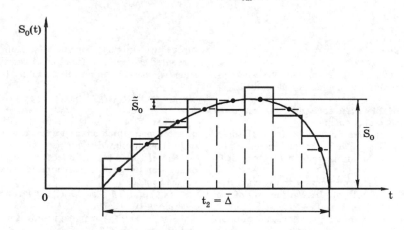

Figure 1   Snow Load on the Ground as a Function of Time:$S_0(t)$ vs. t

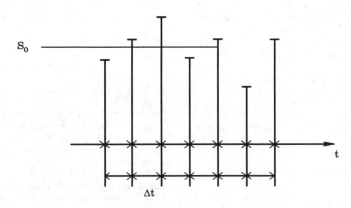

Figure 2   Snow Load So Presented in the Form of a Momentary Impulse with Intermittent Time

α and β differ according to region

In Moscow for example $\alpha = 931$ N/m², $\beta = 365$ N/m², if a building has been designed for n years, it is necessary to find a maximum allowable snow load. S_0 that:

$$P(S_{0n}) = P^n(S_{0m}) \qquad (7)$$

From (7) $V = 1 - P^n(S_{0m})$ and

$$n = \log(1 - V)/\log P(S_{0n}) \qquad (8)$$

and $P(S_{0n}) = \sqrt[n]{1 - V}$

Using equation (4) one will get
$$P(S_{0n}) = \exp(-\exp\alpha_n - S_{0n}/\beta) \qquad (9)$$

where $\alpha_n = \alpha + \beta \ln n$

Parameters of this distribution are written as follows:

$$\bar{S}_{0n} = \bar{S}_{0m} + \beta \ln n \qquad \sigma^2(S_{0n}) = \sigma^2(S_{0m}) \qquad (10)$$

3. ANALYSIS OF RANDOM PROCESS FOR SNOW ACCUMULATION

Random process of snow accumulation must be considered in the case of a functional limit state analysis.

If accumulation processes can be divided into sections, the random function will be substituted with totality of random values.

In each of N sections the numerical parameters of random value S(t) can be calculated:

$$\bar{S}_j = 1/N \sum^N S_{ij} \; ; \quad \sigma^2_j = 1/N - 1\sum_{i \neq 1}^N (S_{ij} - \bar{S}_j)^2 \qquad (11)$$

Distributions in some sections are shown on Figure 3. The analysis makes it clear that asymmetric distributions predominate at the beginning of winter. When transitional autumn period is finished and autumn thaw of snow is stopped, distribution will come close to a normal one. In the end of winter (March, April) one can see an inverted tendency.

Such phenomenon can be explained by the fact that in the beginning and in the end of winter the probability of thaws is large.

For theoretical models of random process of snow load Poisson process can be used. This model supposes that variations of snow weight in winter present a really accountable stream. Change of snow depth is realized with a calculated number of states that is common flaw of random events. This flaw is taken as a stationary and ordinary one.

Data in the form of ten day periods of water equivalents were used from 1947 to 1979 in the north of Russia (9 stations). For each station, 32 processes with 17 sections were taken in the period from November to April.

Accumulation process can be written as follows:

$$\tilde{S}(t) = \sum_{m=1}^{\tilde{n}(t)} F[t, \tilde{\tau}_m, \tilde{V}(t_m)] \qquad (12)$$

t — flowing time, $\tilde{\tau}$-m — time of load variation

$\tilde{V}(tm)$ — random value of load increment in the moment of m bound.

$F[t, \tilde{\tau}_m, \tilde{V}(t_m)]$ — function of snow load increment in m bound.

Random values $n(t_{j+1}) - n(t_j)$ — increment of stream events in a month ($j = 0,1,2,3...$). They have Poisson distribution (Pirson criteria with 5%) with constant values:

$$P[n(t_{j+1}) - n(t_j)] = \Sigma(t_{j+1} - t_j)^n \times$$

$$\times \exp[-\lambda(t_{j+1} - t_j)/n] \qquad (13)$$

When $\lambda = 0,78$, this expression will be the same for each t_j. S_0 that the stream will be stationary, and time of events will have an exponential distribution. $P(\tau) = 1 - \exp(-\lambda\tau)$

4. REPRESENTATION OF SNOW LOAD AS A MARCOV PROCESS

Marcov's process notion enlarges the class of loads used in structural analysis. A process is considered to be Marcov's if connected with the intensity and time when posterior actions and time moments of their appearence depend only on anterior actions and time moments and do not depend upon the loading itself. Intermittent Marcov chain is a random Marcov process and space of state is finite, but a multitude of time indexes T = 0,1,2... coincides with the part of this multitude. Analysis of the Marcov chain is connected with calculation of probability in K-steps.

$p^{(k)} = \| p_{ij}(k) \| p_{ij}^{(k)}$ — probability of transition from i state to j one in K-steps.

System of differential Kolmogorov equations for probabilities of state can be written in following form (Raizer, 1986):

$$\left.\begin{array}{l} dP_0/dt = -\lambda_0 P_0 \\ \\ dP_1/dt = -\lambda_0 - \lambda_1 P_1 \\ \\ dP_k/dt = -\lambda_{k-1} P_{k-1} - \lambda_k P_k \end{array}\right\} \qquad (14)$$

Basic conditions will be:

$$t = 0 \rightarrow P_0 = 1, \; P_1 = P_2 ... = P_m = 0 \qquad (15)$$

Considering the homogeneity of the load process (Smirnov criteria), the preliminary analysis obtains $\lambda = 0,136$, λ — distribution parameter of time to process in state S_k (Weibull distribution)

$$F(x) = 1 - e^{-\lambda t} \qquad (16)$$

271

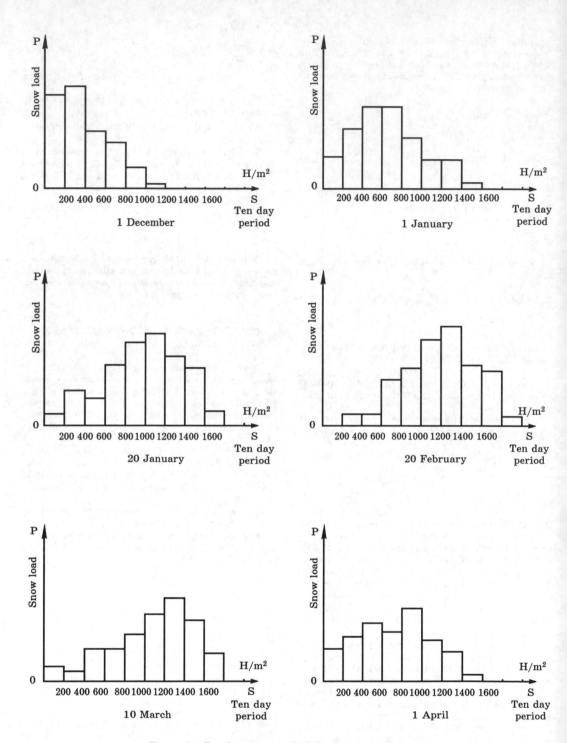

Figure 3   Random process distributions in sections

i	0	1	2	3	4	5	6	7	8
$P_{t,i-1}$	0	0,16	0,11	0,2	0,18	0,14	0,2	0,3	0,32
$P_{v,i}$	0,7	0,57	0,51	0,54	0,55	0,52	0,5	0,2	0,54
$P_{i,i+1}$	0,2	0,18	0,3	0,15	0,12	0,23	0,1	0,3	0

Table 1 Statisticl Values of Transitional Probabilities

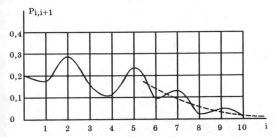

Figure 4 Approximation of transitional probability, $P_{i,i+1}$

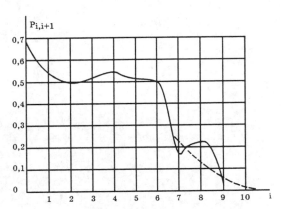

Figure 5 Approximation of transitional probability, $P_{i,i}$

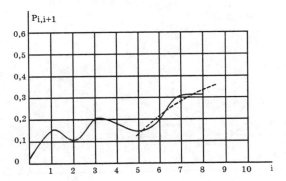

Figure 6 Approximation of transitional probability, $P_{i,i-1}$

Taking into account that the mean time value of the process in state S_k and the distribution parameter are expressed by formula:

$$t_m = 1/\lambda \qquad (17)$$

$t = 7$ days. This means that homogeneous process must be divided in 7 days. Usually the depth of snow is measured every 5 (or 10) days and these observations can be used without change. Steps in snow weight evaluation are 200 N/m². S_0 state S_0 is from 0 to 200 N/m². S_1 — from 200 to 400 N/m². Maximum value is 2200 N/m² (this value has been measured in Moscow region three times only).

5. STATISTICAL VALUES OF TRANSITIONAL PROBABILITIES

See Table 1.
The results are also shown on Figures 4-6. The next important step is the approximation of transitional probabilities in the domain of infrequent events. In Figures 4-6 there are analitical curves shown with dotted lines for k>7.

$$P_{i,t-1} = 0,1\sqrt{4,1i - 25,6}$$

$$P_{i,i} = 4,5/(25i - 161) \qquad (18)$$

$$P_{i,i+1} = 0,7/(3i - 20)$$

6. CONCLUSION

Some rules and methods for statistical base of probabilistic snow load models are given here for estimation of design parameters.

7. REFERENCES

1. Raizer V. Reliability theory methods for design code making procedures. Moscow Strojisdat, 1986 (in Russian).
2. Gumbel E. (1967). Statistics of Extremes. Columbia University Press, New York.

Snow Engineering: Recent Advances, Izumi, Nakamura & Sack (eds) © 1997 Balkema, Rotterdam. ISBN 90 5410 865 7

A survey of snowdrift on the pipeline structures

Naoki Kitamura
Department of Civil Engineering, Maebashi City of Technology, Japan

Kiyomichi Aoyama
Research Institute for Hazards in Snowy Areas, Niigata University, Japan

Teruyasu Nishizawa
Faculty of Economics, Niigata University, Japan

Yukikazu Tsuji
Department of Civil Engineering, Gunma University, Kiryu, Japan

ABSTRACT: Snow loads were observed on prototype pipeline-structures. Little snowfall during the several years of observation so heavey loads could not be measured. Snow dispersion the truss structure and pipe beam was observed. Observations indicated that (1) Snow did not lie uniformly on the structures; (2) Someterminal areas such as truss panel points were heavily cover-ed by snow; and (3) The variation in height of snowdrifts on the pipeline-structures was mainly influenced by temperature.

1 OUTLINE OF OBSERVATIONS

1.1 *Location of observations*

Tests were conducted in Tsugawa town (Figure 1) which is one of the heaviest snowfall arears in Japan. Three meters snowdrifts have occurred in this mountainous place every year prior to 1989.

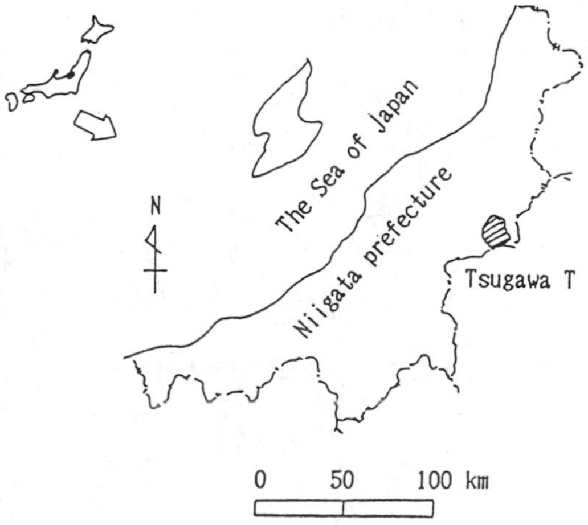

1.2 *Scale of model structures*

The size and shape of these structures are shown in Figures 2 and 3. They are located in a suburb of Tsugawa town. Load cells on the bearings of the truss structure are able to detect snow loads. In order to obtain data of depth of snowdrift on the pipe, the pipe beam is structured there (Figure 3).

2 RESULTS OF OBSERVATION

Rainfall changes to snowfall when the atmospheric temperature is below about 4 °C. Snow began accumulating on the ground and on the structures a little while after snowfall be-

Figure 1 Locations of observations

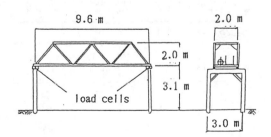

Figure 2 Geometry of truss structure

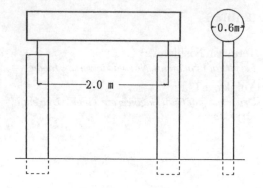

Figure 3 Geometry of pipe beam

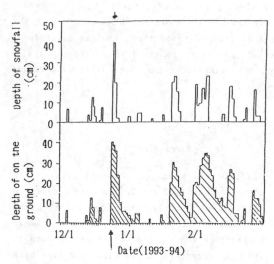

Figure 4 Snowfall and snow on the ground

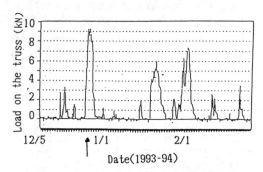

Figure 5 Change in load on the truss

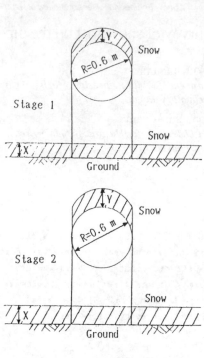

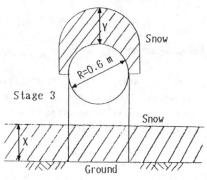

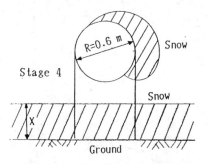

Figure 6 Stages of snow depth

Table 1 Recors of Snowdrifts
(1993 - 1994)   (No. 1)

Date	Demperature	Depth of snowdrift(cm)	
Snowed day	Degree (℃)	on the ground	on the pipe
12/14	2	6	6
15	2	1 3	1 3
16	5	1 3	3
17	5	6	0
18	2	1 0	0
19	2	3	0
24	0	4 0	3 5
25	0	4 6	3 5
26	3	2 8	1 7
27	3	1 9	3
28	4	1 3	0
29	3	7	0
30	5	6	0
1/15	4	1 1	1 1
16	2	4	0
19	1	7	7
20	0	2 6	2 6
21	2	2 5	2 1
22	6	1 9	1 0
23	3	1 6	0
24	5	1 4	0
25	7	1 2	0
26	5	1 0	0
27	8	5	0
28	2	1 3	1 2
29	3	1 8	1 6
30	0	1 5	2
31	1	1 9	5
2/ 1	1	2 3	1 0
2	1	2 5	0
3	7	3 7	0
4	4	1 8	0
5	1 0	1 5	0
6	1 0	1 4	0
7	6	1 2	0
8	1	1 0	0
9	2	2 7	1 5
10	4	3 0	5

Table 1 Recors of Snowdrifts
(1993 - 1994)   (No. 2)

Date	Demperature	Depth of s	wdrift(cm)
Snowed day	Degree (℃)	on the ground	on the pipe
2/11	6	2 4	0
12	7	1 5	0
13	6	1 5	0
14	6	1 5	0
15	6	1 3	0
16	9	9	0
25	3	1 2	0
26	4	8	0
27	7	5	0

gan. Snow on the ground increased little by little with the passage of time. Snow on the structures increased more slowly than on the ground, because some parts of snow on the structures fell off.

2.1 Snowfall and snow on the ground

A change of snowfall and snow on the ground during the 1993-1994 winter is shown in Figure 4. As shown in Figure 4, the maximum depth of snow on the ground within the period of the observation was only 46 cm. This value was less than 12 % of the mamimum depth of snow on the ground past in Tsugawa.

2.2 Load on the truss

Changes in the load on the truss are shown in Figure 5. The maximum load is 9.04 kN. From Figure 4 and Figure 5, Changes in the load on the truss are similar that of snow depth on the ground. But the strict relationship between load and snow depth can not be analyzed as simple problem, so that this observation will continue in the future.

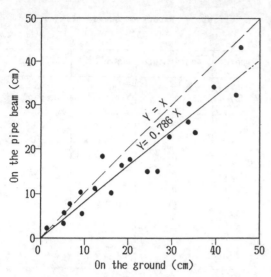

Figure 7 Relationship between Y and X

2.3 *The relationship between the depth of snow on the ground and that on the pipe beam*

This relationship is shown in Figure 6. In this figure, respective symbols are showing the followings:

Stage 1 – Case of a light snowfall,
Stage 2 – Case of a middle snowfall,
Stage 3 – Case of a heavy snowfall,
Stage 4 – Case of fallen snow,
X – Snow depth on the ground, Y – on the pipe. The depth of snow on the pipe beam was less than on the ground.

3 CONSIDERATIONS

The lower diagram in Figure 4 expresses a change of snow depth. Figure 5 shows a change of snow load detected by cells. The arrow in Figure 5 points out to the maximum value within the observation period. Comparing of the two figures, reveals the following two points:
1) Changes in snow load are changes in depth.
2) The arrows in both figures correspond to the same data.
From Figure 6, the relationship between Y and

X in Fifure 7 was estimated as the following:
$$Y = 0.786 X,$$
Fifure 7 was drawn from data based on Table 1.

4 CONCLUSION

From these observations the following results were obtained:
1) Snow did not lie uniformly on the structure.
2) Some terminal areas such as truss panel points were heavily covered by snow.
3) The variation in height of snow on the pipeline-structures was mainly influenced by temperature.
4) The depth of snow on the horizontal pipe beam averaged about 79 % that on ground.
This experiment will continue with the hope that more snow is encountered in the future.

REFERENCES

Boris Bresler & T.Y. Lin 1960. *Design of Steel Structures.* John Wiley & Songs, INC: London.
Japan Road Society 1980. Specifications for Highway Bridges, Part 1 Common Items.
Smith, D.W. 1997. Why Do Bridges Fail. ASCE, No. 462/18:121–130.
Y. Fujita, Y. Honma & N. Kitamura 1994. Suveying of Snowdrift on the Pipeline Structures. *Proceedings of Japan Society for Snow Engineering Voll.11:219–224.*

Snow Engineering: Recent Advances, Izumi, Nakamura & Sack (eds) © 1997 Balkema, Rotterdam. ISBN 90 5410 865 7

Heating characteristics of self temperature control heater

Yasuhiko Hori, Tetsuo Ito & Takeshi Taniguchi
Electrical Engineering Department, CRIEPI, Yokosuka, Japan

Toshiro Yamada & Kenji Shinojima
Tohoku Electric Power Co., Inc., Sendai, Japan

ABSTRACT: This paper describes the heating characteristics of a self temperature control heater (STCH) which is supposed to be used in a melting system for roof snow. The characteristics were considered to be influenced mainly by ambient temperature, wind and snowfall conditions. An experimental roof snow thawing model using an STCH was made, to measure the change of the thermal resistance between a heating element in the STCH and the open air for various conditions. The experiments were carried out in an environmental room, in which ambient temperature, wind and snowfall conditions could be controlled. It was found that the STCH consumed more than twice the electric power when covered with snow, compared to when it was not covered with snow, and that the electric power consumed by the STCH was hardly influenced by ambient temperature with snow cover. Furthermore the electric power consumption decreased about 20% when the STCH surface was half covered with snow, and about 15% when there were some air gaps between the STCH surface and the snow.

1 INTRODUCTION

A self temperature control heater (STCH) is mainly made of fluorocarbon polymers containing electrical conductive carbon grains. The electric resistance of the STCH tends to increase as its temperature increases. Heat expands the STCH making the electrical conductive carbon grains less densely dispersed, thereby increasing the electric resistance of the STCH. Since the electric power consumption decreases, the STCH is not overheated. So, the STCH is one of the candidates for use in a melting system for roof snow. However, it is difficult to design a melting system for roof snow using the STCH, since the characteristics of the STCH are influenced by ambient temperature, wind and snowfall conditions.

This paper describes the heating characteristics of an STCH placed in an environmental room, in which the ambient temperature, wind and snowfall conditions could be changed. Based on the experimental data, an equivalent thermal circuit is worked out, to be used for designing melting systems for roof snow, and a simplified calculation method is proposed.

2 STRUCTURE OF STCH AND EXPERIMENTAL CONDITIONS

The structure of the STCH is shown in Figure 1(a). In the STCH, the heating element is covered with a polyester film for electrical insulation. An aluminum sheet is laminated on the polyester film, to ensure uniform heat supply. The STCH model intended to be used in a melting system for roof snow is shown in Figure 1(b). The aluminum plate functions as a house roof. The characteristics, such as the STCH temperature vs. resistance relation, the thermal resistance between the heating element and the open air, and also the influence of ambient temperature, wind and snowfall condition were clarified.

The experiments were carried out in an environmental room, in which the ambient temperature could be changed from -20°C to 50°C and in which wind and snowfall conditions could also be controlled. Temperatures of the surface of the polyester film, the surface of the aluminum plate and the air in the room were measured. Temperatures of the snow on the aluminum plate and of falling snow were also measured.

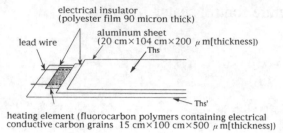

electrical insulator
(polyester film 90 micron thick)

lead wire

aluminum sheet
(20 cm×104 cm×200 μm[thickness])
Ths

Ths'

heating element (fluorocarbon polymers containing electrical
conductive carbon grains 15 cm×100 cm×500 μm[thickness])

aluminum plate
(25 cm×115 cm×1 mm[thickness])

Tsurf

thermal insulator

heating element
(15 cm×100 cm×500 μm[thickness])

(a) Structure of STCH

(b) Experimental roof snow thawing model

Figure 1. STCH components and an experimental roof snow thawing model.

3 RESULTS AND DISCUSSION

3.1 Heating characteristics of STCH

The dependence of electric resistance on the STCH surface temperature is shown in Figure 2. The electric resistance remained in a low range from $200\,\Omega$ to $250\,\Omega$, when the STCH temperature was in a range from -10℃ to 10℃. However, when the STCH temperature was higher than 40℃, the electric resistance increased remarkably. Thus, the STCH changed from $200\,\Omega$ to $1500\,\Omega$ in electric resistance. The STCH surface temperature was affected by ambient temperature, the thermal resistance between the heating element and the open air, and the time after turning on the electric power. Consequently the quantity of heat generated in the STCH (electric power consumption), as the heating area is 1500 cm2, ranged widely from 25W to 200W when A.C. 200V was supplied.

Figure 3 shows the relation among the generated heat flux Q, STCH surface temperature Tsurf, and wind speed without snow. Dividing the electric power consumption of STCH by the heating element area, the generated heat flux Q was obtained. At every ambient temperature, the generated heat flux Q tended to increase and the STCH surface temperature Tsurf tended to decrease with the increase of wind speed, but the increase of generated heat flux tended to level off even if the wind speed increased further. The STCH consumed more than twice the electric power in a wind speed range from 2 to 3 m/s, compared to a state where the STCH was allowed to stand without wind.

Figure 4 shows the heating characteristics with and without snow, as relationships between the generated heat flux and the ambient temperature. Also shown is a case where the STCH surface had snow on only one half of the heating area. The electric power consumed by STCH was hardly influenced by ambient temperature with snow cover.

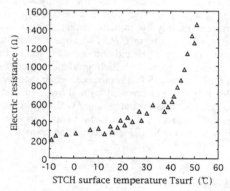

Figure 2. Dependence of electric resistance on STCH surface temperature.

Experimental Conditions

—■— Q at ambient temperature of 20℃
—●— Q at ambient temperature of 0℃
--□-- Tsurf at ambient temperature of 20℃
--○-- Tsurf at ambient temperature of 0℃

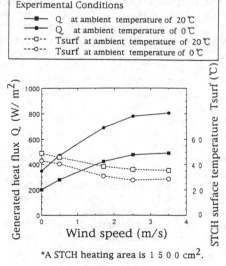

*A STCH heating area is 1 5 0 0 cm^2.

Figure 3. Relationships between generated heat flux, STCH surface temperature, and wind speed without snow.

Experimental Conditions

o with snow cover
o with snow cover placed on only
    one half of the heating area
• without snow cover
** with a gap between the STCH
    surface and the snow

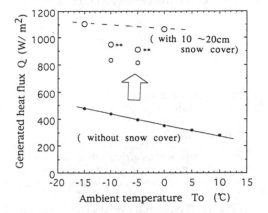

*A STCH heating area is 1 5 0 0 cm^2.
Figure 4. Heating characteristics with snow
cover and without snow cover.

The STCH covered with snow consumed about 1100 W/m2, which was more than twice that of the STCH without snow. Furthermore the electric power consumption decreased about 20% when the STCH surface was half covered with snow. The electric power consumption decreased about 15% when the STCH surface was fully covered with snow, with air gaps between the STCH surface and the snow. Thus, the electric power consumed by the STCH was affected by the snow condition over the STCH surface.

3.2 Simplified calculation method for melting systems for roof snow using STCH

A thermal circuit of a melting system for roof snow using the STCH is shown in figure 5. Generated heat quantity W (electric power consumption) can be divided into the heat quantity W1 and W2, which are defined as radiating heat into air or the snow through the radiating aluminum plate and radiating heat into air through the lower thermal insulation board respectively. The heat quantity W1 radiating through the radiating aluminum plate can be expressed by equations (1) to (4).

$$W1 = \frac{(R_{th1}+R_{th4}) \times W}{\{(R_{th1}+R_{th2}+R_{th3})+(R_{th1}+R_{th4})\}} \qquad (1)$$

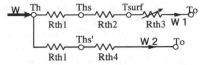

Th : heating element temperature
Ths : surface temperature of heating element
      on aluminum sheet (upper side)
Ths ' : surface temperature of heating element
      (lower side)
Tsurf : STCH surface temperature on aluminum plate
To : ambient temperature
Rth : overall thermal resistance
Rth1 : thermal resistance with a polyester film
Rth2 : thermal resistance between surface of heating
      element and STCH surface
Rth3 : thermal resistance between STCH surface and
      open air
Rth4 : thermal resistance between surface of electrical
      insulator and open air
W (=W1+W2) : generated heat quantity

Figure 5. Thermal circuit of a melting system
for roof snow using the STCH.

$$R_{th} = \frac{(Th - To)}{W} \qquad (2)$$

$$R_{th1}+R_{th4} = \frac{t_1}{\lambda_1 \times S_1} + \frac{t_4}{\lambda_4 \times S_4} \qquad (3)$$

$$R_{th1}+R_{th2}+R_{th3} = \frac{1}{\frac{1}{R_{th}} - \frac{1}{R_{th1}+R_{th4}}} \qquad (4)$$

$\lambda 1$: thermal conductivity of electrical insulator
$t1 (=90 \ \mu m)$; thickness of electrical insulator
$S1 (=1500 \ cm^2)$: area of electrical insulator
$\lambda 4$: thermal conductivity of thermal insulator
$t4 (=6 \ cm)$: thickness of thermal insulator
$S4 (=1500 \ cm^2)$: area of thermal insulator

The generated heat quantity (W), heating element temperature (Th), and ambient temperature (To) were obtained by measurement. By using W, Th and To, the overall thermal resistance (Rth) was calculated from equation (2). It is supposed that the thermal conductivity of the electrical insulator (polyester) $\lambda 1$ is 0.27 to 0.37 W/m K (The Society of Polymer Science 1986), and that the thermal conductivity of the thermal insulator $\lambda 4$ is 0.033 W/m K (The research committee of Heating Technology 1991, Thermal Properties Handbook editorial committee 1990). By using these thermal conductivities, Rth1+Rth4 was calculated from equation (3). Rth1+Rth2+Rth3 was transformed by equation (4). Consequently, W1 which radiated into air or snow through the aluminum plate with a size of 25 cm×115 cm × 1mm (thickness) could be calculated from equation (1).

As the result, in a wind speed range from 0 to 4m/s without snow at 0℃, the percentage of W1 to the overall heating power (W) became from 92% to 98%. With snow on the STCH surface, the percentage of W1 became over 98%. With snow cover, W could be expressed by equation (5) approximately.

$$W = (T_h - T_o) \times (\frac{1}{R_{th1} + R_{th2} + R_{th3}}) \quad (5)$$

Rth3 could be calculated by using the STCH surface temperature (Tsurf), ambient temperature (To) and the heat (W1) obtained above. Rth3 was compared with the reference data (The research committee of Heating Technology 1991) used for road heating systems in a temperature range from -10℃ to 0℃. Even though the reference data did not show such conditions as the emissivity of the surface road, surface temperature and ambient temperature, the values of Rth3 agreed well with the reference data.

When the STCH is covered with snow, Rth3 is equal to 0, since the STCH surface temperature (Tsurf) and ambient temperature (To) are equal, being about 0℃ respectively. In this case, the overall heating power (W) can be obtained from an approximate equation (6) :

$$W = (T_h - T_o) \times (\frac{1}{R_{th1} + R_{th2}}) \quad (6)$$
$$*T_o = 0℃$$

The dependence of the STCH surface temperature on the ambient temperature is shown in Figure 6.

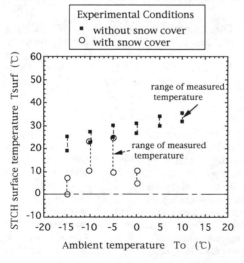

Figure 6. Dependence of STCH surface temperature on ambient temperature.

In some cases, although the STCH surface was covered with snow, the STCH surface temperature ranged from 10℃ to 25℃. This suggests that there were some air gaps between the STCH surface and the snow. In these cases, the temperature of the snow which is going to be thawed can be assumed to be about 0℃. So, the overall heating power (W) can be obtained from an approximate equation (7) :

$$W = (T_h - T_o) \times (\frac{1}{R_{th1} + R_{th2} + R_{th3}}) \quad (7)$$
$$*T_o = 0℃$$

The experimental value of Rth3 became in these cases were from 200 to 500 K cm2/W .

4 CONCLUSION

Experiments were carried out in an artificial environment room, in which ambient temperature, wind and snowfall conditions could be controlled.

(1) The STCH consumed more than twice the electric power in a wind speed range from 2 to 3 m/s, compared to a state without wind.
(2) The STCH consumed more than twice the electric power when it was covered with snow, compared to a condition where it was not covered with snow. The electric power consumption decreased about 20% when the STCH was covered with snow by half. And it decreased about 15% when the STCH surface was fully covered with snow, with existence of air gaps between the STCH surface and snow.
(3) A simplified calculation method was proposed to design melting systems for roof snow using the STCH.

REFERENCES

The Society of Polymer Science. 1986. Polymer Science Data Handbook. BAIFUUKAN LTD., TOKYO. (in Japanese)
The research committee of Heating Technology. 1991. Manual of Heating Technology for design. OHMSYA LTD., TOKYO. (in Japanese)
Thermal Properties Handbook editorial committee. 1990. Thermal Properties Handbook. YOKENDO LTD., TOKYO. (in Japanese)

Snow Engineering: Recent Advances, Izumi, Nakamura & Sack (eds) © 1997 Balkema, Rotterdam. ISBN 90 5410 865 7

Demonstration of an electric heater that prevents icing at eaves

Shinya Hirama
Research & Development Center, Tohoku Electric Power Co., Inc., Sendai, Japan

Hiroaki Wakasa
Heating & Engineering Division, Kitanihon Electric Cable Co., Ltd, Sendai, Japan

ABSTRACT: An electric heater was developed as an aid in preventing icicles from forming on residential eaves in cold and snowy areas. Winter testing resulted in a reduced number of icicles forming but did not eliminate them. Thick snow accumulations caused snow to creep off the roof edges, and icicles formed on the outer edge of such accumulations.

1 OBJECTIVE

In cold and snowy areas, electric panel heaters are installed under residential metal roofs to prevent icicle formations. This method, however, has high initial and operating costs. Our objective was to develop an electric heater that prevents icing at eaves.

2 ELECTRIC HEATER THAT PREVENTS ICING AT EAVES

Figure 1 shows the electric heater installed on an eave. Figure 2 is a cross-section of the heater. The heater is 900 mm long, 50 mm wide, 11 mm thick and installed under the eave. It uses an internal temperature sensor with ON-OFF control.

3 HISTORY

In January and February, 1994, tests were conducted on an initial model of the heater and its effectiveness was confirmed. This model was improved and subjected to further testing. The major improvements were: (1) an outside temperature control was removed leaving an inside temperature sensor as the only control; (2) the inside temperature sensor was located close to the surface of the equipment; (3) an insulating panel was included in the casing.

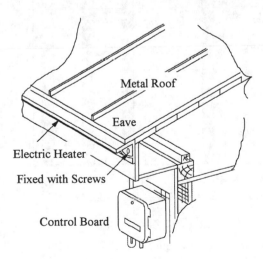

Figure 1 Electric heater installed on an eave.

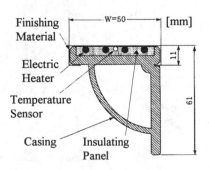

Figure 2   Cross-section.

4 LABORATORY TEST

Prior to outdoor testing, laboratory tests were done to determine the required heat capacity. In the environmental laboratory, at -15℃, artificial snow was precipitated on a model roof surface which had a heater installed. Heating the roof surface and sprinkling with water produced conditions suitable for icicle formation. Test results indicated the lowest values of heater width and heat capacity to be 50 mm and 30 W/m, respectively.

5 OUTDOOR TEST CONDITIONS

Tests were made at eight locations in Aomori City during the winter. Figure 3 shows one of the locations. Equipment consisted of three electric heaters with a total length of 2.7m. Power to the equipment was 30 and 40 W/m. Control conditions were 0℃ for ON and 5, 10, and 15℃ for OFF.

Figure 3   Testing (right and left: installed, center: not installed).

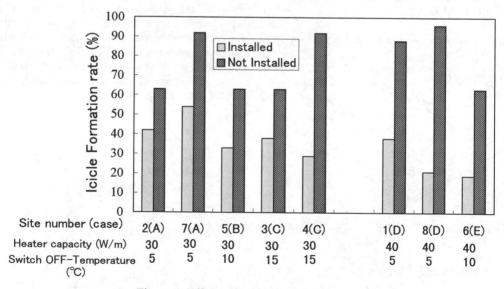

Figure 4 Effect of heaters on icicle formation.

6 OUTDOOR TEST RESULTS

Figure 4 shows the results of the icicle prevention test conducted from January 15 to February 7, 1995. We defined icicle formation rate(%) as (icicle formation days / observation days) × 100. Mean temperature was -1.8℃ and snow accumulation was 215 cm during the test period.

7 CONSIDERATIONS

7.1 *Icicle prevention*

Average icicle formation was 80.3% on the eave without an electric heater and 32.9% with an electric heater. This confirms the effect of the electric heater in reducing icicle formation. With higher power and OFF temperatures, icicle formation decreased. Icicle formation did not drop to zero, for the following reasons:

a. Thick snow accumulation caused snow to creep off roof edges, and icicles began to form at the outside edge of the snow accumulation (Figure 5). Heater configurations cannot solve this.

b. ON-OFF control sometimes leads to a melting-freezing cycle resulting in icicle formation. The heater requires further improvement to prevent this. Figure 6 shows temperature changes under operation.

Figure 5　Snow creep.

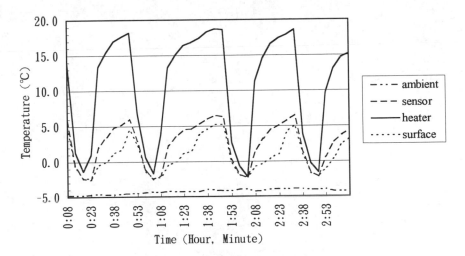

Figure 6　Temperature Changes

285

7.2 *Other evaluations*

The electric heater can be installed without difficulty. The simple configuration and easy installation contribute to costs lower than conventional systems. Running costs are lower than conventional systems because of its smaller capacity.

8 CONCLUSIONS

The electric heater is low cost and simple. However its effectiveness has not yet been completely studied. When used alone without other snow melting measures, the heater cannot afford very severe climatic conditions. The heater should be used with other snow melting systems in very cold climates.

REFERENCES

Hirama, S. , Yoshida, N. , Kisara, A. , Wakasa, H. 1994. Study of electric heater to prevent the edge of eaves frozen. *Proceedings of Japan Society for Snow Engineering,*Vol.11, Nagaoka, Japan.

Snow Engineering: Recent Advances, Izumi, Nakamura & Sack (eds) © 1997 Balkema, Rotterdam. ISBN 90 5410 865 7

Effect of characteristics of the electric heating cable on snow melting

K. Iwamoto
Hokkaido Snow Melting Institute, Japan

S. Sayama, Y. Nishikawa & M. Yamaguchi
Hokkaido National Industrial Research Institute, Japan

M. Sudo & Y. Sakai
Fujii Co., Ltd, Japan

ABSTRACT: There are three kinds of cables which are used for snow melting system in roads. Those are classified according to the materials of the inner wire and the outer insulator. Nichrome cable is generally used, but recently Carbon and Softelec (Teijin) are used also. In this study thermal properties of heating panels using three kinds of heating cables were tested in a cold chamber. The heating efficiency was graded as Softelec, Carbon, Nichrome, in that order. A model of thermal conduction is also discussed. The effect of the characteristics of a heating cable on snow melting is summarized.

1 INTRODUCTION

In general nichrome cable is able for electric road heating. Recently carbon and Softelec(Teijin) are also used. It is said by users that the characteristics of the electric heating cables on snow melting are different for each cable; however, the electric power per unit area is the same. In this report, two kinds of experiments are described; one is an experiment where the cable was heated directly in air and another is an experiment where the cable was set in a concrete panel which was put into a cold (-1℃ or -5℃) chamber and heated. Through these experiments, the thermal characteristics of these three kinds of heating cables were compared. A theoretical analysis was tried and theoretically calculated values were compared with actual values measured in the experiments.

2 ELECTRIC HEATING CABLES

Before describing the experiments, it is necessary to discuss the characteristics of the three kinds of the electric heating cables (Table 1).
The nichrome's electric resistance is very small, so the cable's diameters does not have to be very big. By Joule's law(1840), the cable's heating energy (cal) is calculated using equation (1).

$$Q = 0.24 \ I^2 \cdot R \cdot t = 0.24 \ W \qquad (1)$$

Because the total heating energy is determined by the electric power per unit area (W/ m^2), it is

theoretically correct that the performance of road heating systems are not different for a common electric power supply; however, the materials of cables are different. To understand these differences, an experiment using the heating cables listed in Table 1 was tried.

Table 1. Characteristics of the electric heating cables.

Type	Nichrome	Carbon	Softelec (Teijin)
Inner wire	An alloy of Copper, Nickel, Chrome Φ=1.5 mm	A Compound of Carbon and Plastic Φ=6.0 mm	A Compound of Stainless and Fiber Φ=5.0 mm
Outer cover	PVC t = 1.9 mm	PVC t= 1.5 mm	PVC t = 0.5 mm
Cable's diameter	Φ=5.3 mm	Φ=9.0 mm	Φ=0.5 mm
Electric Resistance	R = 0.372 Ω/m	R = 560 Ω/m	R = 858 Ω/m

* PVC Polyvinyl chloride

3 EXPERIMENT

To determine if the materials of a heating cable influences the performance of road heating, two types of experiments were tried. Three kinds of heating cables (Table 1) were used in each experiment.

3.1 Heating test of cables in air

The three kinds of heating cables were stretched horizontally in air. The length of each was 1 m, and the air temperature was about 15 ℃. Common electric power of 12 W/m was supplied to each cable, and the surface temperature(℃) of heating cable elements stretched in air was measured. As shown in Table 1, the cable's electric resistance was 0.372 Ω /m (Nichrome), 560 Ω/m (Carbon), 858 Ω/m (Softelec(Teijin)).

The temperature was measured at two points. One was the surface of the bared cable element, and another was the insulated cable element (Fig.1) Results are shown in Table 2. The temperature difference between the two points was largest for the Nichrome cable. With reference to this result, the structure of the three cables is shown in Figure 2.

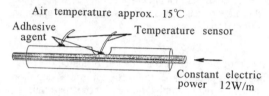

Figure 1. Heating test of the electric heating cables in air.

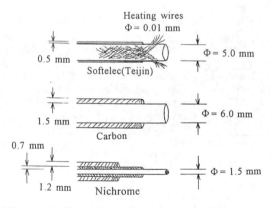

Figure 2. Structure of the three cables

3.2 Concrete block test

Next the temperature of a concrete block (panel) was measured in a cold chamber. The concrete block was 300 mm × 300 mm in area, 50 mm in thickness, and a 20 mm insulation of polystyrene was attached to its bottom. The electric heating cables (Table 1) were placed in the block and heated by constant electric power of 200 W/m². The cable's length was

Table 2. Surface temperature (℃) of heating elements stretched in air.

Type	Nichrome	Carbon	Softelec (Teijin)
Surface temperature (bared)	34.7 ℃	26.2 ℃	21.7 ℃
Surface temperature (insulated)	26.7 ℃	24.2 ℃	21.0 ℃
Measured △T	8.0 ℃	2.0 ℃	0.7 ℃
Calculated △T	8.1 ℃	2.5 ℃	1.2 ℃

Table 3. Temperature (℃) of concrete blocks heated at a constant power of 200 W/ m² in an temperature chamber (after 5 hours).

Type	Nichrome	Carbon	Softelec (Teijin)
Inner temp. Chamber at -1℃	11.5 ℃	13.8 ℃	17.8 ℃
Surface temp. Chamber at -1℃	8.5 ℃	10.8 ℃	13.0 ℃
Inner temp. Chamber at -5℃	9.1 ℃	11.2 ℃	15.2 ℃
Surface temp. Chamber at -5℃	6.6 ℃	8.9 ℃	10.6 ℃

2m (Nichrome, pitch = 5 cm), 1 m (Carbon and Softelec, pitch = 10 cm). The temperature in the cold chamber was set at -1 ℃ and -5 ℃, and after heating for 5hours, temperatures at three points were measured. The points were the surface, bottom, and interior of the block. (Fig. 5)

The results of this experiment are shown in Figure 3 (at -5 ℃) and Figure 4 (at -1 ℃) , and Table 3. Surface temperature of Softelec is highest, and that of Nichrome is lowest. The difference in surface temperature between the highest and the lowest was about 4 ℃ (Table 3) . After 5 hours all temperatures reached constant values.

4 THEORETICAL CALCULATION

There is a certain relationship between total heating energy (Q) and difference of temperature (△T) as shown in equation (2).

$$Q = \frac{2 \pi \lambda}{\ln (r_2/ r_1)} \cdot \triangle T \qquad (2)$$

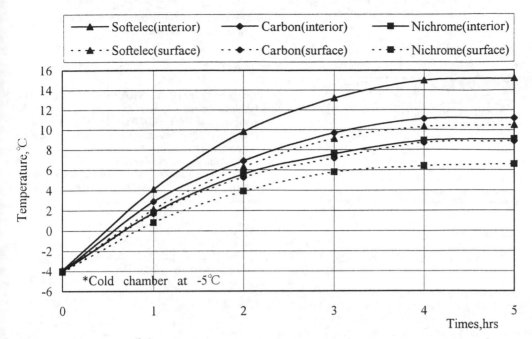

Figure 3. Temperature (℃) of concrete block heated by constant electric power of 200W/m²

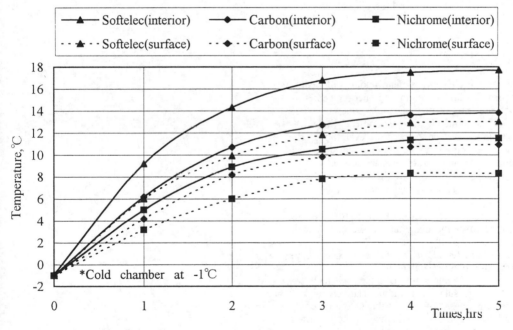

Figure 4. Temperature (℃) of concrete block heated by constant electric power of 200W/m²

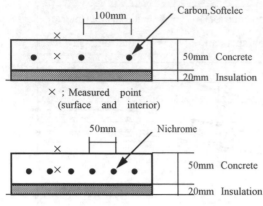

×; Measured point
(surface and interior)

×; Measured point
(surface and interior)

Figure 5. Structure of the concrete block

r_2 is the diameter of inner wire and r_1 is the total diameter (involved outer cover). The difference of temperature ($\triangle T$) is the difference between the surface temperature of the bared cable (inner wire) and that of the insulator (outer cover). λ is the thermal conductivity of cable's cover. In this calculation λ is fixed at 0.3 W/m K.

Calculated values are shown in Table 2 and compared with measured values. These are very close.

5 CONCLUSIONS

According to Joule's law, similar amounts of electric power provide similar amounts of energy. Therefore the surface temperature of each concrete block should be the same. In this experiment, there was difference in the surface temperature even though the electric power supply was the same. The difference between softelec and Carbon was about 4 °C.

The following conclusions are drawn when different heaters are provided the same amount of power.

(1) If the electric resistance of an electric heating cable is low, the diameter of the bared cable (inner wire) is small.
(2) If the diameter of the bared cable is small, its surface temperature will be high.
(3) If the surface temperature of a heater is high, the outer cover (insulator) should be relatively thick.
(4) The thickness of the cable's outer cover (insulator) greatly influences the surface temperature of the cable.
(5) The surface temperature of the concrete block was influenced by the surface temperature of the cable.

(6) Therefore it is better to use a cable with a large and a thin outer cover.

REFERENCES

J.P.Holman.1976. HEAT TRANSFER.McGraw-Hill Book Company

Masahiro Shoji.1995. Heat Transfer Textbook.University of Tokyo Series on Advanced Mechanical Engineering 6

Sogo Sayama, Ken-ichi Miura, Masayosi Sudo, Yosio Sakai, Kunio Tanaka. 1992. Test of the electric Heating cables. *Proceedings of Snow and Ice in Hokkaido* No.11

Snow Engineering: Recent Advances, Izumi, Nakamura & Sack (eds) © 1997 Balkema, Rotterdam. ISBN 90 5410 865 7

Observations of snow accumulation on the roof of an actual domed structure

Kenichi Suda & Michinobu Kikuta
Engineering Research Center, Sato Kogyo Co., Ltd, Atsugi, Japan

Hiroshi Maeda
Fukui Technical Institute, Japan

ABSTRACT : The authors carried out continuous observation of snow loads on an actual domed structure with membrane retractable roofs, using a method based on estimations from the increase in strain of the roof membrane. Based on the observation, we investigated the properties of snow accumulation on the Dome. The distribution of the snow load on the Dome and the shape coefficient of the Dome relative to a flat roof model was also determined.

1 INTRODUCTION

In recent years, demand for large domed structures for use as year-round multi-purpose space has grown in areas with heavy snowfall. Since snow load is an important factor in the design of large domed structures in these regions, the snow load used to design them should be established in a logical manner. However, few observations have been made of the snow load on actual domed structures, and there are few observations regarding changes in snow load with time. Since little is known about the properties of snow accumulation on dome-shaped roofs, the authors carried out continuous observations of snow load on the roof of an actual domed structure.

Photo. 1   The Dome observed

2 OUTLINE OF ACTUAL DOMED STRUCTURE OBSERVED

The Dome observed during the study, which has retractable roofs, is located near Toyama city which is in one of the main snow-belt regions of Japan. Because the temperature does not fall very much during snowfalls, the region tends to have moist, heavy snow.

The Dome encloses a circular arena, and consists of two independent movable roof units, known as the inner roof and the outer roof (see Photo. 1 and Figure 1). Each roof unit is part of a concentric spherical surface. The units have a radius of curvature of 22.4 m and 23.8 m and external spans of 33.1 m and 37.2 m, respectively. They are constructed as a row of

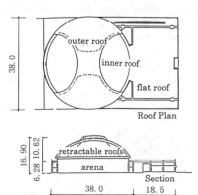

Figure 1   Outline of the Dome

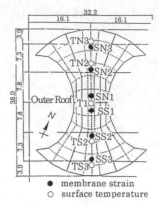

Figure 2   Measuring points of membrane
strain and surface temperature

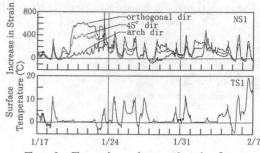

Figure 3   Changes in membrane strain and surface
temperature with time (1/17 -2/7 1993)

arch-shaped trusses connected by subtrusses and stiff
box girders around the edge of the roof units. Two
bogies are installed under each box girder to move
the units. The roof material is a membrane consisting
of Teflon-coated glass fibers. Throughout the
observation period, the roof units did not move and
its snow melting system was not turned on.

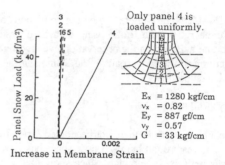

Figure 4   Relationship between snow load and
membrane strain

3   OBSERVATION METHODS AND SNOW
LOAD ESTIMATION METHOD

3.1   *Observation Methods*

Snow observations were made at 9:00 a.m. every day
and membrane strain was monitored continuously.
Daily observations included measurements of the
depth of snow cover, the amount of snowfall, the
density of the snow on the ground, and visual
observation of the state of the snow cover on the
Dome roof.
The continual measurements, which were performed
on the outer roof, included measurements at 30-
minute intervals of membrane strain and roof surface
temperature. As shown in Figure 2, the membrane
strain was measured at six points along the south-
north axis passing through the center of the outer
roof using three-component plastic strain gauges. The
membrane surface temperature was measured at five
points close to the strain measuring points using
thermocouple sheets. The method used to estimate
snow loads from this data is explained in Section 3.2.
The snow load on a flat roof model was also
measured by four load cells at the Dome site. The
model roof, which is the size of 0.9×1.8 meters, was
made of Teflon-coated glass fibers and was set on a
wooden frame 0.9 meters high. Air temperature, solar

radiation, precipitation, wind speed and wind
direction were measured at the same site.

3.2   *Snow Load Estimation Method*

The snow load at each part of the Dome roof was
estimated from the increase in membrane strain,
considering weather data, visual observations and
membrane surface temperatures.
Figure 3 shows the change in membrane strain with
time and nearby membrane surface temperature.
When there is no snow cover, the membrane strain
fluctuates in conformity with the surface temperature.
When there is a snow cover, regardless of changes in
the air temperature, the value of the surface
temperature remains stable at close to 0°C, and the
membrane strain does not change with time. Also, the
increase in membrane strain in the orthogonal
direction of the arch is greater than that in the arch
direction. From this, we can determine if snow
accumulates on the roof of the Dome.
Figure 4 illustrates, based on membrane structural
analysis, the relationship between the increase in
membrane strain (the component in the orthogonal
direction of the arch) and the snow load when only

292

the membrane panel shown in the figure is loaded uniformly. Because on this Dome roof, the warp of the membrane panel is in the orthogonal direction of the arch, and the membrane is supported solely by the arch on both sides and the box girders, the snow load is almost transmitted in orthogonal direction of the arch. This means that, as shown in Figure 4, loading only one membrane panel results in very small strains on the other panels. Consequently, the increase in strain of a panel correlates extremely well with the snow load on that panel.

Because the membrane surface temperature under snow cover remains 0℃, as already stated, it is possible to determine the increase in membrane strain caused by snow load, subtracting that under no snow load at a temperature of 0℃. Then the snow load was estimated using snow load−strain curves determined by membrane structural analysis.

4 OBSERVATION RESULTS

4.1 *Changes in Snow Load with Time*

Figure 5 shows a record of changes in snow load with time on the Dome, for the period with the heaviest snowfall and the longest snow cover on the ground during the observation period. The average of the snow load of six measuring points is shown in the figure. The figure also represents the average snow load on the flat roof model.

While there was snow on the flat roof model throughout this observation period, the snow on the Dome roof disappeared four times in that period. This is assumed to be a result of snow sliding from the roof. Consequently, it is unlikely that there could be continuous snow cover on an actual dome.

Changes in snow load with time on the Dome varied between positions on the roof.

At the lower part of the Dome (SS3 and SN3), the snow load reduced to zero several times during snowfall and did not increase continuously. The snow was more likely to slide off the steeper lower part of the Dome than it was off the middle part (SS2 and SN2) and the upper part (SS1 and SN1).

The snow load on the middle part increased for a while after snowfall began, achieving a higher value than it did on the upper part. However, immediately after snowfall stopped, large snow slides occurred, then the snow load reduced rapidly.

Large snow sliding did not occur on the upper part as it did on the middle part. The snow remained for a relatively long time. But the snow load decreased gradually, finally the snow disappeared even on the upper part without the continuous snow cover seen on the flat roof model.

4.2 *Relationship of Change in Snow Load on the Dome to That on the Flat Roof Model*

Figure 6 represents the correspondence of snow load on the Dome to that on the flat roof model during the period shown in Figure 5. A hysteresis is evident.

On the upper part, the increase in snow load corresponded to about 1/2 that found on the flat roof model. After the maximum load was reached, the snow load began to decrease in conformity with that on the flat roof model, just as it had when the snow load had been increasing. Later, however, it decreased more quickly than the flat roof model and reduced rapidly at the end of this stage. This is believed to be the result of snow sliding off the Dome.

On the middle part, snow load increases

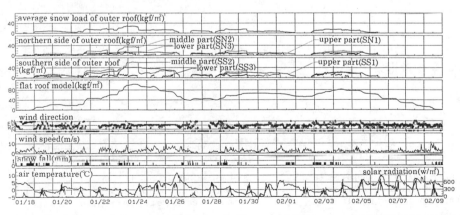

Figure 5   Snow load changes with time on the Dome and flat roof model (1/18 - 2/9 1994)

293

corresponded to those on the flat roof model. For a while but later, its increase declined. Then a sudden reduction of the snow load occasionally happened due to large snow sliding.

On the lower part, snow slid off the Dome whenever the snow load reached a level ranging from 10 to 15 kgf/m².

The snow load averaged over the entire roof displays a hysteresis combining the above results. It did not reduce rapidly to nearly 0, since snow slides over the whole roof did not occur. The snow load of the Dome reached its maximum at almost the same time as the snow load on the flat roof model.

4.3 *Snow Load Distribution over the Dome and Shape Coefficient of the Dome*

Figure 7 represents the change in snow load distribution on the Dome for the period shown in Figure 5. This figure shows the "Shape Coefficient" , which is defined as the ratio of the snow load per unit surface area on the Dome to that of the flat roof model. Because the peak values of snow load for the

flat roof model were frequently obtained between 6:00 and 10:00 a.m., Figure 7 shows the distributions at 8:00 a.m. on each day but on January 27th, its distribution at 3:00 p.m. is shown. Figure 7 also displays the snow load per unit surface area of the flat roof model at the same time and the snow depth on the ground at 9:00 a.m. each day. In Figure 7, snow on the Dome disappeared between periods (3) and (4), periods (4) and (5) and periods (5) and (6).

The snow load distribution was almost uniform for the entire roof at the beginning of snowfall, after which it was replaced by a distribution marked by an extremely low snow load on the lower part, then the distribution gradually changed into a large hill-shaped mass as the snow load on the upper part became relatively high.

When the snow load on the flat roof model was maximum, the Dome snow load distribution featured a slightly higher snow load on the middle part than on the upper part and a very low load on the lower part. Then the snow load on the middle part began to reduce rapidly, resulting in a snow-load distribution characterized by a hill-shaped mass.

During the first snow cover, the shape coefficient of the Dome was large. On the Dome roof snow load repeatedly reduced to zero, the coefficients after new snow accumulated were much smaller than those during the first snow cover, and they decreased further each time the snow cover disappeared. Consequently, the shape coefficients were at their highest value immediately after the first snowfall, but they never exceeded 1.0. Afterwards, when the snow load on the flat roof model was at its maximum value following several snowfalls, the coefficients had decreased to a much lower value.

The snow load averaged over the entire roof of the Dome was at its peak at almost the same time (1/24)

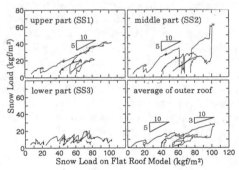

Figure 6   Correspondence of snow load on the Dome to snow load on the flat roof model

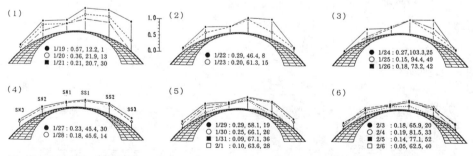

The figure represents "shape coefficient" which is the ratio of snow load on *the Dome* to that on the flat roof model.

The numerals represent date, shape coefficient, snow load on the flat roof model (kgf/m²), and depth of snow cover on the ground (cm) in order.

Figure 7   Changes with day of snow load distribution on the Dome (1/19 - 2/6 1994)

as the maximum snow load of the flat roof model. However, because the snow load on the middle part of the Dome reduced just as the average roof snow load peaked, the period of peak snow load was not very long.

5 CONCLUSION

On an actual domed structure with membrane retractable roofs, changes with time and distribution in snow loads on the roof were investigated using a method based on estimations from the increase in strain of the roof membrane.

The following were confirmed:

[1]The accumulated snow on the Dome disappeared sooner than that deposited on the ground and on the flat roof model. This tendency was particularly marked on the middle and lower parts of the dome, due to frequent snow sliding. Continuous snow cover did not exist on the dome.

[2]Immediately after snowfall started, the distribution of snow load on the dome was relatively uniform over its entire surface, then it gradually changed into a hill-shaped mass by gradual accumulation of snow on the upper part and frequent sudden reductions of snow load on the lower part by sliding. When the snow load was at its maximum value, the load on the middle part was a little larger than that on the upper part, and the load on the lower part was extremely small.

[3]The shape coefficient of the Dome (i.e., the ratio of the average snow load on it to the average load on a flat roof model) was large initially (e.g., 0.57), but was smaller (0.27) when the snow load was at its peak value.

Snow Engineering: Recent Advances, Izumi, Nakamura & Sack (eds) © 1997 Balkema, Rotterdam. ISBN 90 5410 865 7

Observations of snow banks formed under the very large sloped roofs

Osamu Abe

Shinjo Branch of Snow and Ice Studies, NIED, Yamagata, Japan

ABSTRACT: Recently, very large buildings with sloped roofs have been built in snowy areas of Japan. In winter, snow that slides off such roofs forms a snow bank under them. Sometimes an air burst is produced by impact of the falling snow on the ground. We need a safety area on the ground for such buildings. In this study, observations of snow banks are described to establish the characteristics of problems created by very large sloped roofs.

1 INTRODUCTION

Roof snow problems include the following: snow accumulation on the roof, snow sliding on the roof, snow falling from the roof and the formation of snow banks. For a small house, Nakamura (1978-a) proposed an experimental equation to estimate the shape of snow banks from snow that has fallen from the roof. However, that equation should not be used for a large roof since snow sliding down it shoots away from the eaves. We need another idea for the large roof. A theoretical estimate of the shape of the snow bank for a large roof was carried out by Takita & Watanabe (1994-a,b); however, actual data is lacking. In this paper some observations of snow banks from large roofs are shown.

2 OBSERVATION

2.1 *Period*

During the winter of 1988/89 a preliminary study was conducted for a large building. Some problems were pointed out in this study. During the winters of 1989/90 and 1990/91 snow banks were observed at the end of January for three large buildings. Regular observations were carried out during the winter of 1995/96.

2.2 *Object*

The three buildings with very large sloped roofs chosen for this study are located around Shinjo-city in the northern part of Honshu Island. Roof conditions are shown in Table 1. The width of the roofs is from 16.25 m to 29.00 m. The predominant wind direction is WNW in the study area.

2.3 *Method*

Snow bank shape was measured using a carpenter's level and rods. The end of the bank was defined as the limit of impacted snow onto the ground snow cover. Densities and weights of the snow were measured

Table 1. Roof conditions.

Roof	Horizontal width,L(m)	Height of eaves,H(m)	Angle,θ (degree)	Direction	Material	Color	Roofing
A	16.25	11.14	21.8	West	Color paint	Black	Seam joint (Horizontal)
B	29.00	12.15	21.8	West	Polyvinyl fluoride	Silver	Seam joint (Horizontal)
C	21.70	11.875	33.0	South	Color paint	Cream	Folded plate

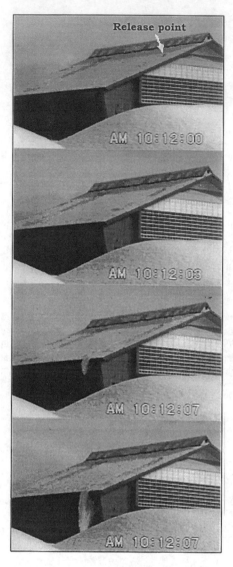

Figure 1. Snow sliding down roof B.

using a snow sampler with an area of 20 cm². The sliding of roof snow was filmed by a video camera.

3 RESULTS

3.1 Sliding phenomenon

Snow sometimes slides down a large roof as an avalanche. Figure 1 shows this phenomenon which occurred on roof B on Jan. 30, 1991. The roof was

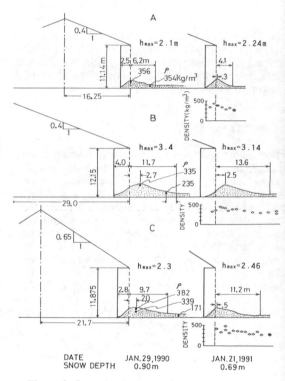

Figure 2. Snow bank shapes during two winters.

Table 2. Height at the peak (h_p), distance to the peak (X_p) and distance to the end (X_e) of snow bank.

Roof	Jan.29,1990			Jan.21,1991			Jan.26,1996		
	h_p(m)	X_p	X_e(m)	h_p(m)	X_p	X_e(m)	X_p(m)	X_p	X_e(m)
A	2.1	0.0	6.2	2.24	0.3	4.1	1.79	0.4	5.2
B	3.4	2.7	11.7	3.14	2.5	13.6	2.22	1.8	9.9
C	2.3	2.0	9.7	2.46	0.5	11.2	1.65	1.1	>12.6
Snow depth(m)	0.90			0.69			0.75		

slippery and the new snow weak . This is characteristic of a large roof. A resident who was beside the
large building said that an air burst is sometimes
produced by the impact of snow on the ground.
Attempts to measure the pressure of the air burst have
not been successful.

3.2 *Variations in snow bank shape*

Figure 2 shows the shape of snow banks at the end of
January during the winters of 1989/90 and 1990/91.
Snow depths at the same day in Shinjo are shown at
the bottom of this figure (Abe et al. 1996). The shape
changes slightly each winter. Some peaks are right
under the eaves but others are some distance away
from them. Numerical descriptions of the snow banks
for three winters are shown in Table 2.

3.3 *Seasonal variation*

In the winter of 1995/96 the area experienced a heavy
snowfall compared to that of the previous seven
winters. Complete data were obtained during this
winter. Meteorological data observed in the Shinjo

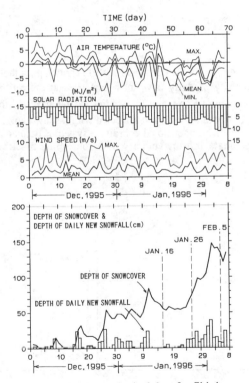

Figure 3. Meteorological data for Shinjo.

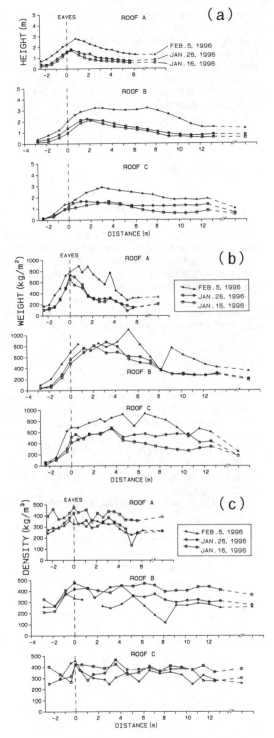

Figure 4. 1996 snow bank measurements.
a: shape, b: weight, and c: density

299

Branch of Snow and Ice Studies, NIED are shown in Figure 3. Figure 4-a,b and c shows seasonal variations in the shape, weight and density on Jan. 16 and 26 and Feb. 5, 1996. The peaks of the height sift toward the right with increasing the snow depth. In the case of roof B, two peaks appeared in the bank on Feb. 5, 1996. The peak to the right had a low density, so it is felt that the second peak was formed by a plowing effect by impacting snow. The highest densities of all snow banks appear near the peaks.

4 DISCUSSION

4.1 Snow accumulation on large roofs

Snow accumulation on large roofs is affected by wind speed and air temperature(Abe & Nakamura 1984, Tomabechi & Hashimoto 1993). For large building, the snow accumulation on the roof is less than that on the ground because of winds. The snow bank below the eaves is formed by snow that accumulated on the roof then slide off. Snow accumulation on the roof is estimated from the snow bank as

$$W_R = (\int_a^b W_B(x)\,dx - W_G b\,)/L \tag{1}$$

where W_R is the weight of snow on a horizontal unit area of the roof, $W_B(x)$ is the weight of snow in the bank along horizontal distance x, W_G is the weight of the snow cover on the ground, L is horizontal width of the roof, a is the negative horizontal distance from the eaves to the wall, and b is the distance from the eaves to the end of the snow bank. The effect of snow melting is ignored.

Figure 5 shows the relationship between W_R and

W_G. The ratio of the weight on the roof to the ground snow cover is calculated from the slope of the line for each roof. The ratios are obtained to 0.45 for A, 0.65 for B and 0.83 for C by the method of least squares. The ratio of roof C is larger than those of roof A and B. It is considered that the accumulation on roof C faced to leeward is larger than those of other two roofs which faced to windward. Nakamura et al.(1984) reported ratios of 0.56 and 0.81 for flat roofs.

4.2 Impact compaction of roof snow

As mentioned above, high density snow is formed by roof snow impact. The density of new snow compacted within snow bank was measured. The fractional volume compaction F (no dimension) of the compacting snow is defined as

$$F = \frac{V_0 - V_1}{V_0 - V_{ice}} \tag{2}$$

where V is the volume of snow, the subscripts 0, 1 and ice denote the initial snow, after impact and ice (terminal). This equation is rewritten for density ρ

$$F = \frac{\rho_{ice}(\rho_1 - \rho_0)}{\rho_1(\rho_{ice} - \rho_0)} \tag{3}$$

where ρ_{ice} is the density of ice($917 kg/m^3$), ρ_0 is the density of daily new snow observed at the Shinjo Branch of Snow and Ice Studies, NIED and ρ_1 is the density of new snow compacted within snow bank.

A two-dimensional coordinate system is defined to estimate impact velocity of a roof snow block as

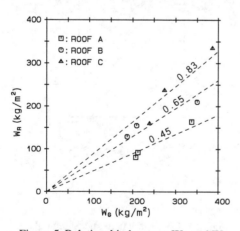

Figure 5. Relationship between W_R and W_G.

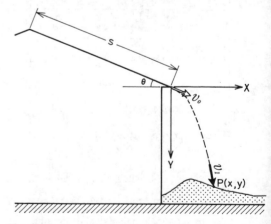

Figure 6. Two dimensional coordinate system of movement of snow block.

shown in Figure 6. Then the initial velocity v_0 at the eaves is divided into the X and Y components as

$$v_{0x} = v_0 \cos\theta \qquad (4)$$

$$v_{0y} = v_0 \sin\theta \qquad (5)$$

where θ is the angle of the roof. If we assume that air resistance can be ignored, the impact point P(x, y) on the bank is described as

$$X_p = v_{0x} \cdot t \qquad (6)$$

$$Y_p = \frac{1}{2}(v_{0y} + v_{1y}) \cdot t \qquad (7)$$

where t is the time from the start to the impact. Impact velocities (v_{1x}, v_{1y}) at P are calculated as

$$v_{1x} = v_{0x} \qquad (8)$$

$$v_{1y} = \sqrt{2gY_p + v_{0y}^2} \qquad (9)$$

where g is the gravity acceleration. From Equations (8) and (9), the absolute impact velocity v_1 is given by

$$v_1 = \sqrt{v_{1x}^2 + v_{1y}^2} \qquad (10)$$

Also, v_0 is calculated by combination of Equations (5), (6), (7) and (9) as

$$v_0 = X_p \sqrt{g / \left\{ 2\cos\theta (Y_p \cos\theta - X_p \sin\theta) \right\}} \ . \qquad (11)$$

To estimate v_1, first v_0 is calculated by Equation (11) with X_p, Y_p and θ obtained by observation.

Figure 7 shows the relationship between F and v_1.

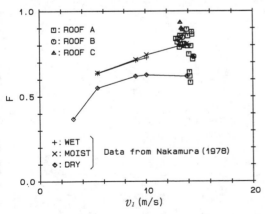

Figure 7. Relationship between the fractional volume compaction F and the impact velocity v_1.

Data by Nakamura (1978-b) are also shown in this figure. It is found that F increases with v_1, and the gradient of F decreases with v_1. In this case, air within the snow affects compaction.

4.3 Possibility of air burst occurrences

It is suggested that an air burst might be produced by the impact of falling snow. Here the possibility of air burst occurrence is described. The porosity S of the snow is defined as

$$S = 1 - \frac{\rho_s}{\rho_{ice}} \qquad (12)$$

where ρ_s is the density of the snow. When the snow is compressed instantly, the compression of the air can be considered to be an adiabatic change by the second law of thermodynamics. In this case the pressure P and volume V are described as

$$PV^\gamma = \text{const.} \qquad (13)$$

where γ is the ratio of specific heat at constant pressure to that at constant volume of air (=1.40). Here a typical case of compression is assumed, namely an initial density of 80 kg/m^3 and a density after the compression of 300 kg/m^3. The unit volume of this snow is compressed to 0.27 its initial size. In the same time Equation (12) indicates S decreases from 0.91 to 0.67, so that the air within the snow is consequently compressed to 0.20 of its initial volume. By Equation (13), the air at 1 atm increases to 9.52 atm. This high pressure might produce an air burst when a large amount of roof snow falls down to the ground.

4.4 Snow sliding on the roof

Snow sliding on the roof determines the shape of the snow bank. Endo et al.(1988) reported the flight distance of snow sliding off a roof. However, little data is available on the movement of snow on large roofs. As mentioned above, sometimes the snow on the roof fractures during sliding (see Fig. 1). We need to know more about this.

5 CONCLUDING REMARKS

Snow banks formed under roofs were observed in this study. The snow bank peak was located beyond the eaves when the snow was deep. Snow accumulation on each roof was estimated from the mass of the snow bank. The ratio of the snow weight on the roof to that on the ground ranged from 0.45 to 0.83. The fractional volume compaction F of impacting snow was related to the impact velocity v_1. It is found that F increases with v_1, and the gradient of F decreases with v_1. Air within the snow is compressed instantly. According to a trial calculation, air at 1 atm increases to 9.52 atm at impact. This high pressure might produce an air burst when a large amount of snow falls.

ACKNOWLEDGMENTS

The author is grateful to the administration section of all buildings studied for their cooperation, and to Mr. Suzuki K., Mr. Otsu M. and the Hutaba Kensetsu Consultant Co. Ltd. for their help. The author also thank Dr. Wayne Tobiasson for his critical reading of this manuscript.

REFERENCES

Abe O. & Nakamura T. 1984. Wind effect on the daily new snow depth on flat roofs. Report of the National Research Center for Disaster Prevention 32: 73-87.

Abe O., Sato T., Sato A., Higashiura M., Numano N., Kosugi K., Nakamura H. & Nakamura T. 1996. Observations of snowcover and new snowfall at the Shinjo Branch of Snow and Ice Studies during the eleven winter periods from 1984/85 to 1994/95. Technical Report of the National Research Institute for Earth Science and Disaster Prevention 175: 1-74.

Endo Y., Ozeki Y., Niwano S., Kobayashi S., Minagawa T., Shinojima K. And Yoshida N. 1988. Conditions of snow sliding on roofs and flight distance. Proc. Cold Region Technology Conference'88 4: 220-225.

Nakamura H. 1978. Shape of the snow bank formed under eaves by the fall of snow on roofs. Seppyo 40: 37-41.

Nakamura H. 1978. Change in density of blocks of new snow before and after the collision against natural snow surface. Report of the National Research Center for Disaster Prevention 19: 239-242.

Nakamura T., Abe O., Nakamura H., Higashiura M. & Numano N. 1984. Comparison of the roof snow depth on three different types of buildings with the ground snow depth. Report of the National Research Center for Disaster Prevention 32: 56-72.

Takita M. & Watanabe M. 1994. Simulation of sliding and accumulation of roof snow. Proc. Japan Society for Snow Engineering 10: 155-158.

Takita M. & Watanabe M. 1994. Simulation of sliding and accumulation of roof snow (Part 2. Effects of viscous drag). Proc. Japan Society for Snow Engineering 11: 39-42.

Tomabechi T. & Hashimoto S. 1993. Considerations on snow drift on roof surface. J. Snow Engineering 9: 92-98.

Snow Engineering: Recent Advances, Izumi, Nakamura & Sack (eds) © 1997 Balkema, Rotterdam. ISBN 90 5410 865 7

Wind tunnel model studies of roof snow loads resulting from multiple snowstorms

N. Isyumov & M. Mikitiuk
Boundary Layer Wind Tunnel Laboratory, The University of Western Ontario, London, Ont., Canada

ABSTRACT: Most physical model studies of roof snow loads deal with the aerodynamic influence of buildings on shaping snow accumulations during particular snowstorms. While the effects of such "single events" can be satisfactorily predicted with physical model studies in wind tunnels and/or water flumes, the evaluation of how weather events combine to produce maximum loads during a particular winter is more difficult. Numerical techniques are used in combination with physical model tests to predict maximum snow accumulations which can occur on roofs. Numerical methods are effective in evaluating the influence of temperature, solar radiation, heat loss from the building and the effects of rain. It is more difficult to use numerical methods to evaluate the accumulation and redistribution of snow on roofs. These depend on the aerodynamics of roofs and are difficult to predict without information from wind tunnel model tests. This paper describes wind tunnel experiments carried out to examine the snow accumulation on roofs which can result due to weather events during cold periods when the roof snow layer remains susceptible to drifting. Results of wind tunnel model studies of such multiple-events and those obtained for a maximum single storm are presented for a large roof.

1 INTRODUCTION

Physical model studies, using wind tunnels and/or water flumes are effective aids for predicting snow accumulations which may occur on roofs and horizontal surfaces of buildings and structures. Details of experimental technique and similarity requirements for such studies can be found in the literature, see Kind (1986), Iversen (1982), Isyumov and Mikitiuk (1992), including the proceedings of ICSE-1 (1988) and ICSE-2 (1992). Unfortunately, it is difficult to achieve full similarity of the behaviour of snow particles in model studies, carried out at a small geometric scale. Model snow particles which have been used in wind tunnels cover a broad range, including heavy granular particles such as sand and glass beads, chemicals like borax and sodium bicarbonate, various natural fibrous materials such as sawdust, ground walnut shells, cracked wheat, bran (finely ground wheat), activated clay and other materials. There is evidence that experimental procedures using some of these materials can provide reasonable agreement with full-scale for situations where the snow is dry or "driftable". An important similarity requirement in such studies is the modelling of the mean and turbulent characteristics of natural wind. This provides similarity of overall building aerodynamics and is an important requirement for achieving similar accumulation of snow, during a particular storm, and its subsequent drifting and redistribution.

Numerical methods are normally used to superimpose snow loads due to individual storms and to keep track of changes in the roof snow accumulation due to other weather events. A mass-balance method, initially proposed at The University of Western Ontario (UWO) by Isyumov (1971) and subsequently extended by Isyumov and Mikitiuk (1977) has been developed to track snow accumulations on roofs during the course of a winter. This method used wind tunnel model data to develop snow load accumulation patterns on particular roofs for different values of wind speed V, wind direction D and snowfall rate s, in combination with empirically-derived wind and thermal ablation functions to estimate what happens to roof snow accumulations between snow storms. In this approach it was assumed that newly-fallen snow remains driftable as long as the air temperature T remains below freezing and there is no rainfall R to wet the snow surface and to inhibit it from drifting. Actually recorded or simulated sequences of weather data were used to track the roof snow load over the course of a winter. This allowed the search for maximum roof snow loads and the identification of important underlying trends. This approach demonstrated the importance of the surrounding terrain, which influences the wind speed at roof level, and allowed the effects of such climatological variables as V, D, T, s and R to be systematically taken into account. These studies clearly showed that the variability of maximum snow accumulations on roofs was substantially greater than that of the maximum ground snow loads. The UWO mass-balance approach kept track of the total amount of snow on the roof but did not provide information on its spatial distribution, as the empirically-derived depletion factors were not able to determine drift patterns on particular roof geometries. This requires specific aerodynamic information. An experimental/numerical procedure, referred to as the finite area element (FAE) technique, has been reported in the literature by members of the firm Rowan Williams Davies and Irwin (RWDI), see Gamble et al. (1991), and Irwin et al. (1992). This technique uses empirical snow-drifting data from field measurements of ground snow transport to evaluate the wind-induced transport of snow on the roof during periods when the roof snow is driftable. The mean wind speeds and directions over the roof, measured in a wind tunnel simulation, serve as the aerodynamic input required to estimate the snow flux carried onto and from particular finite areas of the roof. The mean wind speed measurements are made at a full-scale height of 1 m above the roof surface over a grid covering the entire roof. While this method does allow sheltered or wind-swept areas of the roof to be identified, the technique concentrates on the snow transport or drifting phase of the process and does not recognize the influence of the roof geometry and the wind speed on accumulation patterns during a snowstorm.

Neither the UWO mass-balance method, nor the FAE method of RWDI, recognize that the presence of snow on the roof can influence both the accumulation patterns of new snow and the drifting of snow already on the roof. In short, both approaches assume linear superposition of the effects of individual storms and drifting episodes. This assumption is questionable in the presence of significant snow accumulations. The accumulation of snow in separated flow regions tends to streamline the roof surface and changes the local velocity field over the roof. These nonlinear effects are extremely difficult to track in a numerical simulation.

In wind tunnel model studies, it is common to categorize snowstorms by some characteristic values of V, D and s, which remain constant during a particular wind tunnel experiment. This is not usually the case in full scale, as these variables tend to vary over the course of a snowstorm. Wind tunnel experiments have shown that the variation of V, D

and s during a storm can influence the resulting maximum snow loads, (Kennedy et al. 1992). Depending on the roof geometry and the sequence of a storm, snow accumulated on the roof during initial stages may subsequently be relocated, resulting in more severe maximum local drifts by the end of the same storm. The objectives of this paper are to describe how physical model studies can capture the consequences of snow storms with varying values of V, D and s and the superposition of such events over periods of time during which the air temperature remains below freezing and the snow can be treated as "driftable".

2 FORMATION OF EXTREME SNOW LOADS

The formation of extreme snow loads on roofs strongly depends on local climatic conditions. Climatic studies at UWO have examined the statistics of individual 24-hour snowfalls and the annual maximum ground snow depths, experienced during particular winters at some 30 Canadian locations. Figure 1 summarizes the magnitude of snow loads due to the 30-year return period maximum 24-hour snow storm, denoted as $\hat{S}_i\,(30yr)$ relative to the maximum 30-year return period ground snow load $\hat{S}_g\,(30yr)$. The ratio of these quantities, denoted as R, is plotted in Figure 1 versus the average air temperature during the portion of the winter with significant ground snow accumulations. It is important to note that R approaches 1 for warm climates, where the entire maximum ground snow load tends to be due to a single snowfall. For locations with colder air temperatures, R is substantially less than 1, suggesting that the maximum ground snow load represents the cumulative effects of several snowfalls. Figure 1 also shows a plot of R vs the 30-year return period ground snow depth in meters. These two plots combine to demonstrate the intuitively obvious. Namely, that large snow loads are most likely experienced in relatively cold climates where the maximum snow accumulations and loads tend to be due to the superposition of multiple weather events. Since such locations tend to have low winter temperatures, there are prolonged periods of time during which newly-fallen snow remains driftable. Physical model studies are expected to provide reasonable estimates of both the accumulation and the redistribution and depletion of the snow by wind action during such periods.

The winter of 1995/96 in eastern North America has been relatively cold and has brought significant amounts of snow. In particular, several major snowstorms moved northward along the Atlantic coastline, dumping snow in areas which generally experience little winter snow and producing unusually large snow accumulations in others. Detailed climatic data for the first part of January 1996 were analyzed to examine the "signatures" of extreme snowstorms and the persistence of cold weather conditions, during which newly-fallen snow is likely to remain driftable. This was done in an attempt to examine the extent to which episodes of several snowstorms can be studied in the wind tunnel.

A severe low-pressure system formed in the southern United States and on January 7th moved northward along the Atlantic coastline. The moist air carried by this storm caused extensive precipitation, mainly in the form of snow, as it encountered colder air along its path. This storm pattern has been identified as the causes of many extreme snowfalls in eastern North America, Kocin and Uccellini, (1990). A second, although substantially less-severe storm, followed a similar pattern to the January 7 to 9, 1996 storm only days later. Signatures of the January 7th to 9th snow storm, as well as preceding and immediately-following weather conditions are summarized for New York and Boston in Figure 2. The data come from the Climate Research Division, Scripps Institution of Oceanography, University of California, San Diego. These graphs show the hourly wind speed and wind direction, the air temperature, the precipitation and the observed ground snow depth. In the case of New York, as recorded at JFK International Airport, there was little snow on the ground prior to the January 7th to 9th snowstorm. Air temperatures remained below freezing between January 4th to January 11th. In the case of Boston, as recorded at Logan International Airport, there was considerable snow on the ground at the arrival of the January 7th to 9th storm. Temperatures remained below freezing for approximately the same period of time and a further substantial storm occurred on January 10th and 11th. Throughout this time, there were periods with significant wind speeds coming from different directions, as the storm progressed northward along the Atlantic coastline.

Figure 3 shows contours of periods, measured in days, during which new snow is likely to have been driftable with temperatures remaining below freezing and with no rain to wet the snow surface and inhibit it from drifting. Moving northward into regions with colder winter temperatures, this period is seen to be 10 days and longer. It is during these times that wind tunnel model studies, using a cohesionless model snow, are expected to provide reasonable estimates of the accumulation, redistribution and depletion of snow on the roofs of buildings and structures. To be effective, however, such studies must capture the effects of the substantial changes in the wind speed and wind direction and the snowfall rate which tend to occur during such periods.

3 OVERVIEW OF WIND TUNNEL MODELLING

Details of the similarity requirements and the experimental procedures which are followed in studies of roof snow loads at UWO have been described elsewhere, Mikitiuk and Isyumov (1989), Isyumov et al. (1989), Isyumov and Mikitiuk (1990), Isyumov and Mikitiuk (1992). It is generally accepted that exact similarity requirements are difficult to achieve and that some approximations of the modelling laws are unavoidable. In studies at UWO, it is recognized that the accumulation, transfer and redistribution of snow on roofs and in the vicinity of buildings are dominated by overall and local aerodynamics, including turbulent mixing and convective transport of the flow, as well as the formation of local regions of flow separation and local vorticity. To model this process at a reduced scale in the wind tunnel, it is necessary to achieve a level of similarity of the mean and turbulent flow fields comparable to that required in wind engineering studies of buildings and structures ASCE (1987). In addition, it is necessary to achieve similarity of the trajectories of particles, both while airborne and when drifting along the roof. Acceptable overall similarity of accumulation and redistribution of driftable snow can be achieved by maintaining similarity of the "bulk" hydraulic properties of the snow phase. Details of the similarity requirements in such simulations have been described by Isyumov and Mikitiuk (1992). In summary, these are as follows:

i) Similarity of the mean and turbulent approach flow.

ii) Similarity of local flow features, including regions of flow separation and reattachment, the speed-up and distortion of the mean flow and turbulence created by buildings under study. For sharp-edged bodies this can be achieved by requiring that the Reynolds number, based on building dimensions, exceeds 10,000.

iii) Similarity of the bulk hydraulic properties of the snow phase including the threshold friction velocity u_{*_t} the terminal velocity w_f , and the density ratio ρ_s /ρ , where ρ_s and ρ are the snow phase and air densities.

iv) Requirement that the saltation hop length ℓ of model snow particles is significantly smaller than the overall dimensions of the roof; namely $\ell << H$ and $\ell << B$, where H and B are characteristic roof dimensions.

The above similarity requirements are best achieved in a wind tunnel, with a large

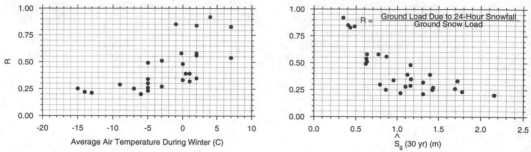

Figure 1. Illustration of the relative importance of single snowstorms at locations with different winter temperatures and different ground snow loads.

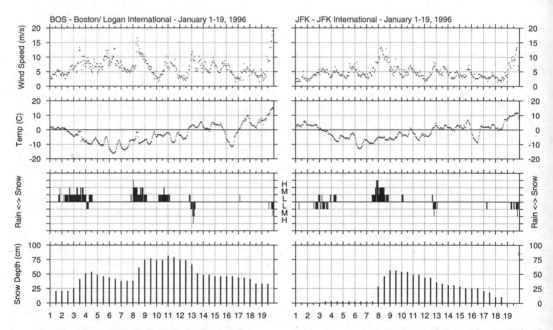

Figure 2. Glimpses of weather conditions at Boston, Massachusetts, and New York, NY, during early January 1996 (Data Source: Climate Research Division, Scripps Institution of Oceanography, University of California, San Diego).

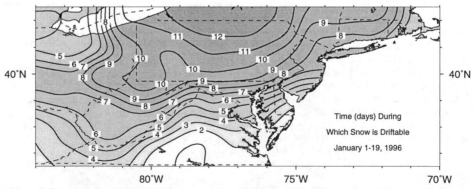

Figure 3. Time periods with "driftable" snow; continuous periods of time with no rain and with below-freezing air temperatures.

cross-section, where experiments can be carried out at as large a geometric scale as possible.

When modelling a particular snow storm in a wind tunnel, it becomes important to simulate the approach flow as characterized by V_H, the mean wind speed at roof level, D, the wind direction, the amount of snowfall during the storm, denoted as Δs, and the duration of the snow storm, Δt. During periods with snowfall, the time scale of the process is determined from the equality of

$$\left(\tfrac{\Delta s}{\Delta t} / V_H\right)_m = \left(\tfrac{\Delta s}{\Delta t} / V_H\right)_p \qquad (1)$$

In the above equation, the subscript m refers to model scale and the subscript p refers to prototype scale.

Correspondingly, the time scale for the snow phase becomes $\lambda_t(snow) = \lambda_t\,(flow)$. This requires that the non-dimensional time for the flow and the snow phase are equal. The time scale of the snow transport during periods of drifting can be based on the similarity of the mass flux of drifting snow. It has been shown by Isyumov and Mikitiuk (1992) that this time scale becomes

$$\left(\frac{u*_t\,\tau}{H}\right)_m \left(\frac{\rho}{\rho_s}\right)_m = \left(\frac{u*_t\,\tau}{H}\right)_p \left(\frac{\rho}{\rho_s}\right)_p \qquad (2)$$

where τ is a characteristic time. This reverts to the time scale of the flow, providing that the density ratio ρ/ρ_s is maintained equal in model and in full scale. Essentially, this implies that the transport of snow particles is comparable to the transport of marked fluid particles. This assumption is approached as snow transport in suspension becomes dominant. The validity of this assumption becomes more approximate at lower wind speeds, when most of the transport is in saltation. Full-scale snow flux measurements over roofs are needed to further develop experimental methods at model scale.

4 SNOW LOADS ON A LARGE ROOF DUE TO SINGLE SNOWSTORMS

Studies of snow loads on large roofs have been carried out at UWO at a geometric scale of 1:100 in the large test section of its boundary layer wind tunnel. An upstream view of this test section, with an experimental setup of a large roof in the foreground, is shown in Figure 4. The width and height of the test section are 4.9 and 3.7 m respectively. A combination of upstream spires and terrain roughness are used to simulate the turbulent boundary layer characteristics of the approach wind. Figure 4 shows an experimental frame mounted over the model of the roof and used to position

displacement sensors which measure changes in the model snow accumulation after particular tests. The frame is not present during the actual wind tunnel experiments. The setup shown in Figure 4 and other large roofs were used in these tests to examine the influence of wind speed and wind direction on snow accumulation patterns on multilevel roofs. Some of these experiments, reported by Kennedy et al. (1992) showed that elongated or "spike" shaped drifts can occur to the lee of a high portion of a roof for skewed winds. The downwind extent of these drifts was observed to be significantly greater than the length of triangular drifts specified by codes at changes in roof elevation.

Figure 5 shows weather data at St. John's, Newfoundland, for a period of several days, ending with a severe winter storm on the 8th of February 1987. That storm dumped 30 cm of snow over a period of 18 hours. The temperature during the storm was below freezing and the hourly wind speed, reported at the St. John's International Airport, reached 18 m/s. The February 8, 1987 storm started with winds from approximately the south, changing counterclockwise to easterly and subsequently to northerly directions as the storm progressed. This shift in wind direction, as the wind speed increased and snow drifting at roof level intensified, was found to have an important influence on the magnitudes and distributions of snow accumulations on the roof, shown in Figure 4. Figure 6 shows model snow accumulation patterns around a high-storey portion of that roof, obtained by approximating the weather events of February 8 and 9, 1987 by four discrete segments or steps. The test was stopped after each segment, the wind direction and speed were changed and the test was resumed. The total experimental sequence continued for a total time which simulated the 18-hour period in full scale.

The high portion of the roof around which snow accumulations are shown in Figure 6, rises one storey above the main roof with its long axis situated approximately northeast to southwest. For skewed wind directions, this high storey tended to act like a barrier or wind deflector which directed the flow parallel to its long walls forming trailing corner drifts to its lee. In the initial stages of the storm, with winds coming from an azimuth of 100°, drift transport occurred parallel to the south wall resulting in elongated drifts at the southwesterly end of the high storey, see Figure 6. As the storm progressed, the wind direction changed to 20° and then to 340° with drifting now occurring parallel to the north wall of the high roof. This resulted in the formation of another trailing drift to the lee of the northwest corner.

The maximum local accumulation was predicted to be about 1.5 m, see Test F9 in Figure 6. This

307

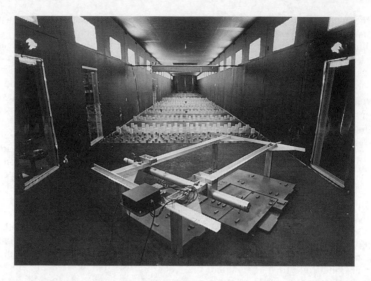

Figure 4. View of Boundary Layer Wind Tunnel at the University of Western Ontario with 1:100 model of a large-area multi-level roof in the foreground.

(Note: Apparatus over model was used to measure snow accumulations after a test. It was not present during test).

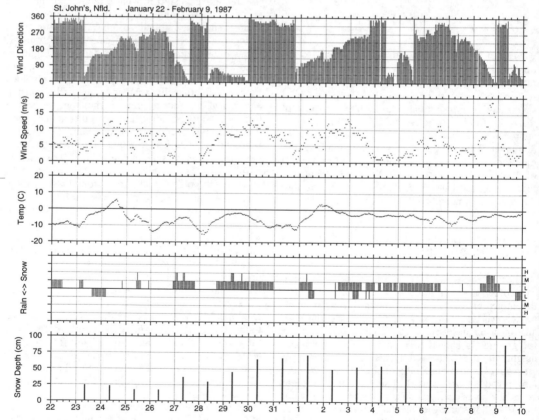

Figure 5. Signature of winter weather at St. John's Newfoundland, with severe winter storm on February 8, 1987.

corresponds to five times the magnitude of the snowfall. Extensive portions of the roof around these trailing corner drifts experienced snow accumulations, which correspond to four times the magnitude of the snowfall. These trailing corner drifts or "spike" drifts, as referred to by Kennedy et al. (1992), have magnitudes and distributions which differ from the triangular drifts suggested by building codes at changes in roof elevation.

5 SNOW LOADS ON A LARGE ROOF DUE TO MULTIPLE SNOWFALL EVENTS

The wind tunnel model studies of the effects of a single extreme snowstorm, see Figure 6, predicted roof snow depths which were lower than the full-scale snow loads measured at locations C and D shown in Figure 6. Upon examination of the weather data in Figure 5, it became apparent that the snowstorm event of February 8, 1987 was preceded by a substantial period during which there were other snowfalls and during which the air temperatures were generally below freezing. Figure 7 summarizes the cumulative snowfall which fell prior to the major snowstorm of February 8, 1987. Temperatures during this time generally remained cold and it is expected that snow falling during this period remained driftable. The cumulative snowfall over this period approached 125 cm which is approximately four times the magnitude of the February 8 storm.

Wind tunnel model studies were carried out to simulate the entire weather sequence for the period of January 25 to February 8, 1987. Best approximations were used to simulate this sequence in a manageable number of steps during the experiment. The wind speed and wind direction for each of these steps was modelled and the snow was allowed to accumulate on the roof. The final snow accumulations near the southwest end of the high portion of the roof, obtained from test sequences G1, G2 and G3, are summarized in Figure 8. The patterns of the snow accumulations differ from those shown in Figure 6 and the maximum drift depth reached 1.8 m and in some cases 2.1 m. The G1, G2 and G3 test sequences represent slightly different approximations of wind speed, wind direction and snowfall, see Figure 7. Wind tunnel sequence G3 used somewhat less snow to allow for a reduction in the amount of driftable snow due to the fixing effects of rain during the time period preceding the February 8 to 9 storm. These wind tunnel tests predicted roof snow depths which were in good agreement with available full-scale measurements taken at locations C and D, shown in Figures 6 and 8. This agreement with full-scale observations provided a valuable confirmation of this modelling approach, which not only included the major

snowstorm of February 8 but also allowed for the contributions of preceding snowfalls, which occurred during low temperatures when new snow was expected to remain driftable.

The snow accumulation patterns shown in Figure 8 differ substantially from those recommended in the National Building Code of Canada (NBCC) for this roof. Figure 9 shows the snow loads along Section A, taken along the centreline of the high roof, and Section B taken along its quarterline. The conversion of snow depths, shown in Figure 8, to snow loads in kPa was made using a snow density determined from snow samples taken from the actual roof. Substantially larger snow loads are seen to occur along Section B. This is attributed to greater drift transport along the north wall of the high storey during the final stages of the February 8 storm. The results shown for sequences G2 and G3 are regarded to provide best approximations of the snow loads which occurred on the roof due to the cumulative effects of weather conditions during the period of January 25 to February 9, 1987. It is important to note that the estimated snow loads along Sections A and B substantially exceeded the provisions of the 1975 edition of the National Building Code of Canada, which was in effect at the time of the construction of this building. The NBCC snow loading provisions changed substantially in 1990. The maximum value of the NBCC 1990 drift snow load at the southwest end wall of the high storey is seen to be comparable to values predicted from the wind tunnel simulations. However, the downstream extent of the code drift load substantially underestimates the snow accumulations observed in the wind tunnel study. It is important to stress that the drift snow loads of the NBCC are based on the assumption that snow is drifted from the high roof and accumulated at the step in roof elevation. The actual phenomena observed in the wind tunnel tests was a different one. The local high accumulations, see Figures 8 and 9, resulted due to the drifting of snow along the main roof and its organization first by the south wall and then, in later stages of the storm, by the north wall of the high storey. The NBCC does not include snow loading provisions for such roof geometries. Clearly this is a situation where wind tunnel model studies become necessary.

6 CONCLUDING REMARKS

This paper has presented information which demonstrates that extreme snow accumulations both on the ground and on roofs are likely to form due to the cumulative effects of several snowstorms. Usually, wind tunnel model studies examine the effects of single storms or events and reliance is made on numerical models to track the cumulative

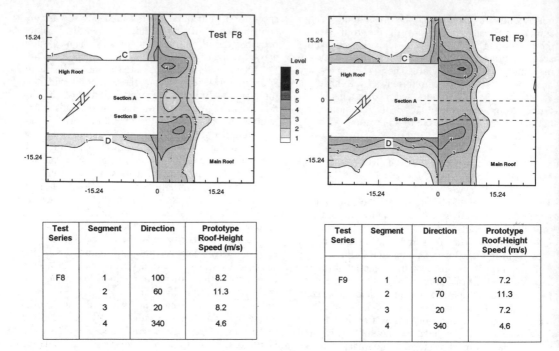

Test Series	Segment	Direction	Prototype Roof-Height Speed (m/s)
F8	1	100	8.2
	2	60	11.3
	3	20	8.2
	4	340	4.6

Test Series	Segment	Direction	Prototype Roof-Height Speed (m/s)
F9	1	100	7.2
	2	70	11.3
	3	20	7.2
	4	340	4.6

Figure 6. Snow accumulation around a high portion of a large roof, resulting from a wind tunnel model simulation of the snowstorm described in Figure 5.
(Note: Locations C and D are full-scale sampling sites. Contour intervals are 30.5 cm).

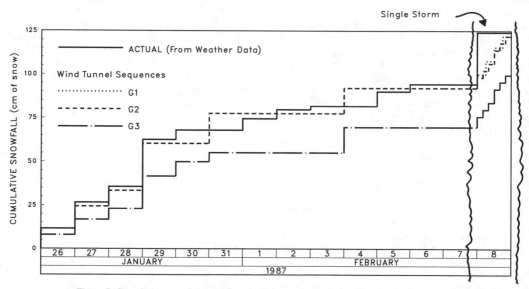

Figure 7. Cumulative snowfall preceding the February 8, 1987 snowstorm, described in Figure 5.

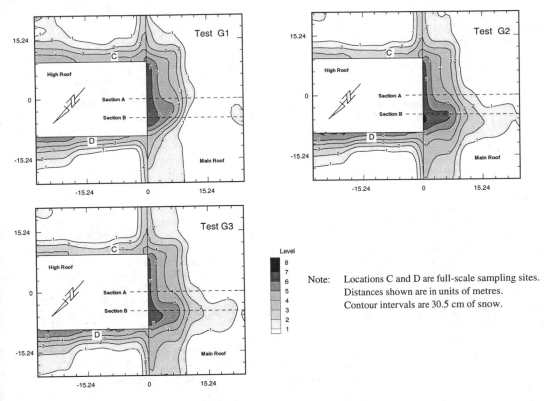

Figure 8. Snow accumulation around a high portion of a large-area roof, resulting from the cumulative effects of several snowfalls during a period of cold temperature, during which the snow was assumed to be driftable.

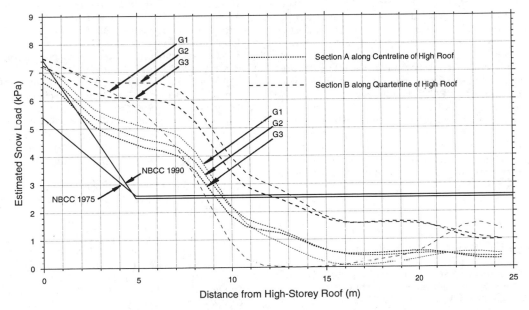

Figure 9. Snow loads predicted from wind tunnel model simulations of the cumulative effects of several snowfalls during cold temperatures and comparisons with snow load requirements of the National Building Code of Canada.

effects of winter weather conditions and to estimate maximum roof snow loads. While numerical models are regularly used to estimate snowmelt, during periods of above freezing temperatures and rainfall, such models have difficulty in allowing for the effects of wind action, which influences both the accumulation of snow and its drifting and redistribution. These processes are strongly influenced by the aerodynamics of the roof which in turn can be modified in the presence of significant snow accumulations. In short, the problem is a nonlinear one and the assumption of linear superposition of individual snowfalls and drifting episodes is an oversimplification. Physical model studies in this regard offer an advantage, as the influence of snow accumulations on the aerodynamics of the roof are taken into account.

Clearly, there are uncertainties in the simulation of the time-varying changes in the material properties of both the prototype and the model snow. Even though both may remain driftable, the mechanical action of drifting fragments and packs both the model and protrotype snow particles. Clearly, further studies are needed to quantify these effects. Nevertheless, the good comparisons with full-scale observations, obtained in the reported studies are encouraging. These results show that maximum snow loads on roofs tend to occur during prolonged periods with cold temperatures, during which a substantial amount of the snow on the roof remains driftable, and can be reshaped by an extreme storm packing winds strong enough to drift the previously-accumulated snow. This pattern of low temperatures clustered around major snowstorms is not uncommon, as demonstrated by the snowstorm sequences experienced in the eastern part of North America during the winter of 1995/96.

This study also demonstrates that triangular drift snow loads, specified by many building codes including the NBCC, do not always capture the magnitudes and shapes of drift snow loads which can occur on unusual roofs. Information on such unusual snow loads can be obtained through physical model studies which, while not exactly simulating the detailed properties of prototype snow, do representatively model the overall aerodynamics of the process.

REFERENCES

American Society of Civil Engineers 1987. *Wind Tunnel Model Studies for Buildings and Structures*, ASCE Manual of Practice No. 67, Published by the ASCE, New York, N.Y.

Gamble, S.L., Kochanski, W.W. and Irwin, P.A. 1991. Finite Area Element Snow Loading Prediction - Applications and Advancements.

Proc. of Eighth International Conference on Wind Engineering. London, Ontario, Canada, Elsevier, Amsterdam.

ICSE-1 1989. Santa Barbara, California, July, 1988, CRREL Special Report 89-6.

ICSE-2 1992. Santa Barbara, California, June, 1992, CRREL Special Report 92-27.

Irwin, P.A., Williams, C.J., Gamble, S.L. and Retziaff, R. 1992. Snow Load Prediction in the Andes Mountains and Downtown Toronto - FAE Simulation Capabilities. ICSE-2: 135-145.

Isyumov, N. 1971, *An Approach to the Prediction of Snow Loads*. Ph.D. Thesis, University of Western Ontario, London, Canada. Published as University of Western Ontario, Engineering Science Research Report BLWT-9-71.

Isyumov, N. and Mikitiuk, M. 1977. Climatology of Snowfall and Related Meteorological Variables with Application to Roof Snow Load Specifications. *Canadian Journal of Civil Engineering*, 4(2):240-256.

Isyumov, N., Mikitiuk, M. and Cookson, P. 1989. Wind Tunnel Modelling of Snow Drifting: Applications to Snow Fences. ICSE-2:210-226.

Isyumov, N. and Mikitiuk, M. 1990. Wind Tunnel Model Tests of Snow Drifting on a Two-Level Flat Roof. *Journal of Wind Engineering and Industrial Aerodynamics*, 36:893-904.

Isyumov, N. and Mikitiuk, M. 1992, Wind Tunnel Modelling of Snow Accumulations on Large-Area Roofs. ICSE-2:181-193.

Iversen, J.D. 1982. Small-Scale Modelling of Snowdrift Phenomena. Proc. Int. Workshop on Wind Tunnel Modelling Criteria in Civil Eng. Applications:V4.1-36, Gaithersburg, Md.

Kennedy, D.J.L., Isyumov, N. and Mikitiuk, M. 1992. The Effectiveness of Code Provisions for Snow Accumulations on Stepped Roofs. ICSE-2:439-452.

Kind, R.J. 1986. Snowdrifting: A Review of Modelling Methods. Cold Regions Science and Technology, 12 (1986), Elsevier, Amsterdam.

Kocin, P.J. and Uccellini, L.W. 1990. *Snowstorms Along the Northeastern Coast of the United States: 1955 to 1985*, Meteorological Monographs, 22(44), American Meteorological Society, Boston, Mass.

Mikitiuk, M. and Isyumov, N. 1989. Variability of Snow Loads on Large-Area Flat Roofs. ICSE-1: 142-157.

National Research Council of Canada 1975. *Supplement No. 4 to the National Building Code of Canada*. Issued by the Associate Committee on the National Building Code, Ottawa, Canada.

National Research Council of Canada 1990. *Supplement to the National Building Code of Canada 1990*. Issued by the Associate Committee on the National Building Code, Ottawa, Canada.

Snow Engineering: Recent Advances, Izumi, Nakamura & Sack (eds) © 1997 Balkema, Rotterdam. ISBN 90 5410 865 7

Laboratory study of snow drifts on gable roofs

Seiji Kamimura
Department of Mechanical Engineering, Nagaoka University of Technology, Japan

Michael J.O'Rourke
Department of Civil and Environmental Engineering, Rensselaer Polytechnic Institute, Troy, N.Y., USA

ABSTRACT: A laboratory investigation of snow drift formation on a gable roof was carried out. Two gable roofs, having slopes of 5° and 10° with wind perpendicular to the ridge line, are considered. Utilizing a water flume and crushed walnut shells to represent snow, transported volume from the windward roof and the surcharged volume on leeward roof were measured at four different wind speeds.

1. INTRODUCTION

Snow drifting on a gable roof causes unbalanced loads by transporting snow from the windward side to the leeward side during and/or after snowfall. The American Society of Civil Engineers Manual 7 (1996) considers unbalanced loads on gable roofs for slopes from 15° to 70°. However there are some building collapse case histories for roof slopes less then 15° (Chin, 1980, DeAngelis, 1988).

In this paper, snow drifts on gable roofs with slopes of 5° and 10° are studied using scale models and a water flume / crushed walnut shell analog. Specifically, the transport rate of snow from the windward roof and the surcharge volume on the leeward roof are investigated.

2. LABORATORY EQUIPMENT

All the laboratory tests were performed in the Rensselaer Polytechnic Institute water flume. The flume is 7.6 m long, 0.76 m wide and 1.22 m deep. There is a metal grate used to smooth the flow at the water inlet. Ten inverted triangular spires and bricks on the flume floor are used to generate appropriate velocity profiles. In this study, velocity profile power law coefficients were 0.08, 0.10, 0.07 and 0.08, which model a rural exposure. The water velocities in the flume were 7.6, 9.1, 10.6 and 12.2 cm/s.

The velocity in the flume is measured using a propeller type velocity meter with an averaging interval of 90 seconds. It is placed 30.5 cm before the front of the building model and 20.3 cm above the flume floor.

Two wooden building models with roof slopes of 5 ° and 10° were constructed of 3/4-inch (19 mm) plywood.

The dimensions of both buildings are as follows: length and width are 60.1 cm, the height of the eave is 20.3 cm. and the heights of the ridge line are about 23.0 cm and 25.7 cm respectively. The scale was 1:30. The model was painted with 2-inch by 2-inch (5 cm × 5 cm) grid line to facilitate data collection.

Crushed walnut shells were used to simulate snow based on similitude of the ratio of threshold shear velocity and terminal fall velocity (O'Rourke and Weitman, 1992).

3. EXPERIMENTAL PROCEDURE

First, a 1.27-cm uniform layer of crushed walnut shells is placed on both the windward and leeward sides of the model roof. The layer corresponds to a prototype snow

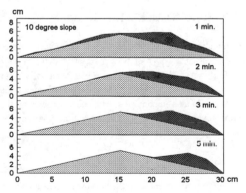

Figure 1. An example of snow drift formation on a gable roof, water velocity = 12.2cm/s.

313

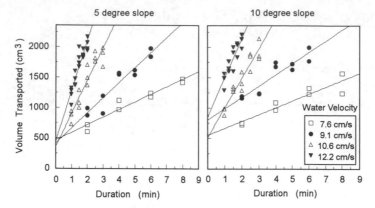

Figure 2. Volume transported from the windward roof.

layer depth of 38.1 cm. A volume of 2360 cm³ is placed on each roof.

The test duration ranged from 1 minute to 12 minutes, dependent upon the water velocity of 7.6, 9.1, 10.6 and 12.2 cm/s (0.25, 0.30, 0.35, 0.40 ft/s). After the test was completed, the flume was drained slowly so as not to disturb the drift formation. With the flume empty of water, the depths of walnut shells at each nodal point on the windward and the leeward sides were manually measured. The remaining shells on both sides were collected separately and set aside to dry. Once the shells were dry, their volumes were measured and recorded. Knowing the initial volume on the windward side and measuring the final volume on the windward side after a test, the volume transported from the windward side was calculated. In addition, the surcharge drift, which is the difference between the initial and final volume on the leeward side, was calculated. The trapping efficiency was then determined as the ratio of the surcharge drift volume divided by the volume transported.

4. EXPERIMENTAL RESULTS

Figure 1 shows the distribution of snow at varies times during a test of a 10° roof with a water velocity of 12.2 cm/s. In this test, the windward side was essentially clear of snow after about 2 minutes, and the total volume on the leeward side was a maximum. After that, particles on the leeward side gradually blew off and the total volume on the leeward side decreased.

4.1 Transport Rate

The volume transported from the windward side at each of the four test velocities is plotted versus the test duration in

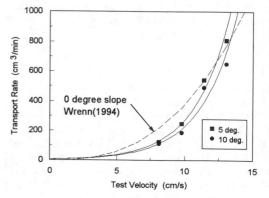

Figure 3. Transport rate from windward roof.

Figure 2. Separate plots for the 5° and 10° roofs are presented. Each data point represents one complete test, whereas all tests at a given velocity composed a data set. Volume transported linearly increased with test duration. The gradient of the least-square line is taken as the volume transport rate, G(t), in cm³ per minute. G(t) for each velocity is shown in Figure 3. The transport rate increases exponentially with test velocity. The rate for the 5° slope is slightly greater than that for the 10° roof. The corresponding transport rate for a flat roof in a rural exposure (Wrenn, 1994) is also shown (dashed line) in Figure 3 as a reference. The curve fit equations from Figure 3 are as follows: for the 5° sloped roof, G(t) = $0.0312V^{4.09}$, for the 10° sloped roof, G(t) = $0.0335V^{3.98}$, with G(t) in cm³/s and V in cm/s.

4.2 Surcharge Volume

Figure 4 in a plot of the surcharge volume as a function of duration. In general, at higher velocities of 10.6 and

314

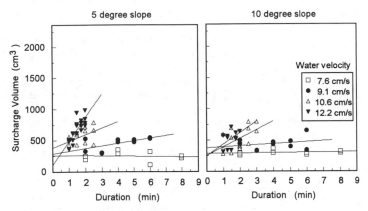

Figure 4. Surcharge Volume on leeward roof.

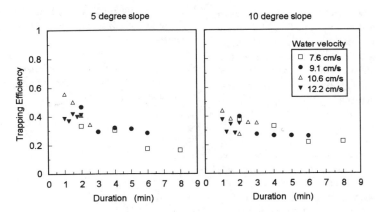

Figure 5. Average trapping efficiency for leeward roof.

12.2 cm/s, surcharge volumes on the leeward side increase with test duration. However at lower velocities of 9.1 and particularly 7.6 cm/s, the surcharge volume was essentially independent of test duration. That is at these lower velocities, the inflow from the windward matches the outflow from the leeward side.

4.3 Trapping Efficiency

The average values of trapping efficiency for each velocity and slope are plotted versus the test duration in Figure 5. The trapping efficiency gradually decreases with test duration but does not appear to be a function of test velocity. The range of trapping efficiency for 5° slope (0.17 to 0.56) is wider than for 10° slope (0.22 to 0.44). That is, the smaller slope would trap more snow on the leeward roof in shorter duration tests but more is blown off in long duration tests. The average values for both model roofs are 0.36 and 0.32 for 5° and 10° slope, respectively.

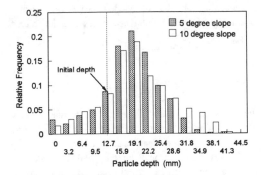

Figure 6. Histogram of particle depth at each nodal point on leeward roof.

4.4 Local Load

A histogram of particle depths at each nodal points on the leeward side from all of the tests is shown in Figure 6. It appears to be a normal distribution with the average

315

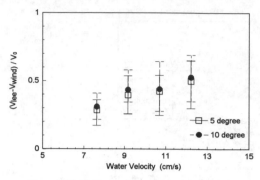

Figure 7. Normalized volume difference between windward and leeward roof. Total initial volume $V_0= 4720$ cm^3

6. BIBLIOGRAPHY

ASCE. 1990. Minimum design loads for buildings and other structures. *ASCE7-88*. New York, NY.

Chin, I., Gouwans, J. & Hanson, M. 1980. Review of Roof failures in the Chicago area under heavy snow loads. *ASCE reprint*. 80-145.

DeAngelis, C. 1993. 1995 revisions of ASCE 7-88. *ASCE 7 Main Committee*. Memo to committee.

O'Rourke, M. & Weitman, N. 1992. Laboratory Studies of Snow Drifts on Multilevel Roofs. *Proceeding of 2nd Int'l conference on Snow Engineering*. Santa Barbara, CA.

Weitman, N. 1992, Laboratory Studies of Snow Drifts on Multilevel Roofs. *Masters Thesis of Rensselaer Polytechnic Institute*. Troy, NY.

Wrenn, P. 1995. Investigation of Triangular Snow Drifts on Multilevel Roofs. *Masters Thesis of Rensselaer Polytechnic Institute*. Troy, NY.

(mean) values of 1.78 and 1.89 cm (11.2/16 and 11.9/16 in.), and the standard deviation of 0.73 and 0.84 cm (4.6/16 and 5.3/16 in.) for 5 ° slope and 10 ° slope, respectively. It should be noted that large depths (i.e. greater than 25 mm) are more common for the 10 ° roof than the 5 ° roof. This may be due to the large aerodynamic shade region with the larger slope.

4.5 Unbalanced Load

As shown in Figure 7, the difference between the volume on the leeward side and the volume on the windward side for all the tests is used to evaluate load unbalance on the roof as a whole. The average (midpoint), maximum (top end of the bar) and minimum (bottom end of bar) values increase with water velocity and the values of 10° sloped roof are slightly greater than 5° sloped roof. It should be noted that although the 10 ° sloped roof have a greater unbalance, the surcharge volumes are less than the volume for the 5° sloped roof.

5. CONCLUSION

A laboratory investigation of snow drift formation on a gable roof was carried out utilizing a water flume and crushed walnut shells to simulate snow in air.

Transported volume from the windward side and surcharged volume on the leeward side were investigated using two model buildings having roof slopes of 5° and 10 °. As a result, it was shown that the 5° sloped roof causes slightly greater transport rate from the windward roof, and also slightly greater volume blown out from the leeward roof. The 10° sloped roof causes slightly greater unbalance of load between the windward and leeward roofs, and yields greater local load on leeward roof.

Snow Engineering: Recent Advances, Izumi, Nakamura & Sack (eds) © 1997 Balkema, Rotterdam. ISBN 90 5410 865 7

Application of wind tunnel modeling to some snow related problems

Toshikazu Nozawa & Jiro Suzuya
Department of Architecture, Tohoku Institute of Technology, Sendai, Japan

Yasushi Uematsu
Department of Architecture, Tohoku University, Sendai, Japan

Yoshiaki Miura
Tohoku Electric Company Ltd, Sendai, Japan

ABSTRACT: To increase the reliability of data obtained in snow wind tunnel tests on roof snow accumulation, it is important to establish similarity conditions. In this paper, we investigate snow accumulation on domed roofs, on the basis of the data obtained in two series of snow wind tunnel tests. Model snow accumulations, similar to the actual snow distribution patterns, were obtained when shifts in wind direction were taken into consideration.

1 INTRODUCTION

Many similarity requirements for snow wind tunnel tests have been proposed to achieve accurate simula--tion of roof snow accumulation. Actually it is di--fficult to satisfy those many similarity requirements simultaneously. The similarity requirements can be roughly classified into three categories: 1) similarity of air flow around the building, 2) similarity of air--borne particles, and 3) similarity of the motion of particles on the surface of the roof snow. More field measurements of roof snow are needed, with results compared to those of snow wind tunnel tests.

We carried out two series of snow wind tunnel tests; one was of a large dome, with results compared with field measurement results by Sakurai et al.(1988), and the other test was of a spherical roof, with test results compared with observed snow accumulation on such a roof.

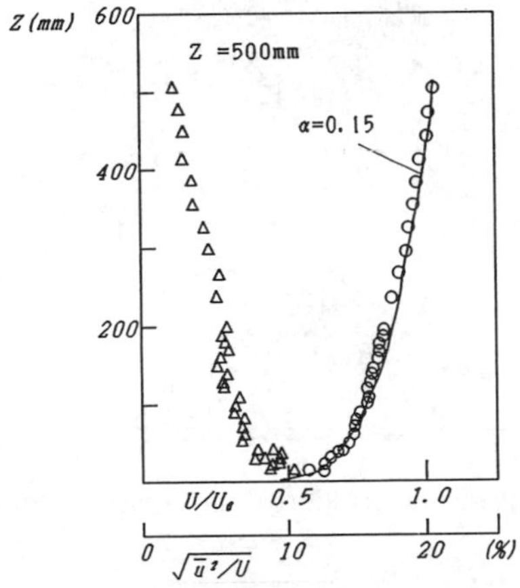

2 WIND TUNNEL TESTS

2.1 *Snow Distribution on Large Dome*

Figure 1 shows the profiles of mean wind velocity and turbulence intensity. Figure 2 shows the roof plan and sections on the large dome. The wind tunnel model was made with a geometric scale of 1/400.

Crushed wheat was used as model snow. It's size distribution is shown in Table 1. The reference wind velocity Ur was measured 50 cm above the wind tunnel floor.

We tests two cases. In the first experiment, the

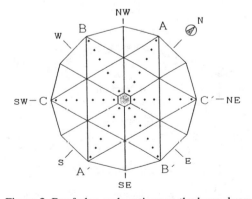

Figure 1. Profiles of mean wind velocity and turbulence intensity.

Figure 2. Roof plan and sections on the large dome.

317

Table 1. Particle size distribution of
the crushed wheat model snow.

Size (μm)	Percentage
500 ~300	39.59
300~150	28.38
150~125	4.53
125~106	9.39
106~75	10.23
75~38	5.59
≦38	2.29

model snow depths, caused by snowfall with wind from W or WNW, were measured by the laser dis--placement sensor(Suzuya et al.). The reference wind velocity in this experiment was 2 m/sec, which was the velocity at the threshold of motion of falling particles. In the second experiment, after snow accumulation caused by snowfall accompanied by the wind from SW, snow accumulation transformed by wind from W was measured.

The velocity of the wind from SW (accompa--nied by snowfall) was 2 m/sec, and that from the W was 2.5 m/sec, which was the velocity at the threshold of motion of particles resting on the model.

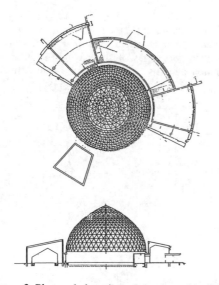

Figure 3. Plan and elevation of the domed building.

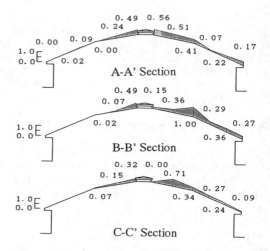

Figure 5a. Model snow accumulation.
(fraction deepest)

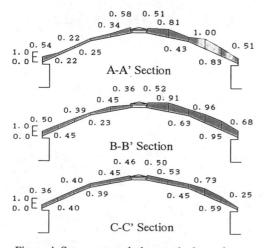

Figure 4. Snow accumulation on the large dome.
(fraction deepest)

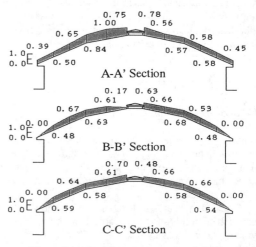

Figure 5b. Model snow accumulation.
(fraction deepest)

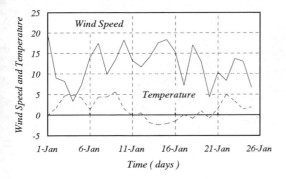

Figure 6. Meteorological records at the site of the domed building in Jan. 1995.

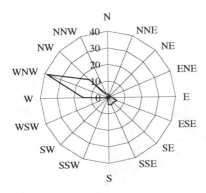

Figure 7. Wind rose for Jan. 1995.

2,2 *Snow Accumulation on Spherical Roof.*

The domed building is located in the coastal area of Noshiro City in Akita Prefecture. The outline of the building is shown in Figure 3. The diameter of the dome is 30 m, and the top of the dome is 22 m high above the ground. Aluminum alloy frames stick out on the surface of the dome. The lowrise buildings with low pitched roof are arranged around the domed building.

Three models made with a geometric scale of 1/400 were tested. The experimental conditions were much the same as that of the above mentioned tests on the large dome. The surface of the three models domes were finished in different manners; that is, smooth, sandblasted and covered with a net of 5 mm mesh.

Snow was accumulated on the models by snowfall in a wind from WNW, then transformed by the winds from NW and W. The reference velocity of the wind from WNW was 2.5 m/sec, and that from NW and W was 3.7 m/sec; the former

corresponded to the threshold velocity for the falling particles and the latter corresponded to that for still particles on the roofs of the surrounding lowrise buildings.

We measured the depths of snow accumulation, first caused by snowfall accompanied the wind and then transformed by wind from the west without snowfall.

3 RESULTS and DISCUSSION

Figure 4 shows the field measurement results for the large dome by Sakurai et al. (1988).

The cross sections of snow accumulations on the model caused by snowfall with wind from W are shown in Figure 5(a). Model snow accumulation caused by snowfall in wind from SW, and transformed by wind from W, is shown in Figure 5(b); the wind directions and cross section lines are shown in Figure 2. The snow depth is normalized by the maximum value on the dome.

The fluctuations of wind speed and temperature, measured at the site of the domed building in Noshiro, in January 1995, are shown in Figure 6.

The wind rose for this period is shown in Figure 7. The distribution of snow on the dome model covered with a net is shown in Figure 8(a). That on the smoothly finished model is shown in Figure 8(b).

The depth of snow in Figure 8 is expressed as the ratio of the measured depth to the maximum depth on the dome.

Snow accumulation on the model dome, as shown in Figure 4(b), has many similarities to the actual snow accumulation shown in Figure 3. In this manner, taking the shift of the wind direction during a snowy season into consideration, we could make snow accumulation on the model similar to actual snow on the large dome. In the test on snow accumu--lation on a spherical dome, the shift of wind direction was considered in the same manner.

Comparing the distribution of snow on three model domes, we found that the pattern shown in Figure 8(b) resembled closely the actual snow accu--mulation pattern.

The third model covered with a net resembled the actual dome in appearance. However, the distribution of snow on this model, as shown in Figure 8(a), did not resemble the actual distribution of snow, in spite of the formal resemblance to a real thing.

4 CONCLUSIONS

Snow accumulation on domed roofs have been in-

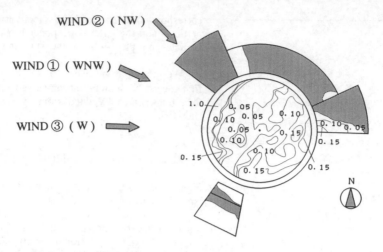

Figure 8a. Model snow depth on the dome model. (on the dome model covered with net.)

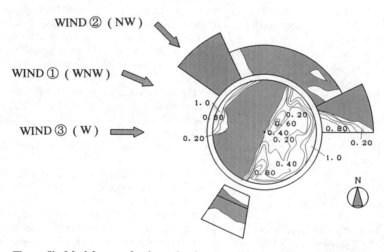

Figure 8b. Model snow depth on the dome model. (on the smoothly finished model.)

-vestigate in a wind tunnel. The results were com-
-pared with those of field measurements.

In wind tunnel tests to determine snow accumu-
-lation on domed buildings, the transformation of
snow on the roof by shifts in wind direction should
be taken into account.

5 REFERENCES

Isyumov, N., Mikitiuk, M. andCookson,1988. Wind
Tunnel Modeling of Snow Drifting, Application
to Snow Fence. *First International Conference
on Snow Engineering*, CRREL.

Petersen, R. L. and Cermak, J. E. 1988. Application
of Physical Modeling for Assessment of Snow
Loading and Drifting. *First International Confe-
-rence on Snow Engineering*, CRREL.

Sakurai, S., Joh, O. and Shibata, T. 1988.Wind
Effect on the Distribution of Snow Depth on a
Large Dome *First International Conference on
Snow Engineering*, CRREL.

Suzuya, J., Uematsu, Y. and Nozawa, T. 1992. Wind
Effects on Snow accumulation on a Flat Roof
*Second International Conference on Snow Eng-
-ineering*, CRREL.

Snow Engineering: Recent Advances, Izumi, Nakamura & Sack (eds) © 1997 Balkema, Rotterdam. ISBN 90 5410 865 7

Water flume evaluation of snowdrift loads on two-level flat roofs

Michael J. O'Rourke
Rensselaer Polytechnic Institute, Troy, N.Y., USA

Peter Wrenn
John G. Waite Associates, Albany, N.Y., USA

ABSTRACT: A laboratory study of snowdrift loads on a roof is presented. In this model study, wind is simulated by the flow of water in a flume while the snow is modeled by crushed walnut shells. Results are presented for a two-level flat roof with wind perpendicular to the roof step, blowing from the upper level towards the lower level roof. The drift formation process is characterized by a transport rate and a trapping efficiency. The transport rate quantifies the amount of snow/walnut shells per unit time which leaves the upper level roof while the trapping efficiency quantifies the percentage of transported snow/walnut shells which eventually becomes part of the drift. The transport rate and trapping efficiency are presented as functions of the fluid velocity, for various upwind terrain roughnesses. Drift volumes from five full-scale case histories are compared to those predicted by the model studies.

1 INTRODUCTION

Snow accumulation on roofs represents an important design load for construction in regions with low winter temperatures. Of particular interest is low-rise construction where the roof system constitutes a significant portion of the total building cost. Hence, accurate predictions of uniform and drifted snow loads are required by the designer. In the past these roof snow loads, particularly drift load, have typically been established by an analysis of existing full-scale case histories in combination with engineering judgment. A new approach is proposed herein for determining roof drift loads. This new approach is based upon the physics of drift formation as evaluated using scale models in a laboratory. The specific roof geometry considered herein is a two-level, nominally flat roof. The following sections discuss similitude relations, laboratory procedures and results. Finally, the relative accuracy of the proposed approach is determined through a comparison with observed drift volumes from five full-scale case histories. For the case of a downwind facing roof step, the accumulation of the redeposited snow often results in a triangular drift (Tabler 1975) on the lower roof along the change in roof elevation. The situation for a two-level flat roof with wind blowing from left to

right with velocity V is shown in Figure 1. In Figure 1, G is the snow transported from the upper level roof, D is the amount of snow which ends up in the drift, E is the trapping efficiency and H_d is the surcharge drift height.

If the wind velocity V is sufficiently large and a snow source is available, upper level roof snow is transported. Some of this transported snow would, in turn, deposit along the change of elevation. The amount transported is quantified as G and the percentage of transported snow which is deposited on the lower roof is quantified as E, where $0 < E < 1$. Since the drift length is linearly proportional to surcharge drift height, O'Rourke and Weitman (1992) proposed the change in the surcharge drift height would be given by

$$\Delta H_d = H_d(t) - H_d(0) \propto \left[\int_0^t G(t) \cdot E(t) dt \right]^{\frac{1}{2}} \quad (1)$$

2 SIMILITUDE REQUIREMENTS

The choice of the model particulate and the correspondence between model and prototype (full scale) results depends upon similitude requirements. Unfortunately, as noted by Iversen (1982), all the

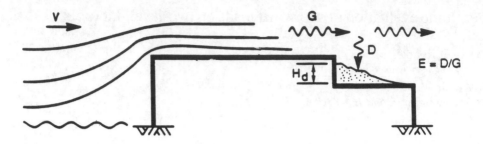

Figure 1. Triangular drift formation at downwind facing roof step.

similitude requirements for modeling drifting snow are not successfully met on a small scale, that is, the large number of modeling parameters cannot be satisfied simultaneously. Herein, similitude requirements are based upon what is considered to be the main transport mechanism.

Snow is transported by wind via three mechanisms: creep, saltation and diffusion (also referred to as suspension). Creep, or the rolling of snow particles along the snow surface, is generally accepted as a minor contributor to the total mass transported. On the other hand, saltation, or the jumping of snow particles over the snow surface, is believed to be the major contributor to snow movement (Kind 1981). The third mechanism of snow transport, diffusion, becomes a significant contributor in turbulent urban exposures (Isyumov et al. 1988). Thus proper similarity of airborne particles and particles near the surface, between the prototype and model, are required. The three primary requirements of similitude are similarity of flow, similarity of airborne particles, and similarity of particles near the surface. Note that in the experiments described herein, snow is modeled as a cohesionless particle, corresponding to dry, loose, fresh snow as defined in the *Handbook of Snow* (Kind 1981).

2.1 Similarity of Flow

In terms of similarity of flow, Isyumov (1971) noted that the terrain roughness in any simulation is extremely important, since the mass flux is dependent on the roughness. Tabler (1975) arrived at similar conclusions, with regard to wind tunnel simulation of snowdrifts.

Due to the no-slip condition at the earth's surface the wind speed is zero along the ground surface. Above the ground surface there is an atmospheric boundary layer in which the wind

speed increases with height, from zero at the ground surface to the geostrophic or gradient wind speed at the top of the atmospheric boundary layer. Proper characterization of the turbulent boundary is important since structures considered herein lay within that layer. The mean velocity profiles can be characterized by a power law (Davenport, 1967; Isyumov, 1971; and Simiu and Scanlan, 1978), which has the form

$$U_z = U_g \left(\frac{z}{z_g} \right)^\alpha \tag{2}$$

where U_z is the wind speed at height z, U_g is the gradient wind speed, z_g is the height of the boundary layer and α is the power law exponent which depends upon the terrain roughness. According to Davenport (1967), average values of α for rural, suburban, and urban exposures are 0.16, 0.28, and 0.40, respectively. As will be shown later, similarity of flow upstream of the structure is accomplished herein by matching the model and prototype power law exponents.

For airborne particles, similarity of the forces acting on the particles and their trajectories are maintained if the ratio of gravitational forces to fluid (aerodynamic) forces, as well as the ratio of inertial (kinematic) forces on the particle to gravitational forces are the same for both the model and the prototype. For the gravitational to fluid force ratio we have

$$\left[\frac{W_f}{U} \right]_m = \left[\frac{W_f}{U} \right]_p \tag{3}$$

where W_f is the terminal fall velocity of the particle and U is the mean velocity of the fluid. The subscripts m and p refer to model and prototype, respectively.

Similarity of the inertial to gravitation force ratio for airborne particles require the densimetric

Froude number be the same in both model and prototype. The densimetric Froude number is a function of the particle mean diameter. Note that Iverson (1982) believes that the densimetric Froude number provides a better correlation of deposition rates than does the conventional Froude number. However, as shown by Isyumov (1971), the Froude number need not be based on the particle's mean diameter, especially when the ratio of the model to prototype particle diameters are much greater than the geometric scale being used. This simplification (i.e., substituting the geometric scale for the particle diameter) does result in a reduction of the particle's ability to follow rapid fluid velocity variations. However, it is felt that this would induce only small errors in tests such as those described herein, where the gradient velocity and velocity profile were relatively constant. The densimetric Froude number based on the geometric scale, is

$$\left[\frac{U^2\rho}{Lg(\rho_s - \rho)}\right]_m = \left[\frac{U^2\rho}{Lg(\rho_s - \rho)}\right]_p \qquad (4)$$

where ρ is the fluid density, L is the length, g is the gravitational acceleration and ρ_s is the particulate density.

2.2 Similarity of Particles Near the Surface

Since saltation is believed to provide the largest portion of the total snow flux for roof drifts, it is necessary to ensure kinematic similarity of the particles' paths in relation to the overall building dimensions.

For these "about to be airborne" particles, similarity requires the fluid forces, expressed as a critical surface shear stress ρu_{*t}^2, acting on particles of diameter d and the restoring forces, due to gravity and proportional to $(\rho_s - \rho) gd^3$, be modeled according to the densimetric Froude number. Here u_{*t} is the threshold shear velocity for the particle, that is, the minimum velocity at which a single particle begins to move. It can be shown that this requirement reduces to

$$\left[\frac{u_{*t}}{U}\right]_m = \left[\frac{u_{*t}}{U}\right]_p \qquad (5)$$

Isyumov (1971) suggested simplifying the similitude requirements by combining Equations (3) and (5) into a single requirement:

$$\left[\frac{W_f}{u_{*t}}\right]_m = \left[\frac{W_f}{u_{*t}}\right]_p \qquad (6)$$

That is, the ratio between terminal fall velocity and the particulate threshold shear velocity should be equivalent for the model and prototype. If this equivalence is maintained, correspondence between model and prototype parameters (specifically, flow velocity and time) are given by Equation (4).

2.3 Properties of Prototype and Model

Based upon experience with water flume snow drift modeling in Canada (personal communication with P. Irwin, 1991), commercially available crushed walnut shells was chosen as the model particulate. The mean model particle diameter of 0.051 cm (0.02 in.) was selected in order to satisfy Equation (6). Table I lists the fluid and particulate properties of the prototype and model. The prototype properties are taken from the *Handbook of Snow* (Kind 1981) for the assumed dry, loose, fresh snow. The model particulate (i.e., crushed walnut shells) are assumed to be spheres and the relation from Randkivi (1976) is used to determine their terminal fall velocity.

Table 1. Properties of prototype and model

Property	Prototype	Model
Diameter, d cm (in)	0.05 (0.0197)	0.051 (0.02)
Particle Density, ρ_s kN/m^3 (pcf)	0.49 (3.12)	11.97 (76.2)
Fluid Density, ρ kN/m^3 (pcf)	0.012 (0.0765)	9.8 (62.4)
Terminal Fall Velocity, W_f cm/s (ft/s)	75 (2.46)	3.11 (0.102)
Threshold Shear Veloctiy, u_{*t} cm/s (ft/s)	15.0 (0.492)	0.66 (0.0218)
W_f/u_{*t}	5.0	4.68

3 EXPERIMENTAL PROCEDURE AND RESULTS

3.1 Laboratory Equipment

All of the laboratory tests described herein were performed in the Rensselaer water flume. The flume has a cross-section of 0.75 m by 1.2 m (2.5 ft by 4 ft) and measures 6.1 m (20 ft) from inlet to outlet. A more detailed description of the Rensselaer flume can be found in Wrenn (1995). The prototype

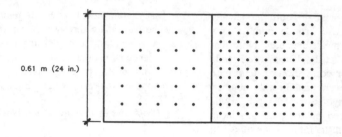

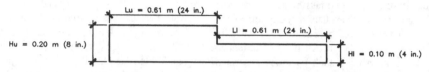

Figure 2 Standard two-level flat roof model with 1:30 scale

modeled in these tests was a two-level, nominally flat roof. A plan and elevation of the scaled model are shown in Figure 2.

3.2 Laboratory Measurements

In all of the tests, a 13 mm (½-in.) dry, uniform layer of crushed walnut shells was placed on the upper and the lower roofs of the model. The 13 mm (½-in.) uniform layer corresponded to a prototype roof snow depth of 380 mm (15 in.).

Crushed walnut shells were not placed on the flume floor, nor was any attempt made to model snowfall while the wind was blowing (i.e., while the water was flowing). Hence, the only "snow source" modeled in these tests was snow already on the model building roofs.

On average, 14 tests of various durations were run for each water velocity and exposure class. The water velocities ranged from 0.06 m/s (0.20 ft/s) to 0.12 m/s (0.40 ft/s), in increments of 0.015 m/s (0.05 ft/s). These velocities were measured at a height of 203 mm (8 in.) above the flume floor and 610 mm (24 in.) upstream of the model. The test durations ranged from 1 minute to 16 minutes: the one minute test duration was considered the shortest to accurately obtain a constant water velocity. The longest duration was determined by shell availability. That is, the longer duration tests were terminated when the upper level roof became free of shells. Note that all tests which exhibited gross duning were neglected. It was felt that this transport mechanism (i.e., creep transport) which was observed in some of the low velocity water

flume tests was not representative of prototype behavior.

After each test, the flume was emptied of water and the depth of walnut shells one the lower roof were manually measured to the nearest 1.5 mm ($^1/_{16}$ in.). Any remaining shells on the upper roof were collected and set aside to air dry. Once the shells were dry, their volumes were measured and recorded. With an initial volume known and a measured volume of remaining shells on the upper roof following each completed test, the volume transport and transport rate (volume transport divided by test duration) for each test were determined.

The surcharge drift volume on the lower roof, which in all tests were roughly triangular in form, was also determined. Having measured the accumulated shell depth at each nodal point on the lower roof, a representative cross-sectional profile was found by averaging the nodal depths parallel to the roof step. The product of the drift area that lay above the initial lower level roof shell depth and the model's width was defined as the surcharge volume, in cubic centimeters. The trapping efficiency for each test was then determined. It is defined as the ratio of the lower level roof surcharge drift volume to the volume transported from the upper level roof.

The model was constructed of wood with a dimensional scale of 1:30. The geometric model scale was chosen such that the eight-inch height of the model corresponded to a typical prototype upper level roof height of 6.1 m (20 ft).

Based upon a suggestion by Irwin (1980), the

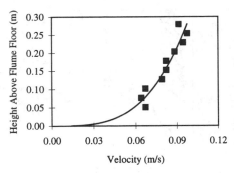

Figure 3   Velocity profile for a suburban exposure at a velocity of 0.09 m/s (0.30 ft/s)

appropriate mean approach velocity profile was attained using ten inverted triangular spires located 0.3 m (1 ft) downstream of the flume inlet area. Surface roughness was modeled using crushed bricks of various heights, intermittently placed on the flume floor upstream of the model. The three main exposure classifications (rural, suburban, and urban) in the water flume were obtained by trial and error. To verify the flume velocity profile, velocity readings were taken 2.4 m (8 ft) downstream of the spires, without the model present. The average velocity, based on the three velocity measurements at each height, was then plotted versus height above the flume floor, and a curve was then fit to the data. Figure 3 shows the resulting velocity profile for a suburban exposure at a velocity of 0.09 m/s (0.30 ft/s).

The power law exponent α in Equation 2 was then back-calculated from the fitted curve. The back-calculated values of α were : 0.159, 0.295 and 0.399 for the simulated rural, suburban and urban exposures, respectively. These model power law exponents agree with the average full-scale values discussed in the previous section.

3.3 Transport Rates

For a given exposure and velocity, the volume transported from the upper level roof was plotted versus the test duration. Each data point represents one complete test, whereas tests at a given velocity and exposure compose a dataset.   A linear regression line was then computed. Figure 4 shows data for a suburban exposure at five water velocities. The slope of each regression line of the volume transport data is defined as the volume transport rate, G, in cm³ per minute. Note that the regression lines do not pass through the origin. This is believed to be due, in part, to the loss of walnut

shells from the upper level roof during initial filling of the flume. Also, the transported volumes shown in Figure 4 are for a fixed length of the upper level roof. As noted in the conclusions, the influence of geometric parameters such as upper roof length is currently unknown.

As one might expect intuitively, the volume transport rate increased with the test velocity. This trend was evident for all three exposures. To determine the influence of terrain roughness, the transport rate was plotted, in Figure 5, versus test velocity for each of the three exposures. Note that the transport rate decreased from rural to suburban to urban exposures. Isyumov's (1971) results also suggested slightly lower than average drift loads (perhaps attributable to lower transport rates) for exposed roofs in an urban exposure as compared to similar roofs in a rural exposure.

For all three exposures, the transport rates were approximately proportional to the velocity raised to the 2.5 power. This agreed with the theoretical transport rate, transport rate approximately proportional to the velocity cubed, as presented in the *Handbook of Snow* (Kind 1981).

3.4 Trapping Efficiency

The trapping efficiency for an individual test was defined as the ratio of the surcharge amount deposited on the lower roof to the amount transported from the upper roof. Thus the trapping efficiency, E, lies between 0 and 1. A general decrease in E with increasing velocity was observed. Intuitively, one would expect the trapping efficiency to decrease as the velocity increases due to a higher percentage of upper roof snow particles jumping over the drift area at higher velocities.

Although there was a fair amount of scatter in

325

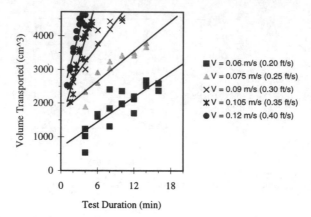

Figure 4. Volume transported from upper roof versus test duration (suburban).

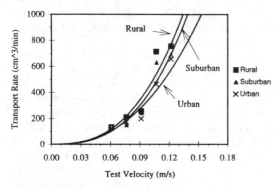

Figure 5 - Transport rate versus test velocity for three exposures

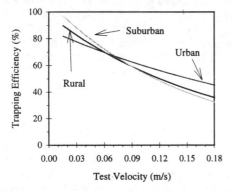

Figure 6. Average trapping efficiency versus test velocity

the trapping efficiency values, on average, there was an observable decrease with velocity. For each exposure, a best fit curve through the average values was determined. These best fit curves are presented in Figure 6. Note that although there are differences in the extrapolated curves at low and high velocities, the influence of exposure on trapping efficiency is small for the middle velocity range where the "actual data" (i.e., water velocities of 0.06 to 0.12 m/sec (0.2 to 0.4 ft/sec)) exists.

4 CASE HISTORY COMPARISONS

In order to gauge the relative effectiveness of the approach described herein, a comparison to actual drifts was made. In all, five full-scale case histories were chosen for this comparison. In selecting the case histories the following geometric criteria were used:

In elevation the structure's geometry should consist of only two roof levels, an upper and a lower, with a difference in roof elevations greater than 1.5 m (5 ft). Each roof level should be nominally flat and have no parapets. The buildings should have, at most, small offsets in plan of the lower roof with respect to the upper roof.

The final criterion involved the snowfall experienced at the case history sites. Snowfall leading to the drift formation should have occurred over a two or three day period. The geometric criterion was intended to insure compatibility between the full scale prototype structures and the relatively simple geometry of the scale models. The short duration snowfall criterion was intended to avoid complicating problems such as crust formation.

The case histories were compiled from insurance claim files and other sources. The scale factor was based upon the case history's upper roof height to the model's upper roof height of 200 mm (8 in). Thus, each case history had a unique scale factor.

All the building's dimensions and snow depths were scaled accordingly.

Wind velocities at the nearest airport, available at three-hour intervals, were taken as representative of conditions at the case history structure in question. The wind velocity was assumed constant over the three-hour interval. Only the component of the wind speed perpendicular to the roof step was considered in the simulation. The scaling of the velocities between the prototype and the model was based on the densimetric Froude number in Equation (4). This assured the same proportion of forces in the model. Since the difference between prototype and model in Equation (6) is less than 7%, as shown in Table I, the time scale was based solely on the geometric scale (Isyumov and Mikitiuk 1990).

In predicting a drift volume, the key parameters are the available snow on the upper roof , the transport rate and the trapping efficiency. Based upon the laboratory results, the following were used to establish the predicted drift size for the scale model for the assumed suburban exposure.

$$G = 235,200.0 \ V^{2.77} \tag{7}$$

$$E = 107.44 \ e^{-6.64V} \tag{8}$$

where V is the flume velocity (m/s) at a height of 20 cm (8 in.).

The prototype velocities were converted to equivalent scale model velocities for each three-hour interval, and then equations (7) and (8) were used to calculate the predicted, scale model drift volumes for each three-hour interval. A comparison between the observed drift volume (i.e., measured value at end of the storm) and predicted drift volumes (i.e., sum of the three-hour drift volume), both at scale model dimension are presented in Table 2. A detailed description of the technique used to account for phenomena such as warm spells and resulting melting is presented in Wrenn (1995).

Table 2. Comparison of observed and predicted drift volumes

Case History	Surcharge Drift Volume		Percent Error (%)
	Observed (cm³)	Predicted (cm³)	
160	1972	4112	+ 109
161	2645	2274	- 14
162	4602	9395	+ 104
167	12011	7232	- 40
234	2404	4092	+ 70

Note that the water flume technique tends to overestimate drift size. That is, the average percent error was + 46%. However, for all five case histories, the predicted values were within a factor of about 2 of the observed values.

5 CONCLUSIONS

The authors are encouraged by the water flume technique results. With further improvements and refinements, it is felt that the technique could be used in developing drift load provisions for building codes and load standards.

Shortcomings with the technique described herein, and by implication areas for future improvements, are listed below.

Flow turbulence was not measured in the laboratory tests. Hence, its impact upon the transport rate and trapping efficiency are unknown.

The geometric characteristics of the laboratory model were fixed. Hence, the influence of parameters, such as the upper roof length, upon the transport rate and trapping efficiency are unknown.

The transport rate and trapping efficiency in the model studies was for flow perpendicular to the roof step. The influence of the angle of incidence is unknown.

6 REFERENCES

Davenport, A.G. (1967) The dependence of wind loads on meteorological parameters, Proceedings of International Seminar on Wind Effects on Buildings and Structures, Ottawa, Canada.

Isyumov, N. (1971) An approach to the prediction of snow loads, Ph.D. Dissertation to the Faculty of Engineering Science, The University of Western Ontario, London, Canada.

Isyumov, N. and M. Mikitiuk (1990) Wind tunnel model tests of snow drifting on a two-level flat roof, Elsevier Science Publishers B.V.

Isyumov, N., M. Mikitiuk and P. Cookson (1988) Wind tunneling modeling of snow drifting— Applications to snow fences, Proceedings of Multidisciplinary Approach to Snow Engineering, Engineering Foundation Conference, Santa Barbara, California, July, pp. 210-226.

Irwin, P. (1980) Personal communication.

Iversen, J.D. (1982) Small-scale modelling of snowdrift phenomena, Proceedings of International Workshop on Wind Tunnel Modeling Criteria in Civil Engineering Applications, Gaithersburg, Maryland, April.

Kind, R.J. (1981). Snow drifting, *Handbook of Snow, Principles, Processes, Management, and Use*, eds. D.M. Gray and D.H. Male, Pergamon Press, Toronto, Canada, pp. 338-359.

O'Rourke M., and N. Weitman (1992) Laboratory studies of snow drifts on multilevel roofs, Proceedings of 2nd International Conference on Snow Engineering, Santa Barbara, California, June, CRREL Special Publication 92-27, pp. 195-206.

Simiu, E., and R.H. Scanlan (1978) Wind effects on structures: An introduction to wind engineering, John Wiley and Sons, New York.

Tabler, R. (1975) Predicted profiles of snowdrifts in topographical catchments, Paper presented at Western Snow Conference, Coronado, California.

Wrenn, P. (1995). Investigation of triangular snow drifts on multilevel roofs, Master of Science Thesis, Rensselaer Polytechnic Institute, Troy, New York, 112 p.

Snow Engineering: Recent Advances, Izumi, Nakamura & Sack (eds) © 1997 Balkema, Rotterdam. ISBN 90 5410 865 7

Snow loads in roof steps – Building code studies

P.A. Irwin
Rowan Williams Davies and Irwin Inc., Guelph, Ont., Canada

ABSTRACT: Snowdrifts in roof steps cause one of the most important roof loading conditions to be accounted for by the snow load provisions in building codes. In the 1995 National Building Code of Canada (NBCC) a new formulation of the roof step loads is used, based on finite area element (FAE) studies and field data. The specified snow loads are given in terms of ground snow load, roof length and width, step height and parapet height on the upper level roof. The FAE method is compared with field data from a roof step. Comparisons are also made between the new NBCC roof step loads, the ASCE 7-95 loads and those recently proposed for the ISO DIS 4355 Standard.

1 INTRODUCTION

The snowdrifts that form in roof steps have been one of the more common causes of roof collapse, as illustrated in Figure 1. The mechanism has been described by Templin and Schriever (1982) and by Taylor (1990). Wind blows snow over the edge of the upper level roof and this snow then becomes trapped in the region of aerodynamic shade on the lower roof behind the step. The form of the drift is illustrated in Figure 2. The snow from the upper level roof forms a triangular shaped drift on top of the snow that is already on the lower roof, and if the upper roof has a large area very high snow loads can build up in the step. Some snow will be held on the upper roof by parapets, if it has any, which tends to reduce the volume of snow in the step.

The above scenario of snow blowing off the upper level roof is the primary reason for the buildup of high loads in steps. Snow can also drift over the lower roof towards the step and this can lead to additional accumulations, but it appears that typically it is the snow off the upper roof that causes most high loads.

The National Building Code of Canada (NBCC) has specified a load distribution to cover the roof step case since 1965. The NBCC load was based on observations on roofs. The recent editions of the NBCC essentially required that the peak load in the drift be taken as three times the 30 year return period ground load or, since there is a maximum volume that the step can contain before it fills up, taken as equal to the step height times snow density, whichever was less. The snow load was reduced linearly with distance from the step out to a distance equal to twice the step height (but not outside the range 3 to 9 m), at which point the drift surcharge load was assumed to have been reduced to zero, leaving only the normal uniform roof load beyond that point. A shortcoming of the NBCC approach was that it had no dependence on roof size. For most normal size roofs it appeared to be adequate but intuitively one would expect the loads to increase with roof size, particularly that of the upper roof, because of the larger reservoir of snow available for drifting. There was concern that with the trend towards building warehouses and industrial buildings with larger and larger roof areas, the existing NBCC provisions may not be adequate.

Therefore studies were undertaken by RWDI and the National Research Council of Canada to improve the snow load provisions, specifically with a view to accounting for roof size. The details of these studies have been given by Irwin et al. (1995). In this paper the results will be summarized and some comparisons with full scale observations made. Also the new 1995 NBCC step load provisions are compared with the American ASCE 7-95 Standard and with the proposed ISO DIS 4355 loads (Apeland 1992) for roof steps.

Figure 1: Example of roof collapse under step snow loads.

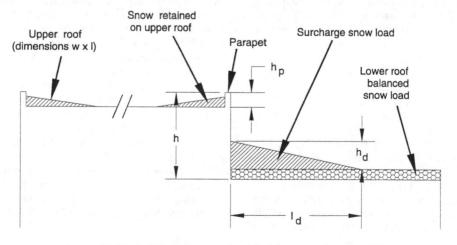

Figure 2: Triangular snow load on a lower roof at a roof step.

2 FINITE AREA ELEMENT (FAE) SNOW-DRIFTING STUDIES

The FAE method has been described in several previous papers: Irwin and Gamble (1988), Gamble et al. (1992), Irwin et al. (1992); and Irwin et al. (1995). Therefore it will only be summarized here.

In the FAE method the roof is divided by a grid pattern into a large number of area elements, as illustrated in Figure 3. The wind velocity pattern one metre above the roof is determined at every grid point either by calculation or by means of wind tunnel tests on a scale model for typically 16 wind directions at 22.5° intervals. In the present parametric studies which were limited to flat roofs the wind velocity one metre above the roof was taken to be equal to that at the same level in the approaching wind.

The wind velocity data are then used by a computer program to compute the drifting rates over the boundaries of the area elements, the drift rate being related to the wind velocity by an empirical expression based on full-scale field measurements such as those described by Kobayashi (1973). The computations proceed in a time step manner in one-hour time steps using hourly records of wind speed and wind direction to set the local wind vectors for each time step. At the end of each time step the excess (or deficit) of the snow mass that has drifted into an area element compared with that which has drifted out gives the net increase (or decrease) in snow mass in the element. This net mass change will be modified by the amount of snow that fell into the area element during the hour as a result of snowfall. Again the snowfall amount is derived from meteorological records. Thus a mass balancing computation is undertaken at each time step.

In addition a heat balance of the snowpack is undertaken so as to determine whether melting will occur and how much snow mass will melt. This part of the computation uses the meteorological records of temperature and cloud cover plus the time of day to evaluate the heat input into or output from the snowpack. If there is significant heat transfer through the roof this additional heat source can be included in this heat balance and melting computation. Meltwater is initially absorbed within the snowpack but eventually a saturation point is reached and the meltwater is allowed to drain out. The absorption of rainfall into the snowpack is treated in a similar manner to the meltwater, the total mass of water in the liquid and solid phases being tracked for each time step. Again the meteorological records are used to provide the amount of rainfall if any for each hour.

It is known that the ability of snow to drift depends on its age and whether it has experienced melting or rainfall (see Kind [1981] for example). In the FAE simulation the drifting rate is essentially set to zero if significant melting or rainfall is experienced and the mass of snow in each element at that time stored. If new snow subsequently arrives it is allowed to drift but the original mass remains fixed in place. For further details of the FAE method the reader is referred to the papers already cited above.

3 FAE STUDIES FOR THE 1995 NBCC

The results of the FAE studies by Irwin et al. (1995) were found to be quite well fitted by the following expression which is now in the 1995 edition of the NBCC:

$$\frac{S_{peak}}{S_s} c_1 = \sqrt{\frac{\gamma 1^{1*}}{S_s} - 6 \left(\frac{\gamma h_p}{S_s}\right)^2} + 0.8 \quad (1)$$

where S_{peak} = peak snow load in the drift

S_s = ground snow load for selected return period

c_1 = empirical constant = 0.35

l^* = effective upper roof length $= 2w - \dfrac{w^2}{l}$

l = longer dimension of upper roof

w = shorter dimension of upper roof

γ = specific weight of snow

h_p = height of parapet on the upper roof

The form of Equation 1 was derived by Irwin et al. (1995) on the basis of simple theoretical arguments with the empirical constant, c_1, left to be determined. The FAE results and Equation 1 were found to agree fairly well for roofs on exposed sites if the constant c_1 was taken to be 0.35 (Irwin et al. 1995). This is illustrated in Figure 4 where the FAE results for several Canadian cities are compared with Equation 1 for the case of zero parapet height ($h_p = 0$).

In the 1995 NBCC if the drift depth predicted by Equation 1 exceeds the height of the step h, then the maximum load is instead taken to be γh. Also, the peak drift load factor S_{peak}/S_s is not allowed to fall

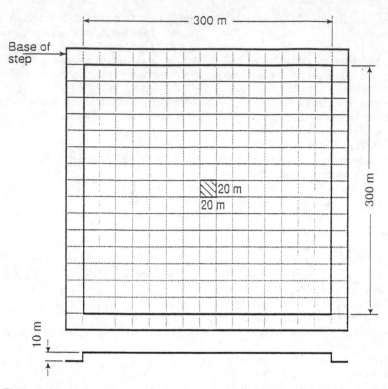

Figure 3: Example of finite area element grid, 300 m square roof with 10 m step height.

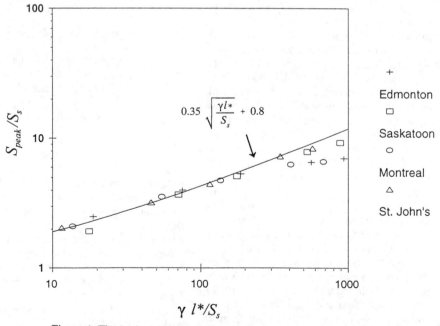

Figure 4: Theoretical fit to FAE results open country, no parapet.

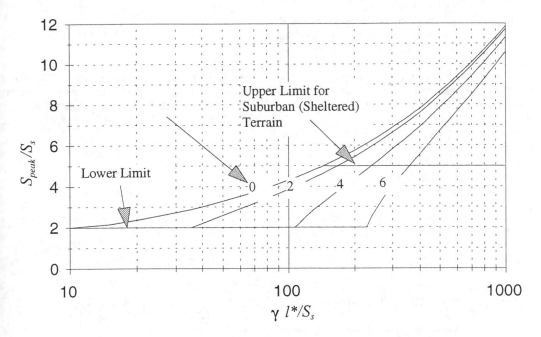

Figure 5: NBCC step snow load factor.

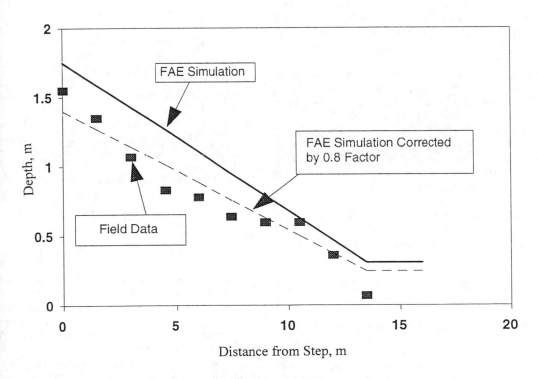

Figure 6: Comparison with field measurements.

below 2. In all cases the snowdrift surface is assumed to have a 1:5 slope. It also should be noted that the NBCC separates out rain load from snow load. Therefore an additional load called S_r is added to all snow loads to allow for rain soaking into the snowpack. S_r is taken to be the maximum 24-hour rain load for the area, nominally for the same return period as S_s.

Figure 5 shows a plot of Equation 1. It was noted in the FAE simulations that for roofs in wind approaching over suburban terrain, as opposed to open terrain, the drift loads in the roof step did not reach as high a value as predicted by Equation 1 for the larger roofs. There was simply not enough wind over the course of the winter to move all the available snow into the step. Therefore an upper limit of 5 for the ratio S_{peak}/S_s is applied for roofs in suburban, i.e., sheltered, terrain.

4 COMPARISON WITH FULL SCALE

Figure 6 shows a comparison of the FAE method with a full scale drift measured on a large area roof in Canada. The simulation was started at the beginning of the winter and was stopped at the point in time when the full scale measurements were made. It can be seen that the FAE prediction was of the right general magnitude but was higher than the actual loads.

Applying a 0.8 correction factor to the FAE load brings it into line with the measurements. Note that in deducing the value of $c_1 = 0.35$ in Equation 1 from the FAE results, a 0.8 factor had been applied to the FAE loads as described by Irwin et al. (1995). It is considered that the primary reason for needing the correction factor is that the FAE method currently assumes all snow transport is by saltation whereas at higher wind speeds some is transported by suspension (Schmidt 1986). Suspension, in which smaller snow particles are picked up and carried long distances, can result in significant loss of snow mass. A method of accounting for this suspension loss during the FAE simulation is currently being developed at RWDI.

5 COMPARISON WITH THE ASCE 7-95 AND ISO STANDARDS

Both the ASCE 7-95 and the proposed ISO DIS 4355 standard have provisions for step loads that depend on upper roof length. The ASCE 7-95 formula for the peak snow load in a step in metric units (newtons and metres) is based on the work of O'Rourke and Wood (1986) and is

$$\frac{S_{peak}}{S_s} = \left(0.0741 l^{1/3}(S_s+479)^{1/4}-0.457\right)\frac{\gamma}{S_s}+\frac{S_b}{S_s} \quad (2)$$

where γ is the specific weight of snow in N/m^3 and S_b is the balanced snow load, taken here as 0.7 times ground load S_s. Both S_b and S_s are in units of N/m^2 and γ is assumed to be given by

$$\gamma = 0.426\ S_s+2200 \quad (3)$$

The slope of the drift surface is given as 1:4. A rain-on-snow load equal to 0.24 kPa is to be added but only in light snow areas where the ground snow load is less than 0.96 kPa.

The recently proposed ISO DIS 4355 maximum load in the step (Apeland 1992) is

$$\frac{S_{peak}}{S_s} = C_e\ C_t\ \mu_b\ (1+\mu_d) \quad (4)$$

where $C_e =$ exposure coefficient

   $C_t =$ thermal coefficient

   $\mu_b =$ balanced snow load coefficient

   $\mu_d =$ drift load coefficient

The drift load coefficient is given by

$$\mu_d = \sqrt{0.5(1-0.95C_e)\frac{\gamma l}{S_s}} \quad (5)$$

The length of the drift l_d is taken as

$$l_d = 4\mu_b\ \mu_d\ \frac{S_s}{\gamma} \quad (6)$$

which implies a 1:4 slope to the drift surface.

Figure 7 compares the NBCC, ASCE and ISO peak step loads as a function of upper roof length for roofs without parapets in an area with 2-kPa ground load. The effect of rain-on-snow was ignored in these comparisons. Also the fact that the NBCC uses a 30 year ground load whereas the other two use a 50-year value was ignored. The ratio S_b/S_s in the ASCE expression was set at 0.7 (the most common value) and in the ISO expression it was assumed that C_e =

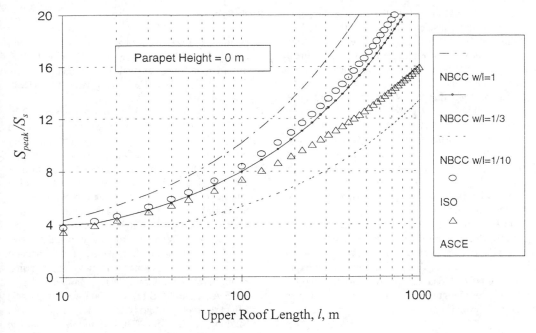

Figure 7: Peak drift load comparisons with ASCE and ISO, 2-kPa ground load.

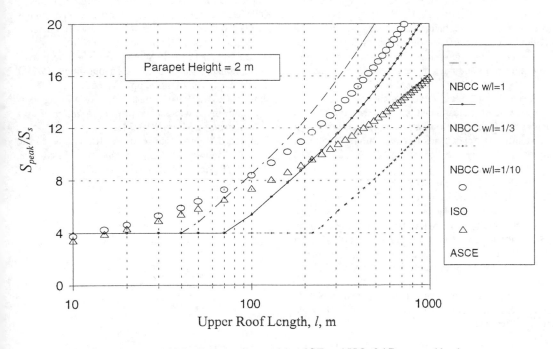

Figure 8: Peak drift load comparisons with ASCE and ISO, 2-kPa ground load.

0.8, $C_t = 1.0$ and $\mu_b = 1.0$. The ASCE and ISO formulae do not consider the effect of the length to width ratio w/l of the roof. Therefore the longer length l was used for the ASCE and ISO curves in Figure 7 with the assumption that it is the direction perpendicular to the step.

It can be seen that the NBCC curves are quite sensitive to the ratio w/l, giving progressively lower loadings for smaller and smaller values of w/l. The ISO provision is in fairly close accord with the new NBCC load for $w/l = 1/3$, which is a fairly common shape for roof plan forms. The ASCE curve, which is based on field observations, is also close to the NBCC curve for $w/l = 1/3$ for roof lengths less than about 100 m, but falls below it for longer roofs.

Figure 8 shows a similar comparison to that in Figure 7 but in this case the parapet height was 2 m. The ASCE and ISO curves remain the same as in Figure 7 because they do not incorporate a parapet effect. The NBCC loads are significantly reduced for the smaller roof lengths for which the parapet effect is much more pronounced.

6 CONCLUSIONS

The new NBCC load provisions, which are based on a combination of FAE studies, theoretical considerations and field observations, quantify the effect of upper roof size on the snow loads in roof steps. They also incorporate the effects of roof width to length ratio and parapet height, both of which can be significant.

Comparing the FAE method with field measurements indicates that typically the current FAE method, which ignores snow losses due to suspension of smaller particles, tends to overestimate step loads by about 20%. The studies for the NBCC incorporated a 20% correction.

In the comparisons with the ASCE 7-95 and proposed ISO provisions, the NBCC loads for zero parapet height are higher for square roofs ($w/l = 1$), are lower for very long narrow roofs, and are similar to the ISO loads for roofs with w/l equal to about 1/3. The NBCC loads for $w/l = 1/3$ and $h_p = 0$ are also similar to the ASCE loads for the more common sizes, i.e., $l < 100$ m.

For upper roofs with large parapets the new NBCC provisions allow significant reductions in step loads for smaller sized roofs.

REFERENCES

Apeland, K. (1992) Standardization of snow loads on roofs - DIS 4355: Revision of ISO Standard 4355, Proc. Second International Conference on Snow Engineering, Santa Barbara, publ. by CRREL, Hanover NH., pp 411-437.

Gamble, S.L., Kochanski, W.K., and Irwin, P.A. (1992) Finite area element snow loading prediction - Applications and advancements, Eighth Int. Conf. on Wind Eng., London, Ontario, pub. Elsevier.

Irwin, P.A., and Gamble, S.L. (1988) Prediction of snow loading on the Toronto SkyDome, Proc. First International Conference on Snow Engineering, Santa Barbara, publ. by CRREL, Hanover, NH.

Irwin, P.A., Gamble, S.L., Retzlaff, R.N., and Taylor, D.A. (1992) Effects of roof size and heat transfer on snow loads on flat roofs, Proc. Second International Conference on Snow Engineering, Santa Barbara, publ. by CRREL, Hanover, NH.

Irwin, P.A., Gamble, S.L., and Taylor, D.A. (1995) Effects of roof size and heat transfer on snow load: Studies for the 1995 NBC, *Canadian Journal of Civil Engineering*, August, 1995.

Kind, R.J. (1981) Snowdrifting, Handbook of Snow, Principles, Processes, Management and Use, Chap. 8, ed. D.M. Gray & D.H. Male, Pergamon Press.

Kobayashi, D. (1973) Studies of snow transport in low level drifting snow, Inst. of Low Temp. Sci., Sapporo, Japan, Report no. A31, pp. 1-58.

O'Rourke, M.J., and Wood, E.(1986) Improved relationship for drift loads on buildings, *Can. J. Civ. Eng.*, vol 13, no.6, pp. 647-652.

Schmidt, R.A. (1986) Transport rate of drifting snow and the mean wind speed profile, *Boundary-Layer Meteorology* 34, pp. 213-241.

Templin, J.T., and Schriever, W.R., Loads due to drifted snow load, *J. Struct. Div.*, ASCE, vol 108, no. ST8, Aug. 1982, pp. 1916-1925.

Snow Engineering: Recent Advances, Izumi, Nakamura & Sack (eds) © 1997 Balkema, Rotterdam. ISBN 90 5410 865 7

Removal of newly deposited snow on a gable roof

Toshiichi Kobayashi
Nagaoka Institute of Snow and Ice Studies, NIED, Niigata, Japan

Motonobu Kumagai
Japan Highway Public Corporation Research Institute, Tokyo, Japan

Yukiko Mizuno
Institute of Low Temperature Science, Hokkaido University, Sapporo, Japan

ABSTRACT : In order to remove deposited snow on a gable roof with a small machine, an automatic roof snow remover was developed to cut soft snow which has just fallen. The remover consists of an electric motor (90 W) and a pair of inverse V-shaped snow cutting plates(0.2 m wide and 3.55 m long each). The plates shuttle between gables from six to seven times every two hours to cut newly deposited snow on each plate, leaving the deposited snow between the plates and the roof.

1 INTRODUCTION

In heavy snowfall areas in Japan such as Niigata Prefecture, removing roof snow is hard and dangerous for the people in the areas. Some kinds of recent houses do not require roof snow removal by manpower. Such houses include those with roof snow melting equipment, those designed for roof snow to slide off and those designed for rood snow to be kept on the roof. However, in residences in heavy snowfall areas remove roof snow by manpower. Residents have been involved in snow accidents by falling from a roof when removing snow. Others have been buried under a roof snow avalanche. Some individuals have been injured, others have been killed. Therefore, new mechanical roof snow removing equipment, which was very simple and used little power, was developed and tested.

2 TEST METHOD

The roof snow removing equipment consists of a 90 W electric motor and a pair of inverse V-shaped stainless steel snow cutting plates each 0.2 m wide and 3.55 m long (Figure 1). The plates shuttle between gables from six to seven times every two hours to cut newly deposited snow. The cut snow slips from the top to the bottom of each plate and falls off the roof. However, the snow between the surface of the roof and the cutting plane of the plates is not removed (Figure 2). The weight of the deposited snow was estimated to be about 200 kg/m² , assuming an average density of 300 kg/m³ . It was equivalent to half of the maximum design snow load of 420 kg/m² constituted by the

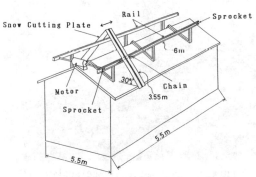

Figure 1. Schematic diagram of the automatic roof snow remover.

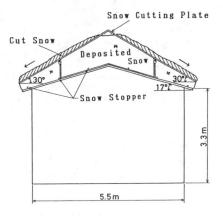

Figure 2. Sectional plan of cut snow.

Ministry of Construction in heavy snowfall areas such as Nagaoka City.

3 RESULTS

The capacity of the roof snow removing equipment obtained from the load curve of the motor was 377 N. When both cutting plates cut 0.06 m deep soft snow which has just fallen (Figure 3), the maximum snow removing force was recorded as 335 N (Figure 4). That was near the limit of the motor. At the beginning of the tests, snow accretion occurred sometimes on the surface of stainless steel snow cutting plates. Then, two kinds of so called accretion resistant films, i.e. , a fluorine resinous film or a polyurethane resinous film were adhered on the surfaces of the plates. After that, the amount of snow accretion on the plate surface decreased considerably.

Figure 3. One example of the sectional
plan after snow cutting.
(1994.1.24, 15:16)

4 DISCUSSION

4. 1 *Snow accretion on the snow cutting plates*

In order to investigate effects of the accretion resistant films on reduction of snow accretion, the adhesive strength between ice and a fluorine resinous film or a polyurethane resinous film was measured at temperatures of 0℃, -0.5℃, -1℃, -3℃ and -5℃.

The experimental apparatus is shown schematically in Figure 5 in which the ice on the sample was prepared by direct freezing of water on the test film at -10℃. The arm with a load cell moves parallel to the sample surface with a speed of 10 mm/min and pushes a ring.

Figure 6 shows the relationship between adhesive strength and temperature. The value at each temperature was an average of three tests at 0℃ and -0.5℃ and of five tests at other temperatures. Although the adhesive strength of ice on either films were almost the same above -1℃, that on a polyurethane film was smaller than on a fluorine resinous film at the lower temperatures.

Separation of ice was always occurred at the interface between ice and a film. Polyurethane resinous film was more effective of preventing snow accretion from the cutting plates.

4. 2 *Operation of the roof snow removing equipment in Nagaoka*

Figure 7 shows the relationship between an accumulated snow depth in every hour and its frequency occurrence observed in Nagaoka from 1983 to 1994. Provided that an accumulation depth in every two hours assumed to be twice as large as that in every

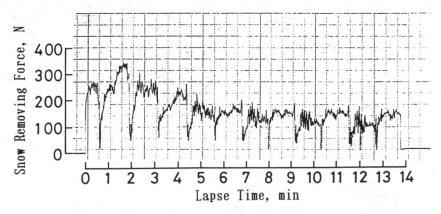

Figure 4. One example of the snow removing force record.
(1994.1.24, 15:00~15:14)

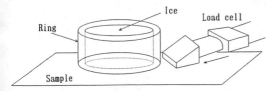

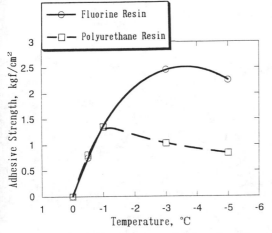

Figure 5. Schematic diagram of the
adhesion tester.

Figure 6. Relation between adhesive strength
and temperature.

hour, the depth in every two hours corresponds to twice of each in Figure 7. Considering that the capacity of the equipment is about 0.06 m, the roof snow removing equipment should be operated under the standard manner as follows :

(1)When an accumulation depth in every hour is less than 0.03 m, the equipment should be operated every two hours interval.

(2)When the accumulation depth is ranging from 0.03 m to 0.06 m, the equipment should be operated every hour interval.

(3)When the accumulation depth exceeds 0.06 m, the equipment must be kept on until the snowfall stops.

In this connection, the number of operation times were estimated to be 140 for every two hours interval, 30 for every hour interval and 4 for continuous operation in an ordinary year in Nagaoka.

5 CONCLUSIONS

In this study, a new type of mechanical roof snow removing equipment was developed and tested. The results of the tests are as follows:

(1)The maximum snow removing force when both snow cutting plates cut 0.06 m deep soft snow which has just fallen was near the limit of the removing capacity of the motor.

(2)In order to reduce snow accretion on the surfaces of the snow cutting plates, various kinds of films were

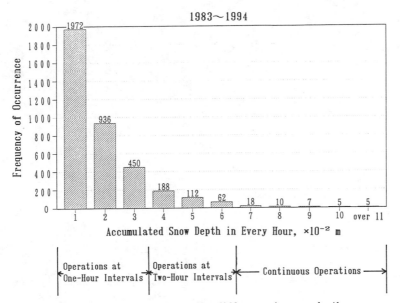

Figure 7. Relation between the difference in snow depths
every hour and the frequency of occurrence.

stuck to them. The adhesive strength of ice onto a polyurethane resinous film was lower than that onto a fluorine resinous film at temperatures from 0 to -5°C.

(3)If the equipment is operated in Nagaoka, the operating time intervals depend on snowfall rate. In an ordinary year, operation at one-hour intervals would be needed 30 times, operation at two-hour intervals would be needed 140 times and continuous operation would be needed 4 times.

REFERENCES

Kobayashi, T. & Kumagai, M. 1995. An automated roof snow remover. *Journal of the Japanese Society of Snow and Ice.* 57:141-148.
Shimizu, M. , Kimura, T. & Nakawo, M. 1993. Meteorological data at the Nagaoka Institute of Snow and Ice Studies (1) (December 1981-April 1991). *Technical Note of the National Research Institute for Earth Science and Disaster Prevention.* 158:1-64.
Yoshida, N. , Nagaoka, M. , Shinojima, K. , Suzuki, Y. & Yoshioka, T. 1989. Study on the wiper-type of roof snow remover. *Proceedings of '89 Cold Region Technology Conference.* 507-512.

Snow Engineering: Recent Advances, Izumi, Nakamura & Sack (eds) © 1997 Balkema, Rotterdam. ISBN 90 5410 865 7

Snow sliding on a membrane roof

Tsukasa Tomabechi, Toshiyuki Ito & Masahiro Takakura
Hokkaido Institute of Technology, Sapporo, Japan

ABSTRACT: Model experiments of snow sliding on a membrane were carried out using artificially prepared model snow in a cold room. A snow survey was also carried out on a roof of membrane in the snowy region by asking managers of a building about snow conditions and sliding. Results obtained are as follows. Deformation of model snow during sliding depends on conditions of fixation.   Observations show that snow slides off smoothly when the slope of roof a V–shaped exceeds 20 ° . Snow remains on gentler slopes near the ridge.

1. INTRODUCTION

It has been shown by studies of snow sliding on roofs that snow on one side of a ridge is balanced by the snow on the another side, and the colder part tends to adfreeze. The temperature around the verge is nearly equal to the air temperature (Tomabechi et al.1995）. Because of these phenomena snow partially slides off frequently as shown in Figure 1. It is necessary for evaluation of snow load and snow removal on a large membrane roof to know the mechanism of partial snow sliding.

Model experiments of snow sliding on a membrane were carried out using artificially prepared model snow in a cold room. A snow survey of the membrane roof of a building in snowy regions was carried out by asking managers of that building about snow conditions and sliding.

2. EXPERIMENTAL METHODS

2.1 *Model snow and sliding experiment*

Model snow particles were made by spraying water into cold air as the same way as Sack (1988) who used fine dusts in the air as freezing nuclei. Water was sprayed at 100 ml/min and 3 to 5 atmospheres in a coldroom at −20 ° C. The tensile strength and shear strength of the model snow were measured by the apparatus shown in Figure 1. The maximum tensile load (Tt) that the snow can sustain prior to is measured by the

Figure 1. Phenomenon of snow partially sliding off which occurs in Sapporo.

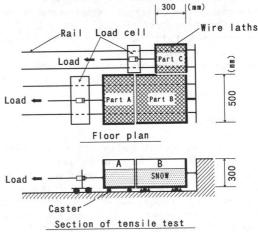

Figure 2. Experimental device used to measure tensile strength and shear strength of the model snow.

attached to part C. The friction resistance (R) at both part A and C was measured after fracture. The tensile strength (*t) and shearing strength (*s) were calculated by equation (1) substituting measured load cell attached to part A in Figure 2. The maximum shear load (Ts) that the snow can sustained prior to fracture is measured by the load cell values;

$$*t,s = (Tt, s - R)/S$$

where S is the cross sectional area.

Sliding experiments were carried out by using the specimens shown in Figure 3. Wire laths were put on the roof to prevent the snow from sliding off. The specimens were prepared to meet the conditions as follows.

Free from fixation: Conditions that no resistant force acts on the sliding snow.

One–point fixation: Conditions that tensile strength acts on the snow at the ridge of the roof.

Two–point Fixation: Conditions that adfreeze force acts on the snow at the verges of the roof.

Three–point Fixation: Conditions that the snow on the roof is acted on by both tensile strength at ridge and adfreeze force at the verges.

The model snow was uniformly 5 cm thick on the membrane for the snow sliding experiment. To observe the development of snow sliding, a 10 cm grid was marked on the surface of the specimens. The specimens were kept at −5 ℃ in a coldroom and the temperature was raised up to +10 ℃ with given temperature gradients. Time–lapse video cameras were used to observe movement of the snow prior to and during slide off. The temperature of the membrane interface was measured by thermocouples every minute during the experiment. The specimens were laid at an angle of 40 ℃ which is most efficient for snow sliding. (Tomabechi et al.1995) .

2.2 Outline of a membrane structure surveyed

The snow survey was carried out at the indoor athletic field (Sundome) in Aomori that is illustrated in Figure 4. The arena part of the Sundome is a big barrel value with a 65 m span, a 24.2 m maximum height and a 102 m long. heating is used to cause roof to slide off. The heating facility is composed of 4 nozzles that blow 50 ℃ hot air against the inner surface of the roof Aomori city recorded a maximum snow cover depth of 209 cm but snow remained on the membrane

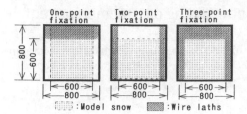

Figure 3. An outline of the specimens using for sliding experiments.

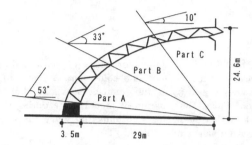

Figure 4. An outline of Sundome in Aomori.

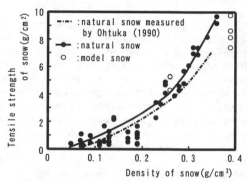

Figure 5. The relationship between the density and the tensile strength of natural snow and model snow.

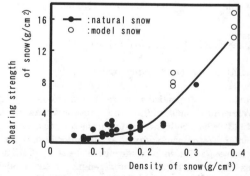

Figure 6. The relationship between the density and the shear strength of natural snow and model snow.

342

roof was quite little. The membrane stretched down between supports in the shape of the letter V. In surveyed and information about snow on the roof was collected from the manager in January 1994. Also local weather data were collected to compare with the survey.

3. SNOW SLIDING OF MODEL EXPERIMENT

The relationship between the density and the tensile strength of model snow is shown in Figure 5. The experimental data of natural snow obtained by Tomabechi et al. (1995) and Honma et al. (1990) are also shown . The tensile strength of the model snow is quite close to that of natural snow. The relationship between the density and the shear strength of the model snow is shown in Figure 6. The values of the model snow are slightly stronger than natural snow. These results suggest that the mechanical properties of the model snow ay the density 0.26 g/cm3 used in this experiment are nearly equal to those of natural snow at the same density.

The process of snow sliding off the roof is shown in Figure 7 for the case of 3−point fixation and a roof slope of 40 ° . As shown in the photograph, the snow on the roof changed remarkably in shape at the downward part. The snow was hanging over and sagging downward. In addition, a crack formed in the middle of the snow. The results of the experiments at a slope of 40 ° are shown in Figure 8 . In the case of 1−point fixation the snow moves downward, except for the fixed part. Meanwhile a linear crack appears between the fixed and moving part, and the snow of the lower part breaks free. In the case of 2−points fixation, the snow is deformed by downward movement of all the snow along its centerline. An upslope crack is formed along the boundary of the fastened and unfastened part. Partial snow sliding starts at the edge of the roof where snow is hanging over. In the case of 3−point fixation, remarkable deformation appears below the unfastened part of the snow. A few cracks were found in the middle part of the roof and snow slid along the cracks as shown in Figure 6. Patterns of snow sliding depend on the type of fixation as mentioned above.

The temperature of inner surface of the materials is around 0 ℃ when deformation becomes great. However, the temperature is different depending on the fixation when snow starts to slide off. It is +3 ℃ with no fixation, around +3 ℃ with 1−point fixation and around +6 ℃ with 2−point and 3−point fixations. There is a tendency to have higher temperatures for sliding as fixation increases. The more fixation

there are, the longer it takes for snow to slide off.

4. RESULTS OF SNOW SURVEY

The depth of snow cover and the air temperature when field surveys were made at Sundome in 1994 are shown in Figures 9 and 10. A maximum snow depth of 96 cm was recorded. It is quite often windy (daily average wind speed >4 m/s) and cold(average temperature <−4 ℃) when daily snow fall exceeds 10 cm. This weather results in blowing snow. The survey was carried out for two days, from 20 to 21 January, when the roof heating system was working. The weather is

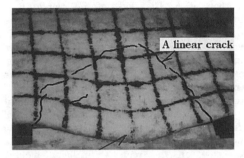

Snow hanging over and extraordinarily twisted

Figure 7. The process of the snow sliding off the roof is shown in the case of 3−points fixation and pitch of roof 40 ° .

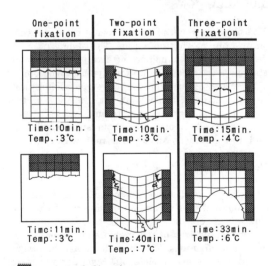

One-point fixation	Two-point fixation	Three-point fixation
Time:10min. Temp.:3℃	Time:10min. Temp.:3℃	Time:15min. Temp.:4℃
Time:11min. Temp.:3℃	Time:40min. Temp.:7℃	Time:33min. Temp.:6℃

▨ : point of fixation

Time : Time after snow slides start

Temp. : Tempreture of the membrance interface

Figure 8. The result of the experiments of the case pitch of roof 40 ° .

shown schematically conditions before and after the survey as follows :

January 19 : Snow depth = 32 cm
             Average temperature = − 5.6 ℃
             Average wind velocity = 6.6 m/s
January 20 : Snow depth = 32 cm
             Average temperature = − 3.8 ℃
             Average wind velocity = 5.1 m/s
January 21: Snow depth = 0 cm
             Average temperature = − 3.0 ℃
             Average wind velocity = 2.4m/s

Snow sliding was analyzed by dividing the roof into three parts as shown in Figure 4. Snow sliding occurs as soon as snow accumulates on part A which has a 50 ° . Snow slides on part B a little later than at A. The slope of the roof at B is around 30 ° . At part C, which has a slope less than 10 ° , snow remained partially as shown in Figure 11; however, the snow slid off the roof during the observation. According to the building managers, there is little snow on part A even when heating is not working. On part B and C there is some snow remaining, and snow does not slide off for about one hour after the start of heating. The passive heating system that blows 50 ℃ hot air at 4 points inside the arena is used to slide snow off the roof while being melted. Managers said that the temperature around the membrane is about 20 ℃ and this system is quite effective for the arena.

5.CONCLUSIONS

The tensile strength and adfreezing act to prevent snow from moving on a roof. The sliding of snow depends on its condition of fixation. The snow slidings were modeled into four types of fixation.

A series of experiments on snow sliding was carried out using model snow and the snow on the roof of a membrane structure in a snowy region was observed. Information about snow conditions and the sliding snow on the roof was collected from managers.

Results obtained are as follows :

1) Deformations of model snow until snow sliding is completed depend on the conditions of fixation. There is a tendency to have high temperatures at the inner surfaces of the membranes according to increasing fixations. The more fixations the longer it takes for snow to slide off. Fixations works to keep snow on the roof.

2) The observations show that snow slides off smoothly if the slope of V−shaped membranes exceeds 20 ° . Snow remains on gentler slopes.

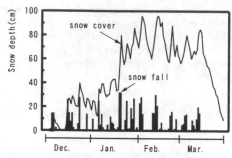

Figure 9. Snow cover and snowfall time histories in Aomori, 1994.

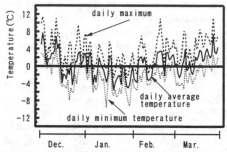

Figure 10. Air temperature time histories in Aomori, 1994.

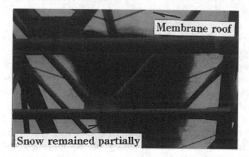

Figure 11. Snow remained partially on the Sundome.

REFERENCES

Sack, R. L. & Baker, R. L. : Cold room studies for sliding snow, *First International Conference on Snow Engineering*, pp.199−209,1988

Ohtsuka,K., Jou,O., Honma,Y., Miyagawa,Y., Masumo, T. & Okada,H. : A study on the removal of snow from membrane structures, *Journal of Membrane structure'90*, No.4, pp.55 − 68, 1990 (in Japanese)

Tomabechi,T., Ito,T. & Takakura, M. : Fundamental study on snow sliding on the roof, *Journal of Snow Engineering of Japan* , Vol.11, No.2, 2−9,Apr.1995 (in Japanese)

3 Infrastructure and transportation

Snow Engineering: Recent Advances, Izumi, Nakamura & Sack (eds) © 1997 Balkema, Rotterdam. ISBN 90 5410 865 7

Snow simulation within the closed space of the Jules Verne Climatic Wind Tunnel

Jacques Gandemer, Pierre Palier & Sophie Boisson-Kouznetzoff
Centre Scientifique et Technique du Bâtiment, Nantes, France

ABSTRACT : The aim of the "Jules Verne" Climatic Wind Tunnel is to reproduce combined climatic parameters at full scale in order to analyse their effects on structures, buildings and vehicles. After a short presentation of the facility, this paper reports on the snow simulation results obtained in the thermal unit. Ambient air, temperature and moisture content are controlled by a device with two heat exchangers (cold and warm) placed in the Wind Tunnel. Three or four snowguns with controlled air/water ratios and pressure allow depositions of 10 cm/h to 20 cm/h of dry or wet snow in the test section. The thickness, density and liquid water content of the produced snow is measured. These measurements are in good agreement with the characteristics of natural snow which has already begun its temporal evolution (half a day) or which has been drifted by wind.

1 INTRODUCTION

The aim of the "Jules Verne" Climatic Wind Tunnel is to study, at full scale, the effects of the climatic parameters, even in extremely harsh conditions, on buildings, vehicles and any materials and machines subjected to one or several climatic parameters. The geometrical characteristics and the performances obtained in the different test chambers for the simulation of combined climatic parameters (wind in association with rain, sand, snow, sun, temperature and moisture content) are given in this paper.

Particularly snow production and different kinds of studies involving snow are presented.

Figure 1. The Jules Verne climatic wind tunnel

2 THE JULES VERNE CLIMATIC WIND TUNNEL

The Jules Verne Climatic Wind Tunnel has been built in order to realise experiments at full scale. We can study the behaviour of structures subjected to the effects of wind combined with other climatic parameters : temperature, snow, rain, frost, sand and sun. Most applications concern :

a) effects of climatic parameters on roofs, walls ...

b) industrial aerodynamics applied to vehicles

c) behaviour and performance of materials and objects subjected to one or more of these parameters.

The Climatic Wind Tunnel has two independent circuits as shown in Figure 2 : the first one (operational since 1990) is called the "dynamic circuit", the second one (operational since march 1995) is called the "thermal circuit". The latter is able to create hot or cold climates (from - 25°C to + 50°C) with snow, sun, rain, freezing rain or frost. The temperature, the relative humidity (from 30% to 95%) and speed of the wind are controlled. Different types of studies can be realised in the various test sections (high speed, environmental, diffuser and thermal) which are complementary. (Figures 3 to 6).

3 SNOW PRODUCTION IN A CLOSED SPACE

One of the interests of the Climatic Wind Tunnel is

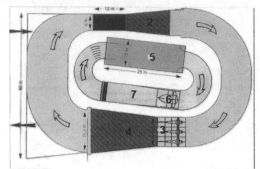

Figure 2. The Jules Verne Climatic Wind Tunnel.

Dynamic circuit
1 High speed tunnel : 300 km/h
2 Diffuser tunnel : 150 km/h
3 Fans : 3200 kW
4 Environment section : 100 km/h
Thermal circuit
5 Experimental chamber : 90-140 km/h, -25°C to +50°C,
humidity control, rain, snow
6 Fan : 1000 kW
7 cold and warm heat exchangers

Figure 3. House under rain in the environment section.

Figure 4. Car in the high speed working section showing the turntable, the boundary layer trap and the force balance.

Fig. 5 House under snow in the thermal section

Figure 6. Car under snow.

its ability to produce snow (Figure 5) within a closed space. The climatic parameters have to be exactly regulated in the wind tunnel in order to produce snow and to control its features (e.g. wettness and density).

The artificial snow obtained in the wind tunnel is similar to natural snow. Snow is produced by "snowguns" (Figures 7 and 8) that project compressed cold air and water into the ambient air. Droplets are formed at the outlet of the gun and have then a ballistic behaviour during which they solidify in a way that depends on climatic conditions. The

ambient air temperature must be negative and the humidity low to improve thermal transfer. More precisely, snowgun control (adjustment of water, air pressure and flow to obtain the required snow quality) depends on the wet bulb temperature.

During their trajectories, water droplets freeze because of thermal transfer with the ambient air and fall down to create a uniform snow cover in the test chamber.

Figure 7. Snowgun.

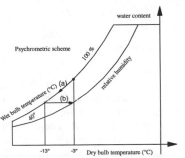

Figure 8. Outlet of the compressed air/water melting.

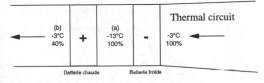

(a) ambient air cooling (with heat exchanger Icing)
(b) ambient air heating (with the moisture content rate decreasing)

Figure 9 : Control of moisture content in the wind tunnel

For the same wet bulb temperature, varying the pressure and the water/air ratio at the inlet of the snowgun produces different "types" of snow : wet snow when this ratio is high and dry snow when it is low.

The control of the moisture content (Figure 9) at a low and constant rate is obtained by using two heat exchangers in the wind tunnel : the first exchanger cools the air in order to take some moisture away (Figure 9a) and the second one warms it again (Figure 9b) to its primary temperature, so that its moisture content diminishes.

4 SNOW CHARACTERISATION

In order to better know and control the characteristics of the snow produced, different studies were realised.

Because of different trajectories times and droplets sizes, the snow characeristics are not expected to be uniform. We are going to describe the snow with 3 criteria : its liquid water content, its density and its distribution on the floor of the test section.

4.1. Liquid water content of snow

Water content defines the proportion of liquid phase of lisuid phase contained in the snow. When snow contains liquid water the temperature is 0°C. When the proportion is great we speak of wet snow, when the proportion is low or when the snow contans no water and has a emperature under 0°C, we speak of "dry snow."

Measurements of water content by calorimeter (Figure 10) and snow temperature by taking samples

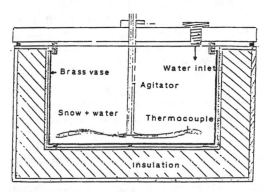

Figure 10. Calorimeter.

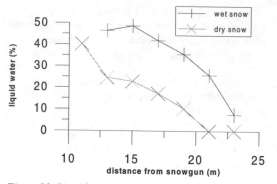

Figure11. Liquid water content

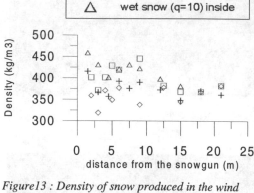

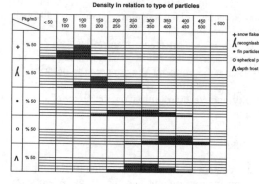

Figure12. Characterisation of natural snow deposits.

Figure13 : Density of snow produced in the wind tunnel

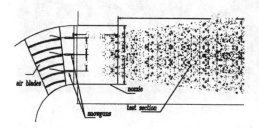

Figure14. Location of snowguns.

in the wind tunnel show that this parameter changes noticeably with distance from the snowgun (Figure 11) and the type of snow produced.

4.2. Density

Density is an important characteristic of snow, particularly for studies concerning snow loads on roofs, avalanche mechanisms and snow transport and accumulation during snowstorms.

Natural snow deposits have densities between 50 and 500 kg/m3 according to their crystalline structures and their temporal evolution (Figure 12).

Snow droplets produced by snowguns are almost spherical with diameters from 0.1 mm to 0.5 mm. Density measurements of snow just produced have been made in the area of the test chamber. Results are given for dry snow and wet snow for samples taken inside and near the surface of the snow deposit (Figure 13) : wet snow named "quality 10" with a low air/water ratio and dry snow "quality -10" for which this ratio is at least four times more important.

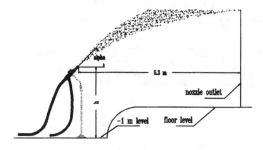

Figure15. Snowgun angle and height.

Density varies from 320 kg/m^3 to 460 kg/m^3 according to the distance to the snowgun. Snow density seems to be lower near the surface due to liquid diffusion within the snow. The snow has a more uniform density of about 370 kg/m^3 in the second part of the test chamber (from 12 m to 25 m) where the snow is thinner (Cf. § 4.3. on the snow distribution). This corresponds to a natural snow which has begun its temporal evolution (half a day)

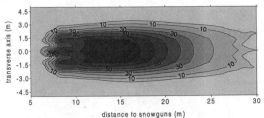

distance between snowguns : 1.5 m
gun's angle : 0° (horizontal)
quality 10 (wet snow))

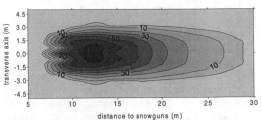

distance between snowguns : 1.5 m
gun's angle : 0° (horizontal)
quality -10 (dry snow)

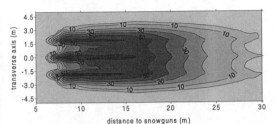

distance between snowguns : 2 m
gun's angle : 0° (horizontal)
quality 10 (wet snow)

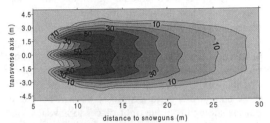

distance between snowguns : 2 m
gun's angle : 0° (horizontal)
quality -10 (dry snow)

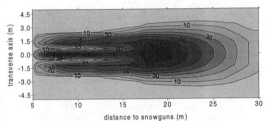

distance between snowguns : 1.5 m
gun's angle : 30°
quality 10 (wet snow)

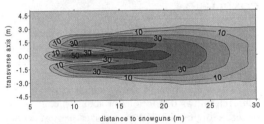

distance between snowguns : 1.5 m
gun's angle : 30°
quality -10 (dry snow)

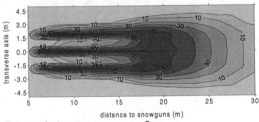

distance between snowguns : 2 m
gun's angle :30°
quality 10 (wet snow)

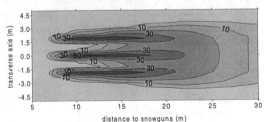

distance between snowguns : 2 m
gun's angle : 30°
quality -10 (dry snow)

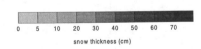

snow thickness (cm)

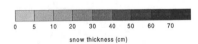

snow thickness (cm)

Figure 16. Snow distribution for "Quality : 10"

Figure 17.Snow distribution for "Quality : -10"

or which has already been drifted by wind causing destruction of the flakes and the crystalline forms.

4.3. Snow distribution

Snow can be obtained using 1, 2, 3 or 4 snowguns. It is obviously interesting to know its uniformity, particularly if, for snow load studies, specific conditions on snow distribution for the testing devices are of interest.

Several geometrical configurations (Figures 14 and 15) have been studied to obtain the most appropriate snow distribution for each case.

Measurements of snow thickness have been made in the all test chamber after two hours of snowgun production (Figures 16 and 17). The different snow mantles obtained define the area where snow distribution is quite uniform and its thickness. With these data (work yet in progress) we will be able to put the snowguns at the right places to have the kind of distribution we want.

CONCLUSION

The "Jules Verne" Climatic Wind Tunnel can simulate an exceptional variety of climatic parameters. Wind is easily reproduced in association with rain, snow, sun, temperature, fog and frosting fog. The simulation of snow, which has now been achieved allows depositions of about 15 cm/h of dry or wet snow on the 250 m² floor of the test section chamber. Snowstorms with windspeeds up to 90 km/h are reproduced, and studies of snow loads on roofs and structures, snow accumulation and snow movement can be realised.

The "Jules Verne" Climatic Wind Tunnel opens new opportunities for study of behaviour of devices subjected to snow and snowstorms.

REFERENCE

"Neige et Avalanches" *n° 25 - avril 1981 - Revue de l'Association Nationale pour l'Etude de la Neige et des Avalanches".*

Snow Engineering: Recent Advances, Izumi, Nakamura & Sack (eds) © 1997 Balkema, Rotterdam. ISBN 90 5410 865 7

Modelling and numerical simulation of snow drift around snow fences

Per-Arne Sundsbø
Narvik Institute of Technology, Department of Building Science, Norway

Ernst W. M. Hansen
SINTEF Energy, Thermal Energy and Hydro Power, Trondheim, Norway

ABSTRACT: A method for computer simulations of snow drift around man made structures is described. Particles are tracked in space and time based on a Lagrangian-Eulerian description of particle and wind velocities, respectively. Calculation of the wind phase is governed by the general Navier-Stokes equations and the proposed equation for particle motion covers both suspension and saltation effects. Saltation is modelled on a statistical basis by assuming random particle distribution in the saltation layer similar to real measurements of vertical mass concentration. Particles are programmed to depart from the surface by saltation provided that the wind is above threshold. Some of the snow particles will then return to the surface while others will remain in suspension in the wind. Computer simulations of snow drift around a two dimensional Wyoming snow fence are performed with a fence height of 2.74 m and a porosity of 50%. The results are in good comparison with the measured equilibrium drift profile typical for such a snow fence.

1 INTRODUCTION

Numerical simulation techniques are becoming more important as an engineering tool and this is also the situation for the field of snow engineering. Computer speed has increased in such a way that fluid flow and limited snow drift problems can be investigated. Routines for computer simulations of snow drift have been developed in order to investigate snow loads on roofs (Irwin and Gamble 1988) and drift around snow fences (Uematsu et al. 1991). The simulations provide rough assumptions of deposition patterns. Simulations of snow drift using a drift-flux model are presented by Bang et al. (1994) where three dimensional snow deposition is investigated. Decker (1991) has reviewed some of the methods in use in modelling of drifting snow and presumably such techniques will be further developed for future snow engineering.

Snow particles will be transported by creep, saltation or by suspension and the amount of transport in each mode is questionable and highly dependent on the actual drift problem. Mellor (1965) assumed that transport by turbulent diffusion is the major factor for drifting around tall buildings, roofs and other engineering structures, where alternative transport modes are hard to explain. Other investigators like Kind (1986) have observed snow drifts and concluded that saltation is the dominant transport mode. This theory is supported by Kobayashi (1972), who claimed that 90% of the total snow movement is within 30 mm of the snow surface at a wind velocity of about 10 m/s. Snow drifting by creep is the only mode which seems to have negligible contribution to the overall amount, regardless of the wind speed at which drifting occurs (Hayes and Tucker 1985).

The distribution in space and size of snow particles in a typical drift profile covers a wide spectra. Particles of all sizes are able to be transported by creep, saltation or by suspension, depending on transport conditions like wind velocity, turbulence level and other fluid and particle properties. This paper considers only particle movements in suspension and by saltation. The described method for simulation of snow drift is tested on a porous fence where there is no separation of the flow, caused by the fence. There is expected to be generated some turbulence by the porous fence, but this paper does not cover the complexity of two-phase turbulence modelling. Storage conditions for snow and effects due to thermal conditions are not evaluated in this work, though they might have a significant effect on the overall drift pattern.

1.1 snow drift physics

A snow particle on the ground is kept on the surface by gravity and cohesive bonds that vary in strength depending on properties like humidity, temperature, ageing, previous influence from wind and other storage conditions. The threshold wind speed is the value above which these bonds are broken and particle motion is initiated. Until this value is reached, no transport occurs. The initial value can be given as a wind speed at a certain height or as a friction velocity based on the surface shear stress at threshold. In general, the threshold friction velocity reflects the capability of the wind to transport a given type of snow particle at all velocities. Threshold friction velocities are reported by Male (1980) in a range from 0.07 m/s to 0.4 m/s which are limits valid for very light dry snow and snow at a wind-hardened surface, respectively. Kind (1986) suggested a threshold friction velocity of 0.15 m/s for dry uncompacted snow where cohesive forces can be neglected. The threshold for wind-transport of snow is dependent on the binding between particles rather than particle size.

2 DRIFT EQUATIONS

2.1 Modelling of friction velocity

The mean vertical velocity profile in a boundary layer is uniquely defined by knowing the friction velocity u_* and the surface roughness z_0. This implies that the friction velocity can be calculated by knowing the surface roughness and the velocity at height z, eq. 1.

$$u_* = u(z)\,\kappa / \ln\!\left(\frac{z}{z_0}\right) \qquad (1)$$

where κ is von Kármán's constant which is approximately equal to 0.4 and z is the vertical distance. The calculated friction velocity is compared to a mean threshold value and consequently the roughness height used in eq. 1 should reflect the conditions at which the threshold friction velocity was measured. Simulations done with different mesh sizes show that the height at which the reference velocity is taken must be considered carefully. It is not convenient to used the velocity in the cell adjacent to the surface because the results seems to be too dependent on mesh size and boundary conditions. A friction velocity which is not representative for the transport region is

determined if this distance is chosen too far from the surface. Choosing a velocity in the middle of a circulation zone will clearly not give representative information about the layer close to the snow surface. In this paper the friction velocity is calculated from a reference velocity 0.25 m above the surface, figure 1, using a roughness height of 0.1 mm for ground drifting, (Norem 1974).

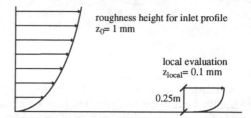

roughness height for inlet profile
$z_0 = 1$ mm

local evaluation
$z_{local} = 0.1$ mm

0.25m

Figure 1. Illustration of the inlet velocity profile and the function applied to calculate the friction velocity based on a local evaluation.

2.2 Particles in a suspension

Snow-particles will be in a suspension if they are small compared to the small-scale turbulence in the flow field. The Magnus lift effect and the lift force caused by simple shear flow are of minor importance in a suspension (White). Effects on suspended particles are governed by the general Basset-Buossinesq-Oseen equation, eq. 2, which is a Lagrangian equation for the particles.

$$\frac{dV_p}{dt} = \frac{3C_d}{4D}\frac{\rho}{\rho_p}\left|V - V_p\right|(V - V_p) - \frac{1}{\rho_p}\frac{\partial p}{\partial r} \qquad (2)$$

$$+ \frac{1}{2}\frac{d}{dt}(V - V_p) + \frac{9}{D\rho_p}\sqrt{\frac{\rho\mu}{\pi}}\int_{t_0}^{t}\frac{\frac{d}{d\tau}(V - V_p)}{\sqrt{t - \tau}}d\tau - g$$

The left side of eq. 2 is the particle acceleration and the terms on the right side is the drag term, pressure gradient term, apparent mass term, Basset term and gravity term.

Since the density of the snow particles is much larger than the density of air, the pressure gradient term, apparent mass term and the Basset term are negligible. Thus the Basset-Buossinesq-Oseen equation will reduce to a simplified expression for transport of suspended particles,

$$\frac{dV_p}{dt} = \frac{3C_d}{4D}\frac{\rho}{\rho_p}\left|V - V_p\right|\left(V - V_p\right) - g \qquad if \quad \rho \ll \rho_p \qquad (3)$$

where $\underline{V}_p$ and $\underline{V}$ are the particle velocity and fluid velocity, respectively, C_d is the Stokesian drag coefficient, D is the particle diameter and $\underline{g}$ includes gravity and other body forces, while ρ and ρ_p are the density of fluid and particle, respectively.

2.3 Modelling of snow particles in motion

A complete numerical model of snow drifting requires specific modelling of all transport modes and for a wide spectra of particle sizes, which will result in a huge set of equations. The large number of equations to be solved will result in large expenses with respect to time consumed during computer simulations. A complete modelling of saltation trajectories will also require a very detailed knowledge about the flow process close to a wall. In modelling of drift-patterns for snow engineering purposes, the physical model domain is normally too big to give detailed information down to the scale needed to fully initiate and describe saltation paths for snow particles. A workable simplification is to assume snow drift modelling by using one mean-particle diameter and occurrence of ground drift in the saltation layer and, to some extent, in the zone above which may be characterized by suspension. The modelling may be performed by applying an equation for particle motion in suspension with an additional lift effect included to initiate and lift particles in a random manner. Utilize a statistical lift approach is in agreement with observations of the chaotic behaviour of snow in saltation. The particle velocity in eq. 3 includes a mean particle velocity and a diffusion velocity to create random particle behaviour: $\underline{V}_p = \underline{V}_{mean} + \underline{V}_{dif}$

Particle diffusion in the flow field may take the form of a Gausian distribution which can be modelled by means of a statistical approach (Barkhudarov and Ditter 1994). It is reasonable to assume that a related distribution occurs for snow particles in a transport mode as the particle concentration increases from the surface. The diffusion approach is therefore applied by using a Monte Carlo technique to compute the vertical distribution of snow in motion along the surface. The diffusion velocity is given as:

$$\underline{V}_{dif} = \sqrt{\frac{4\lambda}{\Delta t}} erf^{-1}(\beta) \qquad where \qquad \beta \in \langle 0, 1 \rangle \qquad (4)$$

where λ is the particle diffusion coefficient, Δt is the time step size and $erf(\beta)$ is the error function with β as a random number between zero and one. This

enlargement in particle velocity is isotropic and for particles near the surface it will have a net effect on the vertical distribution. Concerning the distribution of particles parallel to the surface it will remain the same without the added diffusion effect. The diffusion velocity in eq. 4 will increase with the velocity as the time step size decreases, which is a desirable effect. Particles will be distributed in the saltation layer due to the diffusion effect and the resulting vertical distribution of snow-density is compared to measurements by Mellor and Fellers (1986). Figure 2 shows good similarity between the numerical simulations and measurements of snow transport which supports the use of statistical vertical distribution of particles in motion above the snow surface.

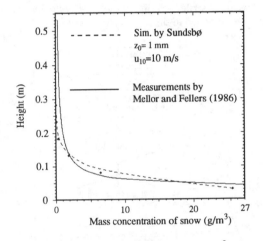

Figure 2. Comparison between field measurements of mass-concentration close to the snow surface and statistical simulations of particle transport.

3 NUMERICAL SOLUTION PROCEDURE

Transport of snow particles in wind is based on a Lagrangian-Eulerian model where the wind velocity field is calculated by solving the well-known Navier-Stokes equations in an Eulerian frame. Particles are moved by the Lagrangian equation of motion, eq. 3, based on Eulerian mean wind velocities. For each time step the particle movements are updated and in this way the particles are tracked through the simulation domain. Particles that deposit on the surface are allowed to form a firm surface which affects the wind flow similar to real conditions.

FLOW-3D, which is a general transient code for fluid flow based on the SOLA algorithm (Hirt 1975), is modified to simulate transport and deposition of snow particles. For each time step the SOLA algorithm calls the routine for particle motions, figure 3 and the routine for evaluation of snow deposition, figure 4.

3.1 Procedure for motion of particles

Equation 3 considers the possibilities of particle saltation and particle suspension. The question is how and when to activate the statistical saltation effect. The particles are moving in a combination of saltation and suspended modes governed by eq. 3 as long as the calculated friction velocity is above the threshold value. As soon as the calculated friction velocity is below the threshold value for particle lift, the diffusion effect is omitted and the particle is allowed to settle with the particle fall velocity and then be attached to the snow surface, see figure 3.

The friction velocity has to increase above the threshold value to release the particle from the attaching bonds.

3.2 Procedure for deposition

Each particle represents a small volume of snow and for each time step the SOLA algorithm calls for an evaluation of the amount of snow in the computational cells. A given cell is considered solid if the amount or volume of the deposited snow is equal to or greater than the cell volume, figure 4. Particles in the saltation mode will bounce and reflect from solidificated cells in a similar manner that they are reflecting the ground boundary surface. Mesh cells will remain solid until the computed threshold friction velocity is reached.

It is impossible to expect that computer simulations of snow drifting for engineering designs can be performed with the same huge number of snow particles that nature can provide. The way of

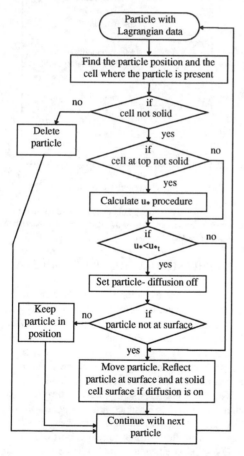

Figure 3. Procedure for statistical-based motion of particles. For each time-step all the particles are treated by the given procedure.

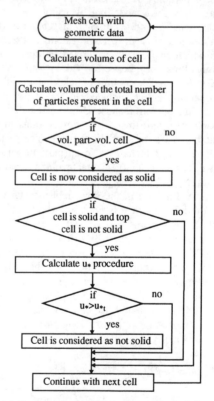

Figure 4. Procedure for deposition of snow. For each time-step the mesh cells are considered as solid if the volume of the particles inside the cell border is equal to or greater than the cell volume.

356

dealing with this problem is to consider the volume of a snow particle attached to the surface to be far greater than the single particle volume of airborne particles. For the purposes of determining deposition rates the deposited particles are considered to be in the order of 10^9 the size of a real snow particle in motion.

4   SIMULATIONS OF SNOW DRIFTS

Computer simulations of snow drifting around a typical Wyoming fence are performed considering the fence has a porosity of 50%, a height of 2.74 m and a bottom gap of approximately 14%. The Wyoming fences have been subjected to thorough investigations and fences of this geometry seem close to optimal regarding storage capacity (Tabler 1988). Figure 6 shows details from the calculated velocity field and snow deposition around a Wyoming fence at two different time steps. The velocities are being retarded rather than forming a recirculation zone downstream of the fence, which is a correct observation according to Perera (1981). He claims that velocities behind fences with a porosity above approximately 0.3 will not form the characteristic recirculating bubble. The zone with retarded velocities is extended further downstream than for solid type fences, which causes a larger deposition area. Figure 6 is also exposing the way the wind is affected by the snow surface. As soon as the bottom gap of the fence is filled with snow, this opening is no longer forcing the snow deposition downstream and consequently the deposition is approaching the fence at leeward side.

Snow particles are released upstream of the fence from particle sources in the inlet velocity profile where they instantaneously will be captured by the flow and drifted further downstream. Particles will

settle down to the surface driven by gravity provided the calculated local friction velocity is less than threshold. In this paper the threshold friction velocity is chosen to be 0.25 m/s and the friction velocity is calculated from a reference height of 0.25 m above the current snow surface using a local roughness height of 10^{-4} m. Snow particles are considered to have a mean density of 700 kg/m^3, a diameter of 100 μm and a terminal fall velocity of 0.4 m/s (Kim et al. 1991).

The deposition in figure 5 is based on a simulation time of 1600 s. This simulation time may seem unrealistic compared to real development over the winter season, but feasible due to the fact that snow particles in deposition are set to be larger than airborne particles.

Simulation of snow drift pattern is compared to an equilibrium profile proposed by Tabler (1989) and found to be acceptable within this profile, see figure 5. Other fence geometries with 50% porosity will, according to Tabler form a deposition similar to this equilibrium profile as long as the fence is not more than half filled with snow.

5   CONCLUDING REMARKS

Snow drifting along the ground is highly dependent on the velocity field close to the surface. Snow particles in saltation and particle behaviour in suspension are modelled in averaging a flow field. A numerical simulation model has been developed and tested to some extent. The particle tracking model includes single particle characteristics and particle movement represents a collection of particles in saltation and in suspension. The threshold velocity for particle saltation is modelled on empirical information of snow drift physics.

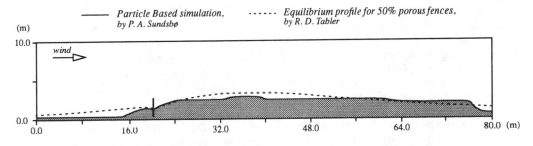

Figure 5. Calculated snow deposition around a 50% porous fence with height of 2.74 m and a bottom gap of 0.4 m. Particle data is as follows: density is 700 kg/m^3, diameter is 100 μm, threshold friction velocity is 0.25 m/s and terminal velocity is 0.4 m/s. Simulation time is 1600 s. Other data is as described for figure 6.

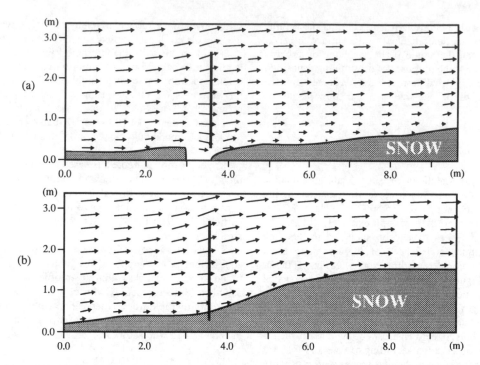

Figure 6. Details from the calculated velocity field and snow deposition around a 50% porous Wyoming fence. The mean inlet velocity profile is given with a roughness height equal to 10^{-3} m and a velocity of 10 m/s at a height of 10 m. Parts (a) and (b) are deposition plots after 600 s and 1000 s, respectively.

The particle simulation model is implemented into the transient fluid flow code FLOW-3D. FLOW-3D is a generalized computer code for the analysis of fluid dynamic and thermal phenomena.

Numerical simulations of snow drift are performed around a typical Wyoming fence. The simulations are performed in two dimensions. Simulated results of snow deposition around the fence are compared with experimental data and found adequate. The model will be extended to three dimensions for snow deposition around complex structures.

Reliable modelling of snow drift and snow deposition are possible provided that weather data and snow properties are well established.

In future studies this model should be extended to incorporate the effects of changes in weather and thermal conditions which are known to occur. Snow drift modelling should also include calculations of a distribution of particle sizes and a two-way coupling between wind and particles. These extensions will clearly increase computer costs but in return give us more accurate results provided the appropriate boundary conditions are given.

Computational fluid dynamics and numerical simulations represent a significant enhancement of design capabilities in snow-engineering.

ACKNOWLEDGEMENT

The authors tanks Mr. B. Bang at Narvik Institute of Technology for his computer assistance during this work.

NOMENCLATURE

Roman letters

C_d	drag coefficient
D	particle diameter
g	gravity
p	pressure
$\underline{r}$	position vector
t	time
u, v, w	components of fluid velocity in x, y, z, direction respectively
$\underline{V}$	fluid velocity vector

$\underline{V}_p$ particle velocity vector
$\underline{V}_{dif}$ diffusion particle velocity vector
$\underline{V}_{mean}$ mean particle velocity vector
u_* friction speed, $= (\tau/\rho)^{1/2}$
u_{*t} friction speed at threshold
x, y, z coordinates of particles position
z_0 roughness height

Greek letters

λ particle diffusion coefficient
β random number
κ von Kármán's constant (~ 0.4)
μ dynamic viscosity
ρ air density
ρ_p particle density
τ time

REFERENCES

Bang, B., Nielsen, A., Sundsbø, P.A. and Wiik, T., 1994, Computer Simulation of Wind Speed, Wind Pressure and Snow Accumulation around Buildings (SNOW-SIM), Energy and Buildings 21, 235-243.

Barkhudarov, M. and Ditter, J.L., 1994, Particle Transport and Diffusion, Flow Science, Inc.

Decker, R., 1991, Multiphase Flows and the Modeling of Drifting Snow, Proceedings of The International Cold Regions Engineering Speciality Conference, Feb. 26-28, p.673-684.

Haehnel, R.B. and Lever, J.H., 1994, Field Measurements of Snowdrifts, Proceedings of ASCE Workshop of the Modeling of Windblown Snow and Sand.

Hayes, W.F. and Tucker, H.G., 1985, Similarity Criteria for Scaling Snow/Wind Interaction Phenomena using Water Flumes and wind Tunnels, American Society of Mechanical Engineers, Fluids Engineering Division, Publ. by ASME, New York.

Hirt, C.W., 1975, SOLA-A Numerical Solution Algorithm for Transient Fluid Flows, Los Alamos Scientific Laboratory Report LA-5852.

Iversen, J.D., 1979, Drifting Snow Similitude, American Society of Civil Engineers, Hydraulics Division, Journal, June, p.737-753.

Irwin, P. A. and Gamble, S. L., 1988, Predicting Snow Loading on the Toronto Skydome, First International Conference on Snow Engineering, Santa Barbara, California.

Kim, D.H., Kwok, K.C.S. and Rohde, H.F., 1991, Similitude Requirements of Snowdrift Modelling for Antarctic Environment, Research Report No. R634, The University of Sydney, Australia.

Kind, R.J., 1986, Snowdrifting: A Review of Modelling Methods, Cold Regions Science and Technology,12, 217-228.

Kobayashi, D., 1972, Studies of Snow Transport in Low-Level Drifting Snow, Contributions No. 1200, Inst. of Low Temp. Sci. Series A, No. 24, Hokkaido University.

Male, D.H., 1980, The Seasonal Snowcover, Dynamics of Snow and Ice Masses.

Mellor, M., 1965, Blowing Snow, Cold Regions Science and Engineering Part III, A3C, U.S. Army Cold Regions Research and Engineering Laboratory, Hampshire.

Mellor, M. and Fellers, G., 1986, Concentration and Flux of Wind-blown Snow, US Army Corps of Engineers, Special Report 86-11.

Norem, H., 1974, Design of Roads in Snow-Drift Areas (in norwegian), The Norwegian Institute of Technology, Dep. of Railways and Roads.

Perera, M.D.A.E.S., 1981, Shelter Behind Two-Dimensional Solid and Porous Fences, Journal of Wind Engineering and Industrial Aerodynamics, 8, 93-104.

Schmidt, R.A., 1980, Threshold Wind-Speeds and Elastic Impact in Snow Transport, Journal of Glaciology, Vol. 26, No. 94.

Tabler, R.D., 1988, Snow Fence Technology, State of The Art, First International Conference on Snow Engineering, Santa Barbara, California.

Tabler, R.D., 1989, Snow Fence Drift Profile Database, American Society of Civil Engineers.

Uematsu, T., Nakata, T., Takeuchi, Y., Arisawa, Y. and Kaneda, Y., 1991, Three-Dimensional Numerical Simulation of Snowdrift, Cold Regions Science and Technology, 20.

White, B.R., Particle Dynamics in Two-Phase Flows, Encyclopedia of Fluid Mechanics, Vol. 4, Chap. 8, 239-282.

Snow Engineering: Recent Advances, Izumi, Nakamura & Sack (eds) © 1997 Balkema, Rotterdam. ISBN 90 5410 865 7

Preventing avalanches at Guozigou in Xinjiang autonomous Uygur Region of China

Changli Li, Lidu Huang & Junchao Li
Xinjiang Science Research Institute of Communication, Urumqi, People's Republic of China

ABSTRACT: Guozigou locate in the western part of Xinjiang Uygur autonomous Region of China. The risk of snow disasters are very serious, and accidents which cost lives often occur. Vehicles are damaged and traffic are cut. Twenty-five people have lost their lives in snow disasters since 1968. The authors and their supporters wish to bring snow disaster under control. After many investigations and studies, we now have full confidence in our "Combined Scheme".This paper introduces the reason, the hazard and on-going of snow disaster prevention methods at Guozigou.

1. INTRODUCTION

Snow disasters at Guozigou are related to nature. They are unacceptable along important national highways such as G312. After the foundation of the People's Republic of China, the Guozigou section of the highway was reconstructed with completion in 1962. Since then there have had three serious snow disasters,which killed 25 people (7 in 1968, 3 in 1988, and 15 in 1994). One third of the dead were of local minority nationality. When the 1988 snow disaster took place, in order to direct the emergency and disaster relief, the secretary of the Party, Zhuang Caiqing and the Director of Xinjiang Communication Department, Zhuma Shabiti spent their Spring Festival at the disaster site. The daughter of the director of Hutubi County Grain Bureau, a recent Uygur college gradute, lost her young life in this disaster.

To reduce such hazards this section of the highway is now being reconstructed.We have carried out many investigations,determined the reason for the hazard and developed rules to prevent snow disasters at Guozigou.

There are two kinds of snow disaster at Guozigou, one is avalanches, which occur on red soil slopes, the other is wind-blown snow which takes place at Songshutou. They endanger the highway in different ways with wind and snow their main and common reason.

2. AVALANCHES

2.1 The hazard and reason for avalanches

The maximum depth of stable snow drift is 120 cm. The wind's maximum speed can reach 24 m/s, and the wind blows large amounts of snow behind and on top of the mountain to the source of an avalanche. This not only enlarges the resource of the avalanche,but also makes snow cornices, which can trigger avalanches.But, when you stand on the road and look up only a high steep cliff can be seen. You will not imagine any possibility of avalanches. However there is a large grass slope out of site above the steep cliff. It extends all the way to the ridge of the mountain and its width is about 200 - 500 m. (See Figure 1).

The grass slope extends along the direction of the mountain ridge for a total length of 2,300 meters. The grade averages between 26 and 39 degrees, which is considered optimum for avalanche occurrence.The slope is covered by thick withered grass (Figure 1 and Figure 2). The withered grass falls down, and becomes a good slip surface. In early spring , the temperature goes up, and the inner part of snow layer changes, loosing its inner friction. An avalanche may be triggered at any moment. If local snow loses its stability, avalanche is triggered, and the chain reaction

Figure 1 The grade of the grass slope is 26-39 degrees

Figure 4 A 40 cm Diameter Tianshan poplar pushed down by an avalanche

Figure 2 Vegetation on the hillside slope

Figure 3   Trees broken by an avalanche

sends a strong snow torrent flooding down, making a violent blast wave. Where the snow torrent goes, all the trees are broken (Figure 3),and the ground surface is eroded. If vehicles are present they will be damaged and people may be killed. The soil, rock, remains of damaged trees and such go with the snow torrent. When blocked, they stop and pile up, cutting traffic and blocking rivers. This happens ofen. For example, on March 10, 1994, a medium-sized bus from Jiuquan of Gansu Province was passing Guozigou, when an avalanche from the opposite mountain crossed the river bed and piled up on the road. It pushed the bus off the road and covered it. But no one knew it at that time. After 5 days and nights the people in the bus were found. Though doctors did their best to save their lives,   only 3 of 17 people survived. Why then do such snow disasters ofen happen at Guozigou? The reason is both of nature and of human being, and the human being playing the leading role.

On slopes which are suitable for an avalanche, even when there is a large amount of snow,an avalanche may not occur. Where he vegetation is complete and trees are dense, there are no avalanches. We discovered that at those places where avalanches re a very serious risk, the vegetation was not depleted before as it is now. There is only grass and no trees now. The trees and egetation were damaged by nature and by humans. In places where the avalanche converged and went through, withered wood nd damaged trees (figure 4) are often seen. It is sufficiently proven that avalanches did not occur until a densly wooded area as changed to only withered grass.

What is frustrating is that there are   still trees being remove unlawfully. The few remaining lonely trees are not strong enough to resist the wind. Some are broken and some

362

Figure 5 A 63cm diameter Tianshan poplar uprooted by an avalanche

are lifted by the root (Figure 5). As wind damage worsens, more avalanches occor. The vicious circle is growing.

2.2 Preventing Method

From the preliminary investigation, we found, that engineering treatment combined with biological control (hereinafter called "Combined Scheme") is the practical way to prevent avalanches at Guozigou.

2.2.1 Outline of "Combined Scheme"

(1) Choose practical avalanche technology for necessary engineering treatment. First protect the highway from the avalanche hazard. Second, protect existing vegetation from avalanches and implement additional vegetion enhancement.

(2) Vegetion enhancement. Plant fast-growing trees by stages, first at the source of avalanches, then at the converging, moving and depositing areas. If the trees are successfully planted, the snow will be more stable at the source ofavalanches.

(3) At the source of avalanches, some special avalanche technology can be done to prevent the formation of snow driffs and cornices. This can reduce the number of avalanches, and elimmate the avalanches friggen.

2.2.2 Keys to the "Combined Scheme"

(1) Choosing the correct shape, distribution and quantity of engineering construction.

(2) Selecting,growing and planting trees which are suitable for the planting area and able to prevent avalanches.

(3) Deciding how to develop the snow shield forest.

2.2.3 Implementation issues

(1) The "Combined Scheme" is a combination of scientific experiment, technical design and engineered construction. Thoughout the implementation period the above three aspects are done in turn over a long period, and much money is needed.Multi-subject (Forestry, Meteorology, Snow Disaster, Highway Engineering) professionals must cooperate closely with each other.In practices,implementation is interfered with by many social aspects. If there is no reasonable policy and support from administrative officials, it will be difficult to succeed.

(2) At the source of an avalanche, and where it converges, moves and deposits, a lot of construction work for stabilizing, slowing, stopping, leading and blocking snow must be done. Most construction sites are above steep cliffs, and have complicated and dangerous steep topography and high elevation. Construction there is difficult. Structures erected to prevent avalanching must be robust.

(3) Engineering treatment is the prerequisite, biological control is the kernel. The implementation of biological engineering can be significantly limited by man and natural factors. Xinjiang is in the middle of Euro-Asia, where drought is the main climate characteristic. Most people are not sensitive to water-needed biological engineering. Also since avalanches are far from people's daily life, it is difficult for them to believe that biological engineering can control avalanches. In that part of Xinjiang it snows a lot. Snow is one-third of the total precipitation). So there are few supporters for biological engineering, and quite a few that oppose it.

(4)The "Combined Scheme" will take up a part of the mountain slope. That will influence some local people's present interests. Because of this, some people will oppose it, and even damage it.

(5) The elevation of Guozigou is very high, the elevation at the source of avalanches is 2,300 meters, and the climate is very bad. it often snows in May and June.This makes the tree species at Guozigou very limited and makes it difficult to introduce desired species.

2.2.4 Advantages of The"Combined Scheme"

(1) Fully utilizes the advantage of local conditions
Engineering construction is mainly earthworks and rockworks using local materials, thus there is little need for

materials bought in from elsewhere.

Trees of selected species planted and protected correctly can grow to become an efficient snow retention device. If they are not destroyed unlawfully, their snow preventing ability will also increase year by year. This will not only control avalanches, but will also bring additional economic profits and ecologic and social benefits. There is no other engineering approach that compares with   it.

(2) Control avalanches and turn the bad to the good

If the 'Combined Scheme" is implemented, the source snow is fixed, and there will be no avalanches. Deposited snow melts in place, and seeps into the ground, moistening plants which include the snow preventing forest, and improves the ecology. Growing plants not only richen the people's material wealth, but they also protect, clear, and increase the freshwater source. This has great present and historical meaning for Xinjiang drought-stricken areas.

(3) A thorough implementation of the"Combined Scheme" will prove its correctness, importance and urgency. Its success will promote biological engineering in other fields where human beings struggle with natural disasters. More and more people will acknowledge, accept, and use it.

3.   WIND-BLOWN SNOW

For about 6 km along the   G312 highway in Songshutou, there are wind-blown snow deposits off and on to a maximum depth of 3 meters. These   drifts blocks vehicles and the snow laden wind obstructs visibility.Traffic most be halfed.

Machine clearing was the usual way of cleaning this snow bot. The speed of that operation is limited by many factors. Some removal requirements can not be satisfied by machine cleaning.

Blowing snow   must be observed at definite spots and at proper times, to   clarify its relation with the topography. Once this relationship is understood,snow engineering works can be set up at various sections of the road according to the wind speed and direction there.

There are many types of snow engineering works that prevent blowing snow. According to our experience, object that block snow work best, with a snow-preventing forest the most effective. The successful planting of a snow-preventing forest can not only eliminate snow deposition on the road, but also clear the air of snow, improving visibility .Further experiments are planned.

Snow Engineering: Recent Advances, Izumi, Nakamura & Sack (eds) © 1997 Balkema, Rotterdam. ISBN 90 5410 865 7

Ammonium carbamate as corrosion inhibiting deicing material

Charles N. Hansen, Hangsil Cho, Heesuk Lee, Hosin Lee & Rand Decker
Center for Advanced Construction Materials, Civil and Environmental Engineering Department, University of Utah, Salt Lake City, Utah, USA

ABSTRACT: Ammonium carbamate is a promising alternative deicing material to salt. It is not corrosive to reinforcing steel in bridges. It inhibits the corrosion of steel and aluminum, and stops the corrosion that the chloride has already started. Ammonium carbamate has a lower eutectic freezing temperature in solution than salt. It produces only minor freeze-thaw damage to concrete, and may also promote plant growth due to its 36% nitrogen content. A mixture of 60% Ammonium carbamate and 40% sodium chloride melts ice faster and reduces both material and environmental costs while effectively preventing the corrosion of steel. For practical applications, a method to pelletize the deicing compound of ammonium carbamate and salt has been researched.

1. INTRODUCTION

Each year about $1.5 billion is spent on highway snow and ice control in the United States. Since 1970, approximately 10 million tons of salt have been used each year on the highways in the U.S.("Highway" 1991). Although salt is an effective and inexpensive deicer, it produces many unintended side effects. These side effects include damage to infrastructure, motor vehicles, and the environment. Therefore, an alternate corrosion inhibiting deicing material is needed to overcome these negative effects of salt. Among other deicing materials, Calcium Magnesium Acetate (CMA) has been one of the widely used deicing materials in the U.S. However, the cost of CMA is approximately twenty times that of salt. Currently, CMA is sold between $600 and $700 per ton delivered while salt is sold at about $30 per ton.

In this paper, the characteristics of a new deicing material, ammonium carbamate, are discussed. The cost of the production of AC is estimated at $160/ton which is much less than that of CMA. The advantages of AC as a potential alternative deicing material are demonstrated in this paper.

2 COMPARISON OF VARIOUS DEICING MATERIALS

Sodium chloride has long been used on the highways and streets as a deicing material. Flake, pellet, and liquid calcium chloride has been also used, usually in combination with rock salt or sand, at the low temperatures. Urea has been also used as deicing compound. It is not considered to be as corrosive to ferrous metal as the deicing salts and it is not as toxic to plants. It has the disadvantage of having a relatively high eutectic temperature of -11.5 °C in a water solution. Other deicing materials commonly manufactured and used are potassium chloride and magnesium chloride. In the past several years, there have been substantial amounts of studies conducted on CMA as a noncorrosive and environmentally safe deicing material ("Handbook" 1992).

A freezing test was conducted on a mixture of 60% ammonium carbamate and 40% salt as shown in Figure 1. The solution "supercooled" before freezing took place. Since supercooling occurred in our experiment, the true freezing point of the solution is estimated at -22.5°C for the salt and ammonium carbamate solution using extrapolation as shown in Figure 1. Freezing temperatures at various concentration levels of the other deicing materials plus ammonium carbamate(NH_4COONH_2) are given in Figure 2.

There are many brands of deicers currently on the market, most of which consist of one or more of the materials shown in the Figure 2. The characteristics of ammonium carbamate are compared with those of other deicing materials in Table 1. based on a literature review and the preliminary test results (Hansen 1987).

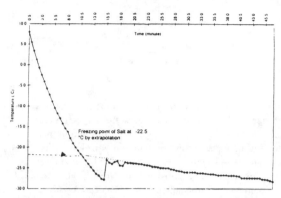

Figure 1 Freezing Temperature of Solution of 60% Ammonium Carbamate and 40% Salt

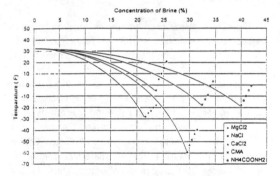

Figure 2 Freezing Points of Deicing Solutions

Table 1. Fundamental characteristics of Deicing Materials

Characteristics of the deicer	Sodium Chloride	Calcium Chloride	CMA	Magnesium Chloride	AC
Eutectic temperature	-21.2°C (-6 °F)	-51°C (-60 °F)	-27.5 °C (-17.5 °F)	-33.3 °C (-28 °F)	-28.9°C (-20°F)
Ice melting speed	High	High	Low	High	Low
Freeze-thaw damage to concrete	Yes	Yes	No	Yes	No
Corrosion of steel	Yes	Yes	No	Yes	No
Environmental cost	High	High	Low	High	Low
Inhibition of chloride corrosion	No	No	N/A	No	Yes

3 CORROSION INHIBITING EFFECT OF AMMONIUM CARBAMATE

3.1 *Laboratory corrosion test*

Many studies have been performed on the effect that chlorides have on the corrosion of steel and reinforcing steel imbedded in concrete. The concrete itself acts as an inhibitor to the corrosion of steel reinforcement (Mozer 1965). The corrosion of the steel in concrete is electrochemical in nature. The area where the metal ion goes into solution becomes the anodic region. If the metal is iron (Fe), it goes into the solutions to form ferrous ions (Fe^{2+}) and two electrons ($2e^-$). To maintain an equilibrium of electrical charges, an equivalent quantity of hydrogen is plated out on adjacent surfaces of the metal, the cathode. The thin film of hydrogen inhibits further corrosion unless the film is removed. Anodic and cathodic reactions may be represented as follows:

Anodic Reaction : $Fe \rightarrow Fe^{2+} + 2e^-$ (1)

Cathodic Reaction: $2H^+ + 2e^- \rightarrow H_2$ (2)

The cathodic reaction is slow in alkaline media because of the low concentration of hydrogen. However, it is accelerated due to the depolarizing action of dissolved oxygen. Therefore, the corrosion rate is proportional to the oxygen concentration as illustrated below:

$2H^+ + 1/2O_2 + 2e^- \rightarrow H_2O$ (3)

$Fe + H_2O + 1/2O_2 \rightarrow Fe(OH)_2$ (4)

As a result, the quantity of electricity flowing through the local cells is equivalent to the amount of corrosion. With the increasing anodic polarization, the corrosion of the metal decreases. In ordinary conditions of steel reinforcement in concrete, where the pH is high, the hydrogen concentration is low, and the oxygen supply is minimal, an anodic coating builds up on the steel. However, when chloride containing deicing materials are applied, the protective anodic film of iron oxide and hydrogen are removed by forming the soluble chloride compounds, thus leaving the steel exposed to further electrochemical attack (Hansen 1995).

Some corrosion tests were performed by using the method of weight loss. The result of the weight loss test showed that AC did not corrode

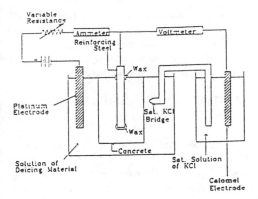

Figure 3. Circuit Diagram for the Corrosion Test.

Table 2a. 2.5% solution of salt.

CONSTANT CURRENT DENSITY (10 ua/cm^2)	
Time (min)	mV
0	-330
5	-180
10	-193
15	-223
20	-248
25	-260
30	-270
35	-280
40	-288
45	-293
50	-295
55	-298
60	-298

Table 2b. 2.5 % solution of Ammonium Carbamate

CONSTANT CURRENT DENSITY (10 ua/cm^2)	
Time (min)	mV
0	-60
5	600
10	630
15	648
20	653
25	663
30	668
35	670
40	673
45	675
50	678
55	680
60	680

steel or aluminum. The other method used for corrosion inhibition evaluation was a modified version of the method used by Gouda (Gouda et al. 1965). The diagram of the circuit used in this corrosion test is shown in Figure 3. The effect of the corrosion inhibitors increases with increasing anodic polarization at relatively low current values. In other words, overall corrosion of the metal diminishes with increasing anodic polarization. Therefore, the corrosion of the metal would be at rest if the potential of the anodes reaches that of the cathodes at infinitesimally small current densities. Polarization current densities commonly used for such tests ranges from 1 to 1000 $\mu A/cm^2$, which presumably approximate the change of values of the actual local cell (Gouda et al. 1965).

In the test using the circuit shown on Figure 3, the potential difference between the calomel electrode and the steel test piece was measured when a positive charge was imposed on the steel test piece to cause a small amount of current to flow. When the corrosion of the steel had been stopped, an anodic coating would build on the test piece and produce a high potential difference between two electrodes. In the cases when the corrosion was not stopped, the test piece becomes negative relative to the calomel electrode.

The corrosion of ferrous-based metals imbedded in concrete that the chloride has already started may also be inhibited by AC. The method is to remove a sufficient amount of chloride ions from the concrete either by immersing the test pieces in the distilled water for two days, or by applying negative voltages to the test pieces while still immersed in distilled water before they are immersed in AC solutions (3 to 6 %) for 1 day to stop the chloride corrosion. The result of the test showed

that AC successfully inhibited chloride corrosion of steel imbedded in concrete (Hansen 1995).

Table 2a shows the voltage levels when deicer composition is sodium chloride. In this test, the test piece is embedded in concrete made with a 2.5% deicer in water solution and the concrete and it is immersed in a 2.5% deicer in water solution. The constant current density of 10 ua/sq. cm for 60 minutes showed that in the presence of a sodium chloride solution, the steel test piece in the concrete had such a strong tendency to corrode that the steel piece developed a voltage, as compared to the calomel electrode, of minus 300 mV as shown in Table 2a.

Table 2b shows the noncorroding effect of ammonium carbamate on steel in concrete. The voltages of the steel test pieces, as compared to the calomel electrode were measured for a constant current density of 10 ua/cm^2. As shown in Table 2b, the high voltage was measured due to the anodic coating on the surface of the test steel, which indicates no corrosion of steel in concrete.

Table 2. Voltages measured between steel and calomel electrode.

3.2 *Advantages of ammonium carbamate as a deicer*

The eutectic freezing temperature of AC in water (-28.9°C) is lower than that of a solution of salt in a water (-21.1°C). AC does not only corrode steel or aluminum, but also inhibits the corrosion of steel which salt has already started. The laboratory test showed that AC stopped the corrosion of steel which is imbedded in concrete (Hansen 1995).

The freeze- thaw damage of AC on concrete was compared with three other deicers in the test performed by the Utah Department of Transportation. The test specimens were from one batch of concrete mixed according to ASTM C-109 specifications, air entrained, using Ideal II cement and Ottawa sand. Test blocks of 5.08 cm^3 were cured for 37 days. Three percent solutions of deicers in distilled water were prepared for calcium chloride, pure sodium chloride, road salt and a AC deicer composition of 95% AC and 5% urea. The minimum and maximum temperatures of the freezing and thawing cycles were at -17.7°C(0° F) and 37.8°C (100° F). The results showed that only minor damage was caused by the solution that was a combination of 2.85% AC, 0.15% urea and 97% distilled water. The damage was about the same as the damage caused by a distilled water control specimen. However, major damages were reported by 3.0% solutions of calcium chloride, road salt and pure sodium chloride.

Because of the small sizes of the sodium and chloride ions, salt will melt ice, at moderate freezing temperature, i.e., -7 °C (+20 °F), faster than any of the other deicers. However, when 60% AC is mixed with 40% salt, the faster melting characteristic of salt is retained while corrosion of steel is inhibited. The combination of deicers (60% AC and 40% salt) will have the ability to melt ice at lower temperatures than a salt type deicer because of the lower freezing point of AC. Salt, which contains about 60% chloride,may produce chloride toxicity to plants. In contrast to salt, AC contains 36% nitrogen which will support plant growth (Hansen 1987). Since toxicity is a function of concentration, a deicer which contains only 40% salt will not be nearly as toxic to plants as one that contains 100% salt.

One main concern for the AC type deicing material is its odor of ammonium at room temperature. Ammonium carbamate slowly decomposes to form two ammonia molecules and one carbon dioxide molecule. The vapor pressure is compared with the vapor pressure of a solution of ammonia in water as shown in Figure 4.

$$H_2O + NH_4OH \leftrightarrow NH_3\uparrow + 2H_2O \qquad (5)$$
$$NH_4COONH_2 \rightarrow 2NH_3 + CO_2 \qquad (6)$$

The vapor pressure of AC (NH_4COONH_2) is approximately 15 mmHg, which is about the same as 5 mole% NH_3 in water near freezing temperature. Since ammonium carbamate yields ammonia vapor, adequate ventilation and respiratory protection are required. However, the application temperature of the deicing material is expected to be around the freezing temperature of the water where the odor is minimal.

4. THE PRODUCTION OF DEICING PELLETS AT LABARATORY

A binding solution was prepared by pulverizing, 0.36-Kg of Surtech L-101 lignite and adding it to one gallon of aqua ammonia. This mixture was then agitated for about 30 minutes and allowed to settle for a day. The solution of humic acid dissolved in aqua ammonia was decanted from the insoluble portion of the Surtech L-101. The 27.2-Kg of AC was pulverized to minus 40 mesh and combined with 18.1-Kg of salt which had also been pulverized to minus 40 mesh. The apparatus for the pelletizing operation were a 41 cm disc pelletizer, a spray bottle

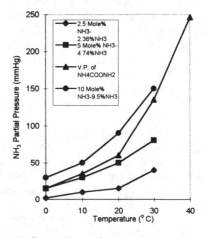

Figure 4. Comparison of vapor pressure of ammonia with that of ammonium carbamate.

to spray the binding solution onto the mixture, and containers. The process of pelletizing was started by turning the disc clockwise rotational speed and slope. In this test, the rotation speed was 24 rpm with a tilt of 60°. The disc was sprayed with the binding solution, and the mixture of AC and salt was added slowly to the disc with a scoop.

The process would be to continuously add the salt with sufficient binding solution. Figure 5 shows a diagram of the pelletizer with three scrapers to produce pellets. A steeper tilt and a faster rotational speed would produce smaller pellets, while a flatter tilt and a slower speed would produce larger pellets. The scrapers of this pelletizer were located in the vertical position, and they direct the first stream or pellets (I), the second stream or growing pellets (II), and the third stream-speed or nuclei (III), as shown in Figure 5. The pellets were collected in an aluminum pan which was placed below the outlet of the pelletizer.

The pellets were generally larger than twice the size of the 4mm pellet desired. At the end of the test, the percentage of binding solution was 17.7%, and the percentage of feed recovered in products was 85.3%. There were several variables in performing a successful pelletizing test. One was controlling the strength of the binding agent and another was the effect of temperature on the pelletizing reaction. The last was the percentage of binding agent needed in order to make strong, stable pellets of the required size.

The test indicated that the hammer mill produced a fine enough grind of the Surtech L-101 to allow the aqua ammonia to dissolve the contained humic acid from the lignite. It appeared that the 62/38 mixture of AC plus salt, wetted more rapidly at low temperatures than it did at higher temperatures. There was no advantage in lowering the temperature

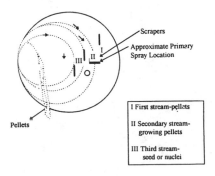

Figure 5. Material Pattern on disc

of the 62/38 AC/salt mixture below 4.4°C (F 40°F). The lower temperature may have slowed the reaction between the binding agent and the AC, such that pellets solidified slowly. The use of a weaker binding solution, produces a slower wetting and setting of the pellets.

At least 25% of the feed to the disc granular should be finer than 200 mesh. In grinding the salts, as they were being prepared for pelletizing, 100% had to pass through a 100 mesh screen. Because of the fact that the two lots of Surtech L-101 varied in strength, some problems were experienced in producing a binding agent that contained a maximum amount of humic acid.

The chemistry reacting carbon dioxide with ammonia to form ammonium carbamate is well-known. A potential method of producing ammonium carbamate is to process an intermediate product in a urea plant, by collecting the ammonium carbamate which is not converted to urea. This may eliminate the need for the recycling of the unconverted ammonium carbamate and increase the total production of the plant. As an intermediate step in the production of urea, ammonium carbamate is produced by reacting two parts of ammonia with one part of carbon dioxide by volume at pressure of 3,000 psi, and a temperature of 187.7°C (370° F). The current intermediate solution from a urea manufacturing plant is composed of 44.5% AC, 3.2% urea, 18.3% ammonia, and 34.0% water. The pH of the water solution, containing 2.37% ammonium carbamate plus 0.133% urea, was 9.13.

In order to handle the high reaction temperature and pressure, the reaction vessel should be fabricated of expensive alloys which will withstand those conditions. The resulting cost of investment for such a plant could be very high. In the urea production plant, in which urea is synthesized from ammonia and carbon dioxide, there are two reactions: the formation of ammonium carbamate, and the conversion of ammonium carbamate into urea. These two reactions are shown as follows:

$$CO_2 + 2NH_3 \rightarrow NH_2CO_2NH_4 \tag{7}$$
$$\text{(ammonium carbamate)}$$
$$NH_2CO_2NH_4 \rightarrow NH_2CONH_2 + H_2O \tag{8}$$
$$\text{(urea)}$$

A solution to such a high investment cost in the production plant is to use a fluidized bed. BASF in Germany has developed a method to use a fluidized bed reactor for producing ammonium carbamate. In the fluidized bed, the reaction is run at

10 to 25 °C under slightly higher than atmospheric pressure. The construction cost of a fluidized bed reactor is less expensive than the corresponding capital cost of the urea plant.

5 SUMMARY AND CONCLUSION

The advantages of ammonium carbamate type deicers include its low eutectic freezing temperature of -30 $^\circ$C (-20 $^\circ$F) in water solution, corrosion inhibiting characteristics, and minor freeze-thaw damage compared to other common deicers. Since ammonium carbamate can be classified as a fertilizer, another objective of the this research paper is to provide deicing compositions based upon ammonium carbamate which are not toxic to plant life. AC mixed with salt would have additional advantages including faster melting speed while effectively inhibiting the corrosion of steel or the steel imbedded in concrete. In addition, the corrosion of the steel reinforcement due to chloride can be stopped by a mixture of ammonium carbamate and salt.

A major effort was made towards pelletizing a mixture of 60% AC and 40% salt. The use of this mixture will further reduce the cost while reducing the negative impacts of the salt. Therefore, if this new deicer works as well as preliminary laboratory tests indicate, it has a potential of being the ideal deicer that will save millions of dollars in maintenance cost of infrastructure while reducing the adverse environmental effects.

REFERENCES

Gouda, Y. K. et al., "A Rapid Method of Study Corrosion Inhibition of Steel in Concrete," Journal of PCA Research & Development Laboratories, Ser. 1175, Sept. 1965, pp.24-31.

"Handbook of Test methods for evaluating Chemical Deicers," SHRP-H-332, Strategic Highway Research Program, National Research Council, Washington D.C., 1992.

Hansen, C.N., "Deicing Compositions," United States Patent No. 4,698,173,1987.

Hansen, C. N., "Method for Inhibiting Corrosion of Metal Imbedded in Concrete," United States Patent No. 5,391,349, 1995.

"Highway Deicing: Comparing Salt and Calcium Magnesium Acetate", TRB Special report 235, Transportation Research Board Committee on the Comparative Costs of Rock Salt and CMA for highway Deicing, Washington D.C., 1991.

Mozer, J. D. et al., "Corrosion of Reinforcing Bars in Concrete," Journal of the American Concrete Institute, Title No. 62-54, Aug. 1965, pp.909-930.

Snow Engineering: Recent Advances, Izumi, Nakamura & Sack (eds) © 1997 Balkema, Rotterdam. ISBN 90 5410 865 7

Field observation of microclimatic and ground thermal regimes during frost penetration beneath highways

T. Ishizaki & M. Fukuda
Institute of Low Temperature Science, Hokkaido University, Sapporo, Japan

N. Mishima & S. Yokota
Research Institute of Japan Highway Public Corporation, Tokyo, Japan

K. Toya
Sapporo Branch of Japan Highway Public Corporation, Japan

ABSTRACT: To understand the ground freezing process beneath highways, we performed field observations of microclimatic and ground thermal regimes during frost penetration. The obtained results show that the surface freezing index is approximately half the air freezing index. The results also revealed that long wave radiation at night plays an important role in advancing frost penetration. A computer model was developed to calculate an equilibrium temperature at the ground surface by taking an energy balance at the ground surface, and to calculate the frost depth. The calculated frost depth coincided well with measured values.

1 INTRODUCTION

In cold regions, when ground freezes, water migrates to the freezing front and segregates as ice lenses. This causes serious damage to highway surfaces. To prevent frost damages, frost-susceptible soil is replaced by non-frost-susceptible material to 80% of the frost depth in Japan using the maximum depth in ten years. For establishing effective measures to prevent frost damage, it is important to know the maximum frost depth.

Up to now, to estimate frost depth, we used the air temperature freezing index. However, it is reported that the frost depth depends on microclimatic condition, and differs greatly even where freezing indexes are similar. Therefore, it is necessary to develop a rational method to predict frost depth more precisely than by conventional methods. For the purpose of establishing a reliable method to predict frost depth, we established full scale highway test sites in Takasu (Asahikawa) and Hokumei (Obihiro). In each test site, we measured frost depth, frost heave amount and the ground temparature profile. In addtion,we measured on weather data such as air temperature, humidity, wind velocity and net radiation balance.

This paper reports the results of field observation performed for 3 years since 1989. We compare measured frost depths with the calculated value to evaluate the validity of this computer program.

2 FIELD OBSERVATION

Figure1 shows a schematic diagram of road structures and locations of the measuring instruments. The soil in this location is clayey sand and it has a high frost susceptibility. The ten-year maximum frost depth is 140 cm in Asahikawa. Three types of road structures were built. In the first type, soil is replaced with a non-frost-susceptible material to a depth of 40 cm, which corresponds to 30% of the ten-year maximum frost depth. In the second and third types, soil is replaced to depths of 70 cm and 100 cm, which correspond to 50% and 70% of the ten-year maximum frost depth, respectively. The size of eash test section is 15 x 15 m wide.

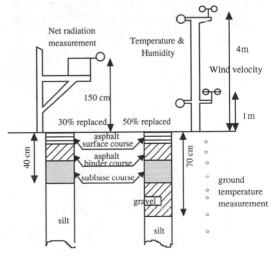

Figure 1. Schematic diagram of road structures and measuring instruments.

To obtain weather data for energy balance analysis, the following measurements were performed:
1) measurements of the temperature profile to a depth of 2 m and measurements of frost depth with a frost tube
2) measurements of temperature, wind velocity and humidity near the ground surface at heights of 1 and 4 m
3) measurements of net long wave radiation and short wave radiation at a he height of 1.5 m. These data were collected by a data logger at 1-hour intervals. The snow cover on the ground surface was removed every morning and evening to expose the black surface. Since a highway surface would have snow removed from it .

3 OBSERVATION RESULTS

3.1 Temperature profile near ground surface

Figure 2 shows typical temperature profiles near the ground surface. The numbers indicate the time of measurement. At 5:00,the ground surface temparature was −16°C and it continues to the air temperature profile. At 11:00,the ground surface temperature increased abruptly to 0°C. We can see a big difference in ground surface temperature and the air temperature. At 15:00, the ground surface temperature was also much higher than the air temparature. These temperature differences were caused by solar radiation.

3.2 Air freezing index and surface freezing index

As shown Figure 2, the surface temperature is usually higher than the air temperature. The freezing index is defined as the cumulative number of degree-days below 0°C for a whole winter. Therefore, the surface freezing index is lower than the air freezing index. The ratio between them are called n-factor. Figure 3 shows the relationships between the surface freezing index and the air freezing index obtained at various places in the central Hokkaido area from 1987 to 1991. The n-factor is approximately 0.5.

3.3 Climate condition when the freezing front advances

Figure 4 shows weather data when the lowest temperature reached -30°C and the freezing front

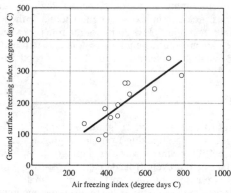

Figure 3. Relationship between the surface freezing index and the air freezing index.

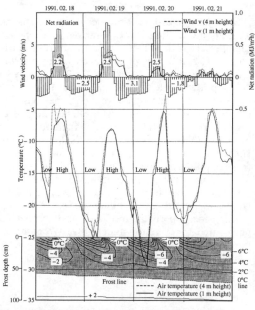

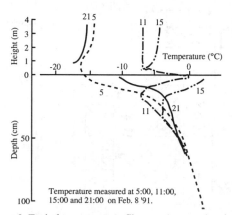

Figure 2. Typical temperature profiles near the ground surface in winter.

Figure 4. Typical weather data near the ground surface during freezing front penetration in winter.

advanced rapidly from Feb. 18, 1991 to Feb. 21, 1991. This figure shows that the ground temperature decreased rapidly with decreasing air temperature. The time lag of temperature change in the ground from air temperature is about 12 hours at depths of 30 to 50 cm. This figure also shows that at night, long wave radiation from the ground surface cools the ground,decreases the ground temperature,and advances the freezing front.

4 FROST DEPTH CALCULATION

4.1 Equilibrium ground surface temperature

Figure 5 shows a schematic diagram of energy balance near the ground surface. The net radiation (Rn) , sensible heat flow (H), latent heat flow (H) and conductive heat flow in the ground (S) are balanced with each other. The energy balance equation is shown by the following equation:

$$Rn + Le + H + S = 0 \qquad (1)$$

In winter, latent heat flow can usually be neglected because the evaporation rate from the frozen ground surface is very low. In the following calculation, the sensible heat flow and latent heat flow are calculated with temperature, wind velocity and humidity profile at heights of 1 and 4 m. The equilibrium ground surface temperature is calculated in the following procedure with equation (1).

4.2 Method of frost depth calculation

Figure 6 shows a flow chart of frost depth calculation following a procedure presented by Outcalt (1975) and Fukuda et al. (1980). At first, the equilibrium ground surface temperature is determined from climate data on the site by repeated calculation of the energy balance at the ground surface. When the equilibrium ground surface temperature is determined, the temperature profile in the ground is obtained by solving a heat transfer equation in which latent heat release is taken into account during ground freezing. The heat transfer equation is solved numerically by the finite difference method. Finally, we obtain the frost depth with the temperature profile in the ground. It should be emphasized that most data needed for this calculation were obtained by various measurements on the test site in our study. In addition to this calculation, we performed an energy balance calculation to determine the equilibrium ground surface temperature by using climate data from a nearby weather station. At first, we compared the climate data,e.g.net radiation, wind velocity and air

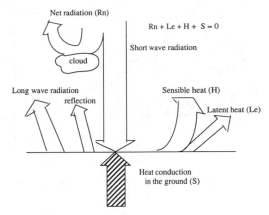

Figure 5. Schematic diagram of energy balance near the ground surface.

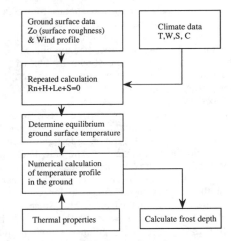

Figure 6. Flow chart of frost depth calculation.

temperature,between the test site and weather station. Then, we obtained a correlation function of climate data between these two points. Finally, we obtained the frost depth by using weather data of a nearby weather station. We called the former method a "frost depth calculation with test site data" and the latter a "frost depth calculation with weather station data".

4.3 Calculated results

Figure 7 shows a typical example of calculated frost depths using test site data together with measured depths. The road structure is of the second type in which a 70-cm soil layer was replaced by non-frost-susceptible soil. The measured maximum frost depth was 78 cm and the calculated maximum frost

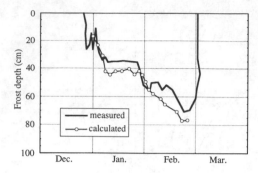

Figure 7. Typical example of frost depth calculated with test site data and measured values.

Table 1. Comparisons between calculated maximum frost depths and the measured maximums.

	Freezing index (Cdays)	Measured frost depth (cm)	Calculated frost depth (cm) with test site data	Calculated frost depth (cm) with weather station data	Calculated frost depth (cm) * Aldrich eq.
(40 cm replacement in Asahikawa)					
1990	645	61	63	65	55 (85)
1991	783	66	73	70	60 (95)
1992	655	61	69	58	58 (85)
(70 cm replacement in Asahikawa)					
1990	645	78	80	78	65 (100)
1991	783	77	85	82	75 (110)
1992	655	72	77	73	65 (100)
(100 cm replacement in Asahikawa)					
1990	645	81	88	82	70 (115)
1991	783	102	101	92	75 (125)
1992	655	100	85	76	70 (115)

* outside ( ) : with ground surface freezing index
  inside ( ) : with air freezing index

depth was 80 cm. The calculated maximum frost depth agreed well with the measured maximum.

We calculated the maximum frost depths for three types of road structure for three years from 1990 to 1991. We also calculated the frost depth using the conventional Aldrich equation,(1956) the air freezing index and the surface freezing index. The calculated results are shown in Table 1. The measured maximum frost depths increased with increasing freezing index and thickness of soil replacement.

The Aldrich equation with air freezing index gives a larger frost depth than the measured one and that with the ground freezing index gives smaller value than the measured one. The calculated frost depth with the energy balance models gives a much closer value to the measured ones. These results show the validity of the energy balance model to calculate the maximum frost depth. Since we can obtain the maximum frost depth with weather station data, this energy balance model is quite effective to calculate the ten-year maximum frost depth for the establishment of counter measures against frost damage to highways.

5 CONCLUSION

To understand the ground freezing process beneath highways, we performed field observation of microclimatic and ground thermal regimes during frost penetration. The obtained results show that the ground surface freezing index is approximately half of air freezing index. The results also revealed that long wave radiation at night plays an important role in advancing frost penetration. We developed a computer model to calculate the equilibrium temperature at the ground surface by taking an energy balance at the ground surface, and to calculate frost depth. The calculated frost depth coincided well with the measured values. This calculation method is quite effective for calculating

the ten-year maximum frost depth for the establishment of countermeasures against frost damage to highways. Since 1995, we have carried out field tests on the highway near Obihiro which opened in 1995. We also plan to compare them with this test site to evaluate the validity of the proposed computer model.

REFERENCES

Fukuda, M. and Ishizaki, T. 1980. A simulation model for frost penetration beneath the ground on the basis of equilibrium surface temperature. *Seppyo*, 42, 2, 71-80.

Outcalt, S. I. 1975. A coupled soil thermal regime surface energy budget simulator. Proc. Conf. on Soil-Water Problems in Cold Regions, Calgary, Alb., 1-32.

Aldrich,H.P.:Frost Penetration Below Highway and Airfield Pavement,Bul.135,H.R.B,

Snow Engineering: Recent Advances, Izumi, Nakamura & Sack (eds) © 1997 Balkema, Rotterdam. ISBN 90 5410 865 7

Experimental studies on the prevention of road freezing using electric heat

S. Kobayashi
Research Institute for Hazards in Snowy Areas, Niigata University, Japan

N. Yoshida & T. Yamada
Applied Technology Research & Development Center, Tohoku Electric Power Co., Inc., Sendai, Japan

ABSTRACT: This paper summarizes results of coldroom freezing protection using road heating and some threshold of road freezing were tested on the various parameters such as air temperature, wind speed, underground heat, solar radiation and road surface conditions.

1 INTRODUCTION

It is very important to protect against road freezing for safe driving in winter, especially after use of studded tires for protection against slipping was abolished in 1990 in Japan. This paper summarizes results of coldroom tests to determine various conditions of road freezing protection using road heating. The most advantageous technique for control is use of electric heat wires, however, it incurs a high running cost. So we expect that the technique will only be used where steep slopes or dangerous curves are present.

To determine the economical running conditions of the electrically heated wire under the road surface, a part of a real road was modeled in a coldroom, and thresholds of road freezing was tested by varying air temperature, wind speed, underground heat, solar radiation and road surface conditions (i.e, amount of snow cover).

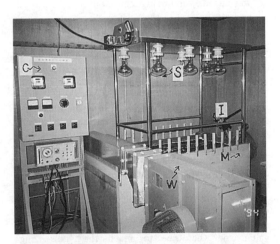

Figure 1 Road model (M) with control panel (C), road heating, underground heat, solar radiation (S), infrared thermometer (I) and blower (W).

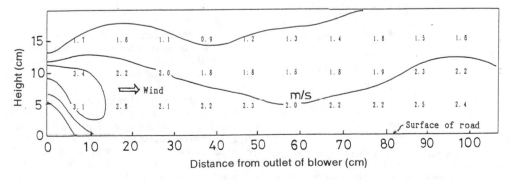

Figure 2 An example of wind distribution above the road surface measured by a small three cup anemometer.

2 METHODS

Figure 1 shows a real road model with 95 ohm electrically heated wires for protection of road freezing and for underground heat, buried at depths from 70 mm to 620 mm from the surface at intervals of 70 mm. The model is 150 mm long, 75 cm wide and 75 cm high. The model is surrounded with 5 cm thick insulation and a wooden plate 2.4 cm thick. To measure temperature profiles, seventy thermocouples were buried in the model. An infrared thermometer and two heat flux meters were used to measure the temperature and heat fluxs at the road surface. Power to the electric wires was controlled automatically. To study wind effect, a blower (200 V, 2.2 kW, 40-80 m³/min) was used. The outlet cross section of the blower was 0.0423 m². An example of wind distribution above the road surface is shown in Figure 2. To research the solar radiation effect, seven infrared lamps (100 V, 500 W) were used as shown in Figure 1.

3 RESULTS

3.1 Relation between air temperature and heating value for protection of road freezing

The temperature of 0°C at the road surface is defined as the "unfreezing" temperature. First, the time to reach 0°C at surface was determined for the case of no snow at the surface and no underground heat. At air temperatures of -2°C, -4°C and -6°C, additional experiments were done with road heating of 100, 150 and 200 W/m², respectively. Figure 3 shows temperature changes at various points in the road for an air temperature of -4°C, heating of 150 W/m², no wind and no snow at the surface. From these experiments the relation between the road heating value and the time to reach the "unfreezing" temperature (0°C) was obtained as shown in Figure 4.

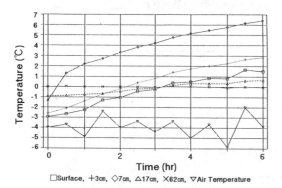

Figure 3 Temperature changes at various points in the road (air temperature, -4°C; road heating, 150 W/m²; no wind; no solar radiation and no snow at the surface).

For example, for an air temperature of -6°C, no wind and no snow above the road, the heat of 200 W/m² was needed during 3.5 hours to protect freezing of water on the surface. With wind blowing above the road, the surface of the road may not reach 0°C. However if there is 3 cm of snow on the surface, the warming rate of the road was somewhat increased. Solar radiation and underground heat act positively to protect against road freezing. In this experiment 1.03 kW/m² of artificial solar radiation was activated for 2 hours under conditions of no wind and wind. Temperature variations without road heating with no wind are shown in Figure 5. Figure 6 shows similar information with a 2 m/s wind. In the case of no wind the surface temperature reached 36°C, with a 2 m/s wind it only reached 22°C. Once the solar heaters were turned off, the surface temperature dropped to 0 °C in 3.5 hours with no wind. That drop only took 1 hour with a 2 m/s wind. Even with 200 W/m² road heating the surface temperature of road decreased with increasing the wind speed.

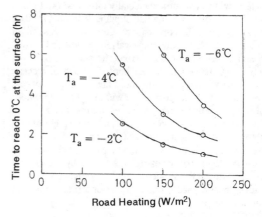

Figure 4 Relation between road heating and time to reach 0°C at the surface for various air temperatures.

3.2 Effects of underground heat and snow

To consider the effect of underground heat, the time to reach 0°C at the surface was determined for underground temperatures of 5°C and 10°C, road heating of 200 W/m², no wind, no solar radiation and no snow on the surface. Results are shown in Figure 7. From Figure 7 the time to reach 0°C at the surface was one hour shorter for an underground temperature of 10°C than an underground temperature of 5°C. When there was about 3.5cm of snow on the surface, the rise of surface temperature was two times faster than with no snow on the surface.

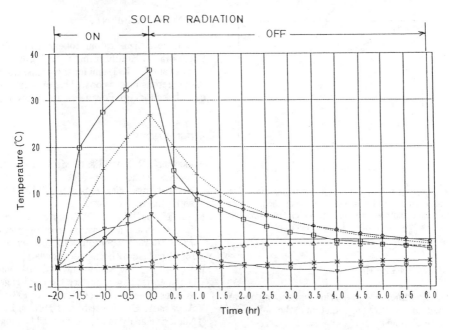

□Surface, +3cm, ◇7cm, △17cm, ×62cm, ▽Air Temperature

Figure 5　Temperature variations without road heating during two hours of solar heating and six hours thereafter, no wind, no snow.

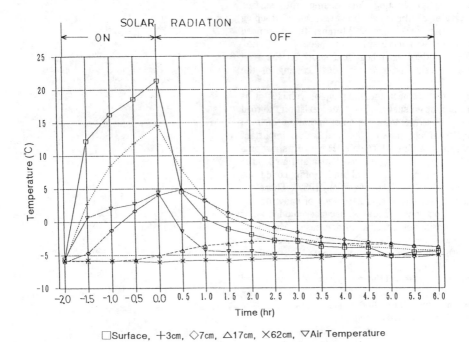

□Surface, +3cm, ◇7cm, △17cm, ×62cm, ▽Air Temperature

Figure 6　Temperature variations without road heating during two hours of solar heating and six hours thereafter, no snow, 2 m/s wind .

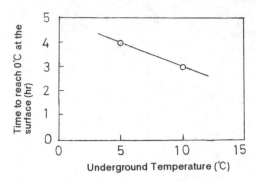

Figure 7 Relation between underground temperature and the time to reach 0℃ at the road surface under conditions of 200 W/m² road heating, no solar radiation, no wind, no snow at the surface and animated air temperature of $-6℃$.

4 CONCLUDING REMARKS

From the experiments with heating and cooling a real road model under various conditions, the following results were obtained:

(1) To protect against freezing, for instance, 200 W/m² of road heating was needed during 3.5 hours for the condition of -6℃ air temperature, no wind, no underground heat and no solar radiation.

(2) Solar radiation was more effective than 200 W/m² of road heating for raising the surface temperature of the road; however, after stopping the radiation, the surface temperature dropped exponentially with increasing wind speed.

(3) Both underground heat and snow on the road provide effective protection against cooling of the road.

To protect against the slipping of vehicles it is important that no snow and no water exist on the road surface, however, for economic operation of road heating, even less snow on the road has an important role of heat loss through the air. However, snow on a road shows a variety of characteristics and structures depending on meteorological and traffic conditions (Maeno et al., 1987; Ishikawa et al., 1987). Therefore we have to study the characteristics of snow on the road under various conditions, because studded snow tires are no longer available to decrease. Moreover, developing instruments such as a motorized snow surface hardness gage is important (Martinelli & Ozment, 1985).

Numerical analysis of these experiments was carried out by Yokoyama et al. (1996) and a satisfactory agreement between the experiments and the numerical analysis was established.

ACKNOWLEDGMENT

The authors thank to our colleagues H. Turumaki, Y. Furukawa, T. Kurebayashi, Y. Ebiko, K. Yamazaki, T. Yakuwa and Dr. K. Izumi for their experimental aid and the numerous discussions. This project was supported by Tohoku Electric Power Co., Inc., Sendai, Japan.

REFERENCES

Ishikawa N., R. Naruse & N. Maeno 1987. Heat balance characteristics of road snow. Low Temperature Science, Ser. A, 46: 151-162 (Japanese with English summary).

Maeno N., H. Narita, K. Nishimura & R. Naruse 1987. Structures and new classification of snow and ice on roads. Low Temperature Science, Ser. A, 46: 119-133 (Japanese with English summary).

Martinelli Jr. M. & A. Ozment 1985. Laboratory tests of a motorized snow surface hardness gage. Cold Regions Science & Technology, 10: 133-140.

Yokoyama T., S. Kobayashi & S. Hirama 1996. Numerical analysis of electric road heating - coupled with full-scaled experiments-. In summary papers of ICSE-3, 145-146.

Snow Engineering: Recent Advances, Izumi, Nakamura & Sack (eds) © 1997 Balkema, Rotterdam. ISBN 90 5410 865 7

Mechanism of ice debonding on an asphalt mixture containing rubber particles

Toyoaki Taniguchi & Yukinori Inaba
Obayashi Road Corporation, Urawa, Japan

Sadanori Murai
Tohoku Institute of Technology, Sendai, Japan

Tatsuo Nishizawa
Ishikawa National College of Technology, Japan

ABSTRACT: A questionnaire survey to roadway agencies showed that ice layers debond from asphalt mixtures containing reclaimed tire rubber particles (AMRP) in moderately cold areas. The mechanism of ice debonding was examined by means of model experiments and analysis by the photoelastic and finite element method (FEM). Model experiments were conducted on enlarged specimens representing an ice layer on the surface of AMRP. The results of the experiment showed that the phenomenon of ice debonding is caused by the existence of rubber particles on the surface. Analysis by the photoelastic method and FEM were conducted to simulate the experiment and calculate the stress in the ice and asphalt layers. The results showed that concentrated stresses, which were not observed without the rubber particle, appeared in the ice layer around the rubber particle. Thus we concluded that the concentrated stresses around the rubber particle initiate radial cracks in the ice and promote debonding of the ice from the asphalt layer. The effects of rubber particle radius, ice layer thickness and density of ice on the concentrated stresses were examined by a parametric study using the FEM model.

1 INTRODUCTION

From a meteorological point of view, about 60% of Japan area is characterized as a so-called "cold and snowy area." For roadway agencies in this region, especially during wintertime, traffic safety has been a major concern for many years. Besides some typical countermeasures, such as a use of de-icing chemicals or installation of heating systems in pavements, asphalt mixtures containing reclaimed tire rubber particles (AMRP) has been practiced successfully in recent years.

It has been observed that ice or compacted snow on a road surface with AMRP easily debonds and can be removed by traffic load applications. Thus, paving with AMRP is considered to be an effective way to provide good safety on wintertime road surfaces. But the mechanism of this phenomenon has not been clearly known and there has been no rational design method for such pavements.

This paper focused on the ice debonding mechanism of AMRP. In order to make the mechanism clear, a condition survey of such pavements, a model experiment, a photoelastic experiment and mathematical analysis by finite element method (FEM) were conducted and the results obtained were discussed.

2 QUESTIONNAIRE SURVEY

2.1 Questionnaire to Roadway Agencies

In order to provide information pertaining to the environmental condition of pavements, the following questions were asked to roadway agencies who have already adopted AMRP. 130 sites were selected from projects completed from 1981 to 1994.
- Traffic volume of heavy vehicles per day in one direction.
- Geographical conditions.
- Average minimum air temperature (the average of daily minimum air temperatures from December 1st to March 31st).
- Maximum depth of snow cover.
- Type of ice and snow deposits on the surface.

Table 1 Result of Multivariable Analysis

Factor	Category	Score	Range	Sample
Traffic volume of heavy vehicles (Daily, one direction)	#1; < 100	0.087	0.26	18
	#2;100<250	0.037		25
	#3;250<1000	0.055		28
	#4;1000<3000	-0.168		24
Average minimum air temperature	#1;+2>-5	-0.638	2.19	55
	#2;-6>-10	0.781		35
	#3;-11>-15	1.549		5
Maximum depth of snow cover	#1;0<9	-0.537	0.98	15
	#2;10<19	-0.127		48
	#3;20<49	0.442		32

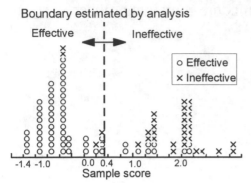

Figure 1. Distribution of sample score

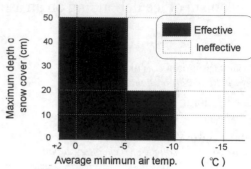

Figure 2. Meteorological conditions in which the ice-removal effect can be expected.

- % of rubber particles mixed in the asphalt mixture.
- Ice debonding effect judged by the agency.

2.2 Analysis of the survey results

Quantification theory was employed to make a statistical analysis of information obtained by the survey. The purpose was to assess factors dominating the ice debonding effect in asphalt pavements with approximately 3% rubber particles. Quantification theory is one of the measures of multivariable analysis and has the benefit of dealing with qualitative factors such as a yes/no judgments of the ice debonding effect.

A relationship between a qualitative factor of the ice debonding effect and the other above-mentioned influencing factors were examined. As a result, it is found that "traffic volume of heavy vehicles," "average minimum air temperature" and "maximum depth of snow cover" have strong influences on the ice debonding effect. When these three factors and the outside criterion, "ice debonding effect," were re-examined, the results shown in Table 1 was obtained. A correlation ratio of 0.562 and a degree of discriminant of 87.4% mean that this analysis has sufficient reliability. Since the value of 1 and 2 is given to quantify "effective" and "ineffective" respectively, a smaller category score means a stronger positive influence on the outside criterion "ice debonding effect." Also a larger value of "range," which is the difference between a maximum and minimum category score, indicates a stronger influence on the criterion. Figure 1 shows a distribution of sample score and actual judgment for each site. From Table 1 and Figure 1, the following results can be drawn.

- The most influencing factor on "ice debonding effect" is "average minimum air temperature". A smaller value of "maximum depth of snow cover" or a larger value of "traffic volume of heavy vehicles" also gives a better circumstances for "ice debonding effect."

- A sample score of 0.4 in Figure 1 is a boundary value of effectiveness estimated from multivariable analysis, and this discriminant boundary shows good agreement with a distribution of actual judgment for each site.
- Using the boundary value of 0.4 and the category scores in Table 1, a combination of factors and categories in which "ice debonding effect" can be expected are illustrated as shown in Figure 2. Meteorological conditions in which the ice debonding can be expected is as follows.
  - ✓ A region with an average minimum air temperature higher than -10 °C and a maximum depth of snow cover of less than 20 cm.
  - ✓ A region with an average minimum air temperature higher than -5 °C and a maximum depth of snow cover of less than 50 cm.
- The traffic volume of heavy vehicles has less influence on the ice debonding effect than the meteorological conditions and is neglected in Figure 2.

3 MODEL EXPERIMENT

3.1 Description of Model

It is believed that the protruding rubber particles

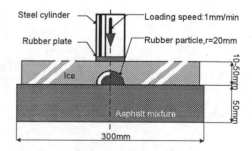

Figure 3. Specimen of the model experiment.

380

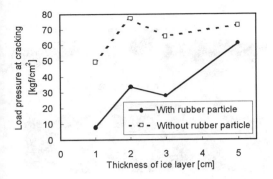

Figure 4. Relationship between the thickness of the ice layer and the load at cracking.

greatly contribute to the ice debonding effect on an actual pavement surface. In order to simplify and observe the pavement surface, rubber particle and ice, a model experiment was conducted using an enlarged specimen. Figure 3 shows the configuration of the enlarged specimen. A rubber particle with a radius of 2 cm was embedded between an asphalt mixture layer and an ice layer. A load was applied at 1 mm/min to the surface of the ice layer, and cracking in the ice layer was observed. The thickness of the ice layer was varied from 10 cm to 50 cm.

3.2 Results

Figure 4 shows the relationship between the load required to initiate cracks in the ice layer and the thickness of the ice layer. It is obvious that a smaller load is required to initiate cracks in a model with the rubber particle than in the model without the rubber particle. The load required to crack thin ice is smaller than that required to crack thicker ice. In a thin 1 cm ice layer, cracks do not reach the outer edge of the ice layer. On the other hand, when the thickness of the ice layer is considerably larger than the radius of the rubber particle, the load required to

crack the ice above the rubber particle is almost identical to that in the model without the rubber particle.

Observation of cracking in the ice layer is summarized in Table 2. In terms of "easy to remove" the ice from the pavement surface, the following characteristics are observed.

- Regardless of the existence of the rubber particle or the thickness of the ice layer, radial cracks occur in the ice layer at peak load.
- In the models with rubber particles, the bonding between the ice layer and the pavement surface is lost after cracks occur and the ice layer can be easily removed. However, in the models without rubber particles, adhesion still exists and the ice can't be removed even after the cracks reach the outer edge.

From the above results, the phenomenon of ice debonding is a function of several factors, such as existence of rubber particles on the pavement surface, size of such particles, thickness of the ice layer and loads on the pavement surface.

4 PHOTOELASTIC EXPERIMENT

4.1 Description of Model

From the results of the model experiment mentioned above, it was recognized that the existence of rubber particles is a main cause of crack initiation in an ice layer on the surface of an asphalt mixture. In order to examine the stress state around the rubber particle, photoelastic experiments were conducted. The photoelastic experiment has the advantage of being able to express the stress concentration visually. Another advantage of this experiment is use of the same rubber particles and epoxy resin with an elastic modulus similar to that of the asphalt pavement.

Figure 5 shows the configuration of the model used in this experiment. The epoxy resin is 10 mm thick. A cylindrical rubber particle with a diameter of 10 mm is embedded in the epoxy resin near the surface. The distance from the surface to the top of

Table 2 Observation of cracking in ice layer

Loading	Particle	hi (cm)	Cracking	Ice debonding or ice removability
Center	Yes	5	5 radial cracks to outer edge	Easy to remove but a portion still bonded
Center	Yes	3	4 radial cracks to outer edge	Easy to remove and all portions de bonded
Center	Yes	2	3-4 radial cracks to outer edge	Half debonded
Center	Yes	1	Ice broken around the particle	Top half of ice broken but interface still bonded
Center	No	5	4 radial cracks to outer edge	All bonded
Center	No	3	2 diametrical cracks to the edge	All bonded
Center	No	2	2 diametrical cracks to the edge	All bonded
Center	No	1	Radial cracks not to the edge	All bonded
Offset	Yes	3	Radial cracks at the bottom	Easy to remove the cracked portion
Offset	Yes	2	Half-round cracks at the bottom	Thin layer peeled off and interface bonded
Offset	Yes	1	Cracks around the loaded area	Thin layer peeled off and interface bonded

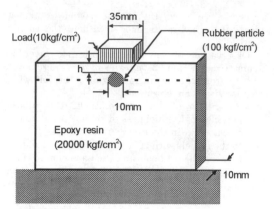

Figure 5. Model used in the photoelastic experiment.

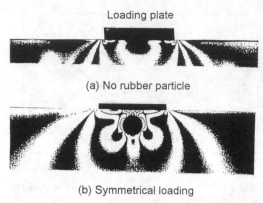

(a) No rubber particle

(b) Symmetrical loading

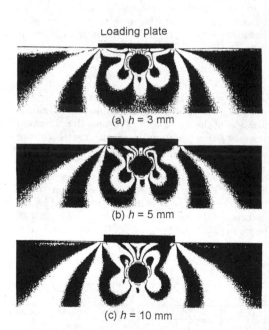

(c) Asymmetrical loading

Photograph 1. Isochromatic lines around a particle.

the rubber particle, *h*, was varied from 3 mm to 5 mm and 10 mm. A distributed load with width of 35 mm was applied on the surface symmetrically and asymmetrically to examine the effect of loading position on the stress states in the pavement and ice layers. The magnitude of the load was almost equal to the tire pressure of commercial vehicles (10 kgf/cm²). The elastic modulus of materials used in this experiment are also presented in the figure. Although the elastic modulus of ice is different from the epoxy resin, we used the same photoelastic materials for the ice and the asphalt mixture to simplify the preparation of the specimen.

4.2 Results

The isochromatic fringe order obtained by photoelastic experiments is proportional to the absolute value of the difference between principal stresses. Accordingly, the fringe pattern of the isochromatic lines expresses the distribution of shear stress. This relationship can be expressed by the following equation.

$$N = \alpha \cdot d \left| \sigma_1 - \sigma_2 \right|$$

N :isochromatic fringe number
α :photoelastic sensitivity
d :thickness of epoxy resin
σ_1, σ_2 :principal stresses

Photograph 1 shows isochromatic lines under symmetrical and asymmetrical loading in the models with and without rubber particles. It can be seen that large stress concentrations appear around the rubber particle. The positions of the largest stresses are different for the symmetrical and asymmetrical loading. Under the symmetrical loading, the maximum stresses occur at both sides of the rubber particle. Under the asymmetrical loading, large stresses occur at the lower side and upper side of the rubber particle. This means that the locations

Photograph 2. Effect of *h* on stress states for symmetrical loading

at which the maximum stress occur depend on positions of traffic loading. These large stresses around the particle might initiate cracks and cause breakdown of the ice.

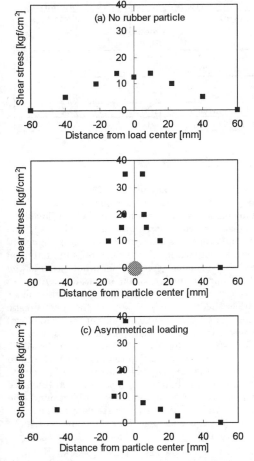

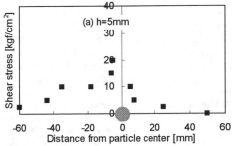

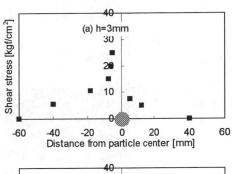

Figure 7. Stress distributions under asymmetrical loading

Figure 6. Stress distributions along the interface (h = 2 mm)

Figure 6 shows the stress distributions along the horizontal line where the largest stress occurs ($h = 3$ mm, symmetrical loading). In the model with the rubber particle, the stress increase is about 2.5 times

higher than in the model without the rubber particle.

Photograph 2 shows the effect of h on the stress state around the rubber particle under symmetrical loading. The magnitude of stress around the rubber particle changes very little when h changes from 3 mm to 5 mm.

Figure 7 shows the stress distribution along the horizontal line under the asymmetrical loading. The stresses around the rubber particle increase as h decreases.

5 FEM ANALYSIS

5.1 FEM model

In order to investigate stresses around the rubber particle under various conditions, we employed an axi-symmetric finite element model (Owen and Hinton, 1989) as shown in Figure 8. In the model, an ice layer was placed on an asphalt mixture layer with a thickness of 150 mm. The thickness of the ice

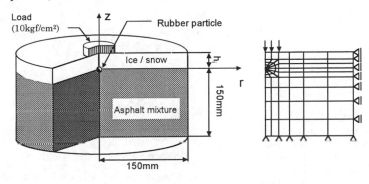

Figure 8. Axisymmetric finite element model.

383

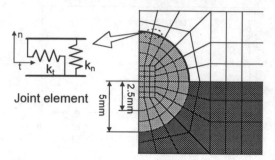

Figure 9. Detailed mesh for a rubber particle.

Table 3. Elastic properties used in FEM analysis.

Material	Young's modulus [kgf/cm²]	Poisson's ratio
Ice/snow (density=500 kg/m³)	20,000	0.35
Ice/snow (density=800 kg/m³)	90,000	0.35
Rubber particle	100	0.49
Asphalt mixture	50,000	0.35

layer, h_i, was varied from 7.5 to 50 mm. A rubber particle was embedded between the ice layer and the asphalt mixture layer. Two sizes of the rubber particle, with radiuses, r, of 2.5 and 5 mm, were assumed.

Nodal points on the cylindrical boundary were radially fixed but vertically free, whereas those on the bottom boundary were radially and vertically fixed. A circular uniform load was applied at the symmetric axis. Its magnitude and radius were 10 kgf/cm² and 20 mm, respectively, corresponding to the photoelastic experiment mentioned above.

Detailed mesh layout around the rubber particle is illustrated in Figure 9. Goodman et al.'s (1968) joint elements were inserted on the surface of the rubber particle to represent the mechanistic interaction between ice and rubber. The joint element has two elastic springs, one a shear spring and the other a normal spring, representing sliding and opening/closing at the interface between the ice and rubber, respectively.

5.2 Properties of materials

In this study, since low temperatures and rapid loading conditions were considered, the ice, rubber and asphalt mixture were assumed as linear elastic materials. Values of elastic constants used in the analysis are summarized in Table 3. Two types of ice were considered. One was an ice with a high density of 800 kg/m³. The another was an ice with a low density of 500 kg/m³ which represents snow compacted by traffic loading. Since the elastic modulus of ice is supposed to depend on its density, the elastic modulus of ice was determined taking the density of ice into account (Maeno and Fukuda 1986).

Two contact conditions on the interface between the ice and rubber were considered. One was a smooth contact condition in which $k_t = 100$ kgf/cm³ and the another was a cohesive contact condition in which $k_t = 10^6$ kgf/cm³. The value of the normal spring stiffness was kept constant at $k_n = 10^6$ kgf/cm³ if in contact and $k_n = 0$ kgf/cm³ if open.

5.3 Stress distributions

To check the accuracy of the FEM model, a stress analysis was carried out for the condition of the photoelastic experiment previously mentioned. A plane-stress condition was assumed and the material properties in Figure 5 were used. Figure 10 shows a comparison of experimental and analytical results of principal stress differentials around the rubber particle under symmetric loading. A stress concentration appears near the surface of the rubber particle in the analysis as in the photoelastic experiment. Thus, it can be said that the FEM model employed in this study has reasonable accuracy.

The stress states around a rubber particle with r = 2.5 mm are shown in Figures 11a and 11b. Figure 11a shows the direction and magnitude of principal stress. From the figure, it can be seen that the large compressive stresses occur vertically around the particle, and radial tensile stresses appear near the top of the rubber particle. Since ice can't withstand excessive tensile stresses, the tensile stress, σ_r, here is the critical stress. Figure 11b shows contours of the maximum shear stress around the rubber particle. A concentration of shear stress can be observed near the right side of the rubber particle in the bottom of the ice layer. The maximum shear stress, τ_{max}, here is another critical stress. Since

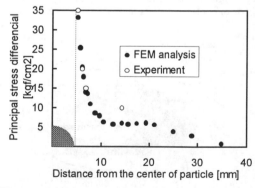

Figure 10. Comparison between the results of the photoelastic experiment and the FEM analysis.

384

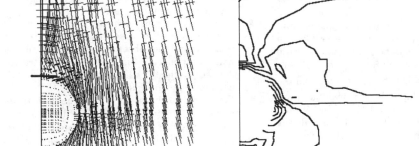

(a) Principalstresses   (b) Shear stress contour

Tensile
Compressive

Figure 11. Stress distribution around the rubber particle(r = 2.5mm)

Figure 12a shows the relationship between the radial stress near the top of the rubber particle, σ_r, and the thickness of the ice layer, h_i. In the case of a thinner ice layer, stresses in the region near the load are generally compressive and thus, σ_r is also compressive. When the ice layer becomes thicker, σ_r turns to be a tensile stress. It reaches a maximum value at $h_i = 30$ mm.

varying the shear stiffness of the joint element made little change in the stress state in the ice layer, the shear spring stiffness was kept constant ($k_t = 100$ kgf/cm³) in the following calculations.

Stresses in the high density ice are greater than those in the low density ice because the high density ice has high elastic modulus. Although stresses in the case of the rubber particle with $r = 5$ mm are greater than the one with $r = 2.5$ mm, the difference

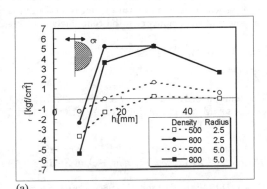

(a)

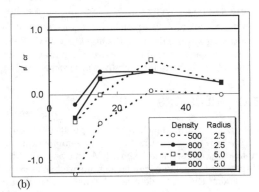

(b)

Figure 12. Relationship between σ_r and thickness of ice layer.

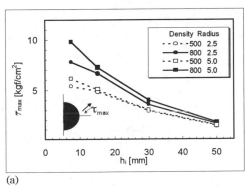

(a)

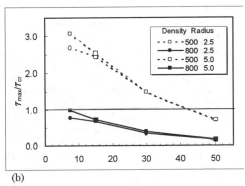

(b)

Figure 13. Relationship between τ_{max} and thickness of ice layer.

is small in high density ice.

Tensile strength, σ_{cr}, of ice depends on the density and varies from 3 kgf/cm² for low density ice to 15 kgf/cm² for high density ice (Maeno and Fukuda 1986). So, the possibility of ice breakdown due to the tensile stress can be evaluated by a ratio of tensile stress to tensile strength, σ_r / σ_{cr}. Figure 12b shows the relationship between σ_r / σ_{cr} and the ice thickness. In the high density ice, the ratio reaches the maximum value of 0.35 at h_i = 15 to 30 mm. For the low density ice, the maximum value is 0.56 at h_i = 30 mm in the case of the rubber particle with r = 5.0 mm.

Figure 13a shows the relationship between the maximum shear stress near the right side of the rubber particle in the bottom of ice layer , τ_{max}, and h_i. It can be seen that τ_{max} decreases proportionally with increasing h_i and that there is little difference due to particle radius.

The shear strength of ice τ_{cr} also depends on the density and varies from 2 kgf/cm² for low density ice to 10 kgf/cm² for high density ice (Maeno and Fukuda 1986). The ratio τ_{max} / τ_{cr} can be an indicator of the possibility of the breakdown of ice due to shear stress. Figure 13b shows the relationship between τ_{max} / τ_{cr} and h_i. In the low density ice, τ_{max} / τ_{cr} exceeds 1.0 in the range of h_i less than 40 mm, and thus, the breakdown can occur in the range. In the high density ice, however, the ratio is less than 1.0 for each thickness considered.

6 CONCLUSIONS

The ice debonding mechanism of asphalt pavements with AMRP was examined by a questionnaire survey, laboratory experiments and FEM analysis. Results obtained in this study can be summarized as follows:
- From the data obtained by field survey, the meteorological conditions in which the ice debonding effect can be expected were derived as follows.

   A region with an average minimum air temperature of higher than -10 ℃ and a maximum depth of snow coverage of less than 20 cm. Also a region with an average minimum air temperature of higher than -5℃ and a maximum depth of snow coverage of less than 50 cm. Thus at relatively lower air temperatures (probably the elastic modulus of ice layer would be higher) the ice debonding effect can be expected for thinner ice layer, whereas at a bit higher temperature (the modulus of ice would be lower) the effect can be expected for thicker layer
- From the model experiment, it was found that the existence of the rubber particle on the

surface of pavement layer plays a very important role since it causes a breakdown of the ice layer on the surface.
- From the results of the photoelastic experiment, it was recognized that stress concentrations occur around the rubber particle, and the degree and position of the stress concentrations depends on the thickness of the ice layer and the position of applied load.
- From the results of the FEM analysis, it was found that the tensile stress near the top of the rubber particle and the shear stress near the right side of the rubber particle are critical stresses. The possibility of ice breakdown due to the tensile stress is maximum at ice thickness of 15 mm to 30 mm. On the other hand, the possibility of the breakdown of ice due to the shear stress is higher in the thin ice layer than in thick one. Breakdown due to shear stress is more likely in the low density ice.

The ice debonding effect of AMRP is due to the existence of the rubber particles on the pavement surface. Therefore, to make the ice debonding more effective, the number of the rubber particles on the pavement surface must be increased. And also, the effectiveness of the ice debonding of AMPR is limited to areas where air temperature is not so low and the thickness of the ice layer on the road surface is less than 40 mm.

REFERENCES

Goodman, R.E, Taylor, R.L. and Brekke, T.L. 1968. A Model for Mechanics of Joint Rock, *Proc. of ASCE*, vol.94, SM3, pp.637-659.

Maeno, K. and Fukuda, M. 1986. *Structure and Characteristics of Ice and Snow*, Kokon-Syoin.

Owen, D.R.J. and Hinton E. 1980. *Finite Elements in Plasticity, Theory and Practice*, Pineridge Press Ltd.

Snow Engineering: Recent Advances, Izumi, Nakamura & Sack (eds) © 1997 Balkema, Rotterdam. ISBN 90 5410 865 7

Transport rate of snow floating in an open channel

Yusuke Fukushima
Nagaoka University of Technology, Niigata, Japan

ABSTRACT: The transport rate of snow floating in a deep open channel is discussed based on theory and experiment. Snow, lighter than water, flows in the upper part of the channel, and water flows in the lower part. A mixture of snow and water occupies the upper part where the velocity is almost uniform. Assuming that the mixture is a Bingham fluid, the velocity profile is derived, and the maximum flow rate of the snow i.e, the transport rate is calculated. The theory and the experimental results are in reasonable agreement. This theory is applicable when the slope of the channel is mild and the water is deep.

1. INTRODUCTION

Heavy snowfall on the Japan Sea side of Honshu Island, Japan, makes snow removal necssary in residential area. Various snow removal systems are widely used in this area, where the weather is relatively mild in winter. Typical snow removal methods are snowplowing, road heating, "Shosetsu pipe" and "Ryusetsuko". The "Shosetsu pipe," which is laid under the surface of roads, is a snow-melting method using a pipe network supplied with warm underground water. The Ryusetsuko is a snow removal system using open channels. The channels look like ditches arranged on sides of road. River water flows in the Ryusetsuko carrying snow lumps with it. All snow removal systems including the Ryusetsuko, have merits and demerits. To evaluate a system, it is necessary to take into consideration the properties of local snowfall, economics, social issues and environmental concerns.

The advantages of the Ryusetsuko are low running costs, the avoidance of environmental problems such as ground sinkage due to the pumping up of underground water, the psychological relief caused by snow removal work at the scene, and the rise of the residents' consciousness through neighborhood collaboration.

On the other hand, several problems related to the Ryusetuko should be solved in order to facilitate its use. The rate at which snow can be removed, should be related to the amount of water supplied in the open channel. This rate will de-pend on snow properties such as density. Effects of a channel slope on the snow discharge should be explained. Also the mechanism of the increase of water depth caused by the snow taransport will be clarified.

In recent years, more Ryusetsukos are being built in mild slope areas. In the past they have been used mostly in areas where the channel slope is steep.

There have been several studies on the Ryusetsuko, which are mainly experimental: for example, Hokuriku Branch of Japan, Ministry of Construction (1983) and Ohkuma et al. (1989). Fukushima et al. (1993) introduced formulae for the snow transport rate of the Ryusetsuko. Also, there are several theoretical investigations concerning the Ryusetsuko, for example, Sato and Shuto (1983), Sasaki et al. (1993) and the author's works (Fukushima, 1990, Fukushima et al., 1992, 1993, 1994). Sato and Shuto (1983) derived the velocity distribution in the Ryusetsuko assuming that the mixture of snow and water is a Bingham fluid, and that the eddy viscosity is uniform in the flow section. The velocity distribution shown in their paper is parabolic in the water part. Sasaki et al. (1993) also derived the velocity distribution in a pipe, assuming that the mixture of water and snow is a dilatant fluid. However, a formula for snow transport rate has not been derived in their studies.

The author (Fukushima et al., 1993) has presented an analytical model of the flow rate of snow

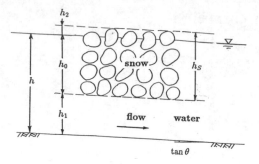

Figure 1: Side view of "Ryusetsuko" and symbols.

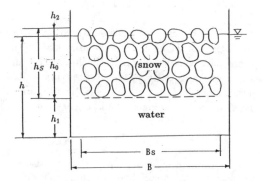

Figure 2: Cross sectional view of "Ryusetsuko" and symbols.

and water in the Ryusetsuko, assuming that the snow chunks touch the bed of the channel. This theory is applicable to the maximum flow rate of snow and the critical conditions of snow transport in the case of a steeply sloped channel. However, the theoretical formulae cannot be applied in the mild slope case because the assumption of snow chunks touching the bed may not be satisfied in this case. The depth of water increases in the case of the mild slope channel with snow chunks floating in the Ryusetsuko.

In this paper, the velocity distribution in a Ryusetsuko will be derived under the assumption that the mixture of snow and water is a Bingham fluid. The formula for the transport rate of snow (maximum flow rate of snow) can be obtained from the velocity distribution and the continuity equation. The characteristics of the present formula are checked by the relationships between the flow rate of snow and the water discharged for a wide range of snow densities, and between the flow rate of snow and the flow depth. Finally, the formula for the transport rate of snow is applied to the experimental data.

2. THEORY OF FLOW RATE OF SNOW

2.1 Distribution of shear stress in the section

Consider a mixture of water and snow, whose constituent are rather small compared with the depth of water, flowing in an open channel. ρ_i denotes the density of the snow pieces (i.e., the density of pure ice,) and ρ_w denotes the density of water. Because ρ_i is smaller than ρ_w, the snow pieces float in the water. Schematic views of the flow in the Ryusetuko are depicted in Figure 1 and Figure 2. As shown in Figure 2, the width of the mixture, B_S, is smaller than the channel width, B. In these figures, h is the depth of water, h_S is the layer thickness of the mixture, h_0 is the snow thickness under the water surface, and h_2 is the snow thickness in the air. Thus, the water depth under the snow layer is expressed by $h_1 = h - h_0$. x denotes the coordinate in the flow direction and z denotes the coordinate perpendicular to x-direction as shown in Figure 3. Here we assume that the flow is uniform in the x-direction. Thus, the discharges of both water and snow are constant and the depths of both water and snow are constant in the x-direction. An additional assumption is that the snow chunks are porous with a uniform porosity λ. The bulk specific gravity of snow, γ_S, is expressed by ρ_i and ρ_w as follows:

$$\gamma_S = \rho_i(1 - \lambda)/\rho_w \qquad (1)$$

Since the snow pieces are in water, we assume that the porous part of the snow will be filled with water. The bulk density of the snow in water is expressed by the following relation:

$$\rho_s = \rho_i(1 - \lambda) + \rho_w\lambda \qquad (2)$$

Sato and Shuto (1983) have assumed that the flow is uniform in the lateral direction, i.e. the flow is two-dimensional, when the mixture of water and snow is a Bingham fluid. The same assumptions are adopted in this paper. The shear stress of the mixture τ, the critical shear stress τ_c, and the Reynolds stress $-\rho_w\overline{u'w'}$ are related as follows:

$$\tau = \tau_c - \rho_w\overline{u'w'} \qquad (3)$$

In the relation above, the shear stress caused by the molecular viscosity is neglected where turbulence is fully developed, and the Reynolds stress is dominant. Under the assumption that the flow is uniform in the flow direction x, the shear stress and the gravitational force are in balance. Thus the distribution of shear stress in the channel (region (a) in Figure 3) is obtained.

The shear stress becomes critical at the bottom of the snow layer. Thus, the critical shear stress is

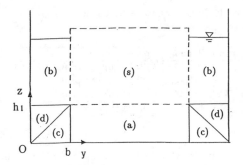

Figure 3: Regions of cross section of Ryusetsuko.

obtained as follows:

$$\tau_c = \rho_w u_{*c}^2 = \rho_w u_{*1}^2 \left\{ \lambda \left(\frac{h}{h_1} - 1 \right) + (1 - \lambda) \frac{\rho_i h_S}{\rho_w h_1} \right\} \tag{4}$$

where u_{*1} is the shear velocity defined by:

$$u_{*1} = (g \sin \theta h_1)^{1/2} \tag{5}$$

The shear stress at the bottom is obtained as follows:

$$\tau_0 = \rho_w u_{*0}^2 = \rho_w (u_{*1}^2 + u_{*c}^2) \tag{6}$$

where u_{*0} is the shear velocity at the bottom. Subtracting Eq. (4) from Eq. (6), the difference between the shear stresses $\tau_o - \tau_c$ is obtained as follows:

$$\tau_o - \tau_c = \rho_w u_{*0}^2 - \rho_w u_{*c}^2 = \rho_w u_{*1}^2 \tag{7}$$

where u_{*c} is the critical shear stress and is expressed by Eq. (4).

2.2 Velocity distribution in the Ryusetsuko

In order to derive the velocity distribution in the Ryusetsuko, the same assumptions and procedures used to derive the velocity distribution in the homogeneous fluid are adopted. First, introducing the mixing length l, l is assumed to be in proportion to the distance from the solid boundary, as follows:

$$l = \kappa z \tag{8}$$

where κ is the von Kármán constant. The shear stress is constant in the water, i.e. a constant stress layer. Thus Eq. (3) becomes:

$$-\rho_w \overline{u'w'} = (\tau - \tau_c) \dot{=} (\tau_o - \tau_c) \tag{9}$$

Using the eddy viscosity K_m, the Renolds stress $-\rho \overline{u'w'}$ is expressed as follows:

$$-\rho_w \overline{u'w'} = \rho_w K_m \frac{\partial u}{\partial z} = \rho_w l^2 \left| \frac{\partial u}{\partial z} \right| \frac{\partial u}{\partial z} \tag{10}$$

Using the relations of Eqs. (8) and (9), Eq.(10) becomes:

$$\frac{du}{dz} = \frac{u_{*1}}{\kappa z} \tag{11}$$

Equation (11) is integrated with z and after some manipulations Eq. (11) becomes:

$$\frac{u}{u_{*1}} = \frac{1}{\kappa} \ln \frac{z}{k_S} + Ar \tag{12}$$

where k_S is the equivalent roughness at the bottom and Ar is a constant. Relation (12) is valid in the range of $k_S \leq z \leq h_1$. Eq. (12) is called the 'log-law' of velocity distribution for region (a) in Fig. 3.

The velocity of the snow layer (region (s) in Figure 3) is expressed by Eq. (12) as $z = h_1$ as follows:

$$u_S = u_{*1} \left(\frac{1}{\kappa} \ln \frac{h_1}{k_S} + Ar \right) \tag{13}$$

This relation is valid in the range of $h_1 \leq z \leq h + h_2$.

A similar log-law is considered applicable to $k_{S2} \leq y \leq b$ (region (b) in Figure 3) as follows:

$$\frac{u}{u_{*2}} = \frac{1}{\kappa} \ln \frac{y}{k_{S2}} + Ar_2 \tag{14}$$

Next, we consider the velocity distribution in the triangular region as shown in Figure 3. The velocity is smoothly connected in regions (a) and (c). Thus, the velocity distribution in region (c) is assumed to be Eq. (12). In a similar manner, the velocity distribution in region (d) is assumed to be Eq. (14). It is noted that Eq. (12) and Eq. (14) are not independent of each other, because the velocity should be continuous on the boundary between region (c) and region (d). Substituting $z = h_1 y / b$ into Eq. (12), the following equation can be obtained:

$$\frac{u}{u_{*1}} = \frac{1}{\kappa} \ln \frac{y}{k_{S2}} + \frac{1}{\kappa} \ln \frac{k_{S2}}{k_S} \frac{h_1}{b} + Ar \tag{15}$$

As Eq. (15) is equal to Eq. (14), the following condition should be satisfied.

$$u_{*1} = u_{*2} \tag{16}$$

$$\frac{1}{\kappa} \ln \frac{k_{S2}}{k_S} \frac{h_1}{b} + Ar = Ar_2 \tag{17}$$

2.3 Discharge of water and snow

The water discharge Q_w and the flow rate of snow Q_{SM} can be derived from the velocity distributions for every regions shown in Figure 3. The water discharges, Q_{w1}, Q_{w2}, Q_{w3}, and Q_{w4} to regions (a), (b), (c) and (d), respectively, are obtained by the integration of the velocity distributions.

$$Q_{w1} = B_S \int_0^{h_1} \left\{ u_{*1} \left(\frac{1}{\kappa} \ln \frac{z}{k_S} + Ar \right) \right\} dz \quad (18)$$

$$Q_{w2} = h_0 \int_0^b \left\{ u_{*2} \left(\frac{1}{\kappa} \ln \frac{y}{k_{S2}} + Ar_2 \right) \right\} dy \quad (19)$$

$$Q_{w3} = \int_0^b \int_0^{h_1 y/b} \left\{ u_{*1} \left(\frac{1}{\kappa} \ln \frac{z}{k_S} + Ar \right) \right\} dz dy \quad (20)$$

$$Q_{w4} = \int_0^{h_1} \int_0^{bz/h_1} \left\{ u_{*2} \left(\frac{1}{\kappa} \ln \frac{y}{k_{S2}} + Ar_2 \right) \right\} dy dz \quad (21)$$

The water discharge Q_{w0} and the flow rate of snow Q_{SM} for region (s) are obtained as follows:

$$Q_{w0} = B_S h_0 u_{*1} \lambda \left(\frac{1}{\kappa} \ln \frac{h_1}{k_S} + Ar \right) \quad (22)$$

$$Q_{SM} = B_S h_S u_{*1} (1 - \lambda) \left(\frac{1}{\kappa} \ln \frac{h_1}{k_S} + Ar \right) \quad (23)$$

The average velocity of the snow layer is expressed by Eq. (13) as follows:

$$u_S = u_{*1} \left(\frac{1}{\kappa} \ln \frac{h_1}{k_S} + Ar \right) = u_{*1} \overline{u_S} \quad (24)$$

where $\overline{u_S}$ is the nondimensional snow layer velocity defined as:

$$\overline{u_S} = \frac{1}{\kappa} \ln \frac{h_1}{k_S} + Ar \quad (25)$$

Using the relation Eq.(25), Eqs. (22) and (23) becomes:

$$Q_{w0} = u_{*1} B_S h_0 \lambda \overline{u_S} \quad (26)$$

$$Q_{SM} = u_{*1} B_S h_S (1 - \lambda) \overline{u_S} \quad (27)$$

The total water discharge is given by the summation of the water discharges of regions (s), (a), (b), (c) and (d) as follows:

$$Q_w = Q_{w0} + Q_{w1} + 2(Q_{w2} + Q_{w3} + Q_{w4})$$

$$= u_{*1} \overline{u_S} (B_S h_1 + 2bh + B_S h_0 \lambda)$$

$$- u_{*1} \frac{1}{\kappa} (B_S h_1 + 3bh_1 + 2bh_0) \quad (28)$$

Multiplying Eq. (27) by the density ρ_i, the mass flow rate of the snow, Q_{SM}, is obtained. It is convenient to use the volume flow rate of snow, Q_s, instead of Q_{SM}. Q_s is defined by Q_{SM} divided by $(1 - \lambda)$ as follows:

$$Q_s = u_{*1} B_S h_S \overline{u_S} \quad (29)$$

Table 1: Experimental conditions. (a) new snow (b) compacted snow (c) granular snow.

run	Slope i_0	Specific gravity γ_S
1	1/500	(a)0.18 (b)0.38 (c)0.55, 0.57
2	1/1000	(a)0.17 (b)0.45 (c)0.55
3	1/2000	(a)0.22 (b)0.41 (c)0.53
4	1/1000	(a)0.28 (b)0.38

The ratio of the flow rate of snow to the water discharge is derived from Eqs. (28) and (29) as follows:

$$Q_s/Q_w = \overline{u_S} B_S h_S / \{ \overline{u_S} (B_S h_1 + \lambda B_S h_0 + 2bh)$$

$$- 1/\kappa (B_S h_1 + 3bh_1 + 2bh_0) \} \quad (30)$$

Finally, the water discharge Q_w can be calculated from Eq. (28), and the flow rate of snow Q_s can be calculated from Eq. (29). It is noted that the bed slope $\sin \theta$ is included only in the shear velocity u_{*1}. Thus, the slope angle θ is not included in the ratio of the flow rate of snow to the water discharge, Q_s/Q_w, which is the function of h_1, k_S, B_S, b and λ .

3. EXPERIMENTS ON THE MAXIMUM FLOW RATE OF SNOW

The experimental apparatuses used in this study are as follows. Channel 1 in Figure 4 is 40 cm wide and 40 cm deep with a valuable slope of from 1/500 to 1/2000. The bottom of channel is covered with sand, and the side walls are made of plastic. Channel 2 in Figure 5 is 25 cm wide and 40 cm deep with a slope of 1/1000. The bottom and side walls of channel 2 are all made of painted veneer plates.

Experiments are carried out as follows. Snow of known volume and weight is thrown into the upstream end of the channel. The snow used in the experiments ranged from (a) new snow ($\gamma_S \approx 0.2$), (b) compacted snow ($\gamma_S \approx 0.4$) to (c)granular snow ($\gamma_S \approx 0.6$) The experimental conditions are shown in Table 1.

4. COMPARISON OF THEORETICAL AND EXPERIMENTAL RESULTS

4.1 *Relationship between the thickness of the snow layer and its flow rate*

In this section, the flow rate of floating snow in the Ryusetsuko is discussed. The flow rate of snow

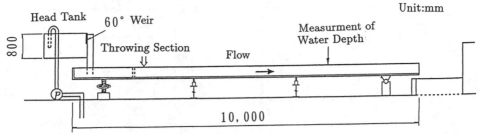

Figure 4: Experimental apparatus (channel 1).

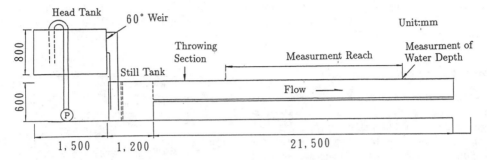

Figure 5: Experimental apparatus (channel 2).

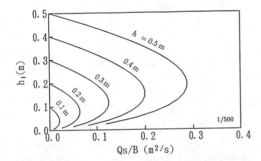

Figure 6: Relationship between water depth and transport rate of snow per unit width ($i_0 = 1/500$, $\beta = 0.8$).

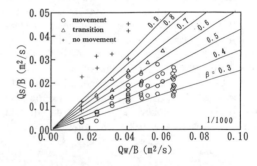

Figure 7: Relationship between the transport rate of snow per unit width and the water discharge per unit width, (channel 2, run 4a.)

varies with the thickness of the snow layer. If the thickness of the snow layer is zero, the flow rate of the snow will be zero, obviously. If the thickness of the snow layer is large, the thickness of the water layer will be small because of the small water velocity. Thus, the flow rate of snow will also become small in this case. Figure 6 shows the variation in the flow rate of snow per unit width in relation to the thickness of the snow layer in channel 1 ($B =25$cm, $i_0 = 1/500$). The flow rate of the snow can be calculated by using the thickness of the snow layer in water as a parameter. In

this calculation, the nondimensional snow width $\beta = (B_S/B)$ is 0.7. It is shown in Figure 6 that the maximum flow rate of snow appears around the water thickness h_1 of 0.3-0.4 of the whole depth h. The relation of Q_s and h_0 is not dependent on the specific gravity of the snow, γ_S, as derived from Eq. (29) and is only a function of the slope angle θ. As previously mentioned, the transport rate of snow is different according to the different snow thicknesses in the case of the same slope.

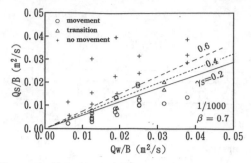

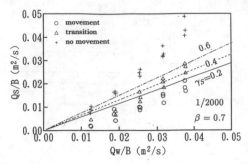

Figure 8: Relationship between the flow rate of snow per unit width and the water discharge per unit width, (channel 1, run 2).

Figure 9: Relationship between the flow rate of snow per unit width and the water discharge per unit width, (channel 1,run 3).

4.2 Snow transport rate in the channel 2

The relationship between the maximum flow rate of snow per unit width, Q_s/B, and the water discharge per unit width, Q_w/B, is shown in Figure 7 with $\gamma_S = 0.7$ according to the experimental condition run 4a. The parameter β is chosen as being in the range of 0.3 to 0.9. The experimental data include the case in which snow lumps are too large to move and thus touch the bottom of the channel. The circle indicates the case in which the snow can move. The plus indicates the case in which the snow cannot move. The triangle indicates the transition. The theoretical line of $\beta = 0.6$ is seen to be best fitted to the experimental data. This line is approximated by the formula by Japan National Railway (Hokuriku Branch of Japan Ministry of Construction, 1983) as follows:

$$Q_w = 2Q_s \qquad (31)$$

4.3 Transport rate of snow in channel 1

As previously mentioned, the ratio of the flow rate of snow to the water discharge is not dependent on the channel slope. In Figure 8 through 13, a specific gravity is chosen as a parameter with $\beta = 0.7$. Therefore, the theoretical lines in Figures 8 and 9, in Figures 11, 12 to 13 are the same line, respectively.

The experimental results of the transport rate of snow per unit width are compared with the theory in which the specific gravity γ_S is chosen as a parameter as shown in Figure 8 ($i_0 = 1/1000$) and Figure 9 ($i_0 = 1/2000$). In these figures, the nondimensional width is set as $\beta = 0.7$. In the experiment, the three values of specific gravity are chosen. The systematic tendency on γ_S

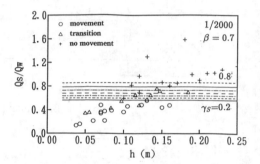

Figure 10: Influence of a specific gravity of snow on the ratio of flow rate of snow to water discharge (run 3).

is not found. As mentioned before, the theoretical lines are not dependent on slope. Small differences in the theoretical lines for different specific gravities of snow appear in Figures 8 and 9. These relations are very close to the formula by Japan National Railway (Eq. (31)) which is not dependent on the specific gravity or on the slope of the channel. The maximum flow rate of snow is half the water discharge in JNR's formula (31). In Ohkuma's (1989) experiments, very similar results are obtained. They pointed out that the experimental results are approximately explained by JNR's equation. The present analysis explains well the experimental results of the other investigators and the empirical formula (31) and finds out that the maximum flow rate of snow is not dependent on the specific gravity of snow. These findings are the important results of the present analysis.

In Figure 10 the relationship between the ratio of flow rate of snow to the water discharge are plotted against the water depth in which the specific gravity is chosen as a parameter. The bed slope

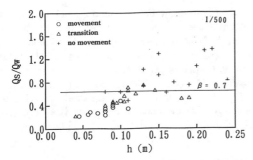

Figure 11: Relationship between the ratio of the flow rate of snow to the water discharge and water depth, (run1).

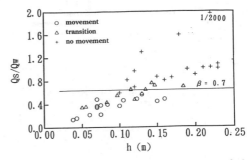

Figure 13: Relationship between the ratio of the flow rate of snow to the water disharge and water depth, (run 3).

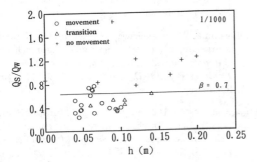

Figure 12: Relationship between the ratio of the flow rate of snow to the water discharge and water depth, (run 2).

is 1/2000 and the experimental condition is run 3. The specific gravity of γ_S=0.2 to 0.8 and $\beta = 0.7$ are used to obtain the theoretical lines. As shown in Figure 10, the theoretical values of the ratio of the flow rate of snow to the water discharge are nearly constant for a wide range of water depths and specific gravity. Experimental results include the case in which the water depth is not so great and the snow lump touches the bed of the channel. Thus, sometimes the snow moves and sometimes it does not. The values of specific gravity are shown in Table 1 as γ_S=0.17 to 0.55. In this range of the specific gravity, the experimental data of Q_s/Q_w vary slightly. This is explained by the theory. It is concluded that the specific gravity does not strongly impact the ratio of water and snow discharge.

The ratios of the flow rate of snow to water discharge based on the experimental data are plotted against the water depth in Figure 11 (run 1), Figure 12 (run 2) and Figure 13 (run 3), in which the slopes are 1/500, 1/1000 and 1/2000, respec-

tively. These figures indicate that Q_s/Q_w is about 0.6 to 0.7 and does not change much with the depth of water. The values of Q_s/Q_w are larger than those obtained by Fukushima et. al (1993) because Fukushima et. al's formula is obtained by assuming that the snow lumps are in contact with the bottom of the channel. This assumption is not satisfied in the case of the large slope. Thus, Q_s/Q_w in the present analysis becomes larger than the values based on the former theory. For these runs of experiments, the snow floats in the channel and the flow rate of the snow becomes greater. From these figures it can be seen that the discharge ratio of snow and water varies little for a wide range of water depths. Therefore, the flow rate of snow is simply in proportion to the amount of the water discharge.

4.4 Increase of water depth caused by snow dumping

The depth of water increases when snow is thrown into the Ryusetsuko compared with a simple water flow. This effect is taken into consideration in the design of the Ryusetsuko. Otherwise, an overflow of water will occur. The increase of water caused by snow is shown in Figure 14 and Figure 15. In these figures, the ordinate indicates the water depth h and the abscissa indicates the flow rate of snow unit width, Q_s/B. The experiment on run 4 is used in these figures. The solid line indicates the water depth with snow, and the broken line indicates the water depth (the normal depth h_{no}). The solid line is located near the transition.

The ratio of the water depth with snow to the water depth without snow is about 1.5. This value is smaller than the observed value of 2.5 (Hokuriku

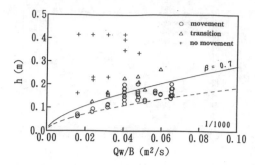

Figure 14: Increase of water depth caused by snow thrown into a Ryusetsuko (run 4a).

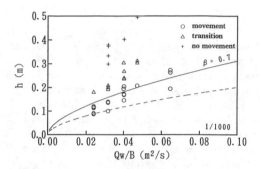

Figure 15: Increase of water depth caused by snow thrown into a Ryusetsuko (run 4b).

Branch of Japan Ministry of Construction, 1983). One of the reasons is that the snow lumps are large enough to block the water, while the present analysis is based on the assumption that snow is fully floated.

5. CONCLUSIONS

This paper examines the maximum flow rate of snow in a mild slope in a "Ryusetsuko" where the water depth is large enough to float the snow layer. This theory assumes that the mixture of snow and water is a Bingham fluid. Then, the velocity distributions in the Ryusetsuko and a formula for the flow rate of snow are derived. The main factors that affect the flow rate of snow are the channel slope, the water depth, the channel width and the water discharge. On the other hand, the specific gravity of snow, γ_S, does not strongly affect it. The empirical formula by JNR clearly explains the experimental results. It is important that the present theory gives a base for this empirical formula for the flow rate of snow.

The present theory is derived from the condition in which the snow floats and does not touch the bottom of the channel. Generally, the snow chunks touch the bottom in a steeply sloped channel. This case is treated by Fukushima et. al (1993). The present theory is applicable to the case in which the slope of a channel is mild, i.e. $i_0 < 1/500$.

A part of this study was supported by a Grant-in-Aid for Developmental Scientific Research (B) (No. 06555147) from the Japanese Ministry of Education, Science, Sports and Culture.

6. REFERENCES

Fukushima, Y. (1990), Simulation of snow motion in "Ryusetsuko", *Proceedings of Annual Meeting of Japan Society of Civil Engineers*, Vol. II, pp. 534-535 (in Japanese).

Fukushima, Y., Hayakawa, N., Okamura, K. and Murakami, M. (1992), Hydraulics approach on the snow remival system, *Procdings of Hydraulic Engineering, JSCE*, Vol. 36, pp.287-291 (in Japanese).

Fukushima, Y., Hayakawa, N. and Murakami, M. (1993), Snow transport in an open channel on mild slope configuration, *Seppyo (J. Japan Society of Snow and Ice)*, Vol. 55-4, pp. 343-351 (in Japanese).

Fukushima, Y., Hayakawa, N. and Murakami, M. (1994), Critical condition for movement of snow chunks in an open channel, *Proc. Japan Society of Civil Engineers*, No. 497/II-28, pp. 51-52 (in Japanese).

Hokuriku Branch of Japan Ministry of Construction, 1983, *Design and Operation Manual of "Ryusetsuko,"* p.35 (in Japanese).

Ohkuma, T., Ohkawa, H., Kandatsu, H., Miya, T., Mizuochi, N. and Nakamura, I., (1989), A fundamental study on standard-design methods of snow conveyance gutter systems and snow melting gutter systems, *Seppyo (J. Japan Society of Snow and Ice*, Vol. 51-4, pp. 239-251 (in Japanese).

Sasaki, M., Takahashi, H. and Kawashima, T., 1993, Velocity distribution of solid particles in solid-water mixture flows in pipelines, *Proceedings of Hydraulic Engineering, JSCE*, Vol. 37, pp. 511-516 (in Japanese).

Sato, T. and Shuto, N., 1983, A velocity formula for the snow and water flow, *Proceedings of 27th Japanese Conference on Hydraulics*, Vol. 27, pp. 801-805 (in Japanese).

Snow Engineering: Recent Advances, Izumi, Nakamura & Sack (eds) © 1997 Balkema, Rotterdam. ISBN 90 5410 865 7

Height and density of new snow on slopes

Yasoichi Endo, Yuji Kominami & Shoji Niwano
Tohkamachi Experiment Station, FFPRI, Niigata, Japan

ABSTRACT: Weights per unit horizontal area, vertical heights and densities were measured in new snow deposited on various slopes of 0° to 75°. Although the weights of new snow on all slopes were nearly the same, the vertical heights of new snow increased and the densities decreased with slope. The difference of heights or densities of new snow due to slope results mainly from the process of deposition of snow crystals and snow-flakes on slopes.

1 INTRODUCTION

When snow is deposited under windless conditions, weights per unit horizontal area, vertical heights and densities of new snow deposited on slopes may be considered in general to be equal to those on a horizontal surface. It was found first by Takahashi and Nogami (1952) and confirmed by Shidei (1952) that vertical heights of new snow on slopes increase with increases in slope and densities decrease with slope. Weights per unit horizontal area are nearly the same. However, these results did not receive much attention and are no longer in people's memory. The dependence of height or density on slope angle is not only essential as basic knowledge, but also may play an important part in avalanche formation. Hence, we re-investigated the dependence in field observations. In this paper, these results are shown and discussed.

2 METHOD OF MEASUREMENTS

To measure weights, vertical heights and densities of new snow deposited on slopes, we placed six wooden slopes 90 cm long and 60 cm wide inclined at 0°, 15°, 30°, 45°, 60° and 75° on flat ground of the Tohkamachi Experiment Station. They were surrounded by a sheet 3 m high to reduce irregular deposition of snow due to strong winds or sunshine. Measurements of weights (HNW) per unit horizontal area and vertical heights (HN) of new snow deposited during a measuring interval of 3 to 24 hours on the slopes were made by using snow samplers and measuring rulers, of which an end is cut at such an angle that its cut end is parallel to the slope when it is inserted vertically into the snow. Mean densities ρ of new snow were calculated by ρ =HNW/HN. Measurements were made during the winters of 1990/91, 91/92 and 92/93.

3 RESULTS OF MEASUREMENTS

To see situations of deposition of snow on slopes, we plotted all weights HNW measured during the three winters against slope angle θ in Figure 1. Open circles connected by a solid line represent weights HNW of new snow deposited during a snowfall on each slope. As shown in this figure, the values of HNW on slopes below 45° do not vary with slope for about 85% of snowfalls. The values of HNW decrease on slopes above 60° because of sliding of snow and rolling of snow particles on the slope.

If snow falls vertically under windless conditions and is deposited on the slopes without slipping down, weights HNW per unit horizontal area of new snow on the slopes should be the same. In order to study the dependence of height or density on slope angle, we need to investigate new snow deposited uniformly on slopes without the influence of winds and sunshine. Hence, we selected new snow for which the ratio of HNW on a slope to HNW on a horizontal surface is within 100±5%. In Figure 2, densities ρ

of new snow selected thus, were plotted against slope angle θ. Here, open circles connected by a solid line represent densities of new snow deposited during a snowfall. This figure shows that the density ρ of new snow on slopes decreases as the slope increases. Consequently, vertical heights HN, being inversely proportional to density ρ for similar weights HNW, increase as the slope increases.

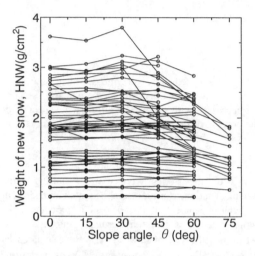

Figure 1. Relation between weight HNW of new snow per unit horizontal area and slope angle θ.

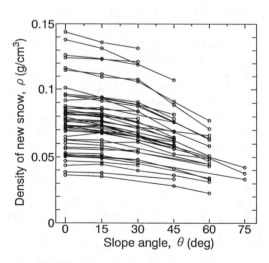

Figure 2. Relation between density ρ of new snow and slope angle θ.

4 RELATION BETWEEN HEIGHT AND DENSITY OF NEW SNOW AND SLOPE

In Figures 3 and 4, vertical heights HN_θ and densities ρ_θ of new snow on a slope of angle θ are plotted respectively against heights HN_0 and densities ρ_0 on a horizontal surface $\theta = 0°$. These figures show that HN_θ and ρ_θ are well approximated respectively by the following straight lines:

$$HN_\theta = (1/B) \cdot HN_0 \quad \text{and} \quad \rho_\theta = B \cdot \rho_0.$$

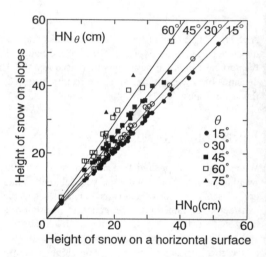

Figure 3. Relation between vertical height HN_θ of new snow on slopes and height HN_0 on a horizontal surface.

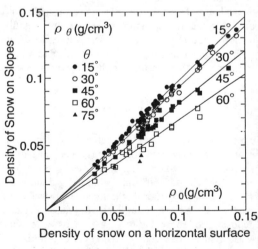

Figure 4. Relation between density ρ_θ of new snow on slopes and density ρ_0 on a horizontal surface.

Here, the mean values of B obtained from the straight lines on Figures 3 and 4 are 0.975 for θ=15°, 0.917 for θ=30°, 0.804 for θ=45° and 0.687 for θ=60°. Then, plotting the obtained values of log B against log(cosθ), we obtained the following relation:

$$B = \rho_\theta / \rho_0 = HN_0 / HN_\theta = (\cos\theta)^\alpha \ ; \ \alpha = 0.57 \quad (1)$$

for new snow deposited during a measuring interval of 3-24 hours. The solid curve in Figure 5 is the plot of B by equation (1), while open circles are the plots of B against θ. Figure 5 shows that the values of B are well approximated by equation (1).

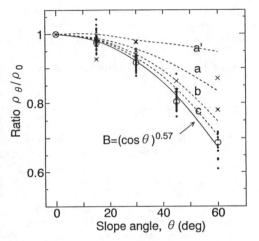

Figure 5. Relation between ratio ρ_θ / ρ_0 and slope angle θ.

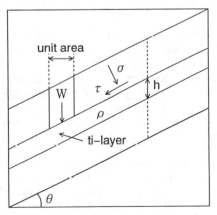

Figure 6. Model of a snow cover on a slope.

To investigate the influence of the length of the interval during which new snow was deposited, we plotted the ratios ρ_θ / ρ_0 given for each snowfall against θ in Figure 5, where crosses are for intervals of 3-7 hours; and solid circles are for intervals of 17-24 hours. Although the data for intervals of 3-7 hours are a few, most plot above the others in Figure 5. That is, the value α for short intervals of 3-7 hours is smaller than the value for 17-24 hours, ranging from 0.2 to 0.5 with a mean of about 0.35. This may suggest that changes in ρ due to θ are caused by the process of densification of snow on slopes.

5 INFLUENCE OF DENSIFICATION OF SNOW

In order to investigate the influence of snow densification, using an equation of snow densification derived theoretically (Endo et al. 1990; Endo 1992), we computed densities of new snow on various slopes, and compared results with measured values.

Now we consider a thin snow layer (called t_i-layer) deposited at time t_i as shown in Figure 6, and assume that the relation between a normal stress σ exerted on the t_i-layer and strain rate ε in the direction normal to the t_i-layer is given by $\sigma = \eta \cdot \varepsilon$, where η is compressive viscosity, expressed by $\eta = C \cdot \rho^n$ with $C \sim 1.11 \times 10^6$ g-wt·hr·cm^{-2}·(g/cm^3)$^{-n}$ and $n \sim 4$ according to Endo et al.(1990).

Then, we have

$$\rho(t_i,t) = \{(n/C) \cdot (\cos\theta)^2 \cdot Q(t_i,t) + (\rho_{\text{init}\,\theta})^n\}^{1/n} \quad (2)$$

where $\rho(t_i,t)$ is a density of t_i-layer at time t ($>t_i$), $\rho_{\text{init}\,\theta}$ the initial density of t_i-layer on a slope of angle θ, and $Q(t_i,t)$ the time-integration of weight W per unit horizontal area of snow lying on t_i-layer from t_i to t:

$$Q(t_i,t) = \int_{t_i}^{t} W(t_i,t)dt = \int_{t_i}^{t} (t-t_s)dW(t_s) = \int_{t_i}^{t} (t-t_s)q(t_s)dt_s$$

where $W(t_i,t)$ is a weight per unit horizontal area of snow lying on t_i-layer at time t, $dW(t_s)=q(t_s)dt_s$ a weight of snow accumulated from time t_s to t_s+dt_s, that is, the weight of t_s-layer, and $q(t_s)$ an accumulation rate at time t_s. Consequently, if hourly precipitation from $t=t_i$ up to t is known, we can compute numerically the density $\rho(t_i,t)$ of t_i-layer at time t from these equations.

Then, since the vertical thickness $h(t_i,t)$ of the t_i-layer at time t is given by $h(t_i,t)=dW(t_i)/\rho(t_i,t)$, the vertical height $HN(t_i,t)$ of new snow deposited from time t_i to t is given by

$$HN(t_i,t)=\int_{t_i}^{t} \{q(t_s)/\rho(t_s,t)\}dt_s$$

and the mean density ρ of the new snow becomes

$$\rho=W(t_i,t)/HN(t_i,t)=\int_{t_i}^{t} q(t_s)dt_s/HN(t_i,t)$$

Now consider the simple case that $Q(t_i,t)$ satisfies the following condition:

$$Q(t_i,t) \gg (C/n)\cdot(\cos\theta)^{-2}\cdot(\rho_{init\theta})^n \qquad (3)$$

Then, equation (2) becomes $\rho(t_i,t)=(\cos\theta)^{2/n}\cdot\{(n/C)\cdot Q(t_i,t)\}$. Hence, if most new snow layers deposited from t_i to t satisfy the condition (3), the ratio ρ_θ/ρ_0 for the snow is given by

$$B=\rho_\theta/\rho_0=HN_0/HN_\theta=(\cos\theta)^{2/n} \qquad (4)$$

with n equal to about 4. This indicates that the value of ρ_θ/ρ_0 approaches equation (4) unrelated to the initial snow density, as time passes after a stop of snowfall or when the measuring interval is long.

Although equation (1) obtained for intervals of 3-24 hours is nearly equal to equation (4), it is doubtful whether our measurements satisfy condition (3). Then, in order to confirm that, we computed mean densities ρ_θ of new snow deposited under constant accumulation rates during a few time intervals on various slopes using the above equations, and compared results with measured values. In these computations, we assumed that the initial density $\rho_{init\theta}$ is related to the initial density ρ_{init0} on a horizontal surface by

$$\rho_{init\theta}/\rho_{init0}=(\cos\theta)^{\alpha0} \qquad (5)$$

and we used the following values: $q(t_s)=0.10$ g·cm^{-2}·hr^{-1}, $\rho_{init0}=0.05$ g·cm^{-3}, $\alpha 0=0$, 0.3, and 0.5, time interval=7 and 24 hrs. The computed ratios ρ_θ/ρ_0 are shown by broken curves in Figure 5, where the curve a' represents the relation for $\alpha 0=0$ for a 7 hour interval and the curves a, b and c represent relations for $\alpha 0=0$, 0.3, and 0.5 for an interval of 24 hours. A comparison between the computed curves and the measured values indicates that the measured dependence of density on slope cannot be explained by the process of densification of snow when the initial densities $\rho_{init\theta}$ on slopes are the same, and that the initial densities $\rho_{init\theta}$ on slopes must be related to ρ_{init0} by equation (5) with $\alpha 0$ equal to about 0.5. According to equation (2), the ratio ρ_θ/ρ_0 in the case of $\alpha 0=0.5$ and n=4.0 is independent of time, being $\rho_{init\theta}/\rho_{init0}=\rho_\theta/\rho_0=(\cos\theta)^{0.5}$. The difference of initial snow density just after snow is deposited on slopes results from the process of deposition of snow crystals and snowflakes, but it is not clear why the initial snow density depends on slope. From the above-mentioned results, we can conclude that differences in density and vertical height of new snow due to slope is due to the process of deposition not densification of snow.

6 CONCLUSIONS

We confirmed that vertical heights of new snow on slopes increase with increases in slope and its density decreases with slope. The weight of new snow per unit horizontal area does not vary much with slope. In addition, we showed that the ratio of height or density on a slope to that on a horizontal surface is expressed as a power function of the cosine of the slope angle, and that differences due to slope result mainly from the process of deposition of snowflakes or snow crystals on slopes.

REFERENCES

Endo, Y., Ohzeki, Y and Niwano, S. 1990. Relation between compressive viscosity and density of low-density snow. *Seppyo* 52: 267-274.

Endo, Y, 1992. Time variation of stability index in new snow on slopes. *Proc. of the Japan-U.S. Workshop on Snow Avalanche, Landslide, Debris Flow Prediction and Control 1991 Tsukuba Japan,* Sponsored by Science and Technology Agency of Japanese Government, 85-94.

Takahashi, K, and Nogami, K. 1952. Metamorphism of snow on slopes. *Yuki* 10: 13-18.

Shidei, T. 1952. Snow cover on slopes. *Bulletin of the Forestry and Forest Products Research Institute* 54: 133-139.

Snow Engineering: Recent Advances, Izumi, Nakamura & Sack (eds) © 1997 Balkema, Rotterdam. ISBN 90 5410 865 7

Snow disasters on roads in Heilongjian Province, China

Eizi Akitaya
Institute of Low Temperature Science, Hokkaido University, Sapporo, Japan

Shunichi Kobayashi
The Research Institute for Hazards in Snowy Areas, Niigata University, Japan

Pengfei Le
Ministry of Communication, Heilongjian Province & Research Group for Snow Disasters in Heilongjian Province, People's Republic of China

ABSTRACT: A study of snow hazards on roads in Heilongjian Province, China, was carried out at the beginning of March 1994. It was found that ice on the roads in flat regions was not slippery because of the soil particles in it. On the other hand, roads in mountainous regions were covered with slippery snow and ice. Although the snow conditions on the roads are not a serious hazard at this time due to the low volume of traffic, measures to prevent accidents will be necessary in the future due to the expected increase in traffic volume .

1 INTRODUCTION

Heilongjian Province, which is major grain producing region, is located in northeast China and covers an area of 710,000 km². The region has moderate temperatures and precipitation in summer, but very cold temperatures and heavy snow drifts in winter. The mean air temperature for January in Harbin is around - 20 ℃ and can fall to as low as - 40 ℃. Visual observations of the snow conditions and snow distribution along roads were carried out by Japanese scientists and Chinese engineers of the Heilongjian Road Office while traveling a distance of 1500 km during a 5-days period.

2 ROUTE AND METHOD OF OBSERVATIONS

Our party traveled east from Harbin along the Songhau river, then north from Roppei along the Heilong river, and south from Jiayin through the forest of Xiao Hinggan Ling, and finally returning to Harbin through Teili (Fig.1). The road was all through farm land or forest, except for ice on frozen rivers at the points indicated by Nos.5, 15 and 21 in Figure 1.

Visual observations were made using data sheets and pictures at 30-minute intervals. The following items were noted: name of the place, snow distribution around the road , snow conditions on the road and possible dangers for the traffic. Snow conditions on the road were classified into snow and

ice based on hardness; snow showed traces of car tires, while ice showed no traces. Friction between tires and a road surface covered with snow or ice is an important factor for the safety of cars.

Temperatures in this region are very low and precipitation is less than 10 mm per month in winter . Snow deposited naturally on the ground becomes a very fragile depth hoar under such conditions. The drifted snow or ice on the road was dirtied by soil blown in the wind from surrounding farm land. Due to the very dry weather, strong winds eroded and blew away the fragile snow and soil particles . The color of snow or ice depended on the amount of soil particles, and was considered to be a simple index of the degree of slipperiness, and thus, danger to traffic. The color changes from white to gray, to dark gray then black in proportion to the increase in impurity.

3 SNOW OBSERVATION USING SNOW PITS

Weather conditions affect the strata, grain size, grain shape, density and hardness of the snow cover. Snow pits were made in naturally deposited and drifted snow around roads. It was found that snow in an open space in forest land (No.31), where snow was deposited calmly without any wind effect on the frozen ground, was approximately 30 cm thick and 150 kg/m³ in density, with a large cup-shaped depth hoar and a ram number smaller than 1 kg. On the other hand, drifted snow in a channel

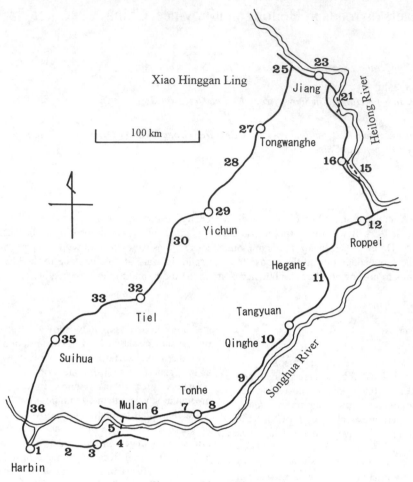

Figure 1. Observation route.

beside a road passing through farm land near No.12 was about 40 cm thick, had a density of 350 kg/m³, a cup-shaped depth hoar including ice sheets in its layer, and a ram number of approximately 50 kg. It is considered from the snow conditions described above that snow which was gently deposited, metamorphosed to depth hoar under very low temperatures then snow drifts were caused by strong winds, eroding the surface of the snow with soil particles and transporting them leeward. Snow and soil particles from farm land transported by wind, deposited behind obstacles such as tree lines, houses and channels. The snow which was deposited densely by blowing snow rapidly became hard.

4 CHARACTERISTICS OF SNOW AND ICE ALONG THE OBSERVATION ROUTE

There was a great variation in the snow conditions and distribution of snow on and in the vicinity of the roads along the 1500 km observation route. The variation in snow conditions is shown in Figure 2. As can be seen from Figure 2, there were some areas along the observation route which had no snow, and in other areas the snow was dirtied by soil, except on roads in forest areas which were covered with white snow. Two other distinctive features were seen along the observation route. One was snowdrifts which were deposited irregularly on the road had changed to black-colored ice by the compression of car tires and the melt-freeze cycle under strong sunshine. The other

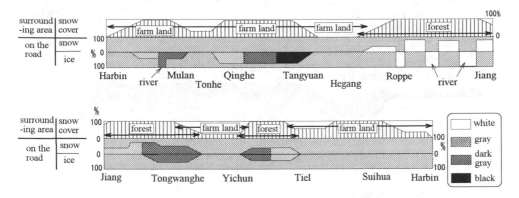

surrounding area snow cover
on the road snow / ice

Harbin river Mulan Tonhe Qinghe Tangyuan Hegang Roppe river Jiang

Jiang Tongwanghe Yichun Tiel Suihua Harbin

white
gray
dark gray
black

Figure 2. Snow conditions on and in the vicinity of roads along the observation route.

feature was an area where all the snow had been blown off the road in the snowy region near Harbin, which seems to have been caused by the geometrical structure of the road .

Figure 3 shows four distinctive types of snow conditions on the road along the observation route. The first is black-colored snow or ice on the road, deposited behind houses or roadside trees by drifting snow which included soil from farm land (10). This type of road was not very slippery due to the soil particles contained in the snow, but the rough surface caused by drifting patterns was often dangerous for traffic.

The second type of snow condition was a thick layer of white snow with deep tire ruts. The surrounding farm land in such regions was completely covered with snow (23). The thick snow drift on such roads is caused by trees planted along the roadside. The third type of snow condition is white, hard and slippery snow or ice on roads surrounded by forest (26). These roads are often dangerous due to steep slopes and curves, and sand was spread on dangerous sections of the road to prevent slipping. The fourth type of snow condition includes two types of snow-free roads in snowy regions: one is a elevated-type of road from surrounding farm land without roadside trees (32), and the other is a elevated road with a deep and wide channel beside the road.

Figure 3. Typical snow conditions on the road.

401

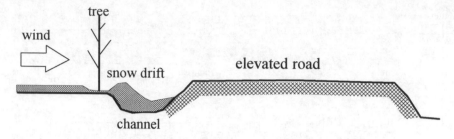

Figure 4. Road structure for reducing snow hazards.

5 A PROPOSAL FOR THE PREVENTION OF SNOW DISASTERS ON ROADS

The traffic along our observation route was not heavy, and there were no snow plows to remove snow from the roads. All of the cars which were seen along the observation route did not have special snow tires. However, despite the bad road conditions, there were no serious traffic accidents during our observation period due to the low volume of traffic. It is expected that traffic accidents will increase in the future as the volume of traffic increases. Therefore, in the future, snow plows will be needed to remove snow drifts and smooth the rough surface on these roads. Another important measure to prevent snow hazards is to reduce the deposition of snow drifts on roads by improving the road structure using a combination of elevated road sections, channels and tree lines outside the channels as shown in Figure 4.

Snow Engineering: Recent Advances, Izumi, Nakamura & Sack (eds) © 1997 Balkema, Rotterdam. ISBN 90 5410 865 7

Research on making indexes of obstruction to traffic in winter

Norio Takahashi
Department of Psychology, Health Sciences
University of Hokkaido, Sapporo, Japan

Fumihiro Hara
Department of Engineering, Hokkaido University,
Sapporo, Japan

Takashi Kawabata
City of Sapporo, Japan

Masaaki Abe
Hokkaido Development and Technology Center, Sapporo,
Japan

ABSTRACT: To ensure secure and comfortable living environments, administrative bodies undertake various measures to deal with snow. Recently, due to urban expansion and resident demand, the snow removal budget has increased. Until very recently, quantitative studies on the effectiveness of municipal snow removal operations have been limited. Here, we conducted questionnaire surveys in Sapporo and Otaru to find the relation between snow and some traffic issues caused by snow. In Sapporo, numerical conversions give us relative ward comparison. We found that snow depth, the ratio of multistory housing developments, subway use and the percentage of aged were major indexes of some traffic issues. Snow removal operations must consider urban and social environments to be effective.

1 INTRODUCTION

The city of Sapporo has conducted systematic snow removal for the last 20 years, but their cost-benefit ratio remains unclear. For an effective use of a limited budget, the following is required: clarifying obstruction; clarifying natural, urban and social conditions related to obstructions; measuring the effectiveness of snow measures; and deciding future budget allocations.

For that purpose, a quantitative analysis of the relation between snow and the living environment is necessary. We focused on traffic obstructions due to snow. We also conducted questionnaire surveys to find the degree of traffic obstructions and then tried to clarify their relation to natural, urban and social conditions.

The indexes we used in this research were: (1) fatigue from snow removal; (2) obstructions to walking; (3) obstructions to driving and taking buses; and (4) obstructions waiting for buses. In addition, we conducted an analysis that gives a relative comparison of each administrative district to express traffic obstructions numerically, as well as regression analyses between subjective environments of nature, urban area and society and obstructed traffic.

In this paper, we report on the results of surveys in Sapporo and Otaru. Consideration was given to the difference in urban environments of these two cities (e.g., slopes).

2 METHODS

2.1 Sapporo

Scope: Men and women over 20 in Sapporo. Stratified two-stage sampling (or sub-stratified sampling). Sample size 1400.
Method: Questionnaires were mailed and collected by visiting respondents two weeks later.
Period of Survey: During the period of snow accumulation.
Percentage of Returns: 46.9 % (656/1400).

2.2 Otaru

Scope: Men and women over 20 in Otaru. Stratified two-stage sampling. Sample size 1500.
Method: Questionnaires were mailed and returned by mail.
Period of Survey: Non-snowy period.
Percentage of Returns: 44 % (653/1500).

3 RESEARCH QUESTIONS

3.1 Sapporo

(1) Four questions were asked about fatigue from removing snow from doorways to roads: (a) physical fatigue from snow removal; (b) undesirable effects on the respondent's work after removing snow; (c) psychological unwillingness to remove snow, and (d) conversion of snow removal

work to physical exercise. Evaluations were made on a 5-point scale and 3-point scale.

(2) Five questions were asked about dissatisfaction with conditions of snow removal: (a) lack of dumping areas; (b) time when snow plows come; (c) snow masses left in front of doorways by snow plows; (d) illegally parked cars, and (e) being unable to drain snowmelt water. Evaluations were made on a 5-point scale.

(3) Twelve questions were asked about obstacles while walking, including: (a) difficulty in walking due to bumpy sidewalks; (b) whether the respondent had a fall on an icy road; (c) whether the respondent met with a traffic accident; (d) detours due to snow; (e) difficulty in crossing a street, and (f) falling snow and ice from roofs. All evaluations were made on a 4-point scale.

(4) Six questions were asked about encountering dangerous walking conditions. Questions included: (a) traffic accidents; (b) falling, and (c) falling snow and ice from roofs. All questions were evaluated on a 4-point scale.

(5) Three questions were asked about driving obstructions: (a) traffic congestion due to snow removal operations; (b) minor collisions with pedestrians due to the narrowed width of roads by piled snow, and (c) delay in arrival due to snow. All questions were evaluated on a 4-point scale.

(6) Questions were asked about obstructions while waiting for buses: (a) harshness of cold; (b) unreliable arrival time, and (c) being left behind because of jammed buses. The questions were evaluated on a 4-point scale.

3.2 Otaru

Questions about fatigue from removing snow and dissatisfaction with snow removal were the same as those for Sapporo. Questions about the danger and obstructions while walking were not separated, while for Sapporo they were separated.

As Otaru has more slopes than Sapporo, 15 questions were asked about walking obstructions on sidewalks, and nine questions were asked about obstructions to driving. Their evaluations were made on a 4-point scale. Five questions about obstructions at bus stops were asked, including one question related to narrowed roads. All questions were evaluated on a 4-point scale.

4 RESULTS AND DISCUSSION

The scores for each obstruction were totaled. The values were calculated for each individual's response to the degree of fatigue, the degree of dissatisfaction with snow removal, the level of obstruction to walking on sidewalks, the level of danger of walking on sidewalks, the level of obstruction to driving and taking buses, and the level of obstruction at bus stops. The conversions of the scores of each obstacle were made, with the highest evaluation given 1 and the lowest 0.

For Sapporo, the converted values were tabulated and normalized for each administrative district (ward) for multiple regression analyses. Thus, the dependent variables represented each obstruction, while the independent variables represented information on the respondent (age, residential form, number of persons the respondent lives with, area of residential lot per home, etc.). For Otaru, as the administrative districts are not clearly separated, a correlation analysis was made using the values for the city as a whole.

4.1 Sapporo

(1) The degree of fatigue from snow removal between doorways and roads
The average score of the converted values was 0.254. Figure 1 shows the normalized scores by ward compared to the seven wards surveyed.

Multiple regression analysis of the normalized scores by ward, considering biological information, gives the regression equation:
$$y = 1.505 - 0.162a + 0.0172b \qquad (1)$$
($R^2 = 0.992$, a is the percentage of the population 60 years and older and b is the snow depth in centimeters.)

This multiple regression analysis shows that fatigue increases as the age increases and as the snow depth increases. This indicates some factors to be considered in designing housing developments in the coming aging society.

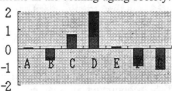

Figure 1. Standardized degree of fatigue from snow removal between doorways and roads.

(2) The degree of dissatisfaction with municipal snow removal operations
The relative value, with the maximum of 1, was 0.507. Figure 2 shows the normalized scores of the seven wards.

Multiple regression analysis shows the regression equation:

y = 0.0308a + 0.0318b + 0.1058c + 1.4935 (2)
(R^2 = 0.9510, a is the area of the residential lot per home in square meters, b is the snow depth in centimeters, c is the percentage of the population 60 years and older.)

This analysis shows that dissatisfaction with snow removal decreases as the residential lot area increases and dissatisfaction increases as the snow depth rises and the proportion of elderly people increases.

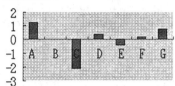

Figure 2. Standardized degree of dissatisfaction with snow removal operation.

(3) The level of obstruction to walking on sidewalks
The score of the converted values was 0.471. Figure 3 shows the normalized score of obstruction to walking on sidewalks according to the seven wards.

Multiple regression analysis shows the regression equation:
y = 6.0581 - 0.099a - 0.013b - 0.14c (3)
(R^2 = 0.871. a is the percentage of using subways, b is snow depth in centimeters, c is the percentage of the population 60 years and older.)

This shows that the levels of obstruction to walking on sidewalks increases as the rate of using subways, snow depth and the proportion of those 60 and over increase.

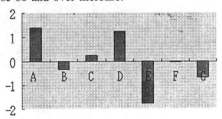

Figure 3. Standardized level of obstruction to walking on sidewalks.

(4) The level of danger of walking on snow-covered sidewalks
The relative score of the converted values was 0.520. Figure 4 shows the level of danger of walking on snow-covered sidewalks according to the seven wards.
Multiple regression analysis shows the regression equation:
y = - 1.68 - 0.084a + 0.02b + 0.131c (4)
(R^2 = 0.832, a is the percentage of using subways, b is snow depth in centimeters, c is the percentage of housing developments.)

This regression equation shows that the danger increases as the rate of using subways declines, and as the snow depth and the rate of housing developments increase.

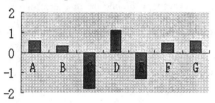

Figure 4. Standardized level of danger of walking on snow-covered sidewalks.

(5) The level of obstruction to driving and to taking buses during the snowy period
The relative score of the converted values was 0.559. Figure 5 shows the normalized score of obstruction to driving and to taking buses according to the seven wards.

Multiple regression analysis shows the regression equation:
y = - 0.0399a - 0.0876b + 0.1177c - 3.6844 (5)
(R^2 = 0.9582, a is the percentage of using subways, b is the percentage of multistory housing developments, c is the percentage of people in their forties and fifties.)

This shows that as the rates of subway use and multistory housing developments increase, obstruction to driving and taking buses decreases, but the obstruction increases as the percentage of people in their forties and fifties increases.

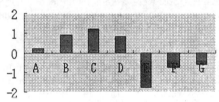

Figure 5. Standardized level of obstruction to driving and to taking buses.

(6) The level of obstruction waiting for buses
The relative score of the converted values was 0.322. Figure 6 shows the level of obstruction at bus stops according to the seven wards.
The multiple regression equation was as follows:
y = - 0.0788a + 0.0174b - 0.0634 (6)
(R^2 = 0.8599, a is the percentage of using subways, b is snow depth in centimeters.)

This shows that the degree of dissatisfaction waiting for a bus lowers as the percentage of subway use rises, and it becomes higher as snow depth becomes greater.

The results of the multiple regression analyses for Sapporo show that snow depth, the rate of using subways, the area of residential lot per home, the percentage of elderly people, the percentage of multistory housing developments and the percentage of people in their forties and fifties are the major indexes to many kinds of traffic obstructions due to snow. However, as the relationships between these variables were not analyzed in our research, we cannot clearly indicate how such obstructions are connected with natural, urban, and social environments.

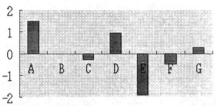

Figure 6. Standardized level of obstruction waiting for buses.

4.2 Otaru

(1) The degree of fatigue from snow removal
Four questions were asked. Evaluations were made on a 5-point scale. The relative value, with the maximum given 1, was 0.4379.

(2) The degree of dissatisfaction with municipal snow removal operations
Five questions were asked, and the evaluations were made on a 5-point scale. The relative value was 0.453, slightly lower than for Sapporo.

(3) The level of obstruction to walking
Unlike Sapporo, questions about obstruction to walking included dangers to pedestrians. Fifteen questions were asked, including questions about slopes and very slippery road surface conditions. Evaluations were made on a 4-point scale. The relative evaluation, with the maximum of 1, was 0.486. This was similar to the value for Sapporo.

(4) The level of obstruction to driving and taking buses
Nine questions, including the conditions of slopes, in addition to the questions for Sapporo, were asked and evaluated on a 4-point scale. The relative evaluation, with a maximum of 1, was 0.505. This was similar to the value for Sapporo.

(5) The level of obstruction waiting for buses
Four questions were asked, including the difficulty in passing due to narrow roads, in addition to the

questions asked in Sapporo. Evaluations were made on a 4-point scale. The relative evaluation, with a maximum of 1, was 0.57. This was considerably higher than for Sapporo.

4.3 Comparison between Sapporo and Otaru

The analysis of variance of differences in the degrees of obstruction, with a maximum of 1, showed that the degree of obstruction waiting for buses in Sapporo was significantly lower than for Otaru (Figures 7 & 8), but significant differences between the other obstructions were not found. Whether this difference was related to the other contents of the questionnaires, correlation with the degrees of obstruction waiting for buses in other cities, was investigated. However, particular tendencies were not found.

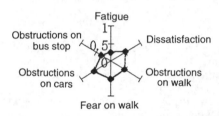

Figure 7. Obstructions by snow (Sapporo).

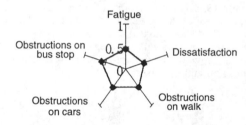

Figure 8. Obstructions by snow (Otaru).

5 CONCLUSION

This study shows that in Sapporo, city planning by ward, natural, social and other conditions are indicators of obstructions to traffic due to snow. The relative degrees of obstruction in the two cities were numerically expressed. However, the causal relationship between the variables and obstructions to traffic remains to be analyzed.

By comparing Otaru and Sapporo, this study shows a large difference in problems with bus stops. However, it remains unclear which variables, and to what degree those variables contribute to this difference.

Snow Engineering: Recent Advances, Izumi, Nakamura & Sack (eds) © 1997 Balkema, Rotterdam. ISBN 90 5410 865 7

Analyses of walking characteristics in winter and falling accidents in Sapporo

Norio Takamiya & Hiroshi Kido
City of Sapporo, Japan

Fumihiro Hara
Department of Engineering, Hokkaido University, Sapporo, Japan

Yoshimitsu Takanishi
Hokkaido Development Engineering Center, Sapporo, Japan

Abstract: The results of measuring walking speed in winter on various road surfaces demonstrate that citizens walk fastest on dry surfaces and slowest on slushy surfaces. Physical hindrance is greater when walking on slush with puddles than on frozen roads. In addition, comparing pedestrian traffic volumes in underground shopping centers, shopping arcades and sidewalks in the summer with those in the winter shows a marked concentration of pedestrians in the underground shopping center in winter. Furthermore, there is a correlation between the number of citizens who have fallen on frozen roads and sustained injuries and the installation rate of studless tires. The findings show that those who have experienced falling accidents include many aged women, and that aged citizens are likely to suffer serious injuries from falling. Also, there is a correlation between gender and which body part sustains an injury from falling. A strong correlation cannot be established between snowfall and falling accidents.

1 INTRODUCTION

The condition of sidewalks in Sapporo in winter becomes extremely poor due to various factors: the width of sidewalks narrowed by snow removed from roadways; frozen roads; snow and ice falling from roofs, etc. Also, countermeasures to remove snow from sidewalks and antifreeze sidewalk surfaces have not been fully developed as compared with those for roadways. We examined walking speed and pedestrians' choices of walking environments as well as analyzed falling accidents on frozen roads in Sapporo in order to understand the present circumstances of walking environments with the aim of improving them. This paper will report the findings.

2 WALKING CHARACTERISTICS IN WINTER

2-1 *Walking speed in winter*

We conducted field surveys on pedestrian traffic volume and walking speed in the central part of the city. For the surveys we designated an 8 to 9 m section and measured walking speeds. The survey was conducted on Feb. 16 (Sun.) and 17 (Mon.), 1992. On the 16th, it was cloudy all day along with snowfall observed between noon and one o'clock in the afternoon, while it was fair throughout the day on the 17th. The sampling size

was 1,721 on the 16th and 1,261 on the 17th. Figure 1 illustrates various walking speeds depending on surface conditions, age and gender. Road surfaces of the sidewalks were classified into the following four categories:

(a) dry
(b) wet A (surface is exposed and wet)
(c) wet B (slushy)
(d) frozen (surface of compacted snow is frozen)

The comparison of average walking speeds for both sexes produced the following findings: both men and women walk fastest on dry surfaces, at average speeds of 5.3 km/h and 4.9 km/h respectively; both men and women walk slowest on slushy surfaces with average speeds of 4.4 km/h and 4.1 km/h respectively. Because slush is very wet, pedestrians take each step very carefully so as not to soak feet, as citizens in Sapporo commonly wear short leather shoes even in the winter. This seems to help explain the decrease in walking speed. Also, it is difficult to step normally on an unstable slushy surface. Aged women walk slowest on frozen surfaces, which differs from other age groups.

2.2 *Choice of walking environment*

Investigation of summer and winter pedestrian volumes was conducted in the city's central area. Field surveys were conducted at 10 locations, among which 6 locations with various walking environments (four sidewalks, one underground shopping center, and one shopping arcade) were selected. Results were collected and compared.

The survey was conducted on July 15 (Wed.) and 19 (Sun.), 1992, and two days (same as the survey of walking speed) in winter. The sampling size was 64,396 on July 15, 91,610 on July 19, 76,291 on Feb. 17, and 107,170 on Feb. 16. Figure 2 illustrates the comparison of traffic volumes depending on various walking environments (sidewalks, underground shopping center, shopping arcade) based on the survey results.

Pedestrian traffic volume in the underground shopping center was higher than other two areas in both summer and winter. In particular, traffic volume in the underground shopping center increased in winter with a decrease in the other two areas. On weekdays, the traffic volume on the sidewalks and in the shopping arcade decreased at almost the same rate. While on weekends, there was a remarkable 6.6 % decrease in the shopping arcade as compared with a 2.4 % decrease on sidewalks. This implies that shoppers tend to concentrate in the underground shopping center during winter.

3 ANALYSIS OF FALLING ACCIDENTS WHILE WALKING IN WINTER

3.1 *Circumstances of falling accidents*

(1) Number and place of accidents
At present it is difficult to fully understand quantitatively the total number of citizens who fell on winter road surfaces. In this paper we will analyze the falling accidents in terms of people who fell and were transported by ambulance (Dec. to Feb.) from data of the Emergency Turnout Report by the Fire Bureau of Sapporo.

Figure 3 shows the number of citizens who fell and sustained injuries each winter season (Dec. to Feb.) from 1983 to 1993, and the trends of installation rates of studded tires as well as developments of regulations on studded tires.

The number of citizens falling and sustaining injuries increased and remained at around 600 since the installation rate of studded tires dropped to 5 % and below. The causes of falling accidents have not yet been clearly determined; however, very slippery road surfaces, appearing more frequently recently, may affect pedestrians.

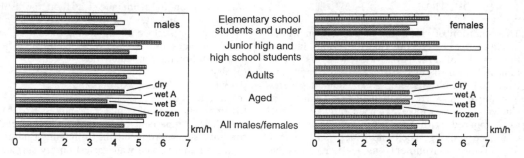

Figure 1. Walking speed by age category.

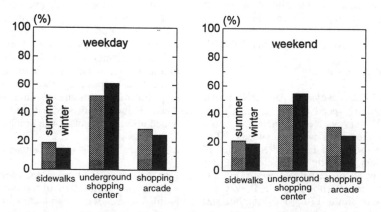

Figure 2. Pedestrian traffic distribution according to walking environment in the city center.

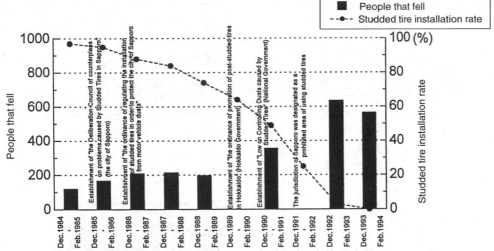

Note 1) The number of citizens who fell and were transported to a hospital by ambulance indicates the seasonal total (Dec. to Feb.). Originally taken from the Emergency Conveyance Record of Sapporo Fire Bureau.
Note 2) The installation rate of studded tires is taken from the results of investigation during the coldest time of winter by the city of Sapporo.
Note 3) 1989 - 1990 and 1991 - 1992 data (people that fell) could not be obtained.

Figure 3. Trends of number of people transported by ambulance and the installation rate of studded tires.

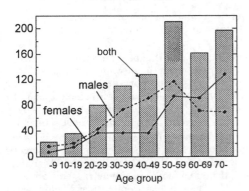

Figure 4. Number of accidents by age group.

(2) Circumstances of falling accidents by age and gender
We analyzed falling accidents in detail, using data of Jan., Feb. and Dec. between 1984 and 1989 focusing on the period just before the introduction of the studded tire regulations.

The number of citizens who fell and sustained injuries increased with age. The number of citizens who fall peaks in the 50-59 age group, then drops (Figure 4). The number of citizens over 65 who fall comprises approximately 30 % of all cases.

The number of men who fall rises linearly up to the fifties then drops after sixty, while the number of women remains low until the forties and then makes a rapid increase after fifty. Women over seventy were the most likely to fall. This can be explained by the difference of leisure time and frequency of going out between men and women depending on their respective age.

(3) Body part and degree of injury
Concerning injured body parts, head injuries are the largest percentage (approx. 43 %) followed by leg (approx. 24 %), hip (11.11 %), arm (6.73 %), and foot (5.1 %) injuries. Head and leg injuries are close to 70 % of all cases. Men are most likely to sustain injuries on the head (50.19 %), 16 % higher than that of women (34.9 %). Women are more likely to sustain injuries on the arms and hips, 10.06 % and 14.56 % respectively, compared with men, 3.70 % and 7.98 % (Figure 5).

Citizens 65 and over have a higher percentage of moderate and serious injuries, 5 % and 12 % higher than those under 65 respectively. Thus, those age 65 and over are more likely to suffer from serious injuries (Figure 6).

3.2 *Correlation between falling accidents and weather*

Pedestrians frequently fell on frozen roads and were taken to hospitals by ambulance between December 1992 and March 1993. An analysis was made of the number of people who have fallen as a function of the weather. Table 1 illustrates the number who fell and sustained injuries as

Female   Male

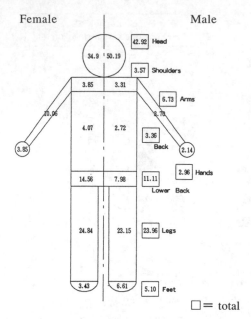

□ = total

Figure 5. Injured body part according to sex.

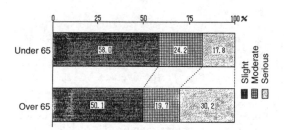

Figure 6. Degree of injuries of older people.

Table 1. The number of citizens who fell.

	days	number of accidents	accidents per day
Dec.1992	31	248	8.0
Jan.1993	31	166	5.4
Feb.1993	28	223	8.0
Mar.1993	31	73	2.4
total	121	710	5.9

investigated by the Fire Bureau.

The total was 710 with an average of 5.9 citizens per day taken to hospitals by ambulance. When citizens fill the streets on Christmas Day, the closing day of government offices for the year, the end of the year, new year office openings, Sapporo Snow Festival, etc., more people are likely to experience falling accidents.

Figure 7 illustrates the relationship between falling accidents and the amount of snowfall.

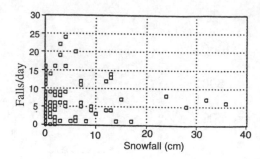

Figure 7. Relationship between falling accidents and snowfall.

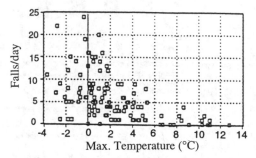

Figure 8. Relationship between falling accidents and maximum temperature.

There is no significant correlation but many citizens fell when it snowed about 3 - 4 cm. However, this does not adequately explain the relationship between snowfall amount and the number of falling accidents because many accidents are reported on days without snowfall.

In terms of the relationship between maximum daytime temperature and the number of citizens who fall, about 2 to 10 people fell on days when the maximum temperature ranged from 2 °C to 5 °C. The number drops to 5 and below when the maximum temperature was 5 °C and warmer (Figure 8).

REFERENCES

Hara, Kawabata and Kobayashi. 1990. On Safety of Winter Walking Environment in the City of Sapporo. Proc. of Cold Region Technology Conference. 151 - 157.

The Road Maintenance Department, Construction Bureau, the City of Sapporo. March, 1993. Report of Fundamental Investigation on Countermeasures against Frozen (Slippery) Road Surfaces.

Study on measures used by women in walking in winter

Fumihiro Hara
Department of Engineering, Hokkaido University, Sapporo, Japan

Chikara Suda
Department of Education, Hokkaido University, Sapporo, Japan

Eiji Akitaya
Low Temperature Scientific Laboratory, Hokkaido University, Sapporo, Japan

Sawako Morita
Sapporo Research Society of Consumers' Life in the Northern Countries, Sapporo Consumers' Association, Japan

ABSTRACT: A questionnaire was sent to women 20 and over living in Sapporo to obtain information on the number of falls they had on winter roads, types of injuries, and types of clothes and shoes worn at the time. Measures were taken from the 1,382 replies. A very high percentage of them experienced falls requiring medical attention, and a relatively high percentage were wearing shoes with anti-slip devices; yet, only a few of which were detachable. Additional research was conducted of ten different types of shoes with anti-slip features to find out comfort levels under varying road conditions. It showed that the anti-slip features were highly effective on roads covered with ice or compacted snow. Most were found to be slippery inside buildings, with only one retaining indoor anti-slip effectiveness. Three types were poor at keeping out the cold. Spikes on the soles of some shoes showed a tendency to collect snow.

1 INTRODUCTION

In Sapporo, heavy snow and cold climate conditions characterize the winter environment. It is not easy to walk on icy or snowy roads; thus, a number of pedestrians fall and get injured. According to a study of out-patients by Ulrik & Soeren (1983) at the Odense University Hospital Orthopedics Department, Denmark, the number of injuries from falling increases more than 14 fold during a winter when Denmark has unusually icy roads, compared to its typical winter. Also, according to an analysis by Gorn Nilsson (1986) of traffic accidents, and the number of injuries and death toll during a year in Ostergotland, Sweden, more than half of the injuries resulted from falls while walking, and two thirds of these falls were caused by slipping. Of the injuries from slipping, two thirds were sustained by women. Those 50 and over were more likely to get injured from slipping than any other age group.

These findings show that falling and getting injured on winter roads is a common problem in the Northern regions, and that women and senior pedestrians are more prone to this problem, often resulting in serious injuries. This report pays special attention to female pedestrians exposed to wintertime risks, and explains the results of a questionnaire. The questionnaire asked women in Sapporo about the winter pedestrian environment and how they responded to these conditions.

Results of the questionnaire highlighted the need to improve winter shoes as part of pedestrian safety measures. In light of this, members of the Sapporo Research Society of Consumers' Life in the Northern Countries, Sapporo Consumers' Association wore ten types of women's winter shoes with anti-slip features for trial use. Their opinions about comfort and improvements required of winter shoes are explained.

2 OPINION SURVEY ON WOMEN ABOUT THE PEDESTRIAN ENVIRONMENT

2.1 Survey methods

In March 1995, a survey was conducted by mailing a questionnaire to female members 20 years and older of the Sapporo Consumers' Association. Inquiries were made in the following areas:
1. Pedestrian environment and occurrence of falls and injuries in winter.
2. Evaluation of winter shoes and effectiveness of anti-slip soles.
3. Walking wear in winter.
4. Safety measures for walking in winter.
Of the reply forms collected, 1,382 were accepted. 75% were in their 40s to 60s and 96.9% had been living in Hokkaido for more than 10 years, thus the survey results represent opinions and experiences of women in their 40s to 60s well accustomed to living in a snowy region.

2.2 Survey results and evaluation

(1) The pedestrian environment and injuries
Over 90% of the respondents felt that sidewalks in Sapporo have become more slippery in the last several winters. Since December 1994, 52.4% have fallen while walking, about 80% had one or two falls and 20% experienced three falls or more.

Figure 1 shows where falls occurred. More than half (54.5%) took place on sidewalks. A rather high percentage, 31.5%, occurred on pedestrian crossings, indicating the influence of frequent occurrences of slippery road surfaces on roadway areas. In 86% of the falls injuries occurred. Body parts injured are shown in Figure 2. The lower back and the knees were most likely to be injured. 6.5% of the injuries were to the head, which can lead to serious ramifications. Most head injuries were sustained to the back of the head, suggesting difficulty in taking a defensive posture when falling backward. Regarding actions following the injuries, 15% of the injured sought medical attention at hospitals, whereas 33.2% administered medical treatment by themselves, indicating that about half of the bruises and injuries required medical treatment.

(2) Evaluation of winter shoes and effectiveness of anti-slip soles
Many women were not fully satisfied with the shoes now using. Most of the respondents rate their shoes as "a little slippery" on slippery roads, 10% think them as "not slippery", while another 10% consider them to be "very slippery" (Figure 3). Figure 4 shows the contributing factors to the "not slippery" evaluations. The most common reasons (38%) for "not slippery" performance are pins, knobs, and ceramics on the sole, followed by natural rubber soles, design, and the shape of grooves on the sole, in that order.

Presently, detachable anti-slip soles are sold and featured in newspapers. 16.3% respondents had detachable anti-slip soles. 32.4% found the soles to be effective; however, 52.0% rated them only as "better to have than not", indicating that these soles are not the ultimate solution. When asked in what areas the detachable soles needed improvement, the respondents cited appearance and ease of attachment/removal.

(3) Walking wear
Winter shoes worn by women in Sapporo are divided as follows: 62.6% wear low-brim winter shoes; 31.0% wear high-brim shoes such as boots; and a very small number continue to wear their summer shoes. Low-brim winter shoes are defined as waterproof with some kind of anti-slip feature on the sole.

If it is snowing, 51.9% wear hats and 43.3% use umbrellas, showing that most use either one or the other (Figure 5). Both hands are free with hats on, while only one hand remains free when

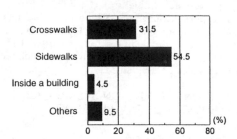

Figure 1. Q. Where did you fall down?

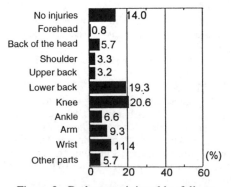

Figure 2. Body parts injured by falls.

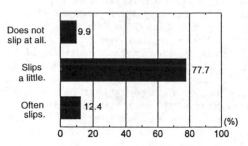

Figure 3. Q. How is the anti-skid performance of the shoes you are wearing?

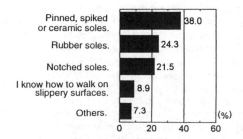

Figure 4. Q. Why don't your shoes slip?

using an umbrella, making it difficult to quickly assume a protective posture during a fall. Because this could lead to serious injury, efforts should be made to provide necessary information on winter lifestyles from the view point of pedestrian safety. When gloves are worn, the body can be supported by one's hands when falling even on slippery roads, and hands and wrists will also be protected. It was found from our survey that most of the respondents wear gloves (Figure 6); thus, few of them put their hands in their pockets when walking.

Walking sticks are another aid. Presently, some are fitted with hardware on the ends to enable penetration into ice on winter roads. Our survey shows, however, that 95.7% do not have walking sticks.

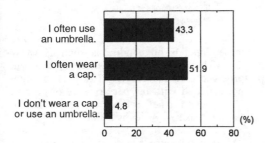

Figure 5. Q. Do you wear a cap or use an umbrella when it is snowing?

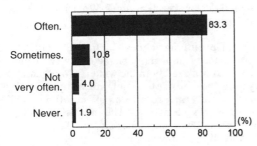

Figure 6. Q. How often do you wear gloves?

(4) Measures to improve pedestrian safety
To improve winter safety for pedestrians, 40.1% of the respondents mentioned the development of slip-proof shoes, 31.4% cited increasing city expenditures on more complete snow removal, and 19.8% think pedestrians should be more careful (Figure 7). The fact that more people prefer the development of slip-proof shoes to more money spent on snow removal may be interpreted that one cannot expect the road maintenance authorities to provide complete snow removal.

Among the measures undertaken by the road maintenance authorities, most respondents chose road heating systems, followed by the spreading of sand and making frozen surfaces coarse.

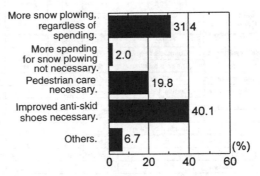

Figure 7. Q. What would be a good measure to improve walking conditions during winter?

3 TRIAL USE OF WINTER SHOES

3-1 *Test method*

The members of the Sapporo Research Society of Consumers' Life in the Northern Countries tested ten kinds of winter shoes currently on the market. Each pair of shoes were used by 3-8 members for 23-54 days to report the level of ease in walking. Seven were low-brim shoes and three were fairly high. Of the ten, seven had anti-slip spikes, including one with rubber-made spikes.

Evaluations included the anti-slip, water-proofing and anti-cold performances. The test was conducted on four environments: icy surface, slush, compacted snow, and inside a building.

3.2 *Test results and considerations*

Table 1 shows the results of the test. In the table, ○ indicates a high evaluation, ● low evaluation and ▲ both high and low evaluations. Generally, anti-slip performance on icy surfaces, compacted snow and inside buildings are important. All of the tested shoes proved to have a certain level of anti-slip effectiveness on compacted snow. Only four pairs of shoes proved to have adequate anti-slip performances on icy surfaces, and only one for inside buildings. However, the evaluation of many of the shoes varied depending on the tester. We also assume that various differences in environmental conditions, and personal differences caused such evaluation variations. A small number of shoes were given a poor evaluation for their anti-slip performances on icy surfaces and compacted snow. The results prove that there is an overall improvement of anti-slip performances. However, the anti-slip performance inside

Table 1. Results of the winter shoes experiment.

	Shoes #	1	2	3	4	5	6	7	8	9	15
	No. of days worn	28	23	50	33	28	43	35	54	38 ...	38
Performance	No. of people wore	4	3	5	4	5	7	5	8	6	4
Skid resistance	Ice	●	●	○	○	○	▲	○	▲	▲	▲
	Compacted snow	○	○	○	○	○	▲	▲	▲	▲	○
	Inside a building	●	●	●	●	▲	▲	○	●	▲	▲
Water resistance	Slush	○	○	○	○	●	●	●	▲	○	▲
Cold resistance	Compacted snow	—	—	—	●	—	—	—	—	—	—
	Slush	—	—	●	—	—	—	●	—	—	—
(*1)	Compacted snow	●	○	○	○	○	●	○	●	○	●
(*2)	Inside a building	▲	○	▲	●	●	○	○	●	○	○
Height of shoes		Low	Low	Low	High	Low	Low	Low	Fairly high	Fairly high	Low
Pins and spikes on soles		Equipped	Equipped	Not equipped	Equipped	Equipped	Not equipped	Not equipped	Equipped	Equipped	Equipped
Sole materials		SR	SR	SRE	SR	SR	SR	SRE	SR	SR	NR

*1 Collection of snow on soles
*2 Noise generated by pins and spikes

NR : Natural rubber
SR : Synthetic rubber
SRE : Synthetic rubber with emery

buildings needs to be improved. Also, many people pointed out the problems caused by side slip of the heel.

Regarding waterproof performance, half of the tested shoes performed well - water did not soak into the shoes, but the remaining shoes performed poorly, with some causing the wearer's socks to become discolored. Because of limited evaluations of cold resistance performance, a clear overall evaluation cannot be given. However, only three pairs were evaluated as poor, thus in general, the cold resistance performance can be considered good. In addition, snow attachments to the bottom of some shoes, especially those with metal spikes, were reported on. Such attachments increased walking difficulty and noise from spikes inside buildings was pointed out as a nuisance of three pairs.

4 CONCLUSIONS

(1) Over 90 % of Sapporo's female citizens state that the winter road surfaces in Sapporo have become much more slippery within the last 2-3 years. More than half have fallen as a result. Women tend to fall on pedestrian sidewalks and crosswalks. Injuries to the lower back and knee constituted a majority of the injuries. More than half of the women who fell required some care, including 15% who went to the hospital. Considering the aging of Japanese society, injuries due to falls on winter roads can be expected to become more serious in the future.

(2) Women in Sapporo, especially in their 40s to 60s, take into consideration anti-slip performance when buying new winter shoes. Also, their purchase depends on conditions such as snow depth. They have a deep interest in the performance of winter shoes. However, as they are not satisfied with the performance of current winter shoes, further follow up research and development of new shoes is needed.

(3) Most people use a hat/cap or an umbrella when it is snowing, and also wear gloves. Few people put their hands in their pockets, limiting their fall protection ability. However, holding an umbrella delays reaction time. Women need to be informed of this danger.

(4) Developing shoes with high anti-slip performance is thought to be the best way to allow people to walk safely in the winter. To organize a system to promote such development by manufacturers, retailers, specialists and consumers is necessary. The areas that need to be improved in winter shoes are:
1. Anti-slip performance in buildings
2. Anti-slip performance on icy surfaces and the prevention of side slip of the heels
3. Waterproof performance
4. Prevention of snow attachment to soles
5. Decrease of noise by spikes inside a building

REFERENCES

Merrild, U. & Bak, S. 1983. An Excess of pedestrian injuries in icy conditions, Accid. Anal. and Prev, Vol. 15, No. 1:, 41-48.

Nilsson, G. 1986. Halkolyckor (Slippery accidents), VTI Rapport 291.

Snow Engineering: Recent Advances, Izumi, Nakamura & Sack (eds) © 1997 Balkema, Rotterdam. ISBN 90 5410 865 7

Variations of road traffic flow according to road surface conditions in a snowy area

Masafumi Horii

Department of Civil Engineering, College of Engineering, Nihon University, Koriyama, Japan

ABSTRACT: This paper discusses the influence of snow conditions on space-mean speed and traffic volume for a two-lane, two-way section of rural highway located in mountainous terrain in Japan. Traffic volume and space-mean speed are reduced under snow or ice conditions, and can be classified according to road surface conditions.

1 INTRODUCTION

Road traffic in snowy areas is frequently paralyzed during the snow season. Expensive snow removal efforts are needed to maintain smooth traffic flow and to maintain social and economic activities during the snow season. Consequently, it is necessary to examine the influence of snow on road traffic flow in order to formulate measures to prevent impairment of the traffic system under snow conditions. However, there are very few studies concerning traffic flow in snowy areas.

This paper presents data on traffic flow under normal conditions and snow conditions, and analyzes the correlation between time-mean speed and space-mean speed. Furthermore, using this correlation, the influence of snow conditions on space-mean speed and traffic volume is examined for a two-lane, two-way section of rural highway located in mountainous terrain in Japan. Multiple regression analysis is applied to explain variations in space-mean speed under snow conditions.

2 TRAFFIC FLOW CHARACTERISTICS

The fundamental characteristics of traffic flow are the traffic volume, q, traffic density, k, and space-mean speed, $\overline{v_s}$. In general, these three characteristics are related as follows:

$$q = k\overline{v_s}. \tag{1}$$

The space-mean speed for a given road section can thus be calculated as follows:

$$\overline{v_s} = \frac{1}{1/m \sum_{i=1}^{m} 1/v_i} \tag{2}$$

where $\overline{v_s}$ is the space-mean speed in kilometers per hour, v_i is the speed of vehicle i and m is the number of observations. As shown in equation (2), space-mean speed is calculated as the harmonic mean of individual speeds, or as the reciprocal of the average travel time rate. On the other hand, when individual speeds are recorded and averaged for vehicles passing a particular point or passing through a short segment over a given time period, the resulting mean speed is referred to as time-mean speed. This is given by the following equation:

$$\overline{v_t} = \frac{1}{n} \sum_{j=1}^{n} v_j \tag{3}$$

where $\overline{v_t}$ is the time-mean speed in kilometers per hour, v_j is the speed of vehicle j and n is the number of observations. The mean speed measured by traffic detectors is usually time-mean speed.

In general, time-mean speed and space-mean speed are related as follows (Wardrop 1952):

$$\overline{v_t} = \overline{v_s} + \frac{\sigma_s^2}{\overline{v_s}} \tag{4}$$

where σ_s^2 is the variance in space-mean speed. In traffic engineering, since space-mean speed is a more important index, it would be desirable if space-mean speed could somehow be estimated from time-mean speed for the road section under consideration. However, there are very few empirical studies concerning this equation or the relation between these two mean speeds (for example, Saito *et al*. 1977 and Horii *et al* 1996) In the present paper, the relation between time-mean speed and space-mean speed is discussed. Furthermore, since road traffic flow during the snow season is affected by several factors, the basic diagram of the traffic flow under snow conditions remains unclear. This relationship is also discussed.

3 DATA COLLECTION

The survey on which this paper is based was carried out in the vicinity of the Tsuchiyu Pass on a road with 5% grade (Route 115). The observation point was equipped with a pair of traffic detectors for measuring speed, occupancy and traffic volume, and with meteorological equipment for measuring temperature, snow depth and road surface conditions. As shown in Table 1, three kinds of data were collected: (1) 5-minute averages of speed, occupancy and traffic volume in both the upgrade and downgrade directions, together with weather conditions; (2) travel times of individual vehicles passing two particular points as determined by a video tape recorder positioned to overlook the points; (3) 60-minute averages of speed, occupancy, traffic volume and weather conditions. Data sets (1) and (2) were used to determine the correlation between time-mean speed and space-mean speed, and then data set (3) was used to examine the influence of snow conditions on road traffic flow.

Table 1. Observed Data

No	Season	Observed from Date	Time	Observed to Date	Time
(1)	No-snow	08/27/90	13:00	08/27/90	16:00
	Snow	01/09/91	13:10	01/09/91	17:10
(2)	No-snow	08/27/90	13:00	08/27/90	16:00
	Snow	01/09/91	13:10	01/09/91	17:10
(3)	No-snow	04/01/90	00:00	04/24/90	11:00
		07/27/90	17:00	09/03/90	13:00
		09/27/90	12:00	10/19/90	02:00
	Snow	03/20/90	12:00	03/31/90	24:00
		11/22/90	11:00	02/04/91	01:10
		03/11/91	11:00	03/28/91	21:00

4 RESULTS

4.1 *Correlation between time-mean speed and space-mean speed*

Figures 1 and 2 show the relation between traffic volume and time-mean speed, space-mean speed on this road section in the upgrade and downgrade directions, respectively. It is apparent from these results that the mean speed decreased as the traffic volume increased. Traffic volume and mean speed were reduced during the snow season, compared with volume and speed under no-snow conditions. During the snow season, traffic volume and mean speed were reduced in the downgrade direction, whereas there was not such a wide difference between traffic flow characteristics in the upgrade and downgrade directions under no-snow conditions. Since the road surface was covered by snow, drivers traveling in the downgrade direction tended to reduce their speed, and this caused traffic volume to drop. As shown in equation (4), time-mean speed was actually greater than space-mean speed. Figures 3 and 4 show the relation between measured values and values of time-mean speed estimated by equation (4). When the 5-minute data set is used for the analysis, a strong correlation is estimated from these results, because the value of correlation coefficient R in the upgrade direction is 0.979, and R in the downgrade direction is 0.942. Therefore, the theoretical equation (4) was empirically confirmed.

In order to estimate space-mean speed from time-mean speed using equation (4), it is necessary to know the variance in space-mean speed. However, calculating the variance for space-mean speed is a time consuming task, and consequently this equation is not useful for estimating space-mean speed. Consequently, these data were analyzed using regression analysis.

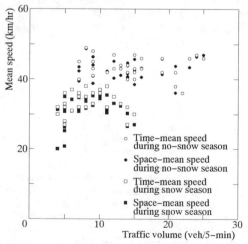

Figure 1. Traffic Volume vs. Mean Speed (Upgrade)

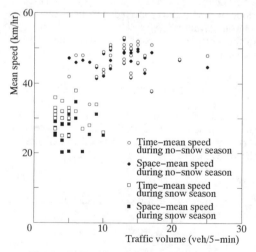

Figure 2. Traffic Volume vs. Mean Speed (Downgrade)

Regression equations (5) and (6) were obtained as follows:

$$\bar{v}_s = -3.783 + 1.064\ \bar{v}_t,\ R = 0.968\ \text{(upgrade)}, \qquad (5)$$
$$\bar{v}_s = -4.441 + 1.071\ \bar{v}_t,\ R = 0.958\ \text{(downgrade)}. \qquad (6)$$

Therefore, it was shown that under certain conditions space-mean speed can be estimated with high precision from time-mean speed on the road section.

4.2 Influence of snow conditions on road traffic flow

First, using equations (5) and (6), 60-minute averages of time-mean speed (data set (3)) were converted to 60-minute averages of space-mean speed. Then the influence of snow was examined on traffic volume and space-mean speed. Figures 5 and 6 show the relation between traffic volume and space-mean speed in the downgrade direction under no-snow and snow conditions, respectively. Under no-snow conditions, the space-mean speed decreases linearly as the traffic volume increases. On the other hand, under snow conditions, the traffic volume and space-mean speed vary widely and are especially reduced when the road surface is covered with snow or ice. In addition, a rather large overlap was observed in the q-v_s distribution during snow and no-snow seasons for dry or wet road conditions. This suggests that traffic volume and space-mean speed show a similar tendency

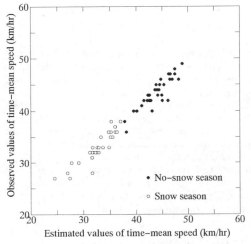

Figure 3. Observed and Estimated values of Time–Mean Speed (Upgrade)

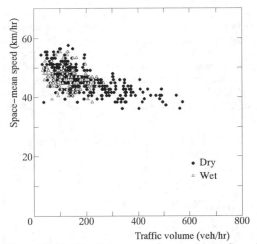

Figure 5. Traffic Volume vs. Space–Mean Speed (Downgrade, No–snow season)

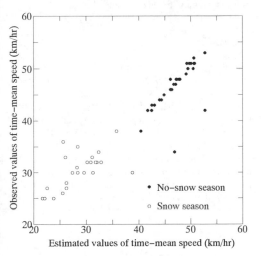

Figure 4. Observed and Estimated Values of Time–Mean Speed (Downgrade)

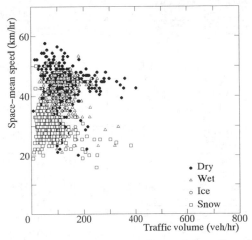

Figure 6. Traffic Volume vs. Space–Mean Speed (Downgrade, snow season)

under dry or wet road conditions during both seasons. Therefore, to examine the influence of snow conditions on the space-mean speed of this road section, multiple regression analysis was applied to these data which include both seasons. Here, the traffic density is estimated from the time-occupancy data, assuming that the mean vehicle length is 5.5 meters. Since the road surface conditions are qualitative variables, three dummy variables were introduced into the regression equation. The variables are (1) traffic density, (2) temperature, (3) road surface dummy 1, which is coded as "1" for a snow-covered road and "0" otherwise, (4) road surface dummy 2, which is coded as "1" for an ice-covered road and "0" otherwise and (5) road surface dummy 3, which is coded as "1" for a road under wet conditions and "0" otherwise. Tables 2 and 3 show the results of the multiple regression analysis for space-mean speed in both directions. Here, s is the standard deviation of residuals. The value of the multiple correlation coefficient R is 0.717 in the upgrade direction and 0.774 in the downgrade direction, so that variations in space-mean speed were fairly well explained by the five variables. From the results of the t-test, all variables were statistically significant.

Table 2. Regression Coefficients and Standard Errors for Space-mean speed (Upgrade)

Variables	Coefficient	Standard error	t value
Traffic Density (veh/km)	-0.770	0.070	-10.955
Temperature ($^\circ$C)	0.312	0.018	17.057
Road Surface Dummy 1	-7.289	0.427	-17.075
Road Surface Dummy 2	-2.598	0.762	-3.410
Road Surface Dummy 3	-1.928	0.288	-6.690
Constant	41.587	0.293	142.032
n=1520	R^2=0.514	R=0.717	s=4.623

Table 3. Regression Coefficients and Standard Errors for Space-mean speed (Downgrade)

Variables	Coefficient	Standard error	t value
Traffic Density (veh/km)	-1.283	0.086	-14.886
Temperature ($^\circ$C)	0.463	0.021	22.405
Road Surface Dummy 1	-8.705	0.479	-18.182
Road Surface Dummy 2	-4.162	0.854	-4.877
Road Surface Dummy 3	-1.669	0.323	-5.170
Constant	44.230	0.341	129.545
n=1520	R^2=0.600	R=0.774	s=5.184

Therefore, efficient regression equations were obtained for space-mean speed .

Based on the regression results, the coefficients of road surface dummy 1 are estimated to be -7.289 (upgrade) and -8.705 (downgrade). That is, the average decrease in space-mean speed under the snow-covered road condition, compared to dry, is estimated to be 7.289 km/hr in the upgrade direction and 8.705 km/hr in the downgrade direction. For road surface dummy 2, space-mean speed under the ice-covered road condition, compared to dry, is estimated to be reduced by 2.598 km/hr and 4.162 km/hr in the upgrade and the downgrade directions, respectively. Thus, the influence of snow on road traffic flow could be quantitatively evaluated.

5 CONCLUSION

Regression analysis was used to analyze the influence of snow conditions on the space-mean speed and traffic volume of a two-lane, two-way section of rural highway located in mountainous terrain in Japan. A strong correlation was observed between space-mean speed and time-mean speed on this road. Furthermore, traffic volume and space-mean speed were reduced under snow- or ice-covered road conditions and could be classified according to road surface conditions. A multiple regression model of space-mean speed that takes into consideration road surface conditions was obtained and variations in space-mean speed could be explained fairly well using this model.

ACKNOWLEDGMENT

The author wishes to thank Dr. Tadashi Fukuda, Professor, Graduate School of Information Sciences, Tohoku University, for his thoughtful suggestions.

REFERENCES

Wardrop J.G. 1952. Some theoretical aspects of road traffic research, *Institute of Civil Engineers, Proceedings*, Part II, Vol. 1: 325-362.
T. Saito and Y. Nishida 1977. Relations between time mean speed and space mean speed, *National Research Institute of Police Science*, Vol. 18, No. 1: 71-86 (in Japanese).
M. Horii and S. Murai 1996. Study on relationship between time-mean-speed and space-mean-speed of road traffic flow in a snowy area, *Tohoku Journal of Natural Disaster Science*, Vol. 32: 193-198 (in Japanese).

Snow Engineering: Recent Advances, Izumi, Nakamura & Sack (eds) © 1997 Balkema, Rotterdam. ISBN 90 5410 865 7

Studies on the influence of accumulated snow affecting route choice behavior

Jinsuke Orita
Department of Urban and Environmental Studies, Akita National College of Technology, Japan

Akira Yuzawa
Department of Urban and Environmental Studies, Nagaoka National College of Technology, Niigata, Japan

ABSTRACT In this paper traffic−route−choice behavior during the snowy season is analyzed. The factors that influence the decision to drive safely on roads having accumulated snow are evaluated using Quantification Theory II. We verify that in the case of professional drivers the most influential factor is the snow−removal factor ; in the case of ordinary drivers traffic−jam conditions are most important. Finally, the disaggregate behavioral model is applied. In this analysis, it is verified that decision making in regard to traffic−route−choice, professional drivers consider all four conditions before driving, while ordinary drivers only consider "necessary time."

1. THE PURPOSE OF THIS RESEARCH

Snow causes a great amount of trouble in local traffic, causing traffic jams, frequent traffic accidents, changes in traffic routes and so on. Furthermore, t sometimes interferes with the planning of travel. Activities and social functions involving economy and industry are also made difficult because of snow. Using a disaggregate behavioral model, this study aims at traffic−route−choice behavior during the snowy season, in hopes of understanding what determines the choice of route that drivers make. In addition to this analysis, Quantification Theory II will be used to evaluate factors that influence the decision to drive safely on roads having accumulated snow.

In this research, truck and bus drivers (defined as professional drivers) and citizens (defined as ordinary drivers) acted as respondents to questions regarding both their awareness of a need to select traffic routes, and the reasons why they chose certain routes. Both factors are analyzed separately.

2. FORMER STUDIES

2. 1 FORMER STUDIES CONCERNING TRAFFIC BEHAVIOR DURING SNOWY SEASON

There are many kinds of research concerned with snowy seasons; for example traffic problems, snow removal and traffic behavior. In his definition and analysis of traffic problems, Yamashita et al. (1972) used the KJ method. These problems have also been systematically analyzed by the DEMATEL method (Takahashi. T and Fukuda. T 1980) in order to quantitatively understand the relationship and influence each problem has on other problems.

In our analysis of traffic behavior, mode choice during summer and snowy seasons have been compared through a field investigation. We have verified that the number of motor vehicle drivers during the snowy season decreases, and that the number of users of public transit (for example, bus and train) increases during the winter months.

Although many useful facts have been reported, research on traffic route behavior under snow conditions, using the disaggregate behavioral model, has never been published.

2. 2 FORMER STUDIES CONCERNING TRAFFIC−ROUTE−CHOICE BEHAVIOR

Much research on traffic behavior using the disaggregate behavioral model has been reported. Several studies focus on traffic−route −choice behavior, for example, Yabe et al. (1994), Yamanaka et al. (1983), and Shimizu et al. (1995).

Yabe at al. analyzed traffic−route−choice behavior using the multinominal logit model and the nested logit model. In their study, RP (Revealed Preference) data were used. Signal density, number of traffic lanes, and necessary time variables were used in these models. Similarly, Yamanaka et al.(1983) studied traffic−route−choice behavior using an ordinary logit model with bus and subway users as subjects. In this research, necessary time, boarding time , waiting time, etc. were generic variables in constructing the model. Age−dummy(60 age over and subway choice = 1) and attending−school−aim−dummy (attending−school−aim and subway choice =1) variables were adopted as alternative specific variables.

In the above−mentioned research, some variables were used to construct models, but none of them involve snowy conditions. The research of Shimizu et al.(1995) follows the same pattern.

3 .INVESTIGATION AND METHOD OF ANALYSIS

Two kinds of data are necessary to construct the disaggregate behavioral model. RP(Revealed Preference) data describe actual traffic behavior, and SP(Stated Preference)data describe the awareness of traffic preference behavior. In this present study, SP data are used, while the conditions of traffic route were established in advance. That is, the routes chosen by our respondents are analyzed. The conditions of each route under various circumstances are defined using design of experiments(Taguchi 1990).

Factors considered necessary when driving safely on snow covered roads are analyzed using Quantification Theory II (Ishihara et al.1990). Traffic−route−choice behavior is analyzed using Equation (1) (Japan society of Traffic Engineers 1993). We constructed a disaggregate behavioral model to understand the factors affecting route choice behavior, and to simulate that behavior under other driving conditions.

4 . ANALYSIS OF FACTORS INFLUENCING THE DECISION FOR SAFE DRIVING ON ROADS WITH ACCUMULATED SNOW

The factors affecting these decisions are analyzed using Quantification Theory II. Established items for analysis are shown in Table 1 and Table 2. Catagories for each item include three grades: " strongly considered," " slightly considered, " and "never considered." External criteria refer to the number of responding groups and the general evaluation of the degree of consideration given to driving on snowy roads. The grades are the same as those in the categories. Table 1 shows the results in the case of professional drivers. Item 1 and Item 2 are excluded because no one responded to the category of "never considered. " According to Table 1 the most

Table1. Analysis of affected using Quantification Theory II (professional drivers)

No.	Items	Categories	Coefficient	Range
2	Quality of accumulated snow	C_{21}	−1.0123	1.0294
		C_{22}	0.0170	
		C_{23}	(0)	
3	Num. of traffic signals	C_{31}	0.9224	1.0895
		C_{32}	−0.1671	
		C_{33}	(0)	
4	Width of traffic lanes	C_{41}	−0.7446	0.7447
		C_{42}	−0.0249	
		C_{43}	(0)	
5	Distance	C_{51}	−0.8952	0.9440
		C_{52}	−0.0249	
		C_{53}	(0)	
6	Cost of fuel	C_{61}	−1.3145	1.3146
		C_{62}	−0.3871	
		C_{63}	(0)	
8	Condition of snow removal	C_{81}	−1.6363	1.6364
		C_{82}	−0.8864	
		C_{83}	(0)	
	Correlation ratio			0.4103

Table2. Analysis of affected using Quantification Theory II (ordinary drivers)

No.	Items	Categories	Coefficient	Range
1	Necessary time	C_{11}	−0.5810	0.8036
		C_{12}	−0.8036	
		C_{13}	(0)	
2	Quality of accumulated snow	C_{21}	0.4150	1.0172
		C_{22}	−0.6022	
		C_{23}	(0)	
3	Num. of traffic signals	C_{31}	−0.3980	0.3980
		C_{32}	−0.0196	
		C_{33}	(0)	
4	Width of traffic lanes	C_{41}	1.0545	1.0545
		C_{42}	0.2097	
		C_{43}	(0)	
5	Distance	C_{51}	0.1353	0.2794
		C_{52}	−0.1441	
		C_{53}	(0)	
6	Cost of fuel	C_{61}	−0.6558	0.9007
		C_{62}	0.2450	
		C_{63}	(0)	
7	Traffic volume	C_{71}	1.4485	1.4485
		C_{72}	0.5350	
		C_{73}	(0)	
8	Condition of snow removal	C_{81}	0.7787	0.8369
		C_{82}	0.8369	
		C_{83}	(0)	
	Correlation ratio			0.2704

influential factor when considering safe driving on accmulated snow is the snow−removal factor (Range=1.63638). Following this item, the cost of fuel and the number of traffic signals have high value; there is little value placed on the width of traffic lane. On the other hand, in the case of ordinary drivers, traffic volume has the highest value in Table−2. The second highest value is the width of the traffic lane, and the third value is the type of snow on the roads. It is important to note that both professional and ordinary drivers consider the type and condition of snow on roads.

5. ANALYSIS OF TRAFFIC ROUTE CHOICE BEHAVIOR USING THE DISAGGREGATE BEHAVIORAL MODEL

On the basis of four conditions, namely, distance, nesessary time, type of snow, and the width of traffic lanes during snowfall, nine kinds of routes can be identified. Each condition has three grades levels. These routes are shown in Table 3. Assuming that the commuter road which the respond−ents usually use become impassable for some reason, the respondents chose the route they want to go from the established nine routes. Table 3 shows the conditions of the usual route and Table 4 shows the results of the chosen routes. Many pro−fessional drivers preferred route D first, and then route A. The most common routes chosen by ordinary drivers were different. Ordinary drivers overwhelmingly preferred only route D. Considering " necessary time," routes D and A have the lowest value of any of the established routes. Thus, both sets of respondents considered " nesessary time " when choosing their routes during snowfall.

The authors guess that traffic jams occur frequently when cars slide because of snow−fall. As a result, it takes more time for drivers to reach their destination. Traffic jams have become worse because of the changeto studless tires.

The models for the analysis of traffic−route − choice behavior have been con−structed by Equation (1), separating pro−fessional drivers from ordinary drivers. The number of samples used to construct the models are , respectively, 92 and 80. Table 5 shows the obtained parameters and the t−value. In the case of professional drivers, all adopted explanatory variables are valid because the t−values are significant when their volume is over 1.980 (i.e., $t_{0.005}(\infty)$)

$$P_{in} = \frac{e^{V_i}}{\sum_j e^{V_j}} \quad \cdots \cdots \quad (1)$$

$$V_i = \theta_1 X_{1i} + \theta_2 X_{2i} + \cdots + \theta_k X_{ki}$$

P_{in} :Choice probability in choosing route " i " by an individual

V_i :Deterministics part of utility by choosing route " i "

X_{ik} :The "k" explanatory variables for choice route " i "

θ_k :Parameter of the "k" explanatory variables

Table3. Choice routes

■ Condition of roads when driving to work

Route	Distance	Necessary time in winter (min.)	Type of snow	Width of traffic lane in snow
(J)	10km	30	Wet snow	3.25m

■ Condition of roads to choose from

Route	Distance	Necessary time in winter (min.)	Type of snow	Width of traffic lane in snow*
(A)	10km	40	Wet snow	Extremely narrow
(B)	10km	50	Packed snow	Slightly narrow
(C)	10km	60	Icy road	No great difference
(D)	15km	40	Packed snow	No great difference
(E)	15km	50	Icy road	Extremely narrow
(F)	15km	60	Wet snow	Slightly narrow
(G)	20km	40	Icy road	Slightly narrow
(H)	20km	50	Wet snow	Extremely narrow
(I)	20km	60	Packed snow	No great difference

*Extremely narrow ··· width is about 2.00m
Slightly narrow ·· width is about 2.50m
No great difference ·· width is about 3.00m

= 1.960 : two− sided probability according to the t−distribution test). On the other hand, data for " necessary time " is only available for ordinary drivers. All positive and negative figures for all parameters are significant. According to this table, the largest t−value is " necessary time " for profes−sional drivers, followed by the value of traffic lane width. Generally, in this case all var−iables influence traffic route behavior, while in the case of ordinary drivers only " necessary time " has influence. Ac−

TAble4. Results of chosen route

Route	Num. of answers (1)	Num. of answers (2)
A	18	10
B	12	10
C	2	1
D	38	49
E	1	0
F	12	5
G	0	0
H	1	0
I	8	5
Total	92	80

Note; (1) :professional drivers
(2) :ordinary drivers

Table5. Obtained parameter and t−value

◆ professional drivers

	Parameter	t-value
Distance	-1.134500E-01	-2.2473
Necessary time	-6.008000E-02	-4.3414
Quality of accumulated snow	-1.480700E+00	-3.7556
Width of traffic lane	2.520300E-02	4.1653

◆ ordinary drivers

	Parameter	t-value
Distance	-3.748636E-02	-0.2866
Necessary time	-8.556916E-02	-3.5478
Quality of accumulated snow	-9.400368E-03	-0.5517
Width of traffic lane	1.896962E-02	1.2349

Table6. Results of choice probability

Route	Choice probabilty (1)	Choice probabilty (2)
A	23.23	16.55
B	10.21	11.35
C	4.49	(−)
D	37.26	57.15
E	0.37	(−)
F	13.96	6.40
G	(−)	(−)
H	4.09	(−)
I	6.35	8.56

(1):professional drivers
(2):ordinary drivers
(−) denotes low response, hence no data

cordingly, it is mentioned that decision −
making in traffic−route−choice among pro−
fessional drivers is done considering four
conditions. Ordinary drivers do not consider
the four conditions before driving, they only
consider the quickest route.

Table 6 shows the obtained choice prob−
ability. In this table, choice probabilty
for routes D and A are shown to have

the highest and second highest number
of responses. In the case of ordinary drivers,
the value of route D has the highest prob−
ability. Thus, the obtained models are
all significant.

6. CONCLUSIONS

Accumulated snow on roads has an in−
fluence on traffic route behavior. We have
analyzed this behavior using the disaggregate
behavioral model and found that the time
necessary to reach a destination greatly
affects the choice of traffic route. The models
used in this study are valuable and can be
used for future study in traffic patterns
and behavior.

REFERENCES

1)Yamashita. H, Shimomura. T and Fukuyama. T
(1972) : Study on the Problems Concerned With
Traffic in areas of High Snowfall Using
KJ Method Proceeding of the 27th Annual
Conference of the Japan Society of Civil
Engineers 4, PP. 15−18

2)Takahashi. K and Fukuda. T (1980) : Consider−
ation on the Awareness of Inhabitants Concerned
With Urban Taffic Problems in Snowy Areas,
Proceeding of the 35th Annual Conference
of the Japan Society of Civil Engineers 4,
PP. 209−210

3)Orita. J, Shimizu. K and Kurita. T (1984):Some
Consideration on Traffic Behavior of Residents
in Poor Transportation Service Areas, City
Planning Reviw No. 19, PP. 25〜30

4)Yabe. Y, Fujii. S and Kitamura. R (1994): Con−
structing the Disaggregate Models to Analyze
the Decision Making Factors of Route Choice,
Proceeding of the 49th Annual Conference of the
Japan Society of Civil Engineers 4, PP. 868−869

5)Yamanaka. H, Amano. M and Kotani. M (1983)
: Studies on Bus Systematic Network Planning
Introducing Disaggregate Route Choice Model,
Papers of the 5th Research Meeting on Civil
Engineering Planning, PP. 462〜467

6)Shimizu. E and Hiratani. K (1995) : A Potential
for Application of Network to Choice Behavior
Model, Proceeding of Infrastructure Planning
NO. 17, PP. 127−138

7)Taguchi. G (1990) : Design of Experiments,
Maruzen, PP. 143−159

8)Ishihara . T, Hasegawa . K and Kawaguchi . T
(1990) : Multivarite Analysis, Kyoritu Syuppan,
PP. 182−197

9)Japan Society of Traffic Engineers (1993) :
Disaggregate Travel Demand Analysis

Snow Engineering: Recent Advances, Izumi, Nakamura & Sack (eds) © 1997 Balkema, Rotterdam. ISBN 90 5410 865 7

Effectiveness of fences in protecting workmen at construction sites from severe wind and snowdrifts

T. Hongo
Kajima Technical Research Institute, Tokyo, Japan

T. Tomabechi
Hokkaido Institute of Technology, Sapporo, Japan

ABSTRACT : When carrying out construction work in winter, workmen must be protected not only from snow but also from strong winds. Large wind and snow fences are employed as a countermeasure. This paper reports on the results of a drifting snow wind tunnel test to predict the effectiveness of wind and snow fences, and the results of field measurements of snowdrift, wind speed and wind direction.

1 INTRODUCTION

The site of the study is an open area with only five buildings located in Rokkasho-mura in Aomori Prefecture in north-eastern Honshu, Japan. Rokkasho-mura is one of a few localities in the country which is noted for large snowdrifts. Once snowdrifting begins, visibility becomes almost nil. As the site is large, about 120 x 130 m, it is most important to protect it from snowdrifts. Even when it is not snowing, the mean wind speed commonly exceeds 10m/sec and the mean temperature is often below 0°C. Furthermore, the wind-chill temperature is often significantly lower than ambient, so it is also important to control strong winds.

In this area, winter winds are predominantly WSW, W, WNW and NW. Therefore, attenuation of these winds is sufficient, and it was decided to employ fences to achieve it. However, the effectiveness of fences in reducing wind speed and snowdrifts is greatly influenced by geographical features and surrounding buildings. Thus it is important to establish the effectiveness of fences for a given situation.

2 DRIFTING SNOW WIND TUNNEL TESTS

2.1 Experimental details

Experiments were carried out using the drifting snow wind tunnel at the Hokkaido Institute of Technology (Toyoda 1992). A area with a diameter of about 480 m was modeled at a length scale of 1:600, to simulate the complex undulations as closely as possible.

In the light of past research on snowdrifts and fences, a 15 m-high main fence with a solidity ratio of 100% (H15) and two 5 m-high auxiliary fences with a solidity ratio of 100% (1H5 and 2H5) were adopted

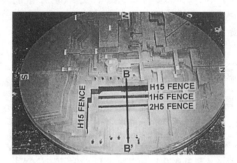

Figure 1. Terrain model and position of model fences.

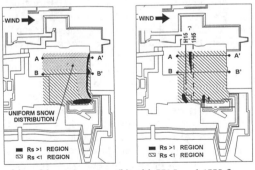

(a) without fences   (b) with H15 and 1H5 fences

Figure 2.   Snowdrift distribution in site area (wind direction : W).

for protection against wind and snow within a distance of approximately 120m. They were installed as shown in Figure.1.

The following three configurations were tested for W and NW wind directions with reference to the

occurrence frequency of wind directions at the site:
1. no fence.
2. main fence (H15) and one auxiliary fence (1H5).
3. main fence (H15) and two auxiliary fences (1H5, 2H5).

Activated clay particles with a moisture content of 8.5 % by weight were used to model the snow. A wind speed of 6 m/sec and a wind tunnel running time of 30 minutes were adopted as a test condition. These test conditions were determined on the basis of snowdrift simulation tests by the authors (Tomabechi 1985, 1986) and the conditions for the occurrence of snowdrifts in Rokkasho-mura. The snowdrift geometry was measured with a laser displacement sensor. The depth of model snow was expressed as a Snowdrift Coefficient Rs as follows:

$$Rs = \frac{\text{Depth of snowdrift at each measurement point}}{\text{Supplied amount of model snow}}$$

The amount of model snow supplied during a 30-minute running period produced a depth of 2.5 mm at the test section without terrain model.

2.2 Experimental results

Figure 2 shows two examples of model snow distribution for the W wind direction. For this wind direction, if a fence is not installed (Figure.2(a)), remarkably large snowdrifts are formed near the retaining wall on the leeward side. The central area of the site is divided into two parts: one showing almost uniform distribution with Rs > 1 where the model snow depth exceeds 2.5 mm and the other a blown-off part with Rs < 1. However, when fences are installed (Figure.2(b)), noticeable snowdrifts are formed between the main fence (H15) and the auxiliary fence (H5). However, the area on the leeward side of the auxiliary fence is a blown-off part, where the snowdrift distribution is uniform at a very small model snow depth.

Figure 3 shows the distribution of snowdrift coefficient Rs along sections A-A' and B-B' in Figure 2. With no fence, the depth of snowdrifts is large all over, and it is clear that the part with the snowdrift coefficient Rs > 1 extends over most of the A-A' section. In particular, on the leeward side, large snowdrifts with Rs of about 3 are formed because of the great influence of the retaining walls. With a main fence (H15) and one auxiliary fence (1H5), a large snowdrift with Rs almost 4 along A-A' is created upwind of the auxiliary fence. However, the coefficient of the snowdrift formed on the leeward side of the auxiliary fence is approximately 0.3, which is very small. It is discerned that in this case, the snow depth is reduced to less than 1/3 of that with no fence. When another auxiliary fence (2H5) is installed in addition to H15 and 1H5, large snowdrifts are formed at the front and rear of each fence. The snowdrift

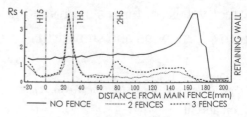

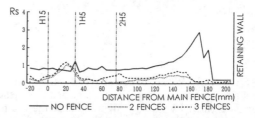

(a) A-A' section (wind direction :W)

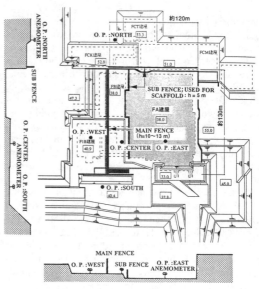

(b) B-B' section (wind direction :W)

Figure 3. Distribution of snowdrift coefficient Rs.

Figure 4. Fence locations and wind observation points.

coefficient on the leeward side of auxiliary fence 2H5 varies from about 0.5~1.0. Adding fence 2H5 causes more snow to de deposited in this area. This may be because the wind speed decreases sharply with the installation of auxiliary fence 2H5 and the snow drifts more easily. Although the results are not shown, the snow depth is also reduced for the NW wind direction.

These results show that the installation of fences resulted in decreasing the snow depths in the working area to 1/3~1/2 the depth with no fences, thus con-

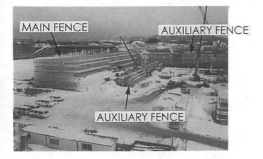

Figure 5. Actual wind and snow fences.

firming the effectiveness of fences against snowdrift.

The design and installation of actual fences were carried out considering the ease of work on the construction site. A panel with a solidity ratio of 100% was fitted to a frame consisting of supports for the main fence, and a 50 cm gap was made between the ground surface and the bottom of panel. Only one auxiliary fence, 1H5, was used. A net with a solidity ratio of 30% was fitted to a scaffold. Since it was ascertained from the wind tunnel test that a main fence in the south could produce undesirable snow-drifting, it was decided not to install one there. Instead, a 5-m high auxiliary fence was installed on the north side of the site to decrease NW wind speeds. Figure 4 shows the locations where fences were installed, and Figure 5 shows the actual fences.

3. FULL-SCALE MEASUREMENTS

Full-scale measurements of wind direction, wind speed and snow depth were carried out at the five points shown in Figure 4. A reference point was established in an open area on the southwest side (on the windward side of the fence) 900 m from the site.

3.1 Comparison of Occurrence Frequency of Wind Direction

Figure 6 shows the variation of wind direction occurrence frequency during both periods, with and without fences. At the reference point, slight differences are discerned between the two periods. However, it can be said that the occurrence frequencies for the wind direction during both periods were nearly same. Therefore, it can be judged that the observation results obtained during both periods with and without fences, may be directly compared. It is assumed that O.P.(observation point) South and O.P.North are not influenced by the presence of the fences. Slight differences can be seen between the two points, but there is no great difference in wind direction occurrence frequency. When there is no fence, a westerly wind at the reference point blows SE, ESE, S or SSW at

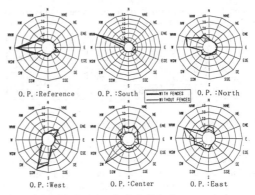

Figure 6. Variation of wind direction occurrence frequency of with and without fences.

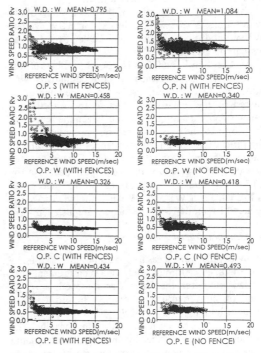

Figure 7. Variation of Rv with and without fences.

O.P.West due to the influence of the 6.4m-high slope to the west of the main fence. When fences are installed, a S or SSW wind blows along the main fence in most cases. At O.P.Center, with no fence, a SW wind blows much of the time. Where fences are installed, no predominant wind direction is discernible. At O.P.East, the wind direction distribution with no fence is similar to that at O.P.North. However, it can be seen that with fences, a W wind is intercepted.

Table 1. Variation of ratio of mean value of Rv with and without fences.

Wind Direction	Ratio of wind speed with/without fences				
	O.P.S	O.P.N	O.P.W	O.P.C	O.P.E
W S W	0.981	1.002	1.441	0.633	0.859
W	1.034	1.022	1.347	0.780	0.880
W N W	1.039	1.051	1.227	0.820	1.004

3.2 Comparison of Wind Speeds

The wind speeds at each of the observation points are indicated by Rv, the ratio of the relevant wind speed to that at the reference point. Figure 7 illustrates examples of the variation of Rv with and without fences for a W wind at the reference point. The abscissa indicates the wind speed at the reference point. Rv varies with wind speed, but tends to gradually approach a constant value (mean value).

Table 1 shows the ratio of the mean value of Rv with fences to that without fences for WSW, W and WNW wind. From Figure 7 and Table 1, the following can be explained:

At O.P.South and O.P.North, the effect of the fence cannot be discerned in the predominant wind direction. At O.P.West, with a fence, the wind intercepted by the fence flows from south to north as a valley wind, through the space between the fence and the slope to the west. Therefore, Rv increases, and the ratio of the mean Rv with fences to that without fences reaches 1 or more. At O.P.Center, located between the two fences, the wind speed decreases by 20%~40% due to the presence of the fences. At O.P.East, for a WSW or W wind, the wind speed is reduced by about 13% when fences are installed.

3.3 Comparison of Snow Depth

Snow depths were measured six times after the fence was installed. Although only one measurement was carried out after the fence was removed, a rough trend of snow depth was obtained. Figure 8 shows the distribution of the full-scale Snowdrift Coefficient Rs with fences when it snowed rather heavily. The full-scale Rs was obtained by dividing the snow depth at each measurement point on the site by the snow depth at the reference point. The measurement points were located along line B-B' in Figure 3. Rs values of 0.3~0.6 were obtained from the full-scale measurements at the most part on the leeward side of auxiliary fence. This approximates the values of 0.3~0.5 obtained in the wind tunnel.

Figure 9 compares snowdrift coefficients Rs with both fences and without the main fence when snow fell almost to the same depth at the reference point. It is clear that with fences, the snow depth on the site was reduced to 1/2~1/5 of that at the reference point. When the main fence was removed, snow drifted to a depth 1.2~1.5 times that at the reference point.

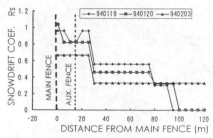

Figure 8. Full scale measurement of Rs.

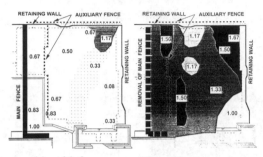

(a) with fences   (b) without main fence

Figure 9. Comparison of full scale Rs with and without fences.

4. CONCLUSION

The effectiveness of a system of fences for protection against wind and snow was investigated. Drifting snow wind tunnel tests and full-scale measurements of wind direction, wind speed and snow depth confirmed their effectiveness.

ACKNOWLEDGMENT

The authors wish to express their gratitude for the cooperation and valuable suggestions given by the staff of Japan Nuclear Fuel Ltd. and all persons concerned.

REFERRENCES

Toyoda, T. & T. Tomabechi 1992. Development of a wind tunnel for the study of snowdrifting. *Second International Conference in Snow Engineering* :207-214.

Tomabechi, T. & A. Endo 1985. The relation between the roof inclination and the forming conditions of the snow depth. *Journal of Structural and Construction Engineering, Trans. of AIJ.* 357:20-28 (in Japanese).

Tomabechi, T. & A. Endo 1986. Snowdrift formation around buildings. *J. of Snow Engineering of Japan.* 1:1-8.

Snow Engineering: Recent Advances, Izumi, Nakamura & Sack (eds) © 1997 Balkema, Rotterdam. ISBN 90 5410 865 7

Strength and frost heave properties of stabilized soils using lime and slag

Shinichiro Kawabata & Mitsuhiko Kamiya
Department of Civil Engineering, Hokkaido Institute of Technology, Sapporo, Japan

ABSTRACT: The replacement method (replacing frost susceptible soils with non-frost-susceptible soils) is simple and widely applied to the control of a frost heave in cold regions. However, a shortage of non-frost-susceptible materials may be anticipated, as the method requires high-quality and coarse-grained soil. To overcome this difficulty, a new method of frost heave restraint is discussed. In this paper, strength and frost heave properties were evaluated in order to investigate the applicability of the lime stabilization. Frost heave properties are influenced by the amount of stabilizer, soil type and curing period. Moreover, frost heave is proportional to the California Bearing Ratio (CBR). It was found that stabilized soil having a value of 100 in the CBR test could restrict frost heave ratio to below 5%. From the test results of the in-situ measurement, it was found that a 10% lime addition was as effective as replacement with sand.

1. INTRODUCTION

Frost heave restraint is one of the major problems for the road design in cold regions. The soil replacement method has been widely applied for frost heave restraint in Japan for a long time. However, as the method requires a large quantity of coarse-grained soil, the lack of such material will become a serious problem in the near future. Therefore, the development of a new method against frost heave has become promising not only for the natural preservation of borrow pits, but the disposal problem of the replaced soil (Kamiya, M et al., 1996).

This paper investigates the effect of lime stabilization on frost heave restraint. Winter observations were made of a lime stabilized soil below a test pavement construction.

2. SOIL SAMPLES AND TESTING METHOD

Three types of soils, A, B and C, were used in this paper. Figure 1 shows the grain size distribution curves. Table 1 summarizes the physical properties of the soils. The stabilizer used in this study is lime. Four types of mixtures of a ground granulated blast-furnace slag were added to increase soil strength. These were lime : slag = 100:0 (L100 series), 75:25 (L75), 50:50 (L50), 25:75 (L25). The amount of stabilizer varied from 3% to 10% of the dry soil weight.

The experimental water content was the natural

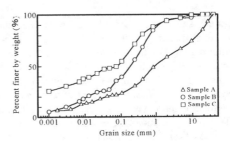

Figure 1. Grain size distribution curves.

Table 1. Physical properties of samples.

	Sample A	Sample B	Sample C
Soil particle density (g/cm³)	2.606	2.710	2.650
Natural water content (%)	17.3	26.8	30.0
Plasticity index	NP	NP	26.0
Optimum water content (%)	16.3	19.1	15.9
Maximum dry density (g/cm³)	1.674	1.708	1.814
Fines (%)	19	36	50

water content (Wn). However, sample A was tested at two conditions of water contents (Wn: natural water content and Wx:Wn+5%), as the natural water content was almost the same as the optimum water content. Hydrated-lime was used with sample A(Wn). Quick-lime was used with samples A (Wx), B and C.

Unconfined compression tests, frost heave tests and CBR tests were conducted after curing period of 7, 14 and 28 days.

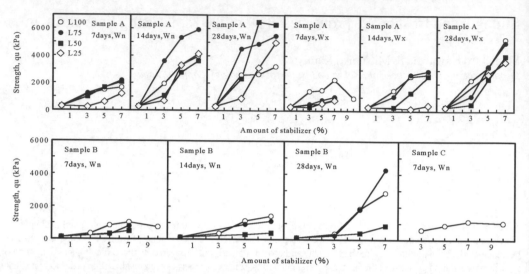

Figure 2. Relationship between unconfined compressive strength (qu) and amount of stabilizer.

3. TEST RESULTS AND DISCUSSION

3. 1 Unconfined compression strength

Figure 2 shows the relationship between unconfined compressive strength and amount of stabilizer at each curing period. For the sample A at its natural water content, the strength increased in proportion to the amount of stabilizer for the relatively short curing period of 7days. When the amount of stabilizer (3%), the effect of curing period on strength was significant. Effect of slag addition on the strength was also found to be significant. Strength increases with increases in curing period (L75 and L50). This indicates that the slag used in this study has a high hydration property. Moreover, the strength increases in proportion to the addition rate of the stabilizer and also to the period of curing in other test series (Sample A(Wx), B). When a high experimental water content occurs such as samples A(Wx), B and C, the lime stabilization is an effective way of increasing strength. This implies that the strength of the soil is influenced by the decrease in water content which is caused by the addition of the stabilizer. Lime significantly increased the strength of soils the L100 and L75 series.

3.2 Frost heave and CBR

The frost heave ratio (frost heave ratio (%) = frost heave height / initial specimen height) of sample A decreases as more lime is added. A higher water content results in greater heave for the same amount of lime (Fig.3). As for sample B whose strength was relatively low (see Fig.2 for 7 days), frost heave

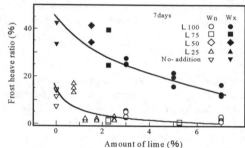

Figure 3. Relationship between frost heave ratio and amount of lime (Sample A).

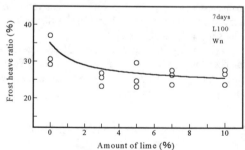

Figure 4. Relationship between frost heave ratio and amount of lime (Sample B).

restraint was found to be low (Fig.4). From the test result of sample C, the effect of the curing period on the frost heave was found to be significant (Fig.5).

From the aforementioned results, frost heave properties are often effected by the amount of stabilizer and the curing period. Therefore, the frost heave properties of stabilization soils have been often

428

studied relative to strength (Ishida, H. 1991, 1993).

Figure 6 shows the relationship between frost heave ratio and CBR. Frost heave is influenced significantly by soil type, water content and curing period when the CBR is low. However, these effects are negligible when the CBR becomes high. It was found that stabilized soil having a CBR of 100 has a frost heave ratio below 5%.

3.3 In situ evaluation

A test pavement was constructed on five types of stabilized conditions as shown in Figure 7. Type I is ordinary sand replacement method. Type II is Sample C soil without any frost heave control. Types III, IV

and V use lime stabilization of Sample C soil. The amounts of stabilizer varied from 5% to 10% of dry soil weight. The test pavement was constructed in the middle of September 1994. Therefore, the curing period was about three months until winter.

Measured frost heave was the order of III > II > IV > V > I (Fig. 8). The maximum frost heave was measured for Type III (5% stabilizer). However, for all other stabilized soils, frost heave decreases with addition stabilizer. In the case of Type V (10% stabilizer), maximum frost heave was about 4 mm, the same value as Type I (sand). Figure 9 shows the relationship between frost heave and frost penetration depth. It was confirmed that frost heave of stabilized soils was concentrated at the top of the frost blanket (350mm in depth). One of the major reasons for concentration, these are probably non-homogeneous mixing. After the site investigation, it was found that the amount of lime in the upper layer was less than that is the lower layer. Proper mixing is important because frost heave of stabilized soil slight where proper mixing occurred. In the other hand, the frost heave of Type II increased monotonously as frost penetrated deeper and deeper. From these results, it was confirmed that frost heave could be restricted in the field conditions as well as in the laboratory by lime stabilization.

Furthermore, the maximum frost penetration depth for lime stabilized fine-grained soil was found to be 200 mm, shallower than in sand (Type I in Fig.10). This implies that the thermal conductivity of soils become lower, as their grain size decreases. Therefore, when stabilized soil is the thickness of the replacement can be reduced.

4. CONCLUSIONS

Lime and slag were added to three types of soils as a stabilizer. The strength and frost heave properties of these soils were evaluated with results summarized as follows;

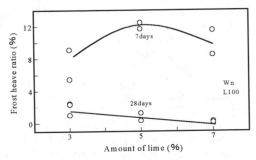

Figure 5. Relationship between frost heave ratio and amount of lime (Sample C).

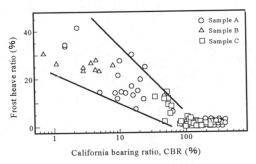

Figure 6. Relationship between frost heave ratio and CBR.

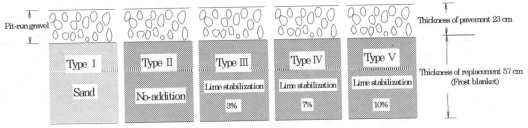

Figure 7. Stabilizer condition of test pavement.

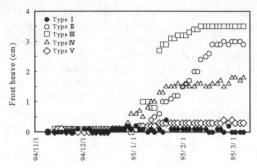

Figure 8. Frost heave in test pavement.

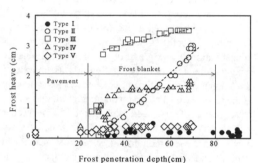

Figure 9. Relationship between frost heave and frost penetration depth.

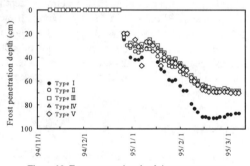

Figure 10. Frost penetration depth in test pavement.

1) When the experimental water content approaches the optimum water content, the long-term strength of stabilization soil with the slag increases significantly.

2) When the experimental water content is relatively high, the lime stabilization method is effective at increasing strength. This implies that the strength of that soil is influenced by the decrease in water content caused by the addition of stabilizers.

3) It was found that lime stabilization could restrict frost heave. The effect of lime stabilization on frost heave was highly significant when the addition rate or curing period increased.

4) Frost heave was proportional to the CBR. It was found that stabilized soil having a CBR of 100 could restrict frost heave ratio below 5%.

5) It was confirmed by the field investigation that stabilization could restrict frost heave. In the case of the additional rate for 10% lime, frost heave was limited to about 4 mm, which was the amount experienced by sand replacement.

6) Frost penetration depth was about 200 mm shallower in stabilized soil than in sand.

7) When stabilized soil is used, the thickness of replacement can be reduced.

REFERENCES

Kamiya, M., Kasahara, A., Hoshi, N. & Watanabe, H. 1996. Utilization of Excavated Soils from Reconstruction of Roads in Sapporo City. *Proc. of 2nd International Congress of Environmental Geotechnics.* (in press)

Ishida, H. 1991. Frost-Heave Prevention for Volcanic Cohesive Soil of High Water Content with Quicklime-Slag Mixture. *Tsuchi-to-kiso.* Vol.39. No.8: 5-10. (in Japanese)

Ishida, H. 1993. Soil Stabilization and Frost Heave Prevention of Various Kinds of Volcanic Cohesive Soils with Quicklime. *Tsuchi-to-kiso.* Vol.41. No.4: 33-38. (in Japanese)

Snow Engineering: Recent Advances, Izumi, Nakamura & Sack (eds) © 1997 Balkema, Rotterdam. ISBN 90 5410 865 7

Numerical analysis of electric road heating coupled with full-scaled experiments

Takao Yokoyama
Department of Mechanical Systems Engineering, Yamagata University, Japan

Shunichi Kobayashi
Research Institute for Hazards in Snowy Areas, Niigata University, Japan

Shinya Hirama
Research Center, Tohoku Electric Co., Ltd, Sendai, Japan

Abstract : A full-scale model of melting snow on a road by Joule heating was constructed and a series of experiments are performed. Temperature response within six hours, which means night preheating, is obtained repeatedly. Numerical analysis was also executed, and satisfactory agreement established. A set of influences under typical variation in the climate are realized both experimentally and theoretically. Finally important thermal behavior is simulated through numerical analysis with the design concept.

1 INTRODUCTORY

Paved roads equipped with snow-melting devices are quite popular in Japan. Wide-spreaded facilities use regional ground water and some use waste heat from hot springs. Electric energy is also attractive especially for focused points or complicated situations, for instance around fire-extinguishers, steps, entrances, etc.

Though electric heating is not uncommon, design methods for various weather situations are not established yet. Therefore a generalized systematic approach is necessary. In this paper we construct a full-scale specimen of a paved road from the thermal point of view. Under different weather conditions, various experiments are carried out. Numerical analysis is constituted to reflect these phenomena practically. Finally other interesting behavior, which are not easily exe-

cuted, are simulated to get useful information for future construction concepts.

2 THEORETICAL TREATMENT

Heat transfer and temperature propagation within solids is given by the Fourie-equation. Thermal properties, however, change due to layer components. Therefore several governmental equations of energy conservation may be connected each other:

$$\vec{q} = -\lambda \nabla \theta, \quad c\rho \frac{\partial \theta}{\partial t} = \nabla \cdot \vec{q} \qquad (1)$$

On the interface between stabilized soil and coarse grained soil, a heating wire is installed 7cm apart. Around the heating wire a specified temperature field occurs. The heat source and thermal resistance by electric insulator are considered in a one-dimension cylinder like as Table 2.

At the surface convective heat transfer is considered, namely,

$$q = h(\theta - \theta_a) \qquad (2)$$

where h W/m²K is the surface heat conductance with air phase. Thermal radiation is also added. The quantity of heat flux onto the surface, however, is determined by experiments, not solved by radiation equation.

2.1 Numerical analysis

Governmental equation (1) is rearranged by the finite-difference method explicitly. Interfaces between layers have the property of continuity of

Table 1 Nomenclature

d	depth	m
$c\rho$	heat capacity	J/m³K
h	surface heat conductance	W/m²K
p	pitch	m
q	heat flux	W/m²
t	time variable	s
w	density of joule heat	W/m
λ	thermal conductivity	W/mK
θ	temperature	°C

subscripts

a	air ambient
u	upper soil
l	lower soil

heat flux. Then the element including the heating wire is modified to have a heat source in addition to continuity of heat flux. Each surface element has not only heat delivery by conduction downward but also heat loss by convection into the air. Furthermore heat excess by thermal radiation is added. Melting, however, is excluded.

Table 2   standard physical properties "S",etc corresponds to Figure 2.

inner radius of wire	mm	2
outer radius of wire	mm	3
heat capacity of wire	J/m³K	3.5×10^6
λ of vinyl	W/mK	0.8
c ρ of "S"	J/m³K	2.0×10^6
c ρ of "C"	J/m³K	3.0×10^6
c ρ of "A"	J/m³K	1.8×10^6
c ρ of "G"	J/m³K	1.8×10^6
c ρ of "I"	J/m³K	3.0×10^6
λ of "S"	W/mK	0.1
λ of "C"	W/mK	0.2
λ of "A"	W/mK	0.7
λ of "G"	W/mK	0.8
λ of "I"	W/mK	1.4

3 EXPERIMENTAL APPARATUS

Figure 1. shows the experimental apparatus made to measure heat conduction under the road surface. Surface conditions such as wind, radiation, etc. were freely controlled. The model pavement consists of five layers surrounded by insulation. Twelve copper-constantan thermocouples are located at A through G and A' through E'. The A, E and E' assembled are used mainly to monitor differences between the wires. Additionally a thermal tracer is used to detect surface temperature. A wind-blower is located at the left side of this model, and solar radiation is delivered by a set of Xenon lamps.

3.1 Method of experiments

Before the experiments, the initial condition was maintained stable. After every experiment, the apparatus was rested. There are four stages of experiments:
1) Stagnant condition-no wind, no radiation
2) Windy condition-windy, no radiation
3) With snow layer-windy,no radiation, snow layer
4) Radiated condition-no wind/windy, with/without snow layer

In each stage three experiments are performed at different heater outputs (100, 150, and 200 W/m²).

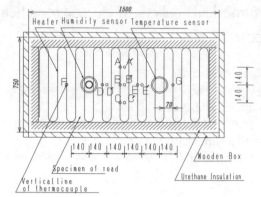

Figure1 Plane view of heated road specimen [mm]

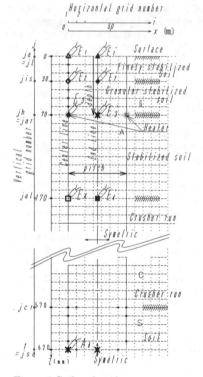

Figure2 Soil and grid structure

3.2 Experimental results

(1) Stagnant condition

Typical experimental results are shown in Figure 3. The highest temperatures appear at E3 and E3'. These are at the interface between G soil (granular stabilized soil) and A soil (stabilized soil) where the road heating-wire was installed. Differences between the wire line and blank line are greatest. E3, however, is accompanied with uncertainty, for it depends on exact location. Surface temperatures of E_1 and E_1' go up just to zero° C, and the difference between them is minimal. Thus, the surface reaches a uniform temperature in spite of line source of heat. The entire surface reaches the melting condition after preheating by night.

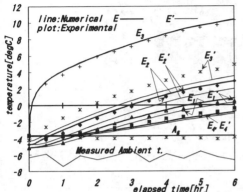

Figure3 Transient temp. at stagnant W=200[W/m²]

(2) With wind

Wind is added to the stagnant condition. Wind velocities were about 2 and 4 m/s. Surface temperature does not rise up so much as for stagnant condition. Additional heater power is required even under attenuated wind. This is observed to deteriorate due to stronger wind of 4 m/s.

(3) With snow layer

Figure 4 shows transient temperatures with 1 cm of snow on the road and a 2 m/s wind. Because of snow layer, the surface temperature does drop much more when snow is absent. Additional snow increases the surface temperature indicating that snow is a good thermal insulator. For that reason pre-heating, for instance at night using reduced prices of electricity, is effective.

(4) Under solar radiation

As solar radiation enter the surface directly, it greatly improves melting. Of course the temperature response at the surface is somewhat different.

3.3 Comparison of numerical results

Numerical results are also presented in Figure 3 and 4. In general, the initial condition was given by measured data, then following temperature propagation were calculated. All the physical properties are given in Table 2. These are normal values, estimated through a series of experiments and from published values of properties. Surface heat conductance h was evaluated, where $Nu \equiv f(\ Pr\ )\ Re^{1/2}$ (Pohlhausen and others 1921). These are Nusselt number, a function of Prandtl number which

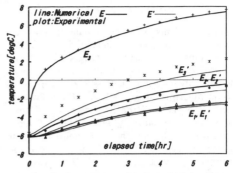

Fig.4 Transient temperature with snow and light wind , W=200 W/m²

depends on laminar or turbulent flow, and the Reynolds number. The simplest evaluation on bare surface under light wind is *4.2 W/m²K*. As turbulent flow is more suitable, it is properly estimated as *10 W/m²K*. At stagnant, natural convection would occur. It is estimated at around 4 W/m²K(McAdams 1954). The most difficult estimation is of snow. It is stored half a year, and becomes granular. This is carefully laid on the bare surface and regulated in height. Its specific density is about *0.2~0.3*, but the thermal conductivity(Maeno,K. etc.) is somewhat high. For a 1-cm snow layer, forced flow could pass through porous granular snow in addition to forming a boundary layer on the surface. This increases thermal conductivity and surface heat conductance.

In Figure 3, Good agreement at E_3 was not expected because of the steep temperature gradient there, for temperature around heating-wire is quite in local. The others gets satisfactory agreement between experiments and numerical results. We expect that thermocouple E_3' was not located properly in detail.

Analysis that of both snow and wind is complicated. When the thickness of the snow was reduced from *10 mm* to *9 mm*, and the thermal conductivity was evaluated a little high (*0.5 W/mK*) as above mentioned, satisfactory agreement is obtained.

Table 3 standard configuration

h	W/m^2K	12
λ_u	W/mK	2
$c\rho_u$	J/m^3K	3×10^6
λ_l	W/mK	0.8
$c\rho_l$	J/m^3K	2×10^6
p	cm	7
d	cm	7
pre-heating time	hour	6
density of heating	W/m	12

4 FURTHER CONSIDERATION

On the simulation model a series of calculations was accomplished as examples for thermal design. From a practical point of view, wire spacing, and depth in addition to soil properties greatly affect the performance of pavement heating system. Figure 5 shows how the time to reach 0 ℃ at the surface varies with the depth of the heating wire. Other components are as listed in Table 3. At least it takes two hours. As the higher thermal conductivity in upper layer becomes, the faster the surface warms. Melting pavement with light load like walkway or steps could operate effectively under thin configuration. Before break-through coming surface is under 0 ℃. Even after the surface above the wire reaches 0 ℃, other portions of the surface might be below freezing if the heating wires are not sufficiently close together.

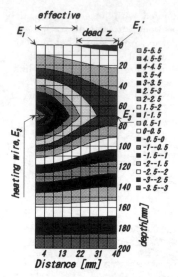

Figure 7 Temperature field after six hours

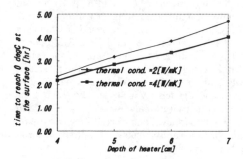

Figure 5 Time to reach 0 ℃ at the surface

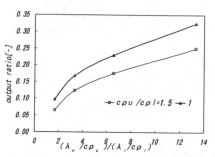

Figure 8 Performance with ratio in the upper to that of the lowwer thermal diffusivity

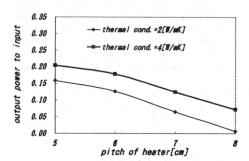

Figure 6 Performance vs. p and λ

Figure 6 indicates how the efficiency ratio of heat-output to heat-delivery drops down as the spacing of heater wires increases. Spacings of *5 to 6 cm* bring much higher performance even for a little additional investment. Temperature profiles in vertical planes are displayed in Figure 7. A dead zone is present on the surface, where the pitch is 8 cm.

Soil layout under the horizontal surface, where heating wire is equipped, should be distinguished from the above. Namely upper layer is to be higher,

but the lower is to be less in thermal diffusivity to control heat flow upward.

Figure 8 suggests the increment of this capability by the growth of ratio of thermal diffusivity in upper layer to the lower.

5 CONCLUDING REMARKS

1. The specimen of pavement works sufficiently and brought important knowledge applicable to practical operation.

2. This numerical model is able to simulate any required configuration of melting components.

6 REFERENCES

Pohlhausen 1921, ZAMM., d.1, SS.115-120

McAdams 1954,Heat Transmission, Mc Graw-Hill,180

Maeno,K. and Fukuda,M. edited 1986, Physical properties and structure of snow, Ko-Kin,173

Snow Engineering: Recent Advances, Izumi, Nakamura & Sack (eds) © 1997 Balkema, Rotterdam. ISBN 90 5410 865 7

A new evaluation of the effectiveness of salt for road anti-icing

Makoto Murakuni
Engineering Department, Head Office, Japan Highway Public Corporation (JH), Tokyo, Japan

ABSTRACT: This paper describes experiments carried out to investigate the phase change of salt solutions on road surfaces at temperatures below freezing, and the effect of this change on the skid-resistance of the road surface.

1 INTRODUCTION

Compared with Europe and America, in Japan there are several characteristics that present particular problems when taking measures to deal with the road traffic conditions in winter. They are:

i) Japan is on the same latitude as the Mediterranean, but the influence of the mountain ranges known as "Japan's Backbone" and the seasonal winds in winter result in heavy snow. In addition, as Japan is at a low latitude, levels of solar radiation, even in winter, cause the snow to melt during the day and then to freeze again at night.

ii) As the Japanese Islands are long and narrow in shape, there are marked regional variations in quantity and type of snow.

iii) Japan does not have natural resources of rock salt, and, despite the fact that the country is surrounded by sea, the high temperature and high humidity climate is unsuitable for the production of natural salt. The salt used for anti-icing of roads is natural salt imported from Mexico or Australia, which leads to higher costs of road management in Japan during the winter season than in Europe or America.

In addition to the above-mentioned conditions, the use of studded tires on Japanese roads was banned in 1991, increasing concern over achieving safe and smooth road transportation. For these reasons, the effective use of salt for anti-icing is a matter of considerable importance to JH.

2 RESEARCH OBJECTIVES

At the present time, the most widely used road anti-icing agent is "salt." The fact that by converting the road surface water to salt water, depending on the concentration, freezing will not occur at 0°C, is of course well known.

However, it is not only the reduction of the freezing point that is relied upon when using salt. If, for example, the temperature one evening was expected to drop as low as −7°C, it is extremely unlikely that salt would be used to provide the concentration of 10% required to prevent freezing (this kind of use will be referred to hereafter as "Freezing Curve Management").

In the case of roads under the management of the Japan Highway Public Corporation (JH), solid salt 30 g/m², or liquid 0.1 ℓ/m² is the standard dosage, which does not necessarily correspond to Freezing Curve Management.

This research was carried out to determine the phase change occurring when road surface water is converted to salt solution of a given concentration and the temperature lowered. Also, the results of tests to evaluate the effect of these changes on the skid-resistance of the road surface are described in this paper.

3 SKID-RESISTANCE OF WET OR FROZEN ROAD SURFACES WITHOUT SALT

First, the variation with falling temperature of skid-resistance of a wet road surface without salt was investigated and the sudden drop in skid-resistance accompanying the change from wet surface to frozen surface was confirmed. It was also verified that a frozen surface covered with a thin film of water provided the slipperiest condition.

3.1 Experimental Method

The skid-resistance was measured in a low temperature test room {temperature range controllable from −30°C to +50°C (±0.5°C), humidity controllable from 40% to 90% RH (accuracy ± 3%)} under the following simulated road surface conditions:

i) Pavement surface covered with water thick enough

435

to completely fill the surface texture, the condition being water just prior to freezing.

ii) Pavement surface covered with ice thick enough to completely fill the surface texture.

The skid-resistance of the surface was measured using a British Portable Skid-Resistance Tester based on the ASTM-E303 test method, and the values obtained were in the form of BPN (British Pendulum Number). The specimens used in the tests were the four types of asphalt mixture surface course shown in Table 1. The asphalt mixture numbers used in the table are from the Manual for Design and Construction of Asphalt Pavement (Japan Road Association), and the maximum aggregate size was 13 mm.

The roughness of the pavement surface (texture), was measured by the sand expansion method using a PWRI (Public Works Research Institute) automatic roughness gauge.

Table 1. Types of asphalt mixture surface course used in the experiments.

Type of Asphalt Mixture	Aggregate Proportions (%)						Asphalt Quantity (%)	Surface Roughness
	No.6 Crushed Stone	No.7 Crushed Stone	Screenings	Coarse Sand	Fine Sand	Mineral Powder		
⑤ Dense and Gap-graded Asphalt Concrete	57	8	13	6	10	6	5.5	0.156
⑥ Dense Grade Asphalt Concrete	40	10	15	15	14	6	6.9	0.244
⑦ Fine Asphalt Concrete	18	10	20	7	35	10	8.6	0.214
⑧ Open-graded Asphalt Concrete	60	17	7	5	5	6	4.3	0.372

3.2 Experimental Results

The results obtained for the two surface conditions described above for a road surface temperature range of 0°C ± 1.0°C, are shown in Figure 1. In Figure 2 are shown the results obtained when the test range was extended to –20°C to +30°C. In these figures, the ▲ symbol indicates results obtained while increasing the temperature, while the ■ symbol represents results obtained while reducing the temperature.

From the experiments, it was confirmed that, in the case of the wet road surface, as 0°C is approached skid-resistance increases, whereas in the case of the completely frozen road surface, the skid-resistance decreases as 0°C is approached. Taking 0°C as the dividing line, depending on whether the surface is wet or frozen, there is a difference in skid-resistance (BPN) value of around 50.

4 SKID-RESISTANCE OF WET OR FROZEN ROAD SURFACES AFTER SALT APPLICATION

4.1 Reasoning from Phase Diagram

With reference to the phase diagram (see Fig. 3) for salt (sodium chloride), the changes occurring as the temperature of a 10% concentration salt solution drops from 10°C can be considered as follows:

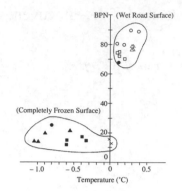

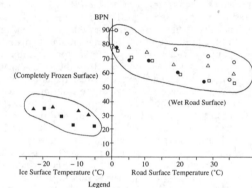

Figure 1. Variation of BPN with road surface temperature for wet and completely frozen surfaces (temperature range 0 ±1.0°C).

Legend
● Dense and Gap-graded Asphalt Concrete
□ Dense Grade Asphalt Concrete
△ Fine Asphalt Concrete
○ Open-graded Asphalt Concrete
▲ Completely Frozen Surface (Temperature Rising)
■ Completely Frozen Surface (Temperature Dropping)
× Completely Frozen Surface Covered with Thin Water Film

Legend
● Dense and Gap-graded Asphalt Concrete
□ Dense Grade Asphalt Concrete
△ Fine Asphalt Concrete
○ Open-graded Asphalt Concrete
▲ Completely Frozen Surface (Temperature Rising)
■ Completely Frozen Surface (Temperature Dropping)

Figure 2. Variation of BPN with road surface temperature for wet and completely frozen surfaces (temperature range –20 to +30°C).

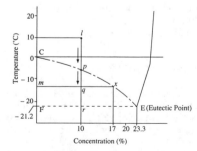

Figure 3. Phase diagram for a salt solution.

436

Condition at ℓ

The 10% salt solution is maintained at 10°C. In changing from ℓ to p, the concentration of 10% remains unchanged as the temperature drops.

Condition at p

Point p is the freezing point (−7°C) for the salt solution with a concentration of 10%, and ice begins to form at this point. If the temperature is further reduced, the amount of pure ice will increase while the remaining salt solution will become more concentrated. The composition (concentration) follows the line towards point E.

Condition at q

If the entire ice and salt solution is considered to be represented by mx (100), the amount of ice will be qx (40), and the amount of salt solution mq (60). The concentrated salt solution remaining will have the concentration at point x (17%).

Condition at r

When the temperature of a salt solution having a concentration of 10% is reduced to the eutectic point (−21.2°C), ice and salt solution of 23.3% concentration will be present in a ratio of $rE:Fr =$ 57:43.

A schematic representation of the above can be seen in Figure 4. In this figure, the x-axis represents the concentration (0 to 23.3%), and the y-axis, the temperature (0 to 21.1°C), while along the z-axis (longitudinal axis), the proportions of ice and salt water are given at various points for the range enclosed by CFE in Figure 3. In Figure 4, values of 5% and 10% are shown as examples of intermediate concentrations. In the case of 10% (Fig. 5), the entire

solution remains in the liquid state until the temperature reaches −7°C. As the temperature continues to drop, the amount of ice continues to increase. On the other hand, the amount of remaining liquid decreases while becoming more concentrated. For example, at −10°C, the ratio of ice to salt solution is 25:75, and the concentration of the salt solution is 13%. The ratio of ice to salt water becomes 50:50 at a temperature of −17°C, at which time the concentration of the salt solution is double, i.e., 20%, as would be expected.

From the above, the points of significance for road transportation in the winter season are as follows:

i) When salt is added to surface water, depending on the concentration, not only is the freezing point lowered but the solution does not change rapidly from the liquid phase to the solid phase as water does.

ii) When salt is added to surface water, even when the temperature is further decreased, the entire solution does not freeze, instead the amount of pure ice continues to increase as far as the eutectic point.

It can therefore be inferred that the application of salt to water on roads will result in a significant difference in the skid-resistance when temperatures drop.

4.2 Relationship Between Salt Solution Phase and Skid-Resistance

An outline of the experiments carried out to verify the above assumptions is as follows:

4.2.1 Experimental Method

Salt solutions having concentrations of 1%, 3%, 5%, and 15% were prepared. Each of the concentrations was applied to the surfaces of paved test specimens in a low temperature test room. The temperature in the low temperature test room was then reduced to set temperatures within the range from 0°C to −20°C. When the test specimens reached the test room temperature, skid-resistance was measured. The asphalt mixture surface course test specimens were made in accordance with JH design guideline Type II, with a maximum aggregate size of 13 mm. The asphalt mixture adopted was that in general use on Expressways in snowy and cold areas.

The salt solution was applied thick enough to completely fill the surface texture (0.5 to 0.75 mm) making the necessary adjustments using a NASA type water film thickness tester. Skid-resistance was evaluated using the mean of 5 BPN values.

4.2.2 Experimental Results

As the skid-resistance was expected to vary depending on the ratio of ice to salt solution, the BPN

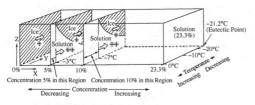

+   Symbol indicates decreasing ice temperature
++ Symbol indicates increasing concentration

Figure 4. 3-Dimensional schematic of variation of salt solution with temperature.

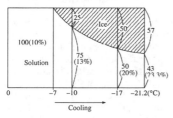

Figure 5. Diagram showing the changes occurring in a 10% salt solution during cooling.

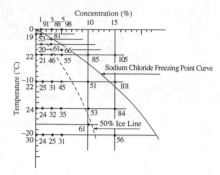

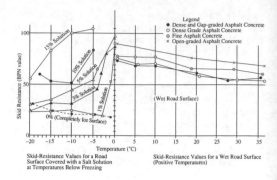

Figure 6. Variation of skid-resistance (BPN) at temperatures below freezing plotted on a phase diagram.

Figure 7. Effect of salt application on improving skid-resistance by preventing the rapid drop in skid-resistance at freezing point and preventing absolute freezing.

values were plotted on a phase diagram as shown in Figure 6. The concentration of 0% in this figure refers to the case of water, that is to say, the values are for the case where no salt is applied to the road surface and the water simply freezes. The broken line in Figure 6 is the line formed by joining the points where the ice and salt solution are in the same ratio i.e. 50:50 (referred to in this paper as the "50% ice line"). As a result of representing the results in this way, the following may be observed:

i) When the ice content is around the 50% mark, the skid-resistance is around BPN 50.

ii) As the amount of ice in the salt solution increases, in other words, the further left of the "50% ice line" the smaller the skid-resistance becomes (BPN<50).

iii)As the amount of ice in the salt solution decreases, in other words, the further right of the "50% ice line" the larger the skid-resistance becomes (BPN>50).

To clearly show the extent to which skid-resistance is improved, Figure 7 was made in the same way as Figure 2. In addition, Figure 8 is a simplified version of Figure 7. From these figures, it can be seen that by the application of salt to the road surface, not only does the freezing point drop with increasing salt concentration, but there is also a corresponding improvement in skid-resistance.

5 CONCLUSIONS

5.1 Without Salt Application

In the case where salt is not applied to the road surface, experimental studies of the relationship between skid-resistance and temperature for the wet road surface or frozen road surface led to the following:

i) It was confirmed that, in the case of the wet road surface, as 0°C is approached skid-resistance increases, and in the case of the completely frozen road surface, the skid-resistance decreases as 0°C is approached. Taking 0°C as the dividing line,

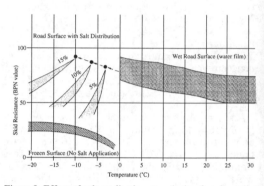

Figure 8. Effect of salt application on reducing freezing point and improving skid-resistance (simplified version of Figure 7).

depending on whether the surface is wet or frozen, there is a difference in skid-resistance (BPN value) of around 50.

ii) The slipperiest surface occurs when a completely frozen surface is covered by a thin film of water, which acts as a lubricant.

5.2 With Salt Application

An investigation of the relationship between the phase change of a salt solution and skid-resistance with reducing temperatures, such as would occur in a cold region with the distribution of salt, led to the following:

i) In the case of a salt solution, there is no sudden drop in skid-resistance when reaching the freezing point as there is in the case of water.

ii) Therefore, in cases where there is a danger of freezing of the road surface in cold conditions, the advance distribution of "salt" has the dual effects of preventing freezing by lowering the freezing point and of preventing absolute freezing of the road surface.

Snow Engineering: Recent Advances, Izumi, Nakamura & Sack (eds) © 1997 Balkema, Rotterdam. ISBN 90 5410 865 7

Study of a snow-melting and antifreeze system for prevention of auto accidents on slippery bridge decks

Hideaki Nakamura
Department of Civil Engineering, Yamaguchi University, Japan

Toshio Tanimoto
Technical Department, Eito Consultants Co., Ltd, Okayama, Japan

Sumio Hamada
Department of Civil Engineering, Yamaguchi University, Japan

ABSTRACT: Car accidents due to slip occur frequently on bridges in the snow season. Melting of snow on the bridge slab is usually later than snow melting on the road, since the road is warmed by the earth. Drivers tend to travel at the same speed on bridges as on roads.

1. INTRODUCTION

Safety of winter traffic is one of the most important policies in the snowy region. Traffic accidents on bridges are a frequent occurrence in such regions, which creates a serious social problem. The cause of these accidents is preferential ice and snow on the bridge due to colder air under the bridge than under the road. A car is driven onto an icy bridge at the same speed as on the snow melted road without notice of the icy or snowy bridge. In order to remove the ice and snow on the bridge, several methods have been proposed and applied. We proposed a pipe heating system as one of the most reasonable procedures. The total cost of the system can be reasonable compared with other methods, although the initial cost of the equipment is not low. The present study aims to obtaining fundamental information related to a pipe heating system for deicing bridge decks. Pipe spacing and water temperature in the pipe are experimentally and theoretical focused. An experiment was conducted outside in the snowy region to prove the theoretical results obtained by the non-steady heat flow finite element method.

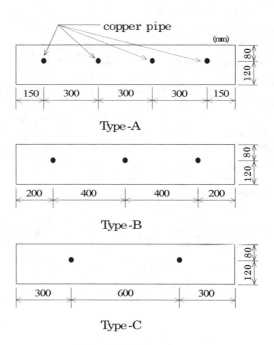

Figure 1. Test specimens.

2. TEST SPECIMENS AND EXPERIMENTAL PROCEDURES

The experiment was conducted on a snowy mountain to obtain information on pipe heating systems and to construct a numerical model of snow melting. The results provide fundamental properties for the thermal analysis of the deck. Pipe spacing, pipe diameter and water temperature to be used can be determined. The test specimens are of a rectangular slab deck ($1200 \times 800 \times 200$ mm) with one to three pipes, as shown in Figure 1. The density of the snow employed in the experiment was 300 kg/m^3. The height of the snow on the deck for the test was 100 mm. Temperature of the water in the pipes was 30 ℃ and 40 ℃. Copper-constantan thermocouples embedded in concrete were used as temperature sensors. Temperatures were measured every 10 minutes. Photographs were taken every hour.

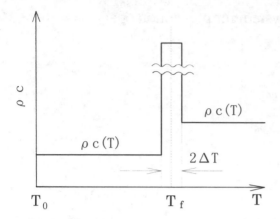

Figure 2. Relationship between material
temperature and heat capacity.

3. ANALYTICAL PROCEDURE FOR DEICING PROCESS

Water has a liquid and solid phases. Change of phase produces heat absorption and emission. In the thermal analysis the heat capacity must be determined by applying latent heat. Figure 2 shows a relationship between the material temperature and the heat capacity ρc. The heat capacity is expressed as Eq.(1) for the present analysis including the phase change. In the figure, T_f is a temperature at the phase change, that is, 0℃ for the snow melting temperature. ΔT is the temperature region at the time when the phase change takes place:

$$\rho c = \rho_i c_i + \rho_i \left(\frac{L}{2\Delta t}\right) \qquad (1)$$

where ρ, c, L and T are the density, specific heat, latent heat and temperature, respectively. Eq.(1) is also expressed by the specific heat as

$$c = c_i + \left(\frac{L}{2\Delta t}\right) \qquad (2)$$

The thermal properties of the snow are considerably influenced by the ratios of ice particle, water and air. The heat capacity of the snow can be expressed as

$$\rho_s c_s = \rho_i c_i (1 - \varepsilon) + \rho_w c_w \varepsilon S + \rho_a c_a (1 - S)\varepsilon \qquad (3)$$

where the subscripts s, i, w and a mean snow, ice, water and air, respectively. The thermal conductivity of the snow has been measured relative to the density of the snow. Devaux[1] proposed a relationship

Table 1. Thermal properties of the materials.

Material	temperature T (K)	density ρ (kg/m^3)	specific heat c (kJ/kgK)	thermal conductivity λ (W/mK)
ice	273	917	2.00	2.2
water	293	998	4.18	0.61
air	293	1.16	1.00	0.026

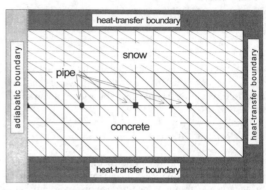

Figure 3. Finite element mesh and boundary conditions.

Table 2. Analytical conditions.
concrete slab

Specific heat of concrete kJ/kgK(kcal/kg℃)	1.30 (0.31)
Density of concrete kg/m^3	2312.0
Thermal conductivity of concrete W/mK(kcal/mh℃)	2.91 (2.5)
Heat-transfer coefficient of concrete W/m^{2}K(kcal/m^{2}h℃)	11.6 (10.0)
Initial temperature ℃	-5.0
Atmospheric temperature ℃	measurement
Heat-transfer coefficient of pipe W/m^{2}K(kcal/m^{2}h℃)	$\alpha_p = 5.52v + 50$ ($\alpha_p = 4.75v + 43.0$) (v : flow rate(cm/s))
Water temperature ℃	measurement
Pipe diameter m	0.015
Flow rate cm/s	360

snow

Initial snow thickness cm	10
Initial snow temperature ℃	-5.0
Thermal conductivity λ_S W/mK	$\lambda_S = 2.93 \times 10^{-2} + 2.93 \times 10^{-6} \rho_S^2$
Heat-transfer coefficient of snow α_S W/m^{2}K(kcal/m^{2}h℃)	85(99)

440

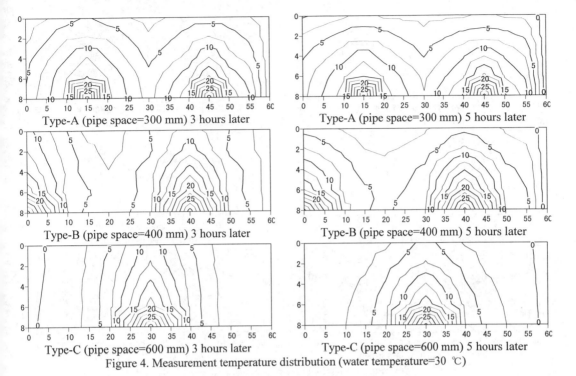

Figure 4. Measurement temperature distribution (water temperature=30 ℃)

between thermal conductivity and density in the range of 100 kg/m³ to 600 kg/m³ as

$$\lambda_S = 2.93 \times 10^{-2} + 2.93 \times 10^{-6} \rho_S^2 \qquad (4)$$

The mean heat-transfer coefficient between the flowing water in the pipe and the concrete has been proposed by Tanabe[2] as a function of the water velocity by

$$\alpha_p = 4.75v + 43.0 \text{ (kcal/m}^2\text{hr}^\circ\text{C)}$$
$$\alpha_p = 5.52v + 50 \text{ (W/mK)} \qquad (5)$$

The transient heat conduction equation discretized by the finite element method can be expressed as below.

$$[K]\{\phi\} + [C]\{\dot{\phi}\} = \{F\} \qquad (6)$$

where $[K], \{\phi\}, [C], \{\dot{\phi}\}$ and $\{F\}$ are the heat conductivity matrix, temperature vector, heat capacity matrix, vector for temperature slope with respect to time, and heat source vector, respectively. Eq. (6) is discretized with respect to space, and not with respect to time. Derivatives with respect to time, $\{\dot{\phi}\}$ is present in Eq. (6). Discretizing this

variables by the Crank-Nicolson method, the temperature at time $t + \dfrac{\Delta t}{2}$ and the derivative of temperature can be expressed respectively as

$$\left\{\phi\left(t + \frac{\Delta t}{2}\right)\right\} = \frac{1}{2}\left(\{\phi(t + \Delta t)\} + \{\phi(t)\}\right) \qquad (7)$$

$$\left\{\dot{\phi}\left(t + \frac{\Delta t}{2}\right)\right\} = \frac{\{\phi(t + \Delta t)\} - \{\phi(t)\}}{\Delta t} \qquad (8)$$

Substituting Eqs. (7) and (8) into Eq. (6), the following equation is obtained:

$$\left(\frac{1}{2}[K] + \frac{1}{\Delta t}[C]\right)\{\phi(t + \Delta t)\}$$
$$= \left(-\frac{1}{2}[K] + \frac{1}{\Delta t}[C]\right)\{\phi(t)\} + \{F\} \qquad (9)$$

Matrices $[K]$ and $[C]$ are non linear functions of unknown nodal temperature, hence Matrices $[K]$ and $[C]$ are corrected until nodal temperatures are sufficiently converged.

441

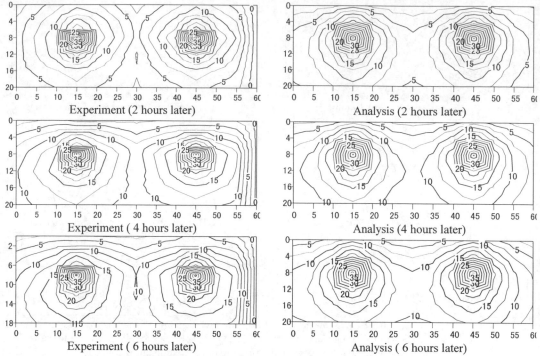

Figure 5. Comparison between experimental and analytical results
(water temperature=40 ℃, pipe space=300 mm)

4. EXPERIMENTAL AND ANALYTICAL RESULTS

The temperature distributions in snow covered concrete section having 300, 400 and 600 mm spaced pipes are given by Figure 4 for a water temperature of 30℃. The mean air temperature was -5℃ during the test. The snow on the surface with 300 mm spacing melted uniformly. That is, the temperature near the surface is rather uniform. On the contrary, the snow on the specimen having 600 mm spacing of pipes melted only above the pipe. However, it may still be possible to make the temperature of the bridge deck with a pipe space of 600 mm higher than that of the road. Figure 5 exhibits the analytical and experimental temperature distributions for the specimen with a 300 mm spacing of pipes and a water temperature of 40℃. The figure indicates good agreement between analytical and experimental results. The experimental temperature after 4 hours dropped compared with the temperature after 2 hours, which may be due to heat transfer of the melted snow. The analysis may provide similar results by employing more accurate properties of ice particles, melted water and air.

5. CONCLUSION

In this study an experiment was conducted to design a pipe heating system and to simulate the analytical model for a numerical analysis. This analysis provides a fairly good simulation to the site experiment, and consequently the analytical method is effective for the design of pipe heating systems.

REFERENCES

Devaux, J. 1933. Annular Physica, 20.
Tanabe, T., Yamakawa, H. and Watanabe, A. 1984. Determination of Convection Coefficient at Cooling Pipe Surface and Analysis of Cooling Effect, *Proc. of JSCE*, No.343, pp.171-179.
Yagawa, G. 1983. *Finite Element Analysis in Fluid Dynamics and Heat Transfer*, Baifukan.

Snow Engineering: Recent Advances, Izumi, Nakamura & Sack (eds) © 1997 Balkema, Rotterdam. ISBN 90 5410 865 7

Development of a narrow, high speed rotary snowplow

Taira Emoto
Tohoku Engineering Office, Tohoku Regional Bureau, Ministry of Construction, Tagajo, Japan

Masaki Ohta
Nihon Josetsuki Seisakusho Co., Ltd, Otaru, Japan

ABSTRACT: As a measure for securing smooth traffic flow in snowy regions, rotary snowplows play an important role.   On the other hand, the operation of these rotary snowplows is one factor contributing to traffic congestion.   Therefore, since 1993, in cooperation with "Tohoku Engineering Office, Tohoku Regional Bureau, Ministry of Construction," we have researched and examined this factor and its countermeasures regarding the 250-HP class rotary snowplow, which is the most popular type in Japan.   The following are the details and results of the research and examination.

1 DETERMINATION OF FACTORS CAUSING TRAFFIC CONGESTION AND EXAMINATION OF SPECIFIC COUNTERMEASURES

1.1 *Snow removal width (vehicle width)*

The snow removal width, 2.6 m of the conventional 250-HP class rotary snowplow seems to be too wide, and when a rotary snowplow of this type removes snow, it is dangerous and difficult for the other vehicles to pass the rotary snowplow within the width of the lane (generally 3.5 m wide).

Therefore, we conducted the following tests to find a better possibility for narrower snow removal width.
(1) The snow removal width of middle- and large-sized rotary snowplows, including the 250-HP class, was determined to be 2.6 m due to the following reasons:

In the past, since there were only a small number of snow removers, the rotary snowplows had to remove snow completely in one pass to allow trucks and other vehicles (max. width: 2.5 m) to travel.

However, as the number of snow removers has increased in recent years, the main assignment of the rotary snowplow has changed from "one-lane snow removal" to "width expansion snow removal."

In this type of snow removal, plow-type snow removal machines, including snowplow trucks, push snow onto the shoulders of the road and thereafter the rotary snowplows remove the snow further off the road.
(2) We collected data on the amount of snowfall and accumulated snow, the width of accumulated snow, from 11 places which are typical roads governed by

"Tohoku Regional Bureau, Ministry of Construction."

The data proved that, as shown in Figure 1, a width of snow removal of 2.6 m was not required.
A width of 2.2 m was enough to accomplish snow removal on 93.6 % of roads governed by the Tohoku Regional Bureau.

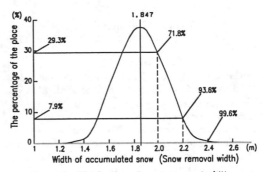

Figure 1. Distribution of snow removal width.

1.2 *Traveling speed*

The traveling speed is 40 km/h for conventional middle- and large-sized rotary snowplows, including the 250-HP class.   Combined with the decreased width of the road and the width of the rotary snowplow, it is difficult and dangerous for the other vehicles to pass the rotary snowplow on downtown roads in heavy traffic.

The following measures to improve traveling performance of the rotary snowplow were examined.

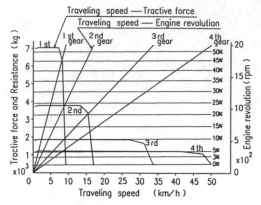

Figure 2. Traveling perfomance curve of newly developed rotary snowplow.

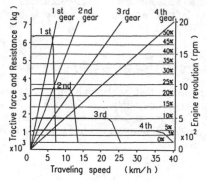

Figure 3. Traveling perfomance curve of conventional rotary snowplow.

(1) The conventional 250-HP rotary snowplow is only providing 220-HP at a traveling speed, 40 km/h. However, engine horsepower of 250-HP, can be utilized for traveling.

(2) As shown in Figure 2 and 3 , by maintaining present climbing ability, the rotary snowplow with 250-HP traveling horsepower, can increase the traveling speed from 40 km/h to 49 km/h.

2   RESULTS OF VARIOUS TESTS

In 1994, based on the above results and the following development concepts and principal specifications, we delivered the newly developed rotary snowplow to "Tohoku Regional Bureau, Ministry of Construction" and conducted tests comparing the newly developed rotary snowplow with the conventional one.

2.1 *Concept of development*

(1) Narrower width of the rotary snowplow to ease traffic congestion during snow removal
(2) Faster traveling speed to ease traffic congestion
(3) Lower noise to reduce environmental pollution
(4) Easier operation and more comfortable cabin
(5) More efficient width expansion snow removal operation by means of relocated blower position.
(6) Improved safety during traveling and snow removal

2.2 *Comparison of principal specifications and appearance*

Table 1. Comparison of principal specifications.

	Unit	Newly developed Rotary snowplow	Conventional Rotary snowplow
Overall length	m	6.95	7.09
Overall width	m	2.2	2.6
Overall height	m	3.6	3.57
Wheel base	m	2.8	2.8
Maximum snow removal capacity	t / h	2,004	2,540
Muximum snow removal capacity for road expansion	t / h	1,990	2,295
Snow removal width	m	2.2	2.6
Total weight of rotary snowplow	kg	14,300	13,350
Engine Rated power	HP/rpm	250/2000	250/2000
Engine Maximum power	HP/rpm	280/1700	280/1700
Engine Maximum torque	kg·m/rpm	130/1400	130/1400
Maximum traveling speed	km/h	49	40

Figure 4. Newly developed rotary snowplow.

2.3 Tests on reduction of traffic congestion

(1) Acceleration performance test
The followings are graphs comparing time to reach maximum speed of the newly developed rotary snowplow and the conventional one, starting in 2nd or 3rd gear.

When the time required to reach 40 km/h is examined, the acceleration performance of the newly developed rotary snowplow is superior to that of the conventional one by 33 to 41%.

Therefore, the newly developed rotary snowplow can contribute to improvement of traffic congestion.

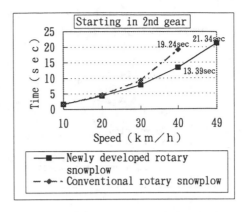

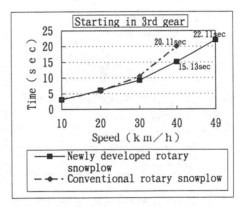

Figure 5. Acceleration performance test.

(2) Tests on reduction of traffic congestion on an actual road
1) Traveling tests were conducted on a 5-km test section of road. The newly developed rotary snowplow and the conventional one traveled at their maximum speeds. Figure 6 shows the number of vehicles which followed the rotary snowplows at the ending point of the test section and which passed the rotary snowplows on the test section.

According to the figure, many vehicles passed the conventional rotary snowplow because of its slower traveling speed, and many vehicles were affected by the rotary snowplow.

On the other hand, only few vehicles passed the newly developed rotary snowplow because of its faster traveling speed. Although many vehicles follwed the newly developed rotary snowplow, they did not try to pass it. Furthermore, it can be seen from the figure that the newly developed rotary snowplow contributed to the improvement of traffic congestion because a smaller propotion of vehicles were affected by it.

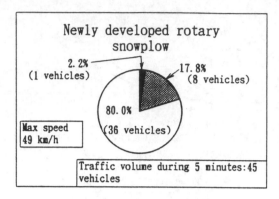

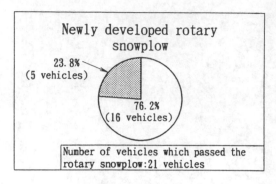

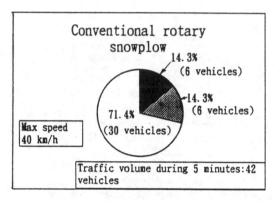

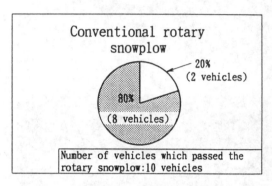

Figure 6. Tests on the reduction of traffic congestion.
(while traveling)

■ Number of vehicles which passed the rotary
snowplow
▨ Number of vehicles which followed the rotary
snowplow
☐ Number of vehicles which were not affected by
the rotary snowplow

2) In snow removal operation work tests, the newly
developed rotary snowplow and the conventional one
traveled on a 500 m test section of an actual road at a
snow removal speed of 3 to 4 km/h.
Figure 7 shows the number of vehicles which passed
and could not pass each rotary snowplow when there
were oncoming vehicles in the opposite lane.
According to the figure, the percentage of following vehicles which could not pass the operating rotary
snowplow due to oncoming vehicles in the opposite
lane is 80% for the conventional rotary snowplow
and 23.8% for the newly developed one, or 1/3 that
of the conventional one. It is apparent in Figure 7
that the smaller overall width of the rotary snowplow
greatly contributed to the improvement of traffic congestion.

Figure 7. Tests on the reduction of traffic congestion.
(during operation)

☐ Number of vehicles which passed the rotary
snowplow
▨ Number of vehicles which could not pass the
rotary snowplow

3 CONCLUSION

According to the results of these traveling test and
snow removal operation work test, it is apparent that
smaller overall width of 2.2 m and faster traveling
speed of 49 km/h can contribute to improve traffic
congestion.
We consider when traffic congestion is increased and
the quality of snow removal will be required more
and more, and then this newly developed rotary
snowplow might contribute to the improvement of
traffic congestion in city area which have narrow road
and heavy traffic.

Snow Engineering: Recent Advances, Izumi, Nakamura & Sack (eds) © 1997 Balkema, Rotterdam. ISBN 90 5410 865 7

Making traffic secure in a heavy snowfall area in winter – Gassan Road information system

Ken Nagata, Takayuki Oba & Asao Kato
Tohoku Regional Bureau, Ministry of Construction, Sendai, Japan

ABSTRACT: Intelligent Transport Systems (ITS) have been introduced to Gassan Road, which runs through one of the heaviest snowfall areas in Japan. The use of information on weather and traffic conditions, together with the image projected by Industrial Television (ITV) cameras, allows for reliable and accurate road management. As a result of the introduction of ITS, Gassan Road has been passable all year and the accident rate has decrease remarkably.

1 INTRODUCTION

1.1 *Intelligent Transport Systems in Japan*

Using the most advanced information technology, research and development of Intelligent Transport Systems (ITS) is being undertaken looking at ways to structure a system incorporating highways, vehicles and people.

The Japanese government announced its "Guidelines on Increasing Use of Information and Communications in the Fields of Roads, Traffic, and Vehicles" in August 1995. The guidelines are aimed at promoting joint development by gov-ernment, business, and academia, and at introducing the various systems into actual use sequentially by the early part of the 21st century.

Eleven different promotional measures are to be undertaken, with development work taking place in nine fields (Fig.1), eight projects are now in progress. ITS in Japan has been sub-jected to field testing and put into practi-cal application in various areas.

The Tohoku Regional Bureau of the Japanese Construction Ministry is actively promoting the development of ITS model developments and field tests in order to cope with harsh natu-ral features including heavy snowfall and traf-fic problems.

Tohoku region is located in the north east-ern part of Japan and covers a wide area with severe circumstances such as heavy snowfall and steep mountains. The region is in habited about 9.8 million, and traffic is required to be open even in the winter period.

1.2 *Gassan Road (National highway Route 112) with heavy snowfall*

Route 112 is 134.5 kilometers long and is the shortest route connecting the inland area of Yamagata Prefecture with the Shonai area.
Since the Tohoku Expressway (Sakata Line) was connected to Route 112 in 1991, it has become an important highway.

Figure 1. ITS development areas

Table 1. Features and functions of Gassan Road

FEATURE/FUNCTION	CHARACTER
Social Economy	·Shortest main road to connect inland and district ·Practical use as a part of Tohoku Expressway (Sakata line)
Tourism	·View line that passes Bandai-Asahi National Park ·Selected are one of the "Best 100 roads in Japan" ·Resort road for skiing ·Pilgrimage passage
Heavy Snowfall Mountain Road	·The heaviest snowfall area in Japan ·Passes through areas of landslides ·High altitude mountain road with long tunnels

Table 1 shows the main functions and features of Gassan Road. Given its geographic position, it continues to increase in importance in economic and social terms.

Gassan Road lies across the mountainous district with a group of long tunnels (Gassan No. tunnel:2620 m, Gassan No.2 tunnel:1530 m, and Yudonosan tunnel:665 m). Like Routes 13 and 17 Gassan Road runs through one of the heaviest snowfall areas in Japan (see Fig.2)

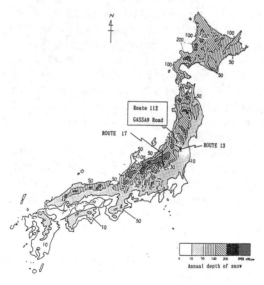

Figure 2. Roads 13, 17 and the Gassan are located in areas that experience heavy snowfall

Northwesterly seasonal winds from the Japan Sea bring a large amount of snow. The maximum depth of cumulative snowfall is about 2,500 cm (Fig.3) The road also runs through Bandai Asahi National Park, for which special measures and controls are required.

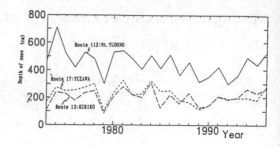

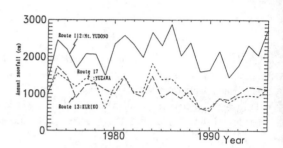

Figure 3. Maximum depth of snow and annual snowfall for Gassan Road and routes 13 and 17

2. GASSAN ROAD INFORMATION SYSTEM

2.1 Outline of the System

In order to make this road passible in winter, great efforts have been made to remove large amounts of snow. Avalanche prevention fences and prevention piles are placed on cut slopes while avalanche retaining walls, snow sheds, similar structures are constructed at the sides of the roadway along with snow protection fences and snow shoulders to prevent, snow drifts and blowing snow that obstructs drivers' vision (Yamatani, 1990). In addition, ITS has been deployed to provide information in real time relating to climate conditions, the road and the tunnels for road management. The system also gives the public information relating to these issues. Three ITS development fields have been introduced:
1) Advanced navigation system
2) Support of driving safety
3) Efficiency in road management

Table 2 shows an outline of the ITS for Gassan Road. The use of information on weather and traffic conditions, together with the image

Table 2. Gassan Road information system

COMPILATION
·Meteorological observation equipment ·TV camera (outside the tunnel) ·TV camera (inside the tunnel) ·Road surface freeze sensor ·Tunnel disaster preventation system ·Road patrol cars ·Snow removing station ·Road information liaison office ·Public telephones

PROVISION/UTILIZATION
1)Gassan Road information terminal 　·Information bulletin board 　·Route map 　·Information board (to post weather/road 　　　　　　　　　　　　　information) 　·Multi-vision (to show road on TV) 2)Gassan Road information subterminal 　·Weather & road information board 　·TV monitor (to show road on TV) 　·Route map 3)Road information board 　·Type A and Type B 4)Roadside boardcasting 5)Emergency alarm board in tunnel 6)Information distribution system by phone 7)Cut-in radio boardcasting system in tunnel
O Construction offices (Yamagata & Sakata) 　Snow removal station 　Yamagata Prefecture Police 　Fire fighting unions 　Road information center

PROCESSING
·Distance observation control system ·Graphic panel ·Keyboard ·TV monitor ·Information board ·Information distribution system by phone ·Radio reboardcasting facility ·Road information by radio ·Mobile wireless telephone system in tunnels ·Gassan Road information network ·Road controlling information network

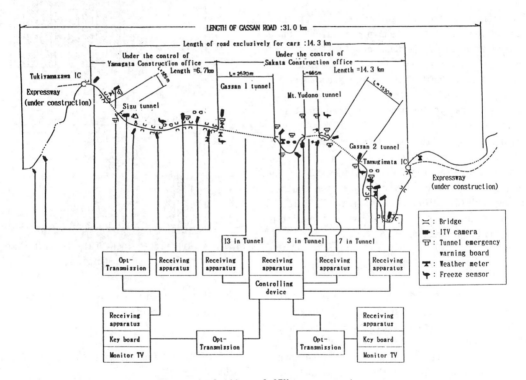

Figure 4. Outline of ITV camera system

projected by a television camera allows for re-liable and accurate road management. The Gassan Road information system has three major components: information compilation, processing and provision, and utilization. The high informa-tion capability and the fast and certain access make optical cable well-suited for the transmis-sion of information between the systems.

Table 3. Major information observation and col
-lection facility in Gassan Road

USE	EQUIPMENT INSTALLED AND QUANTITY

Road observation	ITV Camera (coverage:26.5 km) 19
	  YAMAGATA Side 10
	  SAKATA Side 9

Tunnel observation	ITV Camera (At 3 Tunnels Every 200 m) 23
	  GASSAN 1 Tunnel (2620 m) 13
	  Mt.YUDONO Tunnel (665 m) 3
	  GASSAN 2 Tunnel (1530 m) 7

Tunnel disaster prevention	Emaergency warning system 9 Fire sensor 3

Facility condition observation	Remote controlling device 1
	  Ventilation system in tunnel,
	  Emergency system.
	  Disaster preventionsystem.
	  Observation of communication equipment

Weather condition observation	General meterologocal observation system 13
	  Wind direction, Wind elocity, Humidity
	  Air temperature, Atmospheric pressure,
	  Rate of snowfall, Depth of snow,rainfall
	Road surface observation system 8
	Visibility meter 6

Table 4. Major information supplying facility
in road

PLACE	EQUIPMENT INSTALLED AND QUANTITY
Information terminal	TV monitor (Multi vision) 9 Information board 1 Route map 1 Showcase 1
Information subterminal	TV monitor 3 Radio receivor for road information 2 Atomospheric data indication panel

On the road	Road information board TypeA 3 (S Side)
	                 TypeB 2
	Tunnel emergency warning board in tunnel
	             5 Sets (S Side)
	Roadside boardcasting 3 Points
	          2 km, 2.6 km, 3 km
	Radio re-boardcasting 3 Tunnels
	Wireless 1 Set
	Telephone answering device 1 Set

2.2 *Information collection facilities*

The outline of the information observation and collection facility is shown in Table 3. One characteristic of the Gassan Road ITS is its ability to observe each section of the road by way of a large number of sensors and cameras. This allows weather conditions, the road sur-face, and traffic throughout the road to be ac-curately seen.

Figure 4 shows the outline of the ITV camera systems, a characteristic feature in the ITS of Gassan Road. The image of each section on the road is transmitted to the information control center and the information terminal. Efficiency in removing snow has been increased due to the real-time information.

2.3 *Information provision facilities*

The outline of the information provision system is shown in Table 4. There is an information terminal and an information subterminal near the western entrance of the Gassan Road. Users can watch the current data (weather, road condi-tion) of each section at the terminal and sub-terminal. The radio and signs along the road also provide information on the road.

Road-users as well as managers are able to see current conditions. If drivers "watch" the picture, they can determine "the difficulty" in the mountainous road. By understanding the con-ditions and taking his driving skill and expe-

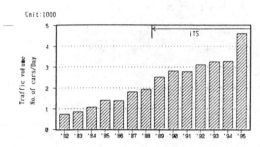

Figure 5. Traffic volume in winter

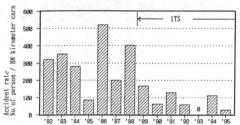

Figure 6. Traffic accident rate in winter at the Sakata side of Gassan Road

rience into account, each driver is able to reach their own decision.

3. THE EFFECT OF THE GASSAN ROAD INFORMATION SYSTEM

Restrictions placed on night traffic on the Gassan Road in winter have been eased as a re-sult of the introduction of ITS. This has made the road passable all year around. As the reli-ability of the road is enhanced, the amount of vehicles using it has increase dramatically

Figure 5 shows the change in number of vehicles on the Gassan Road. It is clear that number of users there was an upward trend before ITS. That trend has notchanged "drastically" increased due to im-proved safety and reliability of the road.

Figure 6 shows the change in traffic accident rate. The accident rate remarkably decresed in spite of an increase in the number of the vehicles on the road. The decrease due to improved road that management provided by implementation of ITS together with additional snow protective structures and increased snow removal capacity. We believe ITS, which provides accurate information to managers and drivers, is playing a major role in providing security of traffic on Gassan Road in winter.

REFERENCE

Yamatani, S (1990) : Situation of countermeasures to avalanches on national roads in Tohoku region (in Japanese), Yuki to doro, 22.

Snow Engineering: Recent Advances, Izumi, Nakamura & Sack (eds) © 1997 Balkema, Rotterdam. ISBN 90 5410 865 7

Improving the efficiency of winter road management: Road surface freezing forecast system, spreading of pre-wetted salt, and continuous salt concentration measuring system

Hiroshi Maeno
Disaster Prevention and Snow Section, Research Institute, Japan Highway Public Corporation, Tokyo, Japan

Tatsumi Suzuki
Sendai Bureau, Japan Highway Public Corporation, Japan

ABSTRACT: This research was carried out to investigate winter road management methods for appropriate and effective use of salt for anti-icing, increasing amounts of which are being used each year. The research was conducted with respect to the following three themes:
i) research on forecasting road surface freezing utilizing a meteorological information system
ii) implementation of a pre-wetted salt spreading method, and
iii) development of a method for measuring the concentration of anti-icing chemicals on the road surface.
Although there are areas remaining for improvement, the basic research on the above produced useful results. It is expected that their integrated adoption will lead to a more effective and efficient spreading of salt for anti-icing.

1 INTRODUCTION

Despite the fact that Japan lies between latitudes of 30° and 45°, the effect of the cold seasonal winds blowing from the Sea of Japan in the north result in about 50% of the country receiving heavy annual snowfalls. In addition, being at a low latitude, there are many cases where the strength and duration of sunshine cause the snow and ice to melt during the day and freeze at night. Snowfall in neighboring regions differs in quantity, quality and precipitation characteristics making ice and snow control difficult.

In addition to these difficult climatic conditions, prohibition of spiked tires in 1990 has led to increasing use of road surface anti-icing chemicals (salts), and their appropriate and efficient use has become a major issue.

For the above reasons, the Japan Highway Public Corporation carried out research on the following three themes in order to achieve more effective and efficient spreading of anti-icing chemicals.
i) research on forecasting road surface freezing utilizing a meteorological information system
ii) implementation of a pre-wetted salt spreading method, and
iii) development of a method for measuring the concentration of anti-icing chemicals on the road surface.

2 ROAD SURFACE FREEZING FORECAST SYSTEM

This method involves statistical analysis of meteorological data for the preceding five years to provide information on road surface temperatures and probabilities of precipitation. The results of the forecast are broadly divided into fixed point information and route direction information.

2.1 Forecast times

From considerations of the times required for anti-icing organization and vehicle dispatch, the system was developed to provide the following two types of forecast:
1) Short Range Forecast: forecast of the conditions from 5:00 in the evening to 9:00 the following morning.
2) Very Short Range Forecast: forecast of the conditions in three hours time.

2.2 Meteorological data used in the forecasts

a) Meteorological observation data
• Road weather observation results: ambient temperature, road surface temperature, wind speed and direction, precipitation, sub-surface temperature (−5 cm and −10 cm).

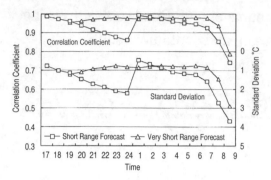

Figure 1. Result of road surface temperature forecast verification test.

• Upper-air observation results: altitude, air temperature, humidity, wind speed and direction for various isobaric altitudes (850, 700, 500 hPa).
• Night-time weather: fine, cloudy, rain, snow.

b) Results of Road Surface Temperature Monitoring
This involves the use of an observation vehicle fitted with an infrared temperature measuring device to measure the temperature of the road surface for the differing road structures and topographical conditions along a specified route to provide basic data for the forecast of route conditions.

2.3 Field verification

An example of the field verification of road surface temperature forecast is given in Figure 1. The reason for the sudden change occurring at 1:00 A.M. is the use of the latest upper-air observation results (announced at 21:00 hours) to provide a modified forecast for the period after 1:00 A.M. Improvements are being made to the accuracy of the forecast.

2.4 Results

Although a drop in accuracy was observed in the field verification for the period between 8:00 and 9:00 A.M., the overall results are considered acceptable. However, as the predicted value is determined using statistical mean values, there are cases where the method is unable to respond to extreme climatic variations. This is believed to be one of the contributory factors in the drop in accuracy for the period between 8:00 and 9:00 A.M. mentioned above.

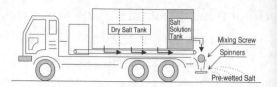

Figure 2. Schematic of pre-wetted salt spreader.

3 SPREADING OF PRE-WETTED SALT

There are two main methods for the spreading of anti-icing chemicals. One is to distribute in the granular form (dry salt spreading), and the other is to distribute the salt dissolved in water (solution spreading). The dry salt spreading tends to be long lasting but has the disadvantage of being easily scattered by the wind or passing vehicles. The solution spreading method has the merit of being fast-acting but tends to flow away easily and is not long lasting. The two methods thus have their own particular advantages and disadvantages.

On the other hand, the combination of salt solution with dry salt immediately before spreading is a method (referred to below as "pre-wetted salt spreading") that is said to mitigate many of the disadvantages associated with the two methods when they are used independently.

For this reason a comparative study was made of the pre-wetted salt spreading method, some of the results of which are presented below.

3.1 Spreading test

A spreading test was performed in which dry salt and pre-wetted salt spreading methods were used and the uniformity of spreading was confirmed for each.

The dry salt was picked up in the wake generated behind the spreading vehicle and left on the road surface in stripes. On the other hand, the pre-wetted salt was uniformly distributed over the road surface even at a speed of 50 km/h. The results of the spreading test are shown in Figures 3 and 4 for the pre-wetted salt spreading (mixing proportion - solution 1: dry salt 2.5) and the dry salt spreading respectively.

3.2 Evaluation of durability

Figure 5 shows the results obtained when the concentration remaining on the road surface after 150 minutes was measured for dry salt (spreading: 30 g/m²) and pre-wetted salt (mixing proportion - solution 1: dry salt 2.5, spreading: 30 g/m² (dry salt

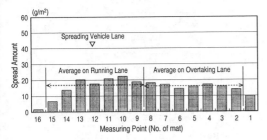

Figure 3. Pre-wetted salt spreading (50 km/h).

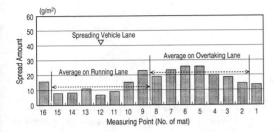

Figure 4. Dry salt spreading (50 km/h).

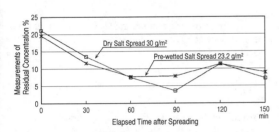

Figure 5. Residual concentration (NaCl).

equivalent: 23.2 g/m²)). From these results it can be seen that, from the time immediately after spreading to the time 30 minutes later, the residual concentration is 2–3% higher in the case of the dry salt. However, after 60 minutes, it can be seen that the pre-wetted salt spreading remains in higher concentrations than the dry salt spreading.

Although the residual concentrations are affected by snowfall and other factors, when the durability of the dry salt and pre-wetted salt spreading methods were evaluated by comparison of the residual concentration, it was found that, despite the fact that the wet spreading was carried out at 23.2 g/m², a little less than the dry spreading of 30 g/m², the results were the same or better for the pre-wetted salt case than for the dry salt case.

3.3 Results

Based on the results of various tests, the results obtained for the pre-wetted salt spreading method in comparison with the other two methods are summarized in Table 1. The pre-wetted salt spreading is relatively unaffected by wind or passing vehicles and has a long lasting effectiveness. In addition, as savings can be made in the amount of salt being distributed and the spreading can be accomplished at higher speeds, the method is clearly efficient.

Table 1. Comparison of spreading methods.

Item	Dry salt	Salt solution	Pre-wetted salt
Durability	Good	Inferior	Good
Immediate effect	Inferior	Good	Good
Ease of spreading	Difficult to spread uniformly	Relatively easy to spread uniformly	Easy to spread uniformly
Effect of cross-grade	Virtually unaffected	Tends to flow away	Unaffected
Effect of wind and traffic	Easily thrown into the shoulder	Unaffected	Unaffected
Standard spread amount	30 g/m²	0.1 ℓ/m² (24 g/m² *20% solution)	30 g/m² (23 g/m² *1:2.5)
Spreading speed	30–40 km/h	40–50 km/h	50–60 km/h

Figure 6. Hand-held refractometer.

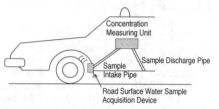

Figure 7. Continuous salt concentration measuring system.

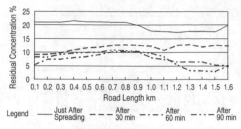

Figure 8. Results obtained using continuous concentration measuring system.

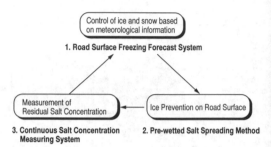

Figure 9. Winter road management system.

4 DEVICE FOR CONTINUOUS MEASUREMENT OF RESIDUAL SALT CONCENTRATIONS

The maintenance of the residual concentration of anti-icing chemicals is carried out using a hand-held portable refractometer of the type indicated in Figure 6. As this gauge can only be used to measure the concentration at one particular location (point), it can only be considered as a guide in the maintenance of road surface anti-icing chemicals concentration. There are therefore difficulties involved when it becomes necessary to maintain the concentration over the entire length of an expressway. As the measurement requires going out onto the expressway the method also involves an element of danger. For these reasons, a device (hereinafter referred to as the "continuous concentration measuring device") was developed, that can continuously measure the concentration along the direction of the highway, in safety, while providing a digital display of the result (Figure 7).

The device is fitted to the front mudguard of an expressway patrol vehicle that takes samples of the up-splash water from the road surface. Results from a concentration measuring unit are then indicated on a digital display located in the front passenger seat location.

4.1 Field tests

Figure 8 shows the results of using the continuous concentration measuring device over a distance of 1.6 km with measurements made at 100-m intervals, and shows the variations of concentration with location and time.

4.2 Results

Providing the road surface is wet, samples can be effectively taken and satisfactory measurement values obtained. However, due to the structure of the device, after prolonged use, particles of road grit and other materials tend to cause blockage of the supply tube that in the worst case can render the device unserviceable. Efforts are currently being made to modify the device to overcome this remaining problem.

5 CONCLUSIONS

In order to best prevent freezing of the road surface, it is necessary to spread the anti-icing chemicals before the road surface water freezes. This "anti-icing," coupled with the full application of the "winter road surface maintenance system" (Figure 9) are of major importance in achieving the most effective and efficient use of anti-icing chemicals.

It would greatly please the authors if this report was put to practical use as a reference on the method of spreading of anti-icing chemicals, a task of great importance in winter road management.

Snow Engineering: Recent Advances, Izumi, Nakamura & Sack (eds) © 1997 Balkema, Rotterdam. ISBN 90 5410 865 7

Problems related to placing snow dumping channels in a flat terrain

Norio Hayakawa & Yusuke Fukushima
Nagaoka University of Technology, Niigata, Japan

Hiroshi Taniuchi
Ebara Engineering Co., Ltd, Tokyo, Japan

ABSTRACT: A pressing need of the snow regions is for effective snow removal devices. One such device is a snow-dumping channel network, which has seldom been installed in a city with flat terrain. This paper first reports an experimental study that shows that snow-dumping channels are possible with a slope as mild as 1 to 1,000. Next, a feasibility study of erecting a snow-dumping channel network is studied with the end result that some kind of measure is needed to solve the problem of downstream condition.

1. INTRODUCTION

The Sea of Japan side of the main island of Japan (The Hokuriku Region) has very heavy snowfall and mild winter temperatures. Consequently, snow is often wet and heavy. Snow removal technologies practiced in this region, therefore, have unusual features. Good examples are the snow-melting pipe system and the snow-dumping channel network. These two technologies have been practiced together with use of snow removal machines, thanks to the mild temperatures and abundant water supply of the region. They are schematically illustrated in Figure 1. All these three technologies have their respective merits and demerits, and the recent trend is to apply two or three of these technologies. Of these three technologies, the snow-dumping channel network has been restricted to areas with abundant supply of water and a sloped terrain in order to wash out dumped snow easily. Recently such channel networks are being considered for city areas with flat terrain. This paper presents a feasibility study of such a system and points out technological problems.

2. SNOW AND CLIMATE OF THE HOKURIKU REGION

Japan's main island is bisected by major mountain ridges. West of the divide lies the Sea of Japan. Seasonal dry wind blows down from Siberia and picks up moisture from the warm water of the Sea of Japan. Consequently, the climate of this Hokuriku region is rather mild and quite humid in winter with a large amount of snowfall annually. Figure 2 shows the data of the annual snowfall and Figure 3, the maximum annual snow depth in the City of Nagaoka which has a population of about 180,000. This feasibility study of a snow dumping channel network was conducted for Nagaoka.

3. SNOW CONVEYING CAPACITY OF MILD SLOPED OPEN CHANNEL

A snow dumping channel is a simple open chan-

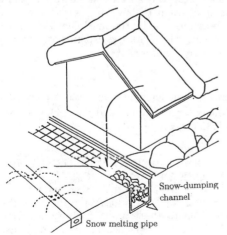

Figure 1: Schematic diagram of snow-dumping channel and snow-melting pipe.

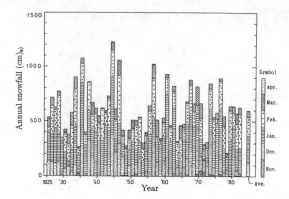

Figure 2: Annual snowfall for Nagaoka, Japan.

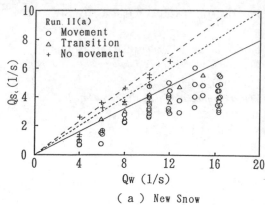

(a) New Snow

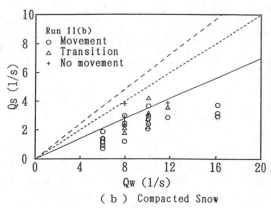

(b) Compacted Snow

Figure 4: Relation between snow and conveying rate of snow and flow rate.

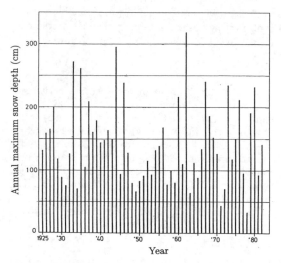

Figure 3: Annual maximum snow depth for Nagaoka, Japan.

"movement" and clogging of a chunk by "no movement". In Figure 4 are drawn a couple of earlier formulae proposed for steeper slopes and the experimental result is shown to conform with them. The best formula for the maximum rate of snow conveyance which fits the experimental data appears to be as follows:

$$Q_s = f Q_w \tag{1}$$

where Q_s is the maximum rate of snow conveyance, Q_w the flow rate of water and $f = 0.4$ for new snow and 0.35 for compacted snow.

4.   MODEL DESIGN OF SNOW DUMPING NETWORK IN NAGAOKA CITY

The city of Nagaoka is located in the snow region of the Niigata Prefecture. The city is currently meshed with snow-melting pipes, and in recent years a plan is being contemplated to erect a snow dumping channel network. At the central part of

nel in which water is supplied at the upstream end and snow is dumped in it in chunks of an oversized shovel size. If a dumped chunk of snow floats over the water surface, the open channel may have unlimited ability to convey snow. In reality, a large-sized chunk of snow touches the channel bottom, thus creating the possibility of clogging. The ability to convey snow chunks has been known and practiced for channels with slopes steeper than 1 to 500 (Hokuriku Branch of Japan Ministry of Construction, 1983). For channels with milder slope, not much has been reported so far. Figure 4 shows our experiment with a wooden open channel with slope of 1 to 1,000 (Fukushima, 1996). The channel has a width of 25 cm and height of 40 cm. Into this channel, chunks of snow are dumped at a set interval and smooth conveyance is marked as

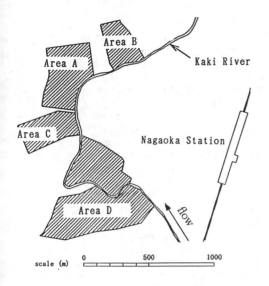

Figure 5: Studied area of Nagaoka City.

the city lies the Kaki River; it is small but may serve as a supply of necessary water. The problem here is that the terrain is very flat necessitating erection of a very mild-sloped channel network. In the following is given a description of the feasibility study of a channel network for this area.

4.1 Climate condition

The snow climate is described in Figures 2 and 3. These data are analyzed statistically and the figures shown in Table 1 are used for model design.

4.2 The studied area

The four areas shown in Figure 5 were selected for study. These areas are heavily congested with houses and the terrain is flat. The Kaki River flows along these areas and flows into the Shinano River, which can supply the necessary water to the Kaki River. Of these four areas, only Area A is studied

Table 1: Statistical data of snowfall of Nagaoka City (cm)

	1/2 year probability	1/10 year probability
Maximum new snow depth	64	77
Annual maximum snow depth	135	235
Cumulative new snow depth	586	983

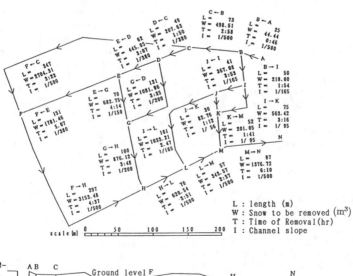

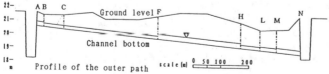

Figure 6: Channel network plan for Area A.

459

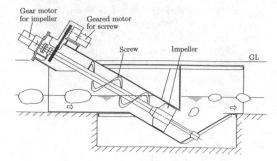

Figure 7: Prototype booster pump.

extensively with the channel network mapped, the amount of snow to be removed calculated and water flow rate determined. The following gives the result of this feasibility design.

4.3 *Outline of model design*

Five steps are followed for design of this study:

(a) Design criteria are fomulated.

(b) Amount of snow to be removed is estimated for road area and roof area.

(c) Necessary amount of water is estimated.

(d) Time shedule for dumping snow is determined.

(e) Water surface profile along the chanel is calculated.

4.4 *Result of Model Design*

(1) *Formulation of design criteria.*
The area covered by the snow-dumping channel network is defined, excluding parks, school grounds, shrines, temples and the river. Snow depth to be removed is defined as the new snow depth of 1/2 year probability for road surfaces and 1.0 m for rooftops.

(2) *Amount of snow to be removed is estimated for road area and roof area.*
This value is calculated for each reach of the channel network, summing up snow on road surfaces and on rooftops which have to be served by this reach. The amount of snow to be dumped per unit of time into a reach of the channel network is determined by the rate of snow volume that can be dumped physically and the distance between each opening for dumping. The ratio of the amount of snow to be removed to the rate of snow that can be dumped is the time schedule to be stipulated. In other words, time schedule, i.e. the frequency at which snow has to be removed, has to be de-

cided in the beginning, size of the channels to be determined thereafter. In this design study, the time necessary to remove snow was decided to be 2 days maximum and the operation of the channel network is limited to 7 hr to 12 hr each day.

(3) *Necessary amount of water is estimated using equation (1).*
The next procedure is to design the channel, especially its slope. In this model study, a slope of 1 to 500 is chosen for the outermost path of the channel network and the slope of the other reach is determined. Figure 6 gives results obtained for Area A, showing, among others, that a time schedule of 5 hours is secured with this plan. The longitudinal profile of the outermost path of this figure indicates, however, that the downstream end of the channel is below the water level of the receiving river. In such a case, a suitable measure has to be implemented at this point to smoothly discharge the snow/water mixture. One possible method is use of a booster pump, which is illustrated in Figure 7.

5. CONCLUSIONS

In view of the pressing needs of the snow region, use of the snow-dumping channel network has been studied. The major technological problems in this case is the possibility of using such a channel in a flat terrain. This paper suggests that such a use is indeed possible with a 1 to 1,000 channel slope. Also, a feasibility study modeling Nagaoka City indicates the need to employ some kind of booster pump to raise the water/snow mixture to be dumped at the downstream end.

6. REFERENCES

Fukushima, Y., 1996, Transport rate of snow floating in an open channel, *Proc. 3rd Internaltional Conference on Snow Engineering.*

Hokuriku Branche of Japan Ministry of Construction, 1983, *Guidlines to Design and Operation of Snow Removal Channels* (in Japanese).

Snow Engineering: Recent Advances, Izumi, Nakamura & Sack (eds) © 1997 Balkema, Rotterdam. ISBN 90 5410 865 7

Stagnation of snow at a bend in a snow drain channel

Fumihiko Imamura & Nobuo Shuto
Disaster Control Research Center, Tohoku University, Sendai, Japan

Jun Yamamoto
National Research Institute of Fisheries Engineering, Japan

Atsushi Kamiyama
Tokyo Construction Consultant Co., Japan

ABSTRACT: Experiments were carried out for different angles of bend, and various water and snow discharge rates to clarify the stagnation mechanism. Stagnation at a bend is associated with three parameters: speed, density , and size of snow lumps. Considering the kinetic energy of broken lumps of snow, we propose a critical condition of stagnation. Given the angle of bend, discharge rate of water, and the slope and width of the channel, the critical speed of the snow and the required distance between the point of dropping and the bend can be estimated to avoid stagnation.

1. INTRODUCTION

Snow drains are widely constructed and used in the heavy snowfall areas of Japan. They are effective at removing snow from urban areas. Snow is dropped into flowing water in the snow drains by manpower or by mechanical power, and is transported by the water. Once stagnation happens somewhere in a snow drain, the system plugs, in the worst case, flooding occurs. Stagnation often occurs at bends, junctions and divergences. Methods of design to avoid stagnation have been also proposed. Some experimental and theoretical studies [JACM, 1988] to remove snow in such channels have been carried out, but they are limited to straight channels. Yamamoto et al. (1992) conducted a field investigation and reported that most stagnation occurs not in straight channels but at bends, junctions, and divergences. Thus, snow removal efficiency is controlled at those points. A new criteria is required.

The aim of this paper is to clarify the stagnation mechanism at a bend theoretically and experimentally. In addition, a numerical simulation of a snow lump in a channel is developed for design of snow removal systems.

2.EXPERIMENTAL STUDY

2.1 Setup and measurement

The bent open-channel of acrylic resin 10 cm wide

shown in Figure 1 was installed in a field where a large quantity of snow of varied density as well as water of low temperature were easily obtained. The angle of the bend was changed from 0° to 90° in 15° increment. Water was supplied and controlled by two pumps. Snow lumps 10 cm high and wide of different lengths and densities were dropped in the channel and their motion near the bend was captured on a video. The definition of stagnation was that a snow lump stopped at the bend over 10 seconds. Water discharge, velocity of the snow lumps approaching the bend, their density, and air and water temperatures were measured. Average water flow velocity was determined from water discharge measurements divided by the sectional area of the channel.

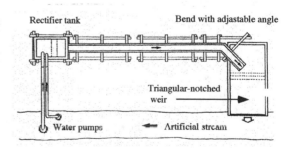

Figure 1 Experimental setup of the open-channel with a bend

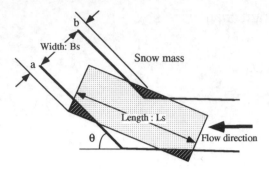

Figure 2 Schematic view of snow passing through a bend

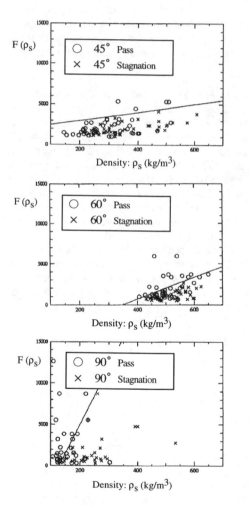

Figure 3 Stagnation condition as a F with different density and a bend angle. The circles and crosses indicate pass and stagnation, respectively.

2.2 Stagnation at a bend

A sketch of a snow lump passing at a bend is shown in Figure 2. If the hatched portion is broken, a large snow lump can pass the bend. The volume of the broken portion would be related to the lumps kinetic energy and their strength. In other words, a fast moving, low densities snow lump can easily pass a smaller angle bend. Among these three factors, the angle of the bend was found to be the most dominant factor to determine the stagnation, as shown in Figure 3.

3. CRITERIA OF SNOW STAGNATION

3.1 Volume of snow mass passing through a bend

Assuming a snow lump of width B_s, height H_s and length L_s is flowing down at a corner of a bend angle θ in a snow drain of the same width B_s, the hatched portion $\triangle V_s$ shown in Figure 2 is easily calculated from the geometry as follows:

$$\Delta V_s = H_s S = H_s (2a^2 + 2b^2) \tag{1}$$

and there is a geometrical relationship between an angle of bend, θ, and a and b as follows:

$$a + B_s + b = B_s \cos\theta + \frac{L_s}{2}\sin\frac{\theta}{2} \tag{2}$$

Substituting Eq.(2) into Eq.(1), we can estimate the minimum value of broken volume to pass, ΔV_s, as

$$\Delta V_s = H_s \sqrt{2}\left\{B_s\left(\cos\frac{\theta}{2} - 1\right) + \frac{L_s}{2}\sin\frac{\theta}{2}\right\}^2 \tag{3}$$

3.2 Criteria of stagnation

A part of the kinetic energy of the snow lump may be converted into the work to break the lump shown in Figure 2. If a linear relationship is assumed between the two, we have

$$\frac{1}{2}\rho_s V_s (v_s)^2 = F \times \Delta V_s \tag{4}$$

where ρ_s is the density of snow, V_s the total volume of the snow lump, v_s the approaching velocity of the snow lump, and F, the fraction of the kinetic energy used to break the lump, which is experimentally determined by the following relation taking the flow direction into account as follows:

$$F = \frac{\frac{1}{2}\rho_s V_s (\sin\theta \times v_s)^2}{\Delta V_s}$$ (5)

Experiment result shows that F at stagnation is related to snow density by the following formula:

$$F = 4\rho_s + 1200$$ (6)

Thus, the critical velocity of stagnation, v_s, can be estimated with Eq. (5) after Eq.(6) is substituted. A snow lump faster than the critical velocity thus determined can pass the bend without stagnating.

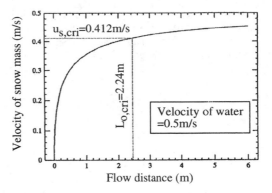

Figure 4 Velocity of a snow mass simulated by Eqs. (7) and (8) with the critical velocity, $u_{s,cri}$, estimated by Eqs.(5) and (6).

4. NUMERICAL MODEL OF SNOW MASS IN A CHANNEL

Without measurements, the velocity of a snow lump should be evaluated by numerical simulation for a design of snow drain channels. Water and a snow lump flow down, mutually exchanging momentum after the snow lump is dropped in a snow drain. Movements of the water and the snow lump are connected through inertia and drag forces as follows:

$$\rho_s L_s H_s \frac{dv_s}{dt} = F_G + F_D + F_I$$ (7)

where F_G is the component of gravity force in the flow direction. F_D is the drag force , and F_I the inertia force, which are respectively defined by:

$$F_D = \frac{1}{2}\rho_f C_D (U - u_s)|U - u_s|A$$

$$F_I = \rho_f C_M V_s \frac{d}{dt}(u_s - U)$$ (8)

where ρ_f is the density of water, U is the velocity of the water, C_D (=2.05) is the coefficient of drag, C_M(=0.25) is the coefficient of inertia force [Kamiyama et al., 1993], and A is the project area of snow mass.

Initially the project area of a snow lump varies because a thrown snow mass oscillates vertically due to buoyancy. The change of its area, A, is included in Eq.(8) by calculating its vertical motion. Parallel calculations of the motion of the water and the snow mass define the position and velocity of the snow mass at any point and time. Figure 4 shows an example. A snow lump is dropped into a flow of 0.5 m/s. It increases in speed as shown by the solid curve. If there is a bend, for which the critical velocity is 0.412 m/s, a distance longer than 2.24 m is required between the point of dropping and the bend for a smooth snow removal operation without stagnation.

5. CONCLUSIONS

From this study, we conclude the following :
(1) Among the factors that influence the stagnation, the angle of bend was found to be the most important.
(2) The critical condition accounting for a kinetic energy and a breaking snow slump to pass through a bend is proposed in Eqs.(5) and (6).
(3) Knowing the angle of bend, the velocity of the water, the slope and width of the channel, the critical speed of the snow and the minimum length between bends in a channel to avoid stagnation can be obtained using proposed numerical simulation and the critical condition.

REFERENCES

Japan Society of Construction and Machinery (1988), Handbook for snow engineering, Morikuta Pub.Co.Ltd, 527p..

Kamiyama, A, M.Noji and F.Imamura (1993), Measurement of snow mass in a channel, Proc. Tech. Meeting of JSCE Tohoku Branch, pp.102-103 (in Japanese).

Yamamoto,J., F.Imamura and N.Shuto (1992), Field
survey and experiments of the blocking in the snow
drain channel, Proc.Japan Society for Snow
Engineering, Vol.18, pp.217-222 (in Japanese).

Snow Engineering: Recent Advances, Izumi, Nakamura & Sack (eds) © 1997 Balkema, Rotterdam. ISBN 90 5410 865 7

A clever device to melt snow using running water

Takeo M.S.Nakagawa
Kanazawa Institute of Technology, Ishikawa, Japan

Ryo Nakagawa Jr
Shisei Institute of Innovative Science, Ishikawa, Japan

ABSTRACT:A clever device to melt snow using running water has been proposed with a clear object of solving an embarrassing problem, viz., how to remove snow in urban areas with high population densities and large snow accumulations. The only requirement when one uses this device is to have a canal or river, in which water is running continuously. It is concluded that the present proposed device is more practical and efficient than any other conventional method for removing snow.

1 INTRODUCTION

Japan has over 260 cities, towns and villages, which are often heavily covered with snow during winter. The common and crucial problem in these areas is how to remove snow from roads, roofs or gardens, in order to make people's daily lives and economic activity possible (Sahara 1983).

In most snow areas of the world, the common strategy might be to carry snow away to some place where it is allowed to melt by the radiant and convective heat exchange processes, instead of trying to melt it before carrying it away. For example, in Japan heavy snow machines as well as underground hot water have been mainly used to cope with this problem, whereas in Germany salt had been used to melt snow on roads. A common disadvantage of these conventional methods is high cost. Furthermore, usage of underground hot water for melting snow is now severely limited, for it may cause ground subsidence. It is, therefore, desirable to develop an innovative device which makes it possible to melt a large amount of snow in a short time without consuming water resources. Indeed, such an attractive device has been developed herein.

In this paper, the principle of the "wire-net method for melting snow" will be introduced. Then, some results of a prototype experiment will be presented and discussed.

2 PRINCIPLE OF THE WIRE-NET METHOD

Figure 1 shows a schematic diagram of the "wire-net method for melting snow"(Kiryu 1983, Nakagawa 1986a,b). The heat capacity of the running water in the water-lane, which is separated from the snow by the wire-net, is conducted to the snow through openings in the wire-net. Thus, snow is melted producing additional water used downstream.

It is necessary for the canal to be separated into a snow loaded part and a water-lane by the wire-net. The water-lane prevents an excessive increase in water depth upstream, which avoids overflow from the canal. At the same time, this ensures a direct contact between the running water and the snow through the openings of the wire-net. Since the temperature of the running water is higher than that of the snow, the heat capacity of the running water is transferred to the snow by conduction. Consequently, the snow will melt, and thus change into the water. The temperature of the running water will be decreased during the heat exchange processes. It is evident that this results in a decrease in heat capacity of the running water for a while. This may not cause any problem in practice, since fresh water with greater heat capacity is supplied continuously from the upper stream.

One important aspect, which must be mentioned here, is another possible heat transfer means from the earth and atmosphere to the snow, as well as to the running water. Especially, the heat transfer from the earth to running water is essential to recover the lost heat capacity of the running water during the snow melting processes.

This is the reason why the running water never gets colder as it flows downstream, but it gets colder while it passes beside the snow.

3 FIELD EXPERIMENT

An experiment was done in the Kuratsuki Canal in Kanazawa, Japan. Figure 2 shows a sketch of the experiment. The wire-net consisted of a diamond pattern(side length of ca. 7 cm) was placed in the canal. The space between the wire-net and right bank was filled with snow, whereas the water flowed mainly in the lane between the left bank and wire-net.

The wire-net covered 15.60 m of the canal. It was fixed to the left bank with supporting rods. The space between the inclined wire-net and right bank was filled with snow of 32.73 m^3, having the density of 0.58 g/cm^3.

The water temperature, water depth and longitudinal flow velocity were measured at five positions $p_1 - p_5$ at 20 min intervals. The water temperature was measured with thermometers 5 cm below the water surface. The water depth was measured with rods, and the longitudinal flow velocity with propeller type flowmeters 10 cm below the water surface. The snow and air temperatures were also measured at 20 min intervals. The snow temperature was measured with a thermometer 30 cm below the surface, while the air temperature near the experimental site was measured with a thermometer.

The water discharge upstream of the experimental site was 0.36 m^3/s. It takes about 20 min to fill the space between the wire-net and right bank with snow, using a bulldozer. Therefore, origin of time was when the space has been completely filled with snow.

4 RESULT

Figure 3 shows how the water temperature at each position in the canal varies with time. In this figure, air and snow temperatures are also plotted for comparison. Once the space is filled completely with snow, water temperatures at p_1 and p_2 start decreasing. The water temperatures at p_1 and p_2 vary with time in a similar way. This suggests that a similar amount of heat is transferred to the snow from the running water at p_1 and p_2, though position p_1 is apart from snow over 2.50 m.

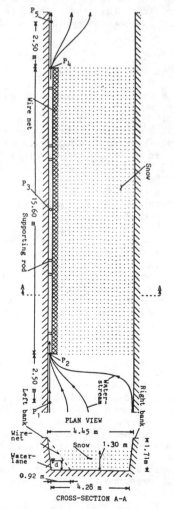

Figure 2 Sketch of the experiment

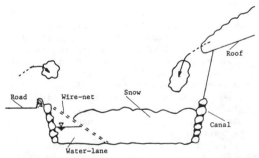

Figure 1 Schematic diagram of the "wire-net method for melting snow".

Water temperatures at p_3, p_4, and p_5 are lower than those at p_1 and p_2, so it is considered that the amount of heat transferred from the running water to the snow at p_3, p_4, and p_5 is greater than that at p_1 and p_2. Note that the water temperature at p_4 is the lowest throughout the experiment.

It can be noted in figure 3 that water temperatures at all of the positions maximize at $t \simeq 37$ min. This is becuase the loaded snow forms a bridge ranging from the wire-net to right bank during the period between $t \simeq 20$ and 35 min, so that the water flows under the snow bridge having lesser surface contact with the snow. The snow bridge breaks at $t \simeq 37$ min, so that the heat transfer to the snow increases suddenly.

Figure 4 shows how the water depth at each position in the canal varies with time. The original water depths at p_1, p_2, p_3, p_4, and p_5 are 34.0, 31.0, 33.8, 30.0, and 26.0 cm, respectively. The water depths at p_1 and p_2 increase rapidly during the period of filling the space between the wire-net and right bank with snow. At $t = 1$ min, the water depths at p_1 and p_2 are their maximum values of 65.0 and 71.0 cm, respectively. Both of these water depths decrease exponentially with time, each approaching the initial value. However, the water depth at p_2 is always greater than that at p_1.

Water depths at p_3, p_4, and p_5 increase less rapidly than those at p_1 and p_2 initially, and they fluctuate with time, reflecting the inhomogeneity of the snow-water mixture in the space between the wire-net and right bank.

Figure 5 shows how the longitudinal flow velocity at each position in the canal varies with time. The original longitudinal flow velocities at p_1, p_2, p_3 p_4, and p_5 are 12, 10, 31, 85, 65 cm/s, respectively. The flow velocity at p_3 decreases exponentially with time and approaches the original value. However, the flow velocities at p_4 and p_5 do not show any clear temporal dependency, reflecting the inhomogeneity of the snow-water mixture in the space between the wire-net and right bank.

5 DISCUSSION

The critical information to assess the performance of the "wire-net method for melting snow" must be the time to melt a given amount of snow. A relevant formula will be, therefore, derived here, and will be discussed in the light of the present

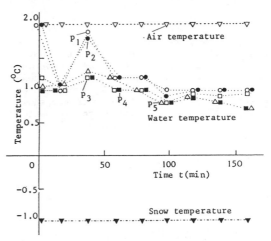

Figure 3 Temperature vs. time for each position in the canal.

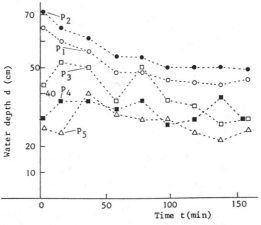

Figure 4 Water depth vs. time for each position in the canal.

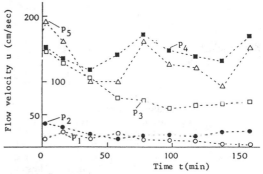

Figure 5 Longitudinal flow velocity vs. time for each position in the canal.

experimental results.

It is assumed that the heat transferred to the snow from the running water, is the same as that used to bring the snow to 0°C and then raising it to the field temperature of the snow-water mixture T_f. Note that the field temperature T_f is not necessarily equal to 0°C, from which the snow starts melting. The above consideration leads immediately to the following relation,

$$\Delta t \cdot \rho_w \cdot Q \cdot c_w (T_{wu} - T_{wd}) = \rho_s \cdot V[L_f + c_s(T_f - T_s)], \quad (1)$$

where Δt is the time to melt the snow, ρ_w and ρ_s are the water and snow densities, respectively, Q the water discharge, c_w and c_s are the specific heats of water and snow, respectively, T_{wu} and T_{wd} are the water temperatures at the upstream and downstream ends of wire-net, respectively, V the snow volume loaded in the space between the wire-net and right bank, L_f the latent heat of snow fusion, and T_s the snow temperature. Hence, equation (1) gives the time to melt the snow as follows,

$$\Delta t = \frac{\rho_s \cdot V[L_f + c_s(T_f - T_s)]}{\rho_w \cdot Q \cdot c_w (T_{wu} - T_{wd})}. \quad (2)$$

It must be noted here that heat transfer to the snow from the air and earth have been neglected, for in this case they are considered to be negligibly small.

Substituting the following physical quantities (Encyclopedia Americana 1968)

$$\rho_s = 0.58 \text{ g/cm}^3, \ \rho_w = 0.9992 \text{ g/cm}^3,$$

$$V = 32.73 \text{ m}^3, \ L_f = 333.9 \text{ J/g},$$

$$c_s = 2.094 \text{ J/g°K}, \ c_w = 4.187 \text{ J/g°K},$$

$$T_f = 1.04 \text{ °C}, \ T_s = -1.0 \text{ °C},$$

$$T_{wu} = 1.24 \text{ °C}, \ T_{wd} = 0.964 \text{ °C},$$

$$Q = 0.36 \text{ m}^3/\text{s},$$

into equation (2), we obtain the time to melt the snow,

$$\Delta t = 257 \text{ min},$$

where T_f is the mean temperature at p_2, p_3 and p_4, T_{wu} the mean temperature at p_1, T_{wd} the mean temperature at p_5. It has been observed that about 80 % of the snow in the space between the wire-net and right bank melts in 196 min. Thus, $0.8 \times \Delta t = 206$ min should be compared with the experimental value of 196 min.

Considering the difficulty to estimate the snow volume accurately, the agreement between the theory and experiment is rather remarkable. Hence, it is suggested that the present proposed formula is useful for the application of the "wire-net method for melting snow".

6 CONCLUSION

The following new knowledge and insights are derived through the present study.

1. This method can melt a large amount of snow in a short time without wasting any water resources, and additional water is produced during the melting process.

2. A formula has been derived to estimate the time to melt the snow for the application of the present method. It has been demonstrated this formula has an engineering value.

3. It is found that the present method is readily applicable to any canal with a different size and shape. When this method is applied to a water canal running through a crowded urban area having only narrow roads, it is especially attractive, for people may be required to carry heavy snow a considerably shorter distance.

4. It is essential that the wire-net is kept clean in order to maintain its effectiveness.

5. The present method has several advantages over the other methods: First, the structure is very simple; second, it is relatively inexpensive; and third, no water overflows from the canal or river, for continuity of the flow is always ensured by the water-lane.

REFERENCES

Encyclopedia Americana 1968. Ice, vol.14: 621. New York:Americana Corporation.
Kiryu, M. 1983. Current state for alleviating snow-disasters at Yonezawa -snow floating and flowing canal, and snow vanishing and melting canal. Development and Conservation, 8:19-25(in Japanese).
Nakagawa, T. 1986a. A device for melting snow. Memoirs of Kanazawa Institute of Technology, A, no.23:11-19.
Nakagawa, T. 1986b. Principle of snow vanishing and melting and its application of compound structures. Proc. of Symposium on Application of Compound Structures, 1:309-316.
Sahara, K. 1983. Snow management and utilization of water resources. J.Japan Sea Society, 6:21-32(in Japanese).

Snow Engineering: Recent Advances, Izumi, Nakamura & Sack (eds) © 1997 Balkema, Rotterdam. ISBN 90 5410 865 7

Rise in stage and limiting conditions for snow transport in elbows of snow drains

Michio Takahashi & Kiyoji Kimura
Department of Civil Engineering, College of Engineering, Nihon University, Koriyama, Japan

Kazue Kitayama
Yamatake-Kitayama Construction Co., Ltd, Kuroishi, Japan

ABSTRACT : A snow drain is one of the most effective facilities to remove snow. The ability to remove snow is remarkably affected by the flow in elbow parts of the drains. This paper treats problems related to the capacity of removing snow by flowing water in the elbows of snow drains. The rise in stage has been analyzed theoretically, and compared with experimental values. Results agreed reasonably well. The limiting conditions for snow transport by flowing water through the elbows of drains have been studied by experimental considerations. As a result, the limiting conditions were estimated. In the experiments, crushed-ice was used instead of snow.

1 INTRODUCTION

A snow drain is one of the most economical and effective facilities to remove snow in the heavy snow cities of Japan. The construction of snow drains has been progressing in these regions. In general, elbows and other rapidly changing parts of snow drains cause the stagnation of snow blocks and the blockade of drains. The ability to remove snow in curving drains is inferior to that in straight drains. As a result, it is necessary to consider the mechanism of flow in those parts. Some consideration about the flow characteristics and the capacity of removing snow in straight snow drains have been made by Sato & Shuto(1983), Watanabe(1986), et al. A few studies on the characteristics of flow in the elbows of snow drains have been reported by Tanaka(1969), Imarura(1992), et al. However, it seems that investigations on the aspect of hydraulic engineering have not been presented sufficiently, and quite a few problems remain unsolved.

Throwing snow into snow drains by means of snow removal equipment is often done to save labor. It is expected that the throwing of snow by equipment will increase from now on. A big lump of snow thrown into a snow drain stays together as the flow goes through the straight drain, And it forms a long, jelly-like, snow block. A laboratory model test was conducted on the flow of a crushed-ice and water mixture using crushed-ice as model snow. This paper describes the rise in stage and the blocking limit of a snow-water mixture in elbows of snow drains, which contain long block of snow.

2 EXPERIMENTAL PROCEDURES

Experiments were conducted in a 4.0 m long, 7.0 cm wide, 10.5 cm deep tilting flume made of acrylic plastic. The length of straight part upstream and its downstream was both 2.0 m. The flume had 90° and 45° elbows in its middle section. It was possible to exchange them. In the experiment, crushed-ice made by an electric ice-slicer was used as model snow. It was thrown into the flume at the upstream end. The crushed-ice was thrown by two methods. In the method A, a crushed-ice block, which was compacted and formed to the required density by manpower, was thrown into the flume. This crushed-ice block was formed in a 60.0 cm long, 6.95 cm wide box. Method A was designed to investigate the blocking limit which varies according to the density and the thickness of crushed-ice. In method B, the crushed-ice was continuously thrown into the flume from two electric ice-slicers which were placed on the upstream end of the flume. The temperature in the laboratory was set as low as possible. Moreover, the water temperature during measuring was kept at about + 2 to 3 °C by throwing great quantities of ice continually into the reservoirs of the upstream and the downstream end. Careful attention was paid so that the crushed-ice would not melt. Three pockets on the side of the flume were used to measure the flow depth. Care was taken so that the measuring instruments would not affect the flow of crushed-ice.

3 RESULTS AND CONSIDERATIONS

3.1 Rise in stage in elbows

The stage in elbows rises as the Froude number of the flow increases. Consequently, it is necessary to consider the change of stage in those parts from the aspect of problems of overflow out of snow drains or estimation of the energy loss of flow in elbows.

To obtain the equation which expresses the change of stage in elbow parts, when the momentum equation is applied within the control sections of I and II shown in Figure 1, the following equation can be produced:

$$\frac{1}{2}\rho gbh_1^{\,2} - \frac{1}{2}\rho gbh_2^{\,2}\cos\theta - F\sin\theta$$
$$= \beta_2\rho Qv_2\cos\theta - \beta_1\rho Qv_1 \tag{1}$$

in which section I = the section before rise in stage, section II = the section after rise in stage, h_1, h_2 = flow depth of I and II, v_1, v_2 = mean velocity of I and II, respectively, Q = rate of discharge, b = flume width, ρ = fluid density, g = gravity acceleration, β = momentum correction factor, and θ =angle of elbows. F, which is the pressure acting on the side wall of flume a-e in elbow part, is assumed to be expressed by following equation:

$$F = \frac{1}{2}\rho gb\tan\left(\frac{\theta}{2}\right)h_e^{\,2} \tag{2}$$

Now, when it is assumed that the flow depth at the point where the stage rises highest in the elbow parts h_e is equal to h_2, and that β_1 and $\beta_2 = 1.0$, the ratio of the change of stage h_e / h_1 in the elbow part is obtained by following equation:

$$\left(\frac{h_e}{h_1}\right)^3 - \left(2F_1^{\,2}+1\right)\left(\frac{h_e}{h_1}\right) + 2F_1^{\,2}\cos\theta = 0 \tag{3}$$

where $F_1 = v_1 / \sqrt{gh_1}$.

In Figure 2, the comparisons of equation (3) with

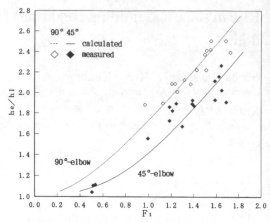

Figure 2.   Rise in stage in elbow parts.

experimental results at elbows of 90° and 45° are shown by the relation between the Froude number F_1 and the relative height of rise in stage h_e / h_1. Both the equation and the experimental results at elbows of 90° and 45° are fairly coincident as a whole, though a little scatter is recognized for the measured values because the fluctuation of stage occurs. Consequently, it is also expected that the change of stage in elbow parts in the mixture flow of crushed-ice and water can be approximately expressed by equation (3)(Noguchi 1992).As a result, it is understood that the change of stage in elbow parts also increases as the Froude number or the angle of the elbow increases.

3.2 Blocking limit of crushed-ice in elbow parts

Flow patterns : Photos 1 and 2 show the flow patterns at elbows of 90° of a formed crushed-ice thrown by means of method A. Photo 1 is the flow pattern where the crushed-ice is plugged. Photo 2 is the flow pattern of non-blocking.

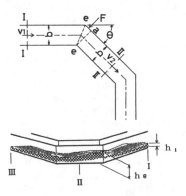

Figure 1.   Schematic description.

Photo 1 . Flow pattern of blocking.

470

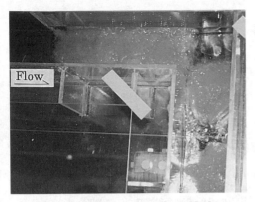

Photo 2 . Flow pattern of non-blocking.

Effects of the thickness and the density of crushed-ice : Figure 3 is some examples of experimental results which show the variations of blocking limits at the elbows of 90° and 45° owing to the differences of the thickness h_S and the density ρ_s of crushed-ice to the Froude number F_{R1}. In this experiment, a formed crushed-ice block was thrown by method A. The Froude number $F_{R1}(= v_1/\sqrt{gR_1})$ is defined by the flow in section I without the crushed-ice, in which R_1 = hydraulic radius in section I. The symbols in those figures indicate as ×=blockade, + =partial blockade, and □ =non-blockade. Those judgments are as follows : blockade indicates that a crushed-ice block was stopped at the elbow more than 5 seconds, non-blockade indicates that the block was not stopped, and partial blockade indicates that a crushed-ice block was stopped at the elbow less than 5 seconds, then

began to move again. It is recognized from Figure 3 that the blocking limits at both elbows are mostly determined by the density of the crushed-ice block, and that they are scarcely affected by the relative thickness of the crushed-ice. However, in Figure 3(d),the blocking limit shifts toward lower values of F_{R1} for thin crushed-ice. It is considered that the thin crushed-ice block presents twists or wave undulations, and goes easily through the elbow.

Blocking limit : Figure 4 shows the relation between the Froude number F_{R1} and the density of crushed-ice ρ_s at the blocking limits obtained from Figure 3. It is expected that the blocking limits in elbows of snow of various densities can be estimated from Figure 4.

And now, the blocking limits in elbows are discussed on the removing snow which goes down attaching to one another and in the condition that it has filled the most of the cross-sectional area of the drain. Figures 5 and 6 show the observed values of blockade and non-blockade at elbows of 90° and 45° , respectively, and the relationship between C_V and the Froude number F_{R1}. C_V is the ratio of the weight of crushed-ice thrown per unit of time to the discharge of flow without crushed-ice. The symbols ◇, ◆ and + are the observed values for crushed-ice thrown into the flume by method A, and symbols ○, ● and × are the observed values by method B. The boundary of blockade and non-blockade is nearly defined by those figures at elbows of 90° and 45° . That boundary is indicated by a line in those figures. As a result, it is expected that the blocking limits of removing snow in elbows which goes through attaching to one another and in the conditions that it has filled the most of the cross-sectional area of snow drains. That is, the weight of snow per unit of time in which it is possible to throw the snow into the drains can be estimated from Figures 5 or 6.

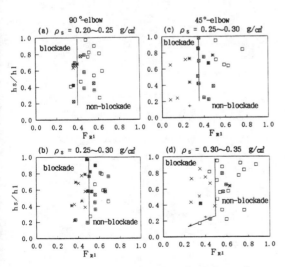

Figure 3 . Variation of blocking limit due to h_S and ρ_s to F_{R1}.

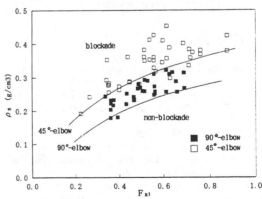

Figure 4 . Blocking limits showing the relation between ρ_s and F_{R1}.

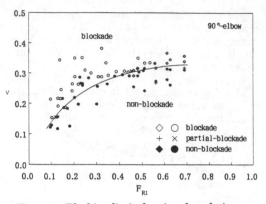

Figure 5 . Blocking limit showing the relation
between C_V and F_{R1} (90° elbow).

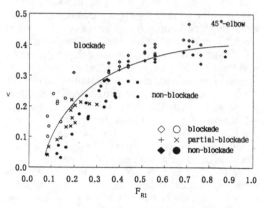

Figure 6 . Blocking limit showing the relation
between C_V and F_{R1} (45° elbow).

In this paper, the crushed-ice floats through at
the blocking limit. In a few cases, the crushed-ice
caused blocking by touching the flume bottom in
streams of small flow depth. Such situations were
excepted from the observed values.

4 CONCLUSIONS

Some considerations were described about the rise
in stage and the blocking limit of blocks of crushed-
ice which goes through snow drain elbows of 90°
and 45° . The results can be summarized as
follows:

1. The equation which was obtained from the
momentum equation expresses the change of stage
in elbow parts, and it agreed with the experimental
results of crushed-ice in water.

2. It was recognized from the results of
experiments in which a formed crushed-ice block
was used that the blocking limits in both elbows are
determined for the most part by the density of the
crushed-ice block, and that they are scarcely
affected by the thickness of the crushed-ice.

3. The blocking limit of removing snow of a block
type which possesses various densities can be
estimated.

4. The rate at which snow can be thrown into the
drains can be approximated knowing the Froude
number.

REFERENCES

Sato, T & N.Shuto 1983. A velocity formula for the
snow and water flow: 801-806. *Proc. of the 27th
Japanese Conference on Hydraulics.*
Watanabe, Z. 1986. Design of a system of snow
removing gutters and the amount of water
required for the snow removal: 141-148. *Jour. Of
the Japanese Society of Snow and Ice,* Vol.48,
No.3.
Tanaka, Y.,et al. 1969. Snow-removal capacity of
gutter systems: 55-66. *Jour. of National Institute
for Disaster Science,* No.3.
Yamamoto, J. & H.Imamura 1992. Fundamental
study on blocking condition and ability of
removing snow in elbows of snow drains: 22-29.
Jour. of Snow Engineering of Japan, Vol.8, No.4.
Noguchi, T., M.Takahashi & K.Kimura 1992.
Hydraulic characteristics on flow of elbow parts of
snow-removing ditches: 293-298. *Proc. of Hydraulic Engineering, JSCE,* Vol.36.

Snow Engineering: Recent Advances, Izumi, Nakamura & Sack (eds) © 1997 Balkema, Rotterdam. ISBN 90 5410 865 7

Characteristics of ice/water mixture flow in a branching pipe and development of an ice fraction control technique

Yoshitaka Kawada
Department of Mechanical Engineering, Nagaoka College of Technology, Niigata, Japan

Shin'ya Takizawa
Shinano Kenshi Corporation, Nagano, Japan

Masataka Shirakashi
Department of Mechanical Engineering, Nagaoka University of Technology, Niigata, Japan

ABSTRACT: A new district cooling system (DCS) utilizing the latent heat of ice is in development to increase the transport capacity of cooling energy supply. In this work, the characteristics of an ice/water mixture flowing in a branching section of a pipeline in a DCS was investigated experimentally and a technique was developed to control the cooling energy distribution through the branch pipe. The branching section consisted of an 80-mm-diameter horizontal main pipe and a 50-mm-diameter branch pipe perpendicular to it. Chip ice of average particle size of 10 mm and natural granulated snow were used as samples. The energy loss of the ice/water mixture flow due to branching is virtually equal to that of water irrespective of branching direction, ice fraction and velocity in the main pipe and particle size. While, the ice fraction in the branch pipe is strongly dependent on these factors. The control equipment devised in this study to hold the ice fraction in the branch pipe at a desired value proved to work successfully for the ice/water mixture under the most unfavorable condition.

1 INTRODUCTION

Recently the ice/water mixture transportation technique has been investigated in many institutes to increase the density of cooling energy through pipelines of a district cooling system (DCS) (Kawada and Hattori 1994). In the pipelines of a DCS, branching sections are used to distribute chilled water to consumers. Hence, it is necessary to clarify the flow characteristics of an ice/water mixture in a branching section for applications of the ice/water mixture transportation technique. Pressure loss at the branching section and the ice fraction in the branch pipe are the most important characteristics from a practical point of view. The former increases the operation power of the DCS and the latter governs the cooling energy supplied through the branch pipe. In spite of its importance, only small number of works under rather limited experimental conditions have been reported on this subject so far (Onojima et al. 1991).

In this study, the effects of the following factors on the energy loss and the ice fraction in the branch pipe were investigated; 1) branching direction, 2) flow conditions, 3) ice particle size. Based on the results, a device was completed to control the flow rate and the ice fraction in the branch pipe at desired values simultaneously, which enables the cooling energy supply through the branch pipe to be controlled to meet the consumption demand.

2 BRANCHING SECTION AND SAMPLE PARTICLES

The branching section used in this study is shown in Figure 1. The nominal diameters of the main horizontal pipe and the branching pipe are 80 mm and 50 mm, respectively. Pressures were measured through 0.8 mm taps at cross sections 1, 2 and 3 located far enough from the branching to allow the flow to become a fully developed circular pipe flow. The branching angle was 90° and the branching direction was either vertically upward or horizontal or vertically downward.

It is expected that the size of ice particles used in the DCS will differ considerably among ice making techniques. Hence, to see the effect of particle size, three types of particles were used as samples: chip ice made by a commercial ice-making machine with average particle size of 10 mm, natural granulated snow

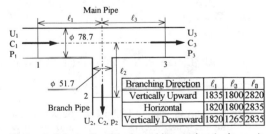

Figure 1. Schematic of the branching section (unit: mm).

snow and polypropylene beads of 0.90 specific gravity and 3.8 mm equivalent diameter.

3 DEFINITION OF BRANCHING LOSS COEFFICIENT

In the branching section shown in Figure 1, the net branching pressure loss between the cross sections 1 and j (= 2 or 3), h_{1j} was calculated by Eq.(1) from the measured pressures by taking into account the pressure losses in the straight pipes.

$$h_{1j} = \frac{p_1 - p_j}{\rho g} - \left(i_1\, \ell_1 + i_j\, \ell_j \right), \qquad (1)$$

Here, p: pressure, ρ: density of water, g: gravitational acceleration, i: pressure gradient of the ice/water mixture in the straight pipe, ℓ: distance from the branching point, subscripts $1, j$: cross sections 1 and j, respectively.
i is expressed as

$$i = i_w + C\phi i_w, \qquad (2)$$

where i_w is the pressure gradient of water, C is the volumetric ice fraction in the pipe and ϕ is the excess pressure loss coefficient. i_w is expressed as

$$i_w = \lambda \frac{1}{d}\frac{U^2}{2g}, \qquad (3)$$

where λ is the pipe friction coefficient, d is the pipe diameter and U is the mean velocity. λ is evaluated by Blasius's formula (4) since the wall of pipes used in this study were hydraulically smooth under the experimental condition. Then,

$$\lambda = 0.3164\left(Ud\,/\,v\right)^{-1/4}, \qquad (4)$$

where v is the kinematic viscosity of water. ϕ is obtained by following empirical formulas (Shirakashi et al. 1995, Kawada et al. 1994).

Horizontal pipe: $\phi = 23Fr^{-0.82} - 0.5$ (5)

Vertical pipe: $\phi = \pm\dfrac{2}{\lambda}Fr^{-1}\ \left(\begin{array}{l}+: \text{downward}\\ -: \text{upward}\end{array}\right)$ (6)

Here Fr is the Froude number defined by

$$Fr = \frac{U^2}{gd\left|1 - S\right|}, \qquad (7)$$

where S is the specific gravity of ice (=0.917). Thus, the branching loss coefficient ζ_{1j} is defined as

$$\zeta_{1j} = \frac{h_{1j}}{U_1^2/2g} \qquad (8)$$

4 EXPERIMENTAL RESULTS AND DISCUSSION

4.1 Branching loss coefficient

Figure 2 shows the relationship between the branching loss coefficients ζ_{12}, ζ_{13} and the flow rate ratio of the branch to the main pipe, V_2/V_1, for upward branching. The branching loss coefficients

of the ice/water mixture almost coincide with those of water shown by the curves in the figure irrespective of the velocity and ice fraction in the main pipe upstream. Similar results were obtained for all the branching directions and particles used in this study. Therefore, the branching loss coefficients of an ice/water mixture are little affected by particle size, and they are virtually equal to those of water.

4.2 Effect of branching direction on ice fraction

Figure 3 shows the relationship between the ice fraction ratio of the branch to the main pipe, C_2/C_1, and flow rate ratio V_2/V_1 for chip ice at U_1=1 m/s and C_1=5 %. When U_1 and C_1 are as low as these values, C_2/C_1-V_2/V_1 curves strongly depend on the branching direction. For upward branching, C_2/C_1 is a little larger than 3 when V_2/V_1 is around 0.1. This means that at low branching velocity the ice fraction in the branch is much higher than in the main pipe upstream. The ice fraction in the upward branching pipe decreases as V_2 increases and, naturally, $C_2=C_1$ at $V_2=V_1$. For downward branching, in contrast, C_2/C_1 is virtually zero at V_2/V_1<0.4 and increases with V_2/V_1 in the region of V_2/V_1>0.4. For the

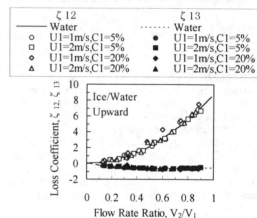

Figure 2. Relationship between the branching loss coefficient and the flow rate ratio.

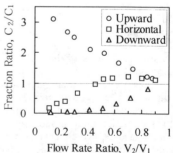

Figure 3. Effect of the branching direction on the ice fraction ratio (chip ice, U_1=1m/s, C_1=5%).

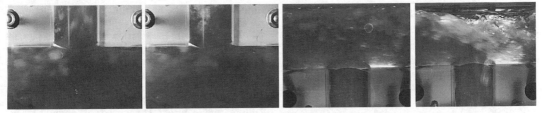

(a) Upward, V_2/V_1=0.14   (b) Upward, V_2/V_1=0.62   (c) Downward, V_2/V_1=0.24   (d) Downward, V_2/V_1=0.62

Figure 4. Flow pattern of ice/water mixture in the branching section (U_1=1 m/s, C_1=5 %).

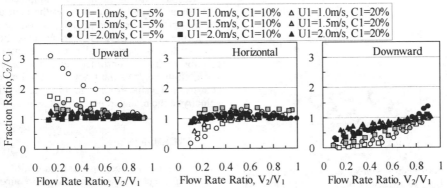

Figure 5. Effects of ice fraction and velocity in the main pipe upstream on the fraction ratio (chip ice).

horizontal branching, C_2/C_1 at a low V_2/V_1 is considerably smaller than unity, increases with V_2/V_1, and virtually equal to unity at V_2/V_1>0.5.

The behaviors of C_2/C_1-V_2/V_1 curves for different branching directions are clearly explained from the flow pattern of ice/water mixture as shown in Figure 4. Ice particles in the pipe adhere to each other and form clusters with a size larger than about one half a pipe diameter at a low ice fraction, say around 5 % (Shirakashi et al. 1995). When U_1 is low, the clusters move along the upper wall of the pipe due to the buoyancy force. Hence, most of the clusters move into the upward branch pipe when V_2/V_1 is low, as seen in Figure 4(a), and an increase in V_2 is mainly supplemented with water, resulting in a decrease of C_2/C_1 with increasing V_2/V_1 (see Figure 4(b)). The effect of buoyancy force on the ice clusters causes the opposite tendency in the C_2/C_1-V_2/V_1 curve for the downward branching as seen in Figures 4(c) and (d). The low value of C_2/C_1 at V_2/V_1<0.3 for the horizontal branching in Figure 3 suggests that the ice clusters have difficulty in following the flow into the branch when V_2 is low because the diameter of the branch pipe is considerably smaller than that of the main pipe.

4.3 Effects of ice fraction and flow velocity in the main pipe

In Figure 5, C_2/C_1-V_2/V_1 relations are compared for different values of U_1 and C_1 to see the effects of ice fraction and flow velocity in the main pipe.

Generally speaking, both C_1 and U_1 have the same effect on the C_2/C_1-V_2/V_1 curve. The C_2/C_1 vs. V_2/V_1 plots collapse to a single curve around C_2/C_1=1 with an increase of either C_1 or U_1 irrespective of the branching direction. However, the mechanisms of these effects are different. When U_1 is higher, mixing in the main pipe is stronger and the distribution of ice particles in the main pipe is more homogeneous (Shirakashi 1994). Hence, C_2 is nearly equal to C_1 even when V_2/V_1 is low. When C_1 is higher, the ice clusters in the main pipe are bigger, almost filling the whole cross section of the pipe (Shirakashi 1994), resulting in nearly homogeneous ice particle distribution, again.

4.4 Effect of size and/or shape of ice particles

In Figure 6, C_2/C_1-V_2/V_1 relations of the three kinds of particles are compared under fixed values of U_1 and C_1. The different behavior among the solid particles is most remarkable for the upward branching. The effect of buoyancy force described in Section 4.2 is seen more clearly with the polypropylene beads for the upward branching. This is because the polypropylene beads do not adhere to each other and do not form clusters. Since particles of granulated snow are finer than the chip ice, the adhesive nature is stronger and it forms clusters more easily than the chip ice. Therefore, the effect of the buoyancy force on snow/water mixture is less remarkable than on ice/water mixture.

4.5 Control of cooling energy distributed to the branch pipe

In operating a DCS, the cooling energy supplied through the branch pipe has to be controlled according to the demand of the consumers connected to it. Hence, we devised the equipment shown in Figure 7 to control the ice fraction and the flow rate in the branch pipe at desired values

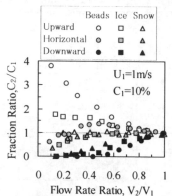

Figure 6. Relationship between the fraction ratio and the flow rate ratio for different solid particles.

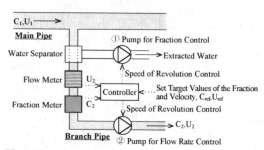

Figure 7. Schematic of the equipment controlling the fraction and the flow rate in the branch pipe.

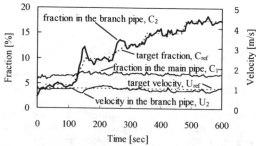

Figure 8. Variations of the ice fraction and the flow rate in the branch pipe with time when the controlling device was applied (downward branching, C_1=7 %, U_1=1.5 m/s).

simultaneously. The basic principles of its operation are as follows:

1) Fraction control: The fraction in the branch pipe is controlled by adjusting the flow rate of water extracted at the water separator. This is attained by controlling the speed of revolution of Pump ① according to the difference between the measured fraction C_2 and the prescribed value C_{ref}.

2) Flow rate control: The speed of revolution of Pump ② is adjusted according to the difference between the measured and the prescribed flow rate to keep the flow rate in the branch pipe constant.

A personal computer was used as the controller, in which a proportional and/or integral control actions were provided. Figure 8 demonstrates how the control device works under a most unfavorable condition. It is confirmed that the equipment successfully controls the fraction in the branch pipe at a desired value from 5 to 17.5 %, holding the velocity in the branch pipe constant.

5 CONCLUSIONS

Characteristics of ice/water mixture flow in the branching section of a horizontal main pipe and a smaller-diameter branch pipe perpendicular to it were investigated experimentally. Based on the results, a device to control the cooling energy supplied through the branch pipe was developed. The conclusions are summarized as follows.

(1) Energy loss due to branching of the ice/water mixture is virtually equal to that of water.

(2) The ice fraction in the branch pipe is affected considerably by the branching direction, the velocity and the fraction in the main pipe, and the particle shape and/or size.

(3) The device to control the flow rate and the ice fraction in the branch pipe worked successfully irrespective of flow conditions in the main pipe.

REFERENCES

Kawada, Y. and Hattori, M. 1994. District cooling system utilizing latent heat of ice, *J. Japanese Society of Snow and Ice*, **56**-2: 169-179.

Kawada, Y., Takahashi, S., Shirakashi, M. and Hayakawa, N. 1994. *Proc. 11th Symp. on Snow Eng. of Japan*: 107-112.

Onojima, H., Takemoto, Y., Fukushima, M. and Takemoto, K. 1991. A study on the hydraulic transportation of ice-water mixture for district cooling systems, *Proc. 1st ASME/JSME Fluids Eng. Conf.*, **FED-Vol.118**: 241-246.

Shirakashi, M., Kawada, Y. and Takahashi, S. 1995. Characteristics of ice/water mixture in horizontal circular pipes, *Trans. Japan Society of Mechanical Engineers*, **61**-585, B: 1632-1639.

Shirakashi, M. 1994. Characteristics of snow/water mixture flow and techniques for its measurement and control, *J. Japanese Society of Snow and Ice*, **56**-2: 159-167.

Snow Engineering: Recent Advances, Izumi, Nakamura & Sack (eds) © 1997 Balkema, Rotterdam. ISBN 90 5410 865 7

Snow melting systems with ground water

Nobuaki Goto, Kiichi Numazawa & Yoshimasa Katsuragi
Japan Ground Water Development Co., Ltd, Yamagata, Japan

ABSTRACT: More than 20% of Japanese live in the snowy area, which occupies about 60% of the country of Japan. Therefore, snow removal and snow control on roads have been essential since early times in Japan. The weather in winter Japan is distinguished from others, because of much snowfall with not so cold atmospheric temperatures, especially in the Northeastern districts. Snow melting systems with ground water have been developed and it has been proven that the temperature of ground water is sufficient for snow melting. It has an economical advantages because it does not need any fuel nor electricity as a heat source. This paper introduces the distinctions of the winter weather, the history of snow removal/control, snow melting systems with ground water and also some considerations on the requirements of heat value for the snow melting system in the Northeastern districts mainly, which seems to be the typical snowy area of Japan.

1 WINTER WEATHER IN JAPAN

Winter weather in Japan is characterized by much snowfall despite moderate temperatures, especially in the Northeastern districts.

Figure 1 shows the relationship between precipitation and average temperature in January in certain cities. It shows that the amount of precipitation of Japanese cities is much more than that of many American and European cities.

It is known that the rain changes to the snow at a ground temperature around 2 to 3℃; therefore the precipitation in the month of January at such Japanese cities is mostly snowfall.

During winter in the Northeastern districts, such as Tohoku and Hokuriku districts, there are many days whose minimum temperatures are below 0℃, but there are a few days whose maximum temperatures are also below 0℃.

Snow on roads melts during daytime with plus temperatures and it freezes again during nighttime with minus temperatures. This is repeated daily or in a short intervals and makes the iced surface irregular.

Such road conditions cause traffic difficulties and many efforts have been made to remove and/or control the snow on the roads in Japan.

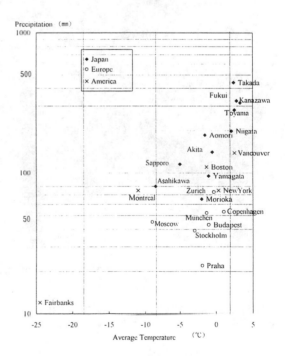

Figure 1. Amount of precipitation and average temperature in January

Table 1. Comparison of initial & running costs for various snow melting systems in Japan.

Snow Melting System	Heat Source	Heating Device	Initial Cost*	Running Cost*	Remarks
Ground Water Without Sprinkling	Earth	Circulation Pipe	50~60	0.5	Preservation of environment
Ground Water With Sprinkling	Earth	Spray Nozzle	15~30	0.45	Risk of land subsidence
Hot Fluid Circulation	Fuel	Circulation Pipe	47	1.0~1.2	Need of facility maintenance
Hot Fluid Circulation	Fuel	Heat Pipe	60	1.2~1.5	Need of facility maintenance
Electric Heating Cable	Electricity	Heating Cable	47	3.5~4.9	Easy to control
Hot Spring Heating	Hot spring	Heat Pipe	30~40	0	Limited to hot spring areas

* cost；1,000¥／㎡.

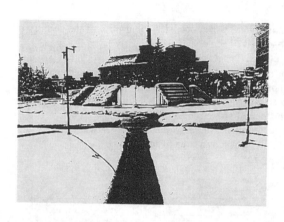

Figure 2. Examples of installations of snow melting systems with ground water.

2 HISTORY OF SNOW REMOVAL/CONTROL IN JAPAN

2.1 Snow removal

Snow removal on roads by civil construction equipment such as graders was started in 1945. We had a very heavy snowfall in 1963 and full-scale snow removal by special snow plows which were developed to suit the snow conditions in Japan started then.

Since the use of snow-tires with studs has been banned since 1990, more applications of effective snow melting facilities, which provide better road conditions, have been desired and installed in the recent years.

478

2.2 Snow melting with sprinkling water

Snow melting systems with sprinkling water are a popular for clearing snow on roads in the snowy area of Japan.

Ground water, river water, sea water, waste water, etc., are sprinkled from nozzles to melt and/or flow out snow on roads by its heat energy and potential energy.

This system was developed in 1961 at Nagaoka-city, and the very heavy snowfall in 1963 made it popular in the Northeastern districts of Japan.

Such systems worked well for a long time, but they are not so convenient for pedestrians and shops beside the roads because of splashing by vehicles passing by.

And it wastes much ground water without recovery, and it some times causes lack of ground water and land subsidence.

2.3 Snow melting without sprinkling water

In 1980, a snow melting system without sprinkling water was developed and demonstrated in Yamagata-city in order to prevent land subsidence and to maintain ground water as a precious resource.

Ground water once pumped up is led to embedded radiation pipes to radiate its heat energy to melt snow on roads and is infiltrated to the ground again, which will be explained in detail below.

3 SNOW MELTING WITHOUT SPRINKLING WATER

This system consists of two wells and a heat radiation area.

One well (the "pumping well") is for pumping ground water and the other (the "recharge well") is for infiltrating used ground water into the aquifer again.

Heat radiation pipes of small diameter are embedded underneath the surface of roads, sidewalks, ramps and parking lots, etc., where snow is to be controlled.

Ground water pumped from the pumping well is led to the radiation pipes and radiates its heat energy to warm the pavements above and melts the snow thereon, and it infiltrated into the aquifer again through the recharge well.

It is said that the temperature of ground water at a certain depth is nearly same as the annual average atmospheric temperature there, so ground water at 10 to 15 ℃ is available in the Northeastern districts of Japan.

Such ground water is not so warm that its heat energy can be utilized for any industrial purpose but it is sufficient as a heat source for snow melting system.

So it can be said that this system utilizes unused energy effectively.

The following are advantages of this system:

1) Since no water is sprinkled on the surface, snow can be cleared from sloped and irregular surface.

2) Since there is no water sprinkled nor splashed on the surface, it is more convenient for pedestrians.

3) Since it utilizes the thermal energy of ground water which is unused, it proves to be energy-saving and has low running costs.

Table 1, prepared by Prof. Zenpachi Watanabe, compares initial costs and running costs for various snow melting systems in Japan.

4) Since it does not waste ground water, it will not cause troubles such as lack of ground water and land subsidence.

4 HEAT VALUE REQUIRED FOR SNOW MELTING

Snow melting systems with heated-slab shall have enough capacity to melt snow and control icing on roads.

Output required for the system can be calculated by the following equation with individual heat values theoretically calculated from weather data:

$$qo = qs + qm + Ar(qe + qh) \qquad (1)$$

where
  qo; output required
  qs; sensible heat transferred to snow
  qm; heat of fusion
  qe; heat of evaporation
  qh; heat transfer by convection and radiation
  Ar; free area ratio

We, in Japan, have used $0.5 \leqq Ar \leqq 1.0$ for free area ratio generally instead of $0 \leqq Ar \leqq 1.0$ and have installed many such systems.

However, it is sometimes reported that systems designed this way have overcapacity especially in the Northeastern districts of Japan.

Thus, we evaluated the theoretical equation again, which supposed that the unit area is partially covered with snow. The area without snow needs the heat of evaporation (qe) and the heat transfer by convection and radiation (qh). The area covered with snow requires the sensible heat (qs) and the heat of fusion (qm).

Considering this, we have derived the theoretical equation below. It is much more applicable to our Northeastern districts.

$$qo = (1 - Ar)(qs + qm) + Ar(qe + qh) \qquad (2)$$

It means that when snow falls (it is permissible to cover the surface with the thin snow layer practically, that is Ar=0), only the sensible heat (qs) and the heat of fusion (qm) need be considered.
When there is no snow there (that is Ar=1.0), only the heat of evaporation (qe) and the heat transfer by convection and radiation (qh) need be considered.

Therefore, the output of the system is sized to cover the heat required to melt snow or the heat required to control icing whichever is larger.

5 CONCLUSION

Snow melting systems with ground water have proved to be effective and have low running costs especially in the Northeastern districts of Japan, where much snowfalls with not so cold atmospheric temperatures.

Output of snow melting systems can be designed to cover the heat required to melt snow or the heat required to control icing whichever is larger.
This approach has proved to be sufficient in the Northeastern districts of Japan.

REFERENCES

ASHRAE. 1980. System Handbook; Chapter 38 "Snow Melting", American Society of Heating, Refrigerating and Air-conditioning Engineer, 38.1-38.15.

Watanabe, Z. 1994. Research on valuation and efficiency of modern snow removal and melting. 59-64. [In Japanese.]

4 Housing and residential planning

Snow Engineering: Recent Advances, Izumi, Nakamura & Sack (eds) © 1997 Balkema, Rotterdam. ISBN 90 5410 865 7

Residential planning under heavy snow conditions in Norway

Kristoffer Apeland
Oslo School of Architecture, Norway

ABSTRACT: Research and development in the field of residential planning in Northern Norway is reported. A method of development is based on a landscape classification, and a developed set of design criteria, depending on the type of landscape, prevailing wind direction and geographical orientation. The methods developed are semiempirical in nature, basing some of the proposed solutions on the building tradition of the Inuit people, as well as old building traditions in Norway and in Central Europe. The R & D-work will hopefully be continued although Professor Anne Brit Börve of the Oslo School of Architecture, who led the development work, unfortunately died from cancer at early age.

1 INTRODUCTION

Norway is a long and narrow country stretching from 58° to 71° N latitude. The area of the country is 324 000 km². One third of this area is situated above the Arctic Circle, and a large number of the settlements are located in regions with very hard climatic conditions.

Due to the very hard climate in the northern region of Norway, the region lends itself to the study of settlement planning in order to create acceptable living conditions for the people over the year, despite the climatic conditions. A number of research and development projects have been performed, in particular from 1976 to 1990.

The initiators of the research projects in Northern Norway have been Professor Arne K. Sterten and Professor Anne Brit Börve, both affiliated with the Oslo School of Architecture.
Professor Sterten is now retired, and Professor Börve died from cancer at a young age. These events have been unfortunate for the continuation of the research. However, young people are eager to go on with the research that was brought up to a professional level by Sterten and Börve.

This paper presents some of the methods developed by Sterten/Börve, and some of the results obtained during the active period 1976-1990.

2 METHOD OF ANALYSIS

The methods developed for the analysis of regions, local areas and buildings, are semiempirical in nature.
Experience is primarily based on the building culture of the Inuits of Baffin land and Hudson Bay (Boas 1901). Planning and development based on the Inuits tradition has been performed by the architect R. Erskine (e.g., Erskine 1968).

Börve (1982) has presented a classification of landscapes. In her doctoral thesis, Börve (1982) has presented design criteria for the planning of homes under arctic conditions.

3 CLASSIFICATION OF LANDSCAPES

Börve's proposal of classification of landscapes, is shown in Figure 1, (Börve 1982)

Landscape A.

Open, plain landscape. Horizon, low and distant.

Open and broad valley floor with low terraces.

Landscape B.

Open, moderately undulating landscape with small hills and ridges. Horizon low and relatively distant.

Landscape C.

High and single sloping landscape forming part of a hillside. Horizon high and near.

Limited sloping landscape rising from sea level into hill plateau.

Landscape D.

"V" og "U" - shaped valley landscapes. High and near horizon on two sides: otherwise distant horizon.

Landscape E.

Heavily undulating landscape with hills and mountains. Dominating high and near horizon.

Figure 1. Classification of landscapes.

4   DESIGN CRITERIA

4.1   *Principal considerations*

Based on litterature reviews, studies of nature and the built landscape, as well as tests and analyses, a number of design criteria have been developed (Börve, 1987). The endeavor has been to establish a set of criteria, that may correspond to the defined landscape classification.

The criteria were developed in a fairly general manner, and may therefore be more a set of guidelines, which require a thorough study of local conditions in order to yield a successful design.

However, the development of criteria has been based on studies of the sun, the wind and the snow as basic parameters. In this connection, the response to climatic parameters observed from vegetation, animals and man has been evaluated.

In addition to the observations made, model tests have been performed. A few tests have been carried out in the wind tunnel at the Norwegian Institute of Technology.

Most of the tests on snow transport, snow drifting and snow accumulation, however, have been carried out using simple models, a small fan and semolina, a method that has proven adequate. (Semolina are the large hard grains left in the bolting machine after the fine flour has been passed through it. Semolina is used in making macaroni, puddings, etc.)

Here, a few examples of the criteria will be referred.

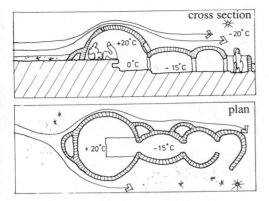

cross section

+20°C —20°C

0°C —15°C

plan

+20°C —15°C

Figure 2. Inuit building, (Boas 1901).

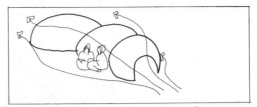

Figure 4. Arctic building for wind, snow, and sun from the same side.

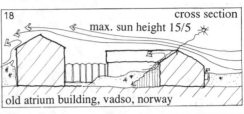

18 cross section

max. sun height 15/5

old atrium building, vadso, norway

plan

Figure 5. Grouping of buildings.

4.2 Building shape in open, weatherexposed flat landscapes.

4.2.1 Strong wind and precipitation/snow drift from one side, sun from the opposite side.

Under such conditions vegetation, animals and people will all turn their back to the heavy wind. The Inuits, native inhabitants of the Arctic, built their dwellings on the basis of this principle (Fig. 2), Börve (1987).
One example of the criteria for building shape based on this principle, is shown in Figure 3.

4.2.2 Strong wind, precipitation/snow drift and sun from the same side.

The Arctic example under these conditions is shown in Figure 4.

For dwellings, and housing developments in the Arctic regions of Norway, this situation has been met, not so much by the shape of the house, but rather with grouping of buildings, see Figure 5. It has thus resulted in various criteria.

When strong winds come from the sunny side, a small roof angle on the windward side will moderate the strain on the building, and the outdoor area. On the other hand, the low outdoor area may result in some snow drift and snow deposit, even though the main part of the snowdrift on the windward side will be deposited some distance from the building as a windward snowdrift, (Figure 6).

4.3 Building shape in an open an sloping landscape

In the development of criteria, a study has been made of how vegetation and birds, in particular sand swallows, have adopted themselves to such conditions, (Fig. 7).
Traditional buildings in mountanious regions of Central Europe, and in Norway have been shaped more or less as shown in Figure 8.

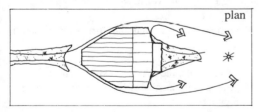

plan

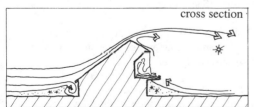

cross section

Figure 3. Building with its back to the wind.

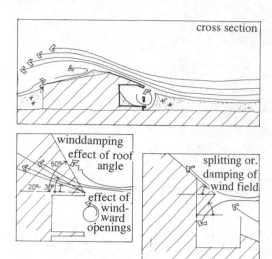

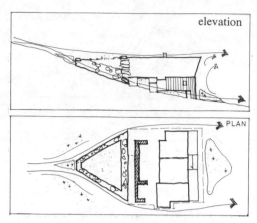

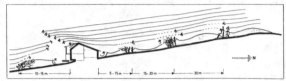

Figure 8. Buildings in the mountains of Central Europe.

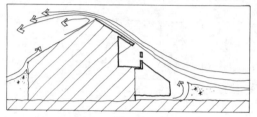

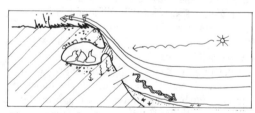

Figure 6. Proposals for buildings when strong winds come from the sunny side.

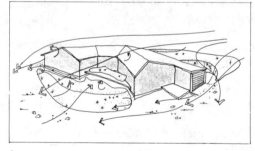

a

Figure 7. Sand swallow's nest facing strong wind.

b

Figure 9. Design of favorable building shape.
a. Cross-section through a lot
b. Recommended main building form - facing south

One of the main conclusions of the studies is that for a single building, one should preferably orient it north/south, and shape it such that a sunny space, free of snow, is formed on the south side, see Figure 9.

5   APPLICATIONS

Professor Börve was active over a number of years in developing settlements and housing designs for regions in Northern Norway, with special regard to local climatic conditions (Börve 1989).

As an example, the development plan for a residential area in the city of Hammerfest is shown in Figures 10 and 11.

Figure 10 shows the area before development.

Figure 11 shows the site plan with synoptic climate charts and an outline of shelter, utility and snow-removal systems.

The experience with this settlement type is very good. This shows that the methodology is an appropriate approach to the problem.

Figure 10. Lot enclosed by dotted line.

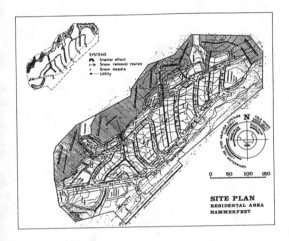

Figure 11. Resulting site plan.

6 FUTURE RESEARCH

The early death of Professor Börve as well as the retirement of Professor Sterten gave a setback to the research that was going on in this field at the Oslo School of Architecture under their inspiring leadership.

However, the seed is sown, and there are a number of people in Norway who are eager to continue the work. We therefore hope that a research group will be built up again in the reasonably near future.

7 CONCLUSIONS

Although modern planning methods and modern construction technology have been greatly improved over the last few decades, the local climate factor has not been included in a sufficient manner in development schemes.

For that reason, a group of planners and research people in Norway initiated research and development projects in order to include the local climate as a main factor in settlement planning in harsh climates. One of the observations was that there was a lot to learn from the way in which vegetation and animals cope with the combination of climatic conditions for specific topographies. Moreover, it turned out to be fruitful to study how the Inuit people of the Arctic had responded to nature in their dwellings and settlements.

Through a systematic study of the landscape in general, and local climate including wind, snow and the sun, a planning tool has been developed. The planning tool includes the following main factors:
- Landscape classification
- Design criteria for various local climatic and topographical conditions
- A general, over-the-year evaluation of solutions, in order to give the inhabitants good living conditions, including reduction of snowdrift problems and snow-removal costs.

When applying these tools for housing developments, the results have been good so far, and it is felt that research and development along these lines should be continued.

8   REFERENCES

Boas, F: "The Eskimo of Baffin land and Hudson Bay". *Bulletin of the American Museum of Natural History*, Vol. 15. 1901.

Erskine, R.: "Architecture and Town Planning in the North". *The PolarRecord*, Vol. 14, No. 89 1968.

Börve, A. B.: "Settlement Planning under Arctic Conditions in Northern Norway", *Energy and Buildings*, 4, 1982.

Börve, A. B.: "Hus og husgrupper i klimautsatte, kalde strok-Utforming og virkemåte", Doctoral thesis, Oslo School of Architecture, 1987.

Börve, A. B.: "Settlement and housing design with special regard to local climatic conditions in cold and polar regions. Examples from Northern Norway", *Energy and Buildings*, 11, 1988.

Snow Engineering: Recent Advances, Izumi, Nakamura & Sack (eds) © 1997 Balkema, Rotterdam. ISBN 90 5410 865 7

The impact of ice dams on buildings in snow country

Ian Mackinlay & Richard S. Flood
Ian Mackinlay Architecture, Inc., San Francisco, Calif., USA

ABSTRACT: Ice dams that adhere to roof edges of snow covered buildings, and elsewhere on roof surfaces, can cause roof loads much greater than either flat roof or ground snow loads. A snow country building design must understand ice dams and deal with them in order to achieve a safe and dry building.

1 DISCUSSION

Snow covered, sloping roofs may be subject to loads greater than either ground snow loads or flat roof load by ice dam formation. These circumstances are created by a combination of site specific environmental factors, the building's own architectural characteristics and its internal temperature. When these design ingredients are ignored or not fully understood, the result is a project where large, heavy ice dams are formed.

In many instances, this mass of ice and the water reservoir backed up behind it, exerts a force on the structure that exceeds the regulatory agency required design values. Ice dams also create hydrostatic pressure at the roof surface which can force water through cracks which otherwise would not leak. This is especially true were buildings have large unheated eaves and building heat melts snow at the roof surface. Snow meltwater runs down under the insulation of the snow blanket, freezes at the eaves, creating an ice dam. This continuing melt/run down/freeze action can build ice dams to substantial height and mass. The deeper the snow on the roof, the greater the potential dam. Ice is heavier than snow and water is heavier than ice. Thus, ice dams produce roof loads far greater than the same volume of drifted snow.

Additional factors which can significantly exacerbate a loading or leakage problem include rain soaked snow pack, snow retentive roofing surface, and snow drifting on roof slopes in the lee of storm winds. Ice dams can prevent snow shedding even from steeply sloped roofs with slippery surfaces

When these design ingredients are ignored or not fully understood, high roof snow loads can occur. The Case Studies will show, that while the architect/engineer cannot change nature's environmental forces, by understanding their capriciousness and characteristics, a satisfactory design solution can be achieved for most conditions.

2 DEFINITIONS

2.1 Ground Snow Load: 2% annual probability of being exceeded (50-year mean recurrence interval, per ASCE 7-95), (100-year per Architectural Institute of Japan) or as published by the regulatory agency. Where only the flat roof snow load is known, the ground snow load is assumed at 142% of the flat roof snow load.

2.2 Flat Roof Snow Load: Basic flat roof design load as published by the regulatory agency or is estimated to be approximately 70% of ground snow load.

2.3 Sloped Roof Snow Load: Regulatory agency (Uniform Building Code) reduction for slope over 20° as outlined in the Case Studies.

3 CASE STUDY ONE

Case Study One illustrates a condition of large ice dam formation. The sloping roof faces away from the sun, is in the lee of the storm winds, and has substantial building heat roof snowmelt. The roof surface, composition shingles, tend to retain the snowpack on the roof. Roof angle 5 in 12 (22.6°), geographical location is 36° N latitude, elevation is 2499m (8200 ft).

This building was constructed in the early 1980's. It is located at Mammoth Lakes, California, USA at the eastern base of Mammoth Mountain (a premier ski area in the Sierra Nevada Mountains). It sits on top of an escarpment with the slopes of its gable roof facing north and south. The south facing roof slope overlooks the valley below and faces the southern storm winds. The north facing roof slope receives the least solar exposure (almost none at midwinter) and is in the lee of the storm winds. Gaps in the roof insulation lead to excessive melting of the snow blanket on the roof. The eaves are unheated. (Figure 1)

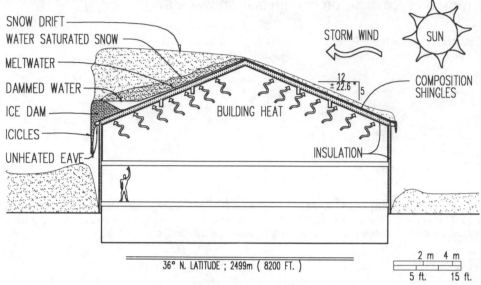

Figure 1 Case study one

The regulatory agency (1976 Uniform Building Code) locally adopted design flat roof snow load was 661 kg/m² (135 lb/ft²), which translates to a ground snow load of 940 kg/m² (192 lb/ft²) for this flat roof load value. Because the roof had a pitch of 5 in 12 (22.6°) the code slope reduction used was:

R_s = S/40 - 1/2
R_s = reduction in lb/sq ft per degree over 20°
S = flat roof snow load in lb/sq ft

This reduction was 14.2 kg/m² (2.91 lb/sq ft per degree) for the 2.6° over 20° for a total reduction of 34 kg/m² (7 lb/sq ft). The sloped roof design snow load was therefore 626 kg/m² (128 lb/sq ft). The structural design was only for a uniform loading condition on both sides of the ridge.

The authors commissioned D.R. Powell to perform an independent evaluation of the site ground snow load conditions. Results of this investigation indicated that the ground snow load should have been in the range of 1273-1400 kg/m² (260-286 lb/sq ft). This would equate to a 5 in 12 pitch roof load of 837-925 kg/m² (171-189 lb/sq ft) minimum. This would be a minimum of 34% to 48% more uniform sloped roof load than originally designed. The investigation also estimated that due to drifts and ice damming, the lee side north roof could readily be 1-1/2 to 3 times the ground snow load. This would put the roof loading on the north facing roof in the range of 1908-3817 kg/m² to 2100-4199 kg/m² (390-780 lb/sq ft to 429-858 lb/sq ft). This is 200% to 570% greater than the original sloped roof design load. At the same time, due to wind stripping and solar radiation, the portion of the roof on the windward side of the ridge could have very little snow cover.

The Architectural Institute of Japan 1996 snow load recommendations indicate that a 22.6° sloped roof (5 in 12) has approximately the highest *basic* shape coefficient. Depending on wind speed, the value ranges from 0.61 at 4.5 m/s (10 mph) to 0.90 at 2.0 m/s (4.5 mph) or less. The Architectural Institute of Japan also states that the shape coefficient for *sliding* snow is determined by the sliding performance of the roofing material for slopes between 10° and 25°. As the roof in this case study has composition shingles, has a 22.6° slope, and assuming a conservative friction value of 0.40 for the composition shingles, the cumulative Architectural Institute of Japan coefficient for *basic* plus *sliding* is 1.01 to 1.30 or 101% to 130% of the ground snow load. Therefore, the ground snow load is exceeded without adding in the *drift* coefficient, which raises the load even further. In the Architectural Institute of Japan's commentary section, it is recommended that the snow eave concentrated loads have a value of 400 kg/m³ (25 lb/cu ft). This is somewhat less than half the density of solid ice. When ice dams are formed, the full density of ice should be used when calculating the ice dam load. This is especially important when the weight of the water behind the ice dam in considered.

This project suffers from significant north roof ice damming and unbalanced loading conditions even in light snow years. North side dormers and chimney projections contribute to the north side snow/ice buildup. Some relief through snow shedding cold have been obtained if the originally designed standing seam metal roof had been used in lieu of the cost saving composition shingles actually installed. Ice dam build-up is magnified by roof surfaces that retain the snow blanket on the roof.

MANUAL ROOF ICE/SNOW REMOVAL

Photo 1 Case study one

Without manual roof snow/ice dam removal multiple times per season, it is estimated that the north side roof would be overstressed structurally most of the time even in moderate snow years. (Photo 1)

4 CASE STUDY TWO

Case Study Two examines ice dam mitigation through the use of a snow arrester/heated gutter combination. (Figure 2) The snow arrester is placed at the edge of the eave and the heated gutter is placed immediately upslope of the arrester. The arrester retains the snow blanket on the roof. The snow blanket insulates the meltwater and keeps it from freezing. The heated gutter intercepts the meltwater prior to its exiting the snow blanket and keeps the intercepted water from freezing as it is drained away, thus reducing ice damming and roof loading. This technique is especially effective where outdoor pedestrian/vehicular circulation space occurs below the eave. Proper design will probably result in not exceeding ground snow load on a roof where no load reductions due to slope are taken (per ASCE 7-95).

In this case study, the snow arrester was designed as a retrofit into an existing structure located at the North Shore of Lake Tahoe, California, USA (38°30' north latitude) at an elevation of approximately 1900 m (6300 ft). (Figure 2)

The regulatory agency (1991 Uniform Building Code) locally adopted design flat roof snow load was 783 kg/m² (160 lb/sq ft), which translates to a ground snow load of 1110 kg/m² (227 lb/sq ft). The existing roof was sloped at a 9 in 12 pitch (about 37°). The code slope reduction used was:

R_s = S/40 - 1/2
R_s = reduction in lb/sq ft per degree over 20°
S = flat roof snow load in lb/sq ft

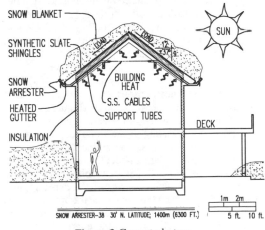

SNOW BLANKET

SYNTHETIC SLATE SHINGLES

SNOW ARRESTER

HEATED GUTTER

INSULATION

LOAD   LOAD

12   9   37°

SUN

BUILDING HEAT

S.S. CABLES

SUPPORT TUBES

DECK

SNOW ARRESTER-38   30' N. LATITUDE; 1400m (6300 FT.)

1m   2m

5 ft.   10 ft.

Figure 2 Case study two

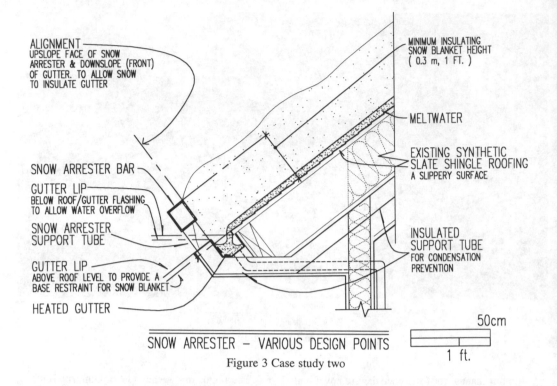

MINIMUM INSULATING
SNOW BLANKET HEIGHT
(0.3 m, 1 FT.)

ALIGNMENT
UPSLOPE FACE OF SNOW
ARRESTER & DOWNSLOPE (FRONT)
OF GUTTER. TO ALLOW SNOW
TO INSULATE GUTTER

MELTWATER

EXISTING SYNTHETIC
SLATE SHINGLE ROOFING
A SLIPPERY SURFACE

SNOW ARRESTER BAR

GUTTER LIP
BELOW ROOF/GUTTER FLASHING
TO ALLOW WATER OVERFLOW

SNOW ARRESTER
SUPPORT TUBE

INSULATED
SUPPORT TUBE
FOR CONDENSATION
PREVENTION

GUTTER LIP
ABOVE ROOF LEVEL TO PROVIDE A
BASE RESTRAINT FOR SNOW BLANKET

HEATED GUTTER

50cm

1 ft.

SNOW ARRESTER – VARIOUS DESIGN POINTS

Figure 3 Case study two

This reduction was 17 kg/m² per degree (3.5 lb/sq ft per degree) for about 17° over 20° for a total reduction of 288 kg/m² (59 lb/sq ft). The sloped roof design snow load was therefore 494 kg/m² (101 lb/sq ft).

If calculated per ASCE 7-95, the C_s factor for a 37° obstructed warm roof would be $C_s = 0.84$. Therefore, the sloped roof load would be 0.84 times the flat roof load or 658 kg/m² (134.4 lb/sq ft). This ASCE 7-95 requirement is 33% more than the 1991 UBC locally adopted requirements.

Synthetic (fiber reinforced cement) slate shingles comprised the sloped roofing surface. Product data for the shingles did not state a coefficient of friction. Empirical field evaluation indicated that, when wet, the shingle surface is very slippery.

Therefore, in designing the snow arrester, no component force reduction was taken due to roofing frictional resistance and the arrester was designed to resist the total snow load up to the ridge. (Figure 2)

The decks below the roofs were to be usable year around so no cascading of snow and ice was permissible. The steel tube arrester bar and support tubes were designed to resist a plane-of-roof load of 2800 kg (6182 lb) over a 1.82 m (6 ft) span, with minimal deflection. Due to the structural capacity of the existing glue-laminated rafters and the need to tie the opposite sides of the roof together, the arrester support tubes were extended about half way into the interior roof rafter space and tied together with stainless steel cables.

(Figure 2) These support tubes were spray foam insulated within the soffit and sheet neoprene insulated within the rafter space to prevent condensation.

The restraint (upslope) face of the arrester bar was positioned to restrain a minimum of a 305-mm (12 in) thick snow blanket. This thickness of snow blanket is sufficient to insulate the protect the roof meltwater from freezing. (Figure 3)

The heated gutter was positioned to align the front (downslope) face nominally in the same plane as the restraint (upslope) face of the arrester bar. This was done to allow the snow blanket to completely cover the gutter and leave only a minimal roof plane surface downslope. (Figure 3)

The gutter profile was dictated by the existing construction and roof slope. The front lip deviates from standard rain gutters in that it extends *above* the plane of the roofing about 29 mm (1 1/8 in.). This high front lip will stop small amounts of snow that slide under the arrester bar and will restrain the bottom of the snow blanket as it tends to migrate downslope. The lip further helps to keep the gutter covered and insulated. The front lip is still below the high point of the gutter back by approximately 25 mm (1 in.) to allow a normal water overflow condition. (Figure 3)

The gutter is heated with two rows of snowmelt cable. The cable is an electrical, self regulating, conductive core type with a maximum power output of 33 watts/m (10 watts/ft) at 0°C (32°F). In this manner the roof meltwater is kept from freezing in the gutter. The

RETROFITTED SNOW ARRESTER

Photo 2 Case study two

heat cable is also installed in the piped/downspout drainage to ensure a free flowing path to on-site drywells. (Figure 3)

Ice dam formation is greatly mitigated because the heated gutter intercepts almost all of the roof meltwater while the arrester keeps the snowpack in place. The sloping roof eave area where ice can form is only 127 mm (5 in.) wide below the gutter. This minimizes the generation of meltwater which in turn severely limits ice dam/icicle formation. (Photo 2)

In heavy snow years, or under certain drifting conditions, the snow blanket may be of such depth that snow cornices will occur over the top of the arrester bar. If the snow cornice seems to become a hazard, it can be manually removed relatively easily using the arrester bar as a guide for the removal tool. Removal of the hazardous snow cornice does not jeopardize retention of the insulating snow blanket.

5   CASE STUDY THREE

Case Study Three reviews the effect of a *cold roof* system to mitigate ice dams. In this *umbrella* design, outside ambient air is allowed to freely circulate in a roof cavity directly below the roofing surface and its supporting substrate. This ventilation cavity is above the primary insulated roof. The ventilation cavity needs to be designed to a depth sufficient to promote an unrestricted flow of outside air. Air volume must be sufficient to absorb building heat and vent it away without contributing to an increase of roof surface temperature. Use of a slippery roofing material will promote shedding of the snow blanket. As a word of caution, the falling roof snow needs a dump area that will not endanger people or property. The architecture and site planning must consider the effects of the snow berm that will be created. (Figure 4)

493

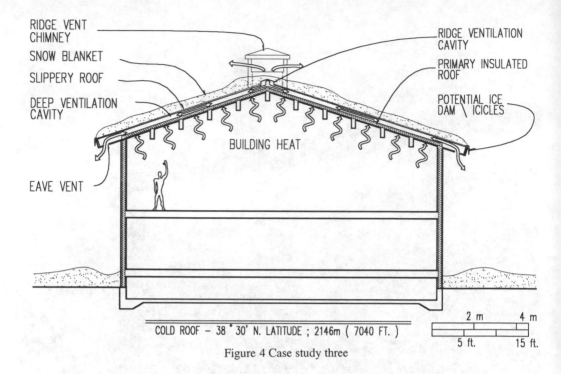

RIDGE VENT
CHIMNEY
SNOW BLANKET
SLIPPERY ROOF
DEEP VENTILATION
CAVITY

RIDGE VENTILATION
CAVITY
PRIMARY INSULATED
ROOF
POTENTIAL ICE
DAM \ ICICLES

BUILDING HEAT

EAVE VENT

COLD ROOF – 38 ° 30' N. LATITUDE ; 2146m (7040 FT.)

2 m   4 m
5 ft.   15 ft.

Figure 4 Case study three

This case study is currently being designed. The project will be located in mountains of the Sierra Nevada, California, USA. The location is about 4 km (2.5 miles) west of the crest of the Sierra Nevada Mountains at an elevation of 2146 m (7040 ft) at 38°30' north latitude. The regulatory agency design ground snow load is 2105 kg/m² (430 lb/sq ft) and the flat roof snow load is 1473 kg/m² (301 lb/sq ft).

The cold roof structure is comprised of 302 mm (11 7/8 inch) deep manufactured wood rafters 610 mm (24 in.) on center. The rafters sit on a vapor retarder covered continuous plane of wood decking, which are structurally supported by glue-laminated beams 1372 mm (4.5 ft) on center. (Figure 5)

Plywood roof sheathing 29 mm (1 1/8 in.) thick covers the rafters and provides support for a standing seam metal roofing system. The plywood forms the top of the deep ventilation cavity. The rafters are laterally braced with metal "X" bracing in lieu of solid blocking at the beam lines to ensure an open ventilation cavity. The bottom half of the rafter space is filled with U = 0.026 (R38) 152 mm (6 in.) thick closed cell polyurethane insulation sealed on the surface. The upper half is left open providing a deep ventilation cavity. (Figure 5)

Structurally, the most challenging issue is the lateral force transfer of the snow loaded roof diaphragm into the exterior shear walls. The lateral design snow load used in this case is 33% of the roof design snow load or 485 kg/m² (99 lb/sq ft). The slippery nature of the metal roof predicated the use of the 33% value, whereas 50% would probably be more appropriate for a nonslippery

roof. Normally, solid wood blocking would be used between the rafters to provide the force path. In this case, solid blocking would negate any airflow into the cold roof cavity. To solve this problem, a steel plate with a hole cut out will be used (ie. a miniature moment frame. (Figure 6)

Pressure equalization between adjacent cavities is achieved by having the wood rafters factory punched with 25 mm (1 in.) diameter holes at 305 mm (12 in.) on center along the upper quarter point of the rafter's plywood web. (Figure 6)

The gable roof thus comprises a series of 610 mm (24 in.) wide by 153 mm (6 in.) ventilation cavities running from eave to ridge and back down to the other eave. The eaves are provided with continuous screened vents with a free area approximately 75% of the cross sectional area of the ventilation cavity. The ventilation cavities will be vented up through the chimney structures at the ridge utilizing the metal roof ridge cap as an airflow duct. (Figure 7)

Cold roofs work best when the outside ambient air temperature remains cold, below freezing, day and night during the snow season. In the Sierra Nevada of California, where the winter air temperature is often above freezing at midday, there will be some melting of the snow blanket, and some ice damming may occur at the eaves. This ice damming will be far less than Case Study One where the building heat significantly contributed to the melting of the snow blanket.

494

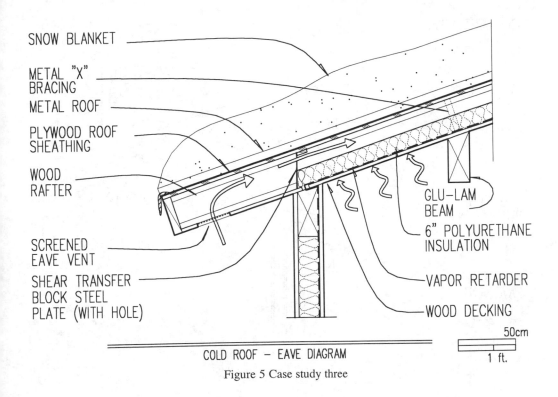

SNOW BLANKET

METAL "X" BRACING

METAL ROOF

PLYWOOD ROOF SHEATHING

WOOD RAFTER

SCREENED EAVE VENT

SHEAR TRANSFER BLOCK STEEL PLATE (WITH HOLE)

GLU-LAM BEAM

6" POLYURETHANE INSULATION

VAPOR RETARDER

WOOD DECKING

50cm
1 ft.

COLD ROOF – EAVE DIAGRAM

Figure 5 Case study three

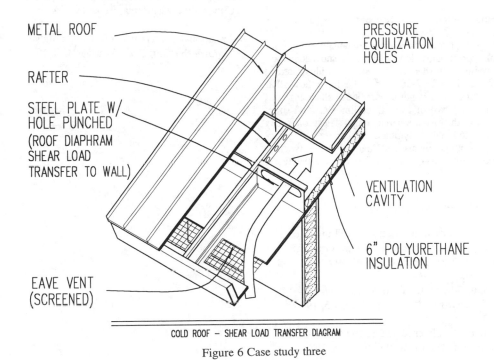

METAL ROOF

RAFTER

STEEL PLATE W/ HOLE PUNCHED (ROOF DIAPHRAM SHEAR LOAD TRANSFER TO WALL)

EAVE VENT (SCREENED)

PRESSURE EQUILIZATION HOLES

VENTILATION CAVITY

6" POLYURETHANE INSULATION

COLD ROOF – SHEAR LOAD TRANSFER DIAGRAM

Figure 6 Case study three

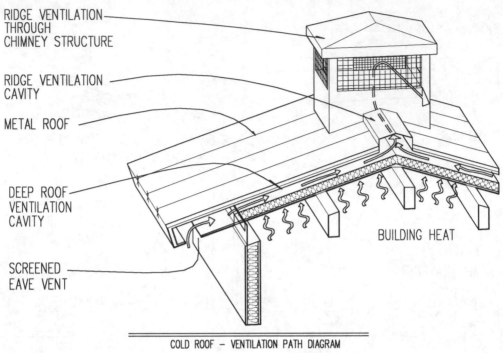

RIDGE VENTILATION THROUGH CHIMNEY STRUCTURE

RIDGE VENTILATION CAVITY

METAL ROOF

DEEP ROOF VENTILATION CAVITY

SCREENED EAVE VENT

BUILDING HEAT

COLD ROOF – VENTILATION PATH DIAGRAM

Figure 7 Case study three

6   CONCLUSION

Ground snow snow loads are generally used as the basis of determining design snow loads for the roofs of buildings. Flat roof snow loads are usually some fraction of ground snow loads. Flat roof snow loads are often reduced further by roof slope, as illustrated in Figure 7-2 of ASCE 7-95. Our studies have shown that these reductions may not be justified in all cases. Roof loads may be two or even three times ground snow loads due to the formation of ice dams. These loads may be distributed highly asymmetrically on the roof structures producing unbalanced loads. Further work is required to quantify the effect of ice dams on structures.

REFERENCES

ASCE 7-95, American Society of Civil Engineers, *Minimum Design Loads for Buildings and Other Structures,* New York, New York 1996.

Architectural Institute of Japan, *Recommendations for Loads on Buildings,* Chapter 5, 1996.

International Conference of Building Officials, *1976 Uniform Building Code,* Whittier, CA, 1976.

Powell, Douglas R. Department of Geography, University of California, Berkeley, CA

International Conference of Building Officials, *1991 Uniform Building Code,* Whittier, CA, 1991.

Snow Engineering: Recent Advances, Izumi, Nakamura & Sack (eds) © 1997 Balkema, Rotterdam. ISBN 90 5410 865 7

A study on the design technique of public housing at snowy area in Japan

Toshiei Tsukidate
Hachinohe Institute of Technology, Aomori, Japan

Takahiro Noguchi
Hokkaido University, Sapporo, Japan

ABSTRACT : We will soon be facing on aging society in Japan. Therefore, I think that public housing to adapt snow to an aging society more than at present. I investigated by questionnaire the inhabitants at Aomori prefecture regarding design techniques against snow damage in public housing. I conclude that many snow damage problems remain.

1. INTRODUCTION

Over 50% of Japan is covered with snow in the winter, and about 25 millions people live in the snowy area. I live in Aomori prefecture. We have an annual snowfall ranging from about 30 cm to about 150 cm here. About 26.6% of the population live in public housing. So, public houses are a very important part of Japanese housing .Many design techniques and countermeasures against snow damage in public housing are not effective. I concluded field surveys at 50 public housing developments in Aomori prefecture.

2. RESULTS OF QUESTIONNAIRE

I sent a questionnaire to 814 inhabitants of 5 low-rise housing developments and 8 medium-rise housing developments in December 1992. The main questions were " Main reasons to select public housing developments, Preparations for snow and cold in winter, and Snow damage " .

2.1 *Main reasons why people select public housing .*

From the questionnaire, I obtained the following results. The main reasons that people select public housing are the convenience of commute (57.3%), the sunshine and function of housing (57.1%) . The countermeasures against snow problems is the sixth (16.2%), but very important. Public housing in which many older people live selected a higher proportion of countermeasures against snow problems since snow shoveling is very hard work for older people.

2.2 *Preparation for winter - snow and cold*

The most important preparation for winter is to prepare snow shoveling tools (52.6%). The second is countermeasures against condensation and the third is freezing of water pipes (19.4%). The fourth is to cover windows and walls to prevent snow damage (9.7%). The inhabitants of medium-rise housing prefer the countermeasures of condensation, and the inhabitants of low-rise housing prefer the countermeasures of water

Table 1. The 13 public housing developments which were sent the questionnaire

| Name of housing development | Housing type (Number of houses) | Site area (m^2) | Land use proportions (%) | | | | | | Area per house (m^2) | Parking area per car (m^2) |
			Building coverage ratio	Road	Park, play lot	Green space	Parking lot	Total		
MT-1	Medium-rise housing (258)	40,559	----	----	----	----	----	----	157.2	----
MT-2	Medium-rise housing (198)	30,726	----	----	----	----	----	----	155.2	----
MT-3	Medium-rise housing (213)	43,091	11.4	13.0	9.2	51.6	14.6	100.0	202.3	29.4
MJ	Medium-rise housing (204)	29,469	25.6	5.0	24.0	29.5	16.5	100.6	144.5	23.9
MA	Medium-rise housing (104)	16,529	25.0	15.4	15.8	34.3	9.5	100.0	158.9	15.2
MS	Medium-rise housing (81)	11,132	20.1	14.4	15.0	33.5	17.0	100.0	137.4	23.5
MI	Medium-rise housing (192)	24,899	19.8	13.1	11.8	41.9	13.5	100.1	129.7	17.4
MF	Medium-rise housing (286)	41,136	20.9	30.5	7.7	37.5	4.4	101.0	143.8	18.2
MS	Medium-rise housing (50)	13,506	26.2	24.8	5.0	38.1	5.9	100.0	270.2	16.0
LYS	Low-rise housing (39)	9,303	18.6	----	----	----	9.7	----	238.5	23.1
LMY	Low-rise housing (16)	4,620	21.3	21.0	----	----	6.1	----	288.8	17.5
LMD	Low-rise housing (19)	3,650	37.2	22.0	----	----	11.9	----	192.1	22.7
LTE	Low-rise housing (20)	3,317	20.7	21.8	----	----	8.1	----	165.8	13.5
Average of over 50 houses (176)		27,894	19.5	14.9	21.6	33.1	11.2	100.2	158.4	20.5
Average of under 50 houses (24)		5,223	24.5	21.6	----	----	9.0	----	218.3	19.2

low-rise housing prefer the countermeasures of water pipes. A maximum of proportion is the inhabitants live in elevated houses (87.5%).

2.3 *Condensation*

There is high of humidity in the winter in Japan, so it is very important to control condensation. The commonest room to grow condensation in is the bedroom (50.4%). The living room is second (38.5%),and the bathroom is third (36.8%). Old medium-rise housings with inadequate thermal insulating material comprise a high percentage of condensation concerns. I think that the reason for bedroom's high condensation is that room is on the north side and has a low temperature. The reason for the living room's condensation is its use to dry washing.

2.4 *Place for snow shoveling tools*

The main snow shoveling tools are the snow shovel , shovel,and snow dump (cf. photo.1). Snow dumps are used in the more snowy area. These tools are storage at the residence (56.4%), in the entrance hall or porch (12.6%), in the vestibule entrance (10.4%) ,or outside of the residence (5.7%). A small number are put inside the " Gangi" which is a Japanese style arcade used in snowy area.

2.5 *Snow damage*

Snow has much influence upon life in public housing developments and give many damages there. Typical troubles are too narrow of parking area (38.7%), sidewalk disturbance (32.9%) and falling snow on access way and sidewalk (22.5%). In a few cases, windows and chimneys are destroyed.

3 . DESIGN TECHNIQUE

3.1 *Methods of snow shoveling in public housing developments*

There are three ways to remove the snow in public housing developments. The first is to shovel it by hand, the second is by snowplow and the third is the street gutter to float snow away. Snow shoveling by hand is the main method everywhere. Snowplows are used to clear roadways and parking areas in large scale medium-rise housing developments. Street gutters are found in some recent housing developments.

3.2 *Snow dump area*

There is no snow dump area in old public housing developments. Thus, snow banks form on both sides of the road and on the sidewalk. Snow on the roadway is removed by snowplow. Parks and play spaces are not good for snow dump areas because of the barriers there.

Recent public housing developments have "Common Spaces" which are multi purpose spaces for parks and play lots in summer and snow dump areas in winter. Many children play there with sleds and skis. It is necessary to link the road with such common spaces.

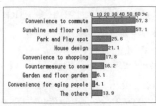
Figure 1. Main reasons people select public housing

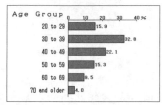
Figure 2. Distribution of inhabitant's age

Figure 3. Preparation for winter - snow and cold

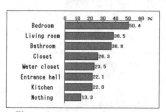
Figure 4. Place of condensation

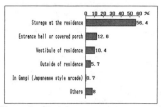
Figure 5. Place for snow shoveling tools

3.3 *Parking area*

The average parking space per car is about 20 m^2 inclusive of car path (Table.1). Because snowplows make snow banks in public housing developments in the winter, parking areas are not constructed enough. At some public housing developments, larger parking space per car (about 30 m^2) are enough allow for snow banks of about 1 m height.

In public housing developments,there are large parks and play lots. They occupy about 20 % of the area. So it is necessary to be able to access them with snow shoveling tools. Recent public has "Common space" where snow is dumped in winter.In summer that space is for parking and play. "Common spaces" must be accessible from roads ,each house and the parking area;to be a good snow dump area and a play lot in winter.

3.4 Roof design

Two styles of sloped roof are in popular in Japan, the first is the gable roof, the second is the hipped roof. Both need to be cleared of snow. Medium-rise residential buildings usually have flat roofs. Gable roofs are increasing because of new designs and their resistance to leaks. Some gable roofs hold snow on them. Snow cornices grow on flat roofs for the wet snow and cold in Aomori prefecture. It is important to keep inhabitants away from such cornices which can fall.

Two new roofs for low-rise housing have been used to avoid falling snow from the roof. At first, Flat roofs in low-rise housing is good allow snow to blow away in dry snow areas. Secondly, the M-type roof with a low parapet is now popular in built-up areas.

3.5 Approach and entrance

(1) Approach

There are two types of approaches, one is the flat access way ,the other is stairs. Though flat access way have few problem, the stair type without a roof becomes with deep snow on the stairs and can be very slippery. Elevated houses with stair access ways without roofs,have deep snow on the stairs .

(2) Entrance and house design

There are two kinds of entrance in low-rise houses. One is a porch , and the other is a covered porch type which we call "Fujoshitsu" . The area of "Fujoshitsu" is only 3.3 m^2 which is not big for several snow shoveling tools but also skis, sleds and such. The entrance hall of medium-rise houses are where many snow shoveling tools are kept. Therefore, in recent medium-rise housings it is larger and lighter than the old ones. But, still there is no special storage space for snow shoveling tools. Elevated low-rise houses and medium-rise houses have a garage under the first floor. These house styles reduce snow removal problems.

3.6 Sidewalk and corridor

There are few sidewalks in low-rise housings developments. The Kinami danchi has Common space, elevated houses and snow fence of larger porch.

Recently,many medium-rise housing developments have "Gangi" which is Japanese style arcade that avoids snow sidewalks. The third Toyama danchi and Josei danchi are constructed with " Gangi " which are

particularly helpful to children and older people. Covered sidewalks are used in Okuno danchi (Aomori,Japan). A sky way which perhaps say an elevated walkways is constructed in Kita 7 Jo danchi (Kuchan ,Hokkaido,Japan). There are some problems on to blow into the corridor that opens on one side.

Photograph 1. Snow dump

Figure 6. Snow problems at public housing developments

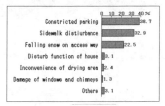

Photograph 2 . Children playing in a "Common Space " (Hachinohe, Aomori, Japan)

Photograph 3. M-type roof of low-rise housing (Sawauchi, Iwate, Japan)

Photograph 4. Snow has fallen from the gable roof in front of these stairs (Inagaki, Aomori, Japan)

Photograph 5. Children playing in a "Gangi"
(Hirosaki, Aomori, Japan)

Photograph 6. Elevated walkways ("Sky-Way")
in KIta 7 Jo danchi (Kuchan, Hokkaido, Japan)

Table 2. Design techniques to reduce snow problems

Item	Design Technique	Result of field survey
Snow dump space	Common space (park and play lot in summer).	Effective snow dump area.
Parking areas	Garage under the 1st floor of he building. Snow shoveling by snow plow	Effective, need to link parking area and common space.
Sidewalk and access balcony	Gangi (Japanese style arcade).Covered side walk. Sky way.	Effective,need to protect against now infiltration.
Bicycle parking lot	Bicycle parking lot from access balcony.	Effective.
Garbage collection lot	Garbage collection lot with roof.	Effective,need to shovel snow.
Roof design	Flat roof and M-type roof with parapet	Effective,need to defend against snow cornices at eaves
Space for snow shoveling tools	Covered porch Storage in entrance hall. Larger entrance hall.	Effective.
Others	Top light for well lighted entrance hall. Sun room using balcony for drying area. Common space as play lot for children.	Effective to communicate Effective to reduce condensation in the house.

3.7 Bicycle parking lot

Bicycles are not useful in snow, so many are kept in bicycle parking lots . For example, some are put in access balconies, landings and in the community hall,too. Bicycle parking lots are of three types. The first is the outside bicycle parking shed, the second is in building with access from the outside, the third is inside on an access balcony. The last one is the newest and most suitable for snowy area.

3.8 Garbage collection lot

Garbage collection lots without roofs are covered with snow and garbage is put on the deep snow. That scene is not beautiful and inhabitants worry about scattering of garbage to surroundings It is good that garbage collection lots with roof are increasing recently. It is necessary that inhabitants take turns shoveling away snow to provide access to garbage collection lots.

4 . CONCLUSIONS

According to the above mentioned results of the questionnaire of public housing developments at Aomori prefecture, 16.2% of inhabitants have chosen the reason of snow control to live in public housing estates.Snow control is very important to older inhabitants. Main problems are constructed parking areas blocked and slippery sidewalks and falling snow. A big problem is that snow shoveling tools block entrances. I made field survey of these problems at a snowy area. The results of this survey are shown as Table 2.

(1) In recently designed public housing , many design techniques are used against snow problems. Common

space " is used as snow dumps in winter and for parking and play space in summer. "Gangi" (Japanese style arcade) are used for sidewalks and sky ways (an elevated walkways) and used in medium-rise public housing. Snow blows inside of "Gangi" it one side without a wall.

(2) Snow in parking area is removed by snowplows at large-scale public housing developments. Because parking areas and roads are not linked to park and play lots, there are no places for snow bank. Elevated houses with garages under their first floor in low-rise housings and medium-rise housing developments are effective in snowy areas. It is necessary to cover entrance stairs .

(3) Because entrance halls of medium-rise housing are larger and lighter than old ones by top light and high-side windows, these spaces are useful for snow shoveling tools and as a communication-place of inhabitants . Entrance halls and porches of low-rise housings are larger,too. Many problems remain. Countermeasures are needed against snow at garbage collection lots and falling snow on approaches and in front of entrance halls.

5 . ACKNOWLEDGMENT

The author thanks all inhabitants in public housing estates that answered the questionnaire and field survey. Aomori local government assistance is greatly appreciated. At last,the author is grateful to all students in Tsukidate eminar,Hachinohe Institute of Technology.

REFERENCE

Communication council of Aomori prefecture public housing.1994.Way of Public housing construction work. AIJ Hokkaido.1995.Public Housing Design List 101

Snow Engineering: Recent Advances, Izumi, Nakamura & Sack (eds) © 1997 Balkema, Rotterdam. ISBN 90 5410 865 7

Role of atrium in Aomori Public College – Design approach to an attractive campus in snowy country

Takashi Sasaki
Akita National College of Technology, Japan

Hideaki Satoh
Institute of New Architecture, Sendai, Japan

ABSTRACT: Sphere of activities are extremely limited in the wintertime in snowy and cold region. It is essential to maintain and rather extend people's activities in all aspects of life through the year. The purpose of design of the campus building is to explore covered open space to provide an protected safe and comfotable space from the outside climate. Covered open spaces(mall, atrium etc) in public facilitiy become effective design tools to create attractive interior spaces for participants in these region. Experiment made on Aomori Public College is to provide an atrium as the infrastructure of the campus building to integrate whole academic functions aming at the activation of campus life through the year.

1 OUTLINE OF AOMORI PUBLIC COLLEGE

Aomori Public College presently offers a diploma program in Management and Economics which is four years in length. The college was founded in April 1993 by the cooperative of Aomori city and the adjacent six towns. It has established an educational program which introduces professional courses from the beginning of the academic year. Therefore the college building should be properly planned and designed to carry out this educational program, considering local climate and site conditions.

2 PROPOSAL OF A "COLLEGE MALL"

2.1 *Design concept of Aomri Public College*

The "Aomori college town plan" is the basic idea of Aomori public college. The college is to be the core of the city, and it activates and promotes the development of the economy of the region. The proposal aims at coexistence of the college and the region. First, the college must be harmonized with the surrounding environment. Second, it must be opened to all citizens. The campus plan must be architecturally so planned and designed

that all citizens have the opportunity to participate in activities inside and outside of the campus in all seasons.

2.2 *Feature of the site and policy of site plan*

1. Climate conditions
Aomori prefecture is known as the most snowy district in Japan. All the activities in daily life and industries are tightly restricted in winter because of the lock of accessibility. So treatment cost of heavy snow becomes a big burden for the local government.

2. Site conditions
The site is about eight km south of the city center on a gently graded hill. It commands a panoramic view of the whole city , the Aomori bay area to the north and the Hakkohda mountains to the south. It is approximately 50 ha in size. The site is surrounded by thickly wooded trees: cedars, red pines, larchs, oaks, beechs and alders etc. These trees and bushes are well maintained, as is the configuration of the land because this area is preserved as the main approch to the Hakkohda–Towada line, it being one of the most famous sightseeing regions in Japan.

3. Site planning policy

Site planning policy is to maintain the nature of the site as much as possible and to harmonize the buildings with nature. Tactics adopted here are, first, to restrict the height of buildings to the top of existing trees on site, and secondly, to match colors of buildings to existing nature.

4. Design tools for interior space

It is essential that interior spaces of the building create a safe and comfortable campus life. Within the complex the atrium is installed named the "College Mall."

3 EVOLUTION OF THE DESIGN CONCEPT

3.1 *Function of the "College Mall"*

Covered open spaces (mall, atrium, portico and so on) are usually used for pedestrian ways. Covers are mostly made of transparent materials such as glass or plastic to admit sunlight into interior spaces. While creating safe and comfortable spaces protected from the outside climate. Since activities are so restricted in winter, inside livable spaces are desperately required here. The atrium named the "College Mall" is proposed as the infrastructure of the campus to integrate academic functions of college, as a spinal column of a living organism. In other words, all functions of the college are connected to the spine and built in one structure. The "College Mall" consists of two key elements:

"Node" and "Path." The "Node" becomes the core to connect different functions of the campus, and it offers a chance for students and staff to participate to various activities and stimulate one another by exchanging conversation and ideas. The "Path" is the channel of activities from one node to another. It becomes an active space and creates an identity of college life to stimulate an awareness of all the creative efforts of students and staff.

Functions of the "Colege Mall" are as follows:

1. Movement

To connect different departments and functions horizontally and vertically, allowing free movement for all partipants like on a street of a city.

2. Junction and buffer

There are three activities in the college: thoes of students, researchers and visitors. The "College Mall" works as a junction to make their different activities into one, and as buffers to avoid disturbance from one another. Also it works as a physical junction to shift the difference of floor levels cuased by the slope of the site.

3. Communication and information

It serves an open space for participants to communicate and exchange ideas with one another to stimulate campus life. Also it serves a place for announcement of information.

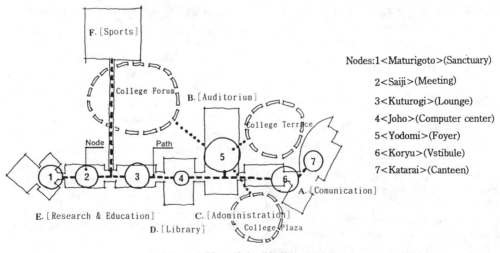

Figure 1 Plan of the "College Mall"

Figure 2 Section of the "College Mall"

3.2 *Structure of the "College Mall"*

1. Plan of the "College Mall"
The "College Mall" is the infrastructure of the campus buildings. It contains two axes to connect the seven nodes with six functional zones. These six zones are: A) Communication, B) Auditorium, C) Administration, D) Library, E) Research and Educatioan, and F) Sports. These six zones are connected to seven nodes to organize themselves as shown on the plan. (Figure 1)
2. Section of the "College Mall"
Three different floor levels are integrated into one building with the "College Mall" as shown on the section. (Figure 2)

3.3 *Design of the "College Mall"*

Various activities will be expected in the "College Mall." Therefore desgin must be considered to create safe and neat interior space so that all participants can enjoy their campus life in all seasons just as on a street in a town.

Design tactics are as follows:
1. To provide open spaces through different levels
The "College Mall" has an open space two or three stories high. This vertical open space creates communication among people on different levels of the building and stimulates their participaton.
2. To create outdoor space inside of the building
The "College Mall" is covered with a transparent roof which allows plentiful sunlight inside the building in winter. It liberates our consciousness to the outside.
3. To provide enjoyable movement space
The "College Mall" is 270 m in length. It is constructed with wide open spaces, narrow aisles, steps, slopes. Also it varries in height and width: sometimes closed with blank walls, open to the outside through glazing. Different shapes are provided in width and height to give various characters to different places. The spaces are cooled in the summertime by moving air and warmed by hot water in the winter. The spaces stimulate students' communication and performances.
4. To characterize nodes
Nodes are characterized and designed according to their different activities: "Matsurigoto" in front of the big lecture room is the sanctuary of the campus, like a temple. (Figure 3) "Saiji" is a multipurpose space. The scaffolds in the middle of this space are used for various events by students. (Figure 4) "Kutsurogi" is a lounge for exchange coversation and ideas where natural white oaks are planted in the middle. (Figure 5) "Koryu" is the main vestibule for all paticipants to contact with. The place is fully openned to the outside through the full glazing wall.

3.4 *Design elements of the building*

The design motif of the college is "Johmon," a life style of very ancient times in Japan. Aomori prefecture is one of the prominent districts where "Johmon" culture still remains. Design features of "Johmon" culture are natural, simple, strong and bold. Materials principally used on this building are rather natural: exposed concrete, earth, ceramic tile, wood, hemp rope and so on. These are applied to floors, walls and ceilings. (Figure 6) Another samples are shown as follows: colomn with henmp rope, (Figure 7) concrete relief at the colomn, (Figure 8) wooden furnitures. (Figure 9)

Figure 3 ‹Maturigoto›(Sanctuary)

Figure 6 Interior space applied Johmon motif

Figure 4 ‹Saiji›(Meeting)

Figure 7 Hemp rope

Figure 5 ‹Kuturogi›(Lounge)

Figure 8 Concrete relief

Figure 9 Wooden furniture

4 EFFECT OF ATRIUM

According to the result of questionnaire sent out to the random samples: students, college staff and some visitors after three years of experience with academic activities, the effect of providing an atrium is conisdered valid. The "College Mall" has been actively utilized in various ways in all seasons so that the campus life is stimulated. Visitors enjoy their participation to the campus life.

Further experiment are to be continued to examine the effect quantitatively for the next step.

Snow Engineering: Recent Advances, Izumi, Nakamura & Sack (eds) © 1997 Balkema, Rotterdam. ISBN 90 5410 865 7

Living style and planning of multifamily housing for snowy, cold regions of Japan

Takahiro Noguchi
Department of Architecture, Hokkaido University, Sapporo, Japan

Toshiei Tsukidate
Hachinohe Institute of Technology, Aomori, Japan

ABSTRACT: The cold climate and heavy snowfall in Hokkaido(located in the northern part of Japan) creates a unique living style for which multifamily housing has great advantages. We propose a new design of apartment house suitable for snowy-cold cities like Sapporo.

1 INTRODUCTION: CLIMATE IN HOKKAIDO AND THE HOUSING SITUATION

1.1 *Purpose and method*

This research is intended to clarify how the cold climate and heavy snowfall in Hokkaido (located in the northern part of Japan) creates a unique living style for which multifamily housing has great advantages. Further, I will examine what multifamily housing should be like in northern areas by investigating the typical living style of detached houses, the present condition of multifamily housing design, and the resident's views of their situation. I have primarily used existing research/study materials and housing surveys, which will appear later.

1.2 *Climate in Hokkaido*

Hokkaido is a prefecture which is located in the northmost part of the Japanese Islands. Its capital is Sapporo, located at Lat.43N, far to the north of Tokyo (Lat.35N). Sapporo is not located at a particularly high latitude compared to other major cities in the frigid zone, though winter conditions are comparable. In Sapporo the average temperature in January is -4.6, the same as in Helsinki and Stockholm (Table 1). High precipitation in winter is characteristic. At over 110 mm, Sapporo's January precipitation is greater than precipitation levels in other major cities in the frigid zone. It has six months of

snowfall days each year, and the maximum snow depth reaches around one meter in a normal year. This climate produces a unique living style which is different from that in temperate zone cities such as Tokyo and Osaka.

1.3 *Diffusion and situation of multi-family housing*

The diffusion rate of multifamily housing is still low. According to 1993 statistics (Table 2), detached houses accounted for about 56% of total dwellings, while multifamily housing, including row houses and apartment houses only acounted for about 43%. Compared with the rest of Japan, the percentage of multifamily housing is slightly higher in Hokkaido, but not significantly.

The statistics clearly show that the ratio of multifamily housing is increasing

Table 1.Mean air temperature and precipitation per month.

	Latitude	Mean air (°c) temperature		Mean (mm) precipitation	
		Jan.	Feb.	Jan.	Feb.
Sapporo	43°N	-4.6	20.2	107.6	68.7
Tokyo	35°N	5.2	25.2	45.1	126.1
Stockholm	59°N	-2.9	17.1	37.4	71.6
Helsinki	60°N	-6.8	16.6	41.4	72.4

Table 2.Housing Types.

	Detached houses	Row-houses	Apart-ments	Others (%)
Hokkaido(total)	55.9	8.4	35.0	0.6
ʺ (~1970)	63.8	19.6	15.5	1.0
ʺ (71~90)	54.6	5.9	38.9	0.5
ʺ (1991~)	53.7	2.6	43.1	0.5
Japan	59.2	5.3	35.0	0.5

*1993 Housing survey of Japan

gradually year by year, but it is unclear
how climate influences these housing
trends. It is not easy to compare these
statistics with other major cities in the
frigid zone because of differences in
history and social customs. However, it is
fair to say that the popularity of
multifamily housing is lower in Hokkaido,
especially in urban areas, than in
comparable cities in the frigid zone. In
part, historical and social factors are
reflected, but the even more basic reason
is due to the quality of the multifamily
housing.

Simply speaking, both the size and the
facilities of apartment houses are not
well appointed for the purpose of
permanent residence. As Table 3 shows, the
floor space of an apartment house is half
or at best 60% of an ordinary detached
house. A private cooperative is 60% to 70%
(70-80m² each home) and public housing
(which provides mainly rental apartments)
is 40% to 50% (50-60m² each home). It is
difficult to provide adequately for lifestyle needs in Hokkaido, which is weighted
much more towards indoor life. This means
that a large living space is much more
important.

Table 3.Total size of floor space per
dwelling.

| | Detached houses | Apartments | | | | (m²) |
| | | Owned | Rented | | Total | Total |
			Public	Private		
Hokkaido	115.3	70.3☆	53.5	46.9	48.0	87.0
Japan	122.3	70.8	48.9	40.3	44.3	91.9

*1993#Housing survey of Japan ☆1988#Housing survey of
Sapporo

2 DETACHED HOUSING IN HOKKAIDO AND ITS LIFE STYLE

2.1 *Indoor-type life style*

In Hokkaido people spend a long time at
home in the wintertime. The average hours
which are spent in the home on holidays in
January are 21~22 hours by householders
and 22~23 hours by housewives (Noguchi
1982). That is to say, people spend most
of their time in the home except to do
some work or to clear snow. It is obvious
according to my research, in comparison to
milder areas such as Kouchi, that people
spend longer hours at home in Hokkaido. It
also means they have to do many things in
the home because they are unable to use
the outside area in wintetime. Most
activities such as children's play,
washing and drying clothes, hobby

carpentry and caring of fixtures, which
can be done outside even in the wintertime
in warm regions, are all brought into the
home in Hokkaido. Families spend a great
deal of time together in the large living
room, reflecting greater emphasis on
indoor life.

Even in Hokkaido, people enjoy gardening
and barbecuing in summer. It is
characteristic that indoor activities are
separated from outdoor ones. In warm
regions, indoor and outdoor activities are
sometimes blended; for example, enjoying
the cool of the evenings in the garden or
doing some housekeeping work outside in
the hot and humid summer. As I will
discuss in greater detail later on, it is
not as easy to do such things in Hokkaido
due to the weather and unsuitable housing
styles. The lifestyle in Hokkaido which
emphasizes indoor living is well suited to
multifamily housing.

2.2 *Construction of living space - simple and compact*

As Figure 1 indicates, there are few
unevenly shaped floor plans. Houses are
more simple and compact than the
complicated-shaped traditional homes in
warm regions. In cold regions, it is
rational to make the living space simple
and compact, and to construct a large open
space (the living room) in the center,
with the other rooms placed around it.
This "living room-centred" home is typical
in Hokkaido. It is also typical that the

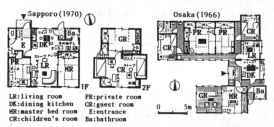

LR:living room   PR:private room
DK:dining kitchen   GR:guest room
MR:master bed room   E:entrance
CR:children's room   Ba:bathroom

Figure 1.Typical detached house plan in Sap-
poro(snowy-cold) and Osaka(warm).

inside of the home is clearly separated
from the outside. Namely, there are many
open places such as the veranda which can
be used together with the garden in a home
in a warm region, but not in Hokkaido. The
comfortable summer of Hokkaido does not
force people onto the veranda as much as
in hotter or more humid climates, and in

the winter such open places are unusable. More important is the airtightness of a home. The logic also applies to the construction of apartment houses.

2.3 *Resident's view of habitability*

Previously, I mentioned that there was a great deal of dissatisfaction with the lack of warmth and sunlight in wintertime (Noguchi 1977). Both factors significantly relate to the habitability of Hokkaido in the wintertime. In recent years these dissatisfactions have been decreasing as the quality of homes improves. At present, residents have many problems related to snow removal (Table 4). There are varoious causes for these, but the problems are all serious, such as "snow dropping nextdoor or on the road", "no place to dump snow", and "no labor to clear snow" (Hoshino 1989). Both in urban centres and in suburbs,housing areas are narrower and smaller. We must seriously consider preserving land adequate for snow removal and providing adequate multifamily housing.

Table 4.Reasons for dissatisfaction with detached houses.

house size	3.9%
room arrangement(plan)	7.6
kitchen,toilet and bathroom facilities	5.9
closet, storeroom, etc. space	8.1
exposure to the sun	6.9
heating	8.1
natural lighting	2.1
ventilation	1.9
durability	6.3
exterior design	3.8
garden area	4.4
parking place in winter	3.6
snow removal around the house	34.8
other	1.2
	(Data:190)

3 LIFE STYLE OF MULTIFAMILY HOUSING AND RESIDENT'S VIEW

3.1 *Living in an apartment house*

Figure 2 indicates the average size of private condominiums. The "living room centered" construction is the same as that of detached houses, but it is a size smaller. Accordingly, the important places (such as the living room, utility room and closet) are very small and inconvenient. The smaller sized public apartments are worse. There is great dissatisfaction regarding their size (Noguchi 1982). Most apartment houses have balconies, but they

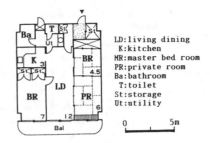

Figure 2.Typical dwelling plan of private apartment house.

tend to be narrow and unusable in the winter. There is an inside-type balcony, but most of them are not used effectively in the winter because they are too cold. Wider inner spaces are also necessary if they are to be useful for housekeeping purposes.

3.2 *Use of common space*

The common space of multifamily housing generally includes the entrance, corridor, elevator hall, storeroom, bicycle storage area, and meeting room. Some new public apartment buildings also feature covered ways (passages covered by a roof which connect one building to another), playrooms, and commons (spaces around the entrance or stairs which can be used for playing or gathering). In northern areas, although these common spaces are important when outdoor life is brought indoors, these areas are not used well (Ekuni 1993). In other words, it is rare to see children playing, neighbors chatting, or groups of residents working in the common entrance, corridor or elevator hall. Children like to go to places where they can play comfortably (i.e., where there is adequate heating). Especially in the wintertime, it is only infrequently that we see children outside playing in the sun. Most prefer to play inside the house (Table 5). The main reason for the unsatisfactory use of these areas is that they are only planned as passageways. Their size and facilities are not suited to playing and resting. For example, most have little or no heating.

3.3 *Outlook of multifamily housing*

The popularity of apartment homes is

Table 5. Places in public housing where children like to play.

	summer	winter
inside the house	48.2%	92.1%
specially designed indoor common space	1.0	13.0
entrance hall,passage,staircase,etc.	4.0	9.2
playroom,meeting place	0	3.0
"Gangi"(covered walkway)	8.1	17.3
play ground,garden within estate	80.0	35.0
parking zone,streets within estate	53.9	20.8
places outside of estate bounds	64.0	38.7
other	10.3	16.0

(Data : 217)

Table 6. Reasons for living in apartment houses.

	Urban residents(98)	Suburban residents(131)
ease of snow removal	78.0%	91.0%
disaster protection	2.0	1.0
quiet neighborhood	17.0	12.0
economical heating cost	18.0	27.0
being secure against crime	24.0	29.0
being full of sunshine	16.0	28.0
conveniences of urban life	34.0	7.0
easy maintenance of house	29.0	22.0
other	1.0	0

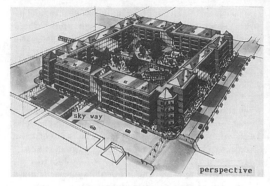

perspective

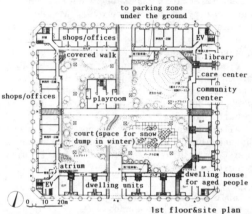

1st floor&site plan

Figure 3. New design of apartment house for snowy, cold regions like Sapporo.

increasing. The survey of 1977 (Noguchi 1977) shows about 70% of residents living in apartments preferred apartment houses if they are to be convenient and large, while 15 years later, the survey of 1992 (Kataoka 1995) shows this percentage increasing to 90%. Besides,the main reason why the respondents preferred apartment houses is that they are free from snow, and house maintenance is easier (Table 6).

Other investigations (Asano 1989) also show a growing number of aged people moving into apartment houses. The main reason for this is that they are free from snow. There is no doubt that it is hard work for residents of detached homes to clear snow.

4 CONCLUSION-PLANNING AND THE FUTURE OF MULTIFAMILY HOUSING

It follows from what has been said that multifamily housing is more suitable in Hokkaido than in milder regions. Multifamily housing is effective as permanent residences considering such factors as the climate of Hokkaido and the point of view of the residents. However, we still have a lot of problems to consider. The size of apartments should be increased to be comparable with detached houses. Facilities, such as closets, utility rooms, and balconies should be improved in size, and common spaces like entrances and corridors could be improved by providing heating. Figure 3 is an example of a multifamily housing complex comprising an entire city block in an

urban area of Sapporo. The buildings of the complex are connected by corridors or covered ways and have shops and public institutions on the ground floor. Thus we see that it is preferable to construct multifamily housing in this northern city.

REFERENCES

Asano,J.,T.Noguchi 1989. "Life Style of the Aged Living in Apartment Houses" Transaction of Hokkaido Branch of AIJ.,No.62.

Ekuni,T.,T.Noguchi 1993. "Common Space of Apartment Houses in Snowy Country(1)" Transaction of Hokkaido Branch of AIJ.,No.66.

Hoshino,N.,T.Noguchi 1989. "Snow Removal around Houses" Transaction of Hokkaido Branch of AIJ.,No.62.

Kataoka,M.,T.Noguchi 1995. "Living Style of Apartments in Urban Area" Transaction of Hokkaido Branch of AIJ.,No.68.

Noguchi,T. 1977. "Residents' View of Houses-Living Style and Houses in Sapporo(9)" Transaction of Hokkaido Branch of AIJ.,No.47.

Noguchi,T.,F.Adachi, et al. 1982. "Living Style and Houses in Hokkaido" Hokkaido University Press.

Snow Engineering: Recent Advances, Izumi, Nakamura & Sack (eds) © 1997 Balkema, Rotterdam. ISBN 90 5410 865 7

Snowdrift control design: Application of CFD simulation techniques

Bill F. Waechter, Raymond J. Sinclair, Glenn D. Schuyler & Colin J. Williams
Rowan Williams Davies & Irwin Inc., Guelph, Ont., Canada

ABSTRACT: Computer modelling techniques, employing computational fluid dynamics (CFD) and a finite area element model (FAE), were used to predict snowdrift deposition patterns around a new building at the South Pole Station, Antarctica. Through interpretation of the wind flow field predicted through CFD, snowdrift prone areas around an existing building, that is raised above the snow surface, were identified. The wind flow field, generated by CFD, was subsequently used as input to an FAE computer snowdrift prediction model. The characteristics of the drift deposition patterns predicted by the computer simulation techniques were in satisfactory agreement with snowdrift patterns measured around the existing reference building at the South Pole. It was concluded that the FAE model predicted realistic snowdrift accumulation patterns when CFD-predicted local wind velocity fields were combined with local meteorological data.

1 INTRODUCTION

Snowdrift formation on and around structures that are located in areas of significant snow accumulation or blowing snow activity is an important design issue. Wind tunnels and water flumes have often been used to predict snowdrift patterns around buildings (Melbourne and Styles 1967, and Irwin and Williams 1983) and in more recent years, computation fluid dynamics (CFD) techniques have been employed (Nakata et al. 1993). A combination of computational fluid dynamics (CFD) and a finite area element (FAE) snowdrifting computer model was used as a design technique to investigate the snowdrift performance of new science facilities planned for construction at the Amundsen-Scott South Pole Station in Antarctica. CFD modelling was used to predict wind velocities around and below the new building, which is elevated above the snow surface. The CFD-predicted velocities near the ground were then used as input to the FAE model to predict snowdrifting. The FAE model was first developed to use measured mean wind speeds from wind tunnel tests. In many cases, however, the CFD method gives a higher resolution of the wind flow field, which was expected to improve the accuracy of the predictions of the snowdrift patterns.

This present study was conducted in two stages. The first was a field comparison stage where a CFD-predicted velocity flow field, specifically low velocity areas, was compared to actual snowdrift patterns recorded around an existing building to assess the accuracy of the velocity predictions. In the second stage, the combined CFD and FAE modelling method was applied to a new building design to predict the extent of snowdrift accumulation that will occur close to the building. If the predicted snow buildup was unsatisfactory, these models were to be used to test the effects on snowdrifting of modifications to the aerodynamic aspects of the new building design.

2 MODELLING METHOD

A three-dimensional CFD model of the actual (full scale) reference building that exists at the South Pole was developed. The computational grid was constructed as a rectangular box and is described as follows. The upwind boundary of the box, inflow boundary into the computational domain, was 10H from the windward edge of the test building, where H is the height of the building, measured from the ground. The downwind, or outflow boundary of the computational domain was located approximately 20H from the end of the building. The side boundaries of the grid were approximately 10H from the lateral edges of the building. The top boundary of the grid was approximately 20H above the roof of the building.

The total number of nodes (calculation points)

Figure 1 Reference building: CFD model and actual.

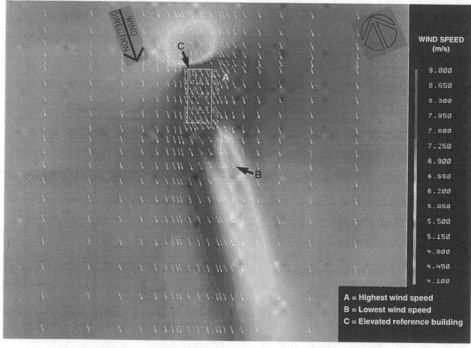

Figure 2 Plan view of wind flow field.

used in the simulation was approximately 70,000. More than 80% of the nodes were concentrated around the building. Beyond the regions of the flow steered by pressure effects of the building and the wake region, the grid resolution was reduced by expanding the node spacing geometrically to the outer boundaries of the grid. The average node spacing around the building was 30 cm to 90 cm in all directions.

A view of the actual (reference) building, which is raised above the snow surface, and a perspective view of the 3-D CFD model are shown in Figure 1. Note the inclusion of the structural steel trusses on the CFD model, since they would retard wind flows beneath the actual reference building and therefore affect snowdrift formation. The reference building is an existing building at the Amundsen-Scott South Pole Station.

Boundary conditions were specified on the various faces of the grid. The upwind, or inflow, boundary was given a uniform wind velocity. The sides of grid were defined as being periodic with each other. This allowed several wind directions to be tested that were not co-incident with the orientation of the grid. The top boundary of the grid had slip-wall conditions applied. The downwind boundary of the grid had fully developed (zero-gradient) outflow conditions applied. The bottom of the computational grid was modelled as a rough wall, characteristic of wind flow over flat snow-covered terrain.

The CFD simulations prepared in this study used the commercial computer code called TASCflow[1]. In all cases, TASCflow was run using the incompressible flow formulation for this forced convection (isothermal) flow problem. The time-averaged Navier-Stokes equations were solved with the standard k-ϵ turbulence model. The discrete equations were a primitive variable, collocated, finite-volume flux-element formulation. This approach combines the geometric flexibility of the finite-element methods with the conservation of transported quantities over each finite volume. The advection terms in the equations were modelled with an accurate upwind differencing method called the mass-weighted-skew (MWS) scheme. This scheme accounts for convection along streamlines of the flow. The solution procedure was iterative, involving solutions of the linearized discrete mass and momentum equations which were solved simultaneously as a coupled set. The equations for k and ϵ were solved in a segregated fashion at each iteration. The iterative process was terminated when the maximum normalized residuals of the mass and momentum equations converged to less than the user specified tolerance of 10^{-3}. Each

simulation required approximately 12 hours of CPU time on an IBM RISC 6000/550 running in core.

The FAE computer model is described in detail in a number of published papers, including: Irwin et al. (1995); Gamble et al. (1991); and, Irwin and Gamble (1988). In simple terms, the FAE model divides a site or building roof into a large number of elemental areas (grid) and computes where snow will be deposited or scoured on an hour-by-hour basis in each elemental area. Statistical methods are then typically used to derive design snow loads; however, in this application the volume of snow "blowing toward" the site in the FAE computer model was determined through application of known snow flux expressions (Kobayashi 1972, and Dyunin 1954) to the meteorological data recorded at the South Pole and the resulting wind speeds predicted by the CFD model. In areas with a climate that is less extreme (cold) than the South Pole, other parameters (e.g., temperature, humidity, rainfall, solar radiation, etc.) that affect snowdrift build-up, erosion, melting, etc., would also have been considered in the FAE simulation.

The meteorological data for the winter of 1992 at the South Pole was reviewed to determine the prevalent wind directions associated with the snowdrift patterns that developed around the reference building. The wind speed and direction recorded on an hourly basis throughout the polar winter were used as inputs to the snow flux expressions to estimate the volume of drifting snow approaching the site for each wind direction. This flux rate was used as an input for the FAE model. The highest flux rate for the winter was associated with wind approaching from the north (360° or 0°), which is measured relative to the station's Grid North (the Greenwich Meridian).

Although snowdrift patterns are the result of drifting events that occur from a variety of wind directions and speeds, a majority of the snow transport winds at the South Pole occur from a limited 30° sector (i.e., 0° - 30°). The CFD model investigation of the reference building therefore focused on the single most dominant wind direction, north, that occurred during the one winter being studied. A more complex simulation would consider all combinations of wind speeds and directions that occurred during the winter and also account for the effects of changes in topography or aerodynamics due to snowdrift deposition and growth. Examination of a single dominant wind direction was considered adequate for a first order comparison to the actual field data (snowdrift patterns) for the one winter.

The wind velocity field from the CFD model was nondimensionalized by the approaching wind speed at 10 m above ground to scale the hourly wind data to

---

[1]   TM of ASC Ltd., Waterloo, Ontario, Canada.

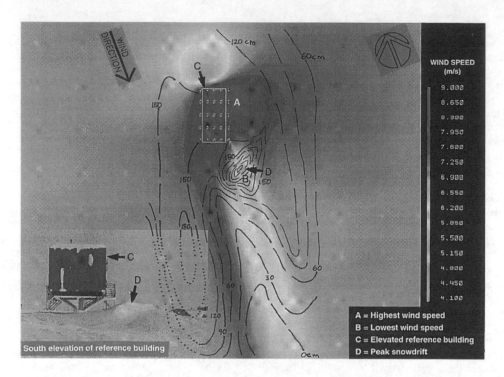

Figure 3 Comparison of CFD wind speeds to snow surface contours.

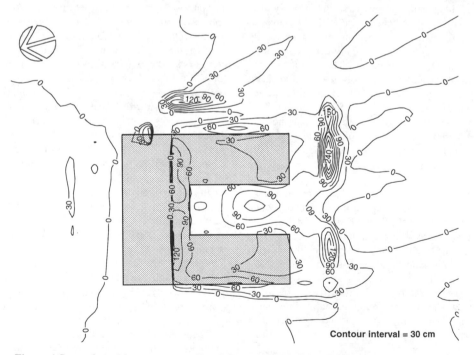

Figure 4 Snow deposition pattern predicted through CFD and an FAE computer snow model.

each grid point in the FAE model. The FAE snowdrifting model used hourly meterological records and a grid system superimposed on the snow surface, dividing the surface into approximately 3500 finite area elements whose horizontal dimensions equalled the CFD grid at a 1.2 m reference height. Values for wind velocities at the reference height at the four corners of each elemental area were used by the FAE model to compute the snowdrift fluxes through the sides of the element. Empirical relationships based on field data are used in the model to relate drift rate to the local mean wind velocity. For this reason, turbulence effects are not explicitly required as input.

The computations were completed for every area element and the change in snow mass contained in each area element during a one-hour time interval was thereby evaluated. This procedure was repeated for each time step covering the entire winter of 1992. The FAE simulation provided a complete history of the snow accumulation in each elemental area for each hour of the year being examined.

3 COMPARISON OF WIND FLOW FIELD TO ACTUAL SNOW DEPOSITION

The velocity patterns predicted through CFD were compared to actual snowdrift patterns measured at the site, around the reference building. The main purpose of this data comparison step was to acquire confidence in the ability of CFD modelling to provide realistic velocity patterns at the reference height used in the FAE model. For this exercise, the seasonal mean wind speed at the South Pole for the winter of 1992 was modelled as the approaching wind in the CFD model.

The snowdrift deposition pattern measured around the reference building as a result of snowdrifting events during the winter of 1992 (March 1992 through October 1992) was used in this study. The reference building is situated in a remote location where surrounding structures would not affect the local drift patterns. Figure 1 shows the reference building and the CFD model. The building itself is elevated approximately 2.4 m above the snow surface. The three-dimensional model of the elevated reference building included the open truss foundation system beneath the building, but not the entry stairs. Based on a review of photographs showing snowdrift conditions around the building, the cumulative effect of the trusses on the wind flows was considered more important than the local effect of the stairs. The CFD-predicted wind velocity field at the reference height of 1.2 m is shown in Figure 2.

A line drawing showing the snow depth contours surveyed around the reference building was superimposed on top of the CFD-predicted wind velocity patterns (Figure 3) for comparison. High wind speeds were predicted through CFD at "A" in the figure with low wind speeds at "B." The upwind corner of the reference building is at "C." The most significant accumulation of snow around the elevated reference building occurred at "D," which is approximately 9 m downwind. The snowdrift developed in an area where the CFD-predicted wind speed was less than approximately 5 m/s. The wind speed "approaching" the CFD model building was the same as the seasonal mean wind speed recorded at the South Pole during the winter of 1992. Other researchers (Melbourne and Styles 1967, Durgin and Floyd 1971, and Jackson and Carroll 1978) support a snowdrift initiation wind speed of approximately 5 m/s. This correlation of a predicted low wind speed zone, based on the seasonal mean wind speed, with the location of an actual major drift accumulation provided sufficient confidence in the validity of the CFD simulation, as it provided a satisfactory representation of the flow field around an elevated building, to use it as input for the FAE model.

A significant difference between the CFD-predicted wind speed contours and the snow surface contours is the elongation of the drift peak perpendicular to the general flow direction. This effect would not be explained by the velocity contours. It was hypothesized that this effect was due to the variability in wind direction over the course of the winter at the site. The actual snow surface contours, shown in the topographic plan, are the result of many different wind directions and wind speeds for the entire winter of 1992, whereas the CFD-predicted contours represent wind speeds for one wind direction only, similar to most wind tunnel and water flume modelling techniques.

Another possible explanation could be the modification of the flow field by the accumulated snowdrift. However, our experience is that this modification would tend to cause accelerated flow at the sides of the snowdrift's peak, that would erode the snow and deposit it in the wake of the peak, thus elongating the drift parallel with the wind direction.

4 APPLICATION

The comparison of the CFD results (i.e., wind speed contours) to the actual snow surface contours, provided the confidence that the CFD simulation provides a satisfactory representation of the flow field around an elevated building. The study was then extended to examine a new building shape, which may be used during future development at the South Pole. The combination of CFD and FAE models, previously described, was used for the new building which is elevated 3 m above the snow surface. In this study of the new building, the CFD-predicted flow fields for the six prevailing wind directions were used to refine the accuracy of the FAE-predicted snow deposition patterns.

The results of the detailed hour-by-hour prediction of snow accumulation around the new building, for the winter of 1992, are presented in Figure 4. Comparison of the drift depth contours predicted for the new building with those actually measured near the reference building (Figure 3) shows a similarity in the drift pattern. The most significant common feature is the broad drift peak running roughly perpendicular to the approaching wind. This similarity lends support to the hypothesis that the broad nature of the peak is due to the wind direction variability. This points to the need to include as many of the prevailing wind directions in an FAE study as is feasible in order to refine the snowdrift definition and thereby increase the accuracy of the snowdrift predictions.

5 CONCLUSIONS

Computer modelling of wind flow patterns around elevated buildings using CFD provides useful and realistic results that assist in locating areas prone to snowdrift deposition. Velocity patterns, predicted by CFD, with regions of reduced wind speed (i.e. ≤ 5 m/s) near the snow surface (grade level), correlate well to actual snowdrift deposition areas measured in the field. The prediction of actual accumulation amounts is a more complicated issue, requiring an hour-by-hour simulation, performed here with the FAE model.

In this study, the FAE model predicted realistic snow drift accumulation patterns when CFD-predicted local wind velocity fields were combined with local meteorological data.

REFERENCES

Durgin, Frank H. and Floyd, Peter (1971) A Study of the drifting of snow under and around raised buildings and building complexes. *Proceedings Third International Conference on Wind Effects on Buildings and Structures,* Tokyo, Japan. pp. 153-165.

Dyunin, A. K. (1954) Vertical distribution of solid flux in a snow-wind flow. National Research Council of Canada, Technical Translation 999.

Gamble, S. L., Kochanski, W. K., and Irwin, P. A. (1991) Finite area element snow loading prediction: Applications and advancements, *Eighth International Conference on Wind Engineering,* Vol. 2, London, Ontario. pp. 1537-1548.

Irwin, P. A. and Gamble, S. L. (1988) Prediction of snow loading on the Toronto SkyDome, *Proceedings of First International Conference on Snow Engineering,* Santa Barbara, publ. by CRREL, Hanover, NH. pp 118-127.

Irwin, P.A., Gamble, S.L. and Taylor, D.A. (1995) Effects of roof size and heat transfer on snow load: Studies for the 1995 NBC, *Canadian Journal of Civil Engineering,* Vol. 22 (1995). pp. 770-784.

Irwin, P.A. and Williams, C.J. (1983) Application of snow simulation model tests to palling and design. *Proceedings, Eastern Snow Conference.* Vol. 28, 40th Annual Meeting, Toronto, Ontario. June 2-3, 1983. pp. 118-130.

Jackson, B. S. and Carroll, J. J. (1978) Aerodynamic roughness as a function of wind direction over asymmetrical surface elements. *Boundary Layer Meteorology,* Vol. 14, pp. 323-330.

Kobayashi, D. (1972) Studies of snow transport in low-level drifting snow. Institute of Low Temperature Science, Sapporo, Japan. Report No. A31, pp. 1-58.

Melbourne, W. H. and Styles, D. F. (1967) Wind tunnel tests on a theory to control Antarctic drift accumulation around buildings. *Proceedings of International Research Seminar on Wind Effects on Buildings and Structures.* Vol. 2. Ottawa, Canada. Paper No. 32. pp. 135-173.

Nakata, T., Uematsu, T., and Kaneda, Y. (1993) Three dimensional numerical simulation of snowdrift. *Journal of Wind Engineering and Industrial Aerodynamics,* Vol. 46 & 47 (1993). pp.741-746.

Snow Engineering: Recent Advances, Izumi, Nakamura & Sack (eds) © 1997 Balkema, Rotterdam. ISBN 90 5410 865 7

Snowdrifting around a tall building in a snowy area

Hiromi Mitsuhashi
Department of Architecture, College of Science and Technology, Nihon University, Tokyo, Japan

ABSTRACT: This paper describes a wind tunnel experiment using model snow which was conducted to predict the patterns of snowdrifts around a proposed 33-story building with balconies planned in Yuzawa-machi, Niigata Prefecture, and differences in snow accumulation on its balconies due to banister shapes. It further clarifies snowdrift accumulation by comparing test results with measurements of snow accumulation at the building after its completion.

1. PRELIMINARY

A large number of high-rise resort condominiums has been constructed in recent years in ski resorts in heavy snowfall regions. Similar to wind damage issues associated with high-rise buildings in urban areas, these condominiums create snow damage problems such as entrances blocked by snowdrift accumulation, and other adverse effects on houses and roads in the neighborhood.

It has become apparent that snow accumulation on balconies, banister design, and other diverse problems should be taken into account at the design stage.

This study examined the properties of snowdrifts around a proposed 33-story building with balconies planned in Yuzawa-machi, Niigata Prefecture, and the conditions of snow accumulation on its balconies by a wind tunnel experiment using model snow.

Measurements of drift accumulation on the completed building were also made, and compared with the results of the experiment for the purpose of providing fundamental information for designing high-rise buildings.

2. WIND TUNNEL EXPERIMENT USING MODEL SNOW

2.1 Outline of the experiment

The subject of the experiment is a 100 meter tall ,33-story building with balconies. The experiment was designed to examine snowdrifts in the vicinity of the building before and after construction from a range of wind directions ,as well as snow accumulation on balconies due to changes in the angle of attack and the shape of banisters.

A whole and two partial models were used in the experiment (Figs. 1 & 2). The whole model consisted of the building and site on a reduced scale of 1 to 600 while four layers of balconies were made into the partial models on a scale of 1 to 100. The wind tunnel was of the Effel type

Figure 1. The whole model.

Figure 2. The partial model.

having a cross section of 300 x 300 mm.

The tests were made at wind velocities of 2.0 m/s and 4.0 m/s, using local wind velocity profiles. Activated clay was used as model snow. The depth of accumulation was measured by laser displacement meter.

Sixteen different tests were conducted on the whole model at four wind directions. Two kinds of the partial model were prepared: one with grating-type banisters and the other with wall-type banisters. Twelve tests were carried out on each of the partial models at three different angles of attack.

2.2 Test results and discussion

See Figures 3 and 4 for an example of measurement results. Results of the tests using the whole model showed that, comparing the site before and after construction of the building for each wind direction, the influence of snowdrifts has become markedly noticeable after the construction. Areas around the building where large drifts will form and areas where snow accumulation was small were clarified. Tests on the partial models indicated

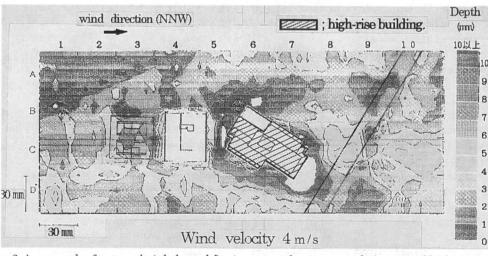

Figure 3. An example of test results(whole model). A contour of snow accumulation around high-rise building.

518

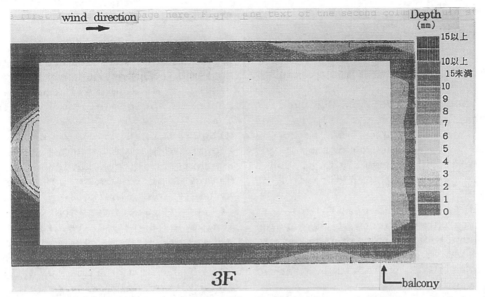

Figure 4. An example of test results(partial model). A contour of snow accumulation on the balcony.

Figure 5. An example of actual measurement.

that the quantity of snow accumulation tended to be larger on lower levels. Large portion of the balconies were affected by snow accumulation for wall type banisters than for grating-type banisters.

3. ACTUAL MEASUREMENT OF SNOW ACCUMULATION ON THE BALCONIES

3.1 Outline of actual measurements

On-site surveys was made in February and December 1995 and January 1996. On the balconies zones covered by snow and snow depths were measured. Points of measurement were chosen on upper, middle and lower stories.

3.2 Results of actual measurements and discussions

See Figure 5 for an example of the results of the actual measurements.

Snow accumulation was observed at balcony corners on each floor. For locations on a given floor, balconies on the north side tended to show larger accumulation, presumably as a result of the prevailing wind direction of NNW.

Almost no snow accumulation was observed on the balcony of the landing on each floor where wire netting was installed to prevent snow from piling up, confirming its effectiveness as a preventive measure.

4. CONCLUSIONS

The following information was obtained from the above results:

1. Properties of snowdrift accumulation around a high-rise building by wind direction, and on balconies by banister shapes were identified from the results of the experiment.

2. Actual measurement revealed properties such as the patterns and quantities of snow accumulation on the balconies.

As outlined above, properties of snowdrift accumulation around a high-rise building, and the conditions of accumulation on its balconies were obtained, providing fundamental information for countermeasures against snowdrifts for high-rise buildings.

On-site measurements will be continued, and resultant data compared further with the results of wind tunnel tests using model snow to examine agreement between them.

REFERENCES

Mitsuhashi, H., 1994. Study on Properties of Snowdrifts around a Tall Building in an Area of Heavy Snowfall, Preprint of the 1994 Conference Japanese Society of Snow and Ice, 166

Mitsuhashi, H., 1995. Study on Preventive Measures of a Tall Building against Snow Accumulation: Actual Measurement of the Conditions of Accumulation on Balconies, Preprint of the 1995 Conference Japanese Society of Snow and Ice, 139

Mitsuhashi, H., 1996. Study on Preventive Measures of a Tall Building against Snow Accumulation: Properties of Snowdrifts as Indicated by Wind Tunnel Tests Using Model Snow, Proceedings of Japan Society for Snow Engineering, vol. 12, 193–196

Investigation of the indoor environment of well-insulated and airtight houses in the Tohoku district

Ken-ichi Hasegawa & Hiroshi Yoshino
Department of Architecture, Faculty of Engineering, Tohoku University, Sendai, Japan

ABSTRACT: This paper describes the thermal environment and indoor air quality of 16 well-insulated and airtight houses in Sendai and Morioka during the winter and summer of 1994. These houses are equipped with mechanical ventilation systems and space heating systems for the entire house. The indoor thermal environment during the winter was comfortable, but some occupants complained of dryness. During the summer, indoor temperatures did not decrease much without air conditioners, and improvement seems possible by cross ventilation or night time ventilation. The concentration of NO_2 was not high. The concentration of NO_2 with an unvented kerosene heater was higher than that without it.

1 INTRODUCTION

Newly constructed detached houses in Japan, especially in the northern portion of Japan, have become more and more airtight and highly insulated due to energy conservation and the demand for thermal comfort. In such houses, it is expected that the quality of the indoor thermal environment will be better than that of existing houses. However, there are many problems related to indoor air quality and humidity in the winter and the thermal indoor environment during the summer. The authors have investigated the indoor thermal environment and indoor air quality in 16 units of highly insulated detached houses in Sendai City and Morioka City. The houses investigated are equipped with me-

chanical ventilation systems and space heating systems for the entire house. Sendai and Morioka are the main cities of the Tohoku District along the Pacific Ocean. The latitudes of Sendai and Morioka are 38°16' and 39°42', respectively. The mean outdoor temperatures of Sendai and Morioka are 1.3°C and -1.9°C in February, and 24°C and 23°C in August, respectively.

2 DESCRIPTION OF HOUSES MEASURED AND THE MEASUREMENT PERIOD

Table 1 describes the houses measured. All houses were built of wood-frame construction method. The floor area of houses was from 105 to 224 m². All houses

Table 1. Description of 16 houses.

No.*	Completion	Floor Area (m²)	Thermal Insulation(mm)			Main Heating System	Supplementary Heating	Cooling System	Family Numbers
			Wall	Floor	Ceiling				
1	1992	182	100	100+30**	200	central / heat pump	electric heater, kotatsu	○	5
2	1989	124	100	150	200	central / circulator	kotatsu	×	4
3	1991	196	100	50+30**	100	central / circulator	electric heater	○	5
4	1989	161	40	30	40	central / circulator	electric heater, unvented kerosene heater	×	3
5	1990	117	40	150	200	central / heating panel	kotatsu	×	5
6	1991	164	100	50+25**	100	electric heater	unvented kerosene heater	○	4
7	1989	155	25	25	50+30***	×	unvented kerosene heater, kotatsu	×	5
8	1992	106	40	40	100	F.F. system	electric carpet	×	4
9	-	-	-	-	-	F.F. system	×	○	6
10	1992	151	50	30+50**	200	central / heating panel	×	○	6
11	1992	206	100	88	200	central / heating panel	electric carpet	○	3
12	1989	116	100	100	200	central / heating panel	×	×	5
13	-	-	-	-	-	F.F. system	×	×	4
14	1992	137	50	30+50**	200	central / heating panel	×	×	4
15	1990	105	100	88	200	central / heating panel	×	×	4
16	1991	224	100	88	200	central / heating panel	×	○	3

* No.1~8 are sited in Sendai City, and No.9~16 in Morioka City. : ** Thermal insulation of foundation : *** Thermal insulation of roof

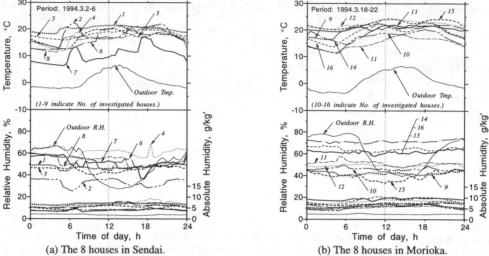

(a) The 8 houses in Sendai.　　　　　　(b) The 8 houses in Morioka.

Figure 1. Temperature and humidity profiles averaged for five days during the winter in Sendai and Morioka.

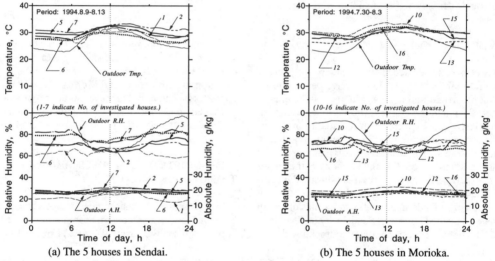

(a) The 5 houses in Sendai.　　　　　　(b) The 5 houses in Morioka.

Figure 2. Temperature and humidity profiles averaged for five days during the summer in Sendai and Morioka.

have thermally insulated walls, ceilings and floors. The thickness of insulation in house no.2, which is the most heavily insulated, is 100 mm for the walls, 150 mm for the ceiling, and 200 mm for the floor. All houses except no.7 have a water heating system with panel radiators or a vented kerosene heater or electric heater. House no.7 has an unvented portable kerosene heater and a "Kotatsu" (a Japanese style electric heater which is mounted under a low table covered with a quilt). A cooling system was present in eight of the sixteen houses. The measurements in winter and summer were taken during February, March and August of 1994.

3 TEMPERATURE AND HUMIDITY IN WINTER AND SUMMER

3.1 Method of measurement

Temperatures at three points and relative humidity at two points in each house were measured continuously for a week by resistance thermopile and data logger. The measuring points were set in the living room (1.1 m and 5 cm above the floor level) and the room with relatively low temperature in the house (1.1 m above the floor level).

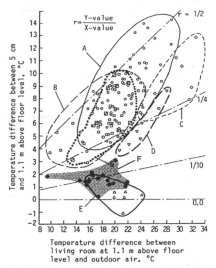

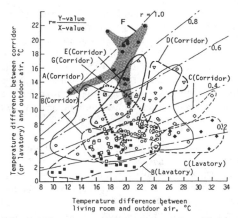

Figure 4 Temperature of living room and corridor (or lavatory or the room with low temperature) during the evening family time.

A: "*PHC Houses*" sold by the public housing corporations in the main 8 cities of the Tohoku District(78 houses, constructed in 1968-1979).

B: "*Rural Houses*" in rural areas of Yamagata Prefecture(30 houses, constructed before 1978).

C: "*Village Houses*" in a rural village of Iwate Prefecture(9 houses, constructed in 1970-1981).

D: "*Insulated Houses*" with a space heater in the city of Sendai(7 houses, constructed in 1982-1984)

E: "*Floor Heated Insulated Houses*" in the city of Sendai(6 houses, constructed in 1982-1984)

F: "*Well-insulated and Airtight Houses*" in the city of Sendai and Morioka(16 houses, constructed in 1989-1992)

G: "*Multi-Family Houses*" constructed of reinforced concrete on the city of Sendai(9 houses, constructed in 1964 and 1971)

Figure 3. Indoor-outdoor temperature difference and vertical temperature difference in living room during the evening family time.

3.2 Daily profiles in living room during winter

Figure 1 shows temperature and humidity profiles of eight houses averaged for 5 days during the winter in Sendai and in Morioka, respectively. Except for house no.7 temperatures varied between 10°C and 24°C. In houses no.1, 5, 12, 13 and 15, the space heating system operated throughout the day. The temperature in these five houses was maintained around 20°C. The temperature of the other houses ranged from 15 to 22°C during the heating time, but after the heater was turned off, the temperature fell and became 5 to 15°C by daybreak. The temperature of houses no.2 and 7 fell rapidly. Except for houses no.2 and 15 the relative humidity was 40~70%. Except for houses no.7, 9 and 10, occupants complained of dryness of the indoor air in winter. Since the relative humidity measured was not very low, it is possible that there are other reasons than humidity related to occupants' feeling dry.

3.3 Daily profiles in living room during summer

Figure 2 shows temperature and humidity profiles of five houses averaged for five days during the summer in Sendai and Morioka, respectively. The temperature of all houses except houses no.1, 6, 15 and 16 were stable compared with changes in outdoor temperature during the day, but indoor temperature did not decrease at night even if the outdoor temperature dropped. In house no.1 the cooling system was operated throughout the whole day, in houses no.6 and 15 it operated intermittently. During operation of the cooling system, temperatures in these three houses were 1 to 2°C lower than the outdoor temperature. In house no.16, occupants depend on more cross ventilation by opening windows, to exhaust the heat. The relative humidity of the measured houses varied between 60% and 80%.

4 COMPARISON OF THERMAL ENVIRONMENT IN VARIOUS HOUSES IN WINTER

The room temperature of 139 houses in the Tohoku District was measured by the authors (Hasegawa and Yoshino, 1987). The thermal environment of the 16 houses reported in this paper was compared to that of the other houses. The mean room temperature during the evening family time after supper was used for comparative analysis of thermal environments. The 155 houses were categorized into seven groups as shown in Figure 3.

4.1 Vertical temperature difference in the living room

Figure 3 shows the relationship between the vertical temperature difference and the indoor-outdoor temperature difference. The vertical temperature difference is the temperature difference between 5cm and 1.1m above the floor level. These temperatures were averaged during the evening family time. The vertical temperature differences are slight in the floor heated insulated houses (Type E) and the measured houses (Type F) . Except for these houses, the vertical temperature difference is between 3°C and 14°C. The ratio of the temperature difference between 5 cm and 1.1 m above the floor to the indoor-outdoor temperature difference (non-dimensional vertical temperature) of the measured houses except no.7 ranges from 0.0 to 0.13.

4.2 Temperatures of lavatory, corridors and the room with relatively low temperature

Figure 4 shows the indoor-outdoor temperature difference and the temperature difference between the living room and the lavatory or the corridor or the room with relatively low temperature (measured houses) averaged during the evening family time. Temperatures in the measured houses except house no.16 were between 13°C and 23°C for the room with relatively low temperature, while that of the living room is 20°C. The distribution of the temperature is due to the differences in the levels of thermal insulation and airtightness, and the types of space heating systems employed.

5 INDOOR AIR QUALITY

5.1 Method of measurement

The concentration of NO_2 was measured by bare detector badges exposed for a measurement period in the living room and the kitchen. This measurement follows the method utilized by Yanagisawa and Nishimura (1980) .

5.2 Results of NO_2 concentration measurements

Figure 5 shows the mean NO_2 concentration for five days in living rooms and kitchens. All concentrations are no very high. The concentrations in houses no.4 and 7 with an unvented kerosene heater was higher than those of other houses. In all of the houses, the concentration in kitchens was higher than that in living rooms due to generation of NO_2 from gas cooking stoves in the kitchen.

6 CONCLUSIONS

1. Indoor environment during the heating season of well-insulated and airtight houses was more thermally comfortable, compared with that of other houses. Some occupants complained of dryness of indoor air in winter. This problem should be investigated in further.
2. During the summer, room temperatures of some measured houses were stable compared with the change of outdoor temperature during the day, but it did not decrease at night even if the outdoor tem-

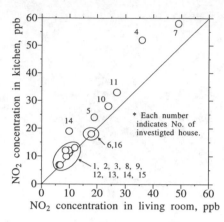

Figure 5. The mean NO_2 concentration.

perature dropped. It seems possible to reduce summer indoor temperatures by cross ventilation with open windows or by night time ventilation.
3. NO_2 concentrations in living rooms and kitchens were not very high. Unvented kerosene heaters caused higher concentrations than with a water heating system or a vented kerosene heater.

REFERENCES

Hasegawa, F. and Yoshino, H. 1987. "Investigation on winter indoor temperature of various types of houses in the Tohoku District." *Journal of Architecture, Planning and Environmental Engineering*. Vol. 371, pp.18-26 (in Japanese)

Yoshino, H., Matsumoto, H., Hasegawa F. and et al. 1990. "Investigation of indoor thermal environment, air quality and energy consumption in new detached houses of wood-frame construction in a small city in Japan." *Environmental International*. Vol.16, pp. 37-52

Yanagisawa, Y and Nishimura, H. 1980. "A personal sampler for measurement of nitrogen dioxide in ambient air." *J. Japan Soc. Air Pollut*. Vol. 15, pp. 316-323 (in Japanese)

Snow Engineering: Recent Advances, Izumi, Nakamura & Sack (eds) © 1997 Balkema, Rotterdam. ISBN 90 5410 865 7

Indoor thermal problems of residential buildings in cold and snowy regions of Japan

Hiroshi Yoshino
Department of Architecture, Faculty of Engineering, Tohoku University, Sendai, Japan

ABSTRACT: Except for Hokkaido District, the quality of the indoor thermal environment of residential buildings in cold and snowy regions is still poor in Japan. The author has investigated the indoor thermal environment of various kinds of residential buildings in the Tohoku district for fifteen years. This paper first describes the climatic conditions of Tohoku district. Secondly, the investigation considers results of building envelope performance, space heating equipment, and room temperature. Thirdly, detailed measurement results of room temperature are demonstrated and the main features are discussed. Lastly, some problems related to indoor thermal environment are pointed out from the background of medical science.

1 INTRODUCTION

Except for Hokkaido District, the quality of indoor thermal environment of residential buildings in cold and snowy regions is still poor in Japan. The author has investigated the indoor thermal environment of various kinds of residential buildings in Tohoku district for fifteen years. This paper reviews the situation for building envelope performance, indoor thermal environment, and life style in the Tohoku District on the basis of investigation results by the author et al. Then the quality of life is discussed from the background of medical science.

2 CLIMATE OF COLD AND SNOWY REGIONS

The main island of Japan is located in the north latitude between 30 and 40°. Although Japan is small, it has various climatic conditions. Figure 1 shows isograms of heating degree-days based on a room temperature of 18 °C and daily mean temperature of 16 °C at which space heating is used (Kimura 1978). Three areas are characterized by more than 3,000°C -days for heating: the northern Tohoku District and Hokkaido, and the mountain areas in the Chubu District in the center of Honshu Island. In the Hokkaido District, some areas have more than 4500 heating degree-days.

As a result of the high mountain ranges in the middle of Honshu Island, there are differences in winter climatic conditions between the area along the Japan Sea and that along the Pacific Ocean. The seasonal northwest wind, including much water vapor from the Japan

Sea, is interrupted by the mountain ranges and brings heavy snow to the area along the Japan Sea. Dry air then passes over the mountain ranges and results in clear, sunny weather in the area along the Pacific Ocean.

In the summer, the cold-climate regions (except for the Hokkaido District) are hot and humid. For example, Akita, a medium-sized city along the Japan Sea with 3000 degree-days, has a mean outdoor temperature of 24.0°C and mean relative humidity of 80% in August. Therefore, not only the winter climate but also the summer climate should be taken into consideration when designing residential buildings in the cold-climatic regions, except for the Hokkaido District.

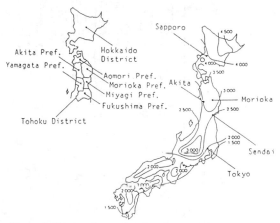

Figure 1. Heating degree days based on a room temperture of 18°C and a daily mean outdoor temperature of 16 °C at which space heating is used(Kimura 1918)

3 SURVEY OF BUILDING COMPONENTS AND SPACE HEATING EQUIPMENT

Building components, space housing equipment, and indoor temperature of the detached houses have been investigated by questionnaire in February, 1992. The 922 investigated houses are located in twelve cities of the Tohoku District, in Sapporo, the capital of the Hokkaido District and in Fuchu city in Tokyo. These houses are selected from the houses of school children in the public schools in each city.

3.1 Building insulation

In Sapporo, all houses are thermally insulated. However in the Tohoku District, some houses have no insulation material. In the cities of the northern portion of the Tohoku District, Aomori, Morioka and Akita, 70 to 90 % of the houses are insulated. In cities in the southern portion of the Tohoku District, Sendai, Fukushima and Iwaki, 30 to 50 % of the houses are insulated. In other cities like Yamagata, Sakata, Yokote, etc., 50 to 70 % of the houses are insulated. This is not only due to outside low temperatures during the winter but to the location of those cities close to the Hokkaido District.

3.2 Window component

The houses in Sapporo have double windows or single windows with double panes. In Aomori, Akita, Morioka and Hachinohe, 50 to 70 % of the houses have such windows. But almost all houses in the other cities have single windows with single panes. In Japanese style rooms of these houses, the Japanese paper window, which is called "shoji", is installed inside the single window.

3.3 Heating equipment

Except for the Hokkaido District, almost houses in Japan are locally and intermittently heated. That means, in many Japanese houses, only the living room is heated in the morning and the evening. In these houses, both kerosene space heaters and "Kotatsu" are used. The "Kotatsu" is a type of electric heater mounted under a low table covered with a cloth. In northern cities like Aomori, Akita, Morioka, Yokote and Hachinohe, more than 50 % of the kerosene space heaters are vented. But in the other cities, most of the space heaters are open-fired and unvented.

3.4 Room temperature

Room temperature is recorded by the occupant's reading of a liquid crystal thermometer. Figure 2 shows the average temperature of the living room and the main bedroom in each city during the evening after the supper. The figure also includes living room temperatures taken in the same manner in 1982. In Sapporo and the northern cities of the Tohoku District, the average temperature of living rooms is more than 20°C, and in the southern cities of the Tohoku district, it is 15 to 20 °C. The bedroom temperature is 10 to 15°C except for Sapporo.

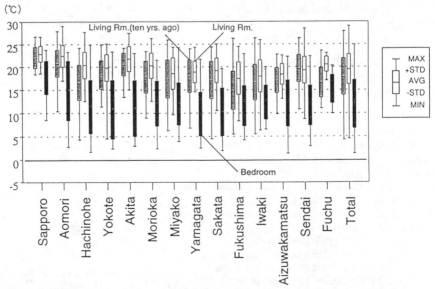

Figure 2. Mean room temperature and the deviation duringthe evening after supper in each city

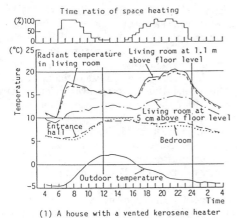

(1) A house with a vented kerosene heater

Figure 3. Temperature profiles of a house in Sendai aver
aged for a week (1985)

4 ROOM TEMPERATURE DETAIL MEASUREMENT

The author has measured room temperatures in 141
houses of various kinds in the Tohoku District since
1979.

4.1 Example of profiles of room temperature

Figure 3 shows an example of average room-tempera-
ture profiles for a house measured in the city of Sendai,
which has 2,800 heating degree-days. Room temperatures, recorded every 30 minutes, were averaged for a
week during the heating season. This house has 5 cm
thick fiberglass insulation for floors and walls, 7.5 cm
thick insulation for ceilings, and airtight sashes with
double panes. A vented space heater using kerosene
was installed in the living room.

The living room temperature at a point 1.1 m above
the floor level of this house was maintained around 20
°C during the evening after supper. After the heater was
turned off, the room temperature fell rapidly and was
10 °C by daybreak. The living room temperature 5 cm
above floor level was 6°C lower than the 1.1 m temperature. There was a high level of temperature stratification in the living room. The radiant temperature was
at most 1 °C lower at the maximum than the dry-bulb
temperature during the heating time. The temperatures
of the entrance hall and the main bedroom were lower
and remained between 5 °C and 10 °C. The high level
of temperature stratification in the living room seemed
to be due to cold-air infiltration from the unheated corridor or the entrance hall.

4.2 Vertical temperature difference in the living room

Figure 4 shows the relationship between the vertical

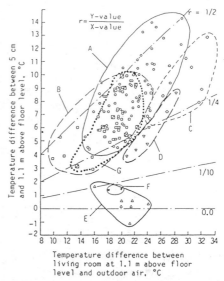

Temperature difference between
living room at 1.1 m above floor
level and outdoor air, °C

A: "PHC Houses" sold by the public housing corporations in the main 8
cities
of the Tohoku District (78 houses, constructed in 1968-1979).
B: "Rural Houses" in rural areas of Yamagata Prefecture (30 houses,
constructed before 1978).
C: "Village Houses" in a rural village of Iwate Prefecture (9 houses,
constructed in 1970-1981 except for one old house).
D: "Insulated Houses" with a space heater in the city of Sendai (7 houses,
constructed in 1981-1984). An example of measured temperature profiles
is shown in Fig.3.
E: "Floor Heated Insulated Houses" with floor heating in the city of
Sendai (6 houses, constructed in 1982-1984).
F: "R-2000 Houses" built in accordance with the Canadian R-2000
Manual in the city of Sendai (2 houses, constructed in 1988).
G: "Multi-Family Houses" constructed of reinforced concrete in the city of
Sendai (9 houses, constructed in 1964 and 1971).

Figure 4. Living room temperature and vertical temperature
difference during the evening after supper

temperature difference and the indoor-outdoor temperature difference in different kinds of houses. The 141
houses were divided into seven groups as shown in the
figure. The vertical temperature difference is the average temperature difference 5 cm and 1.1 m above the
floor level during the evening family time.

Vertical temperature differences are slight in the
floor-heated, insulated houses and the R-2000 houses.
In other houses, the vertical temperature difference is
between 3°C and 14°C and the ratio of the temperature
difference between 5 cm and 1.1 m above the floor to
the indoor-outdoor temperature difference ranges from
0.18 to 0.55. The vertical temperature difference in
some of the PHC houses and the rural houses is large
and that in insulated houses is rather small.

4.3 Temperatures of lavatory and corridors

Figure 5 shows the temperature difference between the
living room and the lavatory or between the living room
and the corridor averaged during the evening after sup-

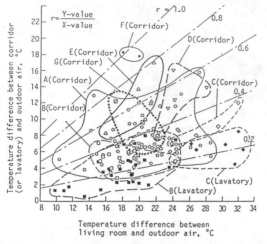

Figure 5. Temperature of living room and corridor (or labotory) during the evening after supper

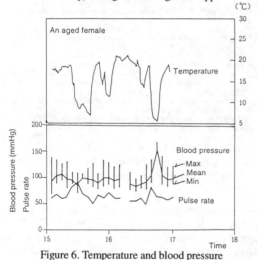

Figure 6. Temperature and blood pressure

space heating has been investigated. Sasaki (1973) on the basis of statistical research from 1957 to 1967 concluded that the mortality from strokes was lower in houses which used space heaters for a long time. Yoshino et al. (1985) investigated the relationship between mortality from strokes and indoor thermal environment during the winter in three towns, in Yamagata prefecture in 1982 and 1983. They reported that room temperatures in bedrooms, bathrooms and kitchens were much lower than the heated living room and pointed out that the great temperature difference between the heated and the other unheated spaces might influence the mortality from strokes. Yoshino et al.(1993) also investigated the relationship between stroke mortality and indoor thermal environment from a study of 161 houses in two towns in Miyagi prefecture. Room temperature in the control households were generally higher than those in the case of household by up to 1.3°C. The thermal conditions of the housing in the case households were little to those in the control households. In addition, they measured the blood pressure pulse rate and the environment temperature of aged occupants every few hours by an automatic recorder. Figure 6 shows an example of blood pressure pulse rate and temperature measurements of an aged female. Her blood pressure rose when she moved from the heated room to an unheated room with low temperature.

6 CONCLUSIONS

Investigation of building thermal performance and room temperature of residential houses in cold and snowy regions of Japan reveals that the quality of life related to the indoor thermal environment is still poor. Improvement in the quality of life of this area should be performed from a hygienic point of view by construction of well-insulated and airtight houses with optimum heating and ventilating systems.

per for the measurement period. Each temperature is indicated as the difference from the outdoor temperature. The lavatory and corridor in the rural and village houses are between 2°C and 5°C for the living room temperature of 20°C. The temperature in the insulated houses, the floor-heated insulated houses, and the R-2000 houses are between 8 and 18 C for a living-room temperature of 20°C. The temperature distribution is due to the difference in the levels of thermal insulation and airtightness and to the types of space-heating systems employed.

5 ROOM TEMPERATURE AND HEALTH

In the past in Japan, the relationship between mortality from cerebrovascular disease (stroke) and residential

REFERENCES

Hasegawa, F., Yoshino, H., Arai, H., Iwasaki, K., Akabayashi, S. and Kikuta, M. 1985. Investigation of the relationship between the indoor thermal environment of houses in the winter and cerebral vascular accident (CVA). Japanese Journal of Public Health Vol 32: 181-193.

Kimura, K. 1978. Weather data for design, Data Book for Building Design, Vol.1: 115.

Sasaki, N. 1973. Lifestyle and stroke mortality, especially space heating, drinking and smoking, Report for 22nd Tohoku Branch Meeting of Public Health, Vol.13, No.13.

Yoshino, H. 1991. Design strategies for houses in cold climate regions of Japan. ASHRAE Transaction: 625-634.

Yoshino, H., Momiyama, M., Sato, T. and Sasaki, K. 1993. Relationship between cerebrovascular disease and indoor thermal environment in two selected towns in Miyagi prefecture, Japan. *Journal of Thermal Biology*. Vol.18, No.5/6:481-486.

5 Development strategy in snow countries

Snow Engineering: Recent Advances, Izumi, Nakamura & Sack (eds) © 1997 Balkema, Rotterdam. ISBN 90 5410 865 7

Principles of avalanche hazard mapping in Switzerland

Bruno Salm

Swiss Federal Institute for Snow and Avalanche Research, Weissfluhjoch/Davos, Switzerland

ABSTRACT: Several Swiss Federal laws state that no buildings shall be erected in areas considerably endangered by avalanches. It has therefore been an essential task for our Institute to quantitatively define avalanche danger and to establish guidelines for hazard mapping. As the life-spans of structures are several decades to centuries, corresponding periods of time have to be considered for hazard assessments. Measures of a potential hazard are the expected frequency and the intensity of a process. In the Swiss Guidelines for the consideration of avalanche hazard with respect to land-use planning, the needed quantities were fixed and have been applied with success.

1 INTRODUCTION

The Swiss Alps are densely populated by permanent inhabitants and - especially in winter season - by tourists. In an alpine region, e.g., the canton of Grisons, with a total surface of 7100 km^2, the permanent population of 182,000 persons is increased in wintertime by 230,000 tourists. Tourism is one of the most important sources of income; facilities such as dwelling houses, hotels, roads and railroads, power lines, ropeways, ski-lifts, etc., have to be maintained to guarantee continuous and safe functioning. The problem is, however, that in mountains potential damages always exist, not only from snow avalanches but also from debris flows, rock falls, landslides, unstable terrain and floods. One may say that no other country has as large a portion of its roads and population threatened by these potentially destructive forces. In Grisons, from 1983 to 1994, average annual damage to buildings by natural hazards amounted to 6.6 million Swiss Francs. After floods, snow avalanches are a major cause (2 million SFr during the same years). Fatalities, injured persons and infrastructural damages are not included in these figures. In Switzerland, on average, 26 persons are killed each year by avalanches. The equivalent "value" placed on a fatality is 1.1 million SFr. However, average values for avalanche hazards are somewhat misleading, because after a long period with no events suddenly a catastrophe occurs. During the extreme winter of 1950/51 total damage in Switzerland amounted to 112 million SFr, including buildings, interior equipment and contents and 98 fatalities (inflation since then not included).

2 JURIDICAL FUNDAMENTALS

The purpose of hazard maps is to consider avalanche danger with respect to land-use planning. Besides of that they are useful instruments for planning avalanche defenses. The principle is that building outside of safe zones shall not be permitted. If the hazard map is not taken into consideration, the Swiss Confederation does not subsidize defense works.

Several legal bases exist in Switzerland to prevent building in hazardous zones. The most general is the so-called "general police clause" which states the right and duty of the communal authority to exercise local police power to guarantee the safety of its inhabitants. The Executive Ordinance of the Federal Forest Law (1992) urges the Cantons to take measures against natural hazards and determines financial support by the Confederation. Finally, the Federal Law for Land-use Planning (1979) requests that building activities shall be restricted to safe areas.

On the basis of the above regulations, the cantons are authorized for the release of restrictions imposed in the public interest.

Two additional remarks have to be made. First, to assure a uniform assessment of danger, the Confederation issues technical documents (e.g. Guidelines by SFISAR). Second, a building prohibition - mostly involving a financial loss for the land owner - never obliges the state for a financial compensation, because natural hazards are considered as a force majeure not originating from human beings.

3 GENERAL HAZARD MAPPING

For avalanches, hazard mapping started as early as 1953 mainly due to the disastrous avalanche winter of 1950/51 the worst in at least 100 years. Afterwards, especially because of the general police clause and the first issue of the Executive Ordinance to the Federal Forest Law in 1965, practically no building in avalanche-prone terrain was legally possible.

Since the fifties the technique of mapping was gradually improved especially through publication of Guidelines for the consideration of avalanche danger for land-use planning (1984 and its 1975 forerunner) and calculation of flowing avalanches (Salm et al. 1990 with different forerunners, e.g., Sommerhalder 1966).

Recently - mainly due to the Federal Forest Law (1992) - comprehensive hazard mapping became a task of first priority in Switzerland. Because avalanche hazard mapping has a relatively long tradition where a lot of practical knowledge could be gathered, the principles for assessment of all other natural hazards were taken from it.

The two quantitites which define danger are
- the expected *frequency*, expressed by return period, and
- the *intensity* of a process.
The considered *return periods* range

between 30 and 300 years, the latter return period being the most extreme phenomenon to be taken into account. Frequent events are considered as dangerous, independent on intensity. So, for instance, even small snow avalanches may endanger people staying in the open.

For *intensity* different quantities are taken for different dangers (Temporary Working Group on Natural Hazards, 1995):
- for snow avalanches, the impact pressure on structures,
- for floods and debris flow, the speed and flow depth
- for rockfalls and landslides, the kinetic energy, and
- for unstable terrain, the depth of the main shear zone in terrain and the speed of displacement.

Hence, for establishing a map one needs dynamical quantitites and consequently a model for calculation. This is described for avalanches in the following sections.

4 DANGER, DAMAGE AND RISK OF AVALANCHES

So far only hazard - or synonymous danger - has been mentioned. The ultimate goal of mapping is however to determine the *risk*. We define risk as the product of the probabilities of danger, damage and presence. The fracture probability of a snow mass on a slope and the subsequent movement until standstill is called *danger* or, more generally a *dangerous process*. Obviously it is a prerequisite that such processes are statistically representative. Secular events occurring in time intervals of several millenia (e.g., huge landslides in geological time scales) are excepted. The *probability of damage* is the possibility that a structure is damaged or destroyed. This depends on the design of the structure (concrete, wooden structures, etc.). For houses in a settlement the *probability of presence* is always one, whereas for persons staying in the open or for motor vehicles it can vary considerably.

The above defines the *real risk* which doesn't describe the risk problem completely. An important, more subjective and irrational component, *aversion,* has to be introduced (Bohnenblust et al, 1987). First of all, one accident with 100 fatalities is perceived more seriously than 100 accidents each with one fatality. Secondly, there seems to be an influence of the type of accident. Accidents due to avalanches are less accepted by the public because they originate from an involuntary accepted danger of nature. In this context the publicity of an avalanche accident may be compared with that of a traffic accident each with one fatality. The real risk multiplied by the so-called *aversion factor* yields the *perceived risk*, which has finally to be introduced into safety planning. Basler and Partner (1985) proposed for avalanches with up to 8 fatalities a factor of 1, from 8 to 60 fatalities a factor of 3 and for more than 60 fatalities, 10. Of course this estimate is a very subjective one and may alter with changing times and with mentality.

A rough estimate shows that the probability of one fatality for a inhabitant of a settlement endangered by avalanches is about 100 times smaller than that of a car driving fatality in the same unit of time. Nevertheless, an avalanche accident is less accepted by the public. This leads to the final problem of the *tolerated risk*, which has to to be compared with the perceived risk. A tolerance of zero can never be required. For example supporting structures, which prevent fracture of unstable snowpacks, reduce the hazard probability considerably, but for snow-mechanical reasons, never to zero.

In conclusion the elaboration of a detailed risk map is difficult although possible. It becomes necessary to calculate benefit-cost ratios for avalanche defense structures. Benefit is taken as the monetary amount of prevented damage, which is much more difficult to determine than the costs of such structures (Altwegg, 1989 and Wilhelm, 1996).

First of all *avalanche hazard and danger maps* are developed by experts, solely based on avalanche dynamics and snow science. Any other aspects are not allowed to be taken into consideration. For instance land owners are mostly interested in using their land for buildings. If an analysis shows a potential hazard for the property, the owner often pretends that in "living memory" no avalanches have been observed on that site. However, the memory of an expert has to be very long, as long as 300 years!

To transform this hazard map into *danger zones* - which are legally binding for land users - risk has to be taken into account in a simplified way (see section 6).

5 AVALANCHE HAZARD MAPS

5.1 Avalanche hazard

5.1.1 Fracture zone
The probability of avalanche formation is given by the following three factors:

Avalanche terrain represents the factor independent of time. Fracture is possible on slopes between 30° and 50°. Generally, the gentler a slope, the larger the released mass, and the more disastrous an avalanche. This is because more overburden snow is necessary to reach shear strength than on steep slopes.

The second factor is given by the *snow conditions* (amount of new snow, temperature, wind, structure of the snowpack) which depend on the development of weather.

The last factor is the *local climate*, remaining unchanged over long time (regional precipitation, prevailing winds and orientation of a slope related to sun-radiation and wind direction).

For hazard assessment we have to consider long intervals of time within which certain extreme situations have to be fixed quantitatively. *Fracture depth* d_0 (thickness of the released snow) here plays a decisive role for the dynamics of potential avalanches and for the determination of potential hazards, runout distances are approximately proportional to it (see paragraph 5.2). According to experience fracture depth is assumed equal to the increase in total snow depth within 3 consecutive days. The necessary data originate from SFISAR observations over 20 to 60 years. With statistics of extremes, the data are then extrapolated to selected return periods (see paragraph 5.2). To get the real fracture depths, two modifications are necessary. First, the observed regional values of the network (valid for about 100 km²) have to be transformed into local ones (an area of about 1 km²), especially taking into consideration wind transported snow. Second, the slope angle restricts the possible amount of deposited new snow because of a given shear strength, so that the real fracture depth decreases with increasing slope (e.g., a reduction of one half for a 45° slope).

Besides fracture depth, the *areal extent* of a fractured zone has to be known. Basically it is given by climatic conditions and topography. Observations showed that this area is often much less than the sum of all slopes inclined more than 30°. A dependence of this area on the return period has to be expected.

5.1.2 Avalanche dynamics

At present, reliable calculation methods only exist for flowing avalanches - those moving close to the ground with high densities - the most frequent and violent ones. To get the same standard for hazard assessment throughout the Swiss Alps we have Guidelines for calculation of flowing avalanches (Salm et al. 1990). The goal of calculations is the determination of avalanche speed, flow depth and runout distance in flat terrain as a function of prescribed return periods. In doing so, two mechanical parameters of the moving snow ("dry" and "turbulent" friction, given by parameters μ and ξ) have to be selected carefully.

It has to be pointed out that calculations are only a part of hazard assessment; inspection of the terrain by experts and consultation of the avalanche cadastre (inventory of all observed events) are mandatory.

5.2 Grades of hazard and probabilities

Commonly 3 grades of hazard are used in maps represented by colours. High hazard (red), moderate hazard (blue) and safe (white). Sometimes a low grade (yellow) is added on for powder snow avalanches, exerting relatively low pressures, or for flowing avalanches with return periods of more than 300 years. Avalanches with a return period T up to 30 years are high hazard independent of pressure. For less frequent events with $30 < T < 300$ years a blue zone is applied up to a pressure of 30 kN/m².

Figure 1 shows the hazard map of the so-called Dorfbach-avalanche in Davos.

Hazard grades are statistical quantities, therefore it is generally impossible to distinguish between absolutely safe and endangered terrain. Probabilities are of great importance in avalanche dynamics. The way statistics is introduced is demonstrated as follows:

It can easily be shown (Salm et al. 1990) that for unchanneled avalanches the runout distance is

$$s = d_0 \, f_{(\mu, \, \xi, \, \psi, \, B)}, \qquad (1)$$

a linear function of fracture depth d_0. Factor f depends on dynamical parameters μ, ξ and on topographical characteristics such as slope angle ψ and width of stream B, with all quantities assumed to be independent of the return period.

The regional values of d_0 depend on T as follows:
The cumulative probability F is (Gumbel, 1958)

$$F = \exp[-\exp(-y)] \qquad (2)$$

where   (3)

$$y = \frac{d_0 - a}{b}$$

with a and b being constants determined by observations in our network.

The return period yields   (4)

$$T = \frac{1}{1-F}$$

which becomes for large periods (F → 1)

$$T = \exp(y) \qquad (5)$$

The density of probability f is generally

$$f = F \exp(-y) \qquad (6)$$

and

$$f = \exp(-y) \qquad (7)$$

for large periods.

In Figure 2 an example is given for a = 76.5 cm; b = 29.4 cm and $f_{(\mu, \, \xi, \, \psi, \, B)} = 200$.

Settlements are endangered by medium to large avalanches which begin with return periods and fracture depths of about 10 years and 1 m respectively.

For periods of several centuries the increase of s is very small:

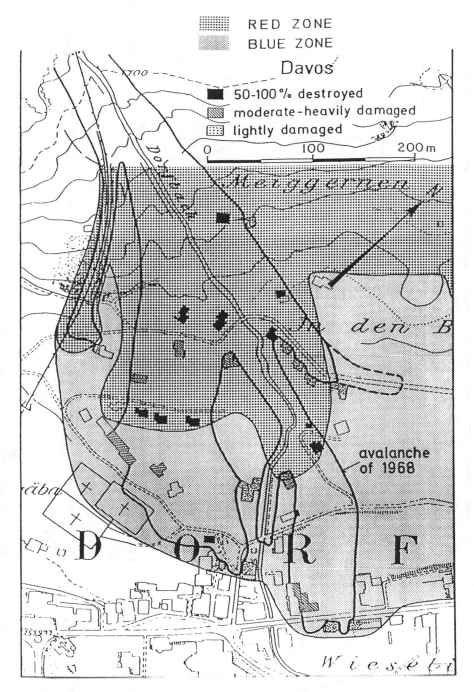

Fig.1 Avalanche hazard map 1991, Dorfbach-avalanche Davos.
Solid lines represent the disastrous avalanche of
27 January, 1968

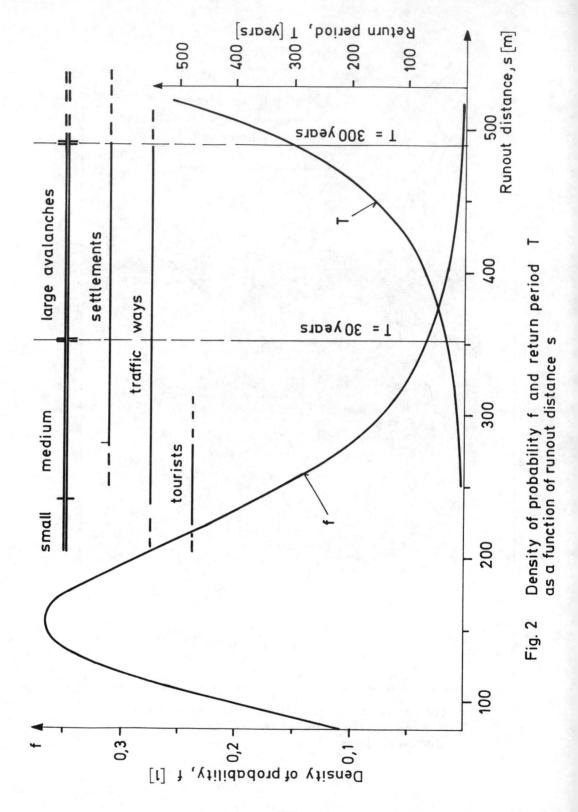

Fig. 2   Density of probability f and return period T
          as a function of runout distance s

T [years]	s [m]
30	350
300	490
500	520
1000	560

Due to the limited data normally available, the determination of a and b in (3) becomes very inaccurate for T > 300 y, i.e., the confidence-interval grows very large and the reliability of the procedure may be exceeded.

Avalanches with a maximum f represent small ones with T on the order of 1 year. They affect tourists and are not considered in our context. Furthermore, the runout of such small masses would be much shorter than in Fig. 1 because of the enlarged μ.

It has been demonstrated that dynamic models not taking probabilities into account would be worthless for hazard mapping.

6 AVALANCHE DANGER ZONES

The transformation of hazard maps into legally binding *danger zones* is dependent on the tolerated risk. As this risk is a rather subjective one, the transformation is principally a political affair. In Switzerland the communities and cantons are responsible for the determination. They are theoretically free to fix their own scales.

In spite of that, the Federal Guidelines (1984) recommend maximum tolerable risks for different grades of hazard. With the limitation to consider events only up to a return period of 300 years, a generally tolerable risk - those with larger periods - is fixed. In practice these recommendations are now widely accepted by the political authorities of communities and cantons.

Principally in red zones building is prohibited. In blue zones building is possible, but with caution and provided that certain safety specifications are met (e.g. reinforcement of walls according to extreme avalanche pressures or the organization of a warning system to evacuate endangered people). The prohibition of hotels, schools or establishments for winter sports (e.g., ice-rinks) reduce the possible extent of damage. Generally, all unusually large assemblages of people have to be avoided.

Serious problems arise when building-sites in a community are rare. The laws do not permit subsidized structural avalanche defenses with the purpose of gaining such sites.

REFERENCES

Altwegg, D. 1989. Die Folgekosten von Waldschäden. Forstwiss. Beiträge des FB Forstökonomie und Forstpolitik, ETH Zürich 1989/8. Unpublished PhD thesis Hochschule St. Gallen.

Basler and Partner 1985. Sicherheitsbeurteilung "Naxberg-Göschenen". SANASIVA-Teilprogramm 8. Zürich, Unpublished consultation, 1985.

Bohnenblust, H.; Troxler, C. 1987. Risk analysis - Is it a useful tool for the politician in making decisions on avalanche safety? Avalanche Formation, Movement and Effects. IAHS Publ. No. 162, 1987.

Calculation of flowing avalanches, Salm et al. 1990. Original title: Salm, B.; Burkard, A.; Gubler, H.U. Berechnung von Fliesslawinen. Eine Anleitung für Praktiker mit Beispielen. Mitt. SLF 47, 1990.

Executive Ordinance of the Federal Forest Law 1992. Original title: Verordnung über den Wald, 30. November 1992.

Federal Law for Land-use Planning 1979. Original title: Bundesgesetz über die Raumplanung, 22. Juni 1979.

Guidelines for the consideration of avalanche danger for land-use planning 1984. Original title: BFL und SLF (Hrsg.). Richtlinien zur Berücksichtigung der Lawinengefahr bei raumwirksamen Tätigkeiten. Bern, EDMZ, 1984.

Gumbel, E.J. 1958. Statistics of Extremes, Columbia University Press, New York.

Sommerhalder, E. 1996. Lawinenkräfte und Objektschutz. Schnee und Lawinen in den Schweizer Alpen Winter 1964/65. Winterbericht SLF Weissfluhjoch/Davos. EDMZ, 3003 Bern.

Temporary working group on natural hazards 1995. Published by Kienzholz, H. 1995. Gefahrenbeurteilung und bewertung - auf dem Weg zu einem Gesamtkonzept. *Schweiz. Zeitschrift für Forstwesen*, 146, 1995, 9: 701-725.

Wilhelm, Ch. 1996. Wirtschaftlichkeit im Lawinenschutz. Unpublished PhD thesis ETH Zürich.

Snow Engineering: Recent Advances, Izumi, Nakamura & Sack (eds) © 1997 Balkema, Rotterdam. ISBN 90 5410 865 7

A snow observation network for a mountainous area

Hideomi Nakamura & Masujiro Shimizu
Nagaoka Institute of Snow and Ice Studies,
NIED, Niigata, Japan

Osamu Abe
Shinjo Branch of Snow and Ices Studies,
NIED, Yamagata, Japan

Tadashi Kimura
Niigata Electric Company, Nagaoka, Japan

Masayoshi Nakawo
Nagoya University, Japan

Tsutomu Nakamura
Iwate University, Morioka, Japan

ABSTRACT: The National Research Institute for Earth Science and Disaster Prevention (NIED) constructed a network of snow observational stations in the snowy country of Japan to investigate annual variations in the snow cover. The network consists of seven pairs of snow observation stations, with each pair of stations consisting of a mountain snow station and a lower altitude snow station at its foot. Meteorological data such as total depth and weight of snow cover, air temperature and solar radiation, etc., are obtained at the each station. From the data measured at Niigata prefecture over six years, it was found that the annual maximum depth of snow cover at the lower altitude stations decreased due to warm winter during the past several years, but the depth of snow cover did not decrease at the mountain stations. (i.e., recent warm winters have not influenced the volume of mountain snow yet).

1 INTRODUCTION

It is very important to monitor mountain snow variations in order to determine if countermeasures to maintain water resources will be needed in the future. But it is almost impossible to get snow data in high mountains, because no stations for snow observation were located there. For this reason, it was necessary to install snow stations in high mountains.

Figure 1. Location of snow stations.

2 AUTOMATIC STATIONS

The National Research Institute for Earth Science and Disaster Prevention (NIED) started to construct a network of snow observational stations in 1989 to investigate annual and spatial variations of the amount of snow cover in Japan. This network is made up of seven pairs of snow observation stations, and each pair of the stations consists of a mountain snow station and a snow station at the foot of the mountain. One pair of snow stations was constructed year by year for seven years. The pairs of snow stations were located at seven places in snowy areas in Japan, as shown in Figure 1. and Table 1. The Okutadami station is shown in Figure 2.

3 MEASUREMENT ITEMS AND DATA OBTAINED

As shown in Table 2, meteorological data such as total depth and weight of snow cover, air temperature, solar radiation, etc., are measured in each station as a rule. The total depth of snow cover is measured mainly by infrared beam snow depth meters, and the snow load is measured by metal wafer meters.

The data measured at nine stations of the fourteen stations can be monitored by telephone, but telephone service is not available at the other stations, so we must go and collect the data recorded by data recorder. Figure 3 shows an example of data obtained.

Table 1. Location of snow stations.

Station	Location		
	Latitude(North)	Longitude(East)	Altitude
Tokachidake	43° 29′ 16″	142° 36′ 17″	520 m
Biei	43° 36′ 13″	142° 29′ 59″	250 m
Iwakisan	40° 39′ 17″	140° 17′ 35″	1238 m
Fujisaki	40° 39′ 08″	140° 29′ 13″	20 m
Gassan	38° 29′ 29″	139° 59′ 56″	710 m
Shinjo	38° 47′ 22″	140° 18′ 46″	127 m
Okutadami	37° 19′ 21″	139° 13′ 35″	1205 m
Nagaoka	37° 25′ 22″	138° 53′ 24″	97 m
Myoko	36° 51′ 55″	138° 04′ 54″	1310 m
Tateyama	36° 34′ 30″	137° 36′ 37″	2450 m
Hakusan	36° 10′ 44″	136° 38′ 20″	825 m
Oguchi	37° 25′ 22″	138° 53′ 24″	97 m
Daisen	35° 20′ 11″	133° 35′ 05″	875 m
Mizoguchi	35° 17′ 49″	133° 24′ 30″	155 m

Sometimes data were not obtained due to machine trouble, etc.. Sensor maintenance is difficult.

4 SNOW DEPTH VARIATION

Annual maximum snow depths at the Okutadami and Nagaoka station are shown in Figure 4. Mean air temperature and total precipitation during December and February of each year at Nagaoka station are also plotted in Figure 4. It is found from the figure that the maximum snow depth at Nagaoka station inverts the phase of mean air temperature as reported by

Figure 2. Okutadami snow station.

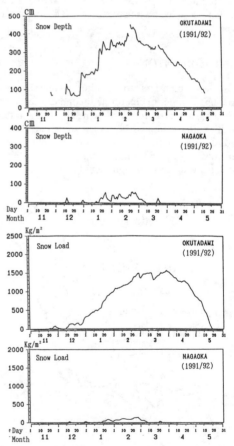

Figure 3. Snow depth and snow load measured at Okutadami and Nagaoka station during the 1991/1992 winter.

Table 2. Measurement items at snow stations. Circles indicate measurements being obtained.

Station	Measurement Items							
	Snow depth	Snow load	Air tem-perature	Soil tem-perature	Global radiation	Albedo	Wind Speed, direction	Precipi-tation
Tokachidake	O	O	O	O	O			
Biei		O	O	O	O			
Iwakisan		O	O	O	O			
Fujisaki	O	O	O	O	O			
Gassan	O	O	O		O			
Shinjo	O	O	O		O		O	O
Okutadami	O	O	O		O	O	O	
Nagaoka	O	O	O	O	O	O	O	O
Myoko	O	O	O			O		O
Tateyama	O		O				O	
Hakusan	O	O	O		O			O
Oguchi	O							
Daisen	O	O	O			O		O
Mizoguchi	O		O		O	O		O

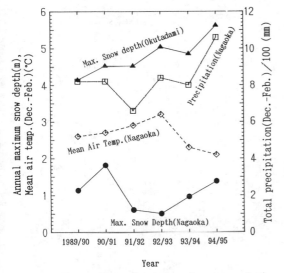

Fig.4. Time variation of annual maximum snow depth, mean air temperature and December-February precipitation.

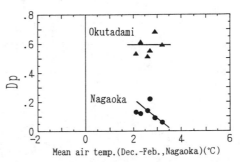

Fig.5. Dependency of maximum snow depth on mean air temperature in Nagaoka.

T.Nakamura and M.Shimizu (1996). On the other hand, the maximum snow depth of Okutadami station seems not to suffer the effect of air temperature variation of Nagaoka station. It seems rather to depend upon the precipitation variation.

Relationships between the mean air temperature at Nagaoka station and the maximum snow depth (in cm) at each station divided by the precipitation (in mm) of Nagaoka station, (Dp), are shown in Figure 5. From the figure, it is clear that Dp at Okutadami is almost independent of the mean temperature at Nagaoka station. At Nagaoka station, however, Dp decreases with the air temperature. The independency of the maximum snow depth at Okutadami station upon the mean air temperature at Nagaoka station is considered due to the elevation difference between Okutadami and Nagaoka stations. This suggests that the effect of recent warm winters have not reduced snow accumulation in the mountains.

REFERENCE

Nakamura,T. & Shimizu,M. 1996.Variation of snow, winter precipitation and winter air temperature during the last century at Nagaoka, Japan. J.Glaciology. Vol.42,No.140:136-140.

Snow Engineering: Recent Advances, Izumi, Nakamura & Sack (eds) © 1997 Balkema, Rotterdam. ISBN 90 5410 865 7

Water balance in a heavy snow region

So Kazama
Institute of Engineering Mechanics, University of Tsukuba, Japan

Masaki Sawamoto
Department of Civil Engineering, Tohoku University, Sendai, Japan

ABSTRACT: The total water balance is discussed in a heavy snow basin. A method for snowfall estimation based on precipitation, the Thornthwaite method of considering evapotranspiration and a simple ground water balance model are used to evaluate each component of the hydrological cycle.

1 INTRODUCTION

Japan was heavy snow district where the maximum snow depth is greater than 5 m. In such a heavy snow region, snow is an important water resource because of its storage nature. Although it is important to evaluate the correct snow volume for water resources planning, snow volume and snowfall in the mountainous areas have never been understood because of the lack of observation points. If many observing stations existed in a basin, it would be easy to analyze various phenomena. However the amount of hydrological data is usually limited except in a few small test basins. Over the past few years a considerable number of studies have been made on water balance in some basins. For example, Hudson and Gilman (1993) investigated long-term variations and Shimotsu and Iwanaga (1993) studied a volcanic basin. The balance with snow was researched on a grassland in New Zealand by Cambell and Murray (1990), and in mountainous Alpine area by Kattelman and Elder (1991). These studies give interesting information on water balance in various types of basins.

In this paper, water balance in a heavy snow region is discussed from the evaluation of various hydrological process using limited data. The results show every hydrological element of the water balance with heavy snow in the case of the Tadami River basin.

2 STUDY BASIN

The basin studied in this paper is the Taki Dam catchment area of 1991 km^2 which is the most upstream portion of the Tadami River basin (Figure 1). The lowest elevation in the basin is at the Tadami Dam site, 335 m, and the highest is at Mt. Hiuchigatake, 2346 m. The basin has two large subbasins. One is the main Tadami River basin which is very steep and located west of the Taki Dam basin and the other is the Ina River basin which is flatter and the east. Ninety-eight percent area of the basin is covered with deciduous forest and the remainder is for human activity. There are five dams in the basin where daily discharge is measured and four AMeDAS(Automated Meteorological Data Acquisition System) observation stations. Three of them are at dam sites and the fourth is at Hinoemata points. In addition, two robot rain gauges work during the summer. Some tunnels divert water between small subbasins for utilization by hydro electric power stations(HPS).

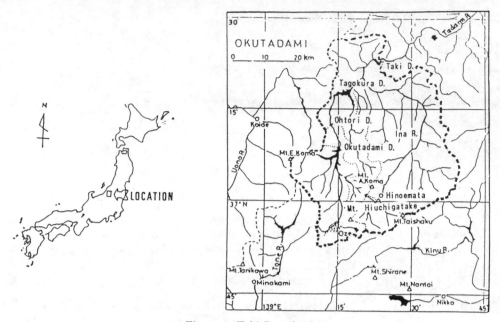

Figure 1: Taki Dam basin

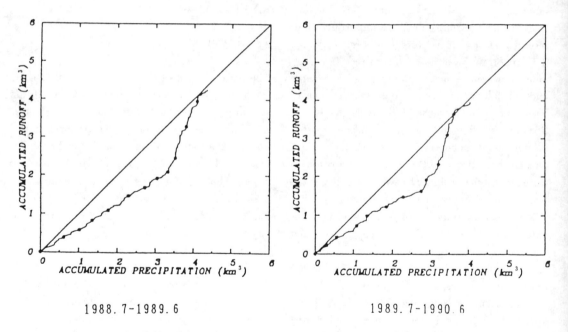

1988. 7-1989. 6   1989. 7-1990. 6

Figure 2: The relationship between runoff and precipitation

3 ESTIMATION OF BASIN MEAN PRECIPITATION

At first, it is shown that observed precipitation data cannot represent basin wide precipitation. A method for estimation of precipitation on the mountain is proposed Runoff from the basin is given by

$$R = \sum_{i=1}^{4}(DamInflow - DamOutflow)_i$$
$$+TakiDamInflow$$
$$+InaHPSInflow, \qquad (1)$$

where i represents the upper four dams. The total runoff from the basin is compared with the total observed precipitation. The result is shown in Figure 2, in which runoff is estimated by the above method and total precipitation is calculated by simple multiplication of each precipitation and its catchment area at four observing stations. Figure 2 concludes that there is no loss in the basin in a year. This result is obviously unrealistic. The reason is that the estimation of basin mean precipitation is not appropriate.

Although it is well known that snowfall increases as elevation increases, little attention has been paid to the distribution of rainfall in a basin. The relationship between elevation and rainfall measured at four observing stations and two robot rain gauges, which work from July to September, is indicated in Figure 3. No correlation is found between rainfall and elevation. Therefore it can be concluded that no systematic data correction is needed for rainfall. On the other hand, snowfall variation due to elevation has to be taken into consideration. The correction used are as follows;

1. Regressed equation between snow depth and elevation is derived from the results of snow surveys by EPDC(the Electric Power Development Company) at about 25 points during snowmelt season.

2. This equation and the area-elevation distribution make it possible to estimate total snow volume in the basin.

3. The mean snow depth in the basin is determined.

4. Correction parameters are calculated from comparisons of the mean snow depth with snow depths at observing stations.

5. Corrected snowfall data is developed by multiplying snowfall at each observing station by the parameter. Here snow density is assumed constant at $0.45(g/cm^3)$.

4 EVAPOTRANSPIRATION

Estimation of evapotranspiration in a basin is usually calculated using the Thornthwaite method, the Hammon method or the Penman method. In this study, the Thornthwaite method is utilized because it was confirmed that this method gives good results on water balance in the snow-free season (Kazama and Sawamoto 1994). Evapotranspiration by the Thornthwaite method is

$$E = 16Do_i(\frac{10T_i}{I})^a, \qquad (2)$$

where

$$a = (0.675I^3 - 77.1I^2 + 17920I + 492390)$$
$$\times 10^{-6}, \qquad (3)$$
$$I = \sum_{i=1}^{12}(\frac{T_i}{5})^{1.514}. \qquad (4)$$

T_i is the monthly mean temperature in cerious degree, the unit of E is mm/month and Do_i is a coefficient to correct for sunshine duration as a function of latitude. This equation gives monthly evapotranspiration. The daily evapotranspiration used in this study is simply transferred from the monthly value by diving E_i by the number of days in the month. The mean temperature T_i in the basin is calculated by multiplying the temperature gradient by the difference between the mean elevation of the basin and the elevation at an observing point.

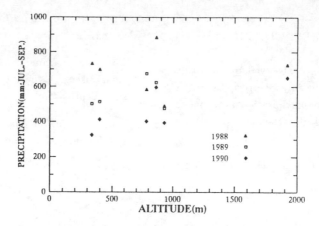

Figure 3: The relationship between rain and elevation in summer

5 ON WATER BALANCE

Figure 4 indicates the relationship between accumulated precipitation and accumulated runoff from August to July obtained by the above mentioned method. From summer to fall, precipitation and "runoff + evapotranspiration" are even balanced. The runoff ratio is almost constant. Runoff ratio decreases from November and becomes minimum in winter. The difference between a line of "runoff" : "precipitation" = 1:1 and a line of "runoff + evapotranspiration" represents snow storage volume in winter. The runoff from the season is supplied from only ground water. This suggests that a simple ground water balance model is applicable; i.e., ground water supplies constant runoff during the snowfall season and is recharged from snowmelt water in the snowmelt season. This ground water balance concept is shown schematically in Figure 5.

From the beginning of March the runoff ratio becomes large due to snowmelt and it continues to the last of June. The runoff volume of snow for this four months is almost equivalent to 50% of the yearly precipitation.

Those results of water balance show 10% error in a year. Some reasons can be considered; the snowfall estimation is still insufficient, snow density is assumed 0.45 but it seems that the density is

lower in the mountains and evapotranspiration and sublimation from a snow surface may not be negligible. Also, the locations of observation points are not so good.

6 DISCUSSION AND CONCLUSION

Shinohara (1965) investigated the distribution of runoff ratio in Japan and reported that his calculation gives unrealistic results in heavy snowfall regions where total runoff becomes greater than precipitation. This is because the estimation of snowfall is difficult. The method presented here can gives a certain solution to this problem because snowfall data are corrected by using a snow depth - elevation relation in mountainous regions. Yearly height of precipitation, runoff, evapotranspiration and snow volume for three years are indicated in Figure 6. Runoff ratio is about 0.70 in this figure. This value is almost the same as that of typical Japanese rivers but it seems that the real runoff ratio is larger than this because accumulated runoff **is estimated 10% less than precipitation on water balance. The reasons for the large runoff ratio are that runoff velocity is large due to steepness of the river and that there is much precipitation in the basin.**

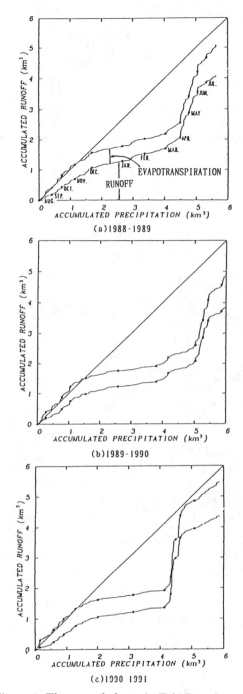

Figure 4: The water balance in Taki Dam basin

The following are concluded.

1. A simple water balance equation for ground water is suggested.

2. Correction for snowfall data is proposed.

3. Validity of the Thornthwaite method is confirmed.

These methods are effective to estimate water balance in large area experiencing snow.

ACKNOWLEDGMENT

The authors are thankful to the Electric Power Development Company and the Tohoku Electric Company for their assistance and permission to use their data. We wish to thank the Foundation of River and Basin Integrated Communications and the Foundation of River and Watershed Environment Management for their generous financial assistance.

REFERENCE

Hudson, J.A. & Gilman 1993. Long-term variability in the water balance of the Plynlimon catchments, *J. Hydrology*, No.143.

Shimotsu, M. & Iwanaga, A. 1993. On the change of hydrological cycle in river basins at the western foot of Aso Volcano, *J. Japan Soc. Hydrol. & Water Resour.*, Vol.6, No.2(Japanese).

Cambell, D.L. & Murray,D.L. 1990. Water balance of snow tussock grassland in New Nealand, *J. Hydrology*, No.118.

Kattelman, R. & Elder,K. 1991. Hydrologic characteristics and water balance of an Alpine basin the Sierra Nevada, *Water Reour. Research*, Vol. 27, No.7.

Kazama, S & Sawamoto, M. 1994. On water balance in a basin with heavy snow area, *Proc. of Hydraulic Eng. JSCE*, Vol.38(Japanese).

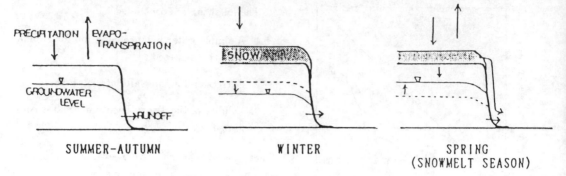

Figure 5: Ground water balance model

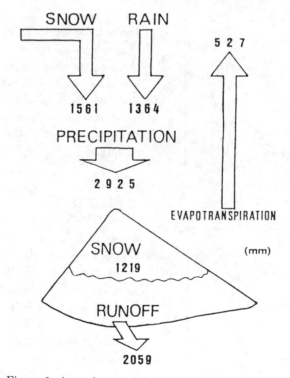

Figure 6: Annual water balance in the Taki Dam basin

Snow Engineering: Recent Advances, Izumi, Nakamura & Sack (eds) © 1997 Balkema, Rotterdam. ISBN 90 5410 865 7

Remaining snow patterns in mountains, Yukigata, as a scene in spring

Yasuaki Nohguchi, Katsuhisa Kawashima, Toshiichi Kobayashi & Yutaka Yamada
Nagaoka Institute of Snow and Ice Studies, NIED, Niigata, Japan

Yasoichi Endo
Tokamachi Experiment Station, FFPRI, Niigata, Japan

Kaoru Izumi
Research Institute for Hazards in Snowy Areas, Niigata University, Japan

ABSTRACT:Remaining snow surrounded by a ground surface and a ground surface surrounded by remaining snow are called *yukigata* in Japan. Some of famous yukigatas have a name and oral literature. There are more than 300 famous yukigatas. The yukigatas have been used mainly as a kind of calendar for agriculture in most of the snowy regions in Japan since long ago. Recently, however, the yukigatas are being forgotten because the primitive ways for agriculture are disappearing with the progress of agricultural technology. On the other hand, the formation of yukigata is governed by the topography of the mountains and snow accumulation and snowmelt. Therefore, yukigatas have information on topographical features, distribution of mountain snow and climate change. Moreover, it is possible to use yukigata as sightseeing resources or material for nature education.

1 INTRODUCTION

When a seasonal snowcover disappears on a plain and a snow line of seasonal snowcover retreats on a mountain in spring, a complex pattern of white domains covered with remaining snow and dark domains of the ground surface appears on a mountainside. Some of the white domains of remaining snow and the dark domains of ground surface are likened to men, animals, tools, etc., and are called *yukigata* in Japanese, which means snow form.

The yukigatas have been used mainly as a kind of calendar for agriculture in most of snowy regions in Japan for many years. Recently, however, the yukigatas are being forgotten because the primitive ways for agriculture are disappearing with the progress of agricultural technology.

For a city near mountains in snowy regions, yukigatas are one of the most impressive scenes in spring or early summer. They are also a kind of cultural inheritance in Japan. In this paper, the yukigatas in Japan will be introduced, and the present-day significance or use of yukigata will be reported.

2 POSITIVE TYPE AND NEGATIVE TYPE

First, we define both the remaining snow surrounded by ground surface and the ground surface surrounded by remaining snow yukigata. Then, the yukigatas are classified into two types; positive and negative, as shown in Figures 1 and 2. In a positive yukigata, a white domain of remaining snow is the figure, and the surrounding dark ground is the background. In a negative yukigata, the dark ground is the figure, and the white snow is the background.

Figure 1. A positive yukigata on Mt. Nokogirisan. It is a Chinese character representing a river.

Figure 2. The negative "Hane-uma"(a jumping horse) yukigata, of Mt.Kannasan. Where is the jumping horse?

Figure 3. Complicated yukigatas in Mt. Awaga-take. ①:"Mamemaki-dori" (a bird) , ②:"Kago-gata"(a basket), ③:"Ushi-gata"(a cow), ④:"Awamaki-baba"(an old woman), ⑤:"Noteware"(straws).

Figure 3 shows a special example of complicated yukigatas in which all of neighboring 3 white domains and 2 dark domains are considered as yukigatas. In this case, a yukigata is a kind of a Rorschach test in psychology (Izumi 1995).

3 MECHANISM OF YUKIGATA FORMATION

The mechanism of yukigata formation is simple, though the pattern is complex. If both accumulation and ablation of snow were spatially uniform, the region covered with snow would change uniformly from white to dark, and yukigatas would not appear on the mountainside. A heterogeneous distribution of snow depth is necessary for the formation of yukigatas.

The heterogeniety depends on the topography of the mountain, snow accumulation and snow melt. Of these three factors, the effect of topographical features on the shape of each yukigata is the most predominant. Vegetation is one of the topographical effects. The complexity of many yukigata patterns originates from that of topographical features with vegetation.

Snow plays a role to visualize the complexity of topographical feature on mountainside (Nohguchi 1995).

Snow accumulation and snow melt affect when yukigatas appear and disappear. Snow accumulation is governed by snowfall, snowdrifting and snow avalanching relating to topographical features.

4 DISTRIBUTION OF YUKIGATA WITH A NAME OR ORAL LITERATURE IN JAPAN

The number of yukigatas is infinite, but only a few yukigatas have a name or oral literature passed from

Table 1. Distribution of famous Yukigata in Japan (Tabuchi 1981).

Prefecture	Positive	Negative	Unknown	Total
Hokkaido	2	0	1	3
Aomori	21	3	14	38
Iwate	12	4	4	20
Akita	4	4	7	15
Miyagi	2	0	2	4
Yamagata	12	1	21	34
Fukushima	9	3	15	27
Niigata	18	10	57	85
Gunma	1	0	2	3
Nagano	22	34	1	57
Yamanashi	2	2	0	4
Toyama	6	4	4	14
Gifu	1	0	0	1
Ishikawa	0	0	1	1
Shizuoka	0	1	1	2
Hyougo	1	0	0	1
Ehime	2	0	0	2
Total	115	66	130	311

generation to generation.

Some conditions for a yukigata to have a name or oral literature are that the yukigata can be seen by many people, and that the yukigata can be recognized as a common shape by them. For these, it is necessary that many people have lived near the mountain for many years. Moreover, it is important for the yukigata to relate to a calender foragriculture or ancient religious sentiments.

The number of famous yukigatas with oral literature in Japan is more than 300 (Table 1, Tabuchi 1981), but the number of the yukigatas that can be confirmed now is not so large, because the collection of natural scientific data on the yukigatas have not been carried out for a long time.

Saito (1988) investigated yukigatas with oral literature only in Niigata prefecture, and reported 124 yukigatas, much more than those that reported by Tabuchi in Table 1. This shows that the number of yukigatas which are not famous, but have names and oral literature, must be incredibly large.

5 NEW YUKIGATA

Currently, most of yukigatas with names and oral literature are being fogotten without any confirmation because their classical utility as a calender of agriculture is no longer needed. Moreover, some

Figure 4. The yukigata "Yuki-usagi"(a snow hare) of Mt. Azumakohuji drawn on a lunch box .

Figure 5. A new positve yukigata, "Mai-hime"(an dancing queen) named by Tabuchi(1981).

yukigatas have vanished by development of ski areas. On the other hand, famous and beautiful yukigatas are sometimes being used as sightseeing resources (e.g., the rabbit in Figure 4).

Tabuchi(1981) introduced a new yukigata with no history which he named"Mai-hime"(a dancing queen in English)(Figure 5). This negative yukigata can be seen only by climbers. A present-day hobby is to find new beautiful yukigatas and to enjoy yukigata watching like bird watching.

6 SCIENTIFIC MEANINGS OF YUKIGATAS

Formation of yukigatas is governed by topography, snow accumulation and snow ablation. Therefore, yukigatas include some scientific informations.

Yukigatas sometimes directly relate to topographical features. Yamada(1995) found that a ring-shaped positive yukigata named "Nichirin" ("the orb of day" in

Figure 6. The "Nichirin"(the orb of day) yukigata of Shimizu.

English) is an outline of a region where a landslide occured (Figure 6). Thus some yukigatas can be used as simple indicators for mass movement, like a landslide, on a mountainside.

The time when a yukigata appears depends on total snowcover in the winter and warming in spring. Kawasima et al. (1995), using the positive yukigata in Figure 1, showed that the dates of its appearance and disappearance can be represented by an accumulated air temperature in spring and maximum water equivalent of snow. This is the reason why yukigatas have been used as a calendar of agriculture. And also this shows that the record of yearly variation in the dates of appearance and disappearance can represent climate change.

Snow distribution on mountains is very complex due to the complex topography. Yukigatas are good for estimating snow distribution on mountains. Endo (1995) estimated the distribution of snow depth on the mountainside of Mt. Hakkai-san where the yukigata "mamemaki-nyudo"(a shaved-head monster) appears (Figure 7) using a method of melting coefficients. In general information on mountain snow is limited because of its complexity and difficulty of observation, so the method of using yukigatas can be a good one.

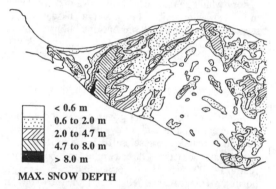

< 0.6 m
0.6 to 2.0 m
2.0 to 4.7 m
4.7 to 8.0 m
> 8.0 m

MAX. SNOW DEPTH

Figure 7. The "Mamemaki-nyudo"(a shaved-head monster) yukigata of Mt. Hakkai-san and the distribution of maximum snow depth.

7 CONCLUSIONS

In this paper we introduced the yukigatas of Japan. Yukigatas with a name and oral literature have been used as a calendar for agriculture, but unfortunately they are being mostly forgotten. For this study, we considered present-day significances of yukigatas and showed that they can provide information about topographical features, distribution of mountain snow and climate change. Moreover, it is possible to use yukigatas as sightseeing resources or for nature education.

In Japan, snow has special meanings in relation to understanding nature. Snow, moon and the flower "setsu-getsu-ka" are the key words to understand Japanese culture. Snow represents the changing of the seasons and the passage of time (Kurita 1987). Paradoxically, snow acts as the harbinger of emerging spring. In this sense, the concept of yukigata corresponds to that of snow.

REFERENCES

Endo, Y. 1995. "Mamemaki-nyuudou" in Mt. Hakkai-san and snow depth destribution (in Japanese). *J. Japanese Soc. of Snow and Ice*, 57, No.4, i.

Izumi, K. 1995. Complicated yukigata (in Japanese). *J. Japanese Soc. of Snow and Ice*, 57, No.4, i.

Kamashima, K. Kobayashi, T. and Nohguchi, Y. 1995. Yearly variation in appearance period of Yukigata "Kawanoji" seen at Nagaoka (in Japanese). *Proceedings of '95 Cold Region Technology Conference*, 264-268.

Kurita, I. 1987. Japanese Identity. *Shouden sha*, 191pp.

Nohguchi, Y. 1995. Spatial patterns of snow areas and no snow areas - snowline and Yukigata - (in Japanese). *Proceedings of '95 Cold Region Technology Conference*, 259-263.

Tabuchi, Y. 1981. Yukigata -a crest of mountain-(in Japanese). *Gakushuu kenkyuu sha.* 371 pp.

Saito, Y. 1988. Yukigatas in Niigata prefecture (in Japanese). History and Folk in Niigata pretesture, *skaiya tosho*, 284-304

Yamada, Y. 1995. On the topographical origin of the Yukigata "Nichirin" (in Japanese). *Proceedings of '95 Cold Region Technology Conference*, 269-275.

Snow Engineering: Recent Advances, Izumi, Nakamura & Sack (eds) © 1997 Balkema, Rotterdam. ISBN 90 5410 865 7

A case study on the utilization of snow and ice as natural cold energy source for low-temperature storage materials

Tetsu Suzuki & Shun'ichi Kobayashi
Niigata University, Japan

Katsutoshi Tsushima
Toyama University, Japan

Shude Shao & Yan Teng
Heilong Jiang Highway Bureau, People's Republic of China

Guozhong Liu
Harbin Highway Engineering Office, People's Republic of China

ABSTRACT: In 1995, the authors conducted low-temperature storage experiments of food, vegetables, etc., with snow throughout the year in a road tunnel which had been put aside as useless in Toyama Prefecture, Japan. Jiayin Town and Harbin City in the Heilong Jiang Province in China are situated in a zone of "seasonally frozen ground". The authors conducted experiments that made it clear that half-underground conservation and year-round utilization of natural ice is possible.

1 EXPERIMENTS ON THE CONSERVATION AND UTILIZATION OF SNOW IN A USED TUNNEL

1.1 *Tunnel and snow dikes for experiments*

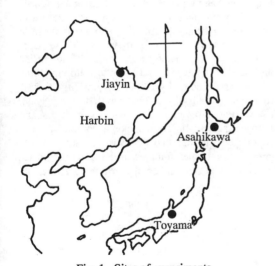

Fig. 1. Sites of experiments.

The road tunnel that was chosen for the experiments is located in a mountainous district in Toyama Prefecture, Japan (Fig.1), and had been put aside as useless. It is 4 m wide, 4.7 m high, 240 m long (east - west), and ~800 m above sea level. With plastic curtains it was separated into 6 rooms (Fig. 2). These rooms were called room 1, room 2, ... room 6 in order from east to west. At the end of April 1995, 91 m³, 180 m³ and 182 m³ of snow were transported into rooms 3-5, respectively, and snow dikes were made in these rooms. The snow dikes were covered with aluminum foil-coated sheets.

1.2 *Observation method*

From May to December, the height and width of the snow dikes were measured 9 times, and the amount of snow thawed was calculated. Temperature of every room was measured 1.5 m above the floor, and the temperature under the sheet in room 4 was measured. At fifteen o'clock on August 1, 1995, the atmospheric temperature outside the east entrance and temperature in each room were 24.1, 10.9, 8.3, 5.5, 4.9, 5.8, and 6.9°C from east to west, respectively. Potatoes, rice, etc., used in the experiment were stored inside and outside the sheet.

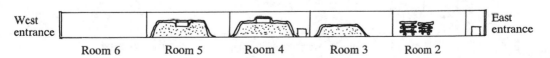

Fig. 2. Cross section of tunnel and 1995 snow dike experiments.

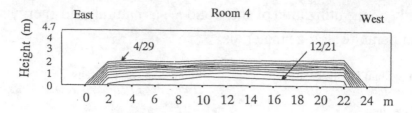

Height of the snow dike measured on 4/29, 5/16, 6/3, 7/1, 7/18, 8/8, 9/10, 10/22 and 12/21 (month/date)

Fig. 3 Height changes of the snow dike in room 4.

1.3 *Changes in the height of the snow dikes and the amount of unmelted snow*

Changes in the height of the snow dike in room 4 are shown in Figure 3. Figure 4 shows the changes of the weight of snow in rooms 3-5. Because the snow dike in room 4 is between rooms 3 and 5, the amount of snow thawed in room 4 is less than that in rooms 3 and 5.

1.4 *Temperatures inside and outside the tunnel*

The atmospheric temperature outside the tunnel and room temperatures within it are shown in Figure 5. Since room 1 is near the eastern entrance of the tunnel and is influenced by the outside temperature, its temperature in summer is high and the temperature curve is arched in shape. The temperature in room 2 is a little lower and it is not arched but straight. Temperatures in rooms 3, 4 and 5 are almost the same (5°C~7°C). Because of the decrement of the volume of snow dikes, the temperatures slightly increase with the increment of time. When the snow dikes exist, the temperature under the sheets is about 1°C.

1.5 *Low-temperature storage of foodstuffs*

1.5.1 Potatoes

In May, potatoes produced in Hokkaido were placed in the low-temperature storage rooms. The potatoes placed outside the sheets (5-6°C) budded, while those placed under the sheets (1-2°C) didn't bud until the end of December.

1.5.2 Rice

Unpolished and unhulled rice (koshihikari) produced in the Toyama plain in 1994 was put in paper sacks and stored in the low-temperature storage rooms from the end of June to the end of August. These paper sacks were placed inside the sheets (1°C) and outside the sheets (5°C), respectively, in the following three conditions: (1) humidity 100%, (2) humidity 56~58% (put in larger polyethylene bags with drying agents), (3) sealing into nylon bags with deoxidizers and nitrogen gas. In order to make a comparison, the unpolished and unhulled rice was also stored in (4) granaries of farmer families near the tunnel, in (5) storehouses at normal atmospheric temperature and

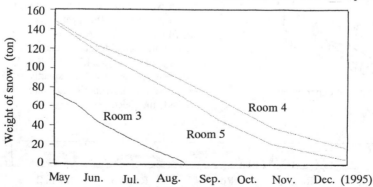

Fig. 4 Weight loss of the snow in rooms 3, 4, and 5.

554

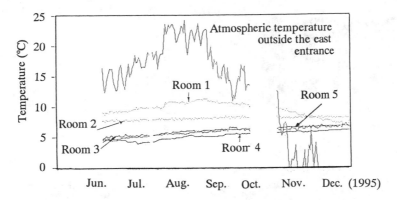

Fig. 5 Temperatures inside and outside the tunnel.

in (6) a low temperature (15°C, humidity 75%) building at the Toyama Agricultural Research Center in the Toyama plain.

The degree of rice spoilage was distinguished by the fatty acid value and the budding rate. If the fatty acid value is higher than 15-20 mg and budding rate is lower than 90%, the rice is considered to have deteriorated.

The unpolished rice used for the experiments had deteriorated somewhat before being used in these tests. After 2 months of the storage, the fatty acid value increased and the budding rate decreased.

Additional spoilage was highest in the case of (4) and (5), a little high at low-temperature storage (6), and low inside the tunnel (1)-(3). For storage inside the tunnel, the spoilage of the unpolished rice was not influenced by changes of storage conditions such as atmospheric temperature (5°C, 1°C), humidity (100%, 56%), or filling with nitrogen gas. The rate of rice spoilage is low at temperatures below 5°C. However at 100% humidity for long periods, rice easily decays.

The above conclusions are also suitable for unhulled rice. As compared with the unpolished rice, the unhulled rice had a lower fatty acid value and higher budding rate, and its spoilage obviously decreased.

1.6 Discussion

On December 21, 1995, about 25 m³ (about 20 tons) of snow remained in the 3 rooms. The unmelted snow tends to vanish in the spring because of the terrestrial heat and radiation of heat from tunnel walls. Since the low-temperature atmosphere outside the tunnel can be drawn into the tunnel, the whole-year low-temperature storage of food, vegetables, etc., in the

tunnel is possible. It is efficient to store potatoes and rice (unpolished and unhulled rice) with snow in the tunnel over a long period of time. But it is necessary to decrease the humidity to prevent decay.

2 EXPERIMENTS ON THE CONSERVATION AND UTILIZATION OF ICE IN HEILONG JIANG PROVINCE, CHINA

In 1992, T. Suzuki conducted low-temperature storage experiments with ice in Asahikawa City, Hokkaido. Asahikawa City is located at 43°46' N and 142°22' E. Its yearly mean atmospheric temperature is 6.3°C and the mean atmospheric temperature in January is -8.5°C. In winter, ice was made in several 1.8 m long ×1.8 m wide×1.8 m high wooden boxes. These boxes were arranged in a rectangular array with an inner space and a space for an entrance or exit. The whole system was covered with an aluminum foil-coated sheet. Experiments using the inner space surrounded by the boxes as a low-temperature storage room were conducted. The ice in the boxes remained until the end of August. Based on these results, in the autumn of 1992, we began experiments on the conservation and utilization of ice in Jiayin Town in Heilong Jiang Province, China, which is colder than Asahikawa City and situated in a "seasonally frozen ground" zone.

2.1 Experiments in Jiayin Town

2.1.1 Situation of experiments and construction of the ice room

Jiayin is located 450 km northeast of Harbin City with coordinates 48°51' North and 130°21' East. The yearly mean atmospheric temperature is about 0°C. Jiayin is a border town along the Heilong Jiang river.

555

Fig. 6 Jiayin experimental site, January 1993.

Fig. 7 Jiayin experimental site, March 1993.

Its opposite bank is Russian territory. In the autumn of 1992, a 4m long × 4m wide × 3.5m deep hole was excavated in a plot of farmland (Fig. 6). In January 1993, water was poured into the hole many times after the bottom and wall of the hole had frozen; finally a 3 m thick ice block was produced. In March, two layers of bags containing coal ashes which served as a heat proofing material were placed on the top of the ice block. Over the whole system an aluminum foil-coated sheet was laid to reflect sunlight (Fig. 7).

The experiments in Jiayin investigated thawing of the ice. No low-temperature storage experiments of vegetables, rice, etc., were conducted.

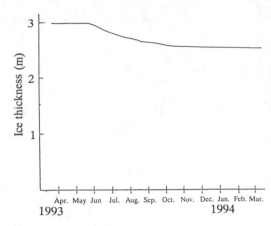

Fig. 8 Decrease in thickness of the Jiayin ice block.

2.1.2 Thickness changes of the ice block

The thickness of the ice block was measured at the beginning of August, 1993. It was found that the ice block which was 3 m thick had melted by 35 cm (Fig. 8). By March, 1994, 45 cm of the year old ice block had melted (i.e., 15% of its thickness). This means that whole-year conservation and utilization of natural ice is possible in Jiayin.

2.2 *Experiments on the conservation and utilization of ice in Harbin City*

2.2.1 Situation of experiments and construction of a low-temperature storage facility

Harbin City is located at 45°41' N and 126°37' E. Its yearly mean atmospheric temperature is 3.5 °C and the mean atmospheric temperature in January is -19.7°C. In the autumn of 1994, a permanent and practical low-temperature storage facility was built of reinforced concrete and brick in a plot of grassland in the suburbs of Harbin City. A front view and floor plan are shown in Figure 9.

Figures 10 and 11 are photographs taken when the facility was under construction and after it was completed, respectively. It is 10m wide × 10m long ×3.5m deep. Low-temperature storage experiments of watermelons and fresh vegetables were conducted in it. As shown in Figure 9, one half of the facility was used for storing ice, the other was used as a storage room in which six 0.8 m wide × 4 m long ×3 m deep shelves were placed. Although a low wall was built between the ice room and the storage room, cold air could flow from the ice room to the storage room.

556

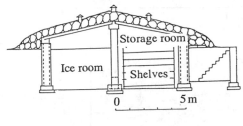

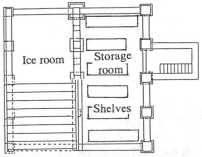

Fig. 9 Facility for conservation and utilization of ice in Harbin (front view and floor plan).

Fig. 10 Harbin facility under construction, November 1994.

Fig. 11 Completed facility at Harbin along National Highway 102 (Peking-Harbin).

2.2.2 Production of the ice block

The ice room was 5m wide × 10m long ×3.5m deep. In February, 1995, water was poured onto crushed ice blocks transported from the nearby Songhua Jiang River, and a 5m wide × 10m long ×3m thick ice block was produced in the ice room.

2.2.3 Atmospheric temperatures outside the facility and storage room temperatures

Atmospheric temperatures outside the facility and storage room temperatures are shown in Figure 12. The atmospheric temperature outside the facility was high from July to August and lower than 0°C after mid-November. The storage room temperature was measured 2m above the floor from the end of July. It was about 8°C until the beginning of November. In order to decrease the humidity in the storage room, the door beside the stairs was opened to ventilate the room, which made the storage room temperature slightly higher from mid-August to the end of September. After mid-November, the storage room temperature decreased. At the beginning of January when the atmospheric temperature outside the facility was -12°C, the storage room temperature was -1°C.

2.2.4 Thickness changes of the ice block

Figure 13 shows the decrease in thickness of the Harbin ice block. The ice block began to melt at the end of May. At the end of August, the ice block which was initially 3m thick melted by 1.5 m. By January, 1996, the thickness of the ice block had decreased by 2m. After that, it is inferred that the ice block will not melt until May because the atmospheric temperature outside the facility is low and outside air is allowed into the facility.

2.2.5 Foodstuffs for low-temperature storage

At the end of August, a large quantity of watermelons and fresh vegetables were carried into the storage room and put on the shelves. In order to decrease the humidity in the storage room, a layer of quick lime was spread on the floor of the storage room. The watermelons were stored until November for more than 2 months. This is the first time that watermelons have been stored for so long a time in northeast China. The agricultural experts set a high value on this kind of storage. The vegetables were stored for 3 months in the storage room.

557

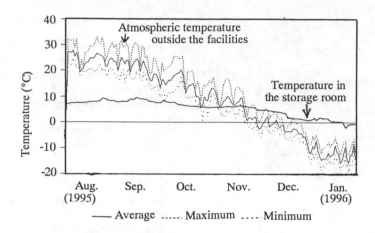

Fig. 12 Atmospheric temperatures outside the Harbin facility and within the storage room
from July 1995 to January 1996.

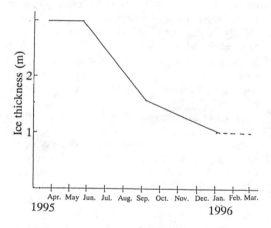

Fig. 13 Decrease in thickness of the
Harbin ice block.

2.2.6 Discussion

The method to decrease the humidity described in
Section 1-6 is also suitable for the case of ice. Instead
of decreasing the humidity in the whole storage room,
airtight spaces with proper volume are made with
polyethylene sheets. The foodstuffs are stored in these
airtight spaces. A appropriate humidity is kept by
placing drying agents in the spaces.

3 CONCLUSIONS

(1).   Year-round   conservation   and   utilization   of
natural snow in a road tunnel is possible.
(2). In the "seasonally frozen ground" zone, the year-
round low-temperature storage of foodstuffs is

possible utilizing ice which is produced by excavating
a half-underground ice room and pouring water onto
crushed natural ice blocks.
(3). In the seasonally frozen ground zone, huge
amounts of natural snow and ice are available.
Utilizing natural snow and ice to store foodstuffs can
save energy sources and promote the development of
the economy.

REFERENCES

H. Kowata, Y. Sato, M. Nara, and S. Katayama, "Ice
    Pond System for Application of Winter Coldness
    to Storage Cooling", Proceedings of the Eleventh
    International Congress on Agricultural Engineering
    (Dublin), 2309-2316,(1989).
K. Tsushima, F.Tanii, "Control of Snow Melting by
    Insulation", J. of Snow Eng. of Japan, 10 (1), 22-31
    (1994).
Kensei Matsumura, "A Method of Snow Storage," J.
    of Hokushin'etsu Branch of Japanese Society of
    Snow and Ice, No.1, 28-34(1988).
Kirkpatrick, D.L., M. Masoero, A. Rabl, C.E.
    Roedder, T.H. Socolow and T.B. Taylor, "The Ice
    Pond - Production and Seasonal Storage Ice for
    Cooling", Solar Energy, 36, 5, 435-445 (1985).
Tetsu Suzuki, "Use of a Cold Energy Element in a
    Low-Temperature   Storage   System",   Second
    International Conference on Snow Engineering,
    Special Report 92-27, pp. 333-339 (1992).
Vigneault, C. and H. Mcnicoll, "Natural Ice Used to
    Refrigerate a Storage Building ", International
    Winter Meeting of American Society of Agri. Eng.,
    Paper No 89-1630 (1989).

Snow Engineering: Recent Advances, Izumi, Nakamura & Sack (eds) © 1997 Balkema, Rotterdam. ISBN 90 5410 865 7

Air-conditioning system by stored snow
Part I: Control system of air temperature/humidity and its performance

M. Kobiyama & A. Wang
Muroran Institute of Technology, Japan

T. Takahashi
Funagata Town Office, Yamagata, Japan

H. Yoshinaga
Sanki Kougyou Co., Ltd, Tokyo, Japan

S. Kawamoto
Hokuyu Kensetu Consultant, Sapporo, Japan

K. Iijima
Sanki Kougyou Co., Ltd, Yamato, Japan

ABSTRACT: In this investigation, the authors propose a new snow air-conditioning system in which hot air is cooled directly on the surfaces of holes bored through a snow pile. A portion of recirculating air is mixed with cold air from the snow storage pit to obtain the prescribed temperature of the air-conditioning area. The characteristics and performance of this system were discussed by a theoretical analysis. From the analytical results, it is clear that the snow air-conditioning system with recirculation is good enough for actual use.

1 INTRODUCTION

It is very important to solve many problems in the fields of energy, food and environment concerned with our livelihood. Snow can contribute not decisively but greatly to solution of such problems.

Snow can be substituted for fossil energy as a source of refrigeration. It has been shown that snow saves a great deal of fossil energy and decreases the peak load of electric power generators when it is used for cooling instead of the electric power refrigerator (Kobiyama 1992). This use of snow is called a snow air-conditioning. This system can contribute to keeping the environment clean, because it does not need any refrigerant and does not exhaust any heat to the surroundings. Moreover, the total cost, i.e., capital cost plus running cost, of snow air-conditioning system is lower than that of electric power refrigerator when the air-conditioning area is larger than 200 m² (Kobiyama et al. 1993).

In this investigation, the authors propose a new snow air-conditioning system in which the hot air is cooled directly on the surface of holes bored through a snow pile. The air cooled by the snow is mixed with outdoor air and recirculating air to adjust the temperature and humidity to prescribed values adequate for an air-conditioned area.

In this report, the characteristics and performance of this system are discussed by a theoretical

analysis based on experimental results reported by the 2nd report of this investigation, in which the heat transfer rate between snow and air was measured.

Analytical results of this system show that the temperature and humidity of cooled air at the entrance of the air-conditioned area can be controlled, and the snow air-conditioning system is good enough for the actual use.

It is too hot during the summer even in snowy country. It will be very comfortable if snow is used for air-conditioning.

2 ALL-AIR TYPE SNOW AIR CONDITIONING SYSTEM WITH β CONTROL

There are many methods of obtaining refrigeration from snow. In the all-air type of snow air-conditioning system, the cold heat is transported by air to avoid use the higher manufactured machinery, such as heat exchangers, etc., and to make a simple control system for temperature and humidity. In the all-air type of snow air-condi-tioning system, when air contacts the surface of snow directly, dirty gas and dust are absorbed on the surface of the snow.

In this system, to suppress the influence of changes in the snow pile and the changes in outdoor conditions on the temperature and the humidity of the area to be cooled, outdoor air and recirculating air are mixed with the cold air from the snow storage pit. Both flow rates are controlled to produce the prescribed values for the air-conditioned area. Use α for the bypass flow rate of the outer air and β for the bypass flow rate of the recirculating air. When both α and β are controlled, we call this system an $\alpha\beta$ control system, if only α or β is controlled, then it is called an α or β control system respectively. Each of these three control systems have special characteristics respectively. We have used one of them on the basis of its characteristics, for example, the accuracy of temperature and humidity control and the simplicity of the system. In this report, the authors investigate the β control system which is utilized widely, because it is very simple. In the β control system, the temperature or humidity of the air-conditioned area can be controlled. In this report, we control β for temperature just as in a general air-condi-tioning system.

2.1 Classification of β control system

In the β control system, a portion of recirculating air β is mixed with the cold air from the snow storage pit, and this flow rate is controlled by a damper automatically to coincide with the temperature prescribed for the air-conditioned area. The flow rate of outdoor air is not controlled automatically.

The β control system is classified according to the mixing point of the outer air into 3 models.

Figure 1 shows model 1 and model 2. In model 3, points 2 and 3 in model 1 are replaced with each other.

In model 1, the air flow rate β is controlled at point 2. Recirculating air is mixed with cold air from the snow storage pit at point 5, and this mixed air is transported into the air-conditioned area by a blower. The remaining recirculating air ($l-\beta$) is mixed with the outer air l_0 at point 3. This mixed air is transported into the snow storage pit, and is cooled by the snow. A part of the air used in the cooling room l_0 is exhausted to the outdoors for ventilation, and the remaining air l is recirculated.

In model 2, all outdoor air for ventilation l_0 is mixed with the cold air from the snow storage pit and the recirculating air β at point 6. This model is very simple and it is easy to use for air ventilation without air cooling.

2.2 Cooling system of air

It is desirable that the air temperature through the snow storage room t_4 is little affected by the amount of snow remaining in the pit or by the air flow rate in order to suppress the β variation. To obtain this character, the hot air is cooled through snow holes. The detail of this cooling system is discussed in the 2nd report.

The melted snow, i.e., the cold water, is exhausted outdoors by a pump. However, we could use this cold water for the individual air-conditioning, special air conditioning and so on.

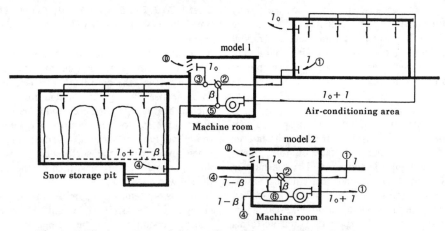

Figure 1 All-air type snow air-conditioning systems with β contr

3 EQUATIONS

Some equations are derived in order to estimate the humidity ϕ_5 or ϕ_6, the air flow rate β and the humidity of the air-conditioned area ϕ_1, under given atmospheric conditions, temperature re-quirements of the air-conditioned area and the amount of snow remaining. Where t is tempe-rature, ϕ relative humidity, x absolute humidity, h enthalpy, Cp_a and Cp_w specific heat of air and steam respectively, and r_0 latent heat of vapori-zation of water.

3.1 x_5 or x_6, β, x_1

3.1.1 Model 1.   From the heat and mass balances at point 5, the following equations are derived:

$$(l_0 + l)h_5 = \beta h_2 + (l_0 + l - \beta)h_4 \qquad (1)$$

$$(l_0 + l)x_5 = \beta x_2 + (l_0 + l - \beta)x_4 \qquad (2)$$

Where the enthalpy h is defined as follows:

$$h_5 = Cp_a t_5 + x_5(Cp_w t_5 + r_0) \qquad (3)$$

β and x_5 are unknown variables in Eqs.(1) and (2). Those values are derived from the simul-taneous equation consisting of both equations, after substitution of Eq.(3) into Eq.(1), as follows:

$$x_5 = \frac{(h_2 - Cp_a t_5)/(h_2 - h_4) - x_4/(x_2 - x_4)}{(Cp_w t_5 + r_0)/(h_2 - h_4) - 1/(x_2 - x_4)} \qquad (4)$$

$$\beta = (l_0 + l)[(x_5 - x_4)/(x_2 - x_4)] \qquad (5)$$

The sensible heat ratio SHF is defined by Eq.(6). The humidity of the air-conditioning area x_1 is estimated by Eq.(7).

$$SHF = Cp_a(t_1 - t_5)/(h_1 - h_5) \qquad (6)$$

$$x_1 = \frac{h_5 - Cp_a t_1 + Cp_a(t_1 - t_5)/SHF}{Cp_w t_2 + r_0} \qquad (7)$$

3.1.2 Model 2.   By the same manner used in 3.1.1 from the heat and mass balances at point 6, the following equations are derived:

$$(l_0 + l)h_6 = (l - \beta)h_4 + \beta h_2 + l_0 h_0 \qquad (8)$$

$$(l_0 + l)x_6 = (l - \beta)x_4 + \beta x_2 + l_0 x_0 \qquad (9)$$

β and x_6 are unknown variables in Eqs.(8) and (9). Those values are derived from the simu-ltaneous equation consisting of both equations after substitution of Eq.(3) into Eq.(8) as follows:

$$x_6 = \{[l_0 h_0 + l h_4 - (l_0 + l)Cp_a t_6]/(h_2 - h_4) - (l_0 x_0 + l x_4)$$
$$/(x_2 - x_4)\}/\{(l_0 + l)[(Cp_w t_6 + r_0)/(h_2 - h_4) - 1/(x_2 - x_4)]\} \qquad (10)$$

$$\beta = [(l_0 + l)x_6 - (l_0 x_0 + l x_4)]/(x_2 - x_4) \qquad (11)$$

The humidity of the air-conditioning area x_1 is estimated by Eq.(7).

3.2 Outlet air temperature, t_4, from the snow storage pit

The initial height of the snow pile is 3 m and the sectional area for each snow hole is 1 m^2. The total running time is 200 hrs and the snow remains 15% at the end of cooling season.

Early in the cooling season, the snow hole grows in both vertical and radial directions up to about h=400 mm and d=500 mm. After this period, the snow melts in the vertical direction with an almost constant hole diameter. From these characteristics of the melting snow hole, we adopted the melting snow model shown in Figure 2. That is, for early melting period up to d=500 mm, the heat transfer rate between snow and air follows the results presented in the 2nd report. After this early period, the heat transfer rate is constant with h=400 mm and d=500 mm. However, the heat transfer area changes as time advances. Then the Nusselt numbers Nu ($= \alpha_H d / \lambda$) are written as follows:

$$d \leq 0.5 : \ Nu = 112\ln[Re^{0.8}(h'/d)^{-1.1}] - 497 \qquad (12)$$

$$d > 0.5 : \ Nu = 112\ln[Re^{0.8}(5/4)^{-1.1}] - 497 \qquad (13)$$

where Re is the Reynolds number defined as Re=Vd/v, V is the air velocity in a snow hole, d is the diameter of the hole and v is the dynamic viscosity of air. The outlet air temperature through the snow hole is expressed as follows:

$$t_4 = t_3 \exp[-\alpha_H \cdot A/(GCp_a)] \qquad (14)$$

561

where α_H is the heat transfer coefficient derived from Nu with the representative length d and thermal conductivity λ, A is the heat transfer area (A=$1^2-\pi$ $d^2/4+\pi$ d h) and G is the mass flow rate of air. The heat transfer coefficients estimated from Eq.(12) or Eq.(13) are higher than the actual ones, because there is not only a converging entrance flow region but also parallel flow regions in the actual snow hole. The maximum difference between the temperature estimated from Eq.(14) and actual one are seemed to be less than 2 to 3 ℃. Within this range of temperature difference, there is no problem to design the actual system, because the temperature t_5 or t_6 is controlled by the air flow rate β.

3.3 Cooling load

Set the design condition of the outer air temperature at t_0=33℃, humidity ϕ_0=63%, temperature of cooling area t_1=27℃, humidity ϕ_1= 55% and specific cooling load for air-conditioned area at 140 kcal/m²h which represents a general office. We assume that the cooling loads for various outdoor temperatures are estimated by the following equation. Where L_0 is the cooling load at t_1=33℃

$$L = L_0(t_0 - 27)/(33 - 27) \tag{15}$$

The air flow rate for ventilation is set at 7.5% of L_0, this quantity corresponds to one air ventilation per an hour. The sensible heat ratio SHF=1/1.3.

A solution is derived by the numerical iterative method for non-linearity of equations.

4 RESULTS AND DISCUSSION

The results of the model 1 are shown in Figure 3, and those of model 2 are shown in Figure 4. In those figures, L_{pit} means the flow rate of air passing through the snow storage pit, that is , in model 1 L_{pit} equals ($l_0+l-\beta$) and in model 2 ($l-\beta$).

The desired temperature for the air-conditioned area is realized for all outdoor air conditions. And this system can easily accommodate higher outdoor air temperatures than that of design point when a larger capacity blower is provided.

In model 1, the humidity of the air-conditioned area is hardly affected by outdoor air variations and is stable throughout the entire cooling period. Since the humidity is rather lower, it is suitable in the high humidity season and region.

In model 2, the desired temperature for the air-conditioned area is realized for all outdoor air conditions same as model 1. The humidity of the air-conditioned area is easily affected by the outdoor air condition, and is higher than in model 1. Sometimes the supplying cool air to the air-conditioning area includes mist. However this condition is very narrow. This model is more compact than the model 1. Because the bypass flow rate of the recirculating air β is smaller than that of the model 1 and the air flow rate to the snow storage pit L_{pit} is almost same.

The characteristics of model 3 are in the middle between those of models 1 and 2.

From those results, the most suitable model in the β control system should be selected based on the needs of the user.

5 CONCLUSION

In this investigation, the authors proposed a new snow air-conditioning system in which hot air was

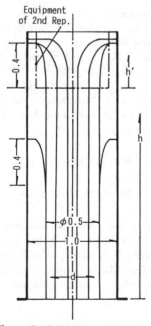

Figure 2　Melting snow model

cooled directly on the surfaces of holes bored through a snow pile. In this report, the authors discuss the β control system, which can be utilized widely because of its simplicity. In this system, a portion of recirculating air is mixed with cold air from the snow storage pit, and the flow rate is controlled by a damper automatically to coincide with the prescribed temperature of the air-conditioned area. The authors classified the β control system into 3 models according to the mixing point of the outdoor air.

The characteristics and performance of this system are discussed by a theoretical analysis. Analytical results show that the temperature and humidity of cooled air at the entrance to the air-conditioned area can be controlled. In model 1, the humidity of the air-conditioned area is not influenced by the humidity of the outdoor air. In model 2, the humidity of the air-conditioned room is easily affected by the humidity of the outdoor

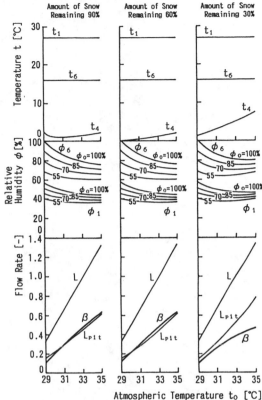

Figure 4　Flow rate, temperature and humidity of Mode 2

air. However, model 2 is very simple. The characteristics of model 3 are in the middle between those of models 1 and 2.

From those results, it is clear that a snow air-conditioning system with β control is good enough for actual use.

REFERENCES

Kobiyama, M. 1992. Oil equivalent of Snow. '92 Cold Region Technology Conference:35-42.

Kobiyama, M. Takahashi T. Wang A. Yoshinaga H.Kawamoto S. and Matumoto H. 1993. All Air Type air-conditioning System Using Snow — Economic Estimation—. '93 Cold Region Technology Conference:64-67.

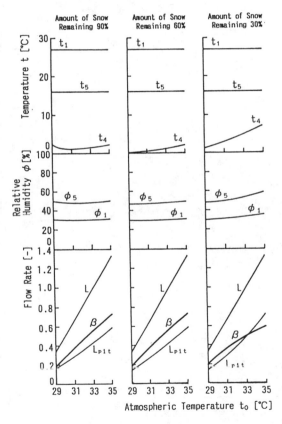

Figure 3　Flow rate, temperature and humidity of Mode 1

Snow Engineering: Recent Advances, Izumi, Nakamura & Sack (eds) © 1997 Balkema, Rotterdam. ISBN 90 5410 865 7

Air-conditioning system by stored snow
Part II: Characteristics of the direct heat exchange between snow and air

A. Wang & M. Kobiyama
Muroran Institute of Technology, Japan

T. Takahashi
Funagata Town Office, Yamagata, Japan

S. Kawamoto
Hokuyu Kensetu Consultant Co., Ltd, Sapporo, Japan

H. Yoshinaga
Sanki Kougyou Co., Ltd, Tokyo, Japan

K. Iijima
Sanki Kougyou Co., Ltd, Yamato, Japan

ABSTRACT: To design an all-air type air-conditioning system using snow, it is desirable that the temperature of air cooled by the snow surface is little affected by the shape or size of the snow. To achieve this, hot air is cooled directly at the snow surface of holes bored through the snow pile. The authors made experiments on direct heat transfer between snow and air. An experimental relationship between Nusselt number, Reynolds number and the height and diameter ratio of a snow hole was deduced.

1 INTRODUCTION

In the case of the design an all-air type air-conditioning system using snow (Kobiyama et al. 1993), it is desirable that the temperature of air cooled by the snow surface is affected little by the shape or size of the snow. To obtain such characteristics, hot air is cooled directly at the snow surface of a hole bored through the snow pile. The phenomenon of direct heat exchange between a snow surface and air is unsteady and complex because snow is melted and its shape changes during the process. Also, the heat transfer rate is affected by many parameters such as air temperature and velocity, and the size and shape of the path through the snow. There are few studies on this type heat transfer except the ablation cooling of a supersonic flighting body (Truitt 1960). Heat transfer between such a snow surface and air has not been discussed still now.

In this report, to provide a series of data for the design of a snow air-conditioning system by direct heat exchange, the authors conducted experiments on the direct heat transfer between snow and air. We determined the relationship among temperature of inlet air, outlet air, area of snow surface, and the size and shape of snow paths. An experimental equation between Nusselt number, Reynolds number and the height and diameter ratio of the snow hole was deduced from these data and the energy balance.

2 EXPERIMENT

Figure 1 shows the experimental apparatus. It consists of a storage box of snow, a blower, an inverter, an electrical heater, an orifice, temperature sensors and a data recorder. Figure 2 shows the storage box of snow. The cross section of this box was square with the inner side of 800 mm, and was made of foam plastic insulation material with the thickness of 100 mm. Two thermocouples were installed at the inlet of the box for measuring the inlet air temperature and two thermocouples were installed at the outlet of the box to measure the outlet temperature. Two straightening plates with a

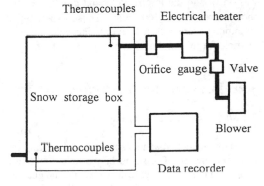

Figure 1   Experimental apparatus.

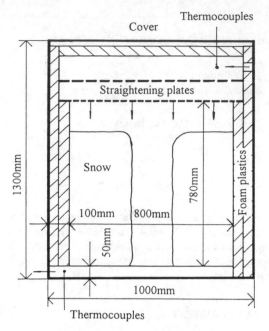

Figure 2   Snow storage box.

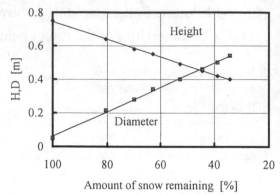

Figure 3   Amount of snow remaining and size of snow path.

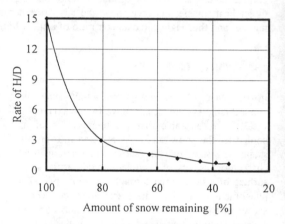

Figure 4   Amount of snow remaining and H/D.

large number of small holes of diameters 7 mm and 9 mm respectively were placed above the snow in order to uniformly distribute the inlet air. The flow rate of the air was adjusted by a valve and the inverter attached to the blower. That rate was measured by the orifice. The inlet air temperature was controlled by the electrical heater with adjustable electric transformer.

The experimental conditions were as follows; the maximum flow rate of air was 0.027 m³/s and the minimum was 0.0054 m³/s, the highest temperature was 35℃ and the lowest 25℃ and densities of the snow were 440 to 460 kg/m³. Those experimental conditions were selected considering actual use in the field for air-conditioning of a general office.

3 RESULTS AND DISCUSSION

3.1 Shape and size of snow path

The snow is melted and its shape changes during the heat exchanging. The shape of the hole was like a convergent nozzle. Figure 3 shows the height and diameter of the snow. It is found that the height and diameter of the snow vary linearly with the amount of snow remaining.

The ratio of height H to diameter D is one of the representative values of this experiment. The relation between the amount of snow remaining and H/D is shown in Figure 4.

3.2 Outlet Air Temperature

Figure 5 shows outlet temperatures for a constant air flow rate of 0.027m³/s and Figure 6 shows outlet temperatures for constant inlet air temperature of 30℃. During the early heat transfer period, when the amount of snow remaining was 100% to 60%, outlet temperatures rise continuously and linearly as the amount of snow decreases. When the amount of snow remaining is between 60% to 30%, outlet air temperatures change gradually. In the early period, changes in the shape and size of snow, the heat

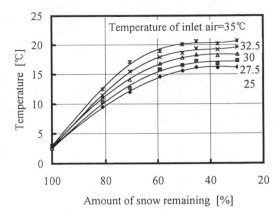

Figure 5   Outlet air temperatures for a constant
air flow rate of 0.027m³/s.

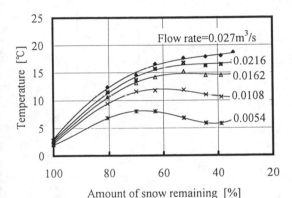

Figure 6   Outlet air temperatures for a constant
inlet air temperature of 30℃.

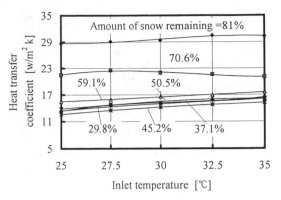

Figure 7   Heat transfer coefficients.

transfer area on the snow surface and the air velocity through the hole were steep. On the other hand, those changes in the middle period were moderate.

3.3 Heat Transfer Coefficient

It is important to generalize the results mentioned above to apply them to actual heat transfer equipment in the air-conditi oning field. Heat transfer coefficients are derived from the experimental results.

The heat transfer rate Q can be expressed by the following equations for energy equilibrium.

$$Q = G \cdot \rho \cdot C_p (t_{in} - t_{out}) \tag{1}$$

$$Q = \alpha \cdot S \cdot \Delta t_m \tag{2}$$

where G is the air flow rate, ρ the density of the air, C_p the specific heat capacity, t_{in} and t_{out} the inlet and outlet temperatures respectively, α the heat transfer coefficient, S the surface of heat exchange and Δt_m the average temperature difference between the air and the snow surface.

From Eqs.(1) and (2), the heat transfer coefficient α is derived as follows,

$$\alpha = G \cdot \rho \cdot C_p \cdot (t_{in} - t_{out})/(S \cdot \Delta t_m) \tag{3}$$

Figure 7 shows heat transfer coefficients. In this figure, it is clear that heat transfer coefficients are little affected by inlet air temperature, however, they increase slightly with the temperature rise.

The Nusselt number and the Reynolds number are defined by the following equations:

$$Nu = \alpha \cdot D/\lambda \tag{4}$$

$$Re = U \cdot D/\nu \tag{5}$$

where D is the measured diameter of the snow hole, and λ the thermal conductivity. The average air velocity, U, through a hole is expressed as $U = G/(\pi D^2/4)$, and ν the viscosity of the air.

The relation between Nu and the values of $Re^{0.8}$ $(H/D)^{-1.1}$ is shown in Figure 8. From this figure, the following experimental equation is deduced:

$$Nu = 112\ln[Re^{0.8} \cdot (H/D)^{-1.1}] - 497 \tag{6}$$

From Eq.(6), the heat transfer coefficient can be estimated when Re and H/D are given. And then the

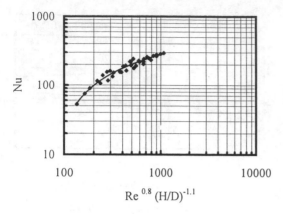

Figure 8   Nu and $Re^{0.8}(H/D)^{-1.1}$

outlet air temperature through the snow hole can be estimated for a given air flow rate and inlet air temperature.

4 CONCLUSIONS

From the experiment on direct heat transfer between the surface of a hole in snow and air passing through the hole, the following conclusions have been derived:

1. During the early heat exchange period, outlet air temperatures rise continuously and linearly with decreasing amount of snow remaining. In the middle period, outlet air temperatures change gradually.

2. Nusselt number Nu, Reynolds number Re and the ratio of height to diameter of the snow hole H/D are related by the following equation:

$$Nu=112\ln[Re^{0.8} \cdot (H/D)^{-1.1}]-497$$

3. Experimental results and the equations in this report are good enough for design of heat transfer equipment for snow air-cooling systems.

REFERENCES

Truitt, R.W. 1960. Fundamentals of Aerodynamic Heating, The Ronald Press Co., N.Y.

Kobiyama M. Wang A. Kawamoto S. Koyama T. Matumoto H. Yamamoto T. Mizino R. and Yoshinaga H. 1993. All-Air Type Air-Conditioning System, 6th Symposium on Engineering in Cold Climate:184-190.

Snow Engineering: Recent Advances, Izumi, Nakamura & Sack (eds) © 1997 Balkema, Rotterdam. ISBN 90 5410 865 7

A cold storage system for food using only natural energy

Ken Okajima & Hiroyuki Nakagawa
Hazama Technical Research Institute, Japan

Syunji Matsuda
Hazama Branch, Tokyo, Japan

Toshihiko Yamasita
Department of Civil Engineering, Faculty of Engineering, Hokkaido University, Sapporo, Japan

ABSTRACT : A cold storage system takes in the severe cold air of winter for cooling a storage room and an ice room. At that time, cold energy transforms water into ice as latent heat in the ice room. The different temperatures and heights existing between the ice room and the storage room cause air flow. Cold air from the ice room is supplied to the storage room from spring to autumn.

1 INTRODUCTION

Supplying farm products all the year around has become popular in Japan. Cold storage of farm products and the like is an important part. From these aspects, cold regions such as Hokkaido are well suited to keep farm products in cold storage. These products can be preserved in a cold storage room by taking advantage of the severe cold as an abundant natural cold energy source.

The present study deals with the efficiency of heat transfer between water (ice) in tanks and the ambient air in the ice room and the ability to send cold air through a duct to the storage room by using air flow due to different temperatures and heights in the ice room and the storage room.

2 EXPERIMENT OF ICE MAKING AND MELTING

2. 1 Outline of the experiment

Figure 1. shows the cold storage system. Table 1. shows the five factors (deepness of water tanks , tank material , kind of water, air velocity around tanks, position of heat transfer) , and the air temperature in the environmental chamber. Figure 2. shows the tanks in the ice room. Eleven tanks were placed in the environmental chamber.

Thermocouples (120) and hygrometers (2) were provided inside and outside the tanks. The thickness of the ice in the tanks was measured at an interval of about ten hours manually during ice-making. The height of the water surface in the tanks was measured manually and the volume of melting water was estimated. The coefficients of heat transfer between the ice tanks and the ambient air were determined from measured values and the model of heat balance shown in Figure 3. We compared the amount of heat transfer by convection to that by evaporation.

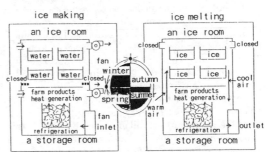

Figure 1. The cold storage system

Table. 1 Experimental variable

	deep-ness of tank	tank material	kind of water	air velocity around tank	position of heat transfer	air temp.
①	540 mm	stainless steel	water	0.2 m/s	water surface	
②	420 mm	stainless steel	water	0.2 m/s	water surface	-5°C during ice making
③	300 mm	"	"	"	"	
④	420 mm	polyethylene	"	"	"	
⑤	"	stainless steel	salt water 3.4 %	"	"	
⑥	"	"	salt water 1.7 %	"	"	
⑦	"	"	water	0.4 m/s	"	
⑧	"	"	"	0.6 m/s	"	5°C during ice melting
⑨	"	"	"	0 m/s	"	
⑩	"	"	"	0.2 m/s	all surface	
⑪	"	"	"	"	Top and bottom	

2. 2 Results of ice making

Figure 4. shows water temperatures insides standard water tank ②. At all the tanks including water tank ②, all water in the tanks was mixed and the water temperature decreased to 4℃(salt water temperature decreased to its melting point). The position of heat transfer is at the water surfaces for tanks ① through ⑨. The temperature there soon became 0℃. And ice began to form. Heat transfer energy from the air to the iced water were spent as ice latent energy after the water at the surface became ice. The speed of ice growth slowed down gradually because thickness of ice is equivalent to thermal resistance. The temperature inside the water tanks decreased rapidly after 100% of the water was ice. Because heat transfer was then only spent as ice sensitive heat. The positions of heat transfer are at the water surface and at the bottom of tank ⑪. Once all the water was 4℃, ice first formed at the water surface and then at the bottom of the tank. Table 2. shows coefficients of heat transfer during the ice making period. The ambient air velocity over was varied ②,⑦,⑧,⑨. The coefficient of heat transfer by convection increased as air velocity increased. The degree of linear slope by the experiment was nearly equal to that obtained by solving for air velocity V in Jurges's formula. Table 3. shows the amount of heat transfer during the ice-making period. Relative humidity was 50% inside the environmental chamber during this period. Heat transfer by evaporation was more than that by convection. In a practical plant, evaporation from the water surface will be more than that in this experiment because cold outdoor air will have a relative humidity below 50%.

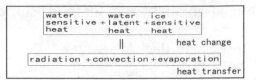

Figure 3. The model of heat balance

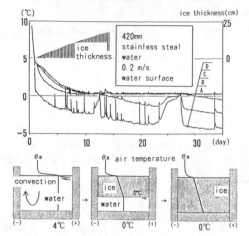

Figure 4. Water temperature of tank ②

Table 2. The coefficients of heat transfer

kcal/hm²℃

	until equal to 4 ℃		after ice formed		ice
	by radiation	by convection	by radiation	by convection	making period
①	3.44	5.10	3.38	4.13	81
②	3.43	4.12	3.38	3.29	50
③	3.42	3.84	3.38	3.22	41
④	3.42	4.72	3.37	2.97	46
⑤	3.35	5.06	3.35	3.75	56
⑥	3.34	3.38	3.36	3.59	66
⑦	3.42	5.41	3.37	4.11	40
⑧	3.42	5.62	3.37	4.73	33
⑨	3.44	3.94	3.38	2.64	72
⑩	3.42		3.37		
	1.81	3.96	1.76	-	-
	1.83		1.76		
⑪	3.42	4.33	3.37	2.17	33
	1.82		1.76		

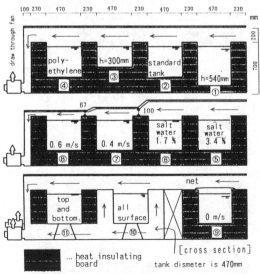

Figure 2. The experimental ice tanks

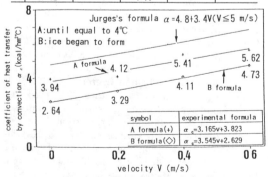

Figure 5. The experimental value and the Jurges's value

2. 3 Results of ice melting

Figure 6. shows water temperatures inside tanks ②,⑩, and ⑪. Ice temperature change inside the water tanks is due to sensitive heat when the temperature changes from -5℃ to the melting point (0℃). The ice temperature increased rapidly. Melted ice water was mixed keeping the condition of 0℃ inside the water tanks until all ice melted. After all ice melted, water temperature change inside the water tanks was again related to sensitive heat during transforming water at the melting point (0℃) to 5℃. Again the water temperature increased rapidly. At tank ⑪ the top and bottom are heat transfer surfaces. At tank ⑩ all surface are heat transfer surfaces and ice was melted at the water surface, at the sides of the tank, and at the bottom of the tank. The ice water temperature was kept of 0℃ at the bottom of the ice(the X layer at Figure 6.). But ice water was relatively cold at the upper part of the tank and relatively warm at the lower part of the tank below the X layer. This occurred because the heaviest specific gravity of water is at 4℃. Since the difference in water temperature between the lower part of the tanks and the ambient air was reduced, heat transfer there decreased. Distribution of water temperature is different due to the position of ice inside the water tanks during the ice melting period. Therefore the volume of heat transfer is different between the water tanks and the ambient air.

3 CONFIRMATION OF CIRCULATION AIR

Figure 7. shows the outline of an experimental model. Table 4. shows experimental parameters influencing air circulation in the system. The model was placed in a thermostatic chamber. Twenty ice pallets(length 0.33m × width 0.198m × height 0.05m) weighting 64 kg which corresponds to about 5100 kcal/h, were set in two rows, two lines, and five layers inside the ice room. The area of heat transfer was 1.31 m² at the water surface, 1.06 m² at the side of all pallets, and 1.31 m² at the bottom of all pallets. A (20 W) heater was set at a height of 150 mm above the storage floor level. This heater corresponds to calorification of farm products. Air temperatures in the thermostatic chamber were defined to change the air temperature to 2℃ in the ice room and to 4℃ in the storage room. Sensors for automatically measuring temperature (34 points) and humidity (2 points) were installed inside the model. Speed of air flow was measured with velocity sensors inside two ducts. Smoke gases were used to visualize air flow motion inside the model.

The warm air in the storage room flowed to the ice room through the upper ducts and collected in the upper part of the ice room. Then the air was mixed

Table 3. Heat transfer by radiation, convection and evaporation kcal/h

	haet trasfer by radiation	heat tranfer by convection	heat transfer by evaporation
①	2. 51	2. 94	3. 77
②	2. 36	2. 46	2. 92
③	2. 36	2. 45	2. 85
④	2. 31	2. 30	2. 60
⑤	1. 86	2. 93	2. 95
⑥	2. 11	2. 81	2. 99
⑦	2. 17	2. 96	3. 11
⑧	2. 33	3. 23	3. 68
⑨	2. 61	1. 81	2. 00
⑩	8. 39	11. 20	2. 20
⑪	3. 76	3. 49	1. 92

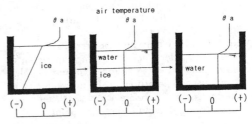

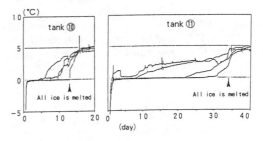

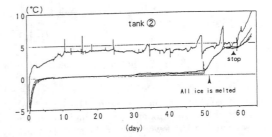

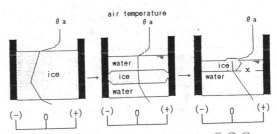

Figure 6. Water temperature of tanks ②,⑩,⑪

571

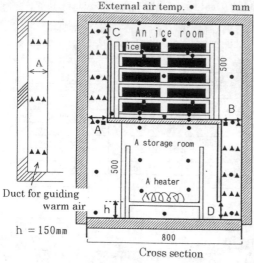
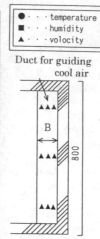

External air temp. ●   mm

● ··· temperature
■ ··· humidity
▲ ··· volocity

Duct for guiding
cool air

Duct for guiding
warm air

h = 150mm

Cross section

Figure 7. The outline of the model

REFERENCES
Nakagawa.H,et al, Japan 1994. Basic Study on The Effective Utilization of Natural Cold Energy in Cold Region. Proceeding of the 10th Cold Region Technology Conference,pp.265-270

Okajima.K,et al, Japan 1994. An Experimental Study on Ice Making and Ice melting Plants for Natural Cold Energy Use vol.2. Proceeding of The Society of Heating, Air-Conditioning and Sanitary Engineers of Japan ,pp.585-588(Ⅱ)

Nakagawa.H,et al, Japan 1994. Cold Storage and Transport System of Food Operated Only by Natural Energy. Proceeding of International Conference on EcoBalance ,pp.102-107

Murakami.S,et al., Weather and Architecture. Asakura Bookstore Corporation.

Table 4. Experimental parameters influencing circulation			
	case 1	case 2	case 3
Area of duct	0. 04	0. 08	0. 12
Width between room and duct	0. 05	0. 10	0. 15
External air temperature	9. 5		

with the cold air in the ice room. The mixed air flowed to the storage room through the lower ducts and collected in the lower part of the storage room. The air was mixed with the warm air in the storage room. The air can't flow backward in either duct.

Figure 8. shows the comparison of the experimental air velocity and the air velocity solved with the formula for stack effect. The experimental air velocity was nearly equal to the calculated value using average temperatures in the ice room and in the storage room. Therefore we can calculate the volume of air circulation in a practical plant.

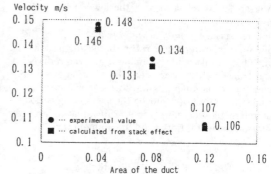

Figure 8. Comparison between experimental velocity and stack effect velocity

4 CONCLUSIONS

In ice making and ice melting using model ice pallets, we have determined the efficiency of heat transfer between the ice pallets and the ambient air. In the design of water tanks, it is important to plan the following;

1. During the ice making period, attention must be paid to the size of the water surface, because the thickness of ice made there causes heat resistance, and evaporation there causes heat transfer.

2. During the ice melting period, the vertical temperature distribution in the water tanks varies with the position of the ice block. Therefore the coefficient of heat transfer varies accordingly.

In the experiment on the volume of air circulation,

3. It was confirmed that the volume of air circulation between the ice room and the storage room is determined by the theory of stack effect, agrees well with that found by the experiment.

Snow Engineering: Recent Advances, Izumi, Nakamura & Sack (eds) © 1997 Balkema, Rotterdam. ISBN 90 5410 865 7

The water source of snow, ice and desertification

Liangwei Wang
Southwest Branch, Chinese Academy of Railway Sciences, People's Republic of China

Tsuneharu Yonetani
National Research Institute for Earth Science and Disaster Prevention, Japan

ABSTRACT: The Source of the Cele River is the northern slope of the Mushi Tage Glacier. It flows into an oasis on the southern margin of the Taklimakan Desert from south to north and disappears into the desert on the northern part of the oasis. The Cele River is supplied by rainfall on the highland region and about 46% of the runoff is from snowmelt. Glaciers, as a source of rivers, are retreating on a large scale. Between a glacier and the desert heat exchange by mountain and valley wind intensifies glacial melt. The price paid for increases in discharge of rivers is decrease of glaciers. Therefore, we suggest that people should plant trees and forests on the middle-mountain region of valleys to reduce the strength of mountain and valley winds to preserve glaciers. This will keep stable the long-time water source and will retard the process of desertification on the oasis.

1. INTRODUCTION

The Cele River is located in the southern margin of the Taklimakan Desert and the drainage basin is mainly between 36 and 37 N, and 80 and 81 E (Fig.1). Figure 2 is longitudinal profile of Cele River. The source of the river is the northern slope of the glacier of Mushi Tage Peak (it is means glacier in Weiwuer language) of Kunlun Mountain. The river flows from south to north into the Taklimakan Desert then disappears into the desert of Northern Cele County. It is the most important water source of the oasis in the Cele Region. This project is a Sino-Japanese joint effort to clarify the desertification mechanism. Meteorological and hydrological data were automatically obtained in Kartashi (which meas "black stone") (2800m a.s.l.) in the middle-mountain area of the Cele River and in Wuku (2500m a.s.l.) on the margin of the desert downstream of the Cele River and in the field of 1480m a.s.l. in northwest Cele County.

According to historical material and practical observation data over four years, the Cele River is almost completely supplied by rainfall from the highland region and about 46 % of the runoff is from snowmelt. Thus it is a typical river supplied by glacier.

2. CHARACTERISTICS OF DISCHARGE IN CELE RIVER BASIN

Figure 3 shows the measurement data of annual runoff process of the Cele River in the Wuku section. The discharge occurs mainly from June to August and is supplied by ground water in winter and spring but

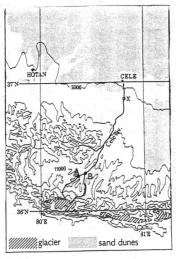

Fig.1
Topographical map around the experimental area.
Contours are drawn in feet and at intervals of 2,000 ft.

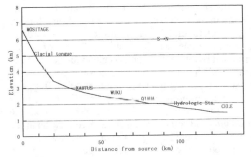

Fig.2 Longitudinal profile of Cele Riv.

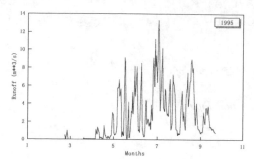

Fig.3 Discharge hydrograph (daily mean)

Table 1. Statistical data for the middle-mountain region.
(Cele River, Elevation 2800m)

1992 yr.	June	July	August	Avg. Value
Air Temp.(℃)	11.0	13.6	12.9	12.5
Earth Temp.(℃)	15.3	18.3	18.3	17.3
Discharge(m³/s)	7.2	3.7	3.7	4.9
Precipitation(mm)	71.5	17.5	12.0	101.0(sum.)

1994 yr.	June	July	August	Avg. Value
Air Temp.(℃)	9.1	13.0	11.7	11.3
Earth Temp.(℃)	17.4	23.4	22.0	20.9
Discharge(m³/s)	7.0	6.4	3.6	5.7
Precipitation(mm)	41.5	7.0	4.0	52.5(sum.)

1995 yr.	June	July	August	Avg. Value
Air Temp.(℃)	10.5	12.0	12.6	11.7
Earth Temp.(℃)	17.4	18.7	21.1	19.1
Discharge(m³/s)	4.2	5.1	3.9	4.4
Precipitation(mm)	45.0	43.0	1.0	89.0(sum.)

not enough to prevent break off of the flow. The precipitation in the Cele River area was very little in 1994 (as shown in Table 1). Precipitation was only 0.5 mm in the oasis of the Cele region from October 1993 to October 1994 and only 67 mm in middle-mountain areas of Kartashi. Information from herdsmen living in the high-mountain areas, indicates that the precipitation was so far below normal on the grassland region of the high-mountains (3800-4500m a.s.l.) that not enough grass used available for their sheep. Herdsmen had to drive their sheep down the mountains and sell them. Thus the precipitation in the whole of the Cele River Basin was quite low in 1994. Compared to the corresponding period in 1992 (a comparison 1994 to 1993 is not possible, because the

hydrological gauge was broken then), the precipitaion was only about 50% during June through August but the discharge of the river increased by 16% and mainly occurred in July (in Apr. to May 1992 the hydrological gauge was broken). However at that moment, there was no snow on the highland. Thus the increase of discharge was due to a great negative balance of the glacier in 1994, in other words, due to an excessive melt of the glacier.

According to the statistic analysis by Mr.Tang Qicheng , in the recent thirty years, the discharge of the river on the Taklimakan Desert has remained constant. But rivers mainly supplied by precipitation are tending to decrease and rivers mainly supplied by glacier are tending to increase.

On the basis of many years of atmospheric temperature data in high-altitude, it can be calculated that the glacier in the river head region was thinned by 130 mm per year. During our investigations on the end of the glacier tongue in early October 1991, we noticed many indications of the retreat of the glacier. The glacier tongue were covered by debrises and stones and silt were mixed in the ice. This proves that the glacier is retreating. The river bed of the Cele River on the middle-mountain area is stable with cobbles firmly inserted there. In the long time scale, the discharge of the Cele River tended to decrease. This correlates with the retreat of the glacier. In 1994, the precipitation decreased but the discharge increased suggesting that the size of the glacier will quickly diminish. Then as a result of the rapid melting of the glacier, after the discharge of river unusually increased for a period of time, it will appear that the discharge will decrease and adjustment function of glacier for conserving water will be weaken on the whole river basin and the distribution of discharge will be uneven in a year. This will further make the water environment worse and residents will have difficulty living here. It is said that the Cele County has moved twice into the south because the sand dunes had moved into the south but this saying is not completely proved. Perhaps the moving was not caused by incursion of sand dunes but to the difficulty of finding water there which forced people to move to the upper reaches of the river.

3. THE ROLE OF MOUNTAIN AND VALLEY WIND IN MELTING OF GLACIER

As shown in Figure 1, the Cele River is almost straight and it joins the glacier and the desert from south to north. The valley of the Cele River is like a long corridor between a satellite hall and a major building of an airport. Here many meteorological and hydrological processes on the middle-mountain region of 2500-2800m a.s.l. have been observed.

In the valley, mountain and valley winds are strong and distinct. By daylight, the hot-air over the desert moves into the valley, then it formed a valley wind. In the evening the cold-air in the highland moved out of the valley and formed a mountain breeze. When weather is clear, the action of the mountain and valley wind is strong. This process had been confirmed by meteorological observations at the mountain pass. This means that the action of the mountain breeze and valley wind proceeded along the entire river valley.

As shown in Table 1, the discharge in summer increased by 16% in 1994 compared to 1992. The mean earth temperature rose by 2.4℃. This was probably caused by the rainfall decrease. The mean air temperature decreased by 1.2℃. The phenomenon of an increase in river discharge but an air temperature decrease in the valley, can only be explained by having the ice and snow on the highland participate in this process.

The action of heat exchange between the mountain breeze and the valley wind is evident in the records of air temperature and discharge in Figure 4. In Figure 5 shows the relationship between discharge and the product of air temperature and wind-speed (the valley wind is plus, the mountain breeze is minus). Evidently, the correlation between the latter and the discharge is better than the former. Although, the product of temperature and wind-speed does not have accurate physical definition, it does indicate heat exchange between the mountain breeze and the valley wind.

Figure 6 shows daily variations in air temperature, wind-speed and wind-direction. The valley-wind is considered positive and the mountain breeze is negative. Just as we noticed, after the valley wind had converted into the mountain breeze, a phenomenon of subsidence air temperature occurred at nightfall. This phenomenon is universal in summer but happens only occasionally in winter. A great quantity of heat was assimilated which made the temperature of the mountain breeze decrease while ice and snow was melting on the mountains in summer. This explains the heat-exchange by the way of mountain and valley wind between the desert and the mountain glacier.

4. AN IDEA ON CONSERVING WATER RESOURCES

In order to make the oasis on the desert last a long time and be a stable source of water, large "Deficits"of the glacier must be eliminated. Mankid cannot change the radiation condition of the sun on control the route of the monsoon. Thus the only feasible way to solve this problem is to place plants for conserving water in the valley on the middle-mountain region and build artificial grasslands on the highland region. This will weaken the action of the mountain and valley winds decreasing heat loss. This will allow the glacier toprogressively return to a

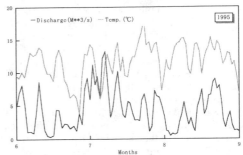

Fig.4 Relation of airtemperature and discharge

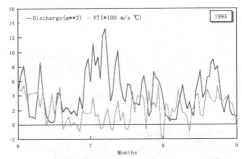

Fig.5 Relation between discharge and
the product of wind-speed and airtemperature

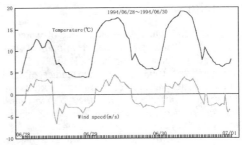

Fig.6 Meteorological process in KARTUS
The valley wind is plus, the mountain wind is minus.

balanced state.

During thirty years, the forest areas were decreased by 13% in the drainage area of SiYang River and the grasslands were destroyed, so that annual mean air temperature in local regions was increased from 0.2 ℃ to 0.4℃ and snow was decreased in Qilian Mountain, Gangsu province. On the basis of measurements, the glacier was retreating at a velocity of 12.5 m to 22.5 m/year. This correlation indicates that forests and grasslands not only are needed to conserve water but they also protect the glacier.

After a forest fire in the Daxianling Northeastward

of China, the annual mean air temperature increased 1.1℃ to 2.1℃. The temperature increase was directly proportion to the range of this forest fire. According to an estimation by professor Xie Zichu, Lanzhou Glaciology and Geocryology, Chinese Academy of Sciences, if the air temperature continues to raise 1℃ in summer, in 150 years the glacier in the Tianshan Mountains will completely disappear.

The water that was expended in the artificial forests and on the artificial grassland never could influence the water balance of the Cele River. It's just that some water evaporated in the desert will only be instead evaporated in middle-mountain region. However, solution of this problem will need to study cooperatively many subjects including hydrology, glaciology, and biology, to implement the government's policy. Protecting the source of water is an action to prevent farther expansion of desertification.

In this research, we were supported and aided financially from the Science and Technology Agency, Japan and Chinese Academy of Sciences. Mr.Tang Xiaosi, Mrs.Tan Jianyu, Mr. Shengying and Mr. Wang Muohai el at., in Lanzhou Glaciology and Geocryology, Chinese Academy of Science and Mr. Kimura have also participated in the study. Professor Xie Zichu has put forward valuable opinions. We deeply appreciate the concern and support of the above persons and organizations and the Xingjiang local government.

REFERENCES

Cheng, L. , & Y.Qu 1994. The source of water in Hexi region and its rational development and application. Sciences Publishing House.

Mikami, M. et al. 1993. Observation studies of land surface-atmosphere interaction in Taklimakan Desert. *Proceedings of the Japan-China international symposium on the study of the mechanism of desertification*:343-348.

Tan, J.Y. 1993. Observation of the water balance in Cele River Basin, Xinjiang, China. *Proceedings of the Japan-China international symposium on the study of the mechanism of desertification* :396-411.

Tang, Q. 1992. Recent tendency analysis of river discharge in Tarim Basin. *J.Desert Research*, 12, No.2:15-20.

Wang, L.W., P.Sun, & T.Yonetani 1994. Hydrological condition in and around the Taklimakan Desert, *Japan Sc. Hydrol & water resources*, 7:223-227.

Wang, L.W., T.Yonetani, et al. 1993. Hydrological characteristics of river on the southern margin of Taklimakan Desert--Cele River as an example, *Proceedings of the Japan-China international symposium on the study of the mechanism of desertification*:461-481.

Yonetani, T, T. Kimura & L.W.Wang 1993. Run-off characteristics in the Cele River Basin--- Preliminary results. *Proceedings of the Japan-China international symposium on the study of the mechanism of desertification*:482-486.

Zhu, Z, & G. Cheng 1994. The desertification of land in China. Sciences Publishing House.

Zhuo, Y. 1994. Change of water and heat state in the frozen after forest-fire on the northern of Daxinan Ling. *Influence and countermeasure of the forest-fire for environment.* Sciences Publishing House.

Snow Engineering: Recent Advances, Izumi, Nakamura & Sack (eds) © 1997 Balkema, Rotterdam. ISBN 90 5410 865 7

The possibility of using snow as a source of active water determined by ^{17}O-NMR analysis

Sankichi Takahashi, Susumu Takahashi & Takeshi Muranaka
Hachinohe Institute of Technology, Aomori, Japan

ABSTRACT: Snow is believed to have great potential as a source of more chemically, or biochemically, active water, the demand for which has recently increased in the semiconductor and biochemical manufacturing industries. The present study has measured the half−width between $278-353$ K, and chemical analysis of 5 snow samples as well as of water collected from two locations in an attempt to clarify the following points.

(1) Establish the dynamic structure of water using ^{17}O − NMR.
(2) Determine influential factors and quantify their effects.
(3) Evaluate the possibility of using snow as a source of active water.

As a result, the following conclusions were derived.

(1) ^{17}O − NMR is effective for analyzing the dynamic structure of water, particularly for chemical analysis of snow and the temperature characteristics of the half−width.
(2) The contributions of structural factors, temperature, water source, and dissolved ion concentration can be quantitatively expressed by applying a polynomial approximation to analyze the temperature characteristics of the half−width.
(3) Snow has great potential to be a source of active water. Therefore, its potential use should be investigated.

1 INTRODUCTION

The recent development of semiconductor and biochemical manufacturing industries has increased the demand for more chemically, or biochemically, active water.

A water unit is composed of both a single H_2O − molecule and an H_2O − aggregate.

This unit is called a cluster. Many clusters can be dynamically bridged by hydrogen bonds to form larger clusters. In general, smaller clusters tend to have higher activity than larger clusters. The basic structure of a cluster consists of a regular tetrahedron with a H_2O − molecule located at its center of gravity and other H_2O − molecules located at each vertex. An ice molecule consists of a hexagonal crystal composed of limitless clusters. Thus, ice tends to be relatively inactive. On the other hand, snow, which is a sublimate of steam consisting of a single H_2O − molecule surrounded by a small aggregate, could provide a source of active water with clusters very close to their basic structure. Recently, ^{17}O − NMR has been used to measure the dynamic structure of water. However, since measurement of the half − width has only been performed at room temperature, the results may not be applicable for practical use.

Therefore, a method which considers such influential factors as the temperature, the origin of the water, and the concentration of dissolved ions must be developed.

The present study measured the half−width at 298 K of untreated, purified, and ultrapurified water collected from two locations. In addition, the half−width of snow samples between $278-353$ K was determined in an attempt to clarify the following points.

(1) Establish the dynamic structure of water using ^{17}O − NMR.
(2) Determine influential factors and quantify their effects.
(3) Quantitatively express differences in the water collected from different locations.
(4) Evaluate the possibility of using snow as a source of active water.

2 SAMPLES USED

The following samples were examined.

(1) Untreated, purified, and ultrapurified water, supplied by two ultrapure water systems for semiconductors, collected from two cities with separate water supplies, Hitachi−city, Ibaraki and Goshogawara−city, Aomori, Japan.

(2) Snow samples.

Thirteen snow samples were collected between Dec., '91 and Mar., '92 in Hachinohe−city, Aomori, Japan from the middle layer of freshly fallen snow immediately after snowfall in order to prevent contamination from external sources such as soil.

Each snow sample was melted at room temperature in a container and the melted water was filtered through a membrane filter with a pore size of 0.1μ m before chemical analysis and measurement of the half−width.

First, pH and electric conductivity of each sample were measured at 298 K.

Since there was no apparent relationship between half−width and electric conductivity, 5 typical samples were selected from the 13 samples, based only on the magnitude of their half−width.

The samples were chemically analysed for the presence of dissolved ions and ^{17}O−NMR analysis was conducted between 278−353 K.

Two samples had a half−width above 100Hz and the other 3 had a half−width between 50 and 80 Hz.

Chemical analysis was performed by ion chromatography in order to determine the ion content of the 5 samples.

Table 1 shows the results of the chemical analysis. Of the ions detected, Na^+, Ca^{2+}, Mg^{2+} were found to be structural ions which promote the formation of bond and the other ions were found to be destructural ions which promote the breakdown of structural integrity.

3 NMR ANALYSIS

3.1 *NMR analyzer used*

Magnetic field strength: 6.7 tesla
Test tube diameter: 10 mm

3.2 *Equations*

The following equations were used to analyze the structure of water based on the half−width of the relaxation curve obtained during ^{17}O − NMR analysis.

- Correlation time/Half−width :
$$\tau_c = 1.44 \pi \, Hw \times 10^{-14} \quad \cdots\cdots(1)$$
- Correlation time/Cluster size :
$$\tau_c = 4 \pi a^3 \mu / 3 \kappa T \quad \cdots\cdots(2)$$
- Correlation time/Activation energy:
$$\tau_c = \tau_0 \exp(E/RT) \quad \cdots\cdots(3)$$

Where, Hw = half−width (Hz)

τ_c = correlation time (s)
τ_0 = initial correlation time (s)
a = radius of cluster (mm)
μ = coefficient of viscosity (Pa•s)
T = temperature (K)
κ = boltzmann constant ($J \cdot K^{-1}$)
R = gas constant ($J \cdot K^{-1} \cdot mol^{-1}$)
E = activation energy ($J \cdot mol^{-1}$)

The correlation time refers to the time required to rotate a cluster, that is, the cluster mobility, which increases as cluster size increases.

Table 1. Chemical analysis of snow samples.

Sample No.	pH (−)	E. Con. (μ S/cm)	Cation (mg/L) / Ion equv. (μ eq/L)						Anion (mg/L)/Ion equv. (μ eq/L)				$T_C + T_A$ (μ eq/L)	T_C / T_A (−)	Cl^- / Na^+ (−)
			Na^+	K^+	Ca^{2+}	Mg^{2+}	NH_4^+	T_C	SO_4^{2-}	NO_3^-	Cl^-	T_A			
1	7.2	59.0	7.65	0.18	0.30	0.93	0.00		1.99	0.30	13.5				
			332.60	4.62	3.75	10.14	0.00	351.1	10.36	4.84	380.3	395.5	746.7	0.89	1.14
2	7.0	37.1	3.95	0.24	1.28	0.72	0.17		1.08	0.54	7.1				
			171.70	6.15	16.00	14.81	9.44	218.1	5.63	8.71	200.0	214.3	432.6	1.02	1.16
3	6.4	22.4	2.02	0.11	0.32	0.35	0.29		1.20	0.49	3.5				
			87.80	2.82	4.00	7.25	6.11	108.1	6.25	7.90	98.6	112.8	220.8	0.96	1.12
7	6.0	29.3	3.41	0.27	0.79	0.36	0.00		0.95	0.13	6.0				
			148.30	6.92	9.88	7.41	0.00	172.5	4.95	2.10	169.0	176.0	350.5	0.99	1.14
12	5.3	37.0	4.15	0.31	0.60	0.56	0.00		1.64	0.18	7.4				
			180.40	7.95	7.50	11.52	0.00	207.4	8.54	2.90	208.5	219.9	432.3	0.97	1.16

It is necessary to determine the temperature dependence of the half−width in order to estimate the cluster size using Eqs. (2),(3).

3.3 *Temperature manipulation*

Few studies have examined the half−width of water at higher temperatures.
Great care must be taken when measuring the half −width at various temperatures in order to ensure stability of the water structure.
Therefore, two methods of temperature manipulation were examined before measurement in order to clarify the influences of temperature manipulation on the stability of the water structure.
One method involved the placement of a test tube in the test location, and then adjusting the test temperature in 2 K steps. The other method involved the transfer of a preheated test tube into the test location at the same test temperature. The test tubes were preheated in a water bath at the test temperature for approximately 2 hours.
Figure 1 shows the results from the same sample measured by both methods.
The first method provided stable values over the entire temperature range, whereas the second method showed substantial fluctuation, particularly between 278−313 K. This instability influenced the accuracy of measurement at higher temperatures. The reasons for the apparent instability may be due to:
(1) Temperature differences between the waterbath and the test location.
(2) Temperature dependence of the water structure.
 Water density is influenced by temperature due to the structure of water.
Figure 2 shows the temperature gradient (d ρ /dT) for water density and the transition zone for water structure between 283 − 313 K. Below this temperature range, the water assumes the structure of ice. However, above 313 K, the structure of water is broken down into smaller units as the hydrogen bonds are broken due to an increase in molecular thermal vibration.
Based on the results from the preliminary study, the temperature was adjusted to the test temperature in 2 K steps for all subsequent tests.

4   RESULTS AND DISCUSSION

4.1 *Influence of water origin on half −width*

Figure 3 shows the relationship between specific resistance and half−width using untreated, purified, and ultrapurified water collected from 2

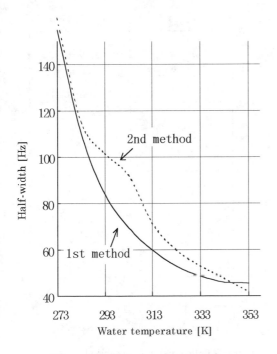

Figure 1. Effect of temperature change method on half-width measured.

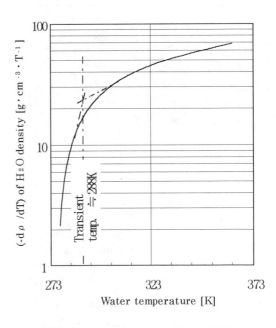

Figure 2. Temperature gradient of H_2O density. (T>277K)

independent sites located approximately 650 km from each other.

Both water sources had a salt concentration estimated to be less than 30 mg/l, consisting primarily of NaCl. The following results are shown in Figure 3.

(1) The influence[11] of dissolved ions in water on the ^{17}O – chemical shift was negligible in the range of the sample concentration because a Na^+ concentration of 1 mol/L is less than ± 1 ppm.

(2) The half – width differed by approximately 15 Hz between the 2 purification systems, regardless of the specific resistance.

These differences may be due to structural differences related to the origin of the water. In the present study, this is referred to as the water difference.

(3) A lower concentration of dissolved salt results in a smaller half – width.

The half – width of the ultrapurified water decreased by approximately 75 Hz to a value that was approximately 50 % of that for untreated water in samples from both locations. This suggests that the concentration of dissolved ions strongly influences the half – width.

Therefore, in order to analyze the dynamic structure of water using ^{17}O – NMR, it is necessary to establish a method which can quantitatively account for structural factors, temperature, water source, and concentration of dissolved ions.

4.2 *Results from snow samples*

Figure 4 shows the relationship between electric conductivity and half – width measured in the 13 samples at 298 K.

The samples were grouped into two ranges according to their Hw, Hw>100 Hz and Hw=50 – 80 Hz. There was no apparent relationship between electric conductivity and Hw.

Thus, the temperature dependence of half – width was determined in 5 of the 13 samples. Samples No.1,3,12 were selected as typical samples of the lower Hw group and samples No.2,7 were selected as typical of the higher Hw group.

Figure 5 shows the results of this analysis. The relationship between temperature and half – width for No.2,7 appeared to be linear, whereas the relationship appeared to be concave for the other samples, composed of two lines which converged at 313 K.

Based on these analyses, it appears that temperature has a greater influence than ion concentration on the structure of water since Hw changed by 100 – 170 Hz from its initial value when the temperature was changed by 75 K(=353 – 278 K).

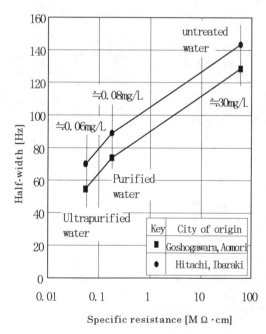

Figure 3. Relationship between half-width and specific resistance.

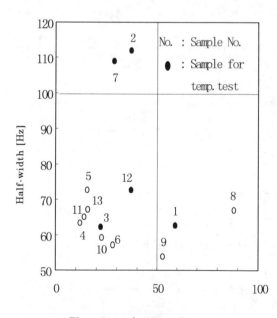

Figure 4. Relationship between half-width and electric conductivity.

580

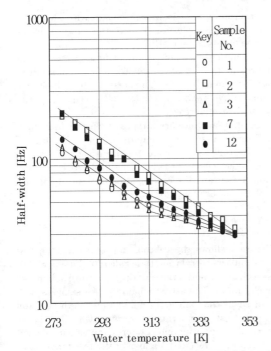

Figure 5. Temperature dependence of
half-width.

5 ANALYSIS

5.1 *Polynomial approximation*

A polynomial approximation is used to quantitatively analyze the effects of each of the structural factors, temperature, water source, and concentration of dissolved ions.

The relationship between temperature and half-width is approximated by eq.(4)

$$Hw(T)=K_0+K_1P_1+K_2P_2+K_3P_3+K_4P_4 \quad \cdots\cdots(4)$$

where, Hw(T)=half-width at a specific temperature (Hz)

$P_1=(T-A_1), P_2=(T-A_2)P_1-B_1$

$P_3=(T-A_3)P_2-B_2P_1, P_4=(T-A_4)P_3-B_3P_2$

T= temperature [K]

$K_0, \cdots K_4$ = constants [Hz·K^{-1}]

$A_0, \cdots A_4, B_0, \cdots B_4$ = constants [K]

Based on the following factors:

(1) The effects of structural factors, temperature, and water source can be quantitatively expressed by Ki, which represents the magnitude of Hw.

(2) The effects of ion concentration can be expressed by Ai,Bi, which represent the slopes of the graphic curves.

5.2 *Effect of water source*

Using eq.(4), the temperature characteristics can be accurately estimated within ± 0.5% of the experimentally observed result.

The constants Ai,Bi are used for all the samples, regardless of the sampling time, as shown below.

$A_1=42, A_2=43.4956, A_3=40.8614, A_4=44.7568$

$B_1=686, B_2=521.037, B_3=343.272$

Since the origin of snow, a sublimate of steam born in the very cold upper atmosphere is believed to be the same water, it should be acceptable to use the same values for A_i, B_i.

5.3 *Effect of dissolved ion concentration*

Table 2 shows the values of K_i, which differ among the samples.

It is noteworthy that the measured Ca^{2+} concentrations shown in Table 1 are in agreement with the absolute values of K_0, K_1.

This suggests that structural ions such as Ca_{2+} may influence the structural composition of water.

Table 3 shows the relationship between K_i and the results from chemical analysis.

This comparison was performed on the basis that the product of Storke's radius and the concentration of a particular ion can express the structural effect since Storke's radius is an index which expresses the structural intensity.

This product is referred to as the structural product (PIR). The order of the absolute values of K_0, K_1 are in agreement with the PIR value for Ca^{2+}.

The relationship between half-width and electric conductivity, as shown in Figure 1, can be explained by the following factors:

Table 2.Numerical values of K_i for each sample.

Sample No.	Constants		
	K0	K1	K2
1	59.110	−1.0487	2.197E−04
2	85.802	−1.9056	1.931E−02
3	60.042	−1.1377	2.652E−02
7	84.162	−1.9081	2.579E−02
12	69.402	−1.5061	3.144E−02

Sample No.	Constants	
	K3	K4
1	−3.694E−04	7.361E−06
2	−3.744E−05	−9.020E−06
3	−4.805E−05	2.683E−06
7	−3.904E−04	9.429E−06
12	−5.619E−04	1.802E−05

Table 3. Effect of ion concentration on H$_2$O clustering.

(mg/1·pm)

Sample No.	1	2	3
$[Na^+]r_{Na}/[Cl^-]r_{Cl}$	1.62	1.59	1.65
$[Mg^+]r_{Mg}/Tor$	1.14	0.84	0.60

Sample No.	4	5	Mean
$[Na^+]r_{Na}/[Cl^-]r_{Cl}$	1.63	1.61	1.62
$[Mg^+]r_{Mg}/Tor$	0.97	1.17	0.94

r_X: Storke's radius of ion

$Tor=[K^+]r_K+[NH_4^+]r_{NH4}+[SO_4^{2-}]r_{SO4}+[NO_3^-]r_{NO3}$

Rain in Japan is strongly influenced by the neighboring seawater which can lead to a high ion equivalent concentration of dissolved salt, particularly Na+ and Cl$^-$.

The lack of equilibrium between the structural effect of Na$^+$ and the destructural effect of Cl$^-$ appears to be related to the mean ratio of $[Na^+]r_{Na}/[Cl^-]r_{Cl}$, which does not equal 1.0. This relationship will continue to exist as long as the ion equivalent concentration ratio of rain is approximately equal to that of seawater.

In the present study, rain water had a ratio of 1.62 compared with the ratio of 1.18 for seawater. Na$^+$ and Cl$^-$ ions do not appear to influence half−width, provided the ratio of concentration between Na+ and Cl$^-$ is similar to that of seawater. There is also an equilibrium between Mg^{2+} and the total of K$^+$, NH^{4+}, SO^{4-}, and NO^{3-}.

Therefore, it appears that only the ion equivalent concentration of Ca^{2+}, which remains free of any interference, is related to K$_0$, and K$_1$ and therefore, influences the half−width.

The constants K$_2$−K$_4$, do not appear to influence the half − width because there is no apparent relationship with the ion equivalent concentration of any of the ions.

As a result, it is possible to express the effect of dissolved ions quantitatively using the constants, K$_0$,K$_1$.

5.4 *Cluster size analysis*

Figure 6 shows the estimated cluster sizes calculated using eq.(2). These ranged in size between $6.3-7.8 \times 10^{-7}$ mm, about 4.6−5.6 times as large as the diameter of a H$_2$O−molecule of 2.6×10^{-7} mm.
This value corresponds to the typical size of an aggregate which is formed by molecules surrounding a centric molecule in approximately double layers.

Furthermore, this is in agreement with the fixed theory that a molecule close to the target nucleus can be observed by NMR.

The cluster size of each sample was set at approximately 7×10^{-7} mm, fixed for a temperature of approximately 333 K. The cluster size of the larger samples No.2 and 7 tended to decrease as the temperature increased.

suggests that there are 2 types of water which have different activation energies at a higher temperature when the cluster size is set at 7×10^{-7} mm, since reducing the cluster size requires greater energy to break the hydrogen bond than increasing the cluster size by hydrogen bonding.

5.5 *Activation energy*

Figure 7 shows the relationship between temperature and the correlation time calculated using eq.(3), as well as the activation energy from the gradient for each sample. The activation energy for samples No.2 and 7 was constant at 17.8 kJ/mol for both over the entire test temperature range. On the other hand,the activation energy for samples No.1,3, and 12 was equal to 17.8 kJ/mol over the test range up to approximately 313 K. Above this temperature, the activation energy was between 14.1 − 15.3 kJ/mol, decreasing as the cluster size decreased. This

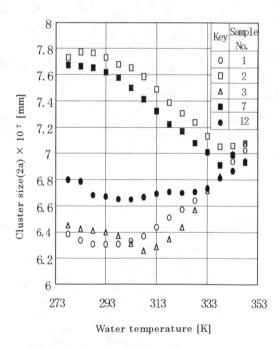

Figure 6. Temperature dependence of cluster size.

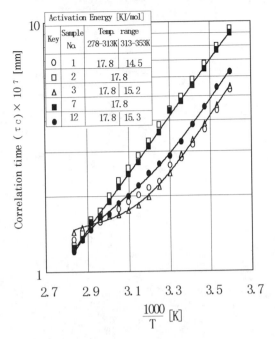

Activation Energy [KJ/mol]			
Key	Sample No.	Temp. range 278–313K	313–353K
○	1	17.8	14.5
□	2	17.8	
△	3	17.8	15.2
■	7	17.8	
●	12	17.8	15.3

$$\frac{1000}{T} \, [K]$$

Figure 7. Temperature dependence
of correlation time.

difference in activation energy is believed to reflect the energy required to break the hydrogen bonds as the temperature rises, since the activation energy was within $8.2-41.8$ kJ/mol, which is the energy associated with the hydrogen bond structuring molecule.

The presence of structural ions would increase the cluster size at a higher temperature.

Krynicki calculated the activation energies shown below from the correlation time of ^{1}H.

For $278-298$ K E=18.48 kJ/mol
For $313-363$ K E=15.49 kJ/mol

These are similar to those reported in the present study.

Based on these results, there appears to be two types of structures with differing activation energies, according to the ion equivalent concentration of Ca^{2+}.

Type 1: Over entire range of tested temperatures:
E=17.8 kJ/mol

Type 2. Range below approximately 313 K.
E=17.8 kJ/mol
Range above approximately 313 K:
E=14.2 - 15.3 kJ/mol

The first type corresponds to water with a cluster size above 7×10^{-7} mm, such as samples No.2 and 7, whereas the other type corresponds to water with a cluster size below 7×10^{-7} mm such as samples No.1,3 and 12.

The existence of volatile water with a small cluster size has been reported.

Small sized water is believed to reduce its latent heat of evaporation due to its low activation energy in a high temperature range.

In addition, this type of water presumably has high biochemical activity, based on its ability to easily evaporate from the leaf of a plant.

This suggests that snow has the potential to be a source of active water. Therefore, its potential use should be investigated further.

6 CONCLUSIONS

The degree to which certain factors influence the dynamic structure of water was investigated for untreated, purified, and ultrapurified water, provided by two ultrapure water systems built at different locations, as well as for 5 snow samples.

The temperature was changed within a range of 278 -353 K.

Based on the results, the following conclusions were derived.

1) The cluster size of water ranges between $6.3-7.8 \times 10^{-7}$ mm, approximately $4.6-5.6$ times as large as the diameter of a H_2O — molecule. This corresponds to the structure of many molecules which surround a molecule in double layers. Furthermore, it suggests that, according to fixed theory, the molecule close to the target nucleus can be observed by NMR.

2) The activation energy determined from the correlation time of ^{17}O is similar to that derived from the correlation time of ^{1}H.

3) ^{17}O — NMR is effective for analyzing the dynamic structure of water, particularly for chemical analysis of rain and the temperature characteristics of the half—width.

4) The contributions of structural factors, temperature, water source, and dissolved ion concentration can be quantitatively expressed by applying a polynomial approximation in order to analyze the temperature characteristics of the half—width.

5) The effect of the water source can be expressed by the polynomial constants, A_i, B_i.

6) The effect of dissolved ions can be expressed by K_0, K_1.

7) Snow has great potential to be a source of active water.

REFERENCES

Arakawa,H. 1989: *Propeties and structure of water and aqueous solution.* Hokkaido univ.:13.

Idem: ibid:15.

Krynicki,K. 1966.: *Physica.*:32,167.

Matsushita,K. 1989.: *Shokuhin to kaihatsu*:24,82.

Matsushita,K. 1990.: *Gekkan hudo kemikaru*:4,42 − 46.

Nakabayashi,H, Konoki,K.1960.: Programming on process design, *Nikkan kogyo sinbunn*:57 − 60.

Nihon denshi Co. NMR group. 1990.: *Nihon denshi news*:31,14 − 16.

Ootaki,H. 1990: *Hydration of ion.* Kyoritu syuppan:55 − 60.

Idem: ibid:61 − 64.

Idem: ibid:67.

Spiegler,K.S. 1962.: *Sea −water purification.* John

Takahashi,S & Koseki,Y. 1991.: *Kagaku kogaku ronbunshu*:17,282 − 283.

Tamaoki,M etal. 1991.: *Nihon kagaku kaishi*:5,667 − 674.

The society of chemical engineer, Japan. 1988: *Handbook of chemical engineering*, p.18, Maruzen, Tokyo(1988).

Watakuki, & Kubota,S. 1992.: *Atarashii mizu no kagaku to riyou gizyutu*, Science forum:103.

Willey & Sons, *Inc.*:138.

Idem: ibid:170.

Snow Engineering: Recent Advances, Izumi, Nakamura & Sack (eds) © 1997 Balkema, Rotterdam. ISBN 90 5410 865 7

Development of Tokamachi-City in a heavy snowfall area: Fighting, utilizing, and enjoying snow

Yoshiro Baba, Kazuaki Sato, Kenji Sudo, Toshiaki Higuchi, Osamu Mori, Shinji Gumizawa
& Katsuya Ikeda
Tokamachi City Municipal Government, Niigata, Japan

ABSTRACT : In the city of Tokamachi, with its heavy snowfall, it is essential to keep the roads and businesses, and thereby the market-based economy, functioning during the winter months. The new Snow Canal System Project will be the most important measure against the snow in the more densely populated areas. This is an introduction to the Snow Canal System and the steps that the city of Tokamachi is taking to cope with snow.

1 INTRODUCTION

Tokamachi is a city of 50,000 people with an area of approximately 212.77 km². It is located in the south of Niigata Prefecture, about 85 km from the prefectural capital, Niigata City. It is known as a leading silk producer and the birthplace of the Snow Festival.

Tokamachi has the most snow of any city of comparable population in the world. In the last decade, the average maximum snowdepth was 194 cm, cumulative snowfall was 982 cm, and average snowfall period was 110 days per year. It is not unheard of to have 80 cm of snow fall overnight. There is a history of struggling with severe snowy weather, and it is our administrative goal to keep traffic going during the winter month.

We have learned much from the endless struggle and experience here in "snow country." In order to effectively deal with the heavy snowfall, the disposal system must be both systematically organized and executed. In crowded residential areas, accumulated snow should be disposed into snow canals, while snow melting pipes take care of falling snow. In more rural sections, snowplows and trucks can keep the roads clean.

Table 1 Snow disposal cost.

Snow disposal by truck	556yen/m³
Snow draining canals	360yen/m³
Snow melting pipes	1,250yen/m³
Road heating	2,471yen/m³

It is necessary to establish such a system which is cost-effective and at the same time saves underground water.

This report introduces the snow canal project as the foundation of the snow removal system.

2 TOKAMACHI CITY SNOW DRAINING CANALS CONSTRUCTION PROJECT

2.1 *Snow Disposal Cost*

As the data in Table 1 demonstrate, the canal system is the best disposal device in terms of cost. In addition, it is impossible to expand the pipe system because of the limitations of the water source. Therefore, Tokamachi City plans to cover roughly 200 ha with canals in order to dispose of excess snow.

2.2 *Basic Requirements on Setting Snow Draining Canals*

Performance of the snow draining canals varies according to the condition of that section of the canal, the shape, amount and speed of waterflow, and so on. Draining canals require the shape and materials to meet the following criteria :

- disposed snow should float away fast.
- falling snow adhering to the canal wall sould not disturb snow removal.
- canal ends should merge into the river smoothly.
- canal structure should be able to endure weight from above.

There are three ways in which snow can be moved into the canals :
by machines, by hand, or a combination of the two.

2.3 *Snow Draining Canals Construction Project*

During this project, the roads under consideration for receiving canals were ranked on the basis of need. Based on research, roads were given one of three priority levels. (Figure 1)
- Priority A : areas with heavy traffic, business districts, national and prefectural roads, areas of dense habitation.
- Priority B : areas with traffic disturbed by snow removed from roofs.
- Priority C : other locations in the designated area.

2.4 *Block Project and Objectives*

Due to limitations on the water source and various geographic reasons, it is difficult to send water to all locations in the project area.
Therefore, the area has been divided into 7 blocks and canals will operate according to a scheduled rotation between these blocks.

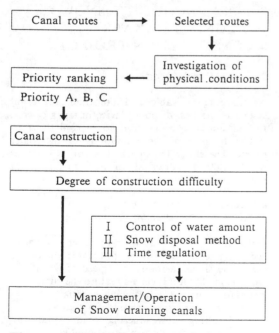

Figure 1　Setting process of snow draining canals.

3 COMPUTATION OF REQUIRED WATER FOR CANAL OPERATION

Figure 2 shows the computation of the flow of water and the length of the period of snow disposal. The amount of snowfall can be calculated in two parts : snow on the roofs of houses, and snow on the roads.　Generally the people of Tokamachi clean the snow off their roofs on weekends or holidays.　Therefore, snow disposal days can be grouped 6 days to 1 cycle.
Snowfall for 1 cycle can than be figured as the maximum snowfall for 6 days during the past 20 years. This amount is approximately 31 cm for 1 day, or 186 cm for 1 cycle.　Snow depth is estimated at 1.2 m for 1 cycle, keeping in mind that snow tends to be less dense during the early period of snowfall and gets progressively denser.
Generally the snow depth on the roads is estimated by the maximum snowfall during a 1 day period over the past 2 years.　This amount has been estimated at 79 cm in Tokamachi.

3.1 *The Amount of Snow Disposed into the Canal*

The amount of snow disposed into the canal can then be calculated using two formulas :
$$M1 = A \cdot H \cdot \alpha \cdot \beta (1+0.1) \times N/S \qquad (1)$$
where, M1 : amount of snow disposed from the roof (m^3)
    A　: snowfall area (m^2)
    H　: snowdepth (1.2m)
    α　: building-to-land ratio
    β　: disposal factor
$$\left(\begin{array}{l} \text{residential/industrial district} \\ \qquad\qquad\qquad\qquad\quad - 60\% \\ \text{business district - 70\%} \end{array} \right)$$
    0.1 : non-disposal factor - 10%
    N　: specific gravity of snow from the roofs - $0.18t/m^3$
    S　: specific gravity of snow in the canals - $0.35t/m^3$

The amount of snow disposed of from the road can be computed as follows :
$$M2 = b \cdot 1 \cdot h \cdot N/S \qquad (2)$$
where, M2 : amount of snow disposed of from the road (m^3)
    b　: road width (m)
    1　: road length (m)
    h　: snow depth (79 cm)
    N　: specific gravity of snow per day - $0.10t/m^3$
    S　: specific gravity of snow in the canals - $0.25/m^3$

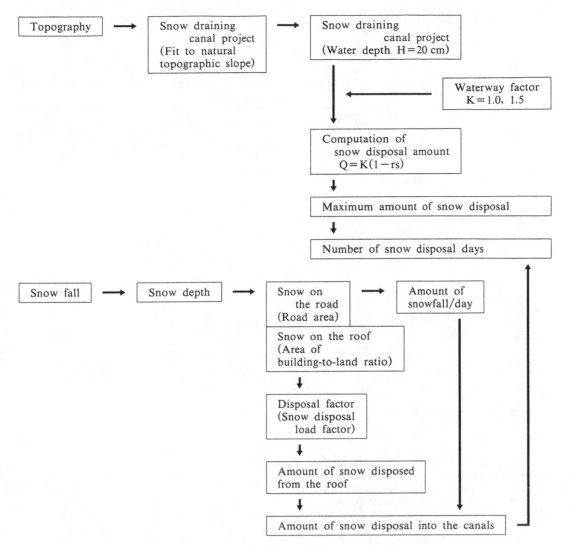

Figure 2   Computation of required water for canal operation.

4 COMPUTATION OF WATER NECESSARY FOR CANAL OPERATION

4.1 *Water Depth in the Canal*

Water depth in the canal is estimated at 25 cm in the Road Structure Ordinance. However, experimental data shows that 20 cm is adequate. In Tokamachi the depth is set at 20 cm taking into account road slope and back current. Historical data proves its applicability.

4.2 *Water Amount in the Canals*

The volume of water in the canals is calculated by using the water depth in the Manning formula:

$$Q = A \cdot V(m^3/s) \tag{3}$$

$$\therefore \quad V = \frac{R^{2/3} \cdot I^{1/2}}{n}(m/s) \tag{4}$$

4.3 *Computation of Maximum Snow Disposal Capacity of Canals*

$$Q = K(1 - rs) \cdot m \tag{5}$$

where,   Q : water flow (m^3/s)

m : volume of snow (m³)

rs : unit weight of snow (0.35)

K : waterway condition factor

$$\begin{pmatrix} \text{good condition} & K=1.0 \\ \text{bad condition} & K=1.5 \end{pmatrix}$$

Thus the capacity of the canals is computed as follows :

When K = 1.0,
$$m = \frac{Q}{K \cdot (1-0.35)} = \frac{Q}{0.65} \quad (6)$$

When K = 1.5,
$$m = \frac{Q}{K \cdot (1-0.35)} = \frac{Q}{0.975} \quad (7)$$

4.4 *Computation of Necessary Water Amount*

As mentioned earlier, the operation of the canals will rotate between 7 blocks. These blocks will receive water according to their priority.

The maximum water flow for each area has been set at Q=2.1m³/s.

5 RESIDENT'S PARTICIPATION AND COOPERATION IN THE CANAL PROJECT

Recently, participation and role allotment is a highlighted subject between citizens and local government. Residents tend to bring all problems to local government, especially in a city like Tokamachi where problems such as snow removal on the roof and flooding caused by melted snow are commonplace.

The city of Tokamachi has set rules regarding the removal of snow and is doing its best to remedy the situation. But it is essential to have local resident's participation and cooperation, but management is left to the resident council.

The canals carry away snow with running water, and can therefore only function when snow is disposed of into the canal. It is residents who do this work, and in exchange they have clear roads. Therefore, the system has a low running cost despite the high initial investment.

Measures against snow must be approached systematically. The city of Tokamachi has spent 5 billion yen during the past 10 years in order to control the snow that falls in this town.

Now the local government is taking the initiative against the elements, acting instead of reacting. This Canal Project is the first step towards an ideal picture of life in "snow country"

6 CONCLUSION

It is said that this is the time of "snow utilization," beyond "snow conquest." There are still many problems to solve in front of us, but we consider that "utilization" and "conquest" are not two different matters.

Utilization is an advanced form of conquest. Tokamachi city has been dealing with snow progressively. We struggled and overcame snow, not hating it, but befriending it. We are sure our snow draining canal system will have a great influence over life and city planning in snowy districts. We developed our own living situation. We will continue to be progressive on challenging projects, and aim high to establish a snow country in which people and nature live harmoniously.

Snow Engineering: Recent Advances, Izumi, Nakamura & Sack (eds) © 1997 Balkema, Rotterdam. ISBN 90 5410 865 7

Quantitative evaluation of walking environment in winter for a snowy city

K. Miyakoshi
Nagaoka College of Technology, Niigata, Japan

K. Kobayashi
Fukui Prefecture Office, Japan

S. Matsumoto
Nagaoka University of Technology, Niigata, Japan

ABSTRACT: The objectives of this paper are to identify the hierarchical structure of inhabitants towards walking environment in winter and to quantitatively evaluate their preferences by applying the method of Analytical Hierarchy Process (AHP). This survey was carried out in the urban areas of Nagaoka. The walking environment of roads can be measured by "a link score", which is mainly dependent on the types of snow removal systems.

1 INTRODUCTION

We have very wet and heavy snowfall in the city of Nagaoka in Niigata prefecture, probably the heaviest snowfall in the world among cities with populations of more than 100,000. Technological development has made it possible to remove and dispose of snow from roads, but walking environment is not improved as compared to driving environment for cars. Further improvement of walking environment in winter has not occured.

Citizens in heavy-snow areas are now asking for higher levels of service than ever before.

The environment of walking has been empirically evaluated, but not by a scientific method.

The objectives of this paper are to identify the hierarchical structure of inhabitants towards walking environment in winter and to quantitatively evaluate their preferences by applying the method of Analytical Hierarchy Process (AHP). A hierarchy comprised of three main factors – convenience, safety and amenity – is specified separately for walking environment.

Modern technology for overcoming snow has developed rapidly over the past thirty years in Japan. We have three basic systems for snow removal from roads : MSR (mechanical snow removal) by bulldozer, motor grader or rotary snow plow, STP (snow-thawing pipe systems) and SCOC (snow-conveying open channels). The same systems are used to remove snow from sidewalks and walking spaces.

Although it is rather difficult to compare the capital cost of facilities (i.e. construction) and operational costs for the three options, we have showed before through a case study that the cost of MSR is rather low and similar to the cost of SCOC (assuming that labor cost of SCOC is excluded), but the cost of STP is much higher than the other two options. One of the technological and institutional problems for local governments is to secure a supply of river water for SCOC and a supply of underground water for STP.

2 ANALYTICAL METHOD

We apply the Analytical Hierarchy Process (AHP) developed by Saaty (1980) to evaluate the preference of inhabitants towards walking environments. The AHP can include and measure all important tangible and intangible, quantitative and qualitative factors. It also allows for differences in opinion and for conflicts, as is the case in the real world. Kinosita (1986) describes one application of the AHP in Japan.

The steps of the AHP for our problem proceed as follows:

(1) The problem is defind as the evaluation of walking environments in winter on a road and side walk where snow removal systems such as MSR, STP, SCOC, or their combinations are furnished.

(2) The criteria for evaluation of the problem are classified into three elements: safety of a road and side walk, amenity for walking, and convenience.

(3) We structure a hierarchy of the criteria, subcriteria and properties of alternative options. Figure 1 shows the hierarchy for walking environments in winter. Figure 2 shows the alternative options of snow removal from a road. We used these options to compute a link score.

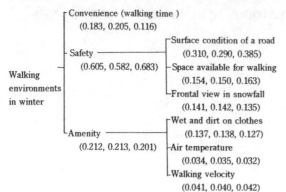

(All, under 60 years of age, over 60 years)

Figure.1. Hierarchy and weight for walking environment in winter

(4) For elements of the lowest level in a hierarchy, we identify a utility function with a value between 0 and 100. The utility function represents the score of an alternative option with respect to an element of the lowest level. This step is our modification and differs from the method proposed by Saaty.

(5) The pairwise elements of the same level are compared in their strength of influence on the next higher level. Then, the matrix of pairwise comparisons is constructed and its eigenvalue is calculated to obtain a set of weights in a hierarchy. For the level which has more than three elements, the consistency index, CI, must be computed to examine the deviation from consistency of pairwise comparisons. The CI can be represented by $(\lambda\ max - n)/(n-1)$, where $\lambda\ max$ is the largest eigenvalue of a matrix of pairwise comparisons, and n is the number of elements of the level in a hierarchy. The CI less than 0.10 is considered acceptable.

(6) We obtain the composite score of an alternative option by summing up the multiplicity of scores of an element and its weight in the lowest level.

Let us denote the weight by Wi and the score of a utility function by Si of element i in the lowest level; then the link score, LS, is

$$LS = \Sigma\ (Wi \times Si),$$

where

$$\Sigma\ Wi = 1.0.$$

3 QUESTIONNAIRE SURVEY AND DATA

We selected the built-up area in the city of Nagaoka for case studies and conducted a questionnaire survey of inhabitants there.

Nagaoka is located in the heavy-snow region, and

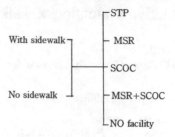

Figure.2. Alternative options of snow removal from a road

the average maximum snow depth is about 1.5 m. In Nagaoka, heavy snowfall has been overcome by snow-thawing pipe systems (STP) and mechanical snow removal (MSR). Snow-conveying open channels (SCOC) have been rarely used until now. SCOC need to be managed by the mutual cooperation of inhabitants.

The questionnaire survey covered both sides of the Shinano River in the built-up area of Nagaoka. The number of valid replies was 1203. The response rate was about 41.9%.

4 RESULTS OF EVALUATION

4.1 *Utility function*

For the lowest elements in the hierarchy, we asked residents to choose one of the alternative situations such that "he/she can accept the situation, but cannot be patient with any worse situation". Then the cumulative distribution curve of the chosen situation yields the utility function of a lowest element, which shows the percentage score of a particular situation of a lowest element. Also, we compared scores of respondents under 60 years of age with those over 60 years of age.

For example, Figure 3 shows the utility function for space available for walking; the score is rather high when people can walk in comfort both on a sidewalk and a road. Figure 4 shows the utility function of the surface condition on a road; the score decreases sharply when the condition changes from 2 to 3 on the horizontal axis. Figures 3 and 4 indicate that there is little difference between respondents under and over 60 years of age.

4.2 *Weight of elements*

Figure 1 also shows the weights of elements in the hierarchy of walking environments, which can be obtained by following the procedure 2 (5) using data of the questionnaire survey. The weight of safety is

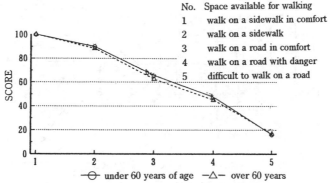

No.	Space available for walking
1	walk on a sidewalk in comfort
2	walk on a sidewalk
3	walk on a road in comfort
4	walk on a road with danger
5	difficult to walk on a road

—○— under 60 years of age   —△— over 60 years

Figure.3. Utility function for space available for walking

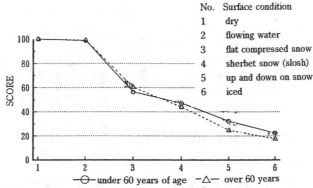

No.	Surface condition
1	dry
2	flowing water
3	flat compressed snow
4	sherbet snow (slosh)
5	up and down on snow
6	iced

—○— under 60 years of age   —△— over 60 years

Figure.4. Utility function of surface condition on a road

greatest (0.605), compared with convenience (0.183), and amenity (0.212). Particularly, for persons over 60 years of age the weight of safety is large. The second level shows that the surface condition of a road (0.310) is more important than the other elements of environment. The space available for walking (Fig.3) has a weight of 0.154. These elements of safety in the second level are important factors.

4.3 Link score for walking environments

We compute the link scores for walking environments under the assumption of several model cases (Fig.2). For the width and structure of a road, we set up several cases; For example, 4 m wide and 6 m wide with or without a sidewalk.

We consider five cases for snow-removal options: MSR only, STP only, SCOC only and the combination of thawing pipe and conveying open channels (STP+SCOC) and no facilities. In the densely built-up area of Nagaoka where houses stand close together and roads are narrow, STP, SCOC or STP + SCOC are primary options. In the outskirts of the area, MSR is the only option for snow-removal and SCOC has not been introduced. Therefor, the combination of MSR and SCOC is not considered in our analysis.

Figure 5 shows the computed link score for some cases. Even under a certain case of snow-removal options and road structure, it is appropriate to assume that the score of elements in the lowest level can have a possible variability or uncertainty. For each case, we compute several link scores (maximum and minimum scores) assuming different and possible conditions of elements; the sensitivity of link score is shown by the arrows in Figure 5.

(1) Given the same road structure, MSR scores lower than STP, and STP + SCOC scores highest. MSR accumulates snow on both sides of the road. In addition, the possible road width and conditions for walking reduce the score of MSR.

(2) STP scores considerably higher than MSR. This is because the surface condition the road environment improves, but on the other hand, the wet and dirt on clothes with STP does not reduce the score .

(3) SCOC scores higher than MSR, because the walking environment improves.

(4) Even under the same method of snow removal, the width of road affects the score considerably, because

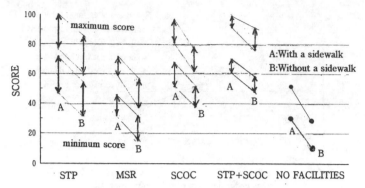

Figure.5. Link score for walking environment by snow removal options

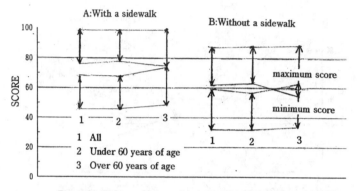

Figure.6. Link score by age for the walking environment with STP

a wider road supplies more space for walking .

(5) As we have already commented, safety is the most important element for walking environment. Where snow is cleared away to both sides of a wide road, a sidewalk of the road is covered with the snow mechanically conveyed from the central part of road, and the sidewalk is no longer recognized to be valuable for pedestrians.

Figure 6 compares scores of under 60 years of age with over 60 years for the walking environment with STP. For persons over 60 years of age, these minimum scores have little difference with those for persons under 60 years.

5 CONCLUSIONS

We applied the method of Analytical Hierarchy Process to evaluate inhabitants' attitudes toward walking environment by snow removal systems, and hence measure the link score of snow-removal options under a variety of conditions. A case study carried out in the city of Nagaoka revealed the following.

(1) The most important factor for evaluation of the walking environment is safety. The surface condition of a road is the major safety issue.

(2) Space available for walking either on a road or a sidewalk, is also rather important.

(3) The link scores for people under 60 years of age are nearly equal to those over 60 years.

(4) Walking environments under the some snow removal options can be evaluated by the link scores.

ACKNOWLEDGEMENTS

This research was supported by the Research and Development Center of Nagaoka University of Technology. We would like to express gratitude to its members, and to the Nagaoka city office members for providing us with materials.

REFERENCES

Kinoshita, E. 1986. The evaluation of travel route choice by the Analytical Hierarchy Process. *Unyu to Keizai* (*Trans. and Econ.*), 46(6), 64–73 (In Japanese).
Miyakosi,K. and S.Matsumoto. 1993 Evaluation the snow-removal options in an urban area based on the preferences of inhabitants. *Annals of Glaciology* 18, 185–189.
Saaty,T.L. 1980. *The Analytical Hierarchy Process.* New York, McGraw-Hill Book Company.

Snow Engineering: Recent Advances, Izumi, Nakamura & Sack (eds) © 1997 Balkema, Rotterdam. ISBN 90 5410 865 7

Intermediate, public-private, territories in snowy cities

Keiji Kitahara
Faculty of Education, Hirosaki University, Aomori, Japan

ABSTRACT: Komise are private territories that have a public character since they are used as pedestrian walkways. Komise are being replaced with arcades in recent years and the spatial meaning of Komise as semi–public space is contributing positiveli to the new style.

1 INTRODUCTORY REMARKS

In snowy cities, we need to consider pedestrian walkways when planning towns. Large mixed–used buildings which cover some blocks and involve public roads in redevelopment projects such as Eaton Center in Toronto, provide examples. In Japan, we do not have many buildings that involve public space except for some experimental projects. It seems reasonable to suppose that we prefer Japanese style structures, which not only border on the road, but also separete the outdoors from the form enclosed by the walls. In our culture, which much use is made of the road and we have a traditional pedestrian way "Gangi" in snow country as I.Mizuno described a few years ago.

– *Each store building is backed away from the street, and the open space is roofed. When each of the roofed spaces along the street are linked together, it becomes a colonnade. Although the arcade in shopping areas, which are very popular in Japan, is similar to Gangi, it is basically different because the Gangi is created on a private lot by each store owner –*

It is popular to regard pedestrian ways such as Gangi, as a design created from life styles in snow country. I consider such space as an intermediate territory that lies between public and private spaces.

2 KOMISE – A BORDER BETWEEN PUBLIC AND PRIVATE

Komise is a traditional architectural form similar to Gangi. From view point of the linguistics, KO means "a little", MI means "watch", and SE means "public". This unique pedestrian way came into existence in the shopping streets in Hirosaki and Kuroishi, called the Tsugaru area in the Edo era.

Gangi is constructed with the clear intention of store owners to link their front open spaces as a continuous pedestrian way like a belt. Komise is generally a part of the architecture, with each shop owner not necessarily intending yo contribute to the creation of a continuous walkway.

Gangi is wider than Komise, and of uniform width store-to-store. Komise is more of a semi–pathway, with various widths (Fig.1).

Komise is basically a private space, but it has a substantial public character. It looks more private than Gangi, but it fulfills its function as a free pathway for pedestrians (Fig.2).

In Japan, open gardens are used in town planning systems, to create public open spaces from private lots. If the building owner ensure the open space with trees and water further the legal standard of the square measure in site, and that space links the front road sufficiently, the government authorizes such open space as an open garden and relaxes restrictions on floor area ratio.

This system is evaluated as a method to produce some comfortable open spaces in urban area, but needs large

Fig.1 Komise with a wide variety of forms in Kuroishi.

Fig.2 A typical Komise as a colonnade in Kuroishi.

Fig.3 An example of a sink provided in a Komise early in the Showa era.

scale lots. So small open spaces which progressive owners deliver are not applicable. If the lot is authorized an opengarden, the owner cannot use such space freely. So there are few cases where open gardens can also serve as walkway or pocket-park. And unattractive open spaces are left over.

Whenever the ambigious intermediate territory between public and private is fixed by ownership or rights of use, the ambigious charm is lost. So the planned development design system that produces open gardens seems risky a method.

But Komise shows the possibility that we can seize intermediate territory not definitely, but probably. It is a space that exists basically as private spot and at the same time has a function as public spot.

But such a intermediate territory on which the ownership is not fixed is forced out of the town planning system. If we want to treat a intermediate territory in public town planning process, we must lose attractive character of ambigious intermediate territory such as the open garden, as I have mentioned before.

So Komise, formed by traditional wisdom as a way to improve daily lives in snow country, has been treated only as a cultural asset not a valuable element of modern town planning.

But we do not need to seize such a traditional space for nostalgia alone since Komise has great value as an active space for realistic town planning. From such points of view, we have some interesting documents on the ownership of Komise in the Edo era.

We have a record that the width and the depth of each lot are described accurately in Kuroishi in 1852. The tax for each building lot was determined according to such data. Komise was not counted as lot area. Even though, columns were placed on the boundary between the lot and the street, and a roof was proveided, the Komise was regarded as a public space and was not a taxable space.

Although Komise was not a taxable space, it does not follow that Komise was a public space. Shop owners displayed their goods in this open space, and, in one case, a sink was proveided in a Komise (Fig.3). In other words,

Komise is a public space having some private use. The responsibility for the land was with the public, but responsibility for daily maintenance was with the private. It is an intermediate concept of ownership, and indicates semi-privatization where a private use occurs on a public lot.

Penal regulations were applied to a deviation from rules of private use. In the Edo era, a famous merchant who held an important post in Kuroishi, built a gate which cut off the continuity of the Komise. He was dismissed from his post and ordered to be penitent.

There are few severe penal regulations in Japanese town regulations or building agreements. A unique town plan where semi-public representation is incorporated into the plan, exixted in pre-modern Kuroishi. This was prompted by the snowy climate.

3 SOME ASPECTS OF MODERN KOMISE

3.1 The Privatization / Semi-Public / Public Issue

Inspite of the former concept of intermediate ownership and use obtained from documents in the Edo era, that concept does not exist in modern Komise today. Komise is regarded as private land and the fixed space where privatization on the ownership progress.

I found that the cars and the vending machines cut off the continuity for pedestrian in Komise, in my survey on modern Komise in Kuroishi. But Komise is regarded as the private land, so we cannot compel the owners to revise their use (Fig.4).

It is difficult for the administration to support financially on the town planning system to the fixed private territory, in Japan. But, even if Komise is recognized as private space, we need to cause the comfortable illusion for the ownership by the representation toward Semi-public such as unique traditional rules in Kuroishi. So Komise can be authorized that it's necessary for administration to support financially.

We can also see examples of modern Komise that

Fig.4 Cars and vending machines cut off this Komise in Kuroishi

Fig.6 A prototype of the modern Komise(Fig.8).

Fig.5 This progressive shop provides an open space facing the Komise, creating a double facade

improve the space for the owner while retaining the public walkway. Figure 5 shows glass doors inlaid between the outside columns, creating a double facade.

3.2 Incorporation of Komise in Modern Buildings

Figure 6 showes a prototype of the modern Komise. I noted a little earlier that the spatial meaning of Komise has been lost. However there are experimental examples where that unique spatial structure has been applied to the design of modern buildings. A department store on the Dotemachi shopping street in Hirosaki built last year provides an example.

Figure 7 showes that is an attractive form of Komise, with the open space used as a cafe terrace and for parking of bicycles. So we regard this spot as semi-public.

Figure 8 is the approach from a park neighboring a

Fig.7 A clever modern Komise in Hirosaki

Fig 8. Komise built in department store, Hirosaki.

Fig.9 The Komise station built at the site conserved by The Komise Club in Kuroishi.

Fig.10 A new Komise produced in the redevelopment in Hirosaki.

depatment store. The space is in the form of a Komise with a double facade (like as Fig.6). This pathway is free to walk after the closing time of the stores, and the elevator can be used to descend to a lower plaza at the river. The concept of having the private sector manage and take the responsibility for the public Komise in former Kuroishi, has evolved into such new designs where the owners manage and bear responsibility for the private land produced as semi-public space. This is an attempt to revive the original Komise concept.

3.3 The Linkage of Komise and Community Design

Six years ago, one developer planned to build a condominium on Komise street in Kuroishi. Then 20 people each contributed about ¥2,500,000 to purchase the land instead as sort of a Komise Trust.

They organized The Komise Club and took an active part in community design. Their approach was to conserve Komise and also apply it to design of a park in the backyard. They have started to link another community groups (Fig.9).

In Hirosaki, Komise is authorized for public property less than in Kuroishi. The traditional storefront has a hard time of existing in a strong movement for redevelopment. But some store owners are trying using to draft a building agreement for new development to make the most of the traditonal design, Komise (Fig.10).

Since it is hard to widen a street, the charm of Komise as intermediate territory will lose to wider streets. Such Komise will become merely colonnades for snow or rain. So We need to consider new development with Komise's spatial meaning, as pre-modern local resources.

4 CONCLUSION

Komise has a place in the spatial structure of snowy cities. However, its intermediate character, how it is ruled by the community rule and managed are important planning issues to all modern cities. So intermediate legislation governing the ownership and use is necessary to street renewal, such as Komise.

REFERENCES

Mizuno,I. 1992. *Re-Application of Traditional Architectural Schemes in the Snow Country, Special Report of Second International Conference on Snow Engineering, 365-371*

Narumi,S. 1992. *A Study on Komise, Kuroishi, 3-26*

Kitahara,K. & Omi,T. 1992. *Formation of SUS-Territories from View Points of Pedestrian Behavior in SUS, Journal of Archit. Plann. Environ. Engng. AIJ, No.433, 119-127*

Snow Engineering: Recent Advances, Izumi, Nakamura & Sack (eds) © 1997 Balkema, Rotterdam. ISBN 90 5410 865 7

Local color traits in connection with the snowy climate of the Tohoku area

Shoji Iijima
Okayama Shoka University, Japan

ABSTRACT: The aim of this study is to clarify the local color attributes in connection with the nature of the regional climate, and to show basic data for color planning of facilities in the Tohoku area. Color investigations were made in the main streets of 12 cities. As a result, the rates of (B)/(A) were correlated with wall color value (R=0.872), and chroma (R=-0.859) [A:sunshine duration per year (Hr), and B: sunshine duration (Dec.-Mar.)]. The numbers of days covered with snow (0<D, D:depth of snowfall) showed some correlation with wall color value (R=-0.519), and with chroma (R=0.482).

1 INTRODUCTION

People living in heavy snow regions suffer not only from damage to buildings but also from uncomfortable visual conditions. For instance, short sunshine duration in winter and snowy landscape are pointed out as visual discomfort. It is said that snowfall and drifted snow make the appearance of streetscapes monotonous and uncomfortable in a snowy region. For color planning, the effect of snow on the visual environment should be taken into consideration. From our experience, we often perceive that vivid (high chroma) walls are beautiful in snowy scenery, but not in scenery without snow.

The aim of this study is to clarify local color attributes in connection with the nature of regional climates (sunshine duration and snowfall), and to show basic data for color planning in the Tohoku area.

········· Climatic boundary line

Figure 1 *Positions of investigated cities*

2 COLOR INVESTIGATION

2.1 *Investigated cities and their climatic enviroments*

Color investigations were made in the main streets of 12 cities ; Aomori, Hirosaki, Hachinohe, Noshiro, Akita, Morioka, Miyako, Omagari, Hanamaki, Kamaishi, Ichinoseki and Sakata. The locations of these cities are shown in Figure 1.

In Asia, especially in Japan, it is known that the regional weather and climate changes from location to location, and that Japanese weather and climate are varied and changeable even in a small area. The nature of Japanese climate is characterized by monsoon activity and a mountain range that divides the

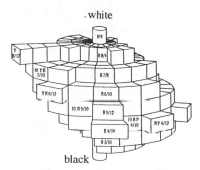

Figure 2 *Munsell color tree*
From Inui (1976)

Japanese island arc into the Omote Nippon facing the Pacific Ocean and the Ura Nippon facing the Sea of Japan (See Figure1). The winter monsoon creates prevailing northwesterly winds and by these winds snowfall and cloudy weather regularly occur in Ura Nippon facing the Sea of Japan (SeeTable 1).

2.2 *Method of color investigation*

In the main street of each city, the wall and signboard colors were measured at around 120 points according to the Munsell book of color based on JIS Z 8721. The measured data consist of Munsell hue, Munsell value and Munsell chroma. Figure 2 shows the Munsell color notation that arranges all colors in a color space. Hue refers to the quality of a color that distinguishes red, blue, green, etc. Hues are arranged around the circumference of the circle. Value shows the lightness of a color and colors vary from dark at the bottom to light at the top. Chroma is the purity or intensity of a color. Chroma increases as it goes outward horizontally from the center.

The investigated streets were chosen for thereason that they are central, prosperous and generally run by local managers. Station squares and the streets adjacent to them were not investigated because nationwide company offices are located there and their signboards tend to be uniform throughout Japan. Wall colors were measured of the colors that covered the largest area of the wall. As to signboards, the representative ones that were larger and self-designed

were measured. When there were more than two signboards per building, a signboard that was considered self-designed was chosen (first standard), and when all were self-designed the larger one was chosen (second standard). On the selected signboards, intensive (high chroma) and light (high value) areas between figures (letters) and the ground were measured. When one of them was achromatic, another chromatic area was measured.

Table 1   Annual average climatic table for the 12 cities*

Station	Temperature (℃)	Sunshine duration (Hr) (A)	Sunshine duration Dec.-Mar.(Hr) (B)	(B)/(A) [%]	The number of days with snowcover ** Snow Depth:D(cm) 0<D
Aomori	9.7	1694.8	328.7	19.4	110.2
Hirosaki	9.8	1965.4	482.3	24.5	108.4
Hachinohe	9.8	1961.8	583.2	29.7	68.8
Noshiro	10.8	1778	373	20.1	114.7
Akita	11.1	1642.2	303.2	18.5	86.6
Morioka	9.8	1815.1	541.9	29.9	88.7
Miyako	10.4	1936.3	656.4	33.9	45.3
Oomagari	10.1	1866.5	432.8	23.2	111.2
Hanamaki	10.3	1730.8	499.5	28.9	89.6
Kamaishi	10.7	1837.4	556.2	30.8	44.5
Ichinoseki	10.6	1728.9	602.2	34.8	71.2
Sakata	12.1	1628.8	281.2	17.3	72.7

* : From The Japan meteorological agency , ;1991,"Climate tables of Japan" , Data observation for 30 years period ,1961-1990
and from The Japan meteorological agency ,;1993,"The monthly normals for AMeDAS stations in Japan(1979-1990)",
** : The Japan meteorogical agency ,;1984," Statistical summary of snow cover in heavy snowfall areas(1954-1982)"

Table 2   Wall colour attributes in the Tohoku area

City	No. of measured points	value mean	value variance	chroma mean	chroma variance	RP	R	YR	Y	GY	G	BG	B	PB	P	2.5YR	5YR	7.5YR	10YR	2.5Y	5Y	7.5Y	10Y
Aomori	176	7.7	2.9	2.1	4.1	0.6	13.1	18.2	36.4	5.7	1.7	0.6	1.7	5.1	0	21.9	15.6	12.5	50	12.5	39.1	23.4	25
Hirosaki	135	7.8	3.1	2.4	6.5	0	8.9	17.8	43	5.2	1.5	0.7	1.5	5.9	0.7	16.7	8.3	12.5	62.5	10.3	37.9	15.5	36.2
Hachinohe	133	8	2.6	1.8	3.9	0.8	8.3	4.5	51.1	3	0.8	0	6	3	0	33.3	50	0	16.7	14.7	23.5	22.1	39.7
Noshiro	100	7.8	1.5	1.8	3	4	9	13	28	3	1	2	12	3	7	7.7	46.2	0	46.2	21.4	39.3	14.3	25
Akita	110	7.3	2.7	2.4	5.9	0	10.9	21.8	37.3	8.2	0.9	0	0	3.6	0	33.3	25	8.3	33.3	4.9	46.3	22	26.8
Morioka	154	8.2	2.2	1.5	3.3	0	9.7	7.1	48.7	1.9	0	0	1.9	4.5	0	0	18.2	0	81.8	9.3	37.3	12	41.3
Miyako	105	8.4	1	1.2	1.1	0	4.8	6.7	50.5	4.8	0	3.8	4.8	1	0	42.8	28.6	0	28.6	11.3	39.6	22.6	26.4
Oomagari	104	8	2	1.8	2.9	1.9	11.5	7.7	47.1	1.9	3.8	0	4.8	1.9	1	12.5	25	12.5	50	6.1	20.4	10.4	63.3
Hanamaki	102	8.6	1	1.6	3.2	0	7.8	4.9	53.9	2.9	0	2.9	2	0	0	40	20	20	20	12.7	29.1	27.3	30.9
Kamaishi	152	8.6	0.9	1.4	3.2	1.3	3.9	13.8	51.3	3.3	0	0	0.7	1.3	0	9.5	42.9	14.3	33.3	5.1	30.8	24.4	39.7
Ichinoseki	100	8.6	0.6	1.2	1.4	0	2	13	43	6	0	0	4	1	0	7.7	46.1	15.4	30.8	7	14	18.6	60.5
Sakata	109	7.7	3.6	2.4	7.2	0.9	11.9	16.5	37.6	4.6	0	0	0.9	2.8	0	44.4	27.8	22.2	5.6	14.6	34.1	12.2	39

* : Total is all measured data.   ** : Total is all chromatic data.
*** : R; red , YR; yellow- red ,Y; yellow ,GY ;green -yellow ,G;green,BG;blue -green,B; blue ,PB;purple- blue, P; purple,RP; red-purple

Table 3   Signboard color attributes in the Tohoku area

City	No. of measured points	value mean	value variance	chroma mean	chroma variance	RP	R	YR	Y	GY	G	BG	B	PB	P
Aomori	176	5.4	4.7	8.4	23.6	4.5	31.8	4.5	6.8	1.1	10.2	1.7	4.5	15.9	0
Hirosaki	135	5.6	4.7	8.1	23.6	2.2	34.7	4.4	4.4	3.7	9.6	0.7	2.2	18.5	1.5
Hachinohe	133	5.2	3.6	8.9	20.4	8.3	32.3	0.8	4.5	3	9	1.5	13.5	9.8	2.3
Noshiro	100	5.3	5	7.6	21.2	3	24	6	5	2	9	1	7	22	2
Akita	110	5	4.3	9.3	20.7	1.8	36.4	6.4	7.3	4.5	4.5	0.9	7.3	15.5	0.9
Morioka	154	5.3	3.9	9.8	17.5	5.2	37	3.9	9.7	3.2	7.1	1.3	4.5	14.9	2.6
Miyako	105	5.6	3.9	9.1	16.2	4.8	36.2	5.7	11.4	1	9.5	1	4.8	16.2	1.9
Oomagari	104	5.5	4.2	7.8	21.6	2.9	36.5	3.8	4.8	1	11.5	0	6.7	10.6	2.9
Hanamaki	102	5.4	4	8.5	17	2.9	32.4	1	13.7	2.9	5.9	1	5.9	19.6	2
Kamaishi	152	5.6	3.9	8.8	19.9	5.9	28.3	6.6	13.2	1.3	5.3	2	7.9	16.4	0
Ichinoseki	100	5.4	4	8	18.1	7	28	2	8	1	11	4	9	14	5
Sakata	109	4.6	4	8.3	20	2.8	23.9	11.9	2.8	1.8	6.4	3.7	8.3	24.8	0.9

* : Total is all chromatic data.
** : R; red , YR; yellow- red ,Y; yellow ,GY ;green -yellow ,G;green,BG;blue -green,B; blue ,PB;purple- blue, P; purple,RP; red-purple

3 RESULTS OF COLOR INVESTIGATION

3.1 Color attributes of each city

Through these investigations, the color attributes of the cities in the Tohoku area were analyzed and identified.

(1) Wall color

Table 2 shows wall color attributes. In the value column, the values in Ura Nippon (Aomori, Sakata etc.) are smaller than those in Omote Nippon (Kamaishi, Hanamaki etc.). As to hues, R (red) and YR (yellow- red) hues predominate in Aomori, Hirosaki, Akita and Sakata. Regarding accessory hues, the color attributes of Aomori, Hirosaki, Akita and Sakata (Ura-Nippon) were characterized by a predominance of GY (Green-yellow), and the color attributes of Miyako, Omagari and Hanamaki were characterized by a predominance of B (Blue).

(2) Signboard color

Table 3 shows signboard color attributes. Yellow hue was predominant in Omote Nippon where there is long sunshine duration. The column for each R hue showed clear tendencies. In Omote nippon, the 2.5R (close to red- purple) hue is higher than in Ura Nippon. In Ura Nippon the10R (close to yellow-red) hue is higher than in Omote Nippon.

3.2 The relation between streetscape color attributes and climatic elements

Correlation analyses were made between climatic elements and color attributes. Regression analyses were done to explain relationships that show high correlation coefficients.

3.2.1 Results of correlation analysis

Correlation analyses were made between climatic elements and color attributes.

(1) Wall color

Table 4 shows the correlation coefficients between wall color attributes and climatic elements. The rates of (B)/(A) were highly correlated with wall color value (R=0.872) and chroma (R=-0.859) [A: sunshine duration in hours per year, B: sunshine duration in hours during Dec.-Mar.]. The numbers of days covered with snow (0<D) showed some correlation with wall color value (R=-0.519), and with chroma (R=0.482).

(2) Signboard color

Table 5 shows the correlation coefficients between signboard color attributes and climatic elements. Sunshine duration (Hr) with signboard value had a correlation of R=0.642. The ratio (B)/(A) had a correlation of R=0.574 with value. The number of days covered with snow (0<D) had a correlation with chroma (R=-0.476).

Table 4   Correlation matrix of wall color

	Sunshine duration (Hr) (A)	Sunshine duration Dec.-Mar.(Hr) (B)	Rate of (B)/(A) [%]	The number of daysr with snowcove Snow Depth:D(cm) 0<D
value	0.305	0.806	0.872	-0.519
chroma	-0.297	-0.808	-0.859	0.482
rate of YR hue	-0.508	-0.667	-0.628	0.233
rate of Y hue	0.501	0.736	0.739	-0.536

Table 5   Correlation matrix of signboard color

	Temperature (℃)	Sunshine duration (Hr) (A)	Sunshine duration (Hr) Dec.-Mar. (B)	Rate of (B)/(A) [%]	The number of days with snowcover Snow Depth:D(cm) 0<D
value	-0.704	0.642	0.628	0.574	-0.003
chroma	-0.134	0.046	0.264	0.285	-0.476
rate of R hue(%)	-0.598	0.396	0.275	0.22	0.076
rate of YR hue(%)	0.773	-0.412	-0.563	-0.558	-0.147
rate of Y hue(%)	-0.156	0.022	0.52	0.614	-0.493
rate of B hue(%)	0.256	-0.038	0.118	0.136	-0.385
rate ofPB hue(%)	0.631	-0.461	-0.47	-0.433	0.09
rate of 2.5R hue	-0.072	0.357	0.797	0.815	-0.693
rate of 5R hue	-0.468	0.448	0.771	0.775	-0.276
rate of 7.5R hue	-0.07	-0.301	-0.657	-0.665	0.58
rate of 10R hue	0.583	-0.473	-0.75	-0.762	0.276

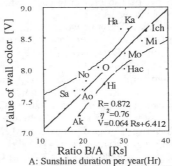

REGEND
(Fig.3 ~6)
Ao :Aomori
Hi :Hirosaki
Hac :Hachinohe
No :Noshiro
Ak :Akita
Mo :Morioka
Mi :Miyako
O :Omagari
Ha :Hanamaki
Ka :Kamaishi
Ich :Ichinoseki
Sa :Sakata

R= 0.872
η^2=0.76
V=0.064 Rs+6.412

Ratio B/A [Rs]
A: Sunshine duration per year(Hr)
B: Sunshine duration (Dec.-Mar.) (Hr)
[Regression Line and 95% Confidence Interval]

Figure 3
Relation between ratio A/B and value of wall color.

3.2.2 Results of regression analysis

The regression analyses were done in order to explain the relation between color attributes and climatic elements.

Figure 3 shows the relation between the ratio (B)/(A) (independent variable) and the value of wall color (dependent variable) by the linear regression equation [1]:

$$V = 0.064 Rs + 6.412 \quad (R = 0.872) \quad [1]$$

where V is the wall color value, and Rs is the ratio (B)/(A). The longer the sunshine duration in winter, the larger the wall color value(V). In other words, areas with a long sunshine duration in winter have high value (bright) wall colorings.

599

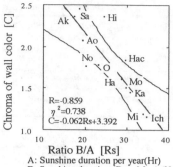

A: Sunshine duration per year(Hr)
B: Sunshine duration (Dec.-Mar.) (Hr)
[Regression Line and 95% Confidence Interval]

Figure 4
*Relation between Ratio A/B
and chroma of wall color.*

Figure 4 shows the relation between the ratio (B)/(A) and chroma of wall color by the linear regression equation [2]:
$$C= -0.062Rs+3.392 \quad (R=-0.859) \qquad [2]$$
where Rs is the ratio (B)/(A) and C is wall color chroma. The shorter the sunshine duration is in winter, the higher the wall color chroma(C). Vivid and chromatic wall colors appear to prevail in cities where having limited sunshine in winter.

Figure 5 shows the relation between the number of days with snow [0<D, D: snow depth] and wall color(value) by linear regression equation [3]:
$$V=-0.01N+8.871 \quad (R=-0.519) \qquad [3]$$
where N is the number of days with snowcover, and V is the wall color value. The smaller the number of days with snowcover, the higher the wall color value.

Figure 6 shows the relation between the number of days with snowcover [0<D , D: snow depth] and chroma by linear regression equation [4] :
$$C=0.009N+1.053 \quad (R=0.482) \qquad [4]$$
where N is the number of days with snowcover, and C is the wall color chroma. The larger the number of days with snowcover, the higher the wall color chroma. In these ways the color attributes of streetscapes in the Tohoku area were made clear.

4 CONCLUSION

In the Tohoku area, the wall color attributes were related the ratio (B)/(A) and the number of days with snowcover. As to color planning, these climatic elements should be paid serious attention with regard to the comfort of streetscape colors.

As to wall color attributes, the ratio (B)/(A) was correlated with wall color value (R=0.872), and chroma (R=-0.859). The numbers of days with snowcover (0<D) showed some correlation with wall color value (R=-0.519), with chroma (R=0.482).

As to signboard color attributes, sunshine duration showed a R=0.642 correlation with signboard value,

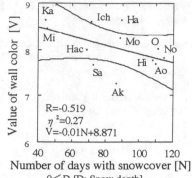

0<D [D: Snow depth]
[Regression Line and 95% Confidence Interval]

Figure 5
*Relation between number of days
covered with snow and value
of wall color.*

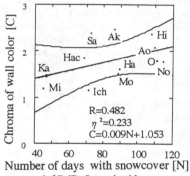

0<D [D: Snow depth]
[Regression Line and 95% Confidence Interval]

Figure 6
*Relation between number of days
covered with snow and chroma
of wall color.*

and the ratio (B)/(A) showed a R=0.628 correlation with signboard value.

REFERENCES

Iijima,Shoji . Study on Relaionship between Local Color Attributes of Streetscape and Climatic Elements.*PAPERS on CITY PLANNING* No. 30. 1995

Inui, Masao. The color planning in architecture. The Kajima press (in Japanses) .1976

The Japan Meteorological Agency. Climate tables of Japan. 1991 (In Japanese)

The Japan Meteorological Agency. Statistical summary of snow cover in heavy snowfall areas. 1984 (In Japanese)

Yoshino, M. Climate in a small area. University of Tokyo press. 1975

Snow Engineering: Recent Advances, Izumi, Nakamura & Sack (eds) © 1997 Balkema, Rotterdam. ISBN 90 5410 865 7

Aufeis growth observed in northeast China

Hideki Narita, Nobuyoshi Ishikawa & Yoshiyuki Ishii
Institute of Low Temperature Science, Hokkaido University, Sapporo, Japan

Shyunichi Kobayashi
Research Institute for Hazard in Snowy Area, Niigata University, Japan

Katutosi Tusima
Faculty of Science, Toyama University, Japan

ABSTRACT: Examination of aufeis was carried out at Heilongjing provice, in northeast China, in 1992 and 1993. Aufeis presents a serious problem for development of roads. Authors observed the process of aufeis growth in situ and considered the growth mechanism.. As a result, it was found that the behavior of water within the aufeis mass and the existence of snow on the surface and air temperature rise are important for aufeis growth.

1 INTRODUCTION

Aufeis distribution was studied in the mountaineous area of Heilongjiang province of northeast China,during the winter. The aufeis was a mass balance problem at McCall Glacier, Alaska (University of Alaska,1972). It is similar to extended ice (Shumskii,1964). In our observed area, underground water overflows the surface or permafrost by breaking the natural balance of the ground water by soring ground freezing. The overflowed water freezes on the surface, and next overflowed water runs on it and also freezes. By repeating the phenomena many times, aufeis grows to a large ice mass.

The aufeises form on road side slopes extend to the road area and obstruct traffic in the mountaineous area of Heilongjing Provice. It is becoming a serious problem for road development. In order to avoid damage from aufeis many ploblems must be solved. The authors examined the process of the aufeis growth during the winters of 1992 and 1993.

2 OBSERVED SITE AND METHODS

The observed aufeis is one of many aufeis which developed by the road side along the Jiayin river in part of the Lesser Khingan range in Heilongjiang Province, northeast China. Figure 1 is a map of the vicinity. Jiayin is a branch of Heilongjiang River (*i.e.,* Amour river).

The observations were carried out during 1992 and1993. Observations during the first year examined mainly the distribution of aufeis and obtained continuous meteorological data. In order to study aufeis growth, some stakes for measuring the thickness of aufeis were set during summer of 1992. In the winter of second year, the stakes were read every day from November, 1992 to March 3, 1993. On March 3, 1993, the length and width of the aufeis was measured and 30 cm deep ice core samples were recovered at three points by core drilling. The samples were brought to Japan and those thin sections were examined for the structural characteristics. Meteorological data (air temperature, solar radiation and wind speed) were obtained at the meteorological station at the mouth of the Jiayin river which is about 4.5 km from the station (Fig.1). They were measured using thermisters, photo-diodes and three-cup anemometers, respectively. Also, the temperature of underground water at the top of the aufeis area was measured with a thermister and recorded by a data-logger.

3 FINDINGS

Figures 2(a) and (b) show the size of the aufeis. The horizontal area was 1550 m^2. The average slope was about $6.5°$. The aufeis began to grow in early October. Its maximum thickness was about 1.3 m on 3 March as shown in Figure. 3. At that time the ice mass was calculated at about 700 m^3. The aufeis

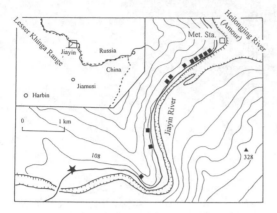

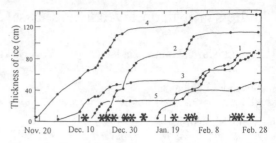

Figure 3　Increase in tickness of aufeis at stakes 1 to 5.

Figure 1　Map of the vicinity of the observed aufeis. Many aufeis formations, which were marked by ■, are distributed along the main road which parallels the Jiayin River. The observed aufeis is shown by the ★. The meteological observation site is shown by the □.

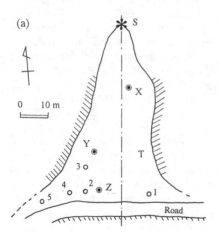

Figure 2(a) and (b)　(a): The plan view and (b): Profile of the aufeis. S and T are points of springing out of ground water. X, Y and Z are ice core drilling sites.

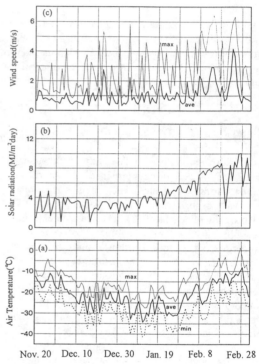

Figure 4(a), (b) and (c)　Air temperature (a), solar radiation (b) and wind speed (c) at the meteorological station during the winter of 1993/1994.

increased in size in steps from early winter to 3 March (Fig. 3). Air temperature, solar radiation and wind speed during the period of growth of the aufeis are shown in Figure 4. The minimum air temperature was -42.5 ℃. The average air temperature was below -10℃ throughout the observational period.

Even during a period of rupid growth of the aufeis, it was below -20℃. The change of solar radiation was very small every day. It was sonny most of the time. The wind speed was very small, also, *i.e.,* the average speed was above 2 m/sec sometimes, but it was below 1 m/sec in most of days.

As mentioned in section 2, in order to examine the structural characteristics of aufeis during the growth of aufeis, thin sections of the ice cores recovered at drilling sites X, Y and Z in Fig.2(a). Figure 5 shows photographs of the thin section of the core drilled at Y. As shown in Figure 5, various stratiform layers which have a different grain shape/size can be found, respectively. The differences are related to the growth conditions of the aufeis.

4 DISCUSSION AND CONCLUDING REMARKS

Aufeises which are distributed in the mountaineous area of Heilongjiang Provice, in northeast China, have formed on roads obstructing traffic; also aufeis which grows on the river has sometimes broken the bridge. In order to understand the damage, it is important to understand the processes of growth and the mechanism of aufeis formation.

The thickness of aufeis increased in steps even where the average air temperature was lower than -10℃ as shown in Figure 3. The fastest rate was commonly in late December and January. The rate was almost constant during middle January and Feburary when the average air temperature was below lower than -20℃. Also, a few increases can be found in late Feburary. It is mentioned above that differences in grain shape and size in various layers are due to the different growth conditions of the aufeis. These layers are classified approximately as follows: (1) a thick layer which has large columnar grains and does not contain many air bubbles, (2) a thin layer which has spherical grains and (3) a fine grained layer which contains many air bubbles. Layer (1) forms after water pools on the surface of aufeis, and freezes slowly (layer A in Fig. 6). Layer (2) forms when water which flows on the surface of the aufeis freezes gradually (layer B in :Fig. 6). Layer (3) forms when new snow covers the aufeis, and water which is flowing on the aufeis infiltrates the snow and freezes (:C-layer in Fig. 6). Mechanis m (1) takes place at warm times. Figure 7 shows the change of thickness of each layer which is measured from Figure 5. The thickest layers contain columnar grains. Coresponding to the layers, heigh air temperature and large solar radiation are recognized

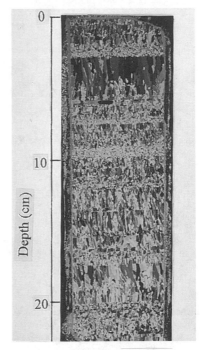

Figure 5　Photograph of a thin section of the site Y aufeis ice core which was taken under cross-polarized light.

Figure 6　Typical layers in the stratigrafical structure of aufeis. (A): Columnar grains. (B) Layer in which water freeze gradually. (C) Layer which froze after water infiltrated into snow on the surface of yhe aufeis.

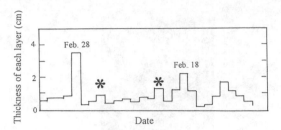

Figure 7 Change of thickness of layers in core-Y. Asteris

REFERENCES

Shumuskii, P.A.(1964): Principles of structural glaciology, New York:Dover Pub..

University of Alaska, (1972) Geophysical Institute, Annual Report 1971-72.

in Figures 4(a) and (b). When the maximum air temperature is in the near of 0℃ and the solar radiation is very large, water on the surface of the aufeis increases abundantly and it pools in some places. This forms thick layer. However, as mentioned before, the period that the growth rate is large also exists when the average temperature it is colder than -20℃. It is thought that this is due to growth by the mechanism (3) which relates to new snow cover. The periods of larger rate agree with periods of snowfall as shown in Figure 4(a). However, for aufeis growth under this condition, water must run on the ice surface or pass throught the ice a long distance before exuding to the surface. Water can not run on the surface without freezing when it is very cold. Flow within the ice will be possible if the temperature within the ice mass is near 0℃. Then, the water can flow in the trifuriate crystal boundaries of the ice. However, the surface temperature was below -10℃, water flow on the surface is hindered. However, as shown in Fig. 4(b) and (c), solar radiation is strong and the wind speed is very small. Since the observed aufeis is located on a southern slope, Intenal melting occurs in the vicinity of the ice surface. As a result, the water which is flowing within the ice will be able to flow out on the surface.

The behavior of water within aufeis ice and the meteorological conditions are very much influence the growth of aufeis.

ACKNOWLEDGMENTS

The authors thank the members of traffic bureau of Heilongjiang, China who give considerable logistical supports to us. This study is conducted as part of a joint study with the traffic bureau of Heilongjing, China. It was funded as an international scientific co-study with Monbu-syou in Japan.

Snow Engineering: Recent Advances, Izumi, Nakamura & Sack (eds) © 1997 Balkema, Rotterdam. ISBN 90 5410 865 7

Preparation of the experimental building for snow and ice disaster prevention

M. Higashiura, O. Abe, T. Sato, N. Numano, A. Sato, H. Yuuki & K. Kosugi
Shinjo Branch of Snow and Ice Studies, NIED, Yamagata, Japan

ABSTRACT: The Experimental Building for Snow and Ice Disaster Prevention is the newest large-scale institution for cooperative use in Japan. The environmental conditions experienced by regions subject to heavy seasonal snowfall will be able to be simulated and various phenomena relating to snow and ice disasters will be studied through experiments conducted there. Construction of the experimental building is expected to take two years and is due to be completed by the end of March 1997.

1 INTRODUCTION

The Experimental Building for Snow and Ice Disaster Prevention (EBSIDP) which was planned by the National Research Institute for Earth Science and Disaster Prevention (NIED), will come under the direction of the NIED's Shinjo Branch of Snow and Ice Studies (Fig. 1). Currently under construction and due to be completed in March, 1997, it is a three-story building with a total floor area of 968 m^2.

In field research, most phenomena related to snow and ice disasters cannot be observed repeatedly. The EBSIDP is designed to simulate in the laboratory, the environmental conditions encountered in regions with heavy seasonal snowfall and to carry out experiments on various phenomena related to snow and ice. The results will be useful in attempts to

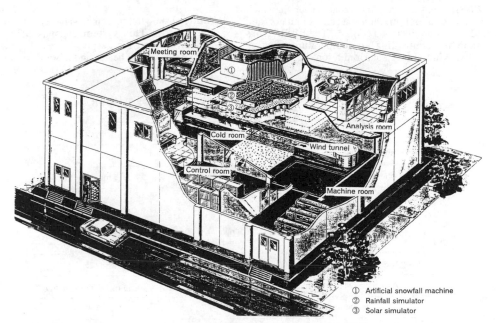

① Artificial snowfall machine
② Rainfall simulator
③ Solar simulator

Figure 1. Schematic plan of the Experimental Building for Snow and Ice Disaster Prevention.

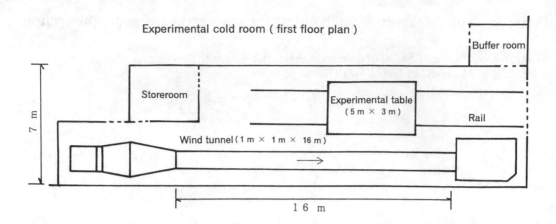

Figure 2. Experimental cold room and equipment.

reduce the impact of disasters caused by snow and ice. The EBSIDP is also expected to make progress in the field of earth science. The facility will also be made available to researchers who wish to cooperate with NIED.

2 MAIN EQUIPMENT AND EXPERIMENTAL LABORATORIES

2.1 Main Equipment

The EBSIDP has an experimental cold room in which snowfall and rainfall machines, a solar simulator, and a wind tunnel are installed (Fig. 2). The EBSIDP has a system for controlling air temperature, relative humidity, wind speed, solar radiation, snowfall and

rainfall. The range of environmental conditions that can be simulated is as follows: air temperature , -30 to +25 °C; relative humidity, 30 to 80%; wind speed up to 20 m/s; solar radiation, 150 to 1000 kW/m^2; snowfall, up to 1 mm/hour; rainfall, up to 2 mm/hour. Wind speed can be controlled in the wind tunnel either by manual operation or automatically by computer. The EBSIDP will be used widely in the research fields of deposited snow, avalanches, snow melt and remote sensing among others. Through wind tunnel experiments, saltation and suspension of blowing snow can be observed in the boundary layer of turbulent flow.

1) Artificial snowfall machine (Table 1)
 System A: This system consists of a cooled adhesive organ for gasphase crystal growth, an evaporator with a heater and a blower to circulate

Table 1. Specifications of the two snowmaking systems.

Specifications	System−A	System−B
Room temperatuare	−2 ℃ ～ −20 ℃	−2 ℃ ～ −20 ℃
Snowfall rate	1 mm/hour (Max)	5 mm/hour (Max)
Areal variation of deposited snow depth	± 10 % >	± 10 % >
Crystal shape of snow	Single crystal (dendrite)	Fine ice particle (sphere)
Size of single crystal	0.5 ～ 5mm	nearly 25 μ m
Area of snowfall	15 m^2 (3m × 5 m)	15 m^2 (3m × 5 m)
Maximum running time	72 hours	72 hours

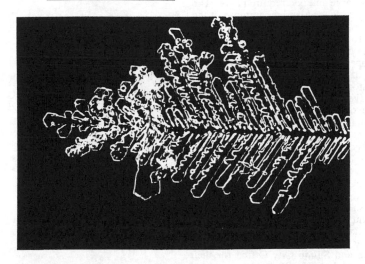

Figure 3. Snow crystal made from the experimental snowfall machine which is planned
(air temperature; -14.7 °C).

air (Nakamura, 1978; Seki, 1996). Frost grown on the adhesive organ is scraped off regularly. The size of crystals produced ranges from approximately 0.5 to 5 mm (Fig. 3).

System B: Compressed water is sprinkled in a cold room. According to adiabatic expansion, fine ice particles 20 to 30 μm in diameter are produced.

2) Wind tunnel

The wind tunnel is a closed circulating type with its own air-conditioning system. It has a collector for blowing snow particles.

3) Solar simulator

The solar radiation produced by the simulator is nearly equal in intensity to the sun's rays. The radiation is variable at the rate of 50 W/m2. Time simulation is controlled by computer.

4) Rain machine

The machine functions in air temperatures greater than 3 °C producing raindrops up to a diameter of approximately 0.5 mm..

2.1.1 How to Use the System

One example of how to use the system,

1) Snow, which is made by the Artificial Snowfall Machine, is deposited on the experimental table. The table measures 3 m x 5 m and can be inclined.

2) After the temperature and humidity in the room have been controlled, solar radiation or rainfall are simulated.

3) Metamorphism of snow can then be studied under the conditions created.

2.2 Experimental Laboratories

The EBSIDP includes an experimental cold room, a buffer room, a low-temperature storeroom, a control room, an analysis room, a telemeter room, machine rooms, a general storeroom, a library, a meeting room with a seating capacity of 50, study rooms and a visiting researcher's room .

3 DISTINCTIVE FEATURES OF MANAGEMENT AND MAINTENANCE

In the planning of the EBSIDP, energy efficiency and unmanned running of the plant were given top priority.

1) In the experimental plant, the biggest power consumption comes from the motors of the refrigerators. As the plant needs many refrigerators, any one compressor may be used for many purposes. For example, one is first used to cool down the experimental room, then later for making artificial snow.

2) The refrigerating compressors have capacity control systems within a 33% to 100% range. The facility has four compressors giving it an overall capacity control system range of 33% to 400%.

3) A hot gas defrost system is used instead of an electrical heating system.

4) The facility will be equipped with two systems for artificial snow-making. System B can produce a large volume of fine ice particles requiring only a

small energy outlay. System A can make crystal snow but has large energy requirements. The choice of system depends on the purpose of the experiment.

5) According to Japanese regulations, when refrigerating capacity exceeds 20 tons, a professional operator is required. In order to meet the objective of unmanned operational capacity, four refrigerators each of which is less than 20 tons in low have been installed, thereby negating the need for an operator.

6) The experimental plant can be operated completely automatically except for the machines which need to be safety-checked on starting.

4 ACKNOWLEDGEMENTS

The authors wish to express our grateful thanks Professor Tsutomu Nakamura of Iwate university for his helpful advice and assistance and also the Japanese society of snow and ice, Yamagata prefecture Office and Shinjo City Office for their support to the plan of the EBSIDP.

5 REFERENCES

Nakamura, H (1978) : Apparatus named "Shimo-bako" to produce a lot of frost. *Seppyo* ,Vol.40, No.1, 31-36.

Seki, M (1996) : Artificial snowfall and snow making machines. *Refrigeration*, Vol.71,No819 , 65-72.

Snow Engineering: Recent Advances, Izumi, Nakamura & Sack (eds) © 1997 Balkema, Rotterdam. ISBN 90 5410 865 7

Density change of ground snow in Hakkoda

M. Sasaki
Hachinohe Institute of Technology, Aomori, Japan

M. Nagao
Ashikaga Institute of Technology, Japan

N. Shuto & M. Sawamoto
Tohoku University, Sendai, Japan

S. Kazama
Tsukuba University, Japan

ABSTRACT: The characteristics of snow depth and density were investigated in this study. The maximum depth of snow is between 3 and 4 m in the Hakkoda Mountains. The depth of snow reaches 3 m early in February, peaks at about 3.5 m, and remains above 3 m until early in April. The depth of snow in mountainous terrain is between 3 and 5 m in the early snowmelt season. The specific gravity of snow is about 0.1 in upper layers at the beginning of winter. The relative density of snow becomes about 0.4 in lower layers early in winter; is above 0.4, except in upper layers, by the middle of winter; and becomes between 0.45 and 0.57 at the end of winter during the snowmelt season. The specific gravity of snow late in winter is above 0.4 in the snow accumulation zone of the Tohoku District in Japan.

1 INTRODUCTION

Disasters caused by floods from snowmelt occur every year in the northern district of Japan. On the other hand, snowmelt runoff from April through July becomes a valuable water resource for irrigation and city water. To forecast stream flow from snowmelt in a particular watershed, one needs to know the depth and density of snow in the mountains. However, those snow conditions have not been shown clearly. We have been observing the density, depth and permeability of snow for 8 years in the Hakkoda Mountains. This paper shows snow densities obtained in the mountainous area, snow depths of the early snowmelt season in the Hakkoda Mountains. The characteristics of ground snow in mountainous areas of the northern districts in Japan are discussed.

2 SNOW DEPTHS IN HAKKODA

Hakkoda have several mountains which are between about 1000 and 1600 m high. The range of the Hakkoda Mountains looms just south of Aomori City located about 700 km north of Tokyo, at 41 ° N, the same latitude as New York, Rome and Madrid.

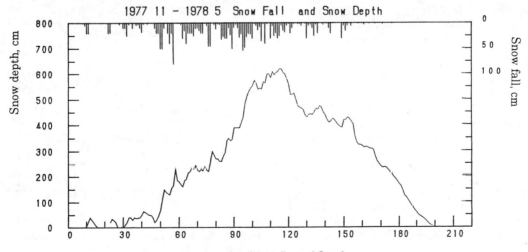

Days after the thirty-first of October

Figure 1. The depth of snow at Sukayu Spa in the Hakkoda Mountains.

The mountains are a short drive or bus-ride from the city (See Appendix Fig. A1).

Snow depth observations have been recorded since 1977 at Sukayu Spa in the Hakkoda Mountains. The maximum depth of snow was recorded at 6.2 m the first year. Figure 1 shows the depth of snow and the snowfall at Sukayu Spa during the 1977-1978 winter. In that figure, the first day of the winter, namely the first of November, corresponds to 1 of the horizontal axis. The depth of snow maximized late in February.

Figure 2 shows the annual maximum depth of snow, hs_{max}, at Sukayu Spa. The figure demonstrates that the maximum depth in 1977 is extremely high. In general, hs_{max} is between 3 and 4 m in the Hakkoda Mountains. As shown in the figure, the maximum depth of snow in Aomori is above 1 m occasionally, and this means that heavy snowfalls occur in that city.

Figure 3 shows the mean depth of snow in the first, middle and last ten days of a month at Sukayu Spa. The mean values are given by averaging 18 years from 1977. The figure suggests that the depth of snow reaches 3 m early in February, peaks at about 3.5 m, and remains above 3 m until early in April. Increases and decreases in snow depth are caused by changes in temperature. Comparing Figures 3 and 4, one can understand the correlation.

Figure 5 is an example of snow depth observations in the Hakkoda Mountains. The observations have been carried out late in March from 1984 along the road from 1040 m-high Kasamatu Pass, through Yachi Spa at 721 m altitude, to Tashiro Gunnkai at 821 m altitude. In the figure, the higher depth of snow in 1985 and the lower depth of snow in 1989 are showing the snow depths in snowy and dry winters, respectively. The middle

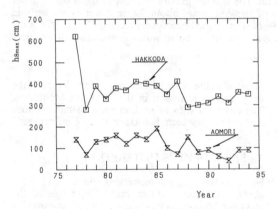

Figure 2. The annual maximum depth of snow.

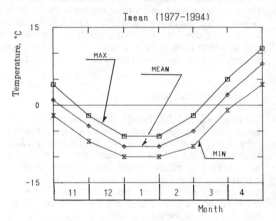

Figure 4. The mean Temperature.

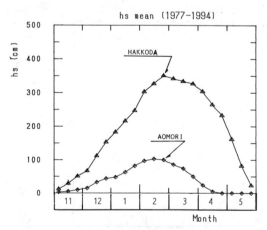

Figure 3. The mean depth of snow.

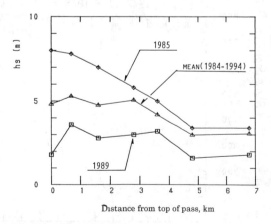

Figure 5. Snow depths in the Hakkoda Mountains.

depth of snow in the figure means the averaged depth for 11 years. As shown in Figure 3, the depth of snow in the mountainous terrain in late in March is lower than that in February. Thence Figure 5 shows that the snow depth is 2 m in a dry winter, and becomes between 3 and 8 m in a snowy winter, and in general is between 3 and 5 m in the early snowmelt season. In Figure 5, the distance in the horizontal axis takes downward from the top of the pass; therefore the higher the area is, the deeper the snow is in the Hakkoda Mountains.

3   CHANGE OF SNOW DENSITY

To forecast volumetric snowmelt, we have been measuring properties of snow, such as depth, density, permeability for 8 years in the Hakkoda Mountains. Table 1 shows the snow density obtained in the snow accumulation zone in December, April and May, where relative densities are shown by values of the specific gravity. As shown in the table, the relative density of snow in Hakkoda is above 0.4 except near the snow surface in the end of winter.

Figure 6 shows an example of the density of snow in the Hakkoda Mountains. In the figure, the vertical distributions of the relative density of snow in December, February and April correspond to early,

Table 1. Specific gravity of snow. The vertical distance, z, is measured from the snow surface.

Date	Specific gravity		
10/12/1985	0.085	at	z = -0.2 m, elevation 1320 m
10/05/1986	0.512 0.525 0.440	at	z = -0.5 m, elevation 925 m
23/04/1988	0.514	at	z = -0.5 m, elevation 960 m
24/04/1989	0.544	at	z = -0.5 m, elevation 970 m
22/04/1990	0.566	at	z = -0.5 m, elevation 1320 m
23/04/1991	0.47	at	z = -0.5 m, elevation 1350 m
	0.54	at	z = -0.5 m, elevation 1380 m
	0.38	at	z = -0.5 m, elevation 1360 m
	0.48	at	z = -0.5 m, elevation 1360 m
	0.575	at	z = -0.5 m, elevation 970 m
16/04/1994	0.45 ~ 0.575	at	z = -0.25 - 2.0 m, elevation 1210 m (See Figure 6)
15/04/1995	0.45	at	z = -0.2 m, elevation 1552 m
	0.48	at	z = -0.5 m, elevation 1552 m
	0.49	at	z = -0.1 m, elevation 925 m
	0.28	at	z = -0.2 m, elevation 925 m

middle and late winter, respectively. The figure suggests that although the specific gravity of snow is about 0.1 in the upper layer, the relative density becomes about 0.4 in lower layers early in winter. In the middle of winter, the relative density is above 0.4 except for the upper layer. The relative density of snow increases to between 0.45 and 0.57 late in winter, in the snowmelt season.

Figure 7 shows vertical profiles of the snow density obtained by one of authors, Sawamoto, at 1440 m altitude in Miyagi Zaoh located about 350 km south of Hakkoda. As shown in the figure, the

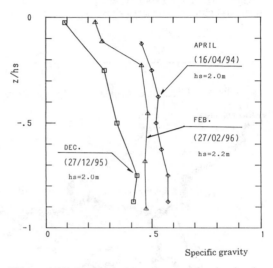

Figure 6. Seasonal change of snow density in the Hakkoda Mountains

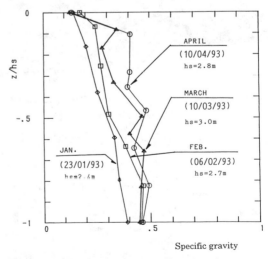

Figure 7. Snow density profiles in the snow accumulation zone in the Tohoku district.

relative density of snow late in winter is above 0.4 in the snow accumulation zone in the Tohoku district of Japan.

4. CONCLUSIONS

The characteristics of snow depth and density were investigated in this study. Within the scope of the present study, the following conclusions were derived:

(1) The maximum depth of snow is between 3 and 4 m in the Hakkoda Mountains. The depth of snow reaches 3 m early in February, and peaks at about 3.5 m late in February, and stays above 3 m until early in April.

(2) The depth of snow in mountainous terrain is between 3 and 5 m early in the snowmelt season. The snow depth is about 2 m in a dry winter, and between 3 and 8 m in a snowy winter. The higher the area is, the deeper the snow is in Hakkoda.

(3) The specific gravity of new snow is about 0.1 in upper layers at the beginning of winter. However, the specific gravity of snow becomes about 0.4 in lower layers early in winter. The relative density is above 0.4, except upper layers, in the middle of winter. The relative density is between 0.45 and 0.57 at the end of winter during snowmelt.

(4) The relative density, namely specific gravity, of snow late in winter is above 0.4 in the snow accumulation zone in the Tohoku District of Japan.

Appendix A.   Location of the Hakkoda Mountains.

Fig. A1.   Location of the Hakkoda Mountains.

Snow Engineering: Recent Advances, Izumi, Nakamura & Sack (eds) © 1997 Balkema, Rotterdam. ISBN 90 5410 865 7

Snow cover conditions in Finnish Lapland

Jussi Hooli
Department of Civil Engineering, University of Oulu, Finland

ABSTRACT: Without exception, the greatest flood of the year in the northernmost part of Finland, Lapland, is caused by melting snow. Because the altitudinal changes are greater than in the rest of the country and the network of hydrological observations is limited, special attention has been given to altitude and areal representativeness of the water equivalent of snow.

1 INTRODUCTION

Finnish Lapland is the northernmost administrative province of Finland. It is about 100,000 km² in size and constitutes almost 30% of the area of Finland (Fig. 1). There are just under 200,000 inhabitants in the area, which is about 4% of the whole population. Southern Lapland is bordered by the flat Northern Ostrobothnia. The more mountainous area begins at the polar circle. Most of Lapland is 200 to 300 m above the sea level. In the eastern and northeastern Lapland, mountains are over 400 meters above sea level and some of the highest peaks are 600 to 800 meters high. In northwestern Lapland, which is geologically connected to the Scandian mountain range, the highest peaks reach up to 1,000 meters. The rest of Lapland is bedrock and the mountaintops are round. The highest mountains are barren even in the southern part of Lapland. The treeline is reached at even lower altitudes in the north. The main mineral soil is moraine and there are boulder fields caused by disintegration at several locations. Swamps amount to over 60% of the area in southern Lapland whereas in the north they amount to less than 20%. Disregarding the northernmost part of Lapland, the waters run mostly to the south, to the watercourses of the River Torniojoki and the River Kemijoki. The climate is continental but the Gulf Stream makes it milder in parts of Lapland.

2 SNOW ACCUMULATION

Snow forms about the same amount of the annual rainfall in all parts of the country, i.e., approximately 200 to 250 mm. Because the annual rainfall diminishes towards the north, the relative proportion of snow is greatest in Lapland where it usually constitutes half of the annual precipitation.

The permanent snow cover falls around Christmas in the southwestern parts of the country and in Lapland in the end of October. The first snow falls in southern Finland 30 to 50 days and in Lapland 15 to 25 days prior to the permanent snow cover.

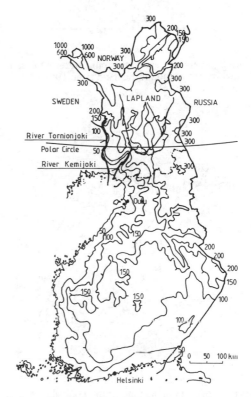

Figure 1. Finland and Lapland; topographic contours above sea level are presented.

The average annual maximum water equivalent of the snow cover is 80 to 140 mm in the southern Finland and in Lapland 160 to 200 mm (Fig. 2). The average date of the maximum water equivalent is 20 March in southern Finland and a month later is the northernmost part of Lapland. The 20 year mean recurrence interval maximum water equivalent of snow is shown in Figure 3. In most parts of southern and central Finland this maximum is 190 to 230 mm. In Lapland, in the areas with most snow, this maximum in nearly 300 mm.

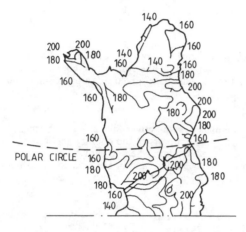

Figure 2. Average annual maximum water equivalent of snow on the ground in Lapland from 1961 to 1975 (Solante 1981) (mm).

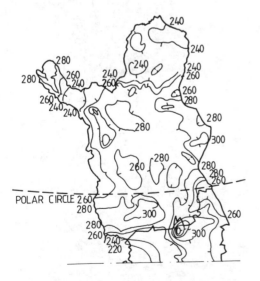

Figure 3. Maximum annual water equivalent (mm) that recur once in 20 years, based on data collected between 1952 and 1984 (Perälä & Reuna 1990).

The average maximum snow depth in southwestern Finland and on the coast is 25 to 45 cm and in eastern and northern Finland 50 to 70 cm. During most snowy winters, there can be as much as one meter of snow on the ground in Lapland.

The snow cover melts from the forests in southwestern Finland in late April-early May and a month later in Lapland. Snow disappears from open fields 8 to 12 days earlier.

3 SNOW SURVEYS AT THE METEOROLOGICAL STATIONS AND ACROSS THIS LAND

Snowfall can be measured with an ordinary precipitation gauge by melting the snow that has accumulated in the container. Precipitation gauges are used in 610 measurement points and the measurements are made once a day. Mainly due to errors caused by wind, these measurements are rather inaccurate. However, corrected results of these measurements are used in real-time hydrological forecasts due to their quick availibility. The regional water equivalents of the snow cover are calculated from the rain observations with the help of a temperature index model. The accuracy of these calculations can be checked by determining the water balance (Vehviläinen, 1994). In addition, the regional values are compared to the values on snow course measurements.

Because snow is subject to melting, evaporation and wind transport, the field measurements aim at observing the properties of the snow pack and its amount. The depth of the snow cover is measured with a snow stake that has a scale in centimeters on it. The stake can either be fixed or transportable. Equipment like a steel cylinder open at one end, a snow balance and a small snow shovel are needed to measure the water equivalent of snow. The cylinder is pressed in the snow. The snow depth can be read from the scale and the water equivalent of snow from a balanced scale.

When moving from one measurement point to another, the snow depth, density and water equivalent vary. In order to get a good representation of the snow conditions in the area, the measurements are made along snow courses. The snow course is located in the terrain so that it represents as well as possible the terrain type and the tree stand, slope and height relations of the area. The course can be in a shape of a square but in mountain areas the course has to run straight across the terrain. On the snow course, snow depths are measured at 50 to 80 points at intervals of 50 m, and the water equivalent of snow is determined at every fifth point in order to determine the snow density. There are a total of 170 snow courses in Finland. the snow courses are measured once or twice a month.

Beginning in 1990, in between the snow course measurements, the areal water equivalent of the snow cover has been estimated by using a network of grid points along with observations on daily rainfall and temperature (Reuna 1994). The procedure of areal water equivalent estimation has been computerized.

The country is covered with a grid having a square size of 10x10 km. Each grid point has given weights in relation to one or several of the 108 drainage basins, for which the areal water equivalents are routinely estimated.

An adjustment factor was calculated for each grid point and for each snow course. This factor takes into account local topography, the proximity to the sea coast and the frequency distribution of wind directions, all of which affect the distribution of snowfall.

The water equivalent at a grid point is estimated by using 1 to 4 of the nearest snow courses. The weight of each snow course is inversely proportional to its distance from the grid point. The adjustment factor ratios of snow courses and grid points are used to correct the unrepresentativity of snow course locations.

4 AEROGAMMASPECTROMETRY

Promising results have been gained by using aerogammaspectrometers to estimate the water equivalent of snow. The method is based on the fact that the gamma radiation emitted by soil is absorbed by the snow cover (Kuittinen & Vironmäki 1980).

$$N = N_0 e^{-(\mu_0 H + \mu_w W_s)} \tag{1}$$

where N = the intensity of the gamma radiation recorded from an airplane; N_0 = the intensity of the gamma radiation emitted by the soil; μ_0 = the absorption coefficient of air to gamma radiation; μ_w = the absorption coefficient of water to gamma radiation; H = flying altitude; W_s = the water equivalent of snow cover.

When the measurement flight is made following the same route during the snow-free season and during snow cover, the only unknown factor in the equation is the water equivalent. In the catchment area of the River Kemijoki, measurement flights have been made during 15 winters, and the measurement results have been used to improve the areal representativeness of the previously described methods, especially as the catchment area has fewer snow courses than usual and variations in altitude are the biggest in the country.

However, if made from an airplane, certain limitations have to be considered as far as flying conditions and radiation are concerned.

Due to absorption of gamma radiation, the flying altitude and the speed have to be rather low. Therefore the plane has to be able to follow the topography of the terrain even at a slow speed. The limited ability of the plane to do this safely thus limits the placement of snow courses in rough terrain.

Some attention has to be given to easy orientation on the snow course. Especially the starting and ending points have to be placed in terrain that is easily recognizable. This is important because the accuracy of water equivalent measurements depends on how well the same route can be followed on each flight. In addition, to minimize costs, the courses should be placed so that moving from one line to another is easy.

Usable snow courses have to have good radiation properties. In order to find areas with intense radiation, good basic geological and geophysical knowledge of the are has to be available. On the other hand, least radiating areas, such as swamps and lakes, can be seen on topographic maps as well. Also, due to the fact that the radiation levels change when the soil moisture changes, the method has been used mostly in northern Lapland where there are less swamps than in the southern part of Lapland.

When using aerogammaspectrometry, the relative error in water equivalent measurements has been somewhere in the region of 5 to 10% of the measured water equivalent (Kuittinen, Autti, Perälä & Vironmäki 1985). This level of accuracy is comparable to course measurements made in the terrain. It has to be kept in mind that if measures are made with the help of the gammaspectrometer, the snow course has to be about 5 km long and background radiation measurements have to be made at one hour intervals and the orienteering error must not be more than 100 m. Nowadays, with better determination of position it is possible to reach even more accurate results by careful navigation. User experiences have shown that this relatively expensive method provides more reliable results before the snowmelt and thus adds to the representativeness of the more conventional methods.

5 THE USE OF SATELLITE IMAGES

The wideness of the snow cover has been observed with the help of satellite images for about ten years. In view of observing the snowmelt in northern Lapland, it has been noticed that the forestation is not too thick to prevent obsevations made by using the NOAA weather satellite. These images are available daily over all of Finland. Based on the gathered material and weather reports from the Meteorological Institute, approximately 2 to 3 satellite images suitable for snow estimates can be obtained each week. A method has been developed to process the satellite images so that the snow lines can easily be seen and thus the broad outline of the snowmelt can be followed even in large catchment areas.

That snow can be seen on satellite images is based on the fact that snow and soil reflect and emit electromagnetic radiation differently. Snow is in fact one of the easiest natural phenomena to observe with satellite images. However, the way the snow reflects radiation changes as the quality of the snow changes. Reflectivity changes with grain size, impurities in the snow, snow depth, water equivalent, free water and the structure of the snow surface.

The use of satellite images allows rough estimates to be made of the water equivalent of snow during the period of snowmelt even when there is no new snow on the ground. When bare spots with no snow appear, it reduces the albedo of the terrain. This can be seen as darkening of the picture.

Images taken by a satellite can easily be processed and different kinds of maps can be added to make the interpretation of information easier.

6 THE SNOW AREAS OF FINLAND

Finland has been divided into different snow areas on the basis of mean water equivalent, duration of the snow cover and thawless period and variation coefficient of the maximum water equivalent:

A. Mean maximum water equivalent of snow on the ground: 1. less than 100 mm; 2. 100 to 180 mm; 3. more than 180 mm.

B. Duration of snow cover in open fields: 1. less than 4 months; 2. 4 to 6 months; 3. more than 6 months.

C. Duration of the thawless period: 1. less than 2 months, no distinct thawless period in mild winter; 2. 2 to 4 months; 3. more than 4 months.

D: Coefficient of variation of the maximum water equivalent in annual series: 1. less than 0.30; 2. 0.30...0.50; 3. more than 0.50.

On the basis of these four characteristics, the regionalization shown in Fig. 4 was obtained. In all, nine different snow regions were distinguished. Their classification codes are as follows:

Regions	A	B	C	D
1	1	1	1	3
2	1	2	1	2
3	2	2	1	2
4	2	2	2	2
5	2	2	2	1
6	2	3	2	1
7	3	3	2	1
8	2	3	3	1
9	3	3	3	1

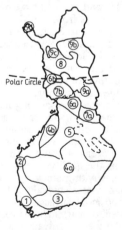

Figure 4. The snow regions of Finland (Kuusisto, 1984).

Regions 4, 6 and 7 each include two separate parts, region 9 includes 4 parts. The sum of the first three digits provides an index of the severity of the winter, while the fourth digit indicates year-to-year-variation.

The dashed line inside region 5 indicates the area, where terrain with higher altitudes belongs to region 7. The Lake Kilpisjärvi regime coincides with snow region 9d. As a rule of thumb, a small basin was considered to belong a "class of a large spring flood" if the mean of two largest monthly runoff values in spring was greater than 30 l/s km^2. This criterion was fulfilled in the region, which roughly coincides with snow regions, 7, 8 and 9.

7 VARIATION IN THE PROPERTIES OF SNOW COVER

Microvariation of water equivalent caused mainly by wind and uneven terrain has been shown to follow normal distribution. Microvariation causes typically a standard deviation of 3 to 15 mm in water equivalent levels.

The mesovariation is caused by altitude, direction of slope and the intensity and duration of storm, wind speed and direction, vegetation (open fields, different forest types) and the process of snowmelt.

On the mountains of Lapland, altitude differences have been observed to have a considerable effect on the water equivalent of snow when the difference in altitude in at least 100 m. When the difference in altitude is 200 m, the water equivalent of snow on the top of the slope can be double to that measured at the bottom (Fig. 5). The increase in water equivalent is at its greatest during the maximum water equivalent, i.e. 35 to 40 mm per elevation of 100 m which is 20 to 30% of the water equivalent of snow courses on the surrounding area. The maximum water equivalent of snow can be measured at tree limit and it reduces above it towards the top. This is mainly due to wind transporting the snow on a treeless slope. Snow travels down the slope with the wind and forms drfits at the treeline where the maximum water equivalent forms (Hooli 1973, Hooli & Ollila 1997, Ollila 1984).

According to five small catchment areas in Lapland, the water equivalent should be raised by 20 to 30% above the value given by the measurement methods normally used in these areas. In the upper part of River Kemijoki catchment area, this increase should be only a few percent. Nowadays this has been taken into consideration when estimating areal water equivalents with the help of the network of grid points mentioned earlier.

The direction of the slope has no significant effect on the maximum annual water equivalent of snow. However, when the snow has begun to melt, the water equivalents on the southern slopes decrease faster than on the northern slopes.

Variation of water equivalent between different terrain types has been researched quite a lot in Finland. In clearings the water equivalent is generally 10 to 20% higher and in a thick coniferous forest 10 to

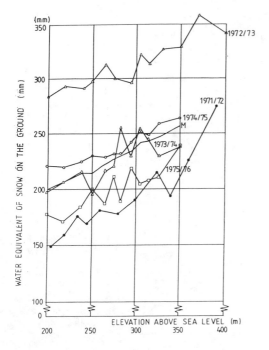

(mm)
350

1972/73

300

WATER EQUIVALENT OF SNOW ON THE GROUND (mm)

1971/72

1974/75
M

250

1973/74

1975/76

200

150

100
0

ELEVATION ABOVE SEA LEVEL (m)
200   250   300   350   400

Figure 5. The effect of elevation on the maximum annual water equivalent of snow on the ground in the mountains of Finnish Lapland.

20% smaller than in open fields (Kaitera, 1949, Mustonen 1965).

8 CROWN SNOW LOAD

Crown snow load means snow and ice that accumulates on trees, masts and other vertical structures during winter. Crown snow load forms on areas where the altitudes is high and where the southwesterly and northeasterly winds bring humid air from the open sea. Areas of extensive crown snow load are situated in the highlands of southern Lapland. Characteristic of these areas is the damage that can be seen in the forests during snowy winters due to uneven snow load.

Crown snow load appears only above a certain altitude. This limit has been calculated by using the equation $y = 0.72x + 55$ m where x is the height of the mountain in meters. According to this equation the limit of the crown snow load on a mountain 300 m high is 271 m and on a mountain 400 m high is 343 m. Crown snow load appears most on eastern slopes and least on western slopes. The effect of the crown snow load on the maximum annual water equivalent of snow on the ground is 5...10% in small catchment areas whereas in larger areas it has no significant effect (Ollila 1984).

9 CONCLUSIONS

Following observations are presented as conclusions:
1. The numerical values of the water equivalent of snow (mm) correspond to ground load caused by snow (kg/m^2). This data is continuously available.
2. The greatest flood in Lapland is always caused by the melting snow.
3. Annual variations of the maximum water equivalent are small in the snow regions of Lapland.
4. Relative altitudes have been found to have an influence on the water equivalent of snow in the mountains of Lapland.

REFERENCES

Hooli, J. 1973. The effect of elevation and slope direction on the water equivalent of the snow cover on mountains in Lapland on the basis of observations taken during the winter of 1971-72. Helsinki University of Technology. Research paper 43. 67 p (Finnish original, English summary).

Hooli, J. and Ollila, M. 1977. Snow Cover Conditions in Mountains Area of Lapland. Helsinki University of Technology. Laboratory for Hydrology and Water Resources Engineering. Reasearch report. 17 p. (Finnish original).

Kaitera, P. 1949. On the melting of snow in springtime and its influence on the discharge maximum in streams and rivers in Finland. Helsinki University of Technology. Research papers 1.

Kuittinen, R. and Vironmäki, J. 1980. Aircraft gamma-ray spectrometry in snow water equivalent measurement Hydrological Sciences Bulletin 25:1, pp. 63-75.

Kuittinen, R., Autti, M., Perälä, J., & Vironmäki, J. 1985. Determination of snow water equivalent with gamma-ray spectrometry and satellite images. Technical Research Centre of Finland. Research report 370. 98 p. (Finnish original).

Kuusisto, E. 1984. Snow accumulation and snowmelt in Finland. Publications of the Water Research Institute. 55. Helsinki. 149 p.

Mustonen, S. 1965. Effect of meteorologic and terrain factors on water equivalent of snow cover and on frost depth. Acta Forestalia Fennica 79. (Finnish original, English summary).

Ollila, M. 1984. Occurence of crown snow load and the effect of elevation and slope direction on the water equivalent of the snow cover. Helsinki University of Technology. Division of Water Engineering. Report 35. 121 p. (Finnish original).

Perälä, J. & Reuna, M. 1990. The areal and time variability of snow water equivalent in Finland. National Board of Waters and Environment. Publ. 56. Series A. 260 p. (Finnish original, English abstract).

Reuna, M. 1994. An Operational Grid Method for Estimation of the Areal Water Equivalent of Snow. Geophysica 30. 1-2, pp 107-121.

Solantie, R., 1981. Average annual maximum

water equivalent in Finland. Vesitalous 1981:5 p. 21-22. (Finnish original).

Vehviläinen, B. 1992. Snow cover models in operational watershed forecasting. National Borad of Waters and Environment, Finland. Helsinki 1992. Publications of Water and Environment Research Institute. No 11. 112 p.

Vehviläinen, B. 1994. The watershed simulation and forecasting system in the National Board of Water and the Environmental Publications of the Water and Environment Research Institute No 17. pp 3-15.

Perspectives

Snow Engineering: Recent Advances, Izumi, Nakamura & Sack (eds) © 1997 Balkema, Rotterdam. ISBN 90 5410 865 7

Perspective on snow technology and science

B. Salm

Swiss Federal Institute for Snow and Avalanche Research, Weissfluhjoch/Davos, Switzerland

ICSE-3 was a valuable conference. It brought people together, which are often not in contact, snow researchers and consumers as architects or civil engineers. Discussion was around the same material snow, but from different point of views. This seems especially a benefit for research. Snow research is primarily applied science, its direction has therefore to be guided by practical problems. Many snow phenomena play in our life a more important role than snow specialists sometime think.

Metamorphism of snow seems pure science, but is fundamental for practical aspects as e.g. stability evolution of the snowpack (avalanche formation and risk for structures) or texture of snow on roofs (formation of roof avalanches on big buildings) or also oversnow transport and snow removal (processed snow).

Avalanche dynamics are crucial for establishing hazard maps and for determining lateral impact loads on structures or forces acting on protective galleries. Field tests are not easy to perform, but badly needed to get more insight. One paper pointed to this.

Friction of snow on snow or on other materials is a very complex problem for a variety of practical applications, as mentioned in the following.

Calculating of *speed and runout of avalanches* depend largely on this resistance. The costs of avalanche galleries e.g., are mainly given by friction on its roofs (dimension of anchors).

Important practical problems are furthermore friction between *tires and snow surface* (safety of traffic), between *shoes and snow* (safety of pedestrians) and between *roof surfaces* and snow (snow loads and roof avalanches). Different papers gave valuable contributions.

Daily snow precipitation, amount and return periods are fundamental for many tasks of the practitioner. Especially unsatisfactory up to date is the prediction of the amount. For prediction of avalanche occurrence especially a bad fact. It should also be better known for traffic (behavior of car-drivers) and organization of snow removal. One would be happy if a precision could be reached as described in a Japanese paper.

To get trustworthly *return periods* for daily, monthly and yearly values, relatively long observation periods are mandatory. For snow loads on roofs or avalanche defense work several decades.

Snow drift is a central problem for formation of avalanches (how much snow is where deposited) and for snow deposition on or around buildings. First of all one should know what is "driftable snow". Besides of the friction velocity it depends very sensitively on snowpack properties (cohesion, friction, grain diameter, temperature). So far no method exists to predict the amount of driftable snow.

Snow drift can be *investigated* by physical models in reduced scales, by numerical simulation or by field tests in scale 1:1 or occasionally also in reduced scales. Each method has specific advantages, but also difficulties.

Tests with *flumes* can easily be performed with different geometries, speeds and densities. But there is the problem of similarity (geometry, time, particle diameter and density etc.). In nature density of ice is about 1000 times higher than that of air, whereas in a flume particle density and water density are similar. The free fall velocity of particle has to be considered. In a 2-phase flow the densometric Froude number depends on density of flow and with this similarity may change form place to place.

Testing in *wind tunnels* would be ideal if tests could be performed with real atmospheric snow. To get a logarithmic profile of the inflow seems problematic too. Promising is certainly the big wind tunnel "Jules Verne" recently built in France.

Numerical simulation is nowadays placed in the foreground due to increased computer capacities. This powerful tool has problems too. A realistic modeling of saltation and the transition to diffusion is needed. The reliability of turbulence models should be tested by physical models. Finally to model snow deposits quantitatively, probably cohesion or even electric charges of particles are to be considered (formation of cornices).
Field tests seem the best way, but parameters can not be changed arbitrarily as in a laboratory.

One may *conclude* that a *combination* of different methods seems the most succesful way.

Recommendations for future Snow Engineering Conferences. A continuation of such valuable conferences is recommended. To have more time for discussion and to reduce parallel session, only papers should be accepted which can be clearly classified into the (well selected) topics I to V. Probably more invited papers on problems of specific interest for snow engineering should be sought, to concentrate on important problems. Finally, I congratulate to this well organized and successful meeting.

Snow Engineering: Recent Advances, Izumi, Nakamura & Sack (eds) © 1997 Balkema, Rotterdam. ISBN 90 5410 865 7

Perspective on building and construction engineering

K. Apeland
Oslo School of Architecture, Norway

INTRODUCTION

The building and construction engineering is concerned primarily with the effect of snow on the buildings, in particular snow load on roofs of various shapes, side pressure from snow on building walls, snow sliding etc.

From owners of buildings, and the building industry, i.e. architects, engineers, contractors, there has been, and still is, a great demand for reliable estimates of the snow actions. So far, this demand has been met with load standards and codes of practice, which have been prepared on the basis of the available knowledge at the time when the code has been prepared.

In addition to withstanding the snow actions, a building must be designed in order to cope with ice dams, leakage problems and other building technology problems, being caused by the existence of snow on or around the building.

When considering the service life of a building and the yearly cost for running it, the detail design, and thus avoidance of faults, may be more important than the construction cost of the building.

STATE OF THE ART

Most countries in which snow is a design concern, have codes of practice in which snow load actions are prescribed.

In 1981, ISO issued its first standard on snow loads, ISO 4355. The ISO standard has been the basis for the code prescriptions in many countries. A revised version of the old ISO 4355 has been prepared, and was adopted in 1995. However, it is not published yet.

In the Third International Conference on Snow

Engineering, Sendai, May 1996, two new codes were presented i.e.:

- The AIJ recommendations for loads on buildings, Japan.
- The Eurocode snow load standard, CEN ENV 1991-2-3 Actions on structures-Snow loads

When considering these new "codes" and comparing them with the new ISO Standard, ISO 4355, considerable differences are observed. Therefore, it is obvious that more research and development is needed in this field in order to establish a scientific basis for the codes.

A secondary goal is to obtain a harmonization of the standards for snow loads. Since determination of snow loads is a worldwide problem, at least in the Northern hemisphere, it is felt that ISO should be the most suitable organization for carrying out such harmonization efforts.

CEN, representing at least Western Europe, and being in close cooperation with ISO, also has a responsibility in this connection.

PERSPECTIVE

Snow loads on roofs.

Research and development seem to be needed in the following specific fields:

- Further studies on ground snow load and conversion factors for the determination of the corresponding snow load on roofs.
- Establishment of statistical data of heavy short term snow falls, which may be applied for defining defining snow loads on roofs as it is proposed in the AIJ recommendations,

and to establish what amount of snow that must be taken into account in the design of glass roofs, for which a continuous melting of snow occurs.
- Studies of the effect of exposure.
- Load surveys on buildings to form a basis for load specifications.
- Evaluation of causes of building damages under snow storms.
- Model studies for the establishment of snow load distribution on various roof shapes.

Design of buildings to resist snow actions.

Although continuous efforts are made by architects and engineers in order to develop robust and wellbehaving buildings, it is felt that this field has been more or less ignored over the years.

Since the life cycle cost of buildings has been brought into focus, as far as building economy is concerned, it is felt that this field should be given considerable attention, also in the field of snow actions on buildings.

CONCLUSIONS

In the field of building and construction engineering, it is felt that there is a substantial lack of knowledge about the snow loads on roofs, exemplified by considerable differences in various codes for snow loads on roofs.

Therefore, there still is a great demand for research in this field.

In view of the fact that life cycle cost is becoming an important issue of building economy, systematic studies should be initiated on the effect of snow actions on buildings in order to keep maintenance costs low.

Snow Engineering: Recent Advances, Izumi, Nakamura & Sack (eds) © 1997 Balkema, Rotterdam. ISBN 90 5410 865 7

Perspective on infrastructure and transportation

N. Hayakawa
Nagaoka University of Technology, Niigata, Japan

There are a variety of topics covered under this theme. They represent importance of advancing snow engineering technology related to building the sound infrastructure, the necessary assets of today's modern society.

Six papers (a02, b03, b04, c06, 04, 07) treated various aspects of road engineering. They include observation of winter road problems in the Heilongjian Province of China (a02), anti- and de- icing techniques (b04, 04), and other snow or ice treating techniques (b03, c06, 07). They include use of road heating, groundwater and rubber chips. These six papers report development of new materials, advanced numerical analysis, and sophisticated field testing or observations. Most, it should be noted, rely heavily on field-based data, expressing need to accumulate good-quality field knowledge.

Somewhat close to this topic ,but appeared to draw a strong attention of researchers, is frost control of the ground (b02, 05). This seems to be the area that is sure to draw strong attention and progress in the future.

Another group of 6 papers (a03, a04, a05, a06, a07, b01) dealt with human behavior, which would fall in the realm of social science. In these works, one collects data how inhabitants, mostly city dwellers, react to the impact caused by snow. Inhabitants in this case could be pedestrians, women, workmen or drivers. This is an important topic for planners of city, road, or any construction works or, in some cases, to everyday life. All these papers are presented by Japanese authors and one wonders if the approach of this kind is uniquely Japanese. And if so, why ? It appears that a variety of problems are hidden under this category. There will be more to be studied as the way of living changes as the age unfolds. One then wonders what is the direction that all these researches lead to. It can probably

be reflected in the planning stage of infrastructure, but this stage in many cases remains untouched in the presentation.

There is yet another group of researches which are definitely called uniquely Japanese. That is a problem of "Ryu-setsu-ko" in Japanese and there are five papers on this topic (09, c01, c02, c03, c04). Ryu-setsu-ko is a road side ditch specially designed so that people can dump snow in. This technology was born in Japan under a number of favorable conditions, namely abundance of water supply to wash snow out, relatively mild winter temperature for free of freezing, abundance of snow to be removed and city life to be maintained. It has been called by Japanese in a variety of terms in English which include snow drain, snow gutter, snow ditch and snow dumping channel. And I am happy to announce that Dr. McKinley christened it in one of the session as "snow removal channel".

Papers presented in this topic deal with a variety of problems related to snow removal channel. They are mostly in the field of hydraulic engineering as most so far are built in the region of mild winter temperature. However, there is a sign that this technology is gaining a support in the region of cold winter temperature. Because of this, there will be a need to study more as the ice and snow problems and there probably is a possibility that this technology can find application in other parts of the world.

Two papers covered snow drift (01, 02), although the Jules Verne (01) appears to hold a more promising future. Snow drift is a big topic and has enjoyed a long history of active research works. There is also one more paper which covers this topic under a different theme (I04). Still, snow drift research, be it numerical, of physical model, of field observation, appears to bear a number of unsolved problems related to the turbulent boundary layer and complicated mechanism of saltation/suspension. Each work is based upon one or two assumptions which are very often left unverified or un-compared paper-to-paper. It will be nice and meritorious if a symposium exclusively devoted to this topic be held some time.

There are two papers which do not fall under any of these categories. Paper a01 dealt with snow depth and snow density on the slope and the paper c05 dealt with the possibility of use of ice and water mixture for air condition in the area of city blocks. The former can be classified as a fundamental study and yet finds its importance in the application related to this theme, whereas the latter concerns itself with a new kind of application of the utilization of snow, representing a sizable number of research which happens to have escaped the attention of researchers on this conference or have found its place of attention in the theme V.

Overall, many problems are treated in the papers of this theme. The methods that have been used include advanced numerical analysis, sophisticated experiment in the field as well as laboratory, advanced statistical analysis of the data and field observation of the snow problems. They undoubtedly reflect the advancement being achieved by today's colleagues of the world. It is probably safe to point out the snow problems of new kind are sprouting and field data cannot be over-emphasized in its importance.

In this theme there are problems, or in the approach, uniquely Japanese. They are unique because of climatic or geographical reasons, or social background in some cases. Possibility in future if they can be proliferated to the other parts of the world is a thing to be watched.

Snow Engineering: Recent Advances, Izumi, Nakamura & Sack (eds) © 1997 Balkema, Rotterdam. ISBN 90 5410 865 7

Perspective on housing and residential planning

I. Mackinlay
Ian Mackinlay Architecture, Inc., San Francisco, Calif., USA

1 DISCUSSION

The major focus of ICSE-3 was engineering design and technical investigation of the properties and characteristics of snow and ice. Only a few papers dealt with the human response to snow and cold. Such investigation is more architectural than engineering, but there is a broad overlap. Perhaps ICSE-4 in Norway should devote a larger section of time to human factors. However, several good points were considered in the Sendai conference:

1. As the population ages, more attention needs to be given to providing old people with a safe snow country environment. Extra special planning and detail to accommodate a less than fully ambulatory populace and people with much less tolerance to environmental extremes and/or sudden changes. Snow and ice should not slip from roofs onto their heads. Icy stairs and walkways must be avoided. Transitional spaces with modulated temperature must translate from inside to outside. Properly conditioned interior spaces must be provided where fresh air and moisture are supplied, not just heat.

2. Special attention must be given to recreation during cold conditions. Skiing and ice skating are not available to the very young, the very old, or the infirm, but these people still need exercise and excitement. They can not spend all winter inside with nothing to do. Attractive exercise rooms are essential to good health.

3. Temperate transition spaces linking various elements of the facilities need to be planned and implemented. It is not enough to provide comfortable interiors, the connections between outside and inside environments must be considered. Besides serving as sheltered circulation arteries, these connecting spaces will allow the older population to walk, browse, shop, sightsee and generally conduct their lives as they would in a warm weather environment.

4. One approach is to provide spring-like interior conditions to a large complex of buildings, such as a school, and interconnect each element of the complex with moderately heated corridors. This approach has been taken for an entire downtown district such as Minneapolis and St. Paul, Minnesota in the USA where many city blocks are interconnected with heated sky bridges that span city streets. (Figure 1) One can move from heated house, to heated automobile, to heated downtown garage, to the heated sky bridge, to the central heated mall, to a conditioned office or department store and not be required to wear more than a light shirt and shorts, even when the outside temperature is well below freezing and the wind chill would be deadly to one so dressed. The risk, of course, is that if the heating fails at any point in this system, the lightly dressed individual may freeze before he can reach a place of refuge. Backup clothing needs to be available.

A second approach would be to provide enclosed arcades along existing public/private right-of-ways. In Japan, these spaces are called "Gangi" (public owned) or "Komise" (privately owned, semi-public). These spaces usually shelter a pedestrian from sun, wind, rain and snow but, in most instances, are not temperature controlled, so they are not as suitable to the elderly or infirm.

5. The snow and cold raises special problems of fire protection. Fire fighting and rescue equipment may not be able to reach a fire through deep snow drifts and/or icy streets. More attention needs to be given to internal fire suppression in Cold Country. While water is the most common media for fire suppression, its freezing potential is a serious problem. Economical, non-toxic, non-staining and odorless anti-freeze additives need to be developed to mitigate freezing and permit safe snow country fire fighting.

Furthermore, development of low cost and low maintenance, self-contained fire suppression systems will also mitigate freezing as they would not be dependent upon public water supply systems.

6. The point of maximum risk is the junction between cold and warm areas. This is where ice forms, roofs slip their snow loads, and hazards

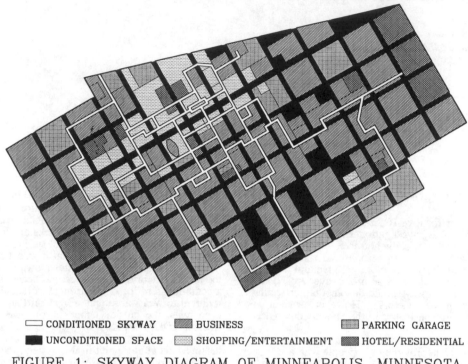

CONDITIONED SKYWAY   BUSINESS   PARKING GARAGE

UNCONDITIONED SPACE   SHOPPING/ENTERTAINMENT   HOTEL/RESIDENTIAL

FIGURE 1: SKYWAY DIAGRAM OF MINNEAPOLIS, MINNESOTA

occur. Special design attention must be given by both architects and engineers to these dangerous interfaces. Most ice dams are created by the freeze-thaw cycle and they are amplified by interior building heat. These are life threatening problems that must be addressed by both architects and engineers.

2 SUMMARY

The underlying principles of Snow County Design must be more widely understood by architects and engineers who work in conditions of snow and cold. It is not possible to rely solely on codes and standards. The snow is too crafty an adversary. The geometry of the structures, the nature of the wind and sun at the site must be analyzed, the history of snow fall and temperature must be understood, the ways the building complex is to be used must be studied. Earthquake forces must be considered. Without this careful investigation, the occupants may be subject to at least discomfort and often to serious risk. The new fallen snow is beautiful and downy, but it can be deadly if its more lethal nature is disregarded.

Snow Engineering: Recent Advances, Izumi, Nakamura & Sack (eds) © 1997 Balkema, Rotterdam. ISBN 90 5410 865 7

Perspective on development strategy in snow countries

M. Kobiyama
Muroran Institute of Technology, Japan

1 Papers presented in Session V

In Session V, the themes presented and discussed were divided into 3 groups. The theme of 1st group was about the technology to conquest snow, 2nd to be familiar with snow and 3rd to utilize snow.

18 papers were presented in this session. 9 papers were concerning with conquest snow, 4 papers were with being familiar with snow and 7 papers were for utilization of snow.

In the papers concerning the conquest snow, avalanche, melting water balance, and flood caused by melting snow were reported and discussed. The discussion on the walking environment was especially interesting. This paper taught us how to estimate the livelihood level in snow country.

In the papers concerning with being familiar with snow, there were many interesting and thoughtful papers, for example, remaining snow pattern, a border between public and private territory, and local color trains under snowy climate. These paper gives us much information that snow country is very interesting, has a long term history and we ought to be proud of our home country, that is, snow country.

In the papers concerning utilizing snow, many examples to utilize snow were reported. Snow can be used for cold heat and water resources. For example, snow and ice stored until summer can be used for storing agricultural products and for air-conditioning instead of an electric air conditioner, that is, refrigerator. A new technology about the temperature and humidity control system of air-conditioning by snow was reported. To use snow in wider fields, this technology is very useful. The technology regarding snow utilization encourages us and will become one of the main themes in the conference which will be held in the future. And an example of international cooperation in snow engineering fields was presented by the experiments of storing snow and ice in China.

In peculiarity, the papers concerning being familiar and utilize snow have been increased. The technology to conquest snow seems to be already matured. This tendency is very important to develop the snow country. This is the time that we have to face snow with a positive attitude to utilize snow or to be familiar with snow.

2 Perspective for development of snow country

As we live in snow country with more severe conditions than that in no snowfall country, we have to consume more energy to keep our life safe and comfortable. However, we must make effort to decrease energy consumption. This problem is very important for us and our children, because if this problem is not solved, we will not be able to live in snow country in the future when the fossil energy will not be in enough supply. Now we must change the life style of higher energy consumption to lower one, and make an effort to use energy more efficiently and to make use of natural and local energy. For example, in winter we live in a town together to decrease energy consumption to use energy more efficiently, in summer we return to our home country to do our individual work. We store hot energy in the soil under the ground or in common materials in summer for cold winter season, and we store cold energy in winter for hot summer season. We have to make clear the value of snow country in the individual state or in the world. For example, as snow country is important for production of many agricultural products, for storage of some fresh foods and so on, we must live there. We engineers who live in snow country have not discussed enough about these problems. We should discuss in future conferences with the standing point of an engineer and a general person living in snow country.

In this conference, the themes were limited, rather they leaned to the hard engineering side. Only a few soft engineering, for example, walking environment, intermediate territory and local color trains, were presented and discussed. Those studies are very important and interesting, because they make our snow country bright. In the same sense, we ought to present and discuss about other soft engineering such as assisting goods to play games, clothes in cold region, and many types of equipment for care the old person and so on.

3 Perspective for the conference

A conference is not a lecture meeting. The discussion in the field of engineering or science is not so important. Rather, the purpose or the aim for developing snow country and how to develop are important, and we should discuss about what we can do for developing snow country and how to live thère.

The most impressive writings in Session V is that by a person who lives in Tokamachi Japan, as follows:
"It is said that this is the time of snow utilization, beyond snow conquest. There are still many problems to solve in front of us, but we consider the utilization and conquest are not two different matters. Utilization is an advanced form of conquest. We will continue to be progressive on challenging projects, and aim high to establish the snow country which people and nature live harmoniously."

Subject index

Author index